Engineering MECHANICS

PROCEEDINGS OF THE 11TH CONFERENCE

SPONSORED BY THE

Engineering Mechanics Division of the American Society of Civil Engineers

AND THE

Florida Atlantic University Center for Applied Stochastic Research

Fort Lauderdale, FL
May 19-22, 1996

Edited by Y.K. Lin and T.C. Su

Volume 1

ASCE 1852

Published by the
American Society of Civil Engineers
345 East 47th Street
New York, New York 10017-2398

ABSTRACT:

This proceedings contains keynote and summary papers presented at the 11th ASCE Engineering Mechanics Conference held in Fort Lauderdale, Florida, May 20-22, 1996. The keynote papers address new trends in biomechanics, basic concepts and application of structural control, and recent advances in environmental fluid mechanics. The summary papers include those of three symposia, poroelasticity, computational geomechanics and smart structures, and the rest of the papers cover such diverse topics as: 1) mechanics in life sciences; 2) turbulence; 3) fluid structure interaction; 4) granular flow; 5) stochastic mechanics; 6) system identification; 7) soil- structure interaction; 8) nondestructive evaluation of bridges and infrastructures; 9) retrofitting and health monitoring of structures; 10) damage and failure of composites; 11) reinforced concrete and other engineering materials; 12) modeling of material with discontinuities and heterogeneities; and 13) mechanics of plates shells and beams.

Library of Congress Cataloging-in-Publication Data

ASCE Engineering Mechanics Conference (11th : 1996 : Fort Lauderdale, Fla.)
Engineering mechanics : proceedings of the 11th Conference / edited by Y. K. Lin and T. C. Su ; sponsored by the Engineering Mechanics Division of the American Society of Civil Engineers and the Florida Atlantic University Center for Applied Stochastic Research
p. cm.
ISBN 0-7844-0172-1
1. Mechanics, Applied—Congresses. I. Lin, Y. K. (Yu-Kweng), 1923- II. Su, T. C. III. American Society of Civil Engineers. Engineering Mechanics Division. IV. Florida Atlantic University. Center for Applied Stochastic Research. V. Title.
TA349.A82 1996 96-16971
620.1—dc20 CIP

Library of Congress Catalog Card No: 96-16971
ISBN 0-7844-0172-1
Manufactured in the United States of America.

FOREWORD

This volume contains summary papers to be presented at the 11th Engineering Mechanics Conference on May 20 through 22, 1996 in Fort Lauderdale, Florida. The conference is organized under the auspices of the Engineering Mechanics Division of ASCE, and is hosted by the Center for Applied Stochastics Research of Florida Atlantic University. The three-day program consists of three plenary key-note sessions and sixty-nine technical sessions. The keynote speakers and their topics are: Professor Y. C. Fung, "New Trends in Biomechanics," Professor T.T. Soong, "Basic Concepts and Applications of Structural Control," and Professor R. L. Street, "Environmental Fluid Mechanics—A Review of Some Recent Results." Each of the summary papers included in the Proceedings has been accepted for publication by the Proceedings Editors. All papers are eligible for discussion in the Journal of Engineering Mechanics. All papers are eligible for ASCE awards.

We would like to thank the key-note speakers for accepting our invitation to address the conference, and all the authors for their valuable contributions. The support and guidance of the EMD Executive Committee, and the help provided by the EMD Technical Committees in organizing many technical sessions, with participation from the Aerospace Division, are gratefully acknowledged. Our special appreciation is due to Mrs. Christine Mikulski of the Center for Applied Stochastics Research for her dedication and hard work in almost all aspects of preparation for the conference.

Y. K. Lin and T. C. Su
Conference Co-Chairs

CONTENTS

KEYNOTE LECTURES

SUBMITTED PAPERS

*Manuscript not received at time of printing.

New Trends in Biomechanics

Y.C. Fung[1]

Abstract

There is a clear trend in the biomechanics community to help solve real problems in biology and medicine, thus, to make a commitment to improve the quality of health care of mankind. Much has been accomplished, but advances were often limited by a lack of real knowledge. The input from biology and medicine has caused a great expansion of the field of mechanics. Living organisms use active transport and nerves and muscles, and can grow and remodel under stress. Hence, mass and geometry are not conserved, the equation of continuity must have a tern of growth. Mass transport across biological membrane is often a movement of matter from the dilute side to the more concentrated side. Stress in connective tissue is largely derived from a decrease of entropy due to an increase of strain. The constitutive equations of materials are not constant: they remodel when stresses deviate from the homeostatic values, and are functions of time. The equation of motion must include the contribution of the muscles, especially the smooth muscles in organs other than the heart and the skeletal muscle. The zero-stress state of a living tissue can change. The boundary of biofluids are soft tissues, so, in few problems in biology and medicine are the motions of fluids and solids uncoupled. In such an expanded domain, there are many interesting and important problems involving cells, tissues, organs, and individuals. Evidences of these features that are relevant to the quality of health care are discussed.

Introduction

Biomechanics seeks to understand how animals and plants work. As a part of man's endeavor to understand oneself, biomechanics is a subject of

[1] Professor Emeritus, Department of Bioengineering, University of California, San Diego, La Jolla, CA. 92093-0412. Von Kármán Medal Awardee, ASCE, 1976.

general interest; and is associated with the names of Huangti (黄帝), Aristotle, Galileo, Harvey, Descartes, Borelli, Hooke, Newton, Euler, Young, Poiseuille, Helmholtz, and others. There are many engineers, biologists, physicians, and surgeons working in the field. There is a trend of commitment to bioengineering and health care. This commitment is based on past record and future prospects. It is fitting for civil engineering, because engineering is science applied for the benefit of man, civil engineering is engineering for the public, and health care is of public interest. I shall focus especially on mechanics. I would like to list some contribution biomechanics has made to the biomedical field; then discuss why mechanics has a major role to play in the study of diseases. The expanding domain of mechanics due to the need to understand physiology and disease is discussed next. Then we shall consider tissue engineering, in which mechanics plays a vital role, and finally, the quality of health care as a measure of future success.

Past Accomplishments of Biomedical Engineering

Today, biomedical engineering has given us the following:

Better understanding of physiology,
Mathematical models, computational methods,
Several artificial internal organs,
Several artificial limbs, joints, and prostheses,
Better understanding of hearing, vision, speaking,
A number of implantable materials,
New biosensors,
New imaging techniques: CT, NMR, PET, etc.,
Clinical devices, instruments, techniques,
Remote sensors, virtual reality,
Minimally invasive surgery techniques, and many more.

New progress is being made, and good news comes frequently. However, every initial success is always stopped by a limitation of knowledge. Most of the successes listed above are still limited in scope and have not achieved their full potential. Real progress will depend on further research.

It is clear that future bioengineering will ride on the momentum of the rapid advances of basic science, including molecular biology, cell biology, neuro-science, immunology, computers, electronics, and mechanics. Bioengineering will develop in the interdisciplinary areas. New devices, instruments, and techniques will be multidisciplinary; and the advance will be accelerated by the society's need for health and economic efficiency.

The Role of Mechanics in Health Care Today

It is easy to see the role played by mechanics in the items listed in the paragraphs above. Artificial heart and heart assist devices are designed to pump blood; they are fluid mechanical devices. Artificial kidney is designed for dialysis; it is a device for mass transport. Artificial limbs are designed to support body weight and perform motion. Artificial heart valves must have high fluid mechanical efficiency, low head loss, long service time, and without calcification and other untoward changes. Pacemaker must induce rhythmic muscle contraction. The role of mechanics in these devices is evident.

New implantable materials must have high strength, low weight, and resistance to corrosion by blood and body fluid. They must be non-carcinogenic and have no undesirable reactions to cells. The surface chemistry and biological effects of implantable materials are often modulated by stress and strain. Mechanics is a principal design factor for new materials.

Biosensors may appear to depend mainly on chemistry and physics; but in reality, fluid mechanics and mass transport analysis often consume most of the developmental time, and success hinges on a thorough understanding of fluid mechanics.

For imaging and clinical devices and virtual reality, mechanics is important not only for the construction of the equipment, but also for the design of experiments and interpretation of the data. For the minimally invasive surgery techniques such as angioplasty, however, biomechanics should play a central role. Angioplasty involves creating an injury and hopes for healing into a normal tissue. Lack of a thorough understanding of the injury and healing processes led to the current difficulties.

Other well-known contributions include sports and sports medicine, orthopedics, treatment of automobile injuries, burn injuries, and equipment for rehabilitation. Biomechanics of hearing, seeing, and speaking are extremely important subjects. Lengthening of bone to solve gait problems is studied in several laboratories.

Clearly, mechanics has become an important knowledge base of health care. (See Fung, 1990, 1993b, 1996 for more detailed discussion of these topics and references to the literature.)

Why Do We Think Stress and Strain are Important in Atherogenesis and Cancer?

Atherosclerosis is one of the most common disease. It occurs only at certain sites in our arteries. The total length of the sites in each of us is no more than one meter. We know that each of us has tens of kilometers of blood vessels. Why should only these short segments be at risk? If you eat a steak, the cholesterol will be circulating in your whole body. All risk factors we know are of the whole body. Why do they cause trouble only at these special sites? Puzzled, a fluid dynamist may question what are the unique features of these sites? Are they stagnation points? Are they the sites of the highest shear stress from the flowing blood? Are they the sites of flow separation, where the streamlines of flow separate from the wall? Are they the sites of transient turbulence? What are the effects of shear stress of the blood on the endothelial cells of the blood vessel wall? What are the effects of turbulence on the endothelial cells? How does LDL get across the endothelial cells into the smooth muscle cells beneath? How do the smooth muscle cells react to the shear stress of blood? Why do the smooth muscle cells transform themselves into foam cells and begin the atherogenesis? How does calcification set in eventually at these sites?

A solid mechanicist may raise different questions. What are the unique features of these atherogenic sites with regard to stress and strain distribution? Are the sites areas of stress concentration? Which stress component matters? Should we be concerned only with stress invariants and strain invariants? If so, which invariant? The mean stress? The principal stresses? The von Mises or octahedral shear? Shall we be concerned with the stress and strain in the endothelial cell membrane? How about the smooth muscle cell membrane? Or should we be concerned with the stress and strain inside these cells? What are these cell's internal machinery that determines gene expression, cell division, cell growth, cell transformation? How does the machinery work? What roles do the ion channels in the cell membranes play? What roles do the actin fibers, microtubules, and other molecules of the cells play? How do various pharmaceuticals and vitamins affect the functions of these ion channels, and internal cellular machinery?

It is clear that we are not going to solve the atherosclerosis problem until these questions are answered. It is clear that the boundaries between fluid mechanics, solid mechanics, cell biology, molecular biology, immunology, and medicine are blurred. It is precise knowledge in these fuzzy borders that is needed. Atherosclerosis is a tough nut to crack. (See Fung, 1996 for references to the literature.)

Similar situation exists in other diseases. Rakesh Jain (1994) laid the tumor's resistance to full penetration by anticancer agents to the high interstitial pressure in the tumor. This is the pressure in the space between capillary blood vessels and the cells of the tumor. Jain has given the following measured values of interstitial pressure in human organs in units of mm Hg relative to the atmospheric pressure which is taken to be zero:

Normal Breast	0.0	Head and Neck Carcinomas	19.0
Normal Skin	0.4	Breast Carcinomas	15.0
Renal Cell Carcinomas	38.0	Metastatic Melanomas	14.3
Cervical Carcinomas	22.8	Lung Carcinomas	10.0
Colorectal Liver Metastases	21.0		

Thus, the interstitial pressure in the cancer mass is almost as high as the capillary blood pressure. In this situation there will be little connective fluid movement from the capillary to the interstitium. Since large molecules of the drug travel mainly by convection, you can see the difficulty of delivering drug molecules to the cancer cells. Jain's observation brings the cancer drug delivery problem to the realm of biomechanics. What can we do about it? How to modify the interstitial pressure? How to grab the drug molecules when they arrive? Many biological and molecular tactics can be suggested. But we need to know much more before the war can be won.

These scenarios are typical. They give us motivation to work harder, to apply biomechanics to disease. If a problem is biomechanical, it must be dealt with biomechanically. No amount of trial in other disciplines will solve the problem.

The Expanding Domain of Biomechanics

I think a major contribution made by biology and medicine to mechanics is the broadening of the scope of classical mechanics. In our bodies, besides gas and urine, all fluids are either non-Newtonian, or viosco-elastic, (see Fung, 1993b for references). The solid materials in our bodies, with the exception of the bone, do not obey Hooke's law. (See Fung, 1993b for references). The bone has very small strains in life, but has a very complex anisotropic structure. Hence, the materials in biomechanics are more complex than those ordinarily encountered in civil engineering. As a result, the first task in biomechanics is to determine the constitutive equations of living tissues whenever the concept of continuum is applicable. The task is big because most biological tissues are nonuniform and multicompartmental. Biofluids are usually mixtures, e.g., normal blood consists of roughly 50% red blood cells, 50% plasma which is itself a mixture. Soft connective tissues are usually a composite of collagen fibers, elastin fibers, ground substances, and fibroblast cells. The living connective tissues usually derive their elasticity through decrease of entropy with increasing strain. This is in contrast to crystalline materials which derive their elasticity by changing internal energy by deformation. Muscle consists of muscle cells embedded in connective tissues. Each cell has a complex specific structure, which is, unfortunately, still largely unknown. It is a common belief in mechanics circles that the mechanical properties of the microscopic elements are presumed known. In biology, we are not so sure. The structure and mechanical properties of a "microscopic" element is as complex and as much unknown as those at a "macroscopic" level. The microscopic elements are only smaller, not simpler. (See Fung, 1993a for a more thorough discussion). Overall, the determination of the constitutive equations of living tissues has been the biggest game in biomechanics for the past two decades. There have been successes; but much remains unknown. (See Fung, 1993b for references).

The imput of a wide variety of materials with non-classical constitutive equations from biology and medicine caused the first broadening of the field of mechanics.

The second broadening of the field of mechanics is forced upon us by the living organism's ability to create, grow, remodel, and die. Hence, if we focus our attention on a given material, e.g., the skin, the blood vessel, the vascular smooth muscle, the collagen, or the calcium ions, then the law of conservation of mass does not apply. We must consider the possibility of addition and subtraction of mass. The equation of continuity must have an additional term describing the rate of growth of the material. What we have learned recently is that in the living tissues, the rate of change of the mass of a given material constituent is modulated by stress and strain. This is equivalent to say that gene expression is modulated by stress and strain.

The third broadening comes with the realization that in addition to the usual passive laws of mass transport, such as Fick's law of diffusion, Starling's law of membrane filtration, and Darcy's law of flow through porous media, living cells usually employ **active transport** mechanisms to move solutes across cell membranes from the less concentrated side to the more concentrated side. Our urine is made in the kidney this way. This kind of mass transport need a supply of energy, (see Fung, 1990, p. 299).

Concomitantly, the equation of energy balance must contain the energy of chemical reactions.

The fourth broadening comes from the realization that the zero-stress state of a living tissue can change with time if the stress and strain deviate from the homeostatic state (a state of stable equilibrium.) If the tissue is a multicompartment multphasic solid, the zero-stress state of each compartment and each phase may change. Recent results show that the changes are stress modulated.

The fifth broadening is that the constitutive equation of a living tissue is not constant, but can vary as a specific function of time when the stress and strain are changed as step functions of time.

The sixth broadening is that we have muscles and nerves in our field, which have no counter part in classical engineering. The equation of motion must include the force and velocity of muscle contraction.

Of course, a field so broadened will have many new boundary-value problems which require new methods for their solution. All these enrichment of mechanics are due to the richness of biology and medicine. With such fundamental broadening we can now commit ourselves to tackle medical problems.

The statements above are too brief. A few details are given below to make them more convincing. I shall take blood vessels as examples because of my limitation. I am more familiar with this organ than with other organs.

Tissue Remodeling of Arteries Under Stress

Figure 1, from Fung and Liu (1991), shows the history of tissue remodeling of rat pulmonary artery when the blood pressure was raised from normal to about 7 mm Hg higher by a step change of the breathing gas from normal sea level air to one with 10% O_2 concentration. The left hand side figure is a sketch of the pulmonary artery of the left lung. The first row shows the morphological changes of the cross-section of the arterial wall in the arch region. It is seen that significant changes occurred already at 2 hours. The intima changed first, followed by the media in which the vascular smooth muscle reside. The adventitia changed slower. The succeeding rows show the changes in the vessel walls of smaller arteries.

Figure 2, from Fung and Liu (1989), shows the change of the zero-stress state of the rat aorta when the aortic blood pressure was changed suddenly by imposing a metal clip to constrict the vessel at the site indicated. Each row shows the history of the shape of the vessel cross-section when it was cut into short segments and then cut radically. The numbers on top are the number of days after surgery.

Figure 3, from Li et al. (1996), shows an example of the indicial functions of the change of the wall thickness of the rat pulmonary artery when the rat breathed 10% O_2 and its lung was hypertensive for 10 days, then returned to normal atmosphere for the number of days indicated in the figure. The indicial functions are plotted for several vessels of decreasing order number. The rat lung has 11 orders of vessels. Order 11 is the largest main pulmonary artery. The diameter ranges of the successive orders of the vessels in the rat are: Order 11, 1750-1292 μm; Order 10, 1291-746 μm; Order 9, 745-498 μm. The calculation of indicial function is explained in Li et al. (1996).

Figure 4, from Liu and Fung (1992), shows how the stress-strain relationship of rat thoracic aorta is changed by tissue remodeling during the development of diabetes following an injection of streptozocin.

Figure 5, from Fung and Liu (1995), shows our current method of determining in vivo the incremental elastic constants of the arterial wall when it is regarded as a two-layered material: an intima-media inner layer, and an adventitia outer layer. (a) refers to the measurement of the vessel diameter when the blood pressure is changed while the length of the vessel is held fixed. (b) refers to changing longitudinal tension due to longitudinal stretch. By varying the length in (a) and the blood pressure in (b), we cover a combined variation of the two variables, length and pressure. (c) refers to bending of the vessel wall with a pair of micropipets of rectangular mouth pushing or pulling diametrically while measuring the blood pressure and the micropipet force and displacement. (d) refers to axial bending with a silk thread, measuring force in the thread and deflection of the vessel. (c) and (f) refer to two ways of imposing shear to the vessel wall with two different distribution of shear strain throughout the wall thickness.

Figure 6, from Fung (1996), shows the length-tension relationship of the vascular smooth muscle in the pig coronary arterioles deduced by the present author (Fung, 1996) from the experimental data of Kuo, Chilian, and Davis (1990, 1991). This is the homeostatic length-tension relationship of the vascular smooth muscle. The relationship is strongly influenced by the shear stress imposed on the blood vessel wall by the flowing blood, τ. The upper panel shows the relationship in vessels without flow. The lower panel shows the effect of flow-induced shear, τ. Note the similarity of these curves to the length-tension curve of the tetanized skeletal muscle, and to the curve of muscle length vs. the peak tension in isometric twitch of heart muscle. The major distinction of the vascular smooth muscle is that it normally works at lengths corresponding to the right leg of an arch shaped curve, whereas the heart and skeletal muscles normally work at lengths corresponding to the left leg of the arch shaped curve. Knowing the curves of Fig. 6, we can now understand the autoregulation of the blood flow; and hyperemia, the increased blood flow in a vessel upon release from compression; and the Bayliss phenomena. Autoregulation is the tendency for blood flow to remain constant in the face of changes in arterial pressure to the organ. It is seen in virtually all organs of the body. It is most pronounced in the brain and kidney. (See Johnson (1978) for detailed discussion and literature.) Each organ has a steady state flow-and-pressure relationship. Suppose in an autoregulated region, say, at a point A in Fig. 6, you made a sudden increase of arterial blood pressure. By the elasticity of the artery the arterial diameter will be increased in response to the sudden increase of pressure; thereby, the resistance to blood flow is transiently reduced and the flow increases. Then the tension in the vascular smooth muscle is increased to a new value above that of the point A in Fig. 7. The curve of Fig. 6 shows that at equilibrium, such an upward movement of tension will have a muscle length which is shorter than that of point A. The smooth muscle shortens to reach the new equilibrium state. The vessel diameter is reduced, the flow resistance increases, and the flow falls back toward the normal value. This is autoregulation, which was used to be considered mysterious. With Fig. 6, we see it as a revelation of the basic feature of the smooth muscle property, and the fact that the normal working condition of the artery corresponds to a point lying on the right leg of the arch-like curve of Fig. 6.

Reactive hyperemia is the period of elevated flow that follows a period of circulatory arrest, e.g., by inflating a cuff on an arm as we use a manometer to measure our blood pressure. On releasing the cuff pressure, the blood flow burst forth far above the normal value. (See Johnson, (1978) for details.) The explanation based on Fig. 6 is that under compression the tension in the smooth muscle is reduced: the condition is represented by a point lower than the normal condition of point A. Hence, the smooth muscle lengthened and the vessel circumference enlarged while the vessel was compressed. Upon release of the compression and the return of the blood flow, the enlarged blood vessel causes a large flow. Later, the adjustments takes place to return to the normal condition. Reactive hyperemia is often used to detect if a vessel has an arteriosclerotic plaque, because the plaque abolishes reactive hyperemia.

Perspective of the Future

The examples above and similar results by other studies clearly show that force and deformation, or stress and strain, play an important role in health and disease. Hence, the recent trend of a commitment of biomechanics to improve the quality of health care is well founded. Since tissues can be remodeled to a certain extent by stress changes, and cells can be directed to proliferate or grow by suitable stresses, it is plausible to assume that tissues can be "engineered" to some extent. In recent years, many biologists are looking into the effect of stress and strain on cell differentiation, proliferation, and growth, see, for example, Ingber (1993). Many chemists and molecular biologists are examining integrins and adhesion molecules that affect gene expressions and cell function due to stress and strain in the extracellular matrix, see Frangos (1993), Resnick et al. (1993), and Reich and Frangos (1993). Dr. Badylak and his associates (1989, 1994) have shown that the intestinal submucosa can be transformed into arteries, veins, or tendons when it was used in the environment and stress field of arteries, veins, or tendons, respectively. It can be anticipated that tissue engineering is not only cell culture and biochemistry, but also biomechanics.

To succeed in their new aspiration, biomechanicists must extend their interest into the neighboring fields of biology, cell culture, molecular biology, as well as physiology and pathology. They must take up the active role to initiate programs.

The field of mechanics has been expanded greatly by the input of biology and medicine. There are many challenges in the expanded domain of biomechanics. At the present time, the most important challenge remains to be the determination of the mechanical properties of the living tissues. The next challenge is to determine the indicial functions of tissue growth or resorption. The third challenge is to determine the zero-stress state of every phase of materials of concern. With the morphometirc data, the constitutive equations, the indicial functions of growth or resorption, and the zero-stress states of the materials as the basic data, we formulate boundary-value problems and solve them. For each boundary-value problem whose solution can be validated by in vivo or in vitro experiments, we have an opportunity to check out the validity of the hypotheses that went into the theory. When every loose ends are fixed and satisfactory agreements are obtained, then we would have obtained a satisfactory working theory. This theory may become a useful tool in medicine, and enrich the basic knowledge.

So, in the future, we must continue to reinforce our expanded field and solve real problems. We want to contribute to the improvement of the quality of health care. As defined by the United States National Academy of Science-Institute of Medicine, **quality of care** is a measure of the degree to which health services for individuals and populations increase the likelihood of desired health outcomes that are consistent with current professional knowledge. Biomechanics' contribution can be measured according to the same definition.

Lewis Thomas (1974) said that there are three kinds of technology in medicine. One is high tech, such as the vaccine for infantile paralysis. The products are highly effective, low cost. The next is the "Half-way" technology: using all kinds of methods to treat a disease that one does not understand. The third is non-tech, all kinds of speculative, do-something methods. He said that the health care cost is so high today mainly because people keep busy on the last tow approaches.

Papers in the field of mechanics have always been judged by their quality. To be judged by the quality of health care may come as a surprise, but it is consistent with our tradition.

References

The literature related to the topics discussed in this paper is discussed in the following:

Badylak, S. F., Lantz, G. C. and Coffey, A. C., "Small Intestine Submucosa as a Large Diameter Vascular Graft in the Dog," J. Surg. Res. Vol. 47, pp. 74-80 (1989).

Badylak, S. F., Coffey, A. C., Lantz, G. C., Tacker, W. A. and Geddes, L. A., "Comparison of the Resistance to Infection of Small Intestinal Submucosa (SIS) vs. Polytetrafluoroethylene Vascular Grafts in a Dog Model," J. Vasc. Surg., pp. 465-472 (March, 1994).

Frangos, J. A. (editor), "Physical Forces and the Mammalian Cell," Academic Press (1993).

Fung, Y. C. "A First Course in Continuum Mechanics, for Physical and Biological Engineers and Scientists, Prentice-Hall, New Jersey, 3rd Edition (1993a).

Fung, Y. C. "Biomechanics: Mechanical Properties of Living Tissues," Springer-Verlag, New York 1st Edition (1981), 2nd Edition (1993b).

Fung, Y. C. "Biodynamics: Circulation," Springer-Verlag, New York, 1st Edition (1984), 2nd Edition (1996).

Fung, Y. C. "Biomechanics: Motion, Flow, Stress, and Growth," Springer-Verlag, New York (1990).

Fung, Y. C. "What Principle Governs the Stress Distribution in Living Organs?" in book "Biomechanics in China, Japan, and USA", Proc. of the Wuhan Conf., Science Press, in May, 1983, Beijing, China (1984).

Fung, Y. C. "Cellular Growth in Soft Tissues Affected by the Stress Level in Service", in "Tissue Engineering," (ed. by R. Skalak and C.F. Fox), Alan Liss, New York, pp. 45-50, (1988).

Fung, Y. C. and Liu, S. Q. "Change of Residual Strains in Arteries Due to Hypertrophy Caused by Aortic Constriction," Circ. Res., 65: 1340-1349 (1989).

Fung, Y. C. and Liu, S. Q. "Changes of Zero-Stress State of Rat Pulmonary Arteries in Hypoxic Hypertension," J. Appl. Physiol., 70: 2455-2470 (1991).

Fung, Y. C. and Liu, S. Q. "Strain Distribution in Small Blood Vessels with Zero-Stress State Taken into Consideration," J. Appl. Physiol., 262: H544-H552 (1992).

Fung, Y. C., Liu, S. Q. and Zhou, J. B. "Remodeling of the Constitutive Equation While a Blood Vessel Remodels Itself Under Stress," J. Biomech. Eng., 115: 453-459 (1993).

Ingber, D. "The Riddle of Morphogenesis: A Question of Solution Chemistry or Molecular Cell Engineering?" Cell, Vol. 75, pp. 1249-1252 (1993).

Jain, R. "Barriers to drug delivery in solid tumors," Scientific American, 271: 58-65 (1994).

Johnson, P. "Peripheral Circulation," New York, Wiley, (1978).

Kuchan, M. J. and Frangos, J. A. "Fluid Flow Regulates Endothelin-1 Release via Protein Kinase C and EDRF in Cultured Endothelial Cells," Am. J. Physiol., 264: H150-H156 (1993).

Kuo, L., Chilian, W.M., and Davis M.J. "Coronary Arteriolar Myogenic Response is Independent of Endothelium," Circ. Res., 66: 860-866 (1990).

Kuo, L., Chilian, W.M., and Davis M.J. "Interaction of Pressure-and Flow-Induced Responses in Porcine Coronary Resistance Vessels," Am. J. Physiol., 261: H1706-H1715 (1991).

Lantz, G. C., Badylak, S. F., Coffey, A. C., Geddes, L. A. and Sandusky, G. E. "Small Intestinal Submucosa as a Superior Vena Cava Graft in the Dog," J. Surg. Res., 53: 175-181 (1992).

Li, Z.J., Jiang, Z.L., Wang, G.R., Liu, S.Q., and Fung, Y.C. "Zero-stress State and Remodeling of the Rat Pulmonary Arteries in Recovery from Hypoxic Hypertension," J. Appl. Physiol., in Press (1996).

Liu, S. Q. and Fung, Y. C. "Changes in Rheological Properties of the Blood Vessels Due to Tissue Remodeling in the Course of the Development of Diabetes," Biorheology, 29: 443-457 (1992).

Reich, K. M. and Frangos, J. A. "Protein Kinase C Mediates Flow-Induced Prostaglandin E_2 Production in Osteoblasts," Calcified Tissue International, 52: 62-66 (1993).

Resnick, N., Collins, T., Atkinson, W., Bonthron, D. T., Dewey Jr., D. F. and Gimbrone Jr., M. A. "Platelet-Derived Growth Factor B Chain Promoter Contains a Cis-Acting Fluid Shear-Stress-Responsive Element," Proc. Nat. Acad. Sic., USA, 90: 4591-4595 (1993).

Thomas, L. "The Technology of Medicine," pp. 31-36 *in* The Lives of a Cell: Notes of a Biology Watcher, New York, Viking (1974).

Vaishnav, R. N. and Vossoughi, J. "Estimation of Residual Strains in Aortic Segment," in Biomedical Engineering, II. Recent Developments (C.W. Hall, ed), Pergamon Press, New York, pp. 330-333 (1983).

Vaishnav, R. and Vossoughi, J. "Residual Stress and Strain in Aortic Segments," J. Biomech., 20: 235-239 (1987).

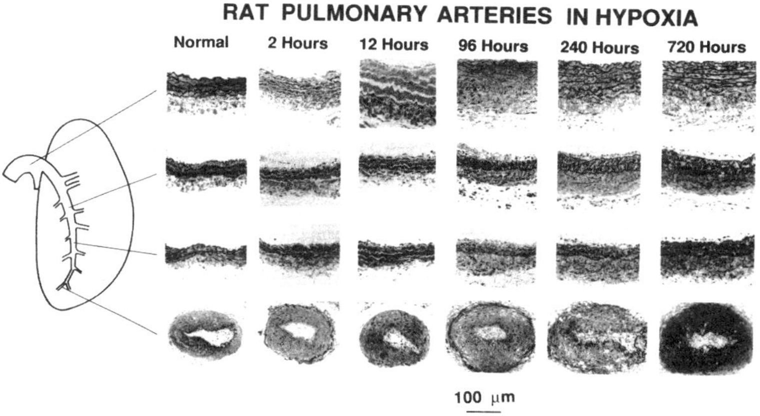

Figure 1. Photographs of histological slides from four regions of the main pulmonary artery of a normal rat and hypertensive rats with different periods of hypoxia. (Reproduced with permission from Fung and Liu (1991).

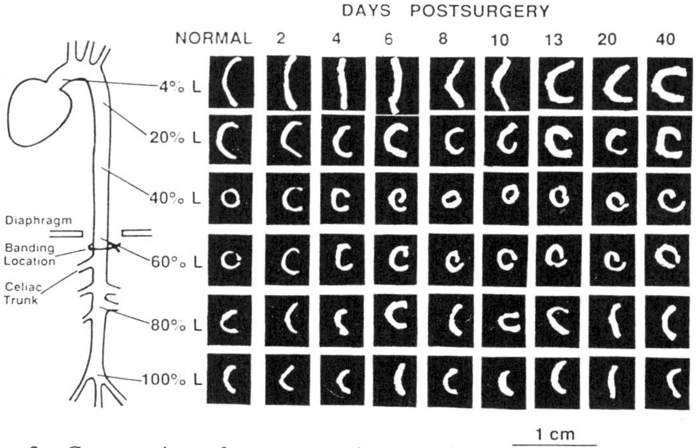

Figure 2. Cross-sections of a rat aorta at the state of zero stress-resultant and zero stress-moment, called zero-stress for short. The first column shows zero-stress state of normal aorta. Other columns show changed zero-stress states a number of days after a sudden onset of hypertension. Successive rows correlate with locations on the aorta expressed in percentage of total length of aorta, L, from the aortic valve. The aortic cross-sectional area was clamped 97% by a metal band below the diaphragm to induce the hypertension. From Liu and Fung (1989) by permission.

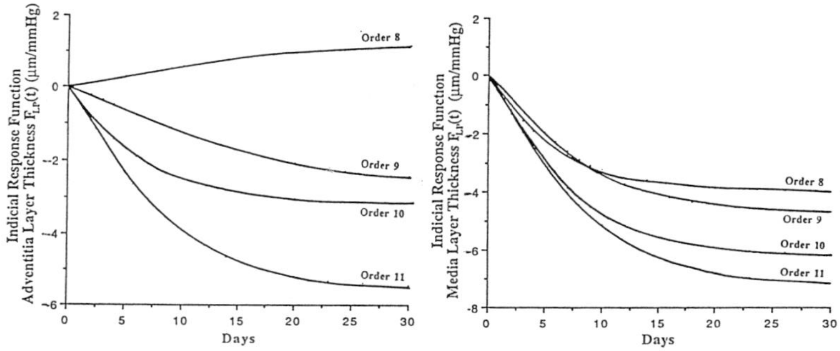

Figure 3. The indicial functions of the thickness of the medial layer of rat pulmonary arteries of orders 8 to 11 in response to a 1 mm Hg decrease of blood pressure after a hypertension period of 10 days. From Li, Jian, Wang, Liu, and Fung (1996), in press.

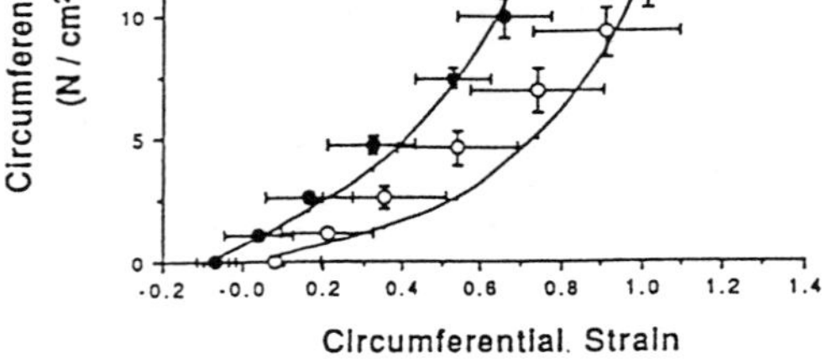

Figure 4. Change of the stress-strain relationship of rat thoracic aorta during the development of diabetes 20 days after an injection of AZT. From Liu and Fung (1992) by permission.

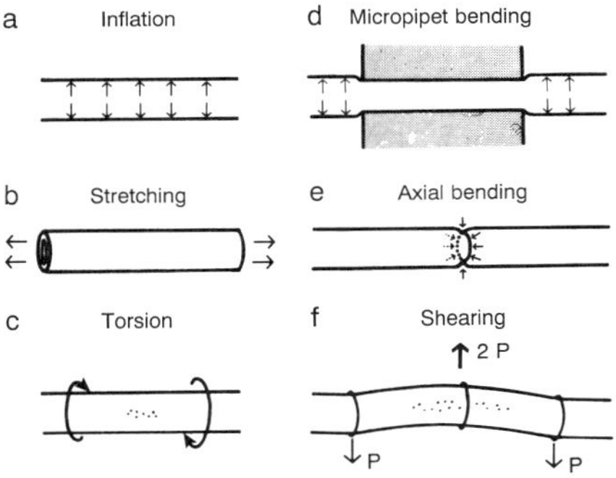

Figure 5. Six types of in vivo experiments that can be used to determine the eight incremental elastic moduli of the two layers of (a) blood vessel at a homeostatic state in vivo. Counting an experiment of type a with variable longitudinal stretch, and another one of type (b) with variable internal pressure as two additional experiments, we have a total of eight experiments for the eight unknowns. From Fung and Liu (1995).

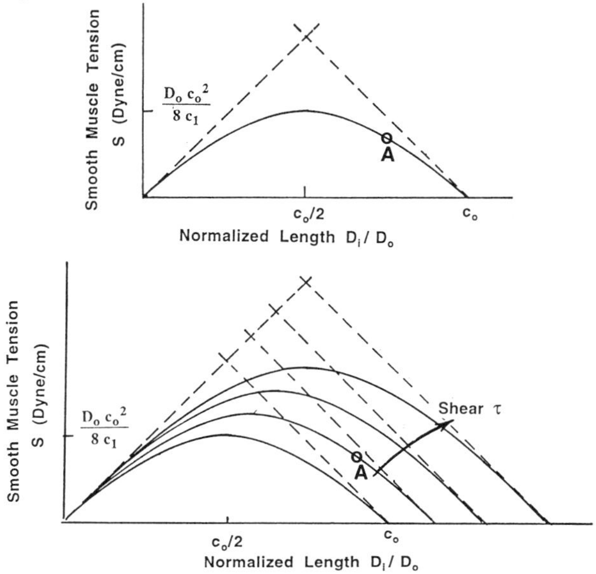

Figure 6. The present author's reduction of the data by Kuo et al. (1990, 1992) showing the length-tension curve of pig coronary arterioles at equilibrium. (A) with blood pressure, without flow. (B) with flow which produces a shear stress of τ on wall. In Fung (1996), in press.

Basic Concepts and Applications of Structural Control

T.T. Soong, ASCE Member[1]

ABSTRACT

In recent years, new and innovative concepts of structural protection have been advanced and are at various stages of development. Modern structural protective systems can be divided into three major groups: seismic isolation, passive energy dissipation, and active control. Passive energy dissipation systems encompass a range of materials and devices for enhancing damping, stiffness and strength, and can be used both for natural hazard mitigation and for rehabilitation of aging or deficient structures. In recent years, serious efforts have been undertaken to develop the concept of energy dissipation, or supplemental damping, into a workable technology, and a number of these devices have been installed in structures throughout the world. Active structural control is an area of structural protection in which the motion of a structure is controlled or modified by means of the action of a control system through some external energy supply, which can be considered as part of the natural evolution of passive control technologies. Considerable attention has also been paid to active structural control research in recent years, with particular emphasis on the alleviation of wind and seismic response. The technology is now at the stage where actual systems have been designed, fabricated and installed in full-scale structures.

The focus of this paper is on active control. It presents an overview of some of the basic active control concepts as applied to civil engineering structures, and provides examples of current full-scale application of the technology. Included in this presentation are some unique requirements and challenges of civil engineering structural control, and some important design considerations related to implementation of this technology to actual structures. An attempt is made to identify areas of future research needs.

1. INTRODUCTION AND BASIC CONCEPTS

The possible use of active control systems and some combinations of passive and active systems, so called hybrid systems, as a means of structural protection against wind and seismic loads has received considerable attention in recent years. Active/hybrid control systems are force delivery devices integrated with real-time processing

[1]Samuel P. Capen Professor, Department of Civil Engineering, State University of New York at Buffalo, Buffalo, New York

evaluators/controllers and sensors within the structure. They react simultaneously with the hazardous excitation to provide enhanced structural behavior for improved service and safety. Remarkable progress has been made over the last twenty years (Soong and Reinhorn, 1993; Soong and Constantinou, 1994). The First World Conference on Structural Control (Housner et al., 1994b) held in Pasadena, California in August 1994, attracted over 300 participants from 35 countries and demonstrated the world-wide interest in structural control. As will be discussed in the following sections, research to date has reached the stage where some active systems have been installed in full-scale structures. Several other active systems have also been used temporarily in construction of bridges or large span structures (e.g., lifelines, roofs) where no other means could provide adequate protection.

An active structural control system has the basic configuration as shown schematically in Fig. 1. It consists of (a) sensors located about the structure to measure either external excitations, or structural response variables, or both; (b) devices to process the measured information and to compute necessary control force needed based on a given control algorithm; and (c) actuators, usually powered by external sources, to produce the required forces.

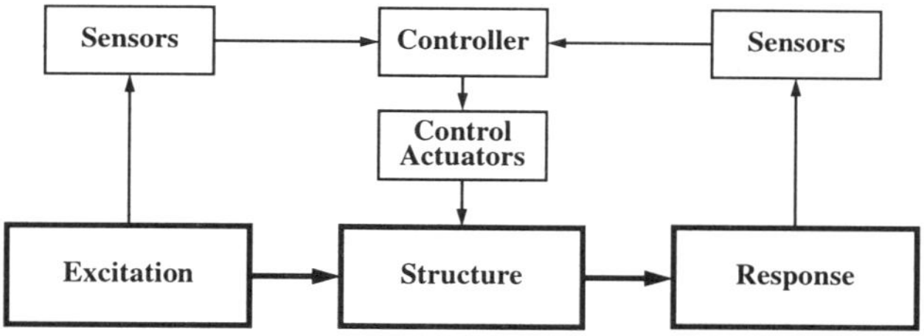

Fig. 1. Block Diagram of Active Structural Control

When only the structural response variables are measured, the control configuration is referred to as feedback control since the structural response is continually monitored and this information is used to make continual corrections to the applied control forces. A feedforward control results when the control forces are regulated only by the measured excitation, which can be achieved, for earthquake inputs, by measuring accelerations at the structural base. In the case where the information on both the response quantities and excitation are utilized for control design, the term feedback-feedforward control is used.

To see the effect of applying such control forces to a linear structure under ideal conditions, consider a building structure modeled by an n-degree-of-freedom lumped mass-spring-dashpot system. The matrix equation of motion of the structural system can be written as

$$\mathbf{M}\ddot{\mathbf{x}}(t) + \mathbf{C}\dot{\mathbf{x}}(t) + \mathbf{K}\mathbf{x}(t) = \mathbf{D}\mathbf{u}(t) + \mathbf{E}\mathbf{f}(t) \tag{1}$$

where $\mathbf{M}$, $\mathbf{C}$ and $\mathbf{K}$ are the $n \times n$ mass, damping and stiffness matrices, respectively, $\mathbf{x}(t)$ is the n-dimensional displacement vector, the m-vector $\mathbf{f}(t)$ represents the applied load or external excitation, and r-vector $\mathbf{u}(t)$ is the applied control force vector. The $n \times r$ matrix $\mathbf{D}$ and the $n \times m$ matrix $\mathbf{E}$ define the locations of the control force vector and the excitation, respectively

Suppose that the feedback-feedforward configuration is used in which the control force $\mathbf{u}(t)$ is designed to be a linear function of measured displacement vector $\mathbf{x}(t)$, velocity vector $\dot{\mathbf{x}}(t)$ and excitation $\mathbf{f}(t)$. The control force vector takes the form

$$\mathbf{u}(t) = \mathbf{G}_x\mathbf{x}(t) + \mathbf{G}_{\dot{x}}\dot{\mathbf{x}}(t) + \mathbf{G}_f\mathbf{f}(t) \tag{2}$$

in which $\mathbf{G}_x$, $\mathbf{G}_{\dot{x}}$ and $\mathbf{G}_f$ are respective control gains which can be time-dependent.

The substitution of Eq. (2) into Eq. (1) yields

$$\mathbf{M}\ddot{\mathbf{x}}(t) + (\mathbf{C} - \mathbf{D}\mathbf{G}_{\dot{x}})\dot{\mathbf{x}}(t) + (\mathbf{K} - \mathbf{D}\mathbf{G}_x)\mathbf{x}(t) = (\mathbf{E} + \mathbf{D}\mathbf{G}_f)\mathbf{f}(t) \tag{3}$$

Comparing Eq. (3) with Eq. (1) in the absence of control, it is seen that the effect of feedback control is to modify the structural parameters (stiffness and damping) so that it can respond more favorably to the external excitation. The effect of the feedforward component is a modification (reduction or total elimination) of the excitation. The choice of the control gain matrices $\mathbf{G}_x$, $\mathbf{G}_{\dot{x}}$ and $\mathbf{G}_f$ depends on the control algorithm selected.

In comparison with passive control systems, a number of advantages associated with active control systems can be cited; among them are (a) enhanced effectiveness in motion control. The degree of effectiveness is, by and large, only limited by the capacity of the control systems; (b) relative insensitivity to site conditions and ground motion; (c) applicability to multi-hazard mitigation situations. An active system can be used, for example, for motion control against both strong wind and earthquakes; and (d) selectivity of control objectives. One may emphasize, for example, human comfort over other aspects of structural motion.

While this description is conceptually in the domain of familiar optimal control theory used in electrical engineering, mechanical engineering, and aerospace engineering, structural control for civil engineering applications has a number of distinctive features, largely due to implementability issues, that set it apart from the general field of feedback control. In particular, when addressing civil engineering structures, there is considerable uncertainty, including nonlinearity, associated with both physical properties and disturbances such as earthquakes and wind, the scale of the forces involved is quite large, there are a limited number of sensors and actuators, the dynamics of the actuators can be quite complex, and the systems must be fail-safe (Soong, 1990; Housner et al., 1994a; Kobori, 1994; Dyke et al., 1994; Dyke et al., 1995).

2. FULL-SCALE APPLICATIONS

As alluded to earlier, the development of active or hybrid control systems has reached the stage of full-scale applications to actual structures. Table 1 lists these installations in building structures in Japan, the only country in which these applications have taken place. In addition, 14 bridge towers have employed active systems during erection (Fujino, 1994). Most of these full scale systems have been subjected to actual wind forces and ground motions and their observed performances provide invaluable information in terms of (i) validating analytical and simulation procedures used to predict actual system performance, (ii) verifying complex electronic-digital-servohydraulic systems under actual loading conditons, and (iii) verifying capability of these systems to operate or shutdown under prescribed conditions.

Table 1. Full Scale Implementations of Active Structural Control

Building	Year Completed	Building Use	Number of Stories	Type of Vibration Control Device
Kyobashi Seiwa	1989	Office	11	Active Mass Damper
Takenaka Corp. Exp. Bldg.	1989	Experimental	6	Active Mass Damper Active Bracing
Kajima Research Lab. No. 21	1990	Office	3	Active Variable Stiffness
Shimizu Tech. Laboratory	1991	Laboratory	7	Active Mass Damper
Toda Corp. Exp. Bldg.	1991	Experimental	6	Hybrid Mass Damper
Okumura Corp. Exp. Building	1991	Experimental	6	Hybrid Mass Damper
Sendagaya INTES Bldg.	1992	Office	11	Hybrid Mass Damper
Elevator Tech. Laboratory	1992	Laboratory	(60 m)	Active Gyro-Stabilizer
Hankyu Chayamachi Bldg.	1992	Hotel & Office	34	Hybrid Mass Damper
Kansai International Airport	1992	Control Tower	(88 m)	Hybrid Mass Damper
Land Mark Tower	1993	Hotel & Office	70	Hybrid Mass Damper
Osaka Resort City 200	1993	Hotel & Office	50	Hybrid Mass Damper
Long Term Credit Bank	1993	Office	21	Hybrid Mass Damper
Ando Nishikicho Bldg.	1993	Office	14	Hybrid Mass Damper
NTT Kuredo Motomachi Bldg.	1993	Office & Hotel	35	Hybrid Mass Damper
Penta-Ocean Exp. Bldg.	1994	Experimental	6	Hybrid Mass Damper
Shinjuku Park Tower	1994	Hotel & Office	52	Hybrid Mass Damper
Dowa Fire & Marine Ins.	1994	Office	29	Hybrid Mass Damper
Porte Kanazawa	1994	Office & Hotel	30	Active Mass Damper
Mitsubishi Heavy Ind. - Yokohama Bldg.	1994	Office	34	Hybrid Mass Damper
Hamamatsu ACT Tower	1994	Office & Hotel	(212 m)	Hybrid Mass Damper
Riverside Sumida	1994	Office	33	Active Mass Damper
J. City	1994	Office	24	Hybrid Mass Damper
Hotel Ocean 45	1994	Hotel	43	Hybrid Mass Damper
Osaka WTC Bldg.	1994	Office	52	Hybrid Mass Damper

Described briefly below are four of these systems and their observed performance, two of which were installed in a full-scale dedicated test structure and two others were installed in actual buildings. The performance of these four systems under recent wind and earthquake episodes is summarized in this section. Also addressed are several practical issues in connection with actual structural applications of these systems.

2.1 Full-scale Test Structure

A dedicated full-scale test structure was erected for performance verification of an active mass damper and an active bracing system under actual seismic ground motions. Located in Tokyo, Japan, the structure is a symmetric two-bay six-story building as shown in Fig. 2. It was constructed of rigidly connected steel frames of rectangular tube columns and W-shaped beams with reinforced concrete slabs at each of the floors. Having rectangular columns, the two orthogonal directions are not structurally identical. Weighing 600 metric tons, the structure was designed as a relatively flexible structure with a fundamental period of 1.1 sec in the strong direction and 1.5 sec in the weak direction in order to simulate a typical high-rise building. The structure was constructed without claddings except for the top story (sixth floor), which housed the active mass damper. Due to lack of cladding and simple connections, the structure has very low damping in the dominant modes (between 0.5% and 1% of critical).

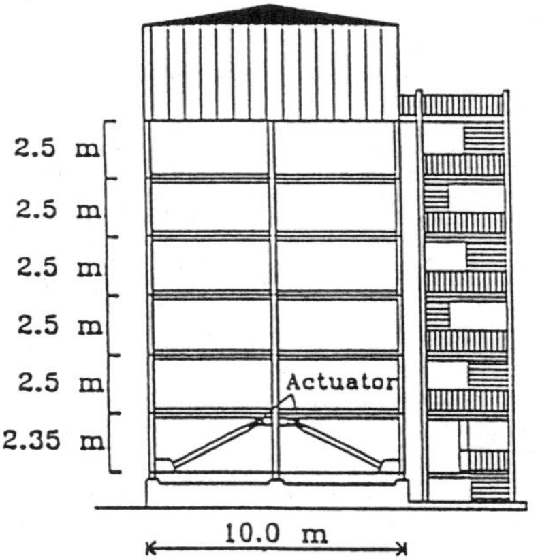

Fig. 2. Full-scale Structure

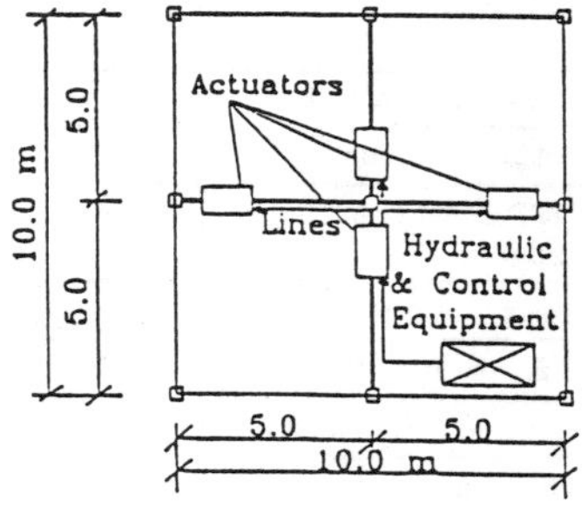

Fig. 3. Top View of ABS

Active Bracing System (ABS). As shown in Figs. 2 and 3, the active bracing system consists of solid diagonal tube braces attached at the first story of the building. The control system enables longitudinal expansion and contraction of the braces by means of hydraulic servocontrolled actuators, inserted between the brace elements and forming an integral part of the bracing system. The control system also includes a hydraulic power supply, an analog and digital controller, and analog sensors. Velocity sensors were located on the ground, at the first, third, and sixth floors of the building, which could also provide acceleration information. Additional transducers were mounted at each floor to monitor building behavior. Each actuator was equipped with a displacement transducer (LVDT) which was used to adjust the length of the brace via the servovalve loop. A more detailed description of the active bracing system can be found in (Soong et al., 1991 and Reinhorn et al., 1993).

Several important issues had to be addressed in order to ensure safe and efficient operation of the system; one of which deals with automated control operation. The hydraulic power for the active bracing system needs to be continuously available, yet it is not practical to have the hydraulic system operating constantly. For this reason, the system had to be designed in such a way that the hydraulic system remained in a ready, but dormant state, with the control software capable of bringing the system to full operation. To accomplish this, the control software had to be able to monitor the status of the control hardware, and adjust the state as necessary. In addition, the hydraulic system had to be capable of almost instantly supplying full power to the active braces.

These requirements led to special modifications of the standard control hardware as well as the addition to the control program of subroutines with the logic for starting, stopping, and monitoring the status of the system. A simplified block diagram of the control system hardware is shown in Fig. 4. The hydraulic power required for rapid starts is stored in the accumulators, which can supply enough power to allow the hydraulic pump to reach full pressure operation. The accumulators can also drive the actuators for approximately one minute, longer than most major earthquakes, in the event of a power failure. The ability of the computer to regulate the system is provided through a series of digital connections between the computer and the analog controller. The digital communications allow the computer to control the hydraulic system, and to monitor the status of the system hardware.

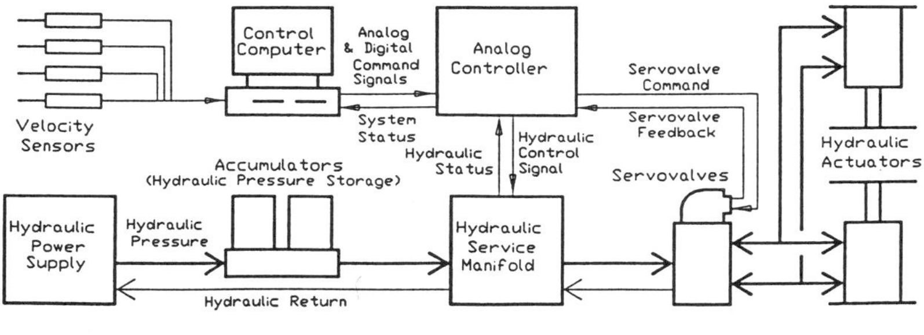

Fig. 4. Block Diagram of ABS

Another important issue is system reliability. In order to properly protect the system and the structure from damage in the event of a full or partial failure of the control system, a number of fail-safes were added to both the hardware and the software. The hardware fail-safes which were installed, will stop the system in a variety of potentially dangerous situations. The control software uses a series of checks and fail-safes to continuously verify its own integrity and the status of most of the system hardware. In addition, the control software contains test routines which can be manually activated in order to verify the efficiency and performance of the entire control system.

The long term maintenance of the system is also an important consideration. If strict tolerances are not met, the continuous wear can lead to degradation in the system performance, and even to failure. The standard maintenance must include manual inspection and verification of the system components on a regular basis.

The structure was subjected to several earthquake motions and its response was recorded while the ABS system was automatically activated during each of these episodes. The ground motion was simultaneously recorded and was used with an analytical model to estimate the probable response of the structure in the uncontrolled mode. The responses at several locations of the structure during two recent earthquakes show that the ABS was able to produce a somewhat uniform reduction of modal responses as indicated in the transfer functions shown in Fig. 5. The controlled response has lower peaks and wider distributions around the peak, indicating damping increase.

Active Mass Damper (AMD). The AMD was housed in the enclosure on the sixth floor. An outline of this device is shown in Fig. 6 which consisted of a suspended 6-ton mass capable of responding instantaneously to structural vibrations induced by wind or earthquakes. A complete description of the AMD system and some observed results are given by Aizawa et al. (1990). The structural response could be controlled in two directions by two electrohydraulic servoactuators, which were installed along orthogonal directions.

While both the ABS and AMD were installed in the same structure, structural control was performed by only one at a time. The control effect of the AMD system was examined during the Izu-Oshima earthquake on October 14, 1989 and on February 20, 1990. An example of the structural response in the frequency domain is shown in Fig. 7, where the transfer function recorded on October 10, 1989, is used as an uncontrolled reference.

Table 2. Comparison of AMD and ABS

	AMD	ABS
RMS Reduction of Peak Response:		
- Displacement	56%	29%
- Acceleration	22%	26%
Maximum Actuator Movement (cm)	10.51	0.194

Summarizing the observation results for both control systems, a rough root-mean-square performance comparison of AMD and ABS is shown in Table 2. Note that the AMD system has a better reduction of displacement response and a lesser reduction of accelerations. This implies a more comfortable response using the ABS and lesser base shear using the AMD. The maximum actuator movement is an indicator of the required input energy. It is large for the AMD system and is quite small for the ABS, indicating substantially less energy needed by the ABS. Comparing the response transfer functions shown in Figs. 5 and 7, it is noted that the AMD reduces the structural response primarily in the lower modes while the ABS system reduces

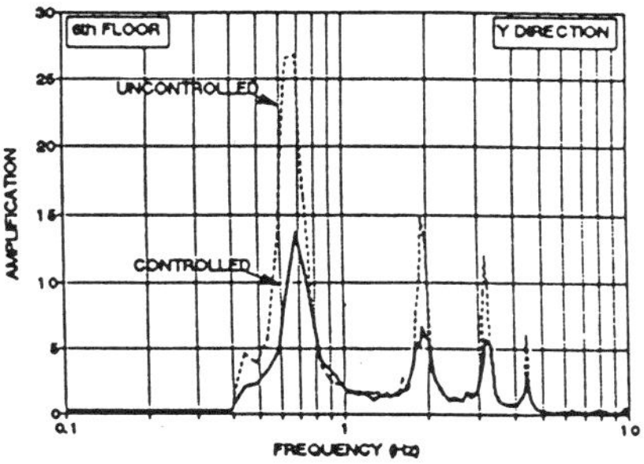

Fig. 5. Acceleration Transfer Function at Sixth Floor

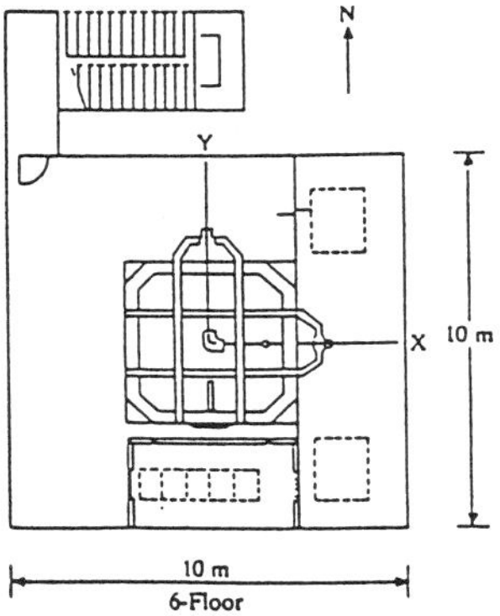

Fig. 6. Top View of AMD

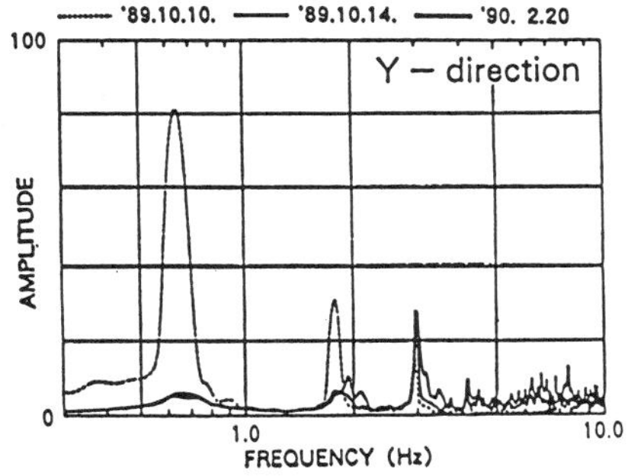

Fig. 7. Acceleration Transfer Function at Sixth Floor

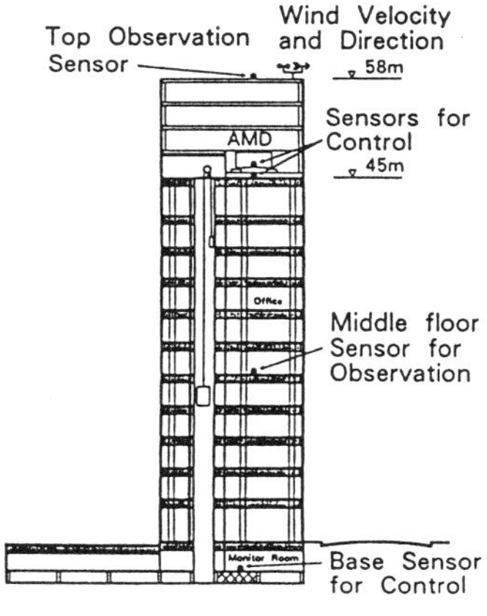

Fig. 8. Sandagaya INTES

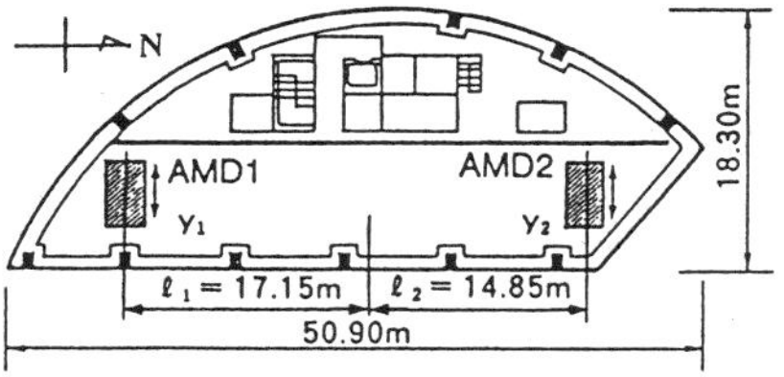

Fig. 9. Top View of AMD

the structural response in all modes. The ABS thus performs better in redistributing earthquake energy in higher modes and suppress their influence.

2.2 Building Applications

Sendagaya INTES Building. An active tuned mass damper system was installed in the Sendagaya INTES building in Tokyo in 1991. As shown in Fig. 8, the AMD was installed atop the 11th floor and consists of two masses to control transverse and torsional motions of the structure while hydraulic actuators provide the active control capabilities. The top view of the control system is shown in Fig. 9 where, as shown in Fig. 10, ice thermal storage tanks are used as mass blocks so that no extra mass is introduced. The masses are supported by multi-stage rubber bearings intended for reducing the control energy consumed in the AMD and for insuring smooth mass movements (Higashino and Aizawa, 1993).

Sufficient data were obtained for evaluation of the AMD performance when the building was subjected to strong wind on March 29, 1993, with peak instantaneous wind speed of 30.6 m/sec. An example of the recorded time histories is shown in Fig. 11, giving both the uncontrolled and controlled states. Their Fourier spectra using samples of 30-second durations are shown in Fig. 12, again showing good performance in the low frequency range but not in the region exceeding 3 Hz. The response at the fundamental mode was reduced by 18% and 28% for translation and torsion, respectively. Similar performance characteristics were observed during a series of earthquakes recorded between May, 1992 and February, 1993.

Hankyu Chayamachi Building. The 160-meter 34-story Hankyu Chayamachi building, as shown in Fig. 13, is located in Osaka, Japan, where an AMD system was installed in 1992 for the primary purpose of occupant comfort control. In this case, the heliport at the roof top is utilized as the moving mass of the AMD, which weighs 480 tons and is about 3.5% of the weight of the tower portion. The layout of AMD devices is shown in Fig. 14. The heliport is supported by six multi-stage rubber bearings. The natural period of rubber and heliport system was set to 3.6 seconds, slightly lower than that of the building (3.8 seconds). The AMD mechanism used here has the same architecture as that of Sendagaya INTES, namely, scheme of the digital controller, servomechanism and the hydraulic design, except that two actuators of 5-ton thrusts are attached in horizontal orthogonal directions. Torsional control is not considered here.

Acceleration time histories during a recent typhoon are shown in Fig. 15 with corresponding Fourier spectra given in Fig. 16. Since the building in this case oscillated primarily in its fundamental mode, significant reductions in acceleration levels were observed.

3. RESEARCH NEEDS

An important observation to be made in the performance observation of control systems such as those described above is that efficient active control systems can be implemented with existing technology under practical constraints such as power requirements and stringent demand of reliability. Thus, significant strides have been made considering that serious implementational efforts began less than ten years ago.

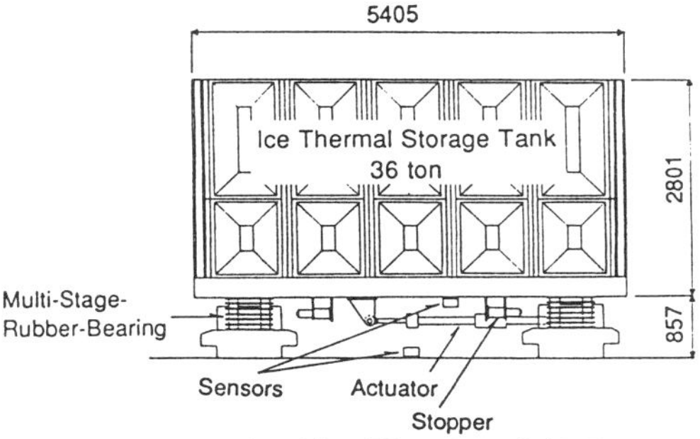

Fig. 10. Elevation of AMD

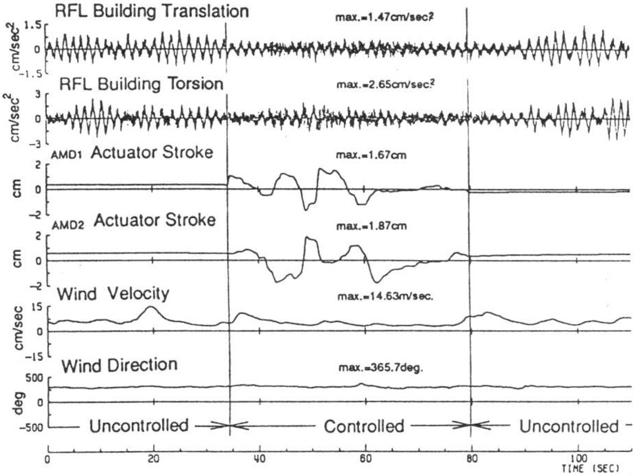

Fig. 11. Response Time Histories (March 29, 1993)

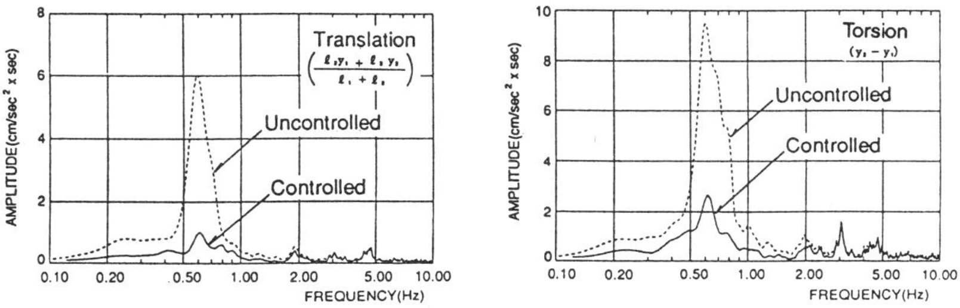

Fig. 12. Response Fourier Spectra (March 29, 1993)

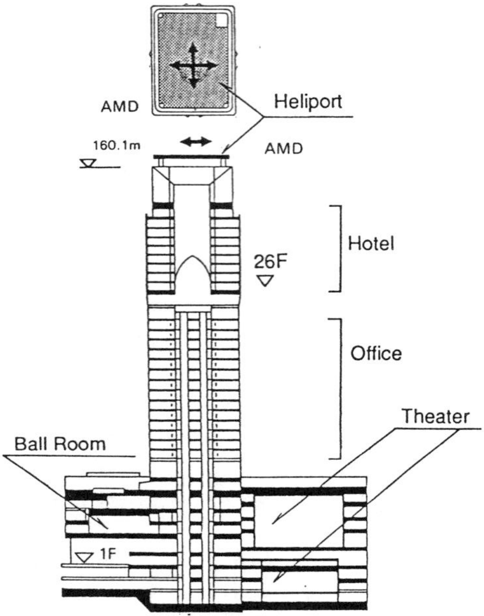

Fig. 13. Hankyu Chayamachi Building

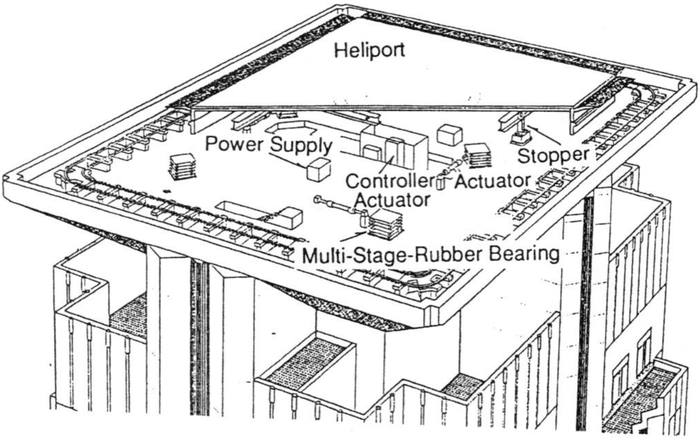

Fig. 14. Layout of AMD

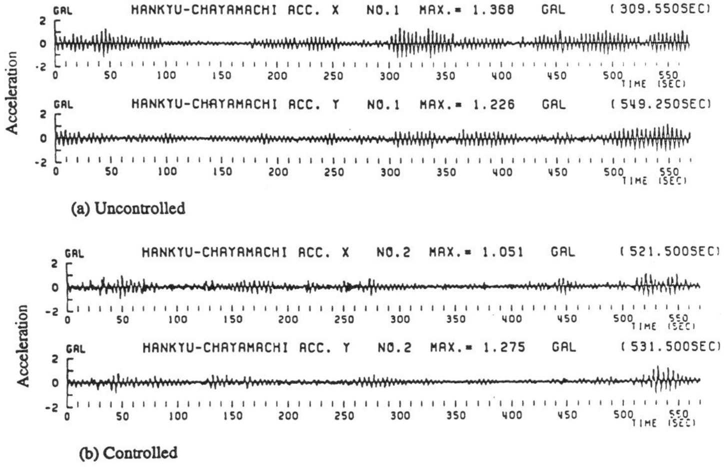

Fig. 15. Acceleration Time Histories

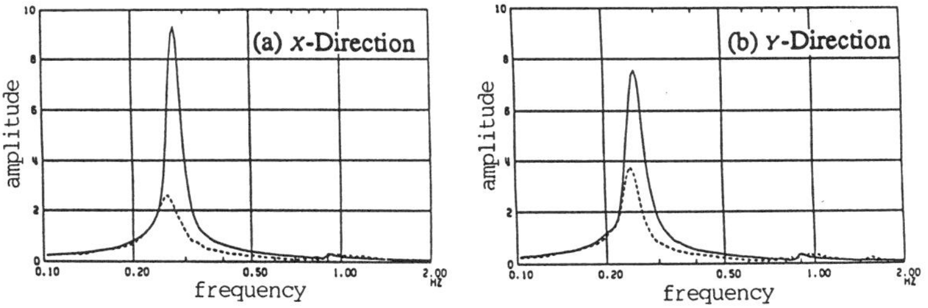

Fig. 16. Acceleration Fourier Spectra

On the other hand, the active mass dampers developed for Sendagaya INTES and Hankyu Chayamachi buildings, as for almost all the buildings listed in Table 1, were designed primarily for response control due to wind and moderate earthquakes. In order to get to the next level in active/hybrid control technology, an outstanding issue that needs to be addressed is whether such systems, with limited control resources and practical constraints such as mass excursions, can be made effective under strong earthquakes. Structural systems are complex combinations of individual structural components. New innovative devices and materials need to be integrated into these complex systems, with realistic evaluation of their performance and impact on the structural system, as well as verify their ability of long-term operation. Some issues of system integration that need more attention are: structure-controller interaction and performance, particularly in highly nonlinear-inelastic structures as used in civil engineering; control strategy and system optimization; integrated control-system design; integrated fail-safe operations; monitoring, self-identification, diagnostics and evaluation; operational issues - hardware/structure/software integration; integrated safety, reliability and maintenance; cost benefit analysis for integrated system on life cycle basis. Effectively addressing integration issues requires a combined experimental, analytical and economical study. Additionally, innovative ideas of devices require exploration through experimentation and adequate basic modeling. A series of standardized benchmark structural models, representing large buildings, bridges, towers, lifelines, etc., with standardized realistic/scaled-down excitations, representing natural hazards, will provide an experimental and analytical test bed for proof-of-concept of existing and new devices.

Although many difficulties still exist in wider application of active control to civil engineering structures, the future appears quite bright. For example, semi-active control strategies appear to be promising in addressing a number of the challenges indicated previously. Such devices typically have low power requirements, which is particularly critical during seismic events when the main power source to the structure may fail. Various semi-active devices have been proposed which utilize forces generated by surface friction or viscous viscoelastic-plastic fluids to dissipate vibratory energy in a structural system. Examples of such devices that have been considered for civil engineering applications include: variable orifice dampers (e.g., Shinozuka et al., 1992; Kawashima et al., 1992; Constantinou and Symans, 1994; Sack et al., 1994; Kurata et al., 1994), friction controllable braces (e.g., Akbay and Aktan, 1991; Dowell and Cheery, 1994), friction controllable isolators (e.g., Feng and Shinozuka, 1990, Kawashima et al., 1992), variable stiffness devices (Kobori et al., 1993; Kobori, 1994), and controllable fluid dampers (e.g., Ehrgott and Masri, 1992,1994; Gavin et al., 1994).

ACKNOWLEDGEMENT

The author wishes to acknowledge the generous support for this work received from the National Center for Earthquake Engineering Research, Buffalo, NY, the Takenaka Corporation of Japan, Kayaba Industries of Japan and MTS Systems Corporation of Minneapolis, MN.

REFERENCES

Akbay, Z. and Aktan, H.M. (1991), "Actively Regulated Friction Slip Devices." *Proc. 6th Canadian Conf. on Earthquake Engrg.*, 367-374.

Aizawa, S., Hayamizu, Y., Higashino, M., Soga, Y. and Yamamoto, M. (1990). "Experimental Study of Dual Axis Active Mass Damper." G.W. Housner and S.F. Masri (eds), *Structural Research*, 68-73, Los Angeles, USC.

Constantinou, M.C. and Symans, M.D. (1994). "Semi-Active Fluid Viscous Dampers for Seismic Response Control." *Proc. 1st World Conf. on Struct. Control*, Pasadena, CA.

Dowell, D.J. and Cherry, S. (1994). "Semi-active Friction Dampers for Seismic Response Control of Structures." *Proc. 5th U.S. Nat. Conf. on Earthquake Engrg.*, 1, 819-828.

Dyke, S.J., Spencer Jr., B.F., Quast, P. and Sain, M.K., Kaspari Jr., D.C. and Soong, T.T. (1994). "Experimental Verification of Acceleration Feedback Control Strategies for an Active Tendon System," *NCEER-94-0024*, National Center for Earthquake Engineering Research, Buffalo, NY.

Dyke, S.J., Spencer Jr., B.F., Quast, P. and Sain, M.K. (1995). "The Role of Control-structure Interaction in Protective System Design." *J. Eng. Mech., ASCE*, 121(2), 322-338.

Ehrgott, R.C. and Masri, S.F. (1994). "Structural Control Applications of an Electrorheological Device." *Proc. Int. Workshop on Struct. Control*, USC Pub. No. CE-9311, 115-129.

Feng, Q. and Shinozuka, M. (1990). "Use of a Variable Damper for Hybrid Control of Bridge Response under Earthquake." *Proc. U.S. Nat. Workshop on Struct. Control Research*, USC Publication No. CE-9013, 107-112.

Fujino, Y. (1994). "Recent Research and Developments on Control of Bridges under Wind and Traffic Excitations in Japan." *Proc. Int. Workshop on Struct. Control*, 144-150.

Gavi, H.P., Hose, Y.D., and Hanson, R.D. (1994). "Design and Control of Electrorheological Dampers." *Proc. 1st World Conf. on Struct. Control*, WP3, 83-92.

Higashino, M. and Aizawa, S. (1993). "Application of Active Mass Damper System in Actual Buildings." G.W. Housner and S.F. Masri (eds), *Proc. Int. Workshop on Structural Control*, 194-205, Los Angeles, USC.

Housner, G., Soong, T.T. and Masri, S.F. (1994a). "Second Generation on Active Structural Control in Civil Engineering." *Proc. 1st World Conf. on Struct. Control*, FA2, 3-18.

Housner, G.W., Masri, S.F. and Chassiakos, A.G. (eds) (1994b). *Proc. 1st World Conf. on Structural Control*, Pasadena, CA, August 3-5.

Kawashima, K., Unjoh, S. and Shimizu, K. (1992). "Experiments on Dynamic Characteristics of Variable Damper." *Proc. of the Japan Nat. Symp. on Struct. Resp. Control*, Tokyo, Japan, 121.

Kobori, T., Takahashi, M., Nasu, T., Niwa, N. and Ogasawara, K. (1993). "Seismic Response Controlled Structure with Active Variable Stiffness System." *Earthquake Engrg. and Struct. Dyn.*, 22, 925-941.

Kobori, T. (1994). "Future Direction on Research and Development of Seismic-Response-Controlled Structure." *Proc. 1st world Conf. on Struct. Control*, Panel, 19-31.

Kurata, N., Kobori, T., Takahashi, M., Niwa, N. and Kurino, H. (1994). "Shaking Table Experiments of Active Variable Damping System." *Proceedings of the 1st World Conf. on Struct. Control*, Pasadena, CA, August 3-5, TP2, 108-117.

Reinhorn, A.M., Soong, T.T., Riley, M.A., Lin, R.C., Aizawa, S. and Higashino, M. (1993). "Full Scale Implementation of Active Control, II." *ASCE, J. Struct. Engrg.*, 119, 1935-1960.

Sack, R.L., Kuo, C.C., Wu, H.C., Liu, L. and Patten, W.N. (1994). "Seismic Motion Control via Semiactive Hydraulic Actuators." *Proc. U.S. 5th Nat. Conf. on Earthquake Engrg.*, Chicago, IL, 2, 311-320.

Shinozuka, M., Constantinou, M.C. and Ghanem, R. (1992). "Passive and Active Fluid Dampers in Structural Applications," *Proc. US/China/Japan Workshop on Struct. Control*, Shanghai, China, 507-516.

Soong, T.T. (1990). *Active Structural Control: Theory and Practice*, Longman, England and Wiley, New York, NY.

Soong, T.T. and Constantinou, M.C. (eds) (1994). *Passive and Active Structural Vibration Control in Civil Engineering*, Springer-Verlag, Wien and New York.

Soong, T.T. and Reinhorn, A.M. (1993). "An Overview of Active and Hybrid Structural Control Research in the U.S." *J. Struct. Design Tall Bldg.*, 2, 193-209.

Soong, T.T., Reinhorn, A.M., Wang, Y.P. and Lin, R.C. (1991). "Full Scale Implementation of Active Control, I." *ASCE, J. Struct. Engrg.*, 117, 3516-3536.

Environmental Fluid Mechanics -- A Review of Some Recent Results

Robert L. Street[1], Member ASCE

Abstract

Hunt (1991) wrote "Environmental fluid mechanics (EFM) refers to fluid motions in the lower atmosphere, in the ground, and in rivers, lakes, and seas that relate in some way to problems connected to human activities within and concern about the environment." Such a scope is too broad for a brief review, so we shall stay above ground, but that shall not limit us too much! The range of EFM includes focus on individual domains such as the near-coastal ocean, lakes, and planetary boundary layers, but it also includes integrative efforts as well. One example of a formal thrust along these lines is the NSF Water and Energy: Atmospheric, Vegetative, and Earth Interactions (WEAVE) program. Its thrust is investigation of the movement of energy and water in the climate defined by the biological, earth, and atmospheric components. In many ways, the WEAVE program's overall objective mirrors the objective of most environmental fluid mechanics research and applications, viz., to gain understanding of the complex interactions and feedbacks among these domains.

To this end this talk examines a number of research studies and applications of EFM. To help focus our time I have constrained my review to the region of California. As well, a majority of the subject matter is based on the efforts in our Environmental Fluid Mechanics Laboratory. Nevertheless we deal with field studies, laboratory experiments, and numerical simulations. Often we shall see a productive

[1] Professor of Fluid Mechanics and Applied Mathematics, Environmental Fluid Mechanics Laboratory, Department of Civil Engineering, Stanford University, Stanford, CA 94305-4020.

interplay between physical (laboratory and field) experiments and numerical experiments (simulations). A sampler of our topics includes:

Atmosphere:

- an Eulerian model for simulation of air pollution in urban and regional areas. The model includes treatment of chemistry, radiation, aerosols, and meteorology. Model results are tested against field data for the Los Angeles basin and some interesting conclusions drawn.
- large-eddy simulation of the evolving stable atmospheric boundary layer over flat terrain, which is intimately connection to the transport and spreading of air pollution.
- an investigation of turbulent, stratified flow over complex terrain and its relationship to pollutant transport.
- a resolved large-eddy simulation of the fluid motion and transport in discrete buoyant plumes in a cross-flowing boundary layer -- an attack on the long-standing "plume-in-grid" problem of air quality modeling.

Estuary:

- the interaction of laboratory and field experiments with numerical modeling in the study of biological and fluid interactions: specifically, clams and phytoplankton in San Francisco Bay.
- modeling of pollution transport in San Francisco Bay, as well as recent field measurements and computations of estuarine tidal flows in San Francisco Bay.

Near-Coastal Ocean:

- transport and mixing in coastal upwelling, with state-of-the-art simulations of laboratory conditions and path-breaking laboratory experiments. This effort is focused on making breakthroughs in the interactive coupling of physical and numerical experiments leading to prediction of geophyical fluid dynamics flows.

Fundamental:

- measurements of turbulence near a water surface, which controls the mixing away from and toward the interface, will be presented to illustrate new experimental methods.
- work on the fundamentals of stratified flows which are an ubiquitous component of EFM flows.

While, as an academic and numerical modeler, I am tempted to focus on research and on whole system modeling, the sample of cases presented here includes examples of where a combination of differing levels of sophistication and type of study or application can yield needed results in EFM. As Hunt says, "In short, there are too many important fluid-flow problems for it to be physically possible to apply the latest expertise and techniques to solving even a minority of them."

Reference

J.C.R. Hunt (1991) "Industrial and environmental fluid mechanics," *Annual Review of Fluid Mechanics*, 23, pp. 1-41.

Engineering Design Considerations for Artificial Lungs

L.F. Mockros and K.E. Cook[1]

Artificial lungs have a variety of uses from the (i) short term (two hours or so), while the natural heart and lungs are bypassed during heart surgery, (ii) intermediate term (days, weeks, or months), during acute, but reversible, respiratory distress, and (iii) long term (months or years), such as a bridge to transplant. Three strategies for use of these devices are developing: the extracorporeal circuit, the intravascular lung assist device, and the intrathoracic artificial lung. Extracorporeal circuits are used for short and intermediate term applications. Intravascular devices are being developed for the intermediate term applications, and intrathoracic devices are intended for intermediate to long term applications.

Engineering design specifications for these devices may be divided into three broad categories: (1) adequate delivery of oxygen to the blood and concomitant removal of carbon dioxide from the blood, (2) limits on the blood trauma, and (3) interactions with the physiology of the natural circulation system. The mechanics of the flow through these devices are important in all three categories. The major device function is to meet required gas transfer rates for various blood flow rates with limits on pressure drop through the device. For both the short term and many intermediate term applications, only basal rates of gas transfer are required. For long term applications, however, the devices must be capable of two or more times basal requirements. Hemolysis was the blood trauma of concern early in the history of artificial organs. That is not of much concern today, but the activation of platelets has long been, and continues to be, a major problem. Of current concern is the more complex production of hormones, autacoids, and cytokines when such devices are used. Interactions between extracorporeal circuits and the physiology

---

[1]Department of Biomedical Engineering, Northwestern University, 2145 Sheridan Road, Evanston, IL 60208

of the natural circulation system is of some concern, but the interactions are much more complex with intravascular and intrathoracic devices. Reduced venous return, due to the blood flow obstruction caused by the device, can be of major concern with an intravascular gas exchanger in the vena cava. Reduced venous return results in reduced cardiac output and, therefore, reduced oxygen delivery to the tissues. The intrathoracic devices presents a different set of issues. Blood flow is provided by the reactive right heart and questions arise as to whether the artificial lung should operate in parallel or series with the natural lung. The essential non-respiratory physiological functions of the lungs are also issues. How much of the blood flow through natural lungs is essential for these functions?

A variety of available models can be used to design devices to meet specifications of gas transfer rates and limits on pressure drops. Quantitatively designing devices to control blood trauma has not been attempted. Most devices are intuitively designed to limit blood trauma based on past experience. The trauma is thought to result from blood contact with the prosthetic materials and the fluid mechanical stresses experienced by the blood as it passes through the device. Surface treatments have been the subject of much research for many years, but it may be necessary to coat the prosthetic materials with endothelial cells to make them truly compatible, at least for the long term. Much is known about relationships between shear stresses and such blood trauma as hemolysis and platelet stimulation, but this information has not been explicitly utilized in designing new devices. Little quantitative information is available relating the blood flow mechanics within the devices and the production of hormones, autacoids, and cytokines. Designing intrathoracic lungs raises additional design considerations. Because of the pulsatile nature of the blood flow input from the right heart, these devices must have a compliance to avoid overloading the right heart. They must have input mechanical impedances similar to the natural pulmonary circulation. These artificial lungs are designed for the gas transfer function. Natural lungs, however, have many other physiological functions that must be considered and the devices may have to function in conjunction with some remaining circulation of the natural lungs.

Biomechanics and Testing of Mechanical Circulatory Support Devices

Harvey Borovetz, James Antaki, William Wagner, Marina Kameneva, John Pristas, Stephen Winowich, William Mandarino, Philip Litwak, Robert Kormos and Bartley Griffith

Since 1985, the University of Pittsburgh (UOP) has maintained one of the world's most active clinical programs in mechanical circulatory support (MCS) as a bridge to cardiac transplantation. As of January 1, 1996 a total of 88 patients (corresponding to more than 4,120 patient days on device) have been "bridged" with either a Symbion /CardioWest total artificial heart (n=21), a Thoratec ventricular assist device (Thoratec Laboratories Corp., n=18), or a Novacor left ventricular assist system [LVAS] (Novacor Division, Baxter Healthcare Corp., n=49). Of the 88 implanted patients, 66 were transplanted (75%); of this number, 53 patients were discharged (80%). An even higher percentage of the Novacor LVAS patients who were transplanted (n=37) were ultimately discharged home (n=33, 89%).

In conjunction with the clinical use of MCS devices at UOP has been an active research program aimed at elucidating the effects of ventricular assist devices on native right and left ventricular function. A second very active effort is investigating the mechanisms underlying thromboembolic complications in patients with implanted MCS systems. The former effort involves an innovative load insensitive measure of ventricular function via pressure-area loops obtained using transesophageal echocardiography and high fidelity pressure manometry. Investigations of device-related thromboembolism have focused on the biocompatibility of the implanted, blood-contacting MCS components, biorheology of the patient's blood, and the micro-flow processes around the inflow/outflow pump valves as visualized using laser-based fluorescent and fiber optic imaging techniques.

Current clinical research efforts at UOP will introduce the use of MCS systems as an alternative to cardiac transplantation.

Artificial Heart & Lung Program, McGowan Center, Department of Surgery, University of Pittsburgh, Pittsburgh, PA, USA

CHANGES IN THE HOST TO BIOMATERIAL SURFACE-BLOOD CONTACT; ACTIVATION OF HOST COMPONENTS

B.G. Hattler, M.D., Ph.D., University of Pittsburgh, Pittsburgh, PA 15213

Cardiopulmonary bypass is known to induce changes in various host components, including the activation of the complement system, increased production and adherence of neutrophils to endothelial surfaces and enhanced detection of pro-inflammatory cytokines in the circulation. Many of these processes are set in motion by contact of the blood and its cellular components with biomaterial surfaces in the cardiopulmonary bypass circuit. Activation of the blood by biomaterial surface interactions has been most extensively studied in the setting of cardiopulmonary bypass. Over 20 different components of the blood, both cellular (i.e., platelets, white cells) and non-cellular (i.e., extrinsic and intrinsic protein regulators of the coagulation cascade), and a series of inflammatory mediators have all been shown to be activated, both in the experimental and clinical setting. We have been particularly interested in the inflammatory mediators Tumor Necrosis Factor Alpha (TNF-α) and Interleukin-6 (IL-6). These pro-inflammatory cytokines have been shown in our laboratory to depress myocardial function in a reversible manner and to be significantly activated during cardiopulmonary bypass [1,2,3]. In a study of over 20 patients undergoing cardiopulmonary bypass, both the expression of message (mRNA) for these two cytokines (i.e., TNF-α, IL-6) and the production of protein (i.e., cytokines detected in the plasma) were evaluated. Message for both cytokines was enhanced 3 to 4 fold in the post-cardiopulmonary bypass interval in all patients studied and free cytokine was elevated 6 to 10 fold in the post-bypass interval [4]. Sequential studies after the return of the patient to the intensive care unit demonstrated the gradual clearance of these two targeted cytokines so that at 48 hours they had returned to baseline levels. Selective area-volume loops performed on these patients by transesophageal echocardiography before and after surgery has revealed that the elevation in cytokine profiles corresponds to a demonstrable depression in myocardial function and that the heart returns to normal baseline function at a time when the cytokines can no longer be detected. These findings are consistent with our in-vitro studies where these same cytokines have been shown to markedly depress isolated muscle strips tested in physiological baths. Once the cytokines are removed from the bath, the muscle returns to a normal functional state. Since the activation of white cells in the blood is critical to cytokine production we tested, in-vitro, white cells removed from the buffy coat of human blood and cultured in media within polyethylene tubes (Table 1). At two hours both message (mRNA for IL-6) and product (IL-6) can be detected in the white cells and media respectively. Similar activation is not demonstrable when white cells are cultured in standard tissue culture plates whose surfaces have been pacified by albumin coating (Table 1). It appears that biomaterials and their effect on the circulating blood play an important role in cytokine expression, synthesis and release. Once expressed, these cytokines have a negative inotropic effect on the myocardium which is actively under investigation.

TABLE 1

White Cell PCR IL-6 mRNA Analysis and IL-6 Production When Cultured In-Vitro on Polyethylene Tubing

	mRNA IL-6 (2 hours Culture; 10^8 WBC; n = 6)	[IL-6] (2 hours Culture; 10^8 WBC; n = 6)
Polyethylene Tubes	**30,133 ± 10,132 cpm**	**310 ± 38 pg/ml**
Culture Flasks	**6,341 ± 2,100 cpm**	**98 ± 28 pg/ml**

References

1. Finkel MS, Oddis CV, Jacob TD, Hattler BG: Negative inotropic effects of cytokines on the heart mediated by NO. Science 1992;257:387-389.
2. Hattler BG, Finkel MS: Interleukin-6, a myocardial depressant factor and its possible role in myocardial reperfusion injury in man. J Cell Biochem Suppl 1991;15F:145.
3. Finkel MS, Hoffman RA, Shen Li, Hattler BG: Interleukin-6 (IL-6) as a mediator of stunned myocardium. Am J. Cardiol 1993;71:1231-1232.
4. Hattler BG, Zeevi A, Oddis CV, Finkel MS: Cytokine induction during cardiac surgery: analysis of TNF-α expression pre and post cardiopulmonary bypass. J Card Surgery 1995; 10:418-422.

PLATELET ACTIVATION IN TIME VARYING SHEAR FLOW FIELD

C. CORNELIUS GLISMANN, Dipl.-Ing., NED H.C. HWANG, Ph. D[1]

Abstract

Samples of human whole blood (WB) were tested in time varying shear and in constant shear flow fields using a rotary viscometer, where the gap can be varied by shifting an outer stationary cylinder relative to an inner rotating cylinder. Platelet aggregation and activation was measured by flow cytometric analysis (FCM) using light scattering and a CD61-FITC and CD62-PE monoclonal antibody respectively. The aggregates were sorted by FCM technique and analyzed by light microscopy.

Introduction

There are three main factors that have profound effects on blood cell function: the material surface in contact with the blood, presence of chemical agonist, and fluid mechanical forces. The latter has been investigated with respect to its effect on blood platelets using constant shear flow in the absence of the other two sources (e.g., Dintenfass et al., 1963; Brown III, C. H. et al., 1975).

In the cardiovascular system (CV) platelet aggregation problems have been reported especially in relation to implanted artificial devices like mechanical heart valves (MHV), vascular grafts, blood pumps, etc. (Koppenstein et al., 1991; Brown, P. et al., 1989). The quantification and correlation of specific flow characteristics and the resulting platelet changes are of major importance in order to improve the design of such devices as well to estimate the risk factors for specific patient groups in their clinical applications.

In this investigation WB was subjected to time varying as well as a time invariant shear flow field in order to determine if there was a measurable difference between the two. The shear flow field was provided by a cylindrical viscometer with a stationary outer cylinder and a rotating inner cylinder which allowed center adjustment to provide either concentric or selected eccentric gap space between the two cylindrical walls. Flow cytometric analysis was used to

[1] Cardiovascular Fluids Laboratory, Department of Biomedical Engineering, University of Miami, Coral Gables, Florida, 33146, USA

evaluate the activation and aggregation of the platelets. The FITC-conjugated monoclonal antibody CD61 was used to discriminate platelets from the rest of the whole blood and measure the changes in aggregation. The PE-conjugated monoclonal antibody CD62 was used to measure the activation status of the platelets. The labeled aggregated platelets were then sorted in the cytometer and analyzed under a light microscope.

Materials and Methods

The principle of the tester is based on a simple cylindrical rotary viscometer where the inner cylinder rotates and the outer cylinder remains stationary. The experimental setup is shown in Figure 1. The transparent Polystyrene material of the test chamber allowed direct visualization of the flow field and laser Doppler anemometer (LDA) mapping using a clear analog fluid. Also the material prevented surface blood interactions due to its high bio-compatibility. In order to create a non uniform gap, the outer cylinder can be adjusted relative to the inner cylinder. The uniform gap width was 1 mm. The maximum shift was 0.8 mm which leads to a minimum gap of 0.2 mm and a maximum gap of 1.8 mm wide (Figure 2). As a result of thereon uniform gap a spatial stationary, but time varying flow field could be generated.

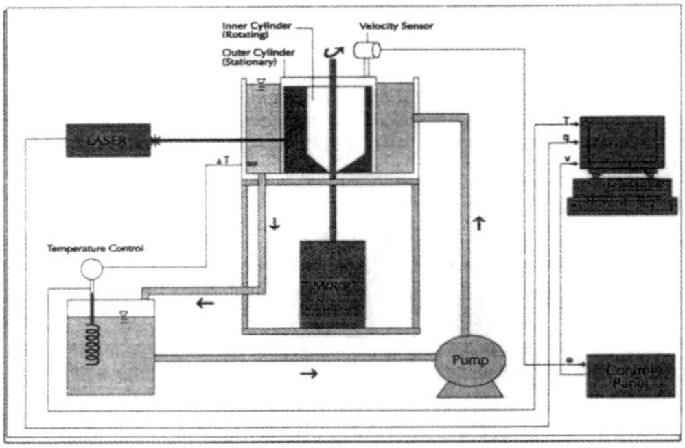

Fig. 1

The measurements were performed for two distinct positions of the tester. Case I, where the gap was uniform (no shift) and case II where the gap was eccentric and the shift was maximum (Figure 2). The shear stresses were in a range of 15 to 50 $dyne/cm^2$ which corresponded to an average shear in large blood vessels. The shear stress-strain relationship of a Newtonian fluid:

$$\tau = \mu * \gamma \quad (1)$$

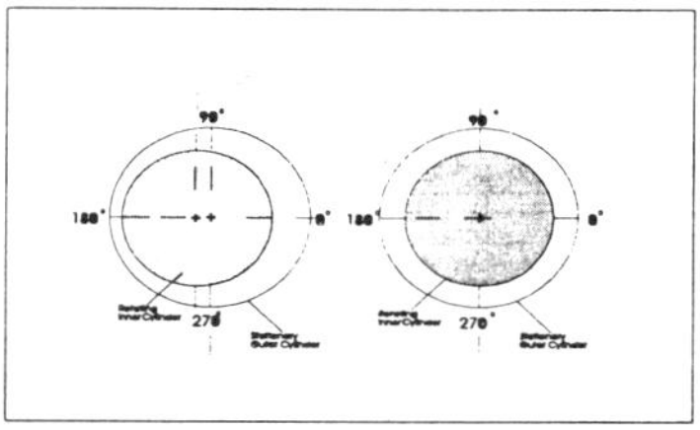

Fig. 2

with τ = shear stress [dyne/cm^2], γ = shear rate [s-1], and μ = viscosity [cp.].

The total volume in the tester was 15 ml. The 100 Ol samples taken for the FCM and microscopic analysis were centerline region of the flow field at different time steps starting from 5 sec up to 600 sec. All experiments were performed at 23°C. All flow cytometry (FCM) measurements were made on an XL-MCL and Elite (Coulter Corp., Miami, FL, USA). The sorted aggregates were monitored using a 100x magnification light microscope (Zeis, Wetzlar, Germany).

Results

The FCM results are illustrated in Figure 3. The first histogram represents the ungated data of the logarithm of forward scatter (FS) and CD61-FITC fluorescence. Region 1 of the first histogram encloses unlabeled platelet-sized and submicron particulate material. Platelets labeled submicron particles (SMP) are in region 2. Large and aggregated platelets are within region 3. In histogram (2:) the ordinate depicts CD61-FITC and the ordinate CD62-PE in logarithmic scale. Finally in histogram (3:) the ordinate represents FS and the abscissa side scatter or 90° scatter light (SS). SS represents the shape and granularity of platelets these are increases in positive axis direction. The population can be divided into three sub populations: single platelets (2), small, fragmented platelets and SMP (1) aggregated and large platelets (3).

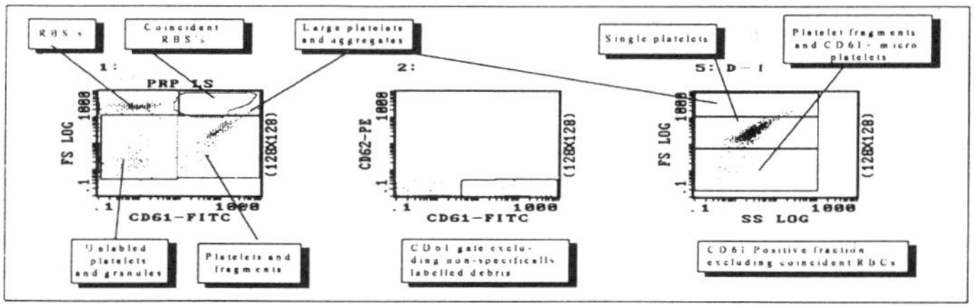

Fig. 3

The aggregated larger lumps of platelets (3) were gated to the sorter and analyzed under the light microscope. They could be determined as true aggregates for the different samples over the full range of 10 min. For large aggregated platelets under constant shear the data show a fairly constant result, while for the time varying case there is an increase of about 20 % of aggregates/large platelets throughout the experiment of 10 minutes. This increase is noticeable considering the low level of shear stress (15 to 50 dyne /cm^2).

Discussion

Large diversities in the threshold of shear induced platelet aggregation as reported by various investigators. This is due to the different method of handling the blood and the specific characteristics the testing devices. A controlled time-varying shear field case was chosen for this study. The time varying results were compared to the that from the constant shear flow.

The resulting 20 % increase in platelet aggregation for the time varying shear stress verse no aggregation in the constant case at this low shear stress level shows that the dynamics plays an important role in shear induced platelet activation. Reports from other investigators confirm that the constant shear stress at this low level does not lead to aggregation (Konstantopolous et al., 1995). The platelet activation results support the findings that there exist a significant increase in platelet activation due to the time varying shear stress.

References

BROWN, III, C. H., LEVERETT, B. L., LEWIS, C. W., ALFREY JR., C. P., and HELLUMS, D. (1975). Morphological, biochemical and functional changes in human platelets subjected to shear stress. *Blood*, 86, 3, 462-471.

BROWN, P., HARRISON, M. J. G. (1989). Changes in blood filterability and platelet aggregability in patients with aortic valve replacement. *Clin Haemorrheol.* 9, 139-147.

DINTENFASS, L., (1963). An application of a cone -in cone viscometer to the study of viscosity, thixotropy and clotting of blood. *Biorheology* 1, 91.

KOPPENSTEINER, R., MORITZ, A., SCHLICK, W., FENZL, G., ROEDLER, S., EHRINGER, H., WOLLNER, E. (1991). Blood rheology after cardiac valve replacement with mechanical prostheses or bioprotheses. *The Amer. Jour. of Cardiology* 67, 79-83.

KONSTANTOPOULOS, K., WU, K.K., UDDEN, M. M., BANEZ, E. I., SHATTIL, S. J. and HELLUMS, J. D. (1995). Flow cytometric studies of platelet responses to shear stress in whole blood. *J. of Biorheologly* 32, No. 1, 73-93.

Key Words

Platelet activation and aggregation, human whole blood, time varying shear flow, flow cytometry analysis, laser Doppler anemometer mapping , flow visualization.

Engineering a Novel Intravenous Oxygenator

William Federspiel[1], Frank Walters[2], Pat Sawzik[2], Gary Reeder[3], Harvey Borovetz[4], and Brack Hattler[5]

The University of Pittsburgh Intravenous Membrane Oxygenator (IMO) is undergoing further engineering development and characterization. The focus of these efforts is an IMO device which can supply up to one-half of basal O_2 consumption and CO_2 elimination rates while residing within the inferior and superior vena cava after peripheral venous insertion. The current IMO design consists of a bundle of hollow fiber membranes potted to manifolds at each end, with an intra-aortic type balloon integrally situated within the fiber bundle. Pulsation of the balloon using helium gas and a balloon pump console promotes fluid and fiber motion and enhances gas exchange. Our goal is to complete engineering development of the IMO during '96-'97 in preparation for formal preclinical animal testing of the device. In the past year over fifteen IMO prototypes have been fabricated and extensively bench-tested to characterize O_2 gas exchange capacity, balloon inflation/deflation over relevant frequency ranges, and the pneumatics of the sweep gas pathway through the device. The results of these performance tests will be presented. The testing has led to several engineering changes, including redesign of the helium and sweep gas pathways within the IMO device. As a result, the maximum rate of balloon pulsation has increased substantially above the previous 70 bpm, and the vacuum required for sufficient sweep gas flow has been reduced. The IMO prototypes have demonstrated an O_2 exchange capacity of up to 90 ml/min/m^2 in water, which appears within 70% of our design goal when extrapolated to scaled-up devices in blood.

The authors appreciate the assistance of the following individuals: J. White, T. Hewitt, M. Hout, and L. Lund. This study was supported by the U.S. Army Medical Research, Development, Acquisition, and Logistics Command under Contract No. DAMD17-94-C-4052. The views, opinions, and/or findings contained in this report are those of the authors and should not be construed as an official Dept. of the Army position, policy, or decision unless so designated by other documentation. The generous support of Medtronics Inc. and the McGowan foundation are also appreciated.

---

[1] Associate Professor, Department of Surgery and of Chemical Engineering, University of Pittsburgh, Pittsburgh, PA 15219

[2] Walters Scientific Instrument Labs, Pittsburgh, PA.

[3] Adjunct Research Assistant Professor, Department of Surgery, University of Pittsburgh, Pittsburgh, PA 15219

[4] Professor, Department of Surgery and of Civil Engineering, University of Pittsburgh, Pittsburgh, PA 15219

[5] Professor, Department of Surgery, University of Pittsburgh, Pittsburgh, PA 15219

Modelling of a Heater Experiment in a Saturated Granite with Thermoporoelastic Source Singularities

Emmanuel Detournay[1], M. ASCE and Ilya Berchenko

Abstract

The paper describes an effort to match the measured pore pressure response induced by a heater located inside a borehole, in an intact saturated granitic rock mass. The effect of the heater is modelled by a combination of heat and fluid sources, which are singular solutions of the linear theory of thermoporoelasticity. It is shown that a single parameter, γ, which lumps several thermoporoelastic constants, controls the magnitude of the pore pressure induced by thermal loading.

Introduction

As part of a multi-year effort to investigate the feasibility of establishing a repository in granitic rock for low-level nuclear waste, a heater experiment was conducted at the Underground Research Laboratory (URL) of Atomic Energy of Canada Limited. In this experiment, heat was produced at a constant rate (about 1000W) for a period of 20 months, from a heating element located inside an emplacement hole drilled in the floor of an underground gallery. Pore pressure changes in the vicinity of the emplacement hole were measured at several piezometers during the heater experiment; in particular, a peak pressure of about 600 kPa was recorded at a distance of approximately 2 m from the heater, 2 months after the heater was turned on, see Fig. 1.

This paper describes an effort to match the measured pore pressure response, by modelling the effect of the heater with a combination of heat and fluid sources. In this approach, the strength of the heat source corresponds to the heat rate produced by the heating element while the strength of the fluid source represents the amount of fluid exchanged between the emplacement hole and the rock mass.

Thermoporoelasticity

Our analysis of the URL heater experiment is carried out within the framework of the linear theory of thermoporoelasticity. The favored formulation of this theory (e.g.,

[1]Department of Civil Engineering, University of Minnesota, Minneapolis, MN 55455.

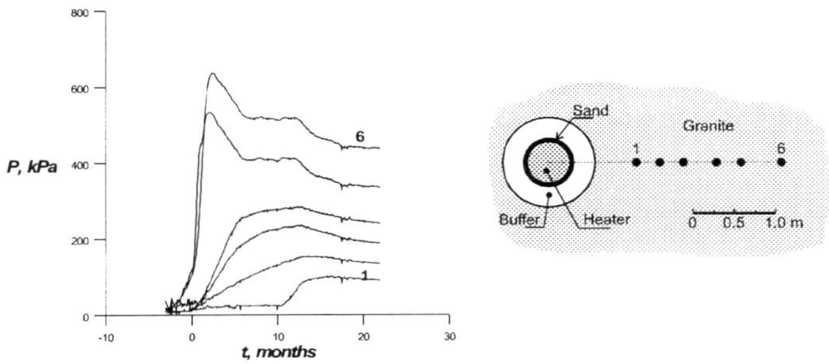

Figure 1: Pore pressure response as measured at piezometers #1 to #6 with origin of time axis corresponding to the date at which the heater was turned on. Plan view sketch of heater and location of piezometers

McTigue 1990, Coussy 1995) is an extension of the classic Biot (1941) theory (see also Detournay and Cheng, 1993). In this formulation, the porous medium is viewed as an open system (Coussy 1995); i.e. an element of the porous medium, identified by an element of the solid skeleton, can gain or lose fluid mass as it deforms.

The essential T-H-M coupled effects are embodied in the volumetric response equations (the deviatoric response of an isotropic material is purely elastic) which can be written as

$$P = -K_u\varepsilon + \alpha M\zeta + \beta_u K_u\theta; \quad p = M(\zeta - \alpha\varepsilon + \beta_m\theta) \tag{1}$$

In the above equations, $P = -\sigma_{kk}/3$ is the total (mean) pressure, p is the pore pressure, ε is the volumetric strain, ζ is the variation of fluid content (Biot 1941) , and θ is the temperature (P, p, and θ are variations with respect to initial values). The constitutive equations can be combined with the balance laws (momentum, mass, energy) and conduction laws (Darcy and Fourier) to yield field equations for the displacement, u_i, pore pressure, p, and temperature, θ. For quasi-static processes of concern here, these equations can be expressed as

$$G\nabla^2 u_i + \frac{1}{3}(G + 3K)\varepsilon_{,i} = \alpha p_{,i} + \beta K\theta_{,i} \tag{2}$$

$$\kappa\nabla^2 p = \frac{1}{M}\frac{\partial p}{\partial t} + \alpha\frac{\partial\varepsilon}{\partial t} - \beta_m\frac{\partial\theta}{\partial t}; \quad c_*\nabla^2\theta = \frac{\partial\theta}{\partial t} \tag{3}$$

Note that (3_b) is a linearized form of the energy balance, in which convected heat flow is neglected. Also the mechanical and hydraulic latent heat coupling terms have been dropped from (3_b) as they can be shown to be small.

The various material constants which appear in these equations can be conveniently organized into 5 groups: (i) *mechanical* (shear modulus, G, drained bulk modulus, K); (ii) *thermal* (drained thermal expansivity, β,volumetric heat capacity, C, thermal conductivity, κ_*); (iii) *hydromechanical* (Biot coefficient, α, Biot modulus,

M, Skempton coefficient, B, undrained bulk modulus, K_u); (iv) *thermohydromechanical* (undrained thermal coefficient, β_u, hydro-thermal coupling coefficient, β_m); (v) *transport* (mobility coefficient, κ). Not all of these coefficients are independent, however, as the following relationships can be established between them: $K_u = K + \alpha^2 M$, $\alpha M = BK_u$, $\beta_u = \beta K/K_u + B\beta_m$. A linear isotropic rock is thus characterized by nine independent thermoporoelastic parameters. Also the thermal and hydraulic diffusivities, c_* and c, can be deduced from the above set of parameters according to

$$c_* = \frac{\kappa_*}{C}; \quad c = \frac{\kappa}{S} \quad \text{with} \quad S = \frac{1}{M} + \frac{3\alpha^2}{3K + 4G} \tag{4}$$

where the storage coefficient S is defined under oedometric conditions (uniaxial strain).

Modelling of the heater problem

Thermal analysis indicates that the exact details of the heater configuration in the emplacement hole are irrelevant at the timescale dictated by the location of the piezometers nearest the heater. Furthermore a two-dimensional analysis can be used to predict the early-time induced pore pressure measured in these piezometers.

An approximate solution to this problem can be obtained by superposing a continuous heat source of strength Σ (the heat rate, W, of the heater scaled by its length) and a time-varying fluid source of strength $\Omega(t)$. The strength of the fluid source is computed by enforcing a zero pore pressure boundary condition at the wall of the emplacement hole (i.e., $p = 0$ at $r = a$), this particular boundary condition being justified as the buffer material is never saturated. Superposition of the heat and fluid source, which actually correspond to singular solutions of the linear theory of thermoporoelasticity (Smith *et al.* 1993), yields

$$\tilde{p}_s = \frac{p_*}{s} \frac{K_0(\omega\xi a)}{2\pi(1-\omega^2)} \left[\frac{K_0(\omega\xi r)}{K_0(\omega\xi a)} - \frac{K_0(\xi r)}{K_0(\xi a)} \right] \quad \text{with} \quad p_* = \gamma\Sigma \tag{5}$$

where the tilde denotes the Laplace transformed quantity, s is the Laplace parameter, $\xi^2 = s/c$, $\omega^2 = c/c_*$, K_0 and K_1 are the modified Bessel functions of order zero and one, respectively. The above equation indicates (through the expression of the 2D characteristic pressure, p_*) that a single parameter, γ, controls the magnitude of the pore pressure induced by thermal loading. This parameter γ ($[\gamma] = TL^{-2}$) lumps several thermoporoelastic constants

$$\gamma = \frac{\beta_o}{\kappa_* S} \quad \text{with} \quad \beta_o = \beta_m - \beta\eta\frac{K}{G} \quad \text{and} \quad \eta = \frac{3\alpha G}{4G + 3K} \tag{6}$$

where β_o is actually the hydrothermal coupling coefficient defined under oedometric condition. Also, the strength of the fluid source is given by

$$\tilde{\Omega}(s) = \frac{c\beta_o\Sigma}{\kappa_*(1-\omega^2)} \left[1 - \frac{K_0(\omega\gamma a)}{K_0(\gamma a)} \right] \tag{7}$$

Note that the fluid source is actually a sink, as Ω has a sign opposite to Σ.

A plot of the variation of dimensionless pore pressure, $\Pi = p/p_*$, with $\rho = r/a$ for the values 1, 5, 10 of the dimensionless time, $\tau = ct/a^2$, is shown in Fig. 2. The profile

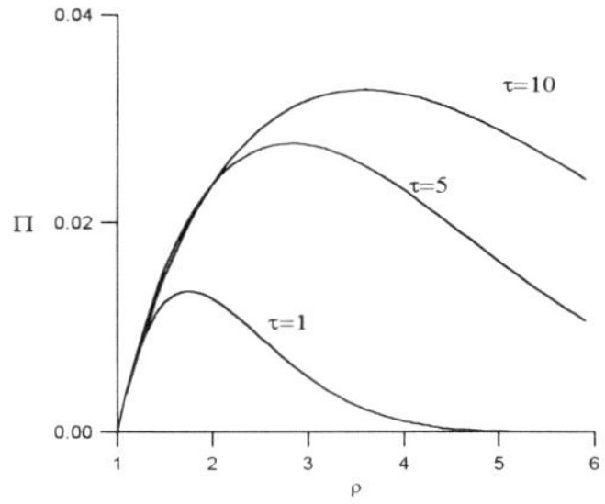

Figure 2: Variation of pore pressure Π versus radius ρ for $\tau = 1, 5, 10$

of pore pressure is characterized by a peak which grows and propagates outward in time. Heating of the borehole thus causes the rock mass to lose fluid, which is collected in the borehole.

Analysis shows that a reasonable match between measured and predicted pore pressure can be achieved with realistic values of the parameters: $\gamma \approx 10^4 \sim 10^5$ s/m^2, $c \approx 10^{-6}$m^2/s, $c_* \approx 10^{-6}$m^2/s, $\kappa_* \simeq 3.5$ W/m ^{o}C, $\beta_o \sim 10^{-6}$ ^{o}C^{-1}. However, the pore pressure at the furthest piezometers can only be predicted by taking into account the three-dimensional nature of the problem.

Acknowledgments

This work has been supported by a grant from AECL, Canada. This support is gratefully acknowledged. The authors would also like to thank Dr. Neil Chandler for his stimulating support.

References

[1] Biot, M. General theory of three-dimensional consolidation. *J. Appl. Phys.*, 12:155–164, 1941.

[2] Coussy, O. *Mechanics of Porous Continua*. Wiley, 1995.

[3] Detournay, E. and Cheng, A. H.-D. *Comprehensive Rock Engineering*, volume 2, chapter 5: Fundamentals of Poroelasticity, pages 113–171. Pergamon, 1993.

[4] McTigue, D. F. Flow to a heated borehole in porous, thermoelastic rock: Analysis. *Water Resour. Res.*, 26(8):1763–1774, August 1990.

[5] Smith, D. W. and Booker, J. R. Green's functions for a fully coupled thermoporoelastic material. *Int. J. Numer. Anal. Methods Geomech.*, 17:139–163, 1993.

Some Thoughts on Thermoporoelastic Coupling

M. Bai[1], Y. Abousleiman[1] and J.-C. Roegiers [1]

Abstract

In order to provide practical solutions for the coupled thermoporoelastic problems without incurring excessive mathematical and/or numerical manipulations while avoiding unreasonable simplifications, a fully coupled thermoporoelastic formulation is presented, and the conditions are discussed with regard to certain geotechnical scenarios where a rigorous coupling should be maintained and, where a partial or full decoupling technique may be acceptable.

Introduction

The study of the thermomechanical response of fluid saturated media appears to be an important task in many research areas such as extraction of geothermal energy, enhanced hydrocarbon recovery using thermal flooding, and reservoir stimulation under nonisothermal conditions. It is understood that these circumstances may result in strong coupling between the heat flow, the interstitial pore fluid flow and the solid matrix deformation. Under specific circumstances, the thermoporoelastic coupling may degenerate into partially-coupled situations where merely thermoelastic or poroelastic effects are manifest. Experience has shown that the mathematical sophistication may not place natural advantages over the coupled approaches. Sound physical implications and significant influences may be the only reasons to render the coupled solutions. In the thermoelastic or poroelastic analyses, justifications must be made to identify major factors, and to incorporate important or exclude trivial coupling mechanisms. This paper examines some geotechnical applications where the coupling should be maintained, and where a partial or full decoupling may be appropriate.

Thermoporoelastic formulation

The chain responses of fluid flow and heat transfer in deformable porous media are demonstrated in a triangular-fashion interplay, where the changes in either solid body strain, pore pressure, or material temperature result in the corresponding unequilibrium of momentum, mass and energy systems. For a linear, quasi-steady elastic system, a fully-coupled thermoporoelastic formulation (Elsworth, 1993; Bai and Roegiers, 1994; Jing et al., 1995) may be described in the sequences of conservations of the momentum, mass and energy as:

[1]Rock Mechanics Institute, The University of Oklahoma, Norman, OK 73019

$$Gu_{i,jj} + (\lambda + G)u_{k,ki} = \alpha p_{,i} + \beta T_{,i} \tag{1}$$

$$\frac{k}{\mu} p_{,kk} = \alpha \dot{\varepsilon}_{kk} - \alpha_h \dot{T} + \alpha^* \dot{p} + Q_f \tag{2}$$

$$K^* T_{,kk} = q_f s T_{,k} + \beta T_0 \dot{\varepsilon}_{kk} + s^* \dot{T} + Q_h \tag{3}$$

where a comma stands for differentiation and summation is implied over the repeated Latin subscripts; a superscripted dot identifies the derivative with respect to time t; u is the displacement, p is the pressure, T is the temperature, ε_{kk} is the total body strain, λ and G are *Lamé*'s constants, α is Biot's coefficient, β is the thermal expansion factor, k is the fluid permeability, μ is the fluid dynamic viscosity, α^* is the lumped compressibility, α_h is the thermal expansion coefficient, K^* is the thermal conductivity, q_f is Darcy's velocity of fluid flow, s is the intrinsic heat capacity for fluid, s^* is the lumped intrinsic heat capacity, T_0 is the reference temperature, Q_f and Q_h are the internal as well as external fluid and heat sources.

In equations (1), (2) and (3), in addition to the traditional quasi-steady mechanical equilibrium, transient fluid flow and heat transport, coupling terms are shown as first two terms on the right-hand side of each equation. They represent, respectively:

- Term (1a) - $\alpha p_{,i}$: modification of the effective stress due to the influence of fluid pressure;
- Term (1b) - $\beta T_{,i}$: adjustment of total strain as a result of thermal expansion;
- Term (2a) - $\alpha \dot{\varepsilon}_{kk}$: mass rate changes expelled by the solid volumetric strain;
- Term (2b) - $\alpha_h \dot{T}$: rate variation as a result of thermal expansion (or contraction) of fluid and solid;
- Term (3a) - $q_f s T_{,k}$: unequilibrium of energy transport due to forced thermal convection in fluid flow;
- Term (3b) - $\beta T_0 \dot{\varepsilon}_{kk}$: energy lost as a result of solid elastic deformation in the form of thermal expansion or contraction.

The interactive thermoporoelastic behavior can be concisely expressed by Figure 1. In fact, any influence of the one mechanism over the other receives retroactive impact from its counterpart. This intricate interaction provides the most convincing evidence for the comprehensive thermal-hydraulic-mechanical coupling.

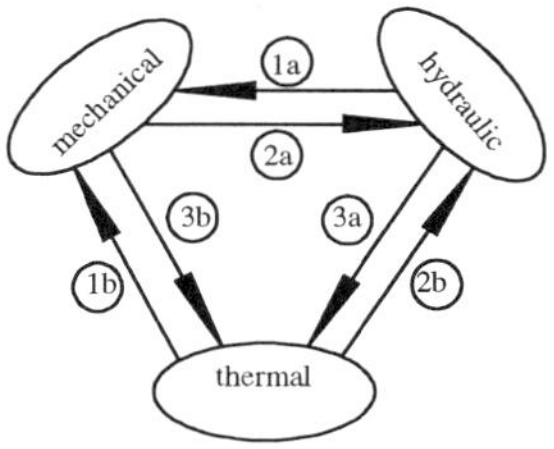

Figure. 1 Relationship of thermoporoelastic coupling

Partially decoupled formulation

Maintaining a fully-coupled thermoporoelastic solution excludes, in most cases, the possibility of using analytical solutions. In practice, partially decoupled formulations are developed to restore the analytical convenience. However, this decoupling requires a justification based on physical intuition, and frequently can be quantified mathematically. The partial decoupling differs from the complete decoupling in its partial replication of one interactive mechanism, instead of ignoring it completely. In any case, the partial decoupling is designed to derive the solutions sequentially. The following provides a couple of decoupling scenarios.

For the stiff material and less significant fluid flow, the effect of elastic deformation and thermal convection may be negligible. The coupled thermoporoelastic formulation is simplified as:

$$\frac{k}{\mu}p_{,kk} = -\alpha_h \dot{T} + \alpha^* \dot{p} + Q_f \tag{4}$$

$$K^* T_{,kk} = s^* \dot{T} + Q_h \tag{5}$$

The temperature T can be derived from equation (5), and then substituted into equation (4) to obtain the pressure p.

For less significant influences of elastic deformation on the changes in both pressure and temperature, and negligible thermal convection, the thermoporoelastic formulation becomes:

$$G u_{i,jj} + (\lambda + G) u_{k,ki} = \alpha p_{,i} + \beta T_{,i} \tag{6}$$

$$\frac{k}{\mu}p_{,kk} = -\alpha_h \dot{T} + \alpha^* \dot{p} + Q_f \tag{7}$$

$$K^* T_{,kk} = s^* \dot{T} + Q_h \tag{8}$$

Solving T from equation (8) and substituting T into equation (7) to obtain p, $\mathbf{u}$ can be finally derived from equation (6), after substituting both T and p. It should be noted that this simplification does not eliminate the inclusion of the impact of both pressure and temperature on the change in solid deformation. This nonretroactive response indeed exists in the real situation.

An illustrative example

The partial decoupling strategy may be demonstrated through examining a simple example in the one-dimensional consolidation. While the detailed formulation and solution technique is presented elsewhere (Bai and Abousleiman, 1996), Figures 2a and 2b represent the results of the analyses for a comprehensive 1-D consolidation under a nonisothermal environment with the negligence of thermal convection. The pressure p, temperature T, and displacement u are normalized with respect to their maximum values. The dimensionless time is defined as: $\tau = c_v t/(h^2)$ where c_v is the coefficient of consolidation, and h is the length of the consolidating column. Figure 2a shows an interactive response of dissipating pressure and temperature to the increase of displacement due to material consolidation. The results of fluid flow with the thermal coupling [with the term 2b in equation (2)] and without the thermal coupling [without the term 2b in equation (2)] are compared in Figure 2b. The coupled solution appears to noticeably delay the pressure dissipation. Under the present circumstance, however, this significant difference may not be observed for the thermal transport where the effect of pressure variation is either coupled or decoupled from the temperature change through the deformation term (3b) in equation (3). This result may provide some avenues for the partially decoupled solutions.

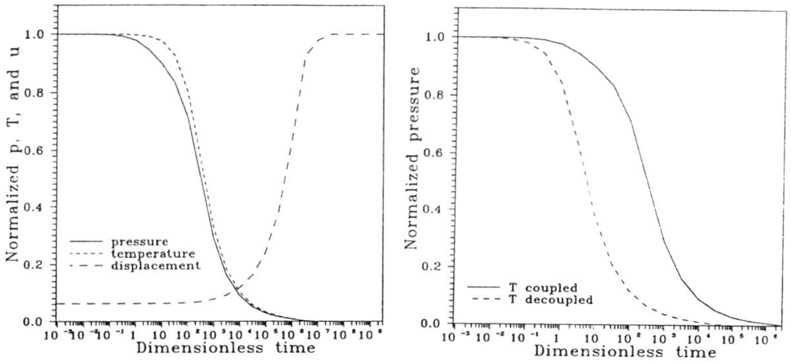

Figure. 2 (a) Thermoporoelastic evolution; (b) Comparison of pressures

Concluding remarks

This paper elicits a statement that the development of a fully coupled thermoporoelastic approach is necessary to minimize potential errors while modeling the behavior of poroelastic media under nonisothermal conditions. However, simplification, such as using partial decoupling, should be attempted when the influence of each coupling term is determined and the omission of the term is justified on physical and analytical ground.

Acknowledgment

Support of the National Science Foundation/State/Industry under the IU-CRC program and under the contract EEC-9209619 is gratefully acknowledged.

References

Bai, M. and Abousleiman, Y. (1996). "Thermoporoelastic coupling with application to consolidation." submitted to *Int. J. Num. Analy. Meth. Geomech.*.

Bai, M. and Roegiers, J.-C. (1994). "Fluid flow and heat flow in deformable fractured porous media." *Int. J. Engng. Sci.*, 32, 10, 1615-1633.

Elsworth, D. (1994). "Computational methods in fluid flow." Comprehensive Rock Engineering, Editor Hudson, *Pergamon Press*, 2, 173-189.

Jing, L., Tsang, C.-F. and Stephansson, O. (1995). "DECOVALEX - An international co-operative research project on mathematical models of coupled THM processes for safety analysis of radioactive waste repositories." *Int. J. Rock Mech. Min. Sci. & Geomech. Abstr.*, 32, 5, 389-398.

Fundamental Thermal Mechanical Modeling of Gas-Filled Porous Composites

N.J. Salamon and Ramnath Ganesan[1]

Abstract

Fundamental modeling issues dealing with the thermal-mechanical decomposition of gas-filled porous polymeric composites are studied using coupled equations for momentum, gas-mass and energy balance and Biot's constitutive relations. Reasonable results are found for closed form solutions to two tests.

Introduction

The objective of this note is to set forth a functional behavior for Biot's pressure-stress coupling factor α in terms of physical parameters for porous polymeric composites. This is important because appropriate experimental data for porosity-related properties does not exist for these materials and their determination must be done indirectly. Hence one must guess the properties and check them through trial and error correlations with data that is available.

The factor α is physically rather vague. Nur and Byerlee (1971) relate it to the effective and pure solid bulk moduli K and K_S, respectively, as $\alpha = 1 - K/K_s$ from which one can argue the bounds: $0 \leq \alpha < 1$. Following Biot and Willis (1957), this can be tightened to $\phi < \alpha \quad 1$ where ϕ is the porosity. It is assumed these arguments can be extended to anisotropic materials and we base our function for α on them.

Biot's theory has been adopted for study of this material by Weiler, F.C. (1992), Sullivan and Salamon (1992), Lee, Salamon and Sullivan (1996), Ganesan (1996) and others referenced there. Using this theory and momentum, mass and energy balance laws to model the physics, the present theory for α is demonstrated for two thermal decomposition tests on carbon phenolic (a rocket insulation material) conducted by Stokes (1987). As the temperature is raised at a constant rate from 290° to 800° K, the FTE test measures free thermal expansion and the RTG test (restrained thermal growth) measures axial stress and lateral strain. The results show reasonable agreement with experiment.

---

1. Professor and Ph.D. Candidate, respectively, Department of Engineering Science and Mechanics, 227 Hammond Bldg., The Pennsylvania State University, University Park, PA 16801

Theory

The α function. It is argued by Ganesan (1996) that the simplest form for the α function is

$$\alpha = \phi^s, \qquad 0 < s = s_0 + C\eta < 1 \tag{1}$$

where s_0 and C are positive constants set according to tests below. The parameter η is given by

$$\eta = (\text{maximum shear stress}) / (\text{shear strength}) \tag{2}$$

which is based upon the hypothesis by Lee, et al. (1996): *If this ratio is high, then pore opening, enhanced by shear stress, increases and the value of α decreases.* Hence s implicitly depends upon pore shape which is affected by shear stress.

One observes that α is an increasing function of φ and a decreasing function of s, hence the bounds cited above are satisfied.

The physical model. An axisymmetric, anisotropic, isothermal, circular cylinder of radius R whose fields vary only radially is treated. The momentum equations reduce to

$$\left(C_{11} - \left.\frac{C_{13}^2}{C_{33}}\right|_{FTE}\right)\left(u_{,rr} + \frac{u_{,r}}{r} - \frac{u}{r^2}\right) = \alpha_{1,r}\, p + \alpha_1\, p_{,r} \tag{3}$$

where the delimited term only applies to the FTE test. The C_{ij} are bulk material stiffnesses, u is the displacement along coordinate r and p is the pore pressure.

The gas-mass balance generally includes gas flow governed by Darcy's law and our finite element models include this, but to achieve a closed form solution we assume no gas flow (hence gas produced is stored) and an ideal gas so that its bulk modulus is p and its thermal expansion coefficient is $1/T$. Then the mass balance is

$$\begin{aligned} &\rho_g\, p\left[\alpha_1\delta_1 + \alpha_2\delta_2 + \alpha_3\delta_3 - \phi\,(\delta_1 + \delta_2 + \delta_3) + \left.\frac{\alpha_3^2}{C_{33}}\right|_{FTE}\right] \\ &+ \rho_g\left(\left(\alpha_1 - \alpha_3\left.\frac{C_{13}}{C_{33}}\right|_{FTE}\right)\left(u_{,r} + \frac{u}{r} - 2\beta_1\Delta T\right) - (\alpha_3\beta_3\Delta T)\,\Big|_{RTG} \right. \\ &\left. + \phi\left(1 - \frac{\Delta T}{T}\right)\Big| \right) = (RF)\,\rho_v\left(\sum_{k=1}^{\text{nreact}} (1 - c_k)\,(W_{ok} - W_{ck})\right) \end{aligned} \tag{4}$$

where ρ_g is the density of the gas, δ_i are Biot's compressibility coefficients, β_i are coefficients of thermal expansion of the solid, T is temperature, and ΔT is the temperature increment during simulation. The right-hand side is gas production modeled after an Arrhenius formulation where RF is the material resin fraction, ρ_v it's virgin density, c_k the degree-of-char, and W_{0k} and W_{ck} are the reactive and residue portions of each reaction k at temperature T.

The energy balance for the isothermal case is trivial. The temperature of the body is uniform and given by $T = T_0 + \dot{Q}t$ where T_0 is the initial temperature, $\dot{Q}$ is the rate of heating and t is time.

Solution

We assume $\alpha_1 = \alpha_2 = \alpha_3 = \alpha$. The material properties, porosity and α are assumed uniform over the section. This and a single integration of (3) yields $\partial u/\partial r + u/r$ independent of r, hence from (4), p is also independent of r and the right-hand side of (3) vanishes.

With boundary conditions $u = 0$ at $r = 0$ and $\tau_{rr} = p_\infty$ at $r = R$ where p_∞ is the ambient pressure, we obtain

$$u = Ar \text{ where}$$

$$A|_{FTE} = \left(\left(\alpha_1 - \alpha_3\frac{C_{13}}{C_{33}}\right)p - p_\infty\right) \Big/ \left(C_{11} + C_{12} - \frac{2C_{13}^2}{C_{33}}\right) + \beta_1 \Delta T \tag{5}$$

$$A|_{RTG} = \frac{\alpha_1 p - p_\infty}{C_{11} + C_{12}} + \beta_1 \Delta T + \frac{C_{13}\beta_1 \Delta T}{C_{11} + C_{12}}$$

With (5) in hand, equation (4) becomes a quadratic equation in p:

$$p^2\left[(\alpha_1\delta_1 + \alpha_2\delta_2 + \alpha_3\delta_3) - \phi(\delta_1 + \delta_2 + \delta_3) + \left.\frac{\alpha_3^2}{C_{33}}\right|_{FTE}\right]$$

$$p\left[\left(\alpha_1 - \alpha_3\left.\frac{C_{13}}{C_{33}}\right|_{FTE}\right)(2A - 2\beta_1\Delta T) - \alpha_3\beta_3\Delta T|_{RTG} + \rho_g\phi(1 - \beta_g\Delta T)\right] \tag{6}$$

$$= \frac{RT}{m_w}(\mathrm{RF})\,\rho_v\left(\sum_{k=1}^{\mathrm{nreact}}(1 - c_k)(W_{ok} - W_{ck})\right)$$

For the FTE solution we plot

$$e_{zz} = \beta_3\Delta T + [\alpha_3 p - 2C_{13}(A|_{FTE} - \beta_1\Delta T)]/C_{33} \tag{7}$$

and for the RTG solution we plot the restraining axial stress and hoop strain:

$$\tau_{zz} = 2C_{13}(A|_{RTG} - \beta_1\Delta T) - C_{33}\beta_3\Delta T - \alpha_3 p \qquad e_{\theta\theta}|_{r=R} = A|_{RTG} \tag{8}$$

Conclusion

The plots below show reasonable agreement with experimental data up to the point of significant gas production by the phenolic, about 750° K, for $1/2 \leq C \leq 1$. Agreement beyond this temperature can be obtained numerically if gas flow is permitted to depressurize the pores. s_0 was set to 0.05 to match an initial value of α determined previously. Future work seeks connections with material morphology.

References

Biot, M.A. and Willis, D.G. (1957), "The Elastic Coefficients of the Theory of Consolidation," ASME J. of Applied Mechanics, Vol. 24, pp. 596-601.

Ganesan, R., (1996), "Fundamental Modeling Studies of the Thermal-Mechanical Response of Gas Filled Porous Decomposing Polymeric Composites," *Ph.D. Dissertation,* The Pennsylvania State University.

Lee, S., Salamon, N.J. and Sullivan, R.M. (1996), "Finite Element Analysis of Poroelastic Composites Undergoing Thermal and Gas Diffusion," AIAA J. of Thermophysics and Heat Transfer (to appear).

Nur, A. and Byerlee, J.D. (1971), "An Exact Effective Stress Law for Elastic Deformation of Rock with Fluids," J. of Geophysical Research, Vol. 76, No. 26, pp. 6414-6419.

Stokes, E.H. (1987), "The Effect of Moisture on the Mechanical and Thermal Response of FM5055 Carbon Phenolic Composites," Southern Research Institute Report No. SRI-EAS-87-1244-6245-6, Southern Research Institute, Birmingham, AL.

Sullivan, R.M., and Salamon, N.J. (1992a), "A Finite Element Method for Thermochemical Decomposition of Polymeric Material - I: Theory, II: Carbon-Phenolic Composites," Int. J. of Engineering Science, Vol. 30, pp. 431-441, Vol. 30, No. 7, pp. 939-951.

Weiler, F.C. (1992), "Fully Coupled Thermo-Poro-Elasto Governing Equations," *Computational Mechanics of Porous Materials and Their Thermal Decomposition*, N.J. Salamon and R.M. Sullivan, Editors, AMD-Vol. 136, ASME Press.

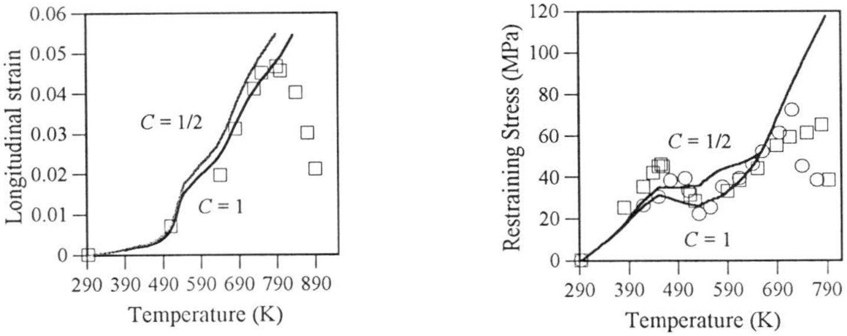

FIGURE 1. Analytical solution vs. experimental data: FTE test (left), RTG test (right).

The Thermodynamic Structure of a Fluid-Saturated Compressible and Incompressible Elastic Porous Solid

Reint de Boer [1]

Abstract

The modern macroscopic porous media theory consists of the mixture theory restricted by the volume fraction concept. The incorporation of this concept leads to serious difficulties in constructing a consistent macrocopic theory.

With the help of the multiplicative decomposition of the deformation gradient and thermodynamic restrictions it will be shown how a consistent phenomenological porous media theory can be developed.

Preliminaries

In order to develop a consistent phenomenological porous media theory many approaches have been carried out in the literature in the past. The crucial point in this theory is concerned with missing field equations due to the introduction of volume fractions. In order to overcome this shortcoming several attempts have been made as for e.g., the introduction of additional balance equations – more or less physically well founded – or additional constitutive equations (see the extensive discussion in de Boer, 1996). Moreover, with the introduction of the volume fraction concept to extend the macroscopic mixture theory the microlevel is touched. In particular, this fact causes serious difficulties in developing a consistent porous media theory. These difficulties can be met by the introduction of a multiplicative decomposition of the deformation gradient.

The aim of this paper is to construct the thermodynamic structure of mathematical models with compressible and incompressible elastic constituents on the basis of reasonable assumptions of the deformations at the microlevel in order to derive restrictions for the constitutive equations. It will be shown that the developed equations are physically well balanced.

---

[1]Professor, Universität Essen, FB 10 – Bauwesen, Institut für Mechanik, D-45117 Essen, Germany

The volume fraction concept

Let $\mathbf{r}$ describe the position of a constituent with volume elements dv_ν in the controlspace and $\mathbf{x}$ the centre of the volume element dv. In order to determine the average volume (realistic) element dv^α of each constituent φ^α, an indicator function

$$\chi^\alpha = \chi^\alpha(\mathbf{r},t) = \quad 1 \text{ for } \mathbf{r} \in dv^\alpha\ , \quad 0 \text{ for } \mathbf{r} \in dv^\beta, \quad \beta \neq \alpha \tag{1}$$

is introduced, where t marks the time. Thus, for the partial volume (realistic) element dv^α with (1) the following relation is obtained:

$$dv^\alpha(\mathbf{x},t) = \int_{dv} \chi^\alpha(\mathbf{r},t)dv_\nu = \int_{dv} \chi^\alpha(\mathbf{x}+\boldsymbol{\xi},t)dv_\nu, \quad \mathbf{r} = \mathbf{x}+\boldsymbol{\xi}\ , \tag{2}$$

where in $(2)_2$ the position vector $\mathbf{r}$ is substituted by $\mathbf{x}+\boldsymbol{\xi}$, and where the integration refers to the microscopic local $\boldsymbol{\xi}$-reference system with the origin $\mathbf{x}$.

With (2) the concept of volume fractions can be formulated as follows

$$n^\alpha(\mathbf{x},t) = \frac{dv^\alpha}{dv} = \frac{1}{dv}\int_{dv} \chi^\alpha(\mathbf{r},t)dv_\nu\ , \quad \sum_{\alpha=1}^{\kappa} n^\alpha = 1\ , \tag{3}$$

whereby $(3)_2$ is known as the volume fraction condition for κ constituents φ^α as can be easily proved.

With the volume fractions n^α the partial density $\rho^\alpha(\mathbf{x},t)$ are related to the realistic density $\rho^{\alpha R}(\mathbf{x},t)$ by (see de Boer, 1996):

$$\rho^\alpha(\mathbf{x},t) = n^\alpha(\mathbf{x},t)\ \rho^{\alpha R}(\mathbf{x},t)\ , \tag{4}$$

where the realistic density $\rho^{\alpha R}(\mathbf{x},t)$ is the average density of the true density $\rho^{\alpha T}(\mathbf{r},t)$.

We arrive at the same results, if an ensemble average mainly used in the suspension theory and the theory of fluidized beds is applied.

The introduction of the volume fractions has the effect that "smeared" continua with reduced densities are created. These partial continua have the same properties as the constituents in an immiscible mixture. In particular, there is an interaction between the individual partial continua due to friction phenomena. Hence, for the "smeared" continua the mixture theory is an adequate tool to describe the thermodynamical behavior of saturated porous solids. In the sequel the mixture theory restricted by the volume fraction concept, will be denoted as the porous media theory.

Kinematics

In this theory it is assumed that the particles X_α of all constituents φ^α, simultaneously occupy at any time t the spatial point $\mathbf{x}$. The particles X_α proceed from different reference positions $\mathbf{X}_\alpha$ at time $t = t_0$. Thus, each constituent is assigned its own independent motion function from which the velocity $\mathbf{x}'_\alpha$ the acceleration $\mathbf{x}''_\alpha$ and the deformation gradient $\mathbf{F}_\alpha$ can be derived:

$$\mathbf{x} = \chi_\alpha(\mathbf{X}_\alpha, t) \ , \ \mathbf{x}'_\alpha = \mathbf{x}'_\alpha(\mathbf{x}, t) \ , \ \mathbf{x}''_\alpha = \mathbf{x}''_\alpha(\mathbf{x}, t) \ , \quad \mathbf{F}_\alpha = \mathrm{Grad}_\alpha\, \chi_\alpha \ . \quad (6)$$

In order to describe some properties at the microscale, as for e.g., compressibility and incompressibility of the true materials, it is convenient to decompose the deformation gradient $\mathbf{F}_\alpha$ multiplicatively into a tensor $\mathbf{F}_{\alpha R}$ which reflects the microscopic deformations of the true material on the macroscale and a tensor $\mathbf{F}_{\alpha N}$ which describes the change of the pores (for more information see Bluhm and de Boer, 1994, and de Boer, 1996).

$$\mathbf{F}_\alpha = \mathbf{F}_{\alpha N}\mathbf{F}_{\alpha R} \ . \quad (7)$$

The entropy inequality

The thermodynamics of the porous media theory consists – as aforementioned – of the volume fraction concept, kinematics, balance equations and the entropy inequality. Due to space limitations the balance equations will not be considered. For a constitutive theory the entropy inequality

$$\sum_{\alpha=1}^{\kappa}[-\rho^\alpha(\psi^\alpha)'_\alpha - \Theta'_\alpha\rho^\alpha\eta^\alpha - \hat{\mathbf{p}}^\alpha\cdot\mathbf{x}'_\alpha + \mathbf{T}^\alpha\cdot\mathbf{D}_\alpha - \frac{1}{\Theta}\mathbf{q}^\alpha\cdot\mathrm{grad}\,\Theta \geq 0 \ , \quad (8)$$

yields important constraints for the constitutive equations. In (8) ψ^α is the free Helmholtz energy, η^α the specific entropy, Θ the absolute temperature, $\hat{\mathbf{p}}^\alpha$ an interaction force, $\mathbf{T}^\alpha$ Cauchy's stress tensor and $\mathbf{D}_\alpha$ the symmetric part of the velocity gradient. Finally, $\mathbf{q}^\alpha$ denotes the external heat flux vector. The model considered is assumed free of mass and moment of momentum exchange.

In the porous media theory for saturated porous solids the entropy inequality must be modified because the volume fraction condition (saturation condition) $(3)_2$ has turned out to be an important constraint (see for e.g., de Boer, 1996). Thus, this constraint, formulated in the rates and provided with a Lagrangian multiplier must be added to the entropy inequality. Moreover, if the constituents are incompressible the incompressibility conditions must be considered in the entropy inequality provided with Lagrangian multipliers. For a binary model consisting of an incompressible liquid and an incompressible porous solid the entropy inequality reads as

$$
\begin{aligned}
&- \rho^S(\psi^S)'_S - \rho^F(\psi^F)'_F + \mathbf{D}_S \cdot (\mathbf{T}^S + n^S\lambda\,\mathbf{I}) + \\
&+ \mathbf{D}_F \cdot (\mathbf{T}^F + n^F\lambda\,\mathbf{I}) + \mathbf{D}_{SR} \cdot (-n^S\lambda + n^S\kappa^{SR})\mathbf{I} + \\
&+ \mathbf{D}_{FR} \cdot (-n^F\lambda + n^F\kappa^{FR})\mathbf{I} - (\hat{\mathbf{p}}^F - \lambda\,\mathrm{grad}\, n^F) \cdot (\mathbf{x}'_F - \mathbf{x}'_S) \geq 0\,,
\end{aligned}
\tag{9}
$$

where the indices S and F denote the solid and fluid phases. $\mathbf{D}_{SR}$ and $\mathbf{D}_{FR}$ are the symmetric parts of the velocity gradients at the microlevel and λ, κ^{SR} and κ^{FR} are the Lagrangian multipliers. In case of compressible constituents the Lagrangian multipliers κ^{SR} and κ^{FR} are zero.

The evaluation of the entropy inequality (9) yields together with reasonable constitutive assumptions important constitutive equations. For an incompressible model in the so-called mixture equilibrium state Cauchy's stress tensor for solid and fluid constituents are represented by

$$
\begin{aligned}
&\mathbf{T}^S = -\, n^S\lambda\,(\mathbf{I} - \mathcal{F}) - n^S\kappa^{SR}\mathcal{F} + \mathbf{T}^S_E\,, \\
&\mathbf{T}^F = -\, n^F\lambda\,\mathbf{I} + \mathbf{T}^F_E, \quad \lambda = \kappa^{FR}\,.
\end{aligned}
\tag{10}
$$

The quantity $\mathcal{F}$ contains constitutive relations of the porous solid at the microscale. The tensors $\mathbf{T}^S_E$ and $\mathbf{T}^F_E$ are the effective stresses governed by the deformations on the macro- and partly on the microlevel.

In the case of a binary model with compressible constituents the Lagrangian multipliers κ^{SR} and κ^{FR} are zero and the tensor function $\mathcal{F}$ and the stresses $\mathbf{T}^S_E$ and $\mathbf{T}^F_E$ alter slightly. For more details refer to de Boer, 1996.

References

Bluhm, J. and de Boer, R.: The volume fraction concept in the porous media theory. Report Mech 94/5, FB10/Mechanik, Universität-GH Essen, submitted for publication.

de Boer, R.: "Highlights in the historical development of the porous media theory – toward a consistent macroscopic theory", to appear in the *Appl. Mech. Rev.* (1996).

Modelling of Thermal-Hydrological-Mechanical Processes in a Discrete Joint

T.S. Nguyen[1], A.P.S. Selvadurai[2] and P.Flavelle[1]

Abstract

This paper presents aspects of a research program which investigates coupled Thermal-Hydrological-Mechanical behaviour of jointed granite. This coupled behaviour has important implications on the safety of concepts proposed for the deep geological disposal of heat-emitting nuclear fuel wastes.

Introduction

The Canadian concept for the disposal of Nuclear Fuel Wastes (NFW) involves the burial of the wastes in a plutonic rock mass of the Canadian Shield, at depths from 500 to 1000 m (AECL, 1994). The wastes will generate heat for thousands of years, creating a thermal pulse that will influence the stability, deformation and water flow characteristics of the rock mass and these in turn have important implications on the rate of contaminant transport from the wastes to the surface. In order to simulate the coupled Thermal-Hydrological-Mechanical (T-H-M) behaviour of the rock mass, the authors have developed a finite element code, FRACON, from a generalization of Biot's theory of consolidation, to include the effects of heat conduction and the presence of rock joints (Nguyen and Selvadurai, 1995). Scoping calculations conducted with the FRACON code in conjunction with experimental investigations on an intact synthetic rock have shown that the pore pressure in the vicinity of the heat source can increase due to the mismatch in the coefficient of thermal expansion and the compressibility characteristics between the pore water and the solid matrix. Hence, the thermal pulse from the disposal of NFW could cause substantial modifications to the stress regime and the flow field, leading to an increase in the rate of contaminant migration from the wastes. Since the natural rock into which a radioactive waste repository is to be excavated is expected to be fractured, a program of experimental and computational research is now being conducted to examine T-H-M processes in both intact and fractured granite.

---

1 . Atomic Energy Control Board, 280 Slater, Ottawa, Ontario, Canada K1P 5S9

2. Dept. Civil Eng. and Applied Mech., McGill University, Montreal, Quebec, Canada H3A 2K6

Experimental setup

An experimental program to assess the combined effects of temperature, hydraulic pressure and ambient rock stress on both fractured and unfractured granite is being conducted at the Department of Civil Engineering and Applied Mechanics of McGill University, Montreal, Canada. The test facility is illustrated in Figure 1. Both intact and artificialy fractured samples of Barre granite from Vermont, USA, are to be tested. The samples have diameters of 457 mm and heights of 508 mm with a cylinderical central cavity of 51 mm. During testing, the cavity will be filled with water, and a pressure transducer will monitor the evolution of the pressure in the borehole. For coupled T-H-M tests, the temperature in the central cavity could be increased either by inserting a heater into the cavity or by heating elements emplaced around the samples. Upon heating, the pressure inside the cavity will increase and gradually decay due to outward drainage into the joint and the rock matrix. The pressure transient will bear the T-H-M signature of the jointed rock sample. It is expected that the main hydraulic conduit will be the joint in the fractured rock samples. Increasing the normal stress on the sample is expected to decrease the aperture of the joint, reducing its permeability and thus influencing the pressure transient in the borehole.

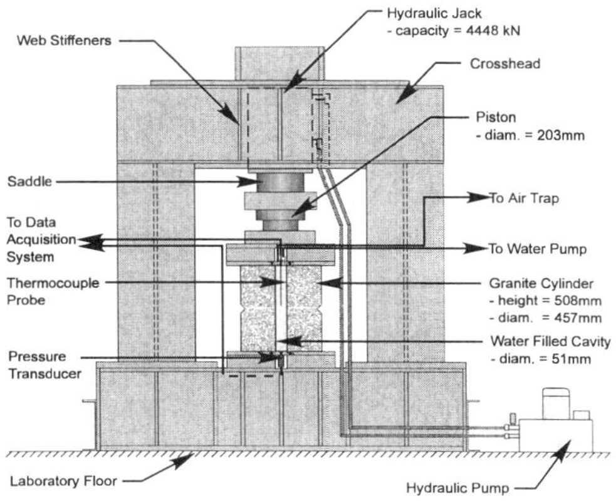

Figure 1. Test Facility for T-H-M modelling of fractured granite

A series of tests under different applied loads will be conducted, and the temperature at several locations of the granite cylinder, the axial compression of the cylinder, the total axial load on the cylinders and the fluid pressure in the borehole will be monitored at prescribed intervals by a computerized data acquisition system. The details of the experimental configurations, theoretical models and computational aspects of the research will be discussed in future publications.

Preliminary computational modelling

In order to finalize the design of the experiments, some preliminary scoping calculations were performed with the FRACON code for T-H-M tests on fractured samples. The finite element mesh with the assumed boundary conditions are shown in Figure 2. Axisymmetric conditions were used in the modelling.

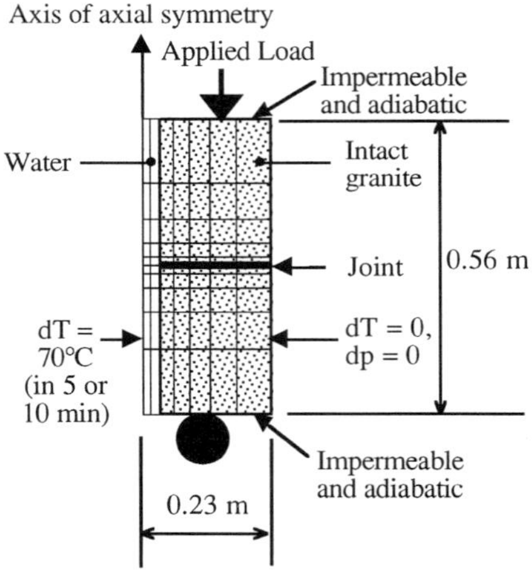

Water:
bulk modulus ≈ 2.2×10^9 Pa; Coefficient of volumetric thermal expansion ≈ 4×10^{-4} °C^{-1}

Intact granite:
Young's modulus E ≈ 60 GPa; Poisson's ratio $\nu \approx 0.1$; porosity n ≈ 0.011; Permeability $k_r \approx 10^{-18}$ - 10^{-20} m^2; Coefficient of volumetric thermal expansion ≈ 0.24×10^{-4} °C^{-1}.

Figure 2. Finite element model for T-H-M test on fractured granite

Fluid elements with zero shear resistance are used to represent the water inside the borehole. Solid poroelastic elements are used to represent the intact granite (Nguyen and Selvadurai, 1995). The T-H-M properties of the intact granite used in the analysis were obtained from independent laboratory tests and are shown in Figure 2. Special joint elements are used to represent the fracture. For all these elements, the primary unknowns are the nodal displacements and pore fluid pressure. The T-H-M properties of the joint under several levels of axial compression are being presently measured. Of most importance is the hydraulic aperture of the joint, e_h. It is assumed that the permeability of the joint is related to its aperture by the relation:

$$k_j = e_h^2/12$$

Four cases of heating the water in the borehole are considered: 1) $k_r=10^{-19}$ m^2, $e_h=10$ μm ; 2) $k_r=10^{-20}$ m^2, $e_h=10$ μm; 3)$k_r=10^{-19}$ m^2, $e_h=5$ μm; 4)$k_r=10^{-19}$ m^2, $e_h=5$ μm For the first three cases,the temperature of the water inside the borehole is increased gradually to 70°C in a period of 5 minutes while for case 4, that heating period is increased to 10 minutes. It is expected that the magnitude of the pressure transient will increase with the rate of heating and decrease with the permeability of the joint. The pressure transients in the borehole predicted by the model are shown in Figure 3. It could be seen that the intact rock permeability does not significantly influence the results, since the results of cases 1 and 2 are very similar. As expected, a decrease in the joint aperture (case 3 vs. case 1) leads to a lower permeability of the joint, and an eight-fold in the magnitude of the maximum water pressure inside the borehole. The slower rate of heating (case 4 vs. case 3) results in a smaller pressure transient, also as expected.

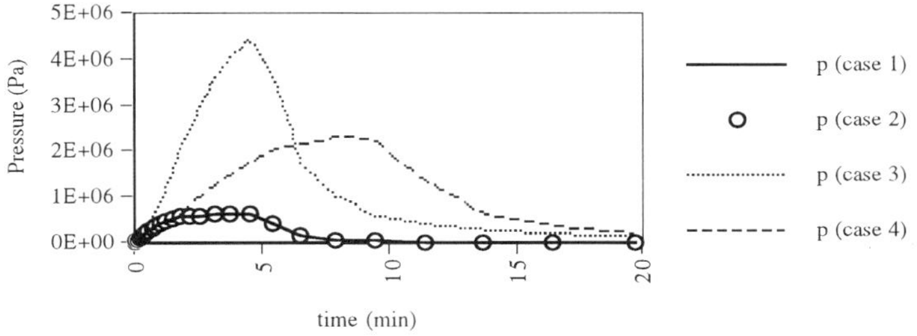

Figure 3. Transient T-H-M tests on fractured samples

Conclusions

Preliminary calculations of the response of fractured granite to internal heating have demonstrated that the T-H-M model predicts the pressure transients as expected: the rate of internal heating controls the buildup of the pressure transients and the permeability of the joint controls the rate of dissipation of the pressure transients. The results of the current measurements of fracture hydraulic apertures as a function of axial compression will be used in the model to design the experimental conditions (rate of heating and applied normal load) to achieve measurable pressure transients.

References

Atomic Energy of Canada Ltd, 1994, 'Environmental Impact statement on the concept for disposal of Canada's nuclear fuel wastes, AECL-10711, COG-93-1.

Nguyen T.S. and A.P.S. Selvadurai, 1995, 'Coupled thermal-mechanical-hydrological behaviour of sparsely fractured rock: implications for nuclear fuel waste disposal', Int. J. Rock Mech. and Min. Sci. Geomech. Abstr., Vol. 32, 465-479.

Optical Fiber Sensors for Advanced Civil Structures

Marten J. de Vries[1], Stephen H. Poland[2], Jennifer L. Grace[2], Vivek Arya[1], Kent A. Murphy[2], Richard O. Claus[1]

Abstract

This paper provides an overview of the design, implementation and performance of optical fiber sensors used for the nondestructive evaluation of civil structures. Optical fiber sensors have been developed during the past twenty years for primary applications in materials and structural analysis. Their advantages for such measurements include an inherent immunity to electromagnetic interference, avoidance of ground loops and sparks, and excellent dynamic range, accuracy and resolution. Although many optical fiber sensors have been studied in the past fifteen years, few experiments have used optical fiber sensors for the structural health monitoring and detection of damage for civil structures. In particular, most of the optical fiber sensors research for civil structure applications performed to date has focused on obtaining qualitative measurements rather than quantitative measurements. In this paper we demonstrate the performance of calibrated optical fiber sensors for the measurement of strain in high performance "smart materials." Assessment of the validity of the measurements was accomplished through direct comparison between the performance of the fiber optic strain sensors and collocated foil strain gages. Quantitative measurements of those strains are presented and future work in this area is discussed.

Introduction

The growth of a civilization as a whole largely depends on the quality of its overall infrastructure. The deterioration of public roads, communication systems, and other critical infrastructure components can lead, and has led, to the downfall of civilizations. History has shown how the decline of a civilization can be accelerated with the decay of its infrastructure.[1] Currently, the United States is seeing a similar trend; the extensive "good" infrastructure that was originally developed and implemented is now rapidly deteriorating with time due to increasing demand, and lack of vital maintenance and repair. For example, according to Federal Highway Administration (FHWA) estimates, 250,000 of the nation's highway bridges have serious structural problems and are in dire need of repair. It is therefore important

[1] Fiber & Electro-Optics Research Center, Bradley Department of Electrical Engineering, Virginia Polytechnic Institute & State University, 106 Plantation Road, Blacksburg, VA 224061-0356.
[2] Fiber and Sensor Technologies, Inc., 2801 Commerce St., Blacksburg, VA 24062.

to increase investment into research for the treatment and repair of the national infrastructure. A growing need is being felt for the design and development of advanced construction materials, and the implementation of electrical and mechanical instrumentation that can perform health monitoring of such infrastructure. Fortunately, recent changes in world politics have refocused part of the effort in such sensor-integrated "smart material and structures" from primarily military uses to civilian infrastructure applications.

To assess the damage and to evaluate the condition of civil structures, novel non-destructive testing and evaluation techniques need to be developed. For this research, optical fiber sensors are being developed and implemented that can be used to quantitatively monitor the performance of bridges, building structures and advanced building materials. In part this requires sensors to monitor various parameters, such as strain and vibration, over extended periods of time. Sensor data obtained from such tests can be used as input to subsequently optimize and improve material and structural design. This data can also suggest possible methods of implementing long-term *in situ* measurement approaches to evaluate the health of constructed structures, and allow the structure to be repaired, restored, or retrofitted. Furthermore, the ability for self-diagnosis will make the "smart" materials of the future structures superior to, and more cost-effective than, the materials used today.

Optical fiber sensors

Optical fiber sensors, unlike conventional electrically-based sensors, are fabricated with high-strength silica which will not corrode or be affected by electromagnetic interference. Their small size and geometric flexibility allows them to be unobtrusively embedded or surface-attached to the host structure. Fiber optic sensors have been used for the non-destructive evaluation of advanced materials and structures for over ten years.[2,3] For example, in composites, fiber sensors have been demonstrated to be feasible for the measurement of internal material changes during fabrication, the in-service lifetime measurement of strain, temperature, vibration, and other physical perturbations, and the eventual detection of damage or property degradation.[4,5] Recent advances in fiber optics communication technology and the widespread use of fiber communications have made available a large selection of reliable and inexpensive opto-electronic devices and components for use in sensors and sensor instrumentation support systems.

For our tests we used extrinsic Fabry-Perot interferometric (EFPI) optical fiber sensors. Murphy *et al.*, first demonstrated the EFPI strain gage on an F-15 fatigue test structure at the Wright Patterson Air Force Base.[6] The EFPI sensor geometry is shown in Figure 1. Here light from a 1300 nm laser light source is coupled into the sensor head which consists of a 126 μm inner diameter hollow core tube into which a 9/125 μm singlemode input fiber and a second fiber, serving solely as a reflector, are positioned. The reflection, approximately 4%, from the first glass-air interface at the end of the input fiber, R_1, and the reflection, R_2, from the air-glass interface at the input to the second fiber interfere and couple back to a detector via a 2x2 coupler. The resulting interference signal varies sinusoidally in response to microdisplacements in the air-gap cavity.

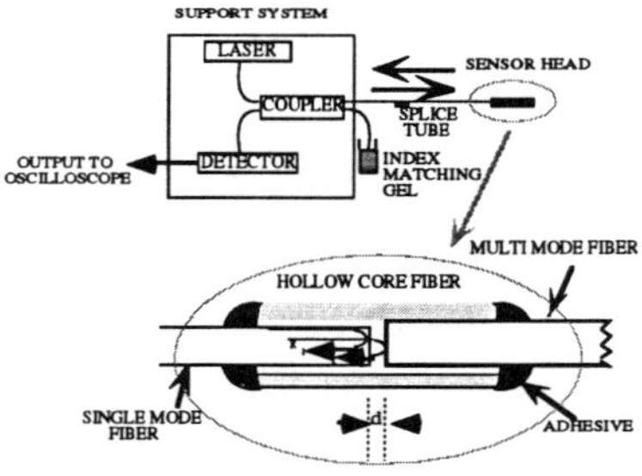

Figure 1. The extrinsic Fabry-Perot interferometric (EFPI) system.

Experiment

Four EFPI sensors were embedded into a prototype composite bridge structure. Figure 2 shows the geometry of the test specimen and the location of the EFPI fiber optic strain sensors. For the tests performed at UCSD's Powell Laboratory, peak-strain measurements were taken during each loading cycle and stored on a LeCroy digital oscilloscope. Two sensors were operated with an F&S FOSS II system and two sensors were operated with an F&S FOSS I system. The FOSS II system provides real-time strain whereas the FOSS I system provides a voltage signal which is proportional to the interferometric optical signal returned from the sensors. Therefore, data from the FOSS I system was post-processed to obtain engineering values.

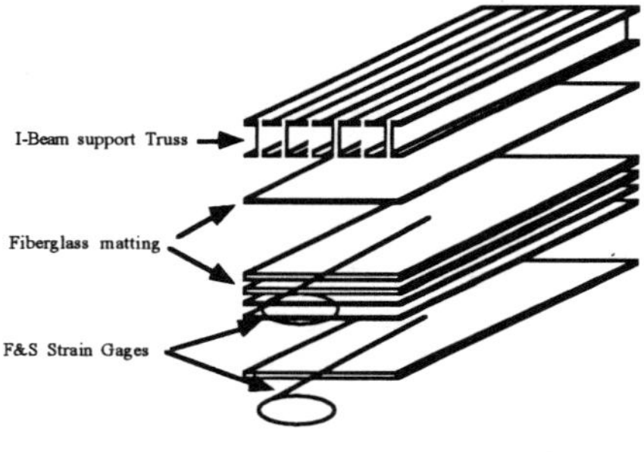

Figure 2. Installation locations of the EFPI strain gages in the prototype composite bridge structure.

Figure 3 shows the results reported from sensors 2, 3, and 4 during the tests. All strains are plotted as strain magnitude, however, sensor 2 was in compression while sensors 3 and 4 were in tension. Also provided in Figure 3 are the least squares-fit to each data set. Sensor 2 was farther away from the neutral axis than were sensors 2 and 3 and this could explain the higher strains experienced by that sensor.

In another experiment, this one performed at the Turner Fairbanks Federal Highway Administration research facility in McLean, Virginia, we attached

Absolute EFPI (AEFPI) sensors on the surface of a composite prestressing strand that is used to prestress concrete. Whereas the EFPI sensor measures relative strain, the AEFPI sensor uses whitelight interferometry to obtain absolute strain information.[7] The load applied to the prestressing tendon was increased in steps of approximately 2000 lbs. The measured strain is plotted versus the applied load in Figure 4. The final value of the AEFPI system was 12,255 με which was within 1% of the value measured by the collocated strain gage.

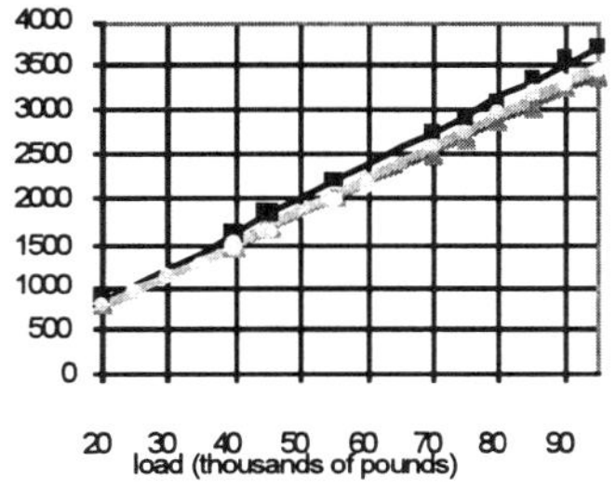

Figure 3. EFPI peak-reported strain during composite specimen loading.

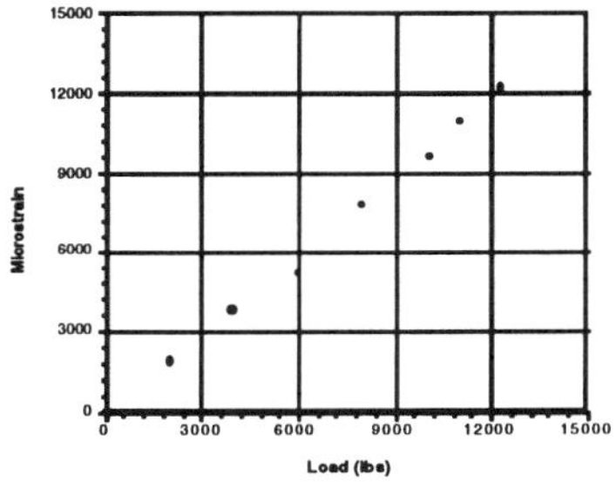

Figure 4. Measured strain versus applied load for composite prestressing strand loading.

Conclusion

We have demonstrated that EFPI sensors can be used to perform quantitative strain measurements. In structures such as buildings and bridges these sensors could be used to provide accurate information regarding the structure's degradation prior to catastrophic failure. In the long term such sensors could reduce the inspection and maintenance costs of such structures.

References

[1] Civil Infrastructure Systems Research: Strategic Issues," A Report of the Civil Infrastructure Systems Task Group, National Science Foundation, January 1993.

[2] J. Dakin, B. Culshaw, *Optical Fiber Sensors: Principles and Components*, Artech House, Boston, MA 1988.

[3] T.G. Giallorenzi, J.A. Bucaro, A. Dandridge, G.H. Sigel, J.H. Cole, S.C. Rashleigh, R.G. Priest, "Optical fiber sensor technology," IEEE J. of Quantum Electronics, vol. 18, pp. 626 - 665, 1982.

[4] E. Udd, *Overview of fiber optic applications to smart structures*, 1988, New York: Plenum.

[5] R.O. Claus, *Fiber Optic Sensor-Based Smart Materials and Structures*, IOP Publishing, Bristol and Philadelphia, April 1992.

[6] K.A. Murphy, M.F. Gunther, A.M. Vengsarkar, R.O. Claus, "Fabry-Perot fiber optic sensors in full-scale fatigue testing on an F-15 aircraft," Applied Optics, 1991.

[7] V. Bhatia, K.A. Murphy, R.O. Claus, T.A. Tran, and J.A. Greene, "Absolute strain and temperature measurements using high finesse EFPI cavities," FEORC Fiber Optics Review Conference, Blacksburg, VA, April 1994.

The Applicability of Neural Network Systems
for
Structural Damage Diagnosis

Chatmongkol Peetathawatchai[1]
Jerome J. Connor, Jr.[2]

Abstract

A general architecture of neural network systems for structural damage diagnosis and a methodology for designing the components of the architecture are developed and evaluated. Important system design issues are the choice of variables to be observed, the methodology for choosing the excitation and type of vibrational signature for the monitored structure, the configuration of the neural networks, and their training algorithm. These design issues are first examined in detail for the case of single-point damage of a multispan beam, and the evaluation is then extended to the case of multiple-point damage.

Introduction

All of the research studies involving damage detection of structures using neural networks (Elkordy, 1993; Rehak, 1992; Wu, 1992; Szewczyk, 1994; Barai, 1995) have been restricted to narrowly defined application areas. Issues related to the architecture and design strategy for a general neural network-based damage diagnosis system have not been addressed. Also, the efforts have focused on very small-scale problems that involve very limited number of possible damage states, and it is not obvious that the methods can be scaled up to deal with larger-scale problems. In most cases, at a particular instant in time, damage is assumed to occur at a single location (single-point damage condition) in order to reduce the complexity of the problem (Wu, 1992; Szewczyk, 1994; Barai, 1995). The treatment of multiple-point damage has been dealt with in a very limited way (Elkordy, 1993), and the methodology for dealing with multiple-point damage has not been adequately developed. These limitations lead to the development of a better approach.

General Architecture of Neural Network-Based Diagnosis System

A general neural network architecture for structural damage diagnosis and a methodology for designing the components of the architecture are formulated and evaluated. As shown in Fig 1, the main components of the

[1]Graduate Student, Civil and Environmental Engineering, MIT, Cambridge, MA 02139
[2]Professor, Civil and Environmental Engineering, MIT, Cambridge, MA 02139

general architecture include i) the physical system of interest and its model ii) the data preprocessing units and iii) a collection of neural networks that operate on the processed data and produce a prediction of the location and magnitude of damage. Important system design issues are the choice of variables to be observed, the methodology for choosing the excitation and type of vibrational signature for the monitored structure, the actual configuration of the neural networks, and their training algorithm. These design issues are first examined in detail for the case of single-point damage condition, and the evaluation is then extended to the case of multiple-point damage.

The evaluation phase is carried out with beam-type structures. As a first step, a single-point damage diagnosis system for a highly-idealized 2-span bending beam model (Fig 2) is developed and evaluated. The simulation model of the beam is a beam model consisting of 22 beam elements, and all damage cases have one damaged beam element. Damage is introduced by reducing the bending stiffness of the corresponding beam element. The numerical model of the beam is formulated using standard matrix procedure. Two choices of excitation and vibrational signature are considered. The first strategy applies ambient excitation to the numerical model of the structure, and then takes the mode shapes as input patterns for the neural networks that estimate the location and magnitude of damage. The second strategy applies a prespecified excitation to generate the model response, and employs the resulting response spectrum as input for the neural network system. Multilayer feedforward networks with back propagation algorithm are employed since the a-priori knowledge of the damage pattern mapping problem is unavailable. The numerical modeling of the structure and the computation of the response are carried out with MATLAB. Simulation studies are performed to evaluate the performance for the different choices of excitation and type of input.

The second step in the evaluation phase considers a 4-span bending beam model (Fig 3) with multiple-point damage. As shown in Fig 4, the diagnosis strategy is based on first identifying which spans are damaged (global structural diagnosis), and then examining independently each individual damaged span to establish the location and extent of damage (local structural diagnosis). Global structural diagnosis requires at least a set of neural networks for predicting which spans are damaged. Local structural diagnosis of each substructure requires at least two sets of neural networks for predicting the locations and extent of damage. The number of local structural diagnosis systems depends on the number of substructures.

Two choices of excitation and vibrational signature are employed for global structural diagnosis. The first choice uses ambient excitation to generate the response of the model, and takes the corresponding mode shapes as the input pattern for the neural networks. The second choice employs a prespecified excitation and uses the corresponding response spectrum as the input pattern. Local structural diagnosis employs a prespecified excitation to create the response of each substructure, and uses the frequency transfer functions of the substructure as input for the neural networks that perform damage diagnosis on the substructure.

Discussion and Conclusion

Observation based on the results of simulation studies indicates that the proposed general architecture has a good potential for any single or multiple-point damage diagnosis problem since the performance of the diagnosis systems based on the architecture is impressive (up to 95% accuracy), and the sensitivity to damage is also adequate (a damage as small as 5% reduction of EI can be effectively detected). The performance of the diagnosis systems always improves with the information incorporated in the input of neural networks until an optimum point is reached. For example, the performance increases with the number of mode shapes used for creating the input for neural networks. The performance also highly depends on the quality of the training data set. The mode shape and the response spectrum approaches perform at similar level, and this suggests the similarity between the two approaches.

References

Barai, S. V., and Pandey, P. C. 'Vibration signature analysis using artificial neural networks.' *Journal of Computing in Civil Engineering*, vol.9, no.4, pp.259-265, 1995.

Elkordy, M. F., Chang, K. C. and Lee, G. C. 'Neural networks trained by analytically simulated damage state.' *Journal of Computing in Civil Engineering*, vol. 7, no. 2, April, 1993.

Rehak, D. R. and Garrett, J. H. 'Strucural System Control using Neural Networks.' In *Proceedings of the 8th Conference on Computing in Civil Engineering*, June 1992.

Szewczyk , Z. P. and Hajela, P. 'Damage detection in structures based on feature-sensitive neural networks.' *Journal of Computing in Civil Engineering*., ASCE, vol.8, no.2, pp. 163-178, 1994.

Wu, X., Ghaboussi, J. and Garrett Jr, J. H. 'Use of neural networks in detection of structural damage.' *Computer and Structure*, 42(4), pp.649-659, 1992.

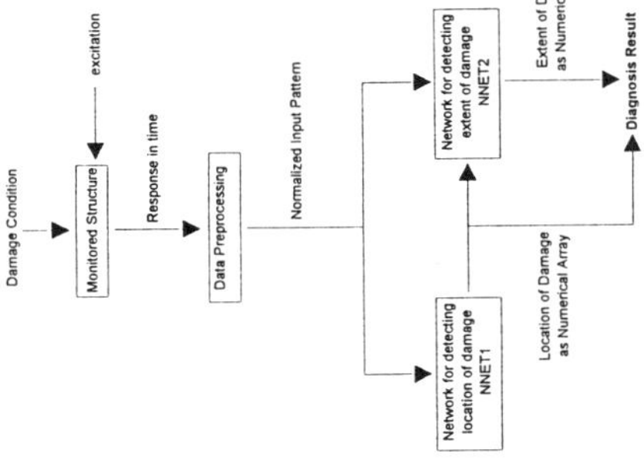

Figure 1

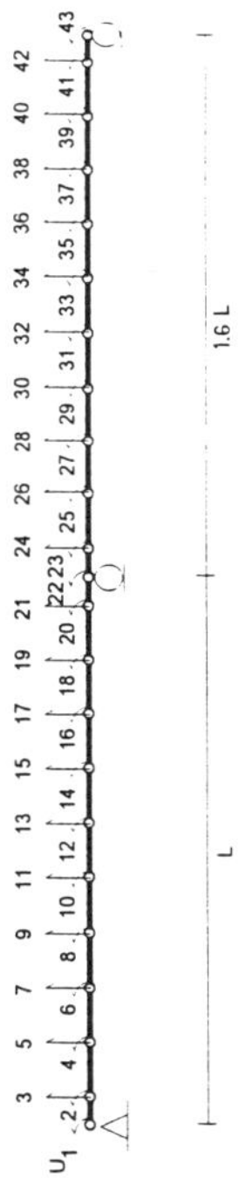

Figure 2

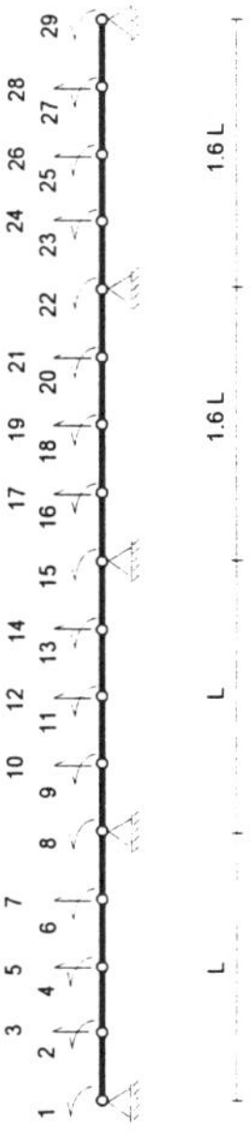

Figure 3

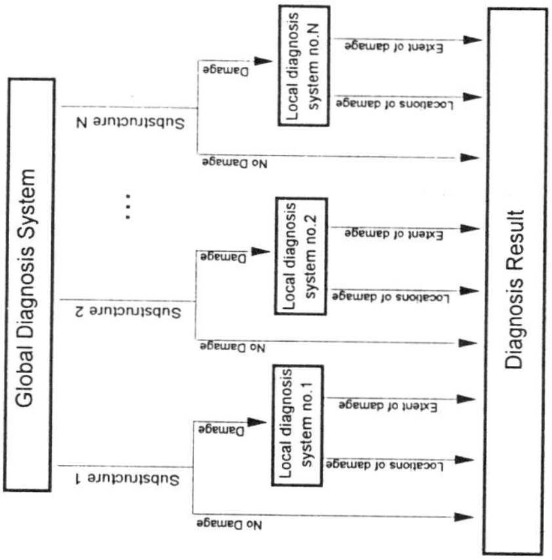

Figure 4

SIMULATION OF NONLINEAR STRUCTURES WITH ARTIFICIAL NEURAL NETWORKS

Thomas L. Paez*, Member ASCE

Abstract

Structural system simulation is important in analysis, design, testing, control, and other areas, but it is particularly difficult when the system under consideration is nonlinear. Artificial neural networks offer a useful tool for the modeling of nonlinear systems, however, such modeling may be inefficient or insufficiently accurate when the system under consideration is complex. This paper shows that there are several transformations that can be used to uncouple and simplify the components of motion of a complex nonlinear system, thereby making its modeling and simulation a much simpler problem. A numerical example is also presented.

Introduction

Artificial neural networks (ANNs) have been applied to the recurrent (autoregressive) modeling of nonlinear systems. Investigations have shown that nonlinear structures can be modeled with ANNs, at least in the case of simple systems. (See, for example, Yamamoto, 1992.) In principle, complicated systems can also be modeled using ANNs. This can be done directly (i.e., without any substantial transformation of the input or output data) using many types of ANNs. As the complexity of the system increases, an ANN that can naturally and efficiently accommodate a large number of inputs must be used for system simulation. When a mechanical system is modeled using a recurrent ANN to directly simulate motions at a large number of degrees of freedom, a very large number of exemplars of motion will be required to train the ANN to accurately represent the system. The reason is that it takes a large number of exemplars to populate a high dimensional input space.

This paper shows how the ANN modeling of nonlinear structures can be made more efficient and accurate when using data measured during experimental vibration. There are a number of operations that can be performed on the data to accomplish these goals. These are: (1) principal component analysis, (2) elimination of statistically dependent components of motion, and (3) transformation of the components of motion to statistically independent, standard normal random signals. These

* Distinguished Member of the Technical Staff, Experimental Structural Dynamics Department, 9741, MS 0557, Sandia National Laboratories, Albuquerque, New Mexico, 87185-0557. This work was sponsored by the Department of Energy under contract No. DE-AC04-94AL85000.

operations are briefly described in the following sections, along with the modeling of the components with an ANN known as the CNLS net. An example is included.

Principal Component Analysis

Principal component analysis of complex structural system motions is aimed at decomposing the motions into their essential constituent parts. A special example of this is the modal decomposition of linear systems, and analogous decompositions can be defined for nonlinear systems using, for example, singular value decomposition (SVD), or a principal component analysis ANN.

SVD is described in detail, for example, in Golub and Van Loan (1983). It can be used to decompose linear or nonlinear structural motions in the following way. Let $\boldsymbol{X}$ be an $N \times n$ matrix representing the motion of a structural system at n transducer locations and at N consecutive times. The form of the SVD is

$$\boldsymbol{X} = \boldsymbol{UWV}^T \cong \boldsymbol{uwv}^T \tag{1}$$

$\boldsymbol{V}$ and $\boldsymbol{W}$ describe characteristic shapes present in $\boldsymbol{X}$ and their corresponding amplitudes. The columns of $\boldsymbol{U}$ are filtered versions of the motions in $\boldsymbol{X}$. The approximate equality on the right indicates that some components of the representation can be eliminated, and still maintain a good approximation to $\boldsymbol{X}$. It is the evolution of the columns in $\boldsymbol{u}$ (a reduced form of $\boldsymbol{U}$) which we seek to simulate.

Elimination of Statistically Dependent Components

Under certain circumstances, some of the components produced during principal component analysis are completely statistically dependent upon others. For the sake of efficiency, we seek to eliminate statistically dependent components from the set to be modeled, then reintroduce these components during physical system simulation. In this way, ANN modeling of structural behavior is simplified.

When a dependency exists, it can be characterized using the conditional expected value of the variables in one column of $\boldsymbol{u}$ given values in another column of $\boldsymbol{u}$. This requires approximation of a joint probability density function (pdf) of the data, and this can be obtained using the kernel density estimator. (See Silverman, 1986.)

The effect of eliminating components of motion that are completely dependent on other components is to eliminate some columns in the matrix $\boldsymbol{u}$. Denote the reduced matrix $\boldsymbol{u}_r$; our objective is to model the evolution of the columns of $\boldsymbol{u}_r$ with an ANN.

Rosenblatt Transform

The previous step produces a description of the motion of a complex structure in terms of a set of components, none of which is completely statistically dependent on others. We can further transform the components, the columns of $\boldsymbol{u}_r$, into signals that are statistically independent with Gaussian distributions. The transformation that accomplishes this is the Rosenblatt transform. (See Rosenblatt, 1952.) The complete form of the transform cannot be written here, but it can be denoted

$$z = \boldsymbol{T}(\boldsymbol{u}_r) \qquad \boldsymbol{u}_r = \boldsymbol{T}^{-1}(z) \tag{2}$$

The second expression indicates that the Rosenblatt transformation is uniquely invertible. The matrix z is the same size as the matrix u_r with the same number of nonzero columns. It is the evolution of the values in these columns that is to be simulated with ANNs. Because the columns in z are statistically independent, we need only to create ANN models for signals in individual columns.

Modeling of Component Motion with the CNLS Net

Our objective at this point is to simulate the components of system motion obtained using the decompositions and transformations described above. An ANN suitable for this simulation is the connectionist normalized linear spline (CNLS) network. We cannot provide the complete form of the ANN here, but that is given in Jones, et.al., (1990). To simulate of a column in z using a CNLS net, we configure the net in a recurrent (autoregressive) framework. This configuration uses as inputs previous response values and the independent excitation, and yields on output, the current response. To specify a simulator CNLS net, we select its parameters so that it simulates the evolution of a column of z with minimum error. A following section provides an example of the use of one of the transformations described above and the modeling of the components of motion with CNLS nets.

Summary

Figure 1 summarizes the decomposition, simulation, and modeling of structural motion described in the previous sections.

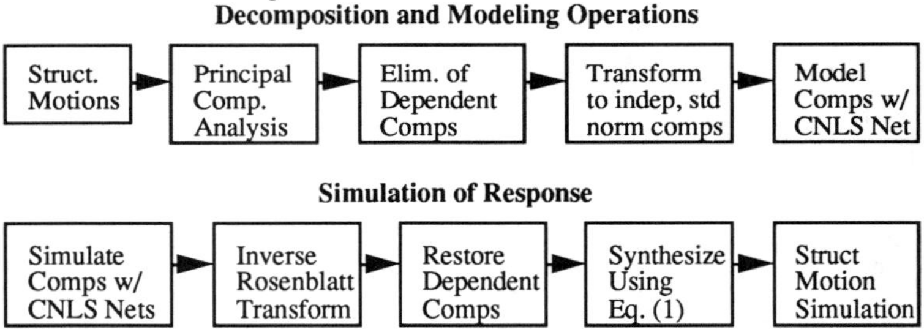

Figure 1.

Numerical Example

We simulate in this example the motion of a simple 10 degree-of-freedom system excited with a Gaussian white noise. Figure 2 shows the system. Training data for the CNLS net were generated by computing response over 6000 time steps; excitation and responses at ten locations were recorded. Figure 3 shows the response at one of the degrees of freedom. The responses were placed in a matrix X as referred to in the previous sections, and its SVD was computed. The singular values of the response indicate that accuracy of about 89% should be achieved by simulating the system response with its first three components. None of the three components were statistically dependent, so none was removed. The system is linear, so its response is Gaussian, therefore, the Rosenblatt transform was not used on the three components in the simulation. (Linearity is of no special benefit in the CNLS net modeling process since the framework for the CNLS net is nonlinear.) The first three components of the response were modeled with CNLS nets. The entire system was tested using data generated over 1000 steps of response computation. The initial conditions and excitation were used to start then execute a response simulation with CNLS nets in the space of $\boldsymbol{u}$. The first column of $\boldsymbol{u}$ from the test data and the first column from the simulated data are compared in Figure 4. This is the dominant component of the response. The match is good, particularly in view of the fact that the simulation is iterated. The simulated response is now reconstructed by substituting into Eq. (1) using the simulated $\boldsymbol{u}$, and the results are compared to the test response at one of the degrees of freedom. This is shown in Figure 5. Of course, the match is quite good since the dominant component is well simulated.

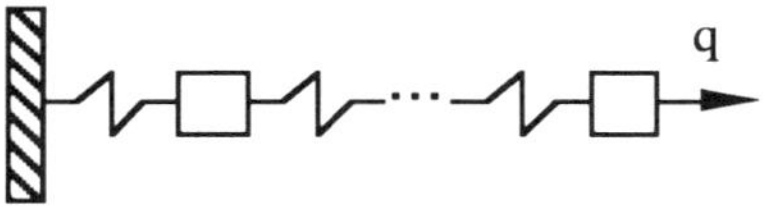

Figure 2. A simple system.

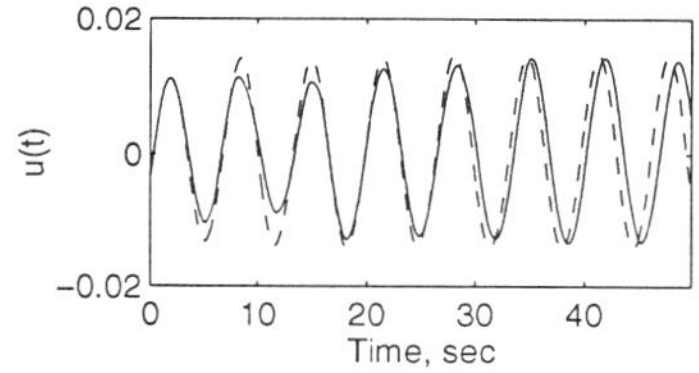

Figure 4. Comparison of test and CNLS simulated responses - Component 1.

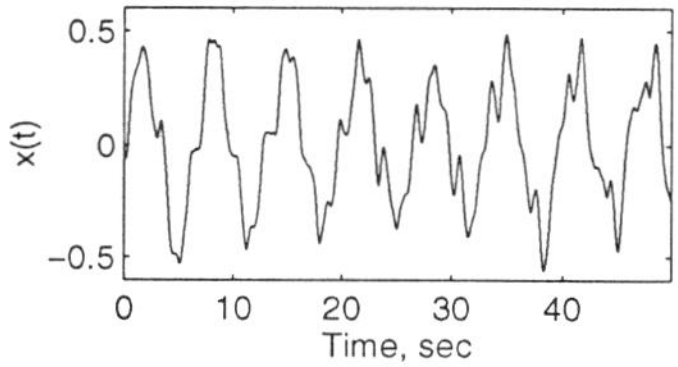

Figure 3. Response to white noise input.

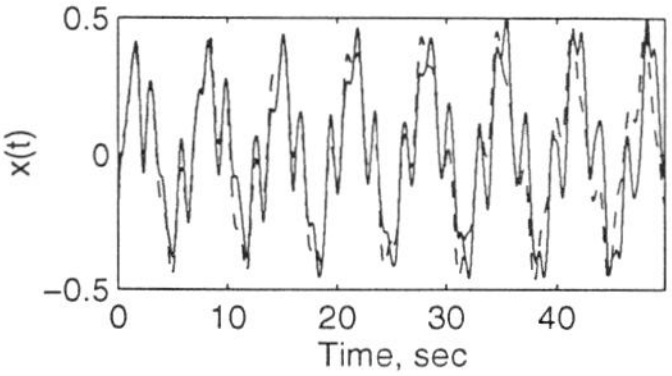

Figure 5. Test and simulated responses at a point in the simple system.

Conclusions

A sequence of operations leading to the simulation of structural system motions with ANNs is described in this paper. It is argued that if the motions can be decomposed and transformed into simple components, then the simulation will be simpler and more accurate. A numerical example confirms that relatively simple motions can, indeed, be modeled with ANNs and the CNLS net in particular.

References

Golub, G. H., Van Loan, C. F., (1983), *Matrix Computations*, Johns Hopkins University Press, Baltimore, Maryland.

Jones, R. D., et. al., (1990), "Nonlinear Adaptive Networks: A Little Theory, A Few Applications," *Cognitive Modeling in System Control*, The Santa Fe Institute.

Rosenblatt, M. (1952), "Remarks on a Multivariate transformation," *Annals of Mathematical Statistics*, **23**, 3, pp. 470-472.

Silverman, B. W.(1986), *Density Estimation for Statistics and Data Analysis*, Chapman and Hall.

Yamamoto, K., (1992), "Modeling of Hysteretic Behavior with Neural Network and its Application to Non-Linear Dynamic Response Analysis," *Applic. Artif. Intell. in Engr., Proc. 7th Conf.*, AING-92, Comp. Mech., UK, pp. 475-486.

Shape Memory Release Device Experiment

Dr. Bernie F. Carpenter[1]

Abstract

Spacecraft require a variety of separation and release devices to accomplish mission related functions. Current off-the-shelf devices such as pyrotechnics, gas-discharge systems, paraffin actuators and other electro-mechanical devices may not be able to meet future design needs. The use of pyrotechnics on advanced lightweight spacecraft, for example, will expose fragile sensors and electronics to high shock levels, and sensitive optics might be subject to contamination. Other areas of consideration include reliability, safety, and cost reduction. Shape memory alloys (SMA) are one class of actuator material that provides a solution to these design problems. Lockheed Martin Astronautics and the Phillips lab have been developing release devices based on actuation and damping properties of shape memory alloys to provide low shock, synchronized deployment of spacecraft. These devices offer "in situ" reset capability allowing the device tested to be the one that is flown. In addition, they eliminate pyrotechnic safety and handling concerns.

Introduction

Spacecraft require a variety of separation and release devices to accomplish mission related functions such as separation from the launch vehicle, deployment of solar arrays and other appendages. Existing technologies may not be able to meet mechanism requirements for future lightweight satellite programs. The use of pyrotechnics, for example, will expose fragile sensors and electronics to high shock levels and sensitive optics might be subject to contamination. A survey of pyroshock flight failures (Moening, 1985) revealed 83 shock related anomalies out of 600 launches with over 50% of these resulting in catastrophic failure. More recently, NASA published a series of lessons learned (Shapio, 1995) and raised concern, for

---

[1] Sr. Staff Engineer, Lockheed Martin Astronautics, P.O. Box 179, MS DC3085, Denver, CO 80201.

fracture of pyrotechnic require special handing and storage procedures significantly increasing integration cost. The technologies described in this paper have broad applicability to Air Force, DoD, and NASA spacecraft.

Release Device Design

In order to demonstrate the effectiveness of shape memory actuators as separation devices, shock output and release time data have been measured for four typical separation nuts: an ordinance device, a nonpyrotechnic device, and two shape memory mechanisms -- the low force nut (LFN) and the two-stage nut (TSN). The pyrotechnic employs NASA Standard Initiators (NSI) that generate gas of sufficient pressure to move a locking sleeve, releasing a set of nut segments and the preloaded bolt. A nonpyrotechnic mechanism in which the preloaded bolt is retained by a locking sleeve and nut segments has also been tested. Operation of this device is accomplished by high temperature failure of a resistive linkwire that retains two spool halves. Parting of the spool halves results in motion of the locking sleeve from an internally preloaded compression spring. Sleeve motion allows the nut segments to separate and the preloaded bolt is released.

A shape memory actuated LFN was developed to provide separation of a 0.64 cm bolt according to specifications provided in Table 1. The LFN shown in Figure 1(a) uses a mechanical advantage system to minimize the initiation force requirements of the shape memory actuation spring. A mechanical ball lock system retains the preloaded bolt. Motion of the plunger (approx. 88.96 N of force is required for a 13,344 N preload) by the resistively heated shape memory initiator spring releases the bolt. Shock generated by the bolt's retaining sleeve and the elastic energy stored in the bolt is dissipated by deformation of the martensitic shape memory damper and reset spring located in the top of the LFN housing. The LFN may be simply reset by heating the reset spring causing it to recover to its elongated austenitic state. The bolt retaining nut segments, ball lock and plunger are automatically reset. Electrical reset features are expected to become increasingly important in future spacecraft as they allow the device that was tested to be the device flown and eliminates the need to remove the nut housing from spacecraft for refurbishment.

Operation of the TSN relies on the reduction of bolt preload prior to actual release of the bolt. Preload requirements are dictated by the launch environment. However, on orbit, these requirements can be relaxed until there is a need for separation. On-orbit relaxation can occur over a long period of time compared to actual separation which must be conducted with 40 msec to avoid "tip off" concerns when using synchronized devices. The TSN shown in Figure 1(b) employs a multistage process to reduce bolt preload using a shape memory cylindrical actuator to transfer the load path from the bolt to the cylinder. This action essentially removes the bolt's stored energy resulting in a shockless device. During the second stage, the bolt is released using helical shape memory actuators integrated into the retaining nut segments.

Since there is no preload, the bolt is released in < 40 msec with minimal power. The TSN is reset by allowing the shape memory actuators to cool by removing the applied power.

Table 1. Shape Memory Release Device Specification

	LFN	TSN
Load		
Working Load (N)	13,344	13,344 to 35,584
Electrical Description		
Voltage Range (Volts)	24 min.	24 min.
Resistance (Ohms)	0.25	Stage 1 15; Stage 2 0.3
Energy Requirements (Watts sec)	90	90
Shock		
Shock Response Spectrum (g)	<500	<200
Temperature (°C)		
Survival	50 to -40	50 to -40
Operations	40 to -20	40 to -20
Other Features		
Shape Memory Technology	Damping	Two-Stage Actuation
Synchronization	Y	Y
Unit Weight (grams)	250	300
Separation Time (ms)	<50	<40
Slow Strain Energy Release	N	Y
Redundancies	Y	Y
Resettable w/o Dissassembly	Y	Y

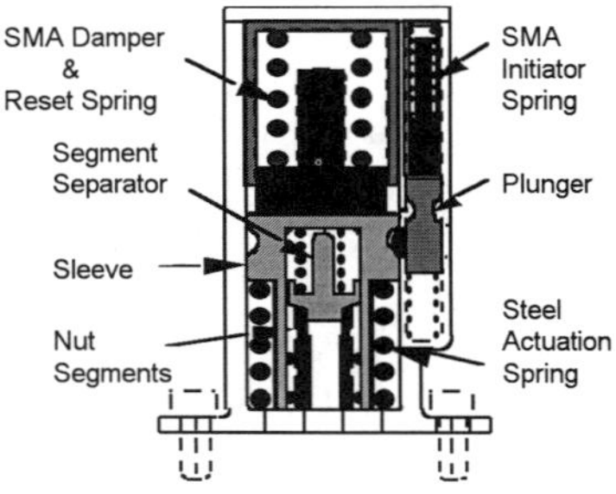

(a) Actuated Low Force Nut

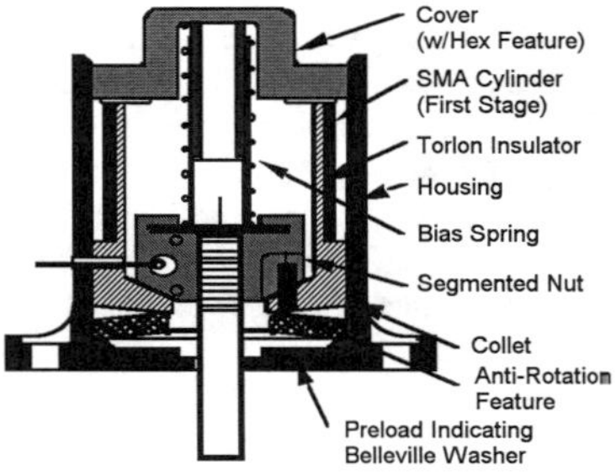

(b) Actuated Two-Stage Nut

Figure 1. Shape Memory Mechanisms.

Shock Response Spectra

Comparative shock spectra (all data was acquired on the same test fixture) have been measured for typical separation devices. These results are provided in Figure 2 along with the spectra for the LFN and TSN. Shock output for the LFN falls below that of both pyrotechnic, and non-pyrotechnic devices is less than the shock design limit of 500 g's. These results demonstrate the exceptional damping capability of shape memory springs while maintaining release times of ≤ 50 msec. Shock output for the TSN is exceptionally low demonstrating the effectiveness of dissipating remaining stored energy from the bolt prior to its release. Separation times of 40 msec have been achieved using less power than the LFN.

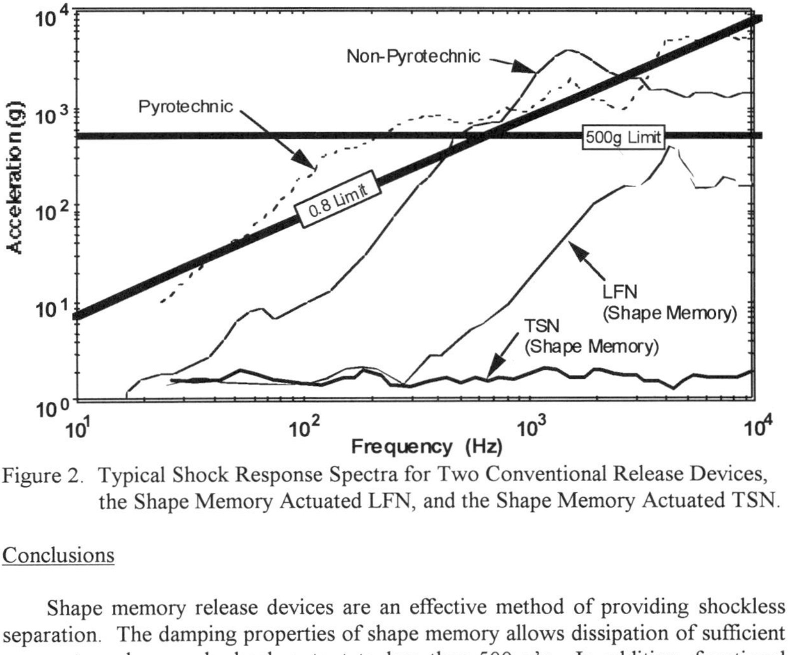

Figure 2. Typical Shock Response Spectra for Two Conventional Release Devices, the Shape Memory Actuated LFN, and the Shape Memory Actuated TSN.

Conclusions

Shape memory release devices are an effective method of providing shockless separation. The damping properties of shape memory allows dissipation of sufficient energy to reduce peak shock output to less than 500 g's. In addition, functional testing has demonstrated separation times of ≤ 50 msec.

References

1. C. J. Moening, "Pyroshock Flight Failures," Proceedings of 31st Annual Technical Meeting of the Institute of Environmental Sciences (IES), 3 May 1985.
2. W. Shapio, F. Murry, R. Howorth, and R. Fusaro, "Space Mechanisms Lessons Learned Study: Volume I – Summary," NASA Technical Memorandum 107046, September 1995.

TIME DOMAIN FUNDAMENTAL SOLUTION TO NONCLASSICAL THERMOELASTICITY WITH ONE RELAXATION TIME
Part I: Three-Dimensional Solution

Jianming Chen* Alexander H.-D. Cheng
Department of Civil Engineering, University of Delaware, Newark, DE 19716 , U.S.A.

ABSTRACT

This paper develops transient fundamental solution for three-dimensional nonclassical thermoelasticity with one relaxation time. The solutions in the Laplace transform domain are presented. The transient solutions have been derived. Lastly, the variations with time of the solid displacements and temperature components for the point loads after the arrival of the waves are studied graphically , the ensuing transient components are compared with Laplace transform domain solutions and are found in excellent agreement.

INTRODUCTION

In recent years, there has been increasing interest in developing a nonclassical thermoelasticity due to its applications in numerous branches of science and engineering, including aerospace engineering and the areas of superconductivity, laser technology and electronic packages *etc.*. In classical thermoelasticity, the displacement field is governed by a wave type equation (hyperbolic), while the temperature field is governed by a diffusion type equation (parabolic), which leads to an infinite propagation speed for finite thermal disturbances, a physically unacceptable situation. To eliminate the paradox, Lord and Shulman (1967), proposed a new theory by incorporating a flux-rate with one relaxation time, t_0 into Fourier's law. Through this relationship the temperature distribution will be described as a wave propagation phenomenan.

REVIEW OF GOVERNING EQUATIONS

Thus we can write the basic equations of the new theory expressing respectively, the first constitutive relation, the second constitutive relation (for the entropy), equilibriom equation, the modified Fourier's Law and the linerlized entropy balance equation, as

$$\sigma_{ij} = \lambda u_{k,k}\delta_{ij} + \mu(u_{i,j} + u_{j,i}) - \gamma_t T\delta_{ij} \tag{1a}$$

$$T_0 S = \gamma_t T_0 u_{k,k} + c_\epsilon T \tag{1b}$$

$$\sigma_{ij,j} + f_i = \rho\ddot{u}_i \tag{1c}$$

$$q_i + t_0\dot{q}_i = -\kappa T_{,i}, \tag{1d}$$

$$T_0\dot{S} + q_{i,i} = \psi \tag{1e}$$

where $i, j = 1, 2, 3$, $c_\epsilon = \rho c$, $\gamma_t = (3\lambda+2\mu)\alpha_t$, λ and μ are the Lame constants, u_i, σ_{ij}, T_0, T, q_i, f_i and ψ are the displacement, stress, reference temperature, temperature difference, heat flux, body force and heat source, and ρ, γ_t, α_t, κ, c_ϵ, and c are the density, stress-temperature modulus, the coefficient of linear volume expansion, conductivity and specific heat at constant strain and the specific heat referred to the unit mass of the body, respectively. S is the entropy per unit volume and unit time.

Eliminating $T_0\dot{S}$ and q_i from eqns (1b,d,e) and $\sigma_{i,j}$ from eqns (1a,c) and taking Laplace transform, we obtain the transform domain motion and energy equations transform domain:

$$\mu\tilde{u}_{i,jj} + (\lambda + \mu)\tilde{u}_{j,ji} - \rho s^2\tilde{u}_i - \gamma_t\tilde{T}_{,i} = -\tilde{f}_i \tag{2a}$$

$$\zeta\tilde{T}_{,jj} - c_\epsilon s\tilde{T} - \gamma_t T_0 s\tilde{u}_{j,j} = -\tilde{\psi} \tag{2b}$$

* **Structural Engineer/STV GROUP 225 Park Ave. South NY NY 10003, U.S.A.**

where $\zeta = \frac{1}{1/\kappa+(t_0/\kappa)s}$, the tilde denotes the Laplace transformation and s is Laplace transform parameter. Eqns (2a,b) can be put in a more convenient form by using nondimensional variables $x_i^* = \sqrt{\frac{\lambda+2\mu}{\rho}}\frac{c_e}{\kappa}$, $\tau = \frac{\lambda+2\mu}{\rho}\frac{c_e}{\kappa}t$, $T^* = \frac{T}{T_0}$, $\sigma_{ij}^* = \frac{\sigma}{(3\lambda+2\mu)\alpha_t T_0}$, $u_i^* = \frac{\lambda+2\mu}{(3\lambda+2\mu)\alpha_t T_0}\sqrt{\frac{\lambda+2\mu}{\rho}}\frac{c_e}{\kappa}u_i$ along with the well known thermomechnical coupling parameter $\delta = \frac{(3\lambda+2\mu)^2\alpha_t^2 T_0}{(\lambda+2\mu)c_e}$ and nondimensional relaxation time $\tau_0 = \frac{\lambda+2\mu}{\rho}\frac{c_e}{\kappa}t_0$.

LAPLACE TRANSFORM DOMAIN FUNDAMENTAL SOLUTION

Following Kupradze (1979) an explicit nondimensional Laplace transform domain fundamental solution, which involves the response of the medium to suddenly applied point forces and a point heat source with Heauiside step functions in time, for eqns(2a,b) can be obtained as follows:

$$\begin{aligned}\tilde{G}_{ij} = &-(A_{ij} + B_{ij}\lambda_1 + C_{ij}\lambda_1^2)\frac{s^2-\lambda_2^2}{\lambda_2^2-\lambda_1^2}\frac{1}{s^3}e^{-\lambda_1 r} + (A_{ij} + B_{ij}\lambda_2 + C_{ij}\lambda_2^2)\frac{s^2-\lambda_1^2}{\lambda_2^2-\lambda_1^2}\frac{1}{s^3}e^{-\lambda_2 r} \\ &-(A_{ij} + B_{ij}\lambda_3 + C_{ij}\lambda_3^2)\frac{1}{s^3}e^{-\lambda_3 r} + D_{ij}\frac{1}{s}e^{-\lambda_3 r}\end{aligned} \tag{3a}$$

$$\tilde{G}_{i4} = \frac{1}{4\pi\zeta s(\lambda_1^2-\lambda_2^2)}\left(-\frac{x_i}{r^3} - \frac{x_i}{r^2}\lambda_1\right)e^{-\lambda_1 r} + \frac{1}{4\pi\zeta s(\lambda_1^2-\lambda_2^2)}\left(\frac{x_i}{r^3} + \frac{x_i}{r^2}\lambda_2\right)e^{-\lambda_2 r} \qquad \tilde{G}_{4i} = \delta s\tilde{G}_{i4} \tag{3b, c}$$

$$\tilde{G}_{44} = \frac{1}{4\pi r\zeta s}\frac{\lambda_1^2-\Lambda^2}{\lambda_1^2-\lambda_2^2}e^{-\lambda_1 r} - \frac{1}{4\pi r\zeta s}\frac{\lambda_2^2-\Lambda^2}{\lambda_1^2-\lambda_2^2}e^{-\lambda_2 r} \tag{3d}$$

$$\begin{aligned}&\lambda_1^2-\lambda_2^2 = \{[s^2 + s(1+\tau_0 s)(1+\delta)]^2 - 4s^3(1+\tau_0 s)\}^{1/2} \qquad \zeta = \frac{1}{1+\tau_o s} \\ &\lambda_{1,2} = \{\frac{1}{2}[s(1+\tau_0 s)(a+\delta) + s^2 \pm (\lambda_1^2-\lambda_2^2)]\}^{1/2} \qquad \lambda_3 = \frac{s}{c_s} \qquad c_s = \sqrt{\mu} \\ &A_{ij} = \frac{1}{4\pi}\left(\frac{3x_ix_j}{r^5} - \frac{\delta_{ij}}{r^3}\right) \quad B_{ij} = \frac{1}{4\pi}\left(\frac{3x_ix_j}{r^4} - \frac{\delta_{ij}}{r^2}\right) \quad C_{ij} = \frac{1}{4\pi}\left(\frac{x_ix_j}{r^3}\right) \quad D_{ij} = \frac{1}{4\pi r\mu}\delta_{ij}\end{aligned} \tag{4}$$

TRANSIENT FUNDAMENTAL SOLUTION

Inspection of eqn(4) shows that $\lambda_1^2 - \lambda_2^2$ can be expressed to the first order of approximation in terms of $\frac{s}{1/\tau_0+s}$ for all time range as long as $\tau_0 > \frac{1}{1+\delta}$ as follows:

$$\lambda_1^2 - \lambda_2^2 \approx cs(s+b) \qquad \lambda_{1,2} = \frac{1}{c_{t,p}}\sqrt{(s+\eta_{t,p})^2 - \xi_{t,p}^2} \tag{5}$$

$$\begin{aligned}&c = \tau_0(1+\delta) + \frac{\delta-1}{1+\delta} \qquad c_p = \sqrt{1+\delta} \qquad \eta_p = \xi_p \approx 0 \\ &b = \frac{(1+\delta)^2}{\tau_0(1+\delta)+\delta-1} \qquad c_t = \sqrt{1+\delta}\sqrt{\frac{1}{\tau_0(1+\delta)^2+\delta}} \qquad \eta_t = \xi_p = \frac{1}{2}\frac{(1+\delta)^2}{\tau_0(1+\delta)^2+\delta}\end{aligned} \tag{6}$$

Armed with approximations for λ_1, λ_2, $\lambda_1^2 - \lambda_2^2$ together with familiar Laplace transform formula, we can invert eqn(3a) to arrive at the following transient solution:

$$\begin{aligned}G_{ij} = &\int_{r/c_t}^{t}\left[P_{11}e^{-b(t-\tau)} + P_{12}\right]e^{-\eta_t\tau}I_0(\xi_t\sqrt{\tau^2 - r^2/c_t^2})d\tau H(t-r/c_t) \\ &+ \left\{P_{21}e^{-a(t-r/c_t)} + P_{22}e^{-b(t-r/c_t)} + P_{23} + P_{24}(t-r/c_t) + P_{25}(t-r/c_t)^2\right\}e^{-\eta_t r/c_t}H(t-r/c_t) \\ &+ \int_{r/c_t}^{t}\left\{P_{21}e^{-a(t-\tau)} + P_{22}e^{-b(t-\tau)} + P_{23} + P_{24}(t-\tau) + P_{25}(t-\tau)^2\right\} \\ &\times e^{-\eta_t\tau}\frac{\xi_t r/c_t}{\sqrt{\tau^2 - r^2/c_t^2}}I_1(\xi_t\sqrt{\tau^2 - r^2/c_t^2})d\tau H(t-r/c_t) \\ &+ \left\{P_{31}e^{-a(t-r/c_p)} + P_{32}e^{-b(t-r/c_p)} + P_{33} + P_{34}(t-r/c_p) + P_{35}(t-r/c_p)^2\right\}H(t-r/c_p) \\ &+ \left\{P_{41} + P_{42}(t-r/c_s) + P_{43}(t-r/c_s)^2\right\}H(t-r/c_s)\end{aligned} \tag{7}$$

$$P_{11}=(-\tfrac{f_9}{b}+f_{10})c_d \quad P_{12}=(\tfrac{f_9}{b}+f_{11})c_d \quad P_{21}=\tfrac{f_1}{a(a-b)}-\tfrac{f_2}{a^2(a-b)}+\tfrac{f_3}{a^3(a-b)}$$

$$P_{22}=-\tfrac{f_1}{b(a-b)}+\tfrac{f_2}{b^2(a-b)}-\tfrac{f_3}{b^3(a-b)}+f_4-\tfrac{f_5}{b}+\tfrac{f_6}{b^2} \quad P_{23}=\tfrac{f_1}{ab}-\tfrac{f_2(a+b)}{a^2b^2}+\tfrac{f_3(a^3-b^3)}{a^3b^3(a-b)}+\tfrac{f_5}{b}-\tfrac{f_6}{b^2}+f_7$$

$$P_{24}=\tfrac{f_2}{ab}-\tfrac{f_3(a+b)}{a^2b^2}+\tfrac{f_6}{b} \quad P_{25}=\tfrac{f_3}{2ab}+\tfrac{f_8}{2} \quad P_{31}=\tfrac{g_1}{a(a-b)}-\tfrac{g_2}{a^2(a-b)}+\tfrac{g_3}{a^3(a-b)}$$

$$P_{32}=-\tfrac{g_1}{b(a-b)}+\tfrac{g_2}{b^2(a-b)}-\tfrac{g_3}{b^3(a-b)}+g_4-\tfrac{g_5}{b}+\tfrac{g_6}{b^2} \quad P_{33}=\tfrac{g_1}{ab}-\tfrac{g_2(a+b)}{a^2b^2}+\tfrac{g_3(a^3-b^3)}{a^3b^3(a-b)}+\tfrac{g_5}{b}-\tfrac{g_6}{b^2}+g_7$$

$$P_{34}=\tfrac{g_2}{ab}-\tfrac{g_3(a+b)}{a^2b^2}+\tfrac{g_6}{b}+g_8 \quad P_{35}=\tfrac{g_3}{2ab}+\tfrac{g_9}{2} \quad P_{41}=D_{ij}-\tfrac{C_{ij}}{\mu} \quad P_{42}=-\tfrac{B_{ij}}{\sqrt{\mu}} \quad P_{43}=-\tfrac{1}{2}A_{ij} \tag{8}$$

$$g_1=\tfrac{A_{ij}d_1}{c} \quad g_2=\tfrac{A_{ij}d_2}{c} \quad g_3=\tfrac{A_{ij}d_3}{c} \quad g_4=\tfrac{C_{ij}e_2}{c} \quad g_5=\tfrac{C_{ij}e_1}{c}+\tfrac{B_{ij}e_2c_p}{c}$$

$$g_6=-\tfrac{A_{ij}}{2c}+\tfrac{B_{ij}e_1c_p}{c} \quad g_7=\tfrac{1}{2}C_{ij} \quad g_8=\tfrac{1}{2}B_{ij}c_p \quad g_9=\tfrac{1}{2}A_{ij}$$

$$f_1=-\tfrac{A_{ij}d_1}{c} \quad f_2=-\tfrac{A_{ij}d_2}{c} \quad f_3=-\tfrac{A_{ij}d_3}{c} \quad f_4=-\tfrac{C_{ij}e_2}{c} \quad f_5=-\tfrac{C_{ij}e_1}{c}$$

$$f_6=\tfrac{A_{ij}}{2c} \quad f_7=\tfrac{1}{2}C_{ij} \quad f_8=\tfrac{1}{2}A_{ij} \quad f_9=-\tfrac{B_{ij}e_1}{c} \quad f_{10}=-\tfrac{B_{ij}e_2}{c} \quad f_{11}=\tfrac{1}{2}B_{ij}$$

$$d_1=\tfrac{\tau_0}{2}(1+\delta) \quad d_2=1+\delta \quad d_3=\tfrac{1}{2\tau_0}(1+\delta) \quad e_1=\tfrac{1}{2}(1-\delta) \quad e_2=\tfrac{1}{2}(\tau_0(1-\delta)-1) \quad a=\tfrac{1}{\tau_0}$$

Similarly we obtain the time domain solution from eqn.(3d) :

$$\begin{aligned} G_{44} =& P_{11}e^{-\eta_t t}\delta(t-r/c_t) + P_{12}e^{-\eta_t t}\frac{1}{\sqrt{t^2-r^2/c_t^2}}I_1(\xi_t\sqrt{t^2-r^2/c_t^2})H(t-r/c_t) \\ &+\int_{r/c_t}^{t}\Big(P_{13}e^{-b(t-\tau)}+P_{14}\Big)e^{-\eta_t\tau}\frac{\xi_t r/c_t}{\sqrt{\tau^2-r^2/c_t^2}}I_1(\xi_t\sqrt{\tau^2-r^2/c_t^2})d\tau H(t-r/c_t) \\ &+\Big(P_{13}e^{-b(t-r/c_t)}+P_{14}\Big)e^{-\eta_t r/c_t}H(t-r/c_t)+P_{21}\delta(t-r/c_p)+\Big(P_{22}e^{-b(t-r/c_p)}+P_{23}\Big)H(t-r/c_p) \end{aligned} \tag{9}$$

$$P_{11}=\frac{\tau_0}{8\pi r}+\frac{d_3}{8\pi rc} \quad P_{12}=(\frac{\tau_0}{8\pi r}+\frac{d_3}{8\pi rc})\xi_d\frac{r}{c_d} \quad P_{13}=\frac{d_2-d_3b}{8\pi rc}-\frac{d_1}{8\pi rbc} \quad P_{14}=\frac{1}{8\pi r}+\frac{d_1}{8\pi rbc}$$

$$P_{21}=\frac{\tau_0}{8\pi r}-\frac{d_3}{8\pi rc} \quad P_{22}=\frac{-d_2+d_3b}{8\pi rc}+\frac{d_1}{8\pi rbc} \quad P_{23}=\frac{1}{8\pi r}-\frac{d_1}{8\pi rbc} \tag{10}$$

$$d_1=1+\delta \quad d_2=2\tau_0(a+\delta)-1 \quad d_3=\tau_0^2-\tau_0+\delta$$

Obviously, the shear wave disappeared in the pressure field due to fluid injection.

Numerical Results

Selected numerical results are presented nondimensionally. The material parameters was selected as follows : $\lambda = 0.5$, $\mu = 0.25$, $\delta = 0.02$, $\tau_0 = 2.0$. The applied force point and the heat source point are located at $(0,0,0)$ and the receiver is chosen at nondimensional coordinate $(1,2,3)$. The nondimensional wave velocities are approximately $c_p = 1.01$ $(Stress Pressure\ Wave)$; $c_t = 0.697$ $(Thermal\ Wave)$; $c_s = 0.5$ $(Shear\ Wave)$, while the nondimensional time required for the three waves to reach the recievers are $t_p = 3.705$, $t_s = 7.483$, and $t_t = 5.37$. In all the Figures, excellent agreement of analytical solution and numerical Laplace inversion is seen.

In figure (4) G_{44}, which is temperature due to thermal injection, clearly demonstrate the existence of two wave fronts i,e, stress pressure wave and thermal wave, which propagate with speeds c_p and c_t, respectively. An interesting feature is the presence of two pulses, which, in the form of the Dirac delta function for the 3-D case, are associated with the arrival of the two dilatational waves.

In figures (1,2,3) G_{ij}, indicate that the displacement of the solid in i-th direction due to point force in j-th direction contains three wave fronts, two dilatational waves and one rotational wave. They propagate with speeds c_p c_t and c_s respectively. Further study reveals that stress wave is true wave, which has negligibly small dissipation factor $\eta_p \approx 0$, while the thermal wave is strongly attenuated with a high dissipative factor $\eta_t = 0.242$ and hence it can be detected only at a close proximity

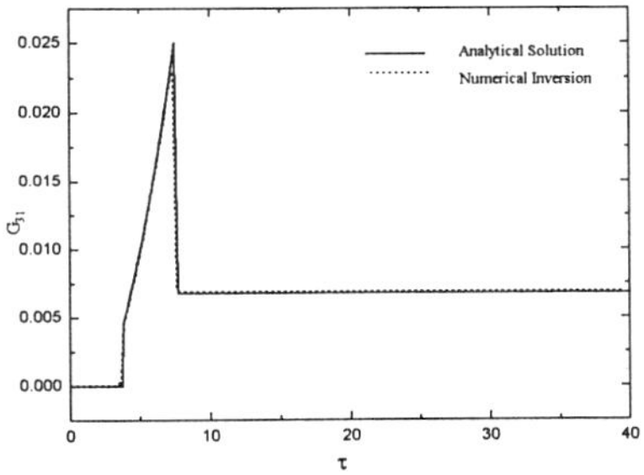

Fig. 1. Displacement time history at (1,2,3) due to point force at (0,0,0)

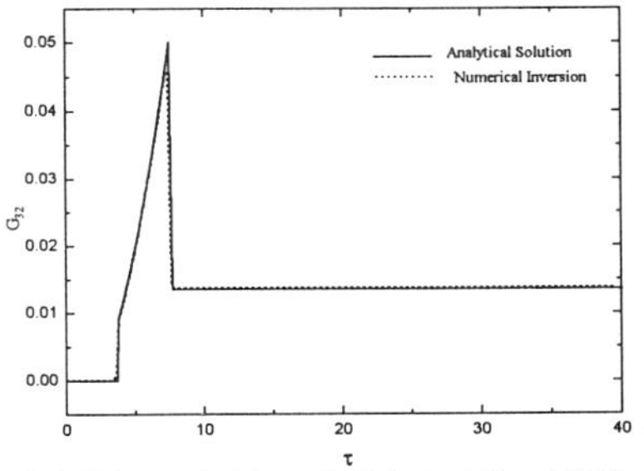

Fig. 2. Displacement time history at (1,2,3) due to point force at (0,0,0)

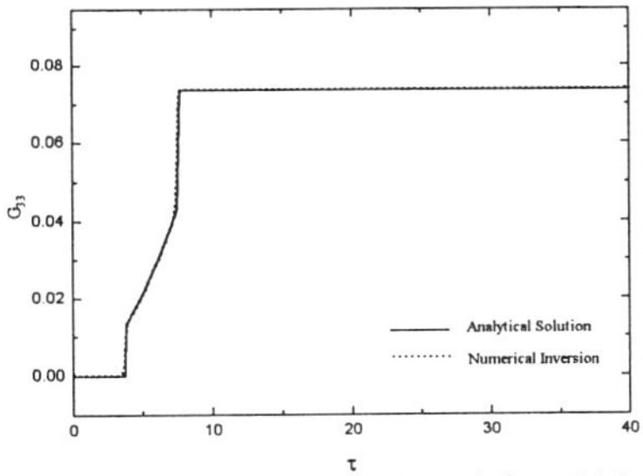

Fig. 3. Displacement time history at (1,2,3) due to point force at (0,0,0)

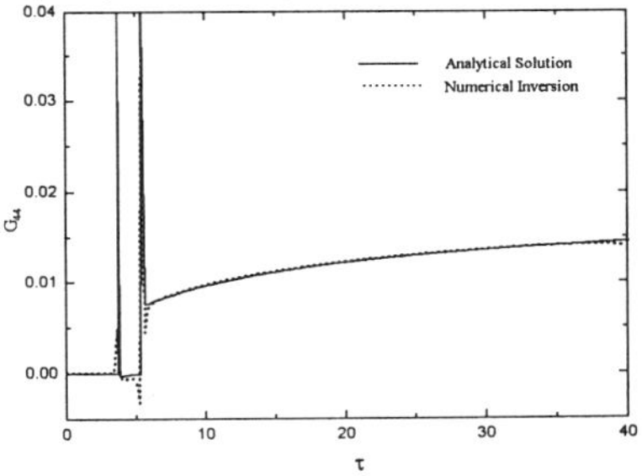

Fig. 4.Temperature time history at (1,2,3) due to heat source at (0,0,0)

of the source. As the disturbance moves into the media, the second wave front slows down and eventually disappears. The dissipative factor for the shear wave is definitely zero.

Conclusion

The transient Green's functions presented in this paper may be used effectively as kernel functions for the time domain boundary element method.

References

Lord, H.W. and Shulman, Y., (1967), 'A Generalized Dynamical Theory of Thermoelasticity', *J. Mech Phys. Solids*, V15, pp. 299-309

Kupradze, V.D., (1979), 'Three-dimensional Problems of the Mathematical Theory of Elasticity and Thermoelasticity', North-Holland

Elastic Moduli of a Bond Model for Reinforced Concrete

James V. Cox[1], Member ASCE

Abstract

This study addresses the elastic moduli of an elastoplastic bond model. The bond model homogenizes the effects of the mechanical interaction between the bar's ribs and adjacent concrete. It couples the tangential and normal response in an interface idealization that can potentially predict both pull-out and splitting failures.

Numerical solutions and limited experimental data suggest that the elastic moduli are significantly affected by the local response of the concrete near the bar, which was not accounted for in the original elastoplastic model. It is postulated that the level of damage in the concrete near the bar is dependent upon the state of the contact between them. An initial attempt to characterize this effect introduces elastoplastic coupling in the bond law. The emphasis of this study is on the radial response; due to the coupled form of the model, the radial response affects many aspects of the model's behavior. Preliminary analyses have produced improved results over the original model, but additional validation studies are needed.

Introduction

Modeling of reinforced concrete requires models for the constituent materials and their interaction, bond behavior. Bond has been modeled at several scales, each of which has well-defined strengths and weaknesses. Phenomenological bond models that are implemented in finite element calculations with interface elements are a compromise of scale. They are not directly useful in modeling large-scale structural systems, and they do not directly represent the underlying mechanics. They represent a homogenization of the underlying mechanics, which occurs in a zone near the bar – the bond zone. Models that couple the response normal and longitudinal to the bar, or *two-dimensional* models, can include the wedging effect of the ribs, thus potentially allowing both pull-out and splitting failures to be predicted in analyses of structural components. De Groot *et al.* (1981) apparently developed the first two-dimensional model; it characterized the response of the concrete between transverse cracks (inclined struts).

---

[1] Assistant Professor, Department of Civil Engineering, Whiting School of Engineering, Johns Hopkins University, Baltimore, MD 21218.

Bond behavior typified by the experiments of Eligehausen *et al.* (1983) motivated the recent development of an elastoplastic bond model (Cox 1994, and Herrmann and Cox 1994) which characterizes the mechanical interaction of steel reinforcing bars with concrete. The generalized stresses for the model (Q_1, Q_2) are the tangential and normal components of the interfacial traction (τ and σ). The generalized strains (q_1, q_2) are the conjugate relative displacements nondimensionalized by the bar diameter (δ_t / D_b and δ_n / D_b). Clearly the effects of the mechanical interaction between the ribs and concrete are not confined to the interface; this is merely an idealization.

Elastoplastic Model

Simplifying assumptions allowed a yield surface and flow rule to be extracted from the experimental data of Malvar (1992). For monotonic response, the evolution of the yield surface and flow rule is characterized by a single internal variable, d, the plastic component of δ_t divided by the rib spacing (a measure of internal length associated with the bar geometry). The bond zone is assumed to be fully-damaged when d equals 1, at which point the model produces a frictional response. The form of the yield surface is given by:

$$\frac{\tau}{f_t} = C(d)\Phi\left(d, \hat{\sigma}(d) - \frac{\sigma}{f_t}\right) \tag{1}$$

where f_t denotes a measure of the concrete tensile strength; C is the isotropic hardening and softening function; and $\hat{\sigma}$ is the kinematic softening function. The form of the flow rule is given by:

$$\left\{\dot{q}^p\right\} = \dot{\lambda}\left\{\begin{array}{c} \text{sgn}(\tau) \\ g(\sigma,d) \end{array}\right\} \tag{2}$$

where $\dot{\lambda}$ is the consistency parameter, and g is the rate of plastic radial dilation with respect to plastic slip.

The "elastic response" deviated from true elastic behavior, since it partially accounted for the underlying contact problem between the ribs and adjacent concrete after the breakdown of adhesion. The incremental moduli used to characterize the elastic response are applicable to monotonic loading and are given by

$$\mathbf{D}^e = E_c\begin{bmatrix} 0.1 & -0.0012\,\text{sgn}(\delta_t) \\ -0.0012\,\text{sgn}(\delta_t) & 0.04 \end{bmatrix} \tag{3}$$

where sgn(0)≡0, and E_c is the elastic modulus of concrete. The sgn(δ_t) factor reflects the contact character of the interface between the inclined rib surface and the adjacent concrete, accounting for the slip direction and the contact symmetry at zero slip (*i.e.*, an incremental contraction of the bond zone does not produce a generalized shear stress when the slip is zero). The implementation accounts for finite precision and treats δ_t as zero if $|\delta_t| < \varepsilon$, where ε denotes a very small tolerance. The small off-diagonal terms reflect the observed lack of significant coupling during the initial elastic response (see *e.g.*, the experimental data of Malvar 1992). Researchers (see *e.g.*, Lutz and Gergely 1967) have concluded that significant wedging action of a rib occurs after crushing of the concrete near the rib face. The model reflects this interpretation of the mechanics since significant radial dilation begins after the initial plastic slip.

Results from some validation problems indicate that the model might be improved by refining its elastic component. In particular, some results may indicate that D^e_{22} should depend on the state of contact between the bar and concrete. The flow rule produces a radial dilation that simulates the wedging action of the ribs, in a homogenized sense. The "elastic moduli" also reflect a homogenization of the mechanical interaction of the ribs, but the above form does not significantly characterize the change in the state of contact between the bar and adjacent concrete and the corresponding damage.

For the radial response, the elastic and plastic components are kinematically opposed. Plastic radial dilation characterizes an average radial displacement of the bond zone due to the mechanical interaction of the ribs. The corresponding elastic radial contraction characterizes the elastic deformation which is concentrated near the regions of contact. Thus, the model's elastic component is important to its prediction of the radial response and the related failure mode – splitting..

Two unpublished tests of Malvar suggest that the elastic radial response becomes more compliant with radial dilation. Increased radial dilation can be attributed to the ribs wedging under the adjacent concrete. This behavior results in a reduced contact area, thus potentially producing increased microcracking that would increase the radial compliance. To account for this observed response, elastoplastic coupling is introduced in the model by expressing the strain energy function as:

$$W\left(\underset{\sim}{q}^{e}, q_2^p\right) = \frac{1}{2}\underset{\sim}{q}^{eT}\mathbf{D}^{e}\left(q_2^p\right)\underset{\sim}{q}^{e} \tag{4}$$

For simplicity, since the off-diagonal moduli are very small, only the diagonal moduli are retained. Including elastoplastic coupling only in the radial component gives:

$$\mathbf{D}^{e}\left(q_2^p\right) = diag\left(D_{11}^{e}, D_{22}^{e}\left(q_2^p\right)\right) \tag{5}$$

The generalized stress components can then be determined from derivatives of W with respect to the corresponding generalized strains resulting in:

$$\underset{\sim}{Q} = \mathbf{D}^{e}\left(q_2^p\right)\underset{\sim}{q}^{e} \tag{6}$$

In this preliminary study, a simple linear relationship is adopted between the elastic compliance modulus for the normal response and q_2^p, *i.e.*,

$$\left(\frac{D_{22}^{e}}{E_c}\right)^{-1} = k_1 + k_2\left(q_2^p\right) \tag{7}$$

where the initial calibration values are k_1= 1/4 and k_2=350. The order of magnitude of k_1 is deduced from a simple elastic analysis. Consider a cylinder of concrete around the bar with a thickness n times the rib height (h_r). Assume that it has numerous longitudinal cracks open such that hoop stresses can be ignored and axisymmetry is a reasonable approximation. For ν=0.17, h_r/D_b=0.07, and n=2, solution of the elasticity equations gives $\left(D_{22}^{e}/E\right)^{-1} \cong 0.12$.

Preliminary Results

Figure 1 presents some preliminary results for the tests of Malvar (1992). In these simple analyses, the effect of the specimen behavior is not addressed. The specimen is relatively small (diameter $\cong 4D_b$) and is pre-split, so omitting a specimen model neglects the initial hoop stresses that occur, thus over predicting the initial magnitude of σ. In addition, ignoring the bar deformation causes the initial elastic stiffness (see the bond stress vs. slip) to be over predicted.

Both the tangent and normal responses of the model reproduce the experimental results with acceptable accuracy. Unlike the experimental results, the initial elastic contraction has not been subtracted. Calibration of the elastoplastic bond model (Cox 1994) also considered this "simplified specimen model," so these results do not validate the bond model. However, the relatively small constant value of D_{22}^{e} (equation 3) led to unrealistic elastic contractions both in the simple model considered here and in axisymmetric finite element models of the specimen.

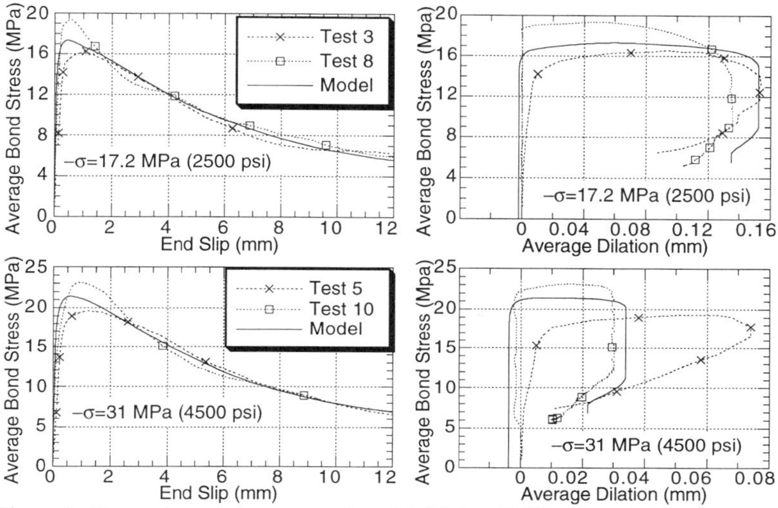

Figure 1. Comparisons between experimental (Malvar 1992) and model results.

Conclusions

Limited experimental data suggests that the elastic modulus D^e_{22} for a two-dimensional bond law is not constant. The model presented here is based on the postulate that this change can be attributed to a change in the state of contact and the resulting damage in the adjacent concrete; furthermore, this behavior is characterized by making the elastic radial compliance a monotonically increasing function (linear in the preliminary study) of the plastic radial dilation.

While additional validation is needed, the model does reduce elastic radial contractions that were excessive in the original model. Additional model development addressing the change in elastic moduli merits additional experimental data.

Appendix I. References

Cox, J.V. (1994). "Development of a plasticity bond model for reinforced concrete – Theory and validation for monotonic applications." *Technical Report TR-2036-SHR*, Naval Facilities Engineering Service Center, Port Hueneme, CA.

De Groot, A.K., Kusters, G.M.A., and Monnier, T. (1981). "Numerical modeling of bond-slip behavior." *Heron, vol 26-1b*, I.B.B.C. Institute TNO, Delft, Netherlands.

Herrmann, L.R., and Cox, J.V. (1994). "Development of a plasticity bond model for reinforced concrete – Preliminary calibration and cyclic applications," *Contract Report CR 94.001-SHR*, Naval Facilities Engineering Service Center, Port Hueneme, CA.

Lutz, L.A., and Gergely, P. (1967). "Mechanics of bond and slip of deformed bars in concrete." *ACI Journal*, 64, 711-721.

Malvar, L.J. (1992). "Bond of Reinforcement under Controlled Confinement." *ACI Materials Journal*, 89(6), 593-601.

TIME DOMAIN FUNDAMENTAL SOLUTION TO NONCLASSICAL THERMOELASTICITY WITH ONE RELAXATION TIME
Part II: Two-Dimensional Solution

Jianming Chen* Alexander H.-D. Cheng
Department of Civil Engineering, University of Delaware, Newark, DE 19716 , U.S.A

ABSTRACT

A transient Green's function for two-dimensional nonclassical thermoelasticity with one relaxation time is obtained. Selected numerical results are presented to demonstrate the features of the waves, and the accuracy of the solution is established by comparing them with Laplace transform domain solutions.

INTRODUCTION

In a companion paper (Chen and Cheng 1996), the authors have presented a set of fundamental solutions for three-dimensional nonclassical dynamic thermoelasticity both in Laplace domain and in time domain. The objective of the present study is to provide its counterpart for the two-dimensional case. In addition to their mathematical attributes and the physical significance in revealing insights into wave propagations in thermoelastic media, the time domain two-dimensional fundamental solution provides a key ingredient, for development of 2-D time domain Boundary Element Method (BEM), which will certainly hold an upper hand over its corresponding Laplace domin BEM (Chen and Dargush 1995) in solving more complicated thermoelastic problems of practical importance including nonlinearity.

GOVERNING EQUATIONS

The two-dimensional governing equations can be derived directly from three-dimensional Eqs.(2a,b) in Part I (Chen and cheng 1996) by assuming displacement components are independent of one of the coordinates, say x_3 and by setting $u_3 = 0$ and $\partial/\partial x_3 = 0$. Thus it leads to the same form as eqns.(2a,b) in Part I except for the indices i, j=1,2.

Laplace Transform Domain Fundamental Solutions

Following the same basic steps as shown in the 3-D case (Chen and Cheng 1996), an explicit nondimensional fundamental solution for 2-D nonclossical thermoelasticity with one relaxation time, which involves a continuous unit line force in the *i-th* direction suddenly applied at the origin *i.e.*, $f_i(\mathrm{x},t) = \delta(x_1)\delta(x_2)H(t)$, and a unit rate of heat source at the origin *i.e.*, $\psi(x,t) = \delta(x_1)\delta(x_2)H(t)$, can be obtained as follows:

$$\begin{aligned}\tilde{G}_{ij} = &-(A_{ij}\lambda_1 K_1(\lambda_1 r) + B_{ij}\lambda_1^2 K_0(\lambda_1 r))\frac{s^2-\lambda_2^2}{\lambda_2^2-\lambda_1^2}\frac{1}{s^3} + (A_{ij}\lambda_2 K_1(\lambda_2 r) + B_{ij}\lambda_2^2 K_0(\lambda_2 r))\frac{s^2-\lambda_1^2}{\lambda_2^2-\lambda_1^2}\frac{1}{s^3} \\ &-(A_{ij}\lambda_3 K_1(\lambda_3 r) + B_{ij}\lambda_3^2 K_0(\lambda_3 r))\frac{1}{s^3} + C_{ij}\frac{1}{s}K_0(\lambda_3 r)\end{aligned} \tag{1a}$$

$$\tilde{G}_{i3} = -\frac{1}{2\pi}\frac{x_i}{\zeta r s}\frac{\lambda_1}{\lambda_1^2-\lambda_2^2}K_1(\lambda_1 r) + \frac{1}{2\pi}\frac{x_i}{\zeta r s}\frac{\lambda_2}{\lambda_1^2-\lambda_2^2}K_1(\lambda_2 r) \qquad \tilde{G}_{3i} = \delta s \tilde{G}_{i3} \tag{1b,c}$$

$$\tilde{G}_{33} = \frac{1}{2\pi\zeta s}\frac{1}{\lambda_1^2-\lambda_2^2}\Big[(\lambda_1^2-s^2)K_0(\lambda_1 r) - (\lambda_2^2-s^2)K_0(\lambda_2 r)\Big] \tag{1d}$$

$$A_{ij} = \frac{1}{2\pi}\Big(\frac{2x_i x_j}{r^3} - \frac{\delta_{ij}}{r}\Big) \qquad B_{ij} = \frac{1}{2\pi}\frac{x_i x_j}{r^2} \qquad C_{ij} = \frac{\delta_{ij}}{2\pi\mu} \tag{2}$$

in which $r^2 = x_i x_i$, $\zeta = \frac{1}{1+\tau_o s}$, $K_0(\lambda_i r)$ and $K_1(\lambda_i r)$ are the modified Bessel functions of the second kind of zero and first order and λ_1, λ_2, $\lambda_3, \lambda_1^2 - \lambda_2^2$ are defined by eqn(4) in Part I of the present paper (Chen and Cheng 1996).

* Structural Engineer/STV GROUP 225 Park Ave. South NY NY 10003, U.S.A.

TRANSIENT FUNDAMENTAL SOLUTION

Following the same argument as presented in Part I, we are justified in expressing $\lambda_1^2 - \lambda_2^2 \approx cs(s+b)$ and $\lambda_{1,2} = \frac{1}{c_{t,p}}\sqrt{(s+\eta_{t,p})^2 - \xi_{t,p}^2}$, which constitute good approximations to the exact result as long as $\tau_0 > \frac{1}{1+\delta}$. In addition we have an exact expression $\lambda_3 = \frac{s}{c_s}$. Obviosly b, c, c_p, c_t, c_s, η_p, η_t, ξ_p, ξ_t are in Part I (Chen and Cheng 1996).

Substitution of the above approximations into eqn.(1b) yield the eqns in terms of know transforms, we can invert it directly to give the following closed form solutions

$$
\begin{aligned}
G_{i3} = & \int_{r/c_t}^{t} (P_{11}e^{-b(t-\tau)} + P_{12})e^{-\eta_t\tau} sinh(\eta_t\sqrt{\tau^2 - r^2/c_t^2})d\tau H(t - r/c_t) \\
& + P_{13}e^{-\eta_t t} sinh(\eta_t\sqrt{t^2 - r^2/c_t^2})H(t - r/c_t) + \int_{r/c_p}^{t} P_{21}e^{-b(t-\tau)}\frac{\tau}{\sqrt{\tau^2 - r^2/c_p^2}}d\tau H(t - r/c_p) \\
& + (P_{22} + P_{23}t)\sqrt{t^2 - r^2/c_p^2}H(t - r/c_p) + P_{24}ln\frac{(t + \sqrt{t^2 - r^2/c_p^2})c_p}{r}H(t - r/c_p)
\end{aligned} \tag{3}
$$

$$
\begin{gathered}
P_{11} = -\frac{f_3}{b}\frac{c_t^2}{\eta_t r} \quad P_{12} = (f_2 + \frac{f_3}{b})\frac{c_t^2}{\eta_t r} \quad P_{13} = f_1\frac{c_t^2}{\eta_t r} \\
P_{21} = \frac{g_3}{b^2}\frac{c_p}{r} \quad P_{22} = (g_1 - \frac{g_3}{b^2})\frac{c_p}{r} \quad P_{23} = \frac{1}{2}(g_2 + \frac{g_3}{b})\frac{c_p}{r} \quad P_{24} = \frac{-r}{2c_p}(g_2 + \frac{g_3}{b}) \\
f_1 = \frac{1}{2\pi}\frac{x_i}{r}(-e_1 - \frac{e_5}{c}) \quad f_2 = \frac{1}{2\pi}\frac{x_i}{r}(-e_2 + \frac{-e_4 + e_5 b}{c}) \quad f_3 = \frac{1}{2\pi}\frac{x_i}{r}\frac{-e_3 + e_4 b - e_5 b^2}{c} \\
g_1 = \frac{1}{2\pi}\frac{x_i}{r}(-e_1 + \frac{e_5}{c})c_p \quad g_2 = \frac{1}{2\pi}\frac{x_i}{r}(-e_2 + \frac{e_4 - e_5 b}{c})c_p \quad g_3 = \frac{1}{2\pi}\frac{x_i}{r}\frac{(e_3 - e_4 b + e_5 b^2)c_p}{c} \\
e_1 = \frac{\tau_0}{2} \quad e_2 = \frac{1}{2} \quad e_3 = \frac{1}{2}(1+\delta) \quad e_4 = \frac{1}{2}\left[2\tau_0(1+\delta) + 1\right] \quad e_5 = \frac{\tau_0}{2}\left[\tau_0(1+\delta) + 1\right]
\end{gathered} \tag{4}
$$

Similarly the desired transient Green's function for eqn(1c) can be easily obtained by applying Laplace inversion.

$$
\begin{aligned}
G_{3i} = & \{P_{11}e^{-\eta_t t} sinh(\eta_t\sqrt{t^2 - r^2/c_t^2}) - P_{11}e^{-\eta_t t} cosh(\eta_t\sqrt{t^2 - r^2/c_t^2})\frac{t}{\sqrt{t^2 - r^2/c_t^2}} \\
& + P_{12}e^{-\eta_t t} sinh(\eta_t\sqrt{t^2 - r^2/c_t^2}) + \int_{r/c_t}^{t} P_{13}e^{-b(t-\tau)}e^{-\eta_t\tau} sinh(\eta_t\sqrt{\tau^2 - r^2/c_t^2})d\tau\}H(t - r/c_t) \\
& + \{\int_{r/c_p}^{t} P_{21}e^{-b(t-\tau)}\frac{\tau}{\sqrt{\tau^2 - r^2/c_p^2}}d\tau + P_{22}\sqrt{t^2 - r^2/c_p^2} + P_{23}\frac{t}{\sqrt{t^2 - r^2/c_p^2}}\}H(t - r/c_p)
\end{aligned} \tag{5}
$$

$$
P_{11} = -f_1\frac{c_t^2}{r} \quad P_{12} = f_2\frac{c_t^2}{\eta_t r} \quad P_{13} = f_3\frac{c_t^2}{\eta_t r} \quad P_{21} = -\frac{g_3}{b}\frac{c_p}{r} \quad P_{22} = (g_2 + \frac{g_3}{b})\frac{c_p}{r} \quad P_{23} = g_1\frac{c_p}{r} \tag{6}
$$

where f_1, f_2, f_3, g_1, g_2, g_3 have the same expressions as in Eq.(4) with the exception that e_1, e_2, e_3, e_4, e_5 should be multiplied by thermal mechanical coupling parameter δ.

Finally, eqn(1d) is inverted to yield

$$G_{33} = \{P_{11}\int_{r/c_p}^{t} e^{-b(t-\tau)}\frac{1}{\sqrt{\tau^2 - r^2/c_p^2}}d\tau + P_{12}ln\frac{(t+\sqrt{t^2-r^2/c_p^2})c_p}{r} + P_{13}\frac{1}{\sqrt{t^2-r^2/c_p^2}}\}H(t-r/c_p)$$

$$+\int_{r/c_t}^{t}(P_{21}e^{-b(t-\tau)} + P_{22})\frac{e^{-\eta_t\tau}}{\sqrt{\tau^2-r^2/c_t^2}}cosh(\eta_t\sqrt{\tau^2-r^2/c_t^2})d\tau H(t-r/c_t) \tag{7}$$

$$+ P_{23}\frac{e^{-\eta_t t}}{\sqrt{t^2-r^2/c_t^2}}cosh(\eta_t\sqrt{t^2-r^2/c_t^2})H(t-r/c_t)$$

$$P_{11} = \frac{1}{2\pi}\frac{-e_4+e_5 b}{c} + \frac{1}{2\pi}\frac{e_3}{bc} \quad P_{12} = \frac{1}{2\pi}e_2 - \frac{1}{2\pi}\frac{e_3}{bc} \quad P_{13} = \frac{1}{2\pi}(e_1 - \frac{e_5}{c})$$

$$P_{21} = \frac{1}{2\pi}\frac{e_4 - e_5 b}{c} - \frac{1}{2\pi}\frac{e_3}{bc} \quad P_{22} = \frac{1}{2\pi}e_2 + \frac{1}{2\pi}\frac{e_3}{bc} \quad P_{23} = \frac{1}{2\pi}(e_1 + \frac{e_5}{c}) \tag{8}$$

$$e_1 = \frac{\tau_0}{2} \quad e_2 = \frac{1}{2} \quad e_3 = \frac{1}{2}(1+\delta) \quad e_4 = \frac{1}{2}\Big[2\tau_0(1+\delta) - 1\Big] \quad e_5 = \frac{\tau_0}{2}\Big[\tau_0(1+\delta) - 1\Big]$$

Numerical Results

The material parameters are presented in nondimensional form as: $\lambda^* = 0.5$, $\mu^* = 0.25$, $\delta = 0.02$, $\tau_0 = 2.0$, In current problem the applied force point (or heat source point) is located at $(0,0)$, the receiver is chosen at nondimensional coordinate $(1,2)$. The nondimensional wave velocities are approximately $c_p = 1.01$ (*Stress Pressure Wave*); $c_t = 0.697$ (*Thermal Wave*); $c_s = 0.5$ (*Shear Wave*). In view of the above, the nondimensional time required for the three waves to reach the recievers are $t_p = 2.214$ (*Stress Pressure Wave*), $t_s = 7.483$ (*Shear Wave*), and $t_t = 3.209$ (*Thermal Wave*). In all the Figures, excellent agreement of analytical solution and numerical Laplace inversion is seen. Two arrival times corresponding to the fast stress pressure wave and second thermal wave can be identified in figure 3 by suddenly changed in displacement due to point heat source *i.e.* G_{13}.

Another interesting feature is the presence of two pulses in G_{31}, G_{32} and G_{33}, which defines the temperature due to line force or line heat source (illustrated Figs. (1,2,4). These pulses, in the form of $\frac{1}{\sqrt{t^2-r^2/c^2}}$ for $2-D$, are associated with the arrival of the two dilatational waves.

The plots show that no diffusion takes place before the arrival of *Diffusive Wave* and it starts right away after its arrival. This is primarily due to the fact that the fast stress wave has practically negligible small dissipation factor $\eta_p = 0$, while the second thermal wave is strongly attenuated with a high dissipative factor $\eta_t = 0.2476$ as the disturbance moves into the media, the difusive wave front slows down and eventually disappears.

These plots also reveal a significant characteristic in the two-dimensional fundamental solutions, *i.e.* the response exhibits a tail immediately behind the wave front when plotted with respect to time. The reason for this phenomena is that the point force or heat source disturbance in two dimensional media is uniformly distributed along the x_3-axis, thus signals continuously arrive at the receiver from all points along the x_3-axis.

CONCLUSION

The availability of a satisfactory transient fundamental solution and an accurate transient boundary integral equation for full dynamic poroelasticity enables the development of an efficient time domain BEM to solve more practical problems including dynamic nonlinear thermoelasticity.

References

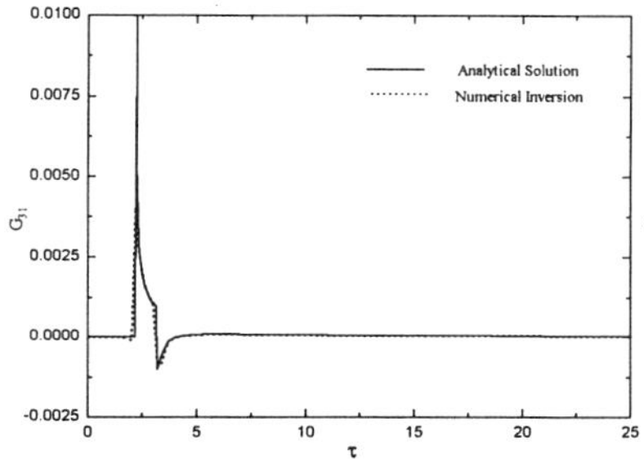

Fig. 1. Temperature time history at (1,2) due to point force at (0,0)

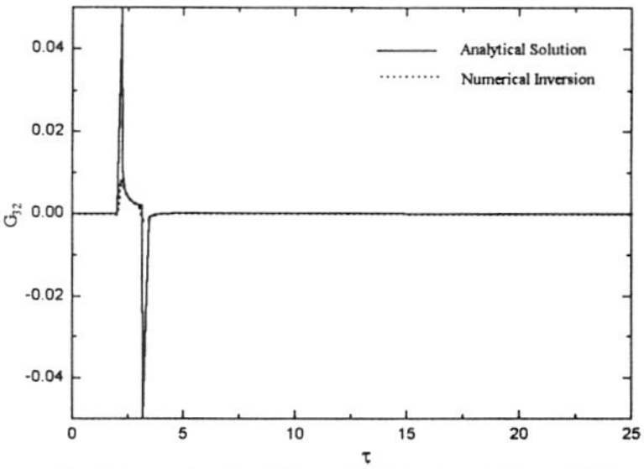

Fig. 2. Temperature time history at (1,2) due to point force at (0,0)

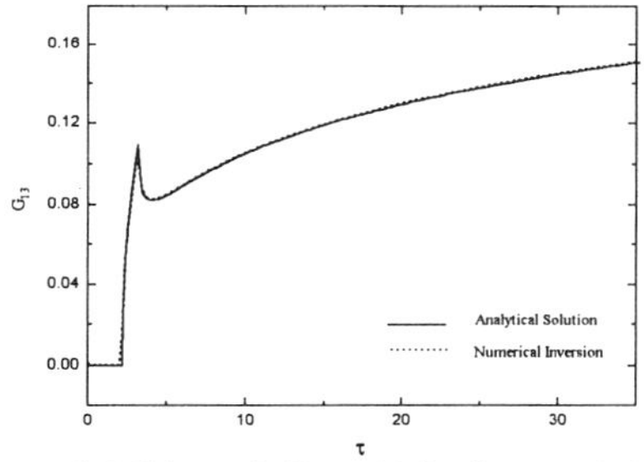

Fig. 3. Displacement time history at (1,2) due to heat source at (0,0)

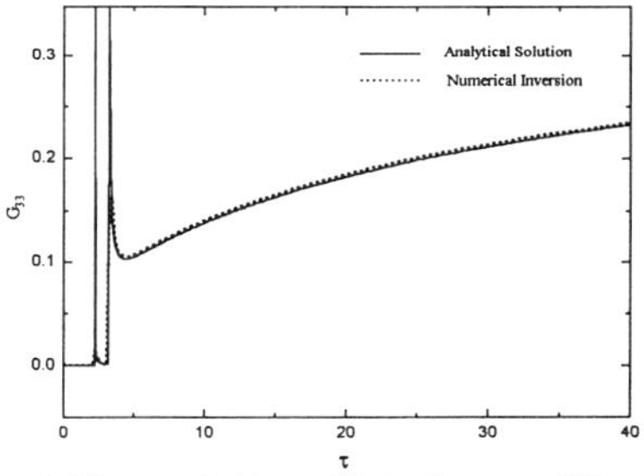

Fig. 4.Temperature time history at (1,2) due to heat source at (0,0)

References

Chen, J. and cheng, A.H.-D. (1996). 'Time Domain Fundamental Solution to Nonclassical Thermoelasticity with One Relaxation Time', 11th ASCE Enginnering Mechanics Conference, Fort Lauderdale, 1996

Chen, J. and Dargush, G.F., (1995), 'Boundary Element Method for Dynamic Poroelastic And Thermoelastic Analyses ', *Int. J. Solids Structures*, Vol. 32, No. 15, pp. 2257-2278.

A Refined Numerical Approach for the Limit-Load Analysis of 3-D Steel Rod Structures

Dr. Norbert Gebbeken[1] (Member ASCE)

Abstract

As far as steel-rod structures are concerned the yield-hinge theory is a very efficient approach of the limit-load theory. Unfortunately, most of the published strategies suffer from considerable deficiencies which depend on two main reasons: firstly, the yield condition is not approximated very well, and secondly, a flow rule is not at all incorporated. This may significantly affect the calculated load-carrying behavior and as a consequence the elasto-plastic failure prediction. In the present paper a closed formulation of a refined numerical method based on the yield-hinge theory is consistently developed from the theory of plasticity.

Introduction

The harmonization of the national and international standards will affect the design of steel structures in the future. Due to the reasons of safety and economy it is advised to apply nonlinear methods. For frame analysis the *yield-hinge theory* has been established. Earlier proposed methods were restricted to the geometrically nonlinear theory of second order or by considering just $P - \Delta$-effects. Moreover, the plastic behavior was only considered for the bending moment. A few authors took the *interaction* of the internal forces in the plastic regime into account. Thus, yielding an inconsistent theory. In order to derive a numerical procedure for the yield-hinge theory the above-mentioned simplifications are not necessary.

Mathematical formulation of the yield-surface

We postulate a function f that defines the *limit-state of elasticity* under any admissible combinations of *limit stress-resultant components (limit internal forces)*. The *yield-criterion* of Huber, v. Mises & Hencky (*J_2-flow theory*) is best suited to

[1]Prof. Dr. Norbert Gebbeken, University of the Federal Armed Forces Munich, Institute of Engineering Mechanics and Structural Mechanics, Werner-Heisenberg-Weg 39, 85579 Neubiberg, Germany

simulate the elastic limit of steel. The problem of formulating yield-functions in the context of beam-theory has been tackled by many scientists in the last three decades (GEBBEKEN 1988). RUBIN derived in 1978 *interaction-functions* which represent the *yield-surface (yield-locus)* (Fig. 1) of double-T cross-sections. The derivations are carried out under consideration of all internal forces, except of the torsional moment. Thus, yielding an exact representation of the yield-surface in the case of plane bending, and a fairly good approximation of the three-dimensional case. For practical purposes

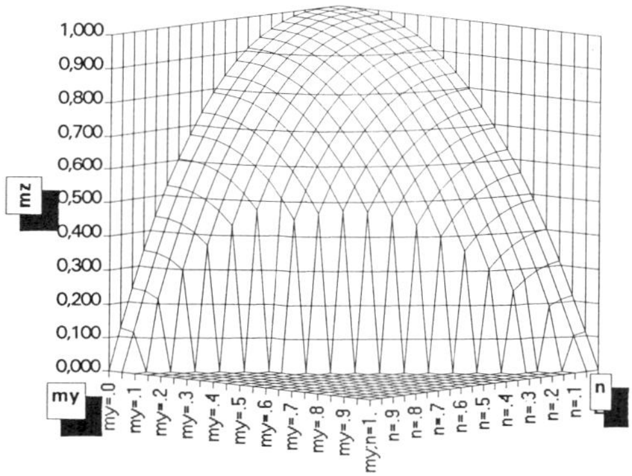

Figure 1: Yield-surface of a HEB 400 according to RUBIN [1978]

simplified interaction-functions have been proposed.

Annotation: *Linearized yield-functions have been introduced to the standards. They are advantageous for graphically based methods. For our purposes it is worthwhile to implement the most efficient ones into the computer program.*

On the yield-hinge concepts

The most simple possibility to represent plastic behavior is the introduction of *concentric yield-hinges*. For frame analysis it is widely accepted to apply concentric yield-hinges associated with the bending moment. *It is worth-mentioning that these concentric yield-hinges are located inevatibly in the centerline of the beam and not in the neutral axis.*

The formulation for generalized yield-hinges is carried out in the framework of a geometrically nonlinear formulation. Because of the incremental, iterative procedure the relations are linearized. Consequently, we can start with the linear relation between the *nodal force vector* and the *nodal displacement vector* in the elastic regime

$$F = k\,v^e\,. \tag{1}$$

For the plastic regime the total incremental displacement vector Δv can be decomposed additively into an elastic part Δv^e and a plastic part Δv^p,

$$\Delta v = \Delta v^e + \Delta v^p \quad . \tag{2}$$

Δv^p can be determined applying the well established *flow-rule* of PRANDTL and REUSS:

$$\Delta v^p = \lambda \frac{\partial f}{\partial F_i} = \lambda \nabla f \quad , \quad \lambda \geq 0 \quad . \tag{3}$$

Annotation: *This is the decisive extension to the concentric yield-hinge concept. With the flow-rule (3) the material formulation is complete, and consistent to the theory of plasticity.*

After some algebra we finally arrive at

$$F = \begin{bmatrix} F_i \\ F_j \end{bmatrix}^{12} = \begin{bmatrix} k_{ii}^p & k_{ij}^p \\ k_{ji}^p & k_{jj}^p \end{bmatrix}^{12} \begin{bmatrix} v_i \\ v_j \end{bmatrix}^{12} \quad , \tag{4}$$

where $[k^p]$ is the *plastic element stiffness matrix*. Applying linear yield-functions it results that the *plastic deformations* of a yield-hinge are approximated by a *fixed kinematic*, i.e. the yield-hinge is blocked. This has severe consequenses. GEBBEKEN 1988 proved that the kinematic constants are neither lower bounds nor upper bounds. Thus, not even in the two-dimensional case we can guarantee a safe solution, unless the magnitude of the present normal force is limited by a certain value of the fraction of the fully plastic normal force N_p. For hot-roled profiles the limits are given in Table 1. For the profile sizes not given here, the limits of n can be determined by interpolation.

Table 1: Normal force limits n

Profile	n_{min}	n_{max}
I	0.33 (I80)	0.39 (I600)
IPE	0.33 (IPE80)	0.38 (IPE600)
HEA	0.20 (HEA100)	0.38 (HEA1000)
HEB	0.20 (HEB100)	0.38 (HEB1000)
HEM	0.22 (HEB100)	0.39 (HEB1000)

Annotations: *Linear yield-functions in conjunction with eccentric yield-hinges result in a wrong simulation of the plastic deformations. If nonlinear yield-functions in connection with eccentrical yield-hinges are used, the alteration of the location of yield-hinges can be considered. It has been proved that the principle of virtual work is locally violated if concentric yield-hinges or linear yield-functions are applied.*

Numerical Example

The frame chosen for analysis is shown in Fig. 2. This example has firstly been investigated by KORN & GALAMBOS. The yield-hinge models are concentric yield-hinges. For the own calculation generalized yield-hinges have been used. In addition, linear yield-functions and "Rubin" are utilized. Large first order sways, slender columns and heavy column compressive axial loads resulted in a 30% sway amplification due to the geometrical nonlinearity. Furthermore, the large deformations cause a tremendeous

Table 2: Limit-load factors and deflections

	κ_u^{II}	κ_{ku}^{I}	u_u^{II}	u_{ku}^{I}
	–	–	m	m
Korn & Galambos	1.283	–	0.853	–
Henning	1.281	2.028	0.857	4.279
Kahn	1.220	1.985	0.840	3.343
Gebbeken nonlin. y.-f.	1.240	2.207	0.914	3.721
Gebbeken lin. y.-f.	1.205	1.950	0.508	2.517

reduction of the magnitude of the load factor when compared to the geometrically linear analysis. Only 60% of the first-order ultimate load was attained. Thus, this frame illustrates an insufficient design, because the system-inherent plastic capacities cannot be activated due to the early loss of elasto-plastic stability. Such frames are called *"Second order frames": Due to second order effects, these frames failed by elasto-plastic instability prior to the formation of a mechanism.*

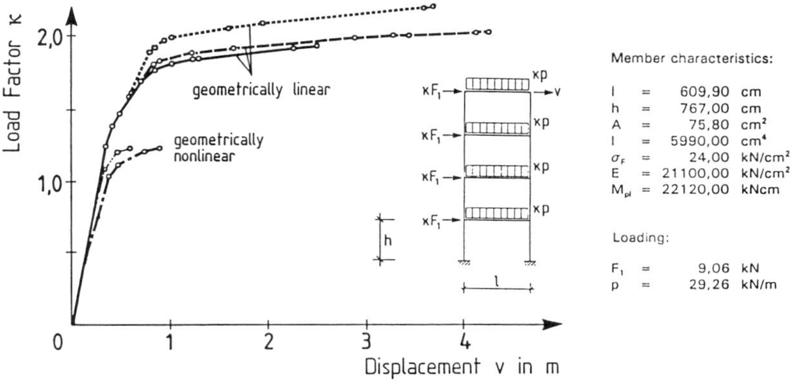

Figure 2: Four-storey one-bay frame

References

Gebbeken, N.: *Eine Fliessgelenktheorie hoeherer Ordnung fuer raeumliche Stabwerke.* Mitt. Nr. 32-88, Inst. fuer Statik, Universitaet Hannover, Stahlbau 57 (1988) 365-372

Gebbeken, N.: *A Refined Numerical Approach for the Limit-Load Analysis of 3-D Steel Rod Structures.* to appear

Korn, A.; Galambos, T.V.: *Behavior of Elastic-Plastic Frames.* Journal of the Structural Division ASCE ST5 (1968) 1119-1142

Rothert, H.; Gebbeken, N.: *On Numerical Results of Reticulated Shell Buckling.* Int. J. of Space Structures 7, No. 4 (1992) 299-319

On Efficiency and Accuracy in Simulations of Granular-type Systems

Yi Sun[1] and Oleg Vinogradov[2]

Abstract

Granular-type systems are treated as systems of coupled particles, i.e. as multibody systems. Different formulations (Lagrangian and Newton-Euler) and different solvers (Runge-Kutta and iterative) are used to study numerically open loop (pendulums with 50, 100, and 150 bodies) and closed-loop (suspended chain with 14 bodies) systems. It is shown that an iterative solver allows to maintain the accuracy of the fourth-order Runge-Kutta solver while reducing significantly the simulation time. This is true for both open and closed-loop systems. It is found also that the time efficiency of the iterative solver grows with the size of the system.

Introduction

In the dynamics of granular-type systems the interaction between the particles leads to a coupled system of governing equations. For the systems of this nature, i.e. for multibody systems, the efficiency and accuracy of simulations are of great importance if any meaningful results are to be expected. Different formulations in multibody dynamics combined with different numerical solvers result in different accuracy and time efficiency. In this paper a representative physical system is studied numerically in order to identify an optimum method-solver combination. Two systems: an open-loop chain (pendulum) and a closed-loop chain are considered. For both systems only the Lagrangian (L) and Newton-Euler (NE) formulations are tested. Also two methods: the fourth order Runge-

---

[1] Graduate student, Dept. of Mech. Engg., University of Calgary, Calgary, Alberta, Canada T2N 1N4

[2] Professor, Dept. of Mech. Engg., University of Calgary, Calgary, Alberta, Canada T2N 1N4

Kutta (RK)(which serves as a benchmark) and an iterative one (I) are used. An initial-value problem for an open chain in the L-formulation is described by a nonlinear system of ODE, while in the NE-formulation by a linear system of ODE and a nonlinear system of constraints equations. In the case of closed-loop chains additional algebraic constraints must be satisfied.

The advantage of the L-formulation is that it leads to a minimum number of equations. Its disadvantage is that the second derivatives are coupled, i.e. are not given explicitly. The NE-formulation, on the other hand, decouples the second derivatives at the expense, however, of the increased number of the equations, since the internal forces (constraints) become additional unknowns. Which formulation is more efficient as a computational tool thus is not self-evident. In addition, the efficiency is method-dependent, and may be problem-dependent.

For an initial-value problem the error accumulation leads to violation of the energy conservation principle and constraints. Any comparison of efficiency should be done if the same error tolerance is maintained. The results presented herein satisfy this requirement.

Methods of Solution and Error Control

A standard RK solver with error control and a variable time step was used. The iterative solver was developed by the authors based on the Gauss-Seidel algorithm. For the L-system the tolerance on positions, and for the NE-system the tolerance on internal forces was prescribed. An iterative method was used to solve the coupled system of NE equations. Among other advantages, the iterative method allows to split the system into one, defining the constraint forces for the given positions and velocities, and another, defining the positions and velocities for the given constraint forces. Both, forces and state coordinates, are updated in the process until a predetermined accuracy is achieved.

The state variables obtained as a result of solving an initial value problem tend to violate the constraints equations due to error accumulation. An error control mechanism built into the iterative scheme allowed to meet the constraints requirements with a prescribed accuracy. However, since the corrected positions and velocities may not satisfy the energy conservation principle, a second corrective step was introduced which allowed to meet the conservation principle "on average".

To increase the efficiency of the iterative solver a variable time step was introduced based on the number of iterations at the previous step, i.e. if this number is outside of the initially prescribed range the time step will go up or down. In this way the smoothness of the function being integrated is taken into account.

Numerical Results

The numerical results are presented for an open chain (pendulum) comprising 50, 100 and 150 equal masses connected by equal links, and for a falling chain fixed at the ends and comprising 14 equal masses also connected by equal links.

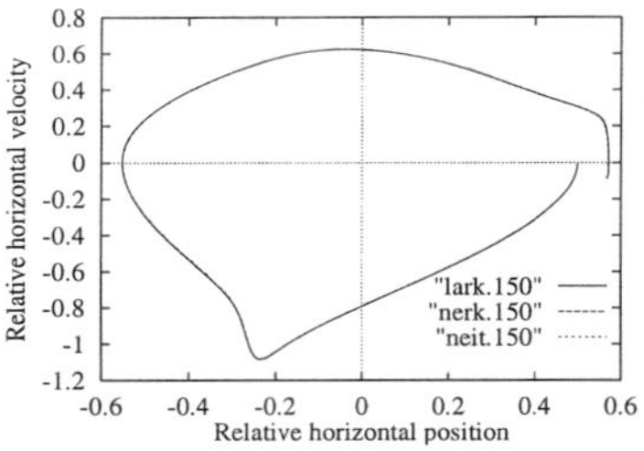

Fig1. Phase diagrams for the last mass in a 150-mass pendulum

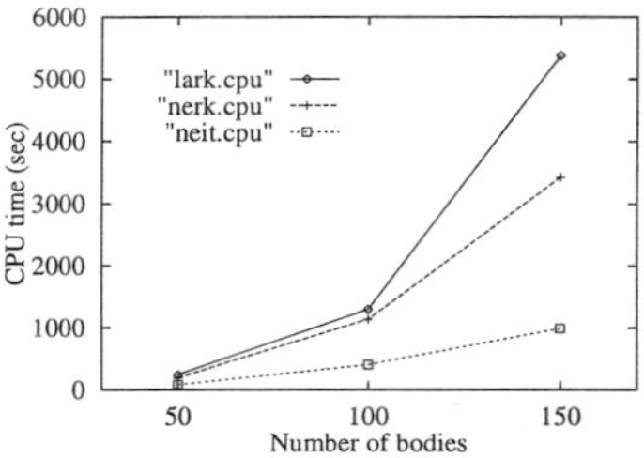

Fig2. CPU times for pendulums with different numbers of bodies

In Fig.1 a phase diagram describing the first cycle completed by the last body in a 150-mass chain is shown for the L-formulation and RK-solver (lark.150), the NE-formulation and RK-solver (nerk.150), and the NE-formulation and I-solver (neit.150). Note that the displacements are normalized with respect to the total pendulum length, while the velocities are normalized with respect to the square root of the gravity constant multiplied by the pendulum length. It is seen that both solvers produce practically identical results for both formulations.

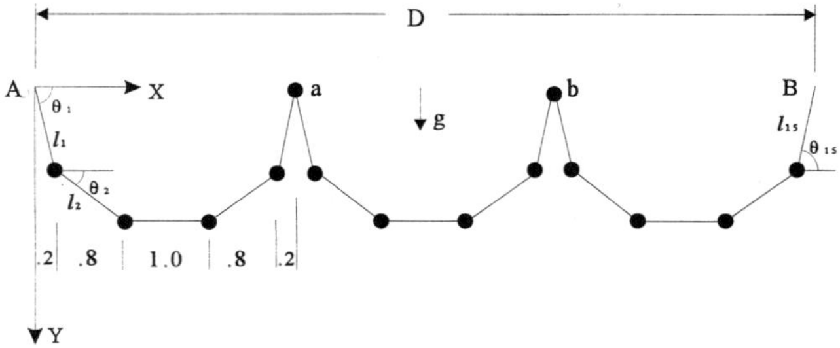

Fig3. Initial geometry of the suspended chain

The CPU time, however, was not identical. In Fig.2 the CPU times for various formulations, solvers and system sizes are shown, and the NE-formulation combined with the I-solver turns out to be the most efficient one.

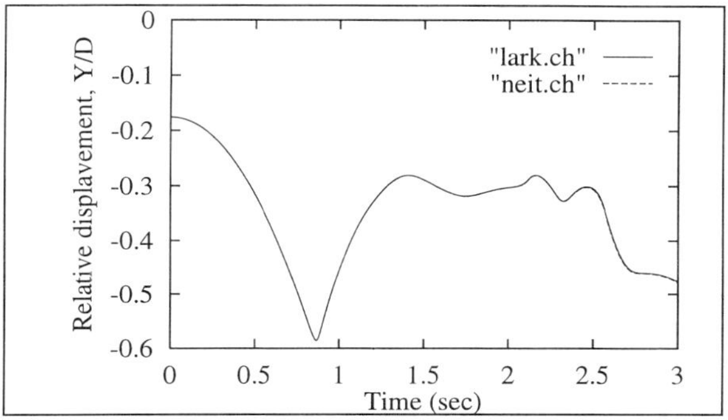

Fig4. Displacement vs. time for the middle point of falling chain

In Fig.3 a 14-mass chain attached initially at four points, A, a, b, B, is shown. At the time t=0 the chain is released at points a and b. The resulting motion of the chain is represented in Fig.4 by the displacement of it's middle point for two formulations (L and NE) and two solvers (RK and I). It is seen that both combinations lead to practically identical results during the first 3 seconds of motion. The NE-I combination requires, however, only 2.1 sec CPU time compared to 13.6 sec for the L-RK combination.

The general conclusion is that for the multibody systems investigated numerically the NE-formulation combined with the I-solver allows to substantially reduce the simulation time while maintaining the accuracy of the fourth order RK method. A comparative investigation of more complex granular-type systems also supports this conclusion.

Acknowledgments

The financial assistance provided by the Natural Sciences and Engineering Research Council of Canada is gratefully acknowledged.

Studying Mixed Granular Flows by Image Analysis

Lennart Gustafsson[1] and Peter Gustafsson[2]

Abstract

Studying mixed granular flows where particles differ in size, density, surface friction, etc. is of importance for such applications as separation and mixing of granular materials. Using the APT (Accumulating Particle Trajectories) method the flow of particles of any one kind can be extracted from the total flow of all particles. It is required only that the different kinds of particles be given different gray values. The flow in a model experiment is shown.

Introduction

The APT (Accumulating Particle Trajectories) method is an image analysis method which was developed for the study of particle flow in mining model experiments (Gustafsson et al., 1994). It has also been used to study the rotational and translational motion of disks (Gustafsson and Gustafsson, 1995) and fine-grain granular flows (Gustafsson and Knutsson, 1994 and Gustafsson and Marklund, 1995). In this presentation the APT method will be shown to be useful to study mixed granular flow where it may be of importance to obtain the flow of each kind of particles separately. The study of mining operations where there may be a high premium on preventing the mixing of ore and waste rock is an example of the potential usefulness of the method. Other applications include operations where particles of different kinds should be separated or mixed.

A brief recapitulation of the functioning principle of the APT method

When the APT method is employed to register the motion of particles which are large compared to a pixel it detects in each pixel the edges of every particle that passes that pixel. That way an associated movement image is built up in which the particle trajectories are accumulated. In the matrix containing the associated movement image each element then contains the number of particle edges which passed each pixel.

In order to detect an edge passage a threshold plus a margin (to reduce the influence of electronic noise) must be passed by the gray level of a pixel. The particle must therefore be distinctly brighter (for practical lighting reasons this is

- - - - - - - - - - - - -

[1]Assoc.Prof., Div. of Ind. Electr., Luleå Univ. of Tech., S-97187 Luleå, Sweden
[2]Grad. Stud., Div. of Mining Engrg., Luleå Univ. of Tech., S-97187 Luleå, Sweden

to be preferred over the alternative of darker particles) than the background (for a formal presentation of the functioning principle of the APT method, see Gustafsson and Gustafsson, 1995).

The APT method employed to study the movement of different kinds of particles.

If different kinds of particles can be painted in different shades of gray, then the analysis of a given video sequence of a flow experiment can be repeated with different threshold levels. That way a sequence of results is obtained and taking the difference of associated movement images will yield the flow of each kind of particle. A simple example will illustrate.

In figure 1.a four disks of two different sizes are shown. They all roll horizontally across the frame. In figure 1.b the movement of all disks is shown. The threshold has been set to detect passages of the gray disks and the passages of the white disks are then also detected. In figure 1.c the threshold has been set to detect only the passages of the white disks. In figure 1.d the difference image shows the movement of the gray disks only.

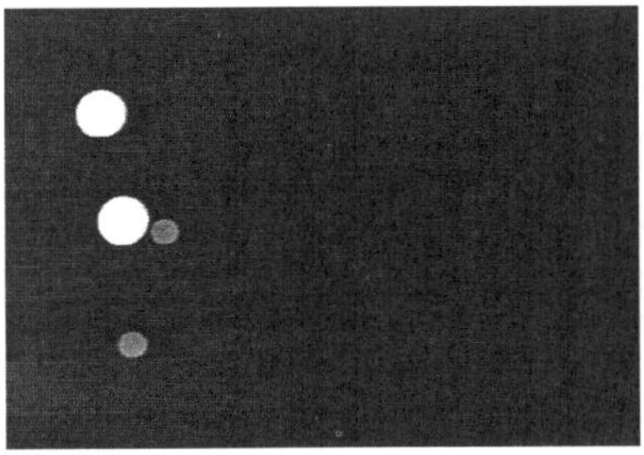

Fig. 1.a Gray and white disks rolling horizontally across the frame

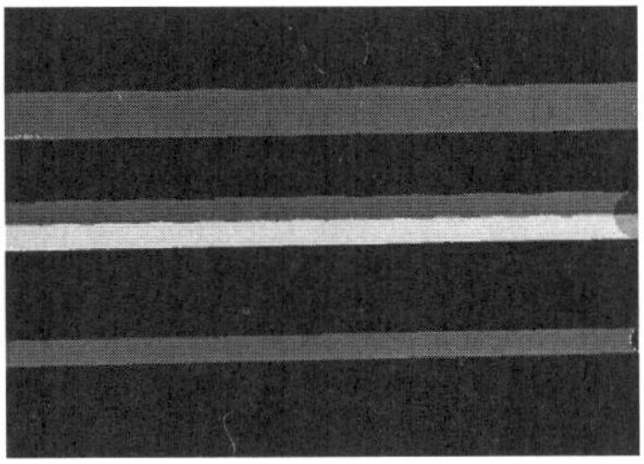

Fig. 1.b Associated movement image showing trajectories of all the disks

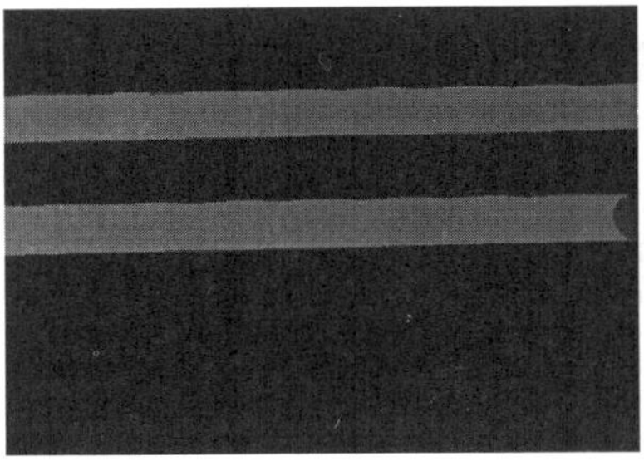

Fig. 1.c Associated movement image showing trajectories of the white disks

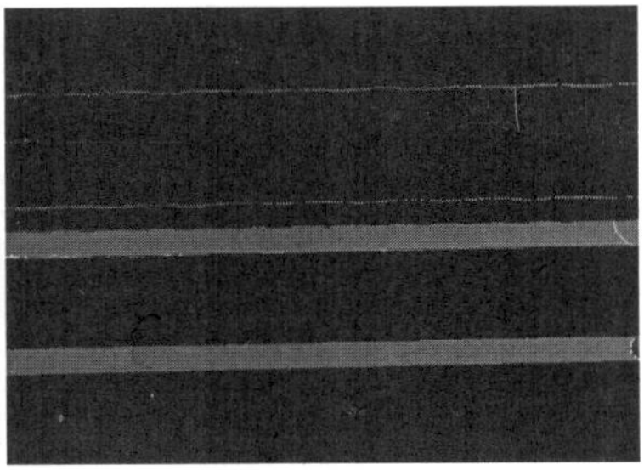

Fig. 1.d Difference image between images in figure 1.b and 1.c, showing the trajectories of the gray disks

There are some traces in figure 1.d which are obviously residues of the movement of the white disks. This is because the trajectory of a disk becomes marginally wider if the threshold is set lower (pixels in which both dark background and white disk are present are more likely to be interpreted as a disk if the threshold is set lower). It is possible to eliminate, at least to a high degree, these residues by widening the trajectories in figure 1.c by one pixel before taking the difference image.

On close scrutiny it is also seen that the trajectories have somewhat wavy edges. This is because the disks are not perfectly circular but slightly ellipsoidal.

<u>An illustration of the use of the APT method</u>

In figure 2.a a slanted circular table, slowly rotating clockwise while vibrating is shown. The disks, 12 mm in diameter, have different frictional properties. The gray disks have well finished sides, whereas the white disks have sides that are coarse. The analysis of 5 seconds of movement, involving 40 video frames, is shown in figures 2.b-2.d.

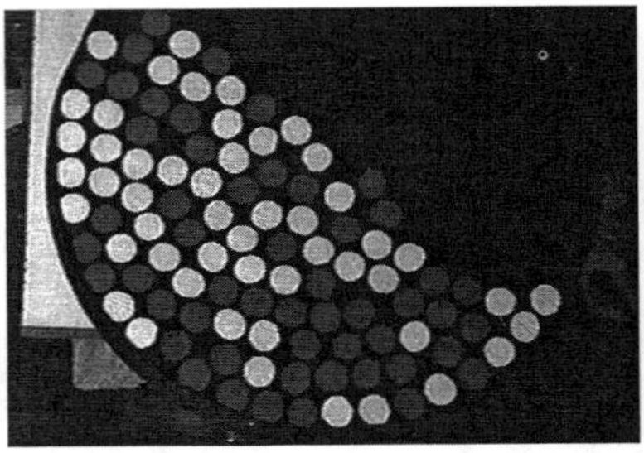

Figure 2.a Rotating table with disks at start of analyzed 5 s interval

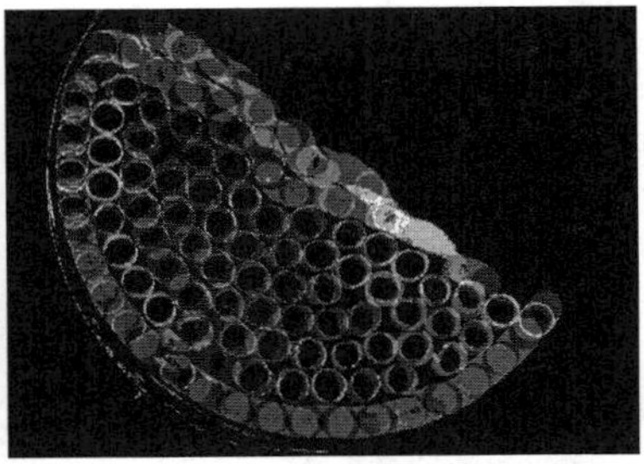

Figure 2.b The associated movement image including the movement of all disks

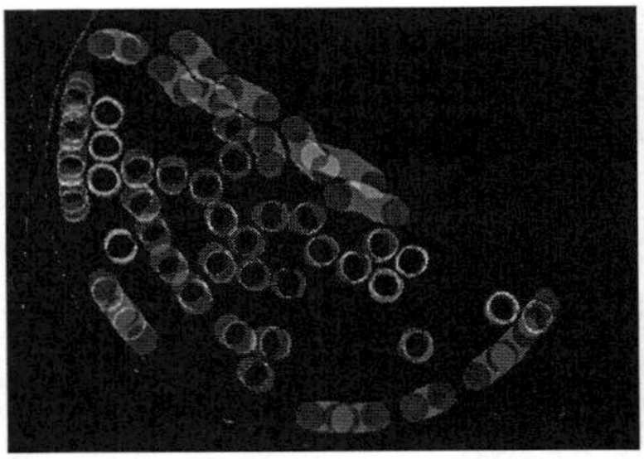

Figure 2.c The associated movement image including the movement of the white disks

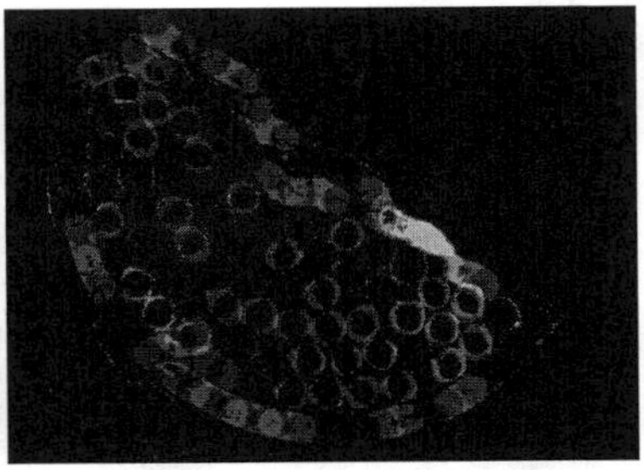

Figure 2.d The associated movement image including the movement of the gray disks (this difference image has been slightly filtered)

It is clearly seen that the disks follow the rotation along the periphery, where a friction band has been mounted, of the circular table. Along the free surface there are rapid movements but in the interior of the "disk area" the disks are slipping and therefore remain fairly immobile.

The number of different disks, the movement of which can be separated

It is advisable to allow for a safety margin of 10 gray levels on each side of the threshold (i.e. an hysteresis band 20 units wide should be employed in the edge detecting operation). Since the number of gray levels allowed in the present APT method system is 256 it should be possible to analyze experiments with 12 different kinds of particles. If colored disks are used this number would theoretically grow to 36. No experiment with such a wide variety of disks has been conducted though and the authors foresee a number of practical problems which would significantly reduce this number.

It is not inconceivable that 8 and 16 different kinds of particles could be allowed in an APT method experiment using black and white and color respectively. This means, e.g., that any combination of 2 different sizes, densities and friction coefficients could be allowed (in the black and white case) in an experiment.

Notes on lighting and vignetting

The illumination must be even across the experimental scene if several different kinds of particles are studied, otherwise the system will make mistakes since the same thresholds are used over the whole frame. This only demands several light sources and some care in the experimental setup. Incandescent lamps should be used to reduce rapid variations in illumination.

Vignetting in the video camera lens should be avoided. The most economical way of doing this is to use a video system camera and mount a lens intended for a 35 mm camera. That way only the center portion of the lens is used and vignetting (the darkening along the edges and especially in the corners) is avoided. If vignetting is present in a previously recorded sequence this can be corrected by real time image filtering equipment which can be programmed to multiply the gray level in each pixel with a correcting factor.

References

Gustafsson, S.L., Gimholt, J., Larsson, R. and S.P. Gustafsson (1994) "Granular Flow Measurement by Image Analysis and a Mining Engineering Application", *CIM Reporter,* Vol. 20, No. 1, pp. 16-18.

Gustafsson, S.L. and S. Knutsson (1994) "An Image Analysis Method for Studying Movements in Granular and Solid Bodies", Geotechnical Testing Journal, ASTM Vol. 17, No. 1, pp. 95-100.

Gustafsson, S.L. and S.P. Gustafsson (1995) "Studying Disk Movements by Novel Image-Analysis Method", *Journal of Engineering Mechanics*, August 1995, pp. 931-934.

Gustafsson, S.L. and O. Marklund (1995) "Image Analysis of Fine-Grain Flow: Conditions for High Quality Measurement Results", *Proceedings of 10th Engineering Mechanics Conference*, pp.614-617.

Centrifugal, Gravity and Side-wall effects in Annular Shear Cells

Cliff K. K. Lun[1]

Abstract

No abrupt stress jump at critical solids volume fraction is found in the annular shear-cell simulation with gravity. Due to the circulatory motion of the solid-walls that formed the trough of the annular shear-cell, a centrifugal force field is induced upon the bulk materials. The granular stresses at the outer side-wall are typically higher than those at the inner one. The top and bottom wall-stresses increase in an oscillatory fashion with increasing radius from the inner side-wall to the outer one, which indicates the existence of internal microstructures in the granular flow medium. Recirculatory secondary-flow motion is clearly evidenced at low to moderate solids fraction, but they become progressively chaotic and the recirculatory speed decrease with further increase in concentration.

Introduction

Annular shear-cells are commonly used to measure the variations of normal and shear stresses with the solids volume fraction and the shear rate developed in the rapid shearing of granular materials (Wang and Campbell 1992; Hanes and Inman 1985; Craig et al. 1986; Savage and Sayed 1984). However, very little is known about the internal flow properties such as the stress distributions, the secondary-flow behavior and the centrifugal effect in such devices.

Numerical Simulation

A simulation program is developed in the present work to study the effects of side walls, centrifugal force and gravity in rapid shearing of bulk materials in annular shear-cells. The computer program is an extension of the one used by Lun and Bent (1994) for gravity-free simple-shear flow to the case of solid wall-bounded annular shear flow in Cylindrical coordinates (r, θ, z).

----

[1]Department of Engineering, Dalhousie University, Halifax, NS, B3H 3J5 Canada

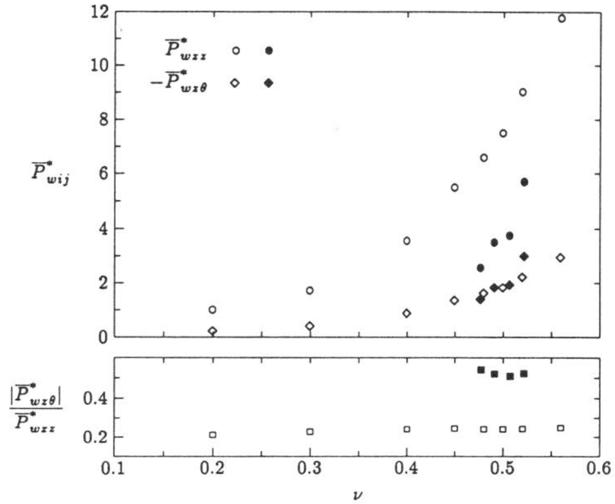

Figure 1. Variations of normal and shear stresses with solids volume fraction; simulation (open symbols); data of Savage and Sayed (1984) (solid symbols).

The computer program is used to simulate the annular shear-cell experiments of Savage and Sayed (1984) for testing 1.8 mm glass beads. The inner and outer radii of the annular trough normalized by the particle diameter were 59.97 and 81.13 respectively. The top and bottom walls were lined with coarse sand papers with grain size close to that of the test particles. In the simulation, hemispheric wall-particles with diameter identical to the flow-particles are fixed in triangular lattice with an area solids fraction of 0.7633 on the top and bottom walls. The coefficients of normal restitution (e_p = 0.97), tangential restitution in sticking contact (β_o = 0.44), and friction (μ_p = 0.092) measured by Foerster et al. (1994) for collisions between glass beads are employed in the simulation for particle-particle collisions, and particles colliding with top and bottom wall-particles. Coefficients e_p = 0.831, β_o = 0.31, and μ_p = 0.125 are used for glass beads colliding with the aluminum side-walls (Foerster et al. 1994).

Result

The variation of dimensionless mean stresses $\overline{P}^*_{wij} = P_{wij}/(\rho_p d^2 U_m^2 / H^2)$ at the top-wall with solids volume fraction ν is presented in figure 1; ρ_p, d, H and U_m are the particle diameter, mass density, shear gap height, and bottom wall speed at mid-trough radius respectively. Reasonable agreement is found between the simulation result and the test data for the shear stress, however it is not so for the normal stress. The rough sand papers boundaries used in the experiments were apparently responsible for the discrepancy in the normal stress. The phenomenon of abrupt stress jump, which was observed by Lun & Bent (1994) in gravity-free simple shearing flows at about 0.50 for inelastic frictional particles, is absent in the flows inside the simulated annular shear-cell with gravity.

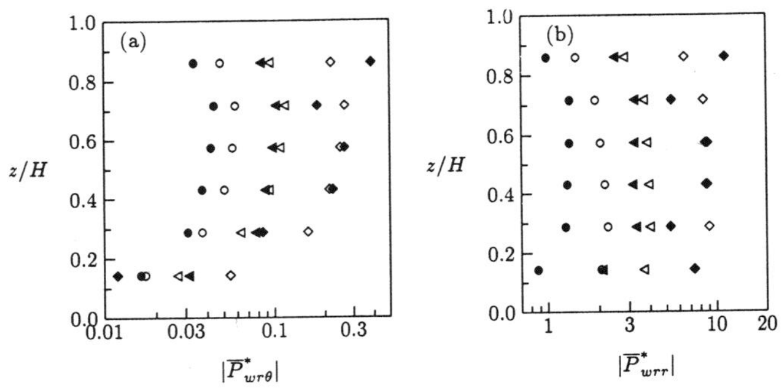

Figure 2. Distributions of (a) shear stress and (b) normal stress on inner (solid symbols) and outer (open symbols) side-walls; ν = 0.30, circle; ν = 0.40, triangle; ν = 0.52, diamond.

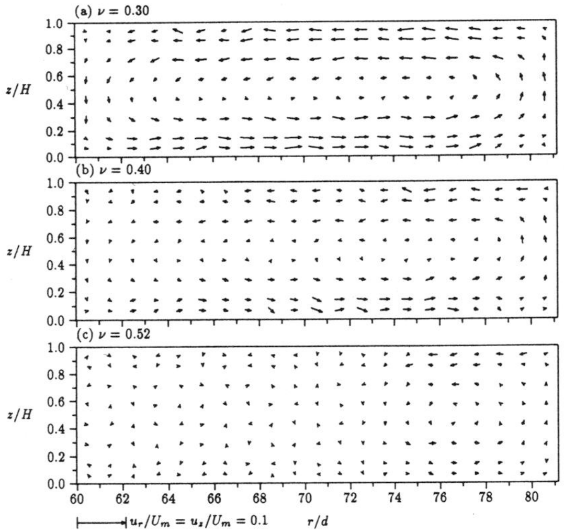

Figure 3. Recirculatory secondary-flow motion at various solid volume fraction.

Figure 2 shows the distributions of shear and normal stresses on inner and outer side-walls at various solids fractions. Due to the induced centrifugal effect, the stresses are in general higher at the outer side-wall than the inner one. The recirculatory motion of the secondary-flow can be seen vividly in figure 3 at solids fraction of 0.30 and 0.40. As the concentration increases further, the secondary-flow pattern become more irregular and the particle recirculatory speed decreases.

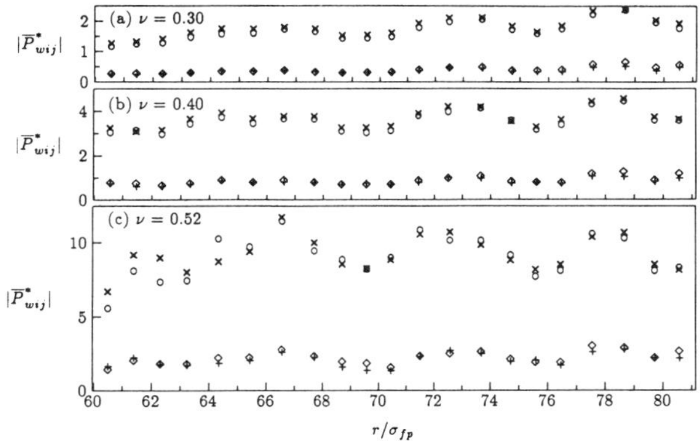

Figure 4. Comparisons of distributions of shear (◊) and normal (o) stresses on the top wall with those (+, ×) of the bottom wall.

The distributions of shear and normal stresses on the top and bottom walls are shown in figure 4. The stresses increase gradually in a sinusoidal manner from the inner side-wall to the outer one. The positions of the peaks are rather consistent in the 3 different solids fractions plotted in the figure which indicate the existence of microstructures in the granular flow medium.

Remark

Work is in progress to simulate other experiments and to compare the numerical result with the measurements.

Acknowledgment

This work was supported by a research grant from **NSERC** of Canada.

References

Craig, K., Buckholz, R. H., and Domoto, G. (1986). "An experimental study of the rapid flow of dry cohesionless metal powders," *Trans. ASME J. Appl. Mech.*, 53, 935-942.

Hanes, D. M., and Inman, D. L. (1985). "Observations of rapidly flowing granular-fluid materials," *J. Fluid Mech.*, 150, 357-380.

Lun, C. K. K., and Bent, A. A. (1994). "Numerical simulation of inelastic, frictional spheres in simple shear flow," *J. Fluid Mech.*, 258, 335-353.

Savage, S. B., and Sayed, M. (1984). "Stresses developed by dry cohesionless granular materials sheared in annular shear cell," *J. Fluid Mech.*, 142, 391-430.

Wang, D., and Campbell, D. S. (1992). "Reynolds analogy for a shearing granular material," *J. Fluid Mech.*, 244, 527-546.

Theory and Simulations of Relaxation and Cyclic Granular Flows

Marijan Babic[1], Associate Member and William J. Bocchieri[2], Student Member

Abstract

Two important types of unsteady granular Couette flows, relaxation flows and cyclic flows, are investigated both theoretically and by the means of the Discrete Element Method (DEM). For both types of flows predictions of the kinetic theory for nearly elastic disks compare favorably with the corresponding DEM simulations.

Introduction

Kinetic theories of rapid granular flows (e.g. Jenkins and Richman 1985) have so far been applied predominantly to relatively simple, steady parallel flows such as the Couette flow, inclined chute flow, and vertical chute flow. In order to develop a better understanding of granular material behavior under more complex conditions as well as to test the performance of the time-dependent kinetic theory it is of interest to investigate unsteady rapid granular flows. A step in this direction was the study of unbounded relaxation shear flows by Louge et al. (1993).

In the present study we consider unsteady wall-bounded Couette flows. Such flows may be generated by equal and opposite time-dependent motion of parallel walls. In particular we consider relaxation flows, in which the wall velocity instantaneously changes from one constant velocity to another, and cyclic flows, in which the wall velocity is a harmonic function of time.

Theory

The granular material is assumed to be composed of uniform circular particles (disks or spheres) of diameter σ. The particles are smooth and nearly elastic, i.e. the coefficient of restitution e is close to unity. The flow is bounded by "bumpy" walls located at $y = \pm L$ and moving with velocities $\pm u_w(t)$. The bumpy walls are created by placing particles of diameter d at a spacing s along the walls. The flow is assumed to be uniform in the x-direction. The governing equations of the kinetic theory reduce to:

$$\frac{\partial c}{\partial t} + v\frac{\partial c}{\partial y} + c\frac{\partial v}{\partial y} = 0, \tag{1}$$

$$c\left(\frac{\partial u}{\partial t} + v\frac{\partial u}{\partial y}\right) = \frac{\partial}{\partial y}\left(f_\mu \sigma w\frac{\partial u}{\partial y}\right), \tag{2}$$

1. Assistant Professor, 2. Undergraduate Student, Department of Civil Engineering and Geological Sciences, University of Notre Dame, Notre Dame, IN 46556

$$c\left(\frac{\partial v}{\partial t}+v\frac{\partial v}{\partial y}\right) = \frac{\partial}{\partial y}\left(-f_\pi w^2+f_\lambda \sigma w\frac{\partial v}{\partial y}\right), \tag{3}$$

$$\frac{n}{2}c\left(\frac{\partial w^2}{\partial t}+v\frac{\partial w^2}{\partial y}\right) = \frac{\partial}{\partial y}\left(f_\kappa \sigma w\frac{\partial w^2}{\partial y}\right)-f_\pi w^2\frac{\partial v}{\partial y}+f_\lambda \sigma w\left(\frac{\partial v}{\partial y}\right)^2$$
$$+f_\mu \sigma w\left(\frac{\partial u}{\partial y}\right)^2-\eta f_\gamma \frac{w^3}{\sigma}, \tag{4}$$

where c is the concentration, u and v are the x- and y- components of the mean velocity, $w = T^{1/2}$ is the mean fluctuation speed, T is the granular temperature, $n = 2$ for disks, $n = 3$ for spheres, $\eta = 1-e$, and f_π, f_λ, f_μ, f_κ, f_γ are functions of c. Due to the flow symmetry about the centerline, the following boundary conditions are imposed at $y = 0$: $u(0,t) = 0$, $v(0,t) = 0$, $(\partial w/\partial y)(0,t) = 0$. The boundary conditions for bumpy walls, developed by Jenkins and Richman (1986), can be reduced to:

$$u_w(t)-u(L,t) = h_1\sigma\left(\frac{\partial u}{\partial y}\right)(L,t), \tag{5}$$

$$\sigma\left(\frac{\partial w^2}{\partial y}\right)(L,t) = \left[h_2\left(\sigma\frac{\partial u}{\partial y}\right)^2-h_3\eta_w w^2\right](L,t), \tag{6}$$

where h_1, h_2 and h_3 are functions of c, d/σ and s/d, $\eta_w = 1-e_w$, and e_w is the wall coefficient of restitution. It is also necessary to impose the no-flow boundary condition at the wall, i.e. $v(L,t) = 0$. The initial conditions for both relaxation and cyclic flows is the solution of the steady flow problem with $u_w = u_w^0$.

Results and Discussion

An implicit finite-difference numerical scheme for the solution of the above problem has been developed. In this scheme the domain $(0, L)$ is divided in N equal intervals. A staggered grid is used so that the discrete values of c are sought at the edges of the intervals, while the discrete values of (u, v, w) are sought at the centers of these intervals. The discretization yields a set of $4N+1$ nonlinear equations for $4N+1$ unknowns that have to be solved each time step. The scheme is first-order accurate in time and appears to be unconditionally stable. It is second-order accurate in space, which allows relatively small values of N to be used.

Numerical simulations of the unsteady Couette flows were performed using the soft-particle discrete element method (DEM). The model tracks the motion of a finite number of particles placed between periodic boundaries at $x = \pm A$ and bumpy walls at $y = \pm(L+(\sigma+d)/2)$. Due to the finite number of particles used in the simulation, the data for the complete unsteady fields (e.g. $w(y,t)$) is quite "noisy", so only the width-averaged data (e.g. $\overline{w}(t)$) is used for comparison with the theory.

Relaxation Flows

For the relaxation flows $u_w(t) = u_w^0 H(-t)+u_w^\infty H(t)$, where $H(t)$ is the Heaviside step function. Relaxation flows are characterized by two time scales: (i) momentum diffusion time scale t_d, which can be estimated by equating the orders of magnitude of the first term on the left-hand side and the term on the right-hand side of the x-momentum equation (2), and (ii) energy relaxation time scale t_r, which can be estimated by equating the orders of magnitude of the unsteady term (the first term on the left-hand side) and the rate of dissipation term (the last term on the right-hand side) in the energy equation (4). Assuming that c, f_μ and f_γ are all of the order of one, $t_d = L^2/(\sigma w_s)$, while $t_r = \sigma/((1-e)w_s)$, where w_s is the fluctuation speed scale. In general it may be stated that the duration of the relaxation flow is of the order of $\max(t_d, t_r)$. An example with $e = 0.96$, $L/\sigma = 5$, $u_w^0 = 5$, $u_w^\infty = 10$, $d/\sigma = 1$, and $s/d = 0$ is shown in Figure 1 along with the width-averaged DEM data (with $A/\sigma = 100$). The comparison is very good. Quantities shown in the figure are nondimensionalized as follows: $t^* = t/t_r$, $y^* = y/L$, $u^* = u/u_s$, $v^* = v/(w_s(L/\sigma)\eta)$, $w^* = w/w_s$, $t_{yx}^* = t_{yx}/(w_s u_s \sigma/L)$, $t_{yy}^* = t_{yy}/w_s^2$, where $w_s = (\sigma/L)u_s/\eta^{1/2}$ and $u_s = u_w^\infty$. For small values of t_d/t_r, the momentum diffusion process is much faster than the energy relaxation process. In this limit the flow becomes similar to the unbounded relaxation flow studied by Louge et al. (1993), i.e. the velocity field is almost instantaneously adjusted to the new wall motion.

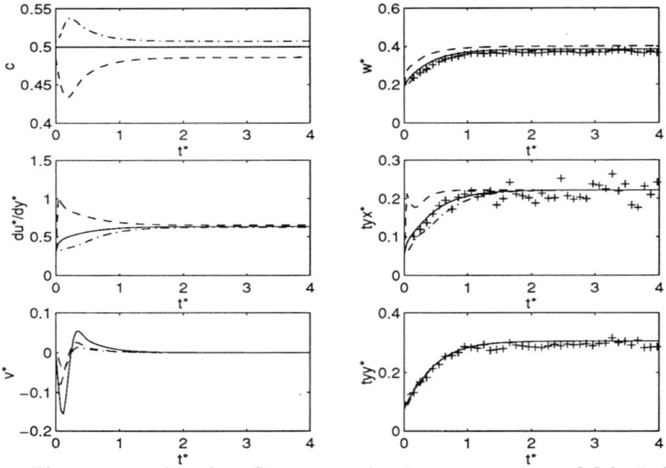

Figure 1: Relaxation flow example. Parameters: $e = 0.96$, $L/\sigma = 5$, $u_w^0 = 5$, $u_w^\infty = 10$, $d/\sigma = 1$, $s/d = 0$. Solid lines: width-averaged values, dashed lines: boundary values, dot-dashed: centerline values, pluses: DEM data.

Cyclic Flows

For the cyclic flows $u_w(t) = u_w^0 H(-t) + \sqrt{2}u_w^{rms}\cos(\Omega t)H(t)$. For sufficiently large times all cyclic flows approach limit cycles that are independent of the initial conditions. The features of the cyclic flow solutions depend upon the relative magnitudes of the energy relaxation time scale t_r, momentum diffusion time scale t_d, and the period of wall oscillations Ω^{-1}. The simplest case occurs for $\Omega^{-1} \gg \max(t_d, t_r)$ (slow oscillations). In this case the flow is quasi-steady, i.e. all instantaneous fields correspond to the steady solutions that would be obtained for the particular instantaneous value of the wall velocity. On the other hand, for $\Omega^{-1} \ll \min(t_d, t_r)$ (fast oscillations) we have a vibrating boundary problem: only the material in the immediate vicinity of the walls is sheared back and forth while the bulk of the material is energized but does not move in the mean sense. In the present study we focus on the generic case $t_d \sim t_r \sim \Omega^{-1}$. An example with $e = 0.96$, $L/\sigma = 5$, $u_w^0 = 5$, $u_w^{rms} = 5$, $\Omega = 0.1$ $d/\sigma = 1$, and $s/d = 0$ is shown in Figure 2 along with the width-averaged DEM data (with $A/\sigma = 100$). The comparison is very good. The nondimensionalization is the same as for the relaxation flow, except that $u_s = u_w^{rms}$. It can be seen that $\partial u/\partial y$ and t_{yx} fields oscillate with a frequency Ω while w, c and t_{yy} fields oscillate with a frequency 2Ω. These quantities are near their maxima when the wall speed is at its maximum. Increasing Ω reduces the amplitudes of the granular temperature and normal stress oscillations. In the vibrating boundary limit these oscillations vanish.

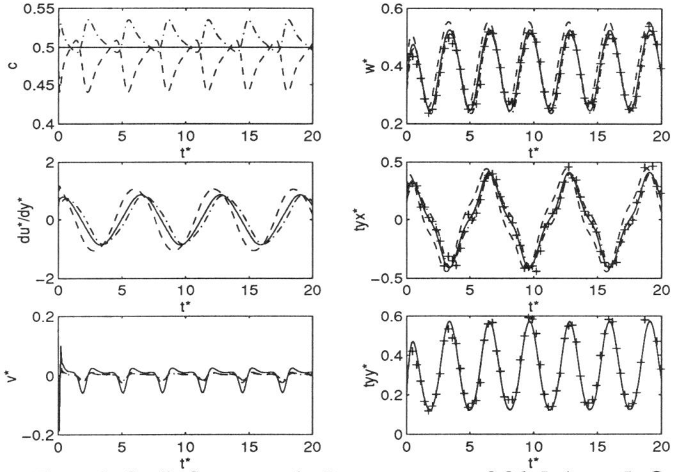

Figure 2: Cyclic flow example. Parameters: $e = 0.96$, $L/\sigma = 5$, $\Omega = 0.2$, $u_w^{rms} = 5$, $u_w^0 = 5$, $u_w^\infty = 10$, $d/\sigma = 1$, $s/d = 0$. Solid lines: width-averaged values, dashed lines: boundary values, dot-dashed: centerline values, pluses: DEM data.

Conclusions

Unsteady Couette granular flows have been investigated. A numerical scheme for solution of the governing equations posed by the kinetic theory has been developed and applied to the relaxation and cyclic flows. The resulting predictions compare favorably with the results of the corresponding DEM simulations.

References

1. Jenkins, J. T. and Richman, M. W., Kinetic theory for plane shear flows of a dense gas of identical, rough, inelastic, circular disks, Phys. Fluids, 28:3485–3494, 1985.
2. Jenkins, J. T. and Richman, M. W., Boundary conditions for plane flows of smooth, nearly elastic, circular disks. J. Fluid Mech., 171:53–69, 1986.
3. Louge, M. Y., Jenkins, J. T. and Hopkins, M. A., The relaxation of the second moments in rapid shear flows of smooth disks, Mech. Materials, 16:199-204, 1993.

Response of an Oscillator with One-Sided Barrier under Non-White Random Excitation

G. Q. Cai[1]

Abstract

A modified version of quasi-conservative averaging is applied to an oscillator with one-sided barrier subjected to a broad-band random excitation . The total energy process is approximated as a Markov process, and its stationary probability density id obtained.

System

Consider an oscillator with one-sided barrier, shown in Fig. 1. The rigid barrier is assumed to be located at the static equilibrium position $X = 0$. The equation of motion of the mass between its impacts against the barrier can be written as

$$\ddot{X} + 2\zeta\omega_0\dot{X} + \frac{1}{2}(1-\nu^2)\dot{X}|\dot{X}|\,\delta(X) + u(X) = \xi(t)\ , \qquad X \geq 0 \qquad (1)$$

where $\delta(X)$ is the Dirac delta function, $u(X)$ is assumed to be a nonlinear odd function of X, $\xi(t)$ is a stationary random excitation with a broad-band spectrum $\Phi(\omega)$, and ν is the restitution coefficient, defined as the ratio of velocities immediately after and before an impact. The third term on the left-hand-side of equation (1) accounts for the energy loss due to impacts. The case of $\nu = 1$ and a linear spring force was investigated by Dimentberg (1995).

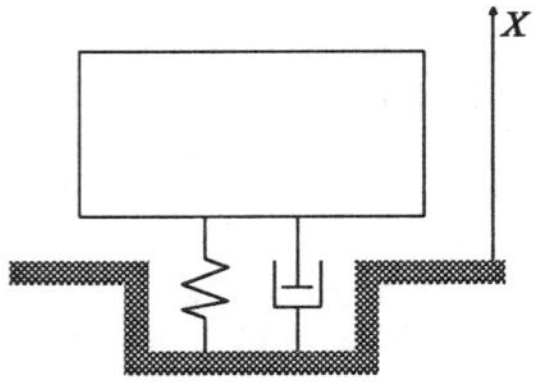

Fig. 1. An oscillator with one-sided barrier

[1]Center for Appl. Stochastics Res., Florida Atlantic Univ., Boca Raton, FL 33431.

Analysis

The potential energy $U(X)$ and the total energy Λ of the system are, respectively,

$$U(X) = \int_0^X u(z)\,dz\,, \qquad \Lambda = \frac{1}{2}\dot{X}^2 + U(X) \tag{2}$$

In the absence of damping and external excitation, the system is conservative, and the amplitude a of the free motion corresponding to an energy level Λ can be obtained from the equation $U(a) = \Lambda$. We assume that this equation has only one positive root. Starting from a maximum displacement a, the mass moves in the negative X-direction until reaching the barrier where an impact occurs. It then reverses the direction and reaches the amplitude a again. The period for such a cycle is

$$T = 2\int_0^a [2\Lambda - 2U(X)]^{-1/2}\,dX \tag{3}$$

It is seen from (3) that T depends on the energy level Λ.

Now return to system (1). Letting

$$\sqrt{U(X)} = \sqrt{\Lambda}\cos\phi\,\mathrm{sgn}(\cos\phi)\,, \qquad \dot{X} = -\sqrt{2\Lambda}\sin\phi\,\mathrm{sgn}(\cos\phi) \tag{4}$$

system (1) is transformed into two first-order differential equations of the energy process $\Lambda(t)$ and the phase process $\phi(t)$ as follows

$$\dot{\Lambda} = -4\zeta\omega_0\Lambda\sin^2\phi - \sqrt{2}(1-\nu^2)\Lambda^{3/2}|\sin\phi|^3\delta(X) - \sqrt{2\Lambda}\sin\phi\,\mathrm{sgn}(\cos\phi)\,\xi(t) \tag{5}$$

$$\dot{\phi} = -2\zeta\omega_0\sin\phi\cos\phi - \sqrt{\Lambda/2}\,(1-\nu^2)\sin\phi\cos\phi\,|\sin\phi|\delta(X) + \frac{u(X)\,\mathrm{sgn}(\cos\phi)}{\sqrt{2\Lambda}\cos\phi} - \frac{\cos\phi\,\mathrm{sgn}(\cos\phi)}{\sqrt{2\Lambda}}\xi(t) \tag{6}$$

where X is interpreted as a function of Λ and ϕ according to transformation (4). Equation (5) show that the energy process $\Lambda(t)$ is slowly varying if ζ and $(1-\nu^2)$ are of order ε, and the excitation $\xi(t)$ is of order $\varepsilon^{1/2}$. Assume, in addition, that the frequency band of the random excitation $\xi(t)$ is sufficiently wide so that its correlation time is much shorter than the relaxation time of the energy process. Then, $\Lambda(t)$ can be approximated as a Markov process by a modified version of quasi-conservative averaging (Roberts, 1982; Cai, 1994) as follows

$$d\Lambda = m(\Lambda)\,dt + \sigma(\Lambda)\,dB(t) \tag{7}$$

where

$$m(\Lambda) = -\frac{4\zeta\omega_0}{T}\int_0^a \sqrt{2\Lambda - 2U(X)}\,dX - \frac{1}{T}(1-\nu^2)\Lambda + \pi c_0^2\Phi(0) + \frac{1}{2}\pi\sum_{n=1}^{\infty}(s_n^2 + c_n^2)\Phi(n\omega_T) \tag{8}$$

$$\sigma^2(\Lambda) = 2\pi\Lambda \sum_{n=1}^{\infty} s_n^2 \Phi(n\omega_T) \tag{9}$$

in which $\omega_T = 2\pi/T$, and coefficients c_0, c_n and s_n are calculated from

$$c_0 = \frac{2}{T\sqrt{\Lambda}} \int_0^{T/2} \sqrt{U[X(t)]}\, dt \tag{10}$$

$$c_n = \frac{4}{T\sqrt{\Lambda}} \int_0^{T/2} \sqrt{U[X(t)]}\, \cos(n\omega_T t)\, dt \tag{11}$$

$$s_n = -\frac{4}{T\sqrt{2\Lambda}} \int_0^{T/2} \dot{X}(t) \sin(n\omega_T t)\, dt \tag{12}$$

For a given energy level Λ, functions $X(t)$ and $\dot{X}(t)$ in (10)-(12) within a half period $[0, T/2]$ may be approximated by those of the undamped free motion.

The stationary probability density of $\Lambda(t)$ is then obtained as

$$p(\lambda) = \frac{C}{\sigma^2(\lambda)} \exp\left[\int \frac{2m(\lambda)}{\sigma^2(\lambda)} d\lambda \right] \tag{13}$$

where λ is the state variable of Λ, and C is a normalization constant.

An Example

The above procedure is illustrated by an example with a nonlinear restoring force

$$u(X) = \omega_0^2 (1 + \alpha X^2) X \tag{14}$$

with $\zeta = 0.02$, $\omega_0 = 1$, and two restitution coefficients $\nu = 1$ and 0.9. The excitation $\xi(t)$ is a filtered noise with a spectral density

$$\Phi(\omega) = \frac{K}{(\omega^2 - \omega_1^2)^2 + 4\zeta_1^2 \omega_1^2 \omega^2} \tag{15}$$

The set of $\zeta_1 = 0.5$, $\omega_1 = 5$, and $K = 30$ is used in the calculation. Figs. 2 and 3 show the stationary probability densities of the energy process with two different levels of nonlinearity $\alpha = 0.1$ and $\alpha = 1$. Also shown in the figures are results obtained from Monte Carlo simulations. For the case of weak nonlinearity ($\alpha = 0.1$), the results obtained from the present method are in good agreement with those from Monte Carlo simulation. When the nonlinearity is strong ($\alpha = 1$), the analytical results deviate from those from Monte Carlo simulations, but still in reasonable accurate range.

Acknowledgement

The work reported in this paper is supported by the U.S. Office of Naval Research under Grant N00014-93-1-0879.

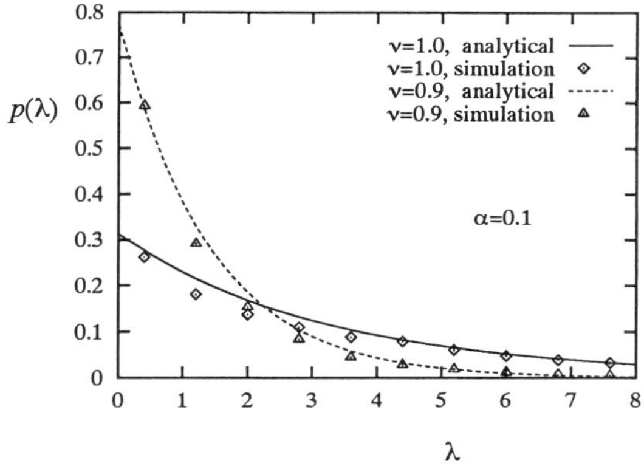

Fig. 2. Stationary probability density of energy process for $\alpha = 0.1$.

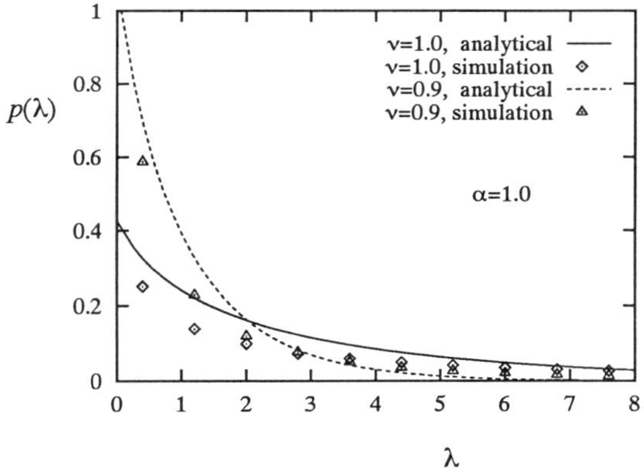

Fig. 3. Stationary probability density of energy process for $\alpha = 1$.

References

Cai, G.Q. (1994). Random vibration of nonlinear systems under non-white excitations. *J. Engrg. Mech.*, 121, 633-639.

Dimentberg, M. (1995). Psudolinear Vibroimpact systems: non-white random excitation and MDOF cases. to appear in *Nonlinear Dynamics.*

Roberts, J.B. (1982). A stochastic theory for nonlinear ship rolling in irregular seas. *J. Ship Res.*, 26, 229-245.

STATISTICS OF FRACTIONAL OCCUPATION TIME FOR NONLINEAR STOCHASTIC RESPONSE

Armen Der Kiureghian[1], Member ASCE, and Chun-Ching Li[2]

Abstract
A new approach is introduced to compute the statistics (mean, variance, probability distribution) of the fractional occupation time, which is the fraction of unit time spent by the response of a system in a specified domain. The approach is based on the first-order reliability method and is applicable to nonlinear and ucertain systems.

Introduction

In some systems, the damage incurred under dynamic loading is related to the time spent by a response quantity outside a specified domain. For example, the cumulative plastic strain in an elasto-plastic system is closely related to the time spent by the response outside the elastic domain. By "fractional occupation time" (FOT), we refer to the fraction of an unit time spent by the system in the specified domain. This paper describes a new method for computing the mean, variance, and probability distribution of FOT for nonlinear systems subjected to stochastic excitation.

Consider a nonlinear system subjected to stochastic excitation whose motion is described by the stochastic differential equation

$$\mathbf{D}[\mathbf{u}(t), \mathbf{v}] = \mathbf{PF}(t) \tag{1}$$

where t denotes time, $\mathbf{D}$ is a differential operator, $\mathbf{u}(t)$ is the response vector (generalized displacement or internal force), $\mathbf{v}$ is a vector of time-invariant system parameters, e.g., mass, damping, restoring force, $\mathbf{P}$ is a load distribution matrix, and $\mathbf{F}(t)$ is a vector of input processes, e.g., components of ground motion. Our interest lies in the excursions of $\mathbf{u}(t)$ into a "failure" domain defined by

$$\mathcal{F} = \{\mathbf{u}, \mathbf{v}\colon\ g(\mathbf{u}, \mathbf{v}) \le 0\} \tag{2}$$

where $g(\mathbf{u}, \mathbf{v})$ is a limit-state function. A realization of the response process in the space of $\mathbf{u}$ and $\mathbf{v}$ may appear as in Fig. 1. The response is seen to make infrequent and brief excursion into $\mathcal{F}$. Let $\eta(t, \Delta t)$ denote the FOT of $\mathbf{u}(t)$ in $\mathcal{F}$ at time t defined by

---

[1] Professor of Civil Engineering, University of California, Berkeley, CA 94720.
[2] Graduate Student, Department of Civil Engineering, University of California, Berkeley, CA 94720.

$$\eta(t,\Delta t) = \frac{1}{\Delta t}\int_{t}^{t+\Delta t} H[-g(\mathbf{u}(t),\mathbf{v})]\mathrm{d}t \tag{3}$$

where Δt is a small interval and $H(.)$ is the Heaviside function, i.e., $H(a)=0$ for $a<0$ and $H(a)=1$ for $a>0$. Using the above definition, one can show that

$$\mathrm{E}[\eta(t,\Delta t)] = \frac{1}{\Delta t}\int_{t}^{t+\Delta t} \mathrm{P}[g(\mathbf{u}(t),\mathbf{v}) \le 0] \tag{4}$$

$$\mathrm{E}[\eta^2(t,\Delta t)] = \frac{1}{\Delta t}\int_{-\Delta t}^{\Delta t} \mathrm{P}[g(\mathbf{u}(t),\mathbf{v}) \le 0 \cap g(\mathbf{u}(t-\tau),\mathbf{v}) \le 0]\left(1-\frac{|\tau|}{\Delta t}\right)d\tau \tag{5}$$

from which the variance is computed as $\mathrm{Var}[\eta(t,\Delta t)] = \mathrm{E}[\eta^2(t,\Delta t)] - \mathrm{E}^2[\eta(t,\Delta t)]$. Furthermore, the probability that $\eta(t)$ exceeds η_0 is given by

$$\mathrm{P}[\eta(t,\Delta t) > \eta_0] \tag{6}$$

In the following, we use the first-order reliability method (FORM) to compute (4)-(6).

Solution by FORM

FORM is an efficient algorithm for computing probability integrals of the form

$$p = \int_{\mathcal{D}} f(\mathbf{x})\mathrm{d}\mathbf{x} \tag{7}$$

where $\mathbf{x}$ is a vector of random variables with PDF $f(\mathbf{x})$, and $\mathcal{D}$ denotes a domain in the outcome space of $\mathbf{x}$. Of interest here are cases where $\mathcal{D} \equiv \{\bigcap_i G_i(\mathbf{x}) \le 0\}$, where $G_i(\mathbf{x})$, i = 1, 2, ..., are a set of limit-state functions. In FORM, an approximations to p is obtained by: (a) transforming the random variables $\mathbf{x}$ into the standard normal space, (b) linearizing each of the limit-state surfaces $G_i(\mathbf{x}) = 0$ at the point of maximum likelihood within $\mathcal{D}$, and (c) using known expressions for the probability content of polyhedral sets in the normal space. This method provides an accurate approximation as long as the limit-state surfaces are not strongly warped. The main computational effort lies in finding the linearization points. This requires the solution of a constrained optimization problem involving the gadient of each $G_i(\mathbf{x})$.

In order to solve the random vibration problem by FORM, it is necessary to represent the input stochastic processes in terms of random variables. For a Gaussian process, one convenient representation of the component $F(t)$ of $\mathbf{F}(t)$ is

$$\hat{F}(t) = \sum_{i=1}^{n} a_i(t) f_i \tag{8}$$

where $a_i(t)$ are deterministic coefficient functions and f_i are zero-mean, statistically independent Gaussian variables. In a recent paper (Li and Der Kiureghian 1995), the authors modeled $\hat{F}(t)$ as the response of a filter to a train of pulses $\mathbf{f} = (f_1, f_2, \cdots)$ at time points t_1, t_2, ... The coefficient functions $a_i(t)$ then are related to the unit impulse response function of the filter. Fig. 2 shows a typical realization of the pulse train and the filter response, where a Kanai-Tajimi filter (Clough and Penzien 1993) with time modulation is used to describe an earthquake-type excitation. The input excitation thus is completely defined by the random variables $\mathbf{f} = (f_1, f_2, \cdots)$ and the set of parameters defining the filter and the modulating function.

Let $\mathbf{x} = (\mathbf{f}, \mathbf{v})$ denote the collection of all random variables describing the problem, where **f** describes the input excitation and **v** describes the system properties. The latter may also include parameters of the filter or the modulating function used in defining the input excitation, if they are uncertain. The response $\mathbf{u}(t)$ clearly is an implicit function **f** and **v**, and hence of **x**. One may write $\mathbf{u}(t) = \mathbf{u}(\mathbf{f}, \mathbf{v}, t) = \mathbf{u}(\mathbf{x}, t)$. It should be clear that the randomness in **u** arises from **x**, whereas t acts as a deterministic parameter. Furthermore, one can write $g(\mathbf{u}(t), \mathbf{v}) = g(\mathbf{x}, t)$. It follows that problems defined in (4)-(6) can be solved by FORM in the space of random variables **x**. Specifically, $\mathrm{E}[\eta(t, \Delta t)]$ is obtained by using the limit-state function $G(\mathbf{x}) = g(\mathbf{x}, t)$; $\mathrm{E}[\eta^2(t, \Delta t)]$ is obtained by using the two limit-state functions $G_1(\mathbf{x}) = g(\mathbf{x}, t)$ and $G_2(\mathbf{x}) = g(\mathbf{x}, t - \tau)$ followed by the integration over τ as in (5); and $\mathrm{P}[\eta(t, \Delta t) > \eta_0]$ is obtained by using the limit-state function $\mathrm{G}(\mathbf{x}) = \eta_0 - \eta(\mathbf{x}, t, \Delta t)$, where $\eta(\mathbf{x}, t, \Delta t)$ is as defined in (3) with $g(\mathbf{u}(t), \mathbf{v})$ replaced by $g(\mathbf{x}, t)$.

Example Application

Consider a single-degree-of-freedom oscillator having the Bouc-Wen hysteretic behavior (Wen 1989) and subjected to the excitation described above. The equation of motion is

$$\ddot{u}(t) + 2\varsigma_0\omega_0\dot{u}(t) + \omega_0^2[\alpha u(t) + (1-\alpha)z(t)] = F(t) \tag{9a}$$

$$\dot{z}(t) = -\gamma|\dot{u}(t)||z(t)|^{n-1}z(t) - \beta\dot{u}(t)|z(t)|^n + A\dot{u}(t) \tag{9b}$$

where $z(t)$ is an auxiliary function, α controls the degree of nonlinearity, and $\omega_0 = 3\pi$ rad/s, $\varsigma_0 = 0.05$, $\gamma = \beta = 0.5/\sigma$, $n = 1$, and $A = 1$ are system parameters, where σ denotes the maximum root-mean-square response of the linear ($\alpha = 1$) oscillator. Consider the FOT spent by the response $u(t)$ above the threshold of 3σ. We have $g(\mathbf{u}(t), \mathbf{v}) = 3\sigma - u(t)$. Since the system parameters are deterministic, **x** = **f**. For the present example t_i are taken at 0.02 s intervals and the size of **x** is 600.

Fig. 3 shows plots of the mean and mean-plus-one-standard-deviation of $\eta(t, \Delta t)$ and the probability $\mathrm{P}[\eta(t, \Delta t) > 0.8]$ for the linear ($\alpha = 1$, dashed lines) and a strongly nonlinear ($\alpha = 0.2$, solid lines) oscillators for $\Delta t = 0.2$ s. For the linear oscillator, $u(t)$ is a linear function of the Gaussian variables **f** and, hence, of **x**. It follows that the limit-state functions used in the FORM solutions of $\mathrm{E}[\eta(t, \Delta t)]$ and $\mathrm{E}[\eta^2(t, \Delta t)]$ for this case are linear and hence these solutions are exact, save for the discretization error. This observation supports the conjecture that the limit-state surfaces for the nonlinear oscillator are not too strongly warped and, therefore, good accuracy is expected for that case as well.

In the above solutions, use is made of the Improved HL-RF algorithm by Zhang and Der Kiureghian (1994) to find the design points. A direct differentiation method (Li and Der Kiureghian 1995) is used to compute the necessary gradients. However, it is noted that the system defined by (9) has discontinuous response gradients at zero crossings of $z(t)$ and $\dot{u}(t)$. Furthermore, the Heaviside function in (3) has a discontinuous gradient at zero crossings of $g(\mathbf{u}(t), \mathbf{v})$. If not handled properly, these discontinuities cause divergence in the optimization algorithm. In the present implementation, these discontinuities are resolved by slightly modifying the relevant equations to provide a smooth ramp in place of a jump. The effect on the limit-state surfaces is only local and does not influence the computed results.

Acknowledgment
The work presented here is supported by a grant from Kajima Corp., Japan, through CUREe. The authors thank Arne B. Huseby for suggesting this problem in a discussion of their earlier paper.

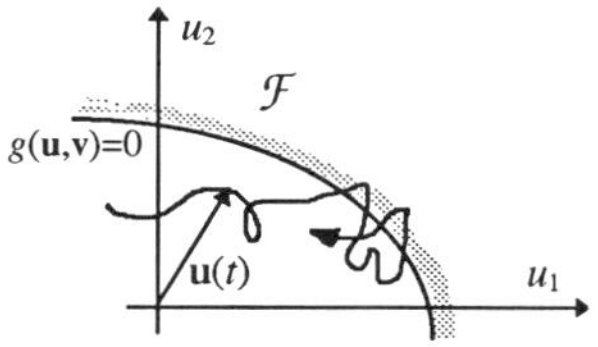

Fig. 1. Problem definition

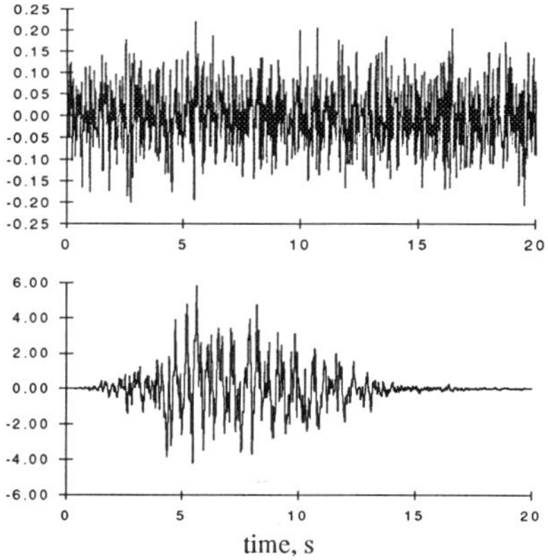

Fig. 2. Sample realization of pulse train (above) and filter response (below)

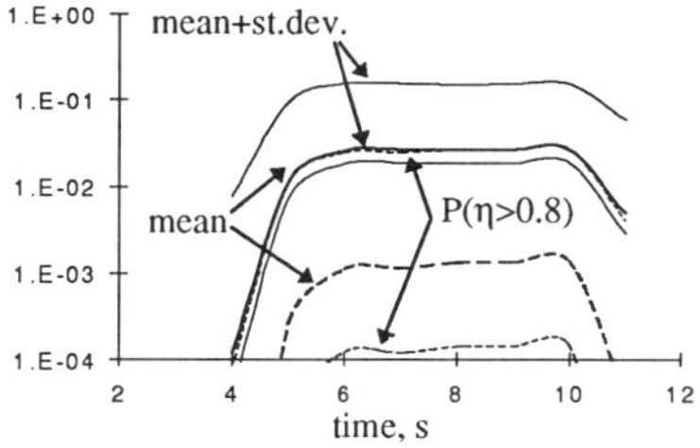

Fig. 3. Statistics of FOT for linear (dashed lines) and nonlinear (solid lines) oscillators

References

1. Clough, R., and J. Penzien (1993). *Dynamics of structures*. McGraw-Hill, NY.
2. Li, C-C., and A. Der Kiureghian (1995). Mean out-crossing rate of nonlinear response to stochastic input. *Proc. ICASP5*, Paris, France, 1, 295-302.
3. Wen, Y.K. (1989). Methods of random vibration for inelastic structures, *Appl. Mech. Revs., ASME*, **42**, 39-52.
4. Zhang, Y., and A. Der Kiureghian (1994). Two improved algorithms for reliability analysis. *Proc. 6th IFIP Working Conference*, Assisi, Italy.

Nonlinear Systems with Poisson White Noise

Mircea Grigoriu[1]

Abstract

A method is developed for calculating statistics of the response of a nonlinear system subjected to Poisson white noise. The method is based on the observation that the system vibrates freely between consecutive pulses of the input. Two algorithms are proposed. The algorithms are based on the definition of the Poisson white noise and a binomial approximation of this noise. The algorithms are applied to find response statistics for two nonlinear systems.

Introduction

The Poisson white noise consists of independent identically distributed pulses arriving at random times according to a Poisson counting process. The process can be used to model traffic loads, seismic ground accelerations, wind gusts, wave forces, forces acting on airplanes, and other actions (Grigoriu 1995; Hu 1995; Köylüoğlu et al. 1995). The determination of statistics of the response $\boldsymbol{X}(t)$ of a nonlinear system subjected to Poisson white noise can be based on generalized versions of the Fokker-Planck-Kolmogorov equations, the Itô differentiation formula, the path integral method, and Monte Carlo simulation (Hu 1995; Köylüoğlu et al. 1995).

This paper presents a new method for calculating statistics of the response of a nonlinear system subjected to Poisson white noise based on the observation that the system exhibits free vibrations between consecutive pulses of the noise. Two algorithms are used for calculations. The first algorithm uses the definition of the Poisson white noise so that the interarrival times between input pulses are random. The second algorithm is based on an approximation of the Poisson white noise referred to as binomial white noise.

---

[1] Professor, School of Civil and Environmental Engineering, Cornell University, Ithaca, NY 14853

Problem definition

Let $\{\boldsymbol{X}(t), t \geq 0\}$ be an $\Re^d$-valued process defined by the stochastic differential equation

$$d\boldsymbol{X}(t) = \boldsymbol{a}(\boldsymbol{X}(t),t)\,dt + \boldsymbol{b}(\boldsymbol{X}(t),t)\,d\boldsymbol{Y}(t), \quad t \geq 0, \tag{1}$$

where the $(d,1)$ and (d,m) matrix functions $\boldsymbol{a}(\boldsymbol{x},t)$ and $\boldsymbol{b}(\boldsymbol{x},t)$ depend on the state vector and time,

$$\boldsymbol{Y}(t) = \begin{cases} 0, & \text{if } N(t) = 0 \\ \sum_{k=1}^{N(t)} \boldsymbol{Y}_k\,\xi(t - S_k), \quad t \geq 0, & \text{if } N(t) > 0 \end{cases} \tag{2}$$

is a compound Poisson process depending on the independent identically distributed vectors $\{\boldsymbol{Y}_k\}$ and a homogeneous Poisson counting process $\{N(t), t \geq 0\}$ with intensity $\nu > 0$ and events at times $\{S_k\}$. The Poisson white noise can be defined by the formal derivative $d\boldsymbol{W}(t)/dt = \boldsymbol{Y}(t)$.

Free vibration based method

Suppose that the system is time invariant, that is $\boldsymbol{a}(\boldsymbol{x},t) = \boldsymbol{a}(\boldsymbol{x})$ and $\boldsymbol{b}(\boldsymbol{x},t) = \boldsymbol{b}(\boldsymbol{x})$. Let $\{\boldsymbol{h}(\boldsymbol{\xi},s), s \geq 0, \boldsymbol{\xi} \in \Re^d\}$ be the homogeneous solutions of Eq. 1 for $\boldsymbol{X}(0) = \boldsymbol{\xi}$. These solutions give the free vibration evolution of the system of Eq. 1 and, generally, need to be determined numerically. The free vibration solutions can be used to develop the mappings

$$\boldsymbol{X}_k^- = \boldsymbol{h}(\boldsymbol{X}_{k-1}^+, T_k) \tag{3}$$

and

$$\boldsymbol{X}(S_{k-1} + s) = \boldsymbol{h}(\boldsymbol{X}_{k-1}^+, s), \quad 0 \leq s < T_k, \tag{4}$$

where $\boldsymbol{X}_k^- = \boldsymbol{X}(S_k-)$, $\boldsymbol{X}_k^+ = \boldsymbol{X}(S_k+)$, and $T_k = S_k - S_{k-1}$ denotes the lag between the arrival times of pulses $(k-1)$ and k. The Euler approximation of Eq. 1 gives

$$\boldsymbol{X}_k^+ = \boldsymbol{X}_k^- + \boldsymbol{b}(\boldsymbol{X}_k^-)\,\boldsymbol{Y}_k. \tag{5}$$

Let $D \subset \Re^d$ be a bounded domain which contains $\boldsymbol{X}(t)$ with nearly unit probability and $\{C_i\}$, $i = 1, 2, \ldots, n_c$, be a partition of D, that is $D = \cup_{i=1}^{n_c} C_i$ and $C_i \cap C_j = \emptyset$, $i \neq j$, where the cells $\{C_i\}$ are small. Define the n_c-dimensional probability vector $\boldsymbol{P}(t) = \{P(\boldsymbol{X}(t+) \in C_1), P(\boldsymbol{X}(t+) \in C_2), \ldots, P(\boldsymbol{X}(t+) \in C_{n_c})\}'$. The probability vector $\boldsymbol{P}(t)$ can be determined from the recurrence formula

$$\boldsymbol{P}(S_k) = \boldsymbol{p}(T_k)\,\boldsymbol{P}(S_{k-1}), \quad k = 1, 2, \ldots, N(t), \tag{6}$$

applied at the Poisson times $0 \leq S_1 < S_2 < \cdots < S_{N(t)} < t$, where $\boldsymbol{P}(0)$ corresponds to the initial state $\boldsymbol{X}(0)$ and $\boldsymbol{p}(T_k)$ is the transition probability matrix for values of $\boldsymbol{X}(t)$ immediately following pulses $\boldsymbol{Y}_{k-1}$ and $\boldsymbol{Y}_k$. The elements of $\boldsymbol{p}(T_k)$ can be approximated by

$$p_{i,j}(T_k) = P(\boldsymbol{X}_k^+ \in C_j \mid \boldsymbol{X}_{k-1}^+ = \boldsymbol{x}_i), \quad i,j = 1,\ldots,n_c, \tag{7}$$

and represent the conditional probability of $\boldsymbol{X}_k^+$ being in cell C_j given that $\boldsymbol{X}_{k-1}^+ = \boldsymbol{x}_i \in C_i$. The probability vector at time t can be found from

$$\boldsymbol{P}(t) = \boldsymbol{p}^* \, \boldsymbol{P}(S_{N(t)}) \tag{8}$$

in which $p^*_{i,j} = \int_{C_i} 1_A(\boldsymbol{x}) \, f(\boldsymbol{x}) \, d\boldsymbol{x} / \int_{C_i} f(\boldsymbol{x}) \, d\boldsymbol{x}$, $f(\boldsymbol{x})$ is the density of $\boldsymbol{X}^+_{N(t)}$, $A = \{\boldsymbol{x} : \boldsymbol{h}(\boldsymbol{x}, t - S_{N(t)}) \in C_j\}$, and $1_A(\boldsymbol{x})$ denotes the indicator of set A.

Statistics of $\boldsymbol{X}(t)$ can be determined from Eqs. 6–8 in three steps. First, n_s samples of the Poisson times $\{S_k\}$ need to be generated. Second, conditional response statistics have to be calculated from Eqs. 6–8 for each realization of the Poisson times. Third, unconditional response statistics can be determined by averaging corresponding conditional response statistics determined in the second step. If $\nu \Delta t \ll 1$, the Poisson white noise can be approximated by the binomial white noise defined as an independent sequence $\{\tilde{\boldsymbol{Y}}_k\}$ with time step $\Delta t > 0$ where $\tilde{\boldsymbol{Y}}_k$ is equal to $\boldsymbol{Y}_k$ and zero with probabilities $1 - \exp(-\nu \, \Delta t)$ and $\exp(-\nu \, \Delta t)$, respectively. The binomial white noise has a fixed time step so that response statistics can be determined directly from Eqs. 6–8.

Numerical example

Consider a nonlinear half oscillator with displacement response $X(t)$ defined by the stochastic differential equation

$$dX(t) = -(\rho \, X(t) + \varepsilon \, X(t)^3) \, dt + dY(t), \quad t \geq 0, \tag{9}$$

with parameters ρ and ε, initial state $X(0) = 0$, and input noise $Y(t)$ given by Eq. 2 with $m = 1$. If the probability density of pulses Y_k is symmetric about zero, the stationary distribution of $X(t)$ is also symmetric about zero because $-X(t)$ is a solution whenever $X(t)$ satisfies Eq. 9. The free vibration solution of this equation is $h(\xi, s) = \xi / \sqrt{[-\varepsilon \, \xi^2 + (\rho + \varepsilon \, \xi^2) \exp(2\rho \, s)]/\rho}$, $s \geq 0$, for $X(0) = \xi$. Because the system is causal, $h(\xi, s)$ vanishes for $s < 0$. If $\{\rho > 0, \varepsilon > 0\}$ and $\{\rho < 0, \varepsilon > 0\}$, the system potential has one and two wells, respectively. The single well is located at $\xi = 0$ and the double wells have the coordinates $\xi = \pm\sqrt{-\rho/\varepsilon}$.

Figures 1($\boldsymbol{a}$) and ($\boldsymbol{b}$) show stationary density and histograms of $X(t)$ for the single and the double-well potential systems with parameters $\{\rho = 1.0, \varepsilon = 0.1\}$ and $\{\rho = -1.0, \varepsilon = 0.1\}$, respectively. The histograms are based on 10,000 independent realizations of $X(t)$. The stationary densities of these figures based on the Poisson and the binomial white noise inputs practically coincide and compare satisfactorily with Monte Carlo simulation results.

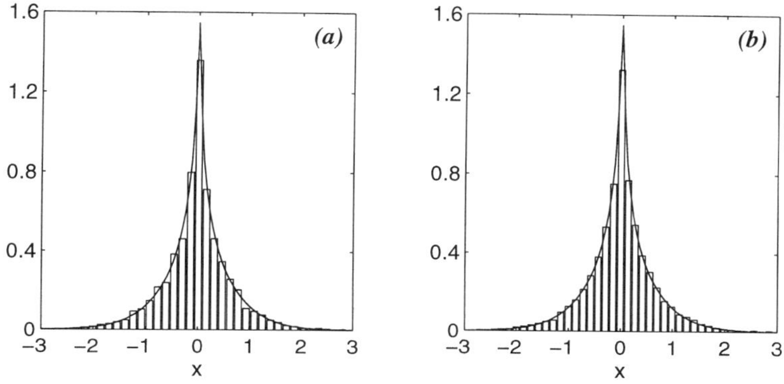

Figure 1: Stationary densities and histograms of $X(t)$

Conclusions

A method was developed for calculating statistics of the response of a nonlinear system subjected to Poisson white noise. The method is based on the observation that the system vibrates freely between consecutive pulses of the input. The free vibration solution can be used to develop a mapping giving the time evolution of the system response. Two algorithms based on the Poisson and the binomial white noises were developed for calculations. The proposed method of analysis was applied to find response statistics for two nonlinear systems characterized by potentials with single and double wells. The accuracy of the proposed method was evaluated by Monte Carlo simulation.

References

Grigoriu, M. (1995). *Applied Non-Gaussian Processes. Examples, Theory, Simulation, Linear Random Vibration, and MATLAB Solutions,* Prentice Hall, Englewood Cliffs, New Jersey.

Hu, S-L. J. (1995). "Response cumulant Equations for Dynamic Systems under Delta-Correlated Processes," *Proceedings of the 10th Engineering Mechanics Conference,* ASCE, Vol. 2, Editor S. Sture, University of Colorado at Boulder, Boulder, Colorado, 2, 1119–1122.

Köylüoğlu H. U., Cakmak, A. S., and Nielsen, S. R. K. (1995). "Fast cell-to-cell mapping (path integral) with probability tails for the stochastic response of non-linear white noise and Poisson driven systems," *Computational Stochastic Mechanics,* Editor P. D. Spanos, A. A. Balkema, Rotterdam, 361–370.

Probabilistic Stability
Analysis of Shallow Arches

R. P. Nordgren,[1] Member, ASCE, and K. S. Hussain[2]

Abstract

The present analysis treats the nonlinear stability of shallow elastic arches with random variation in shape imperfection, load distribution, and elastic stiffness. These variables are represented by one-dimensional random fields with known autocorrelation coefficient functions. In order to employ the computational methods of reliability analysis, the random fields are discretized using various techniques. Computation of the probability distribution for the critical instability loads is carried out by the first- and second-order reliability methods (FORM/SORM) in combination with nonlinear energy minimization. Selected results have been verified using Monte Carlo simulation (MCS).

Introduction

In recent years, extensive developments in reliability methods (Madsen et al., 1986) have enabled the use of versatile and efficient algorithms such as FORM and SORM to estimate the probability of failure of a structure.

The stability of thin, shallow arches has received extensive deterministic analysis in the literature (Gjelsvik et al., 1962). While the behavior can be modeled by reasonably straightforward and uncomplicated nonlinear equations, arches are capable of exhibiting both limit-point and bifurcation-type buckling. The bifurcation buckling behavior is strongly affected by shape imperfections.

In the present study, we first review the deterministic analysis method for the snap-through buckling of shallow arches. The buckling loads for different modes of instability are obtained using nonlinear energy analysis under various loads and boundary conditions. Next, reliability analysis is carried out using FORM, SORM and MCS for various autocorrelation functions for the imperfections. Several methods for random field discretization are compared using numerical results for an example problem.

[1]Prof. of Civil and Mech. Engrg., Rice Univ., Houston, TX 77005
[2]Grad. Student, Dept. of Civil Engrg., Rice Univ., Houston, TX 77005

Deterministic Stability Analysis

The problem considered is either a pinned or clamped shallow circular arch of rectangular cross section subjected to a uniformly distributed or a concentrated radial loading. The total energy H of the system is given by

$$H = \frac{1}{2}\int_{-\beta}^{\beta} \varepsilon^2 d\theta + \frac{t^2}{24R^2}\int_{-\beta}^{\beta} \kappa^2 d\theta - \frac{1}{EtbR}\Omega(P, w) , \tag{1}$$

where κ is the curvature change, ε is the elastic axial strain, t is the thickness and b is the width of the arch, E is Young's modulus, θ is the polar coordinate, β is one-half the included angle of the arch, P is the total radial load acting on the arch, and $\Omega\ (P,w)$ is dependent upon the manner of loading.

After considering equilibrium in both the tangential and radial directions, H can be expressed in terms of the radial displacement w and the initial geometric imperfection $\bar{w}$ only. These quantities can be expanded in Fourier series as

$$w = \sum_{n=1}^{\infty} A_n w_n(\zeta) , \quad \bar{w} = \sum_{n=1}^{\infty} C_n w_n(\zeta) , \quad \zeta = \theta/\beta , \tag{2}$$

where A_n and C_n are the Fourier coefficients for the basis function $w_n(\zeta)$. The series chosen to represent the deflected shape depends upon the end conditions of the arch which $w_n(\zeta)$ should satisfy.

For any position of equilibrium, equating the partial derivative of H with respect to A_n to zero, yields a set of n simultaneous nonlinear algebraic equations in $A_1,..., A_n$ and P. Solution of this set is carried out using the Newton-Raphson method and the load-deflection curve is obtained. Examination of the deflections as P increases determines the value of the load for which either bifurcation instability or limit instability takes place.

Reliability Analysis

The reliability program CALREL (Liu et al., 1989) is used to provide an estimate for the reliability index and the sensitivity measures based on FORM, SORM and MCS. Limit-state functions are formulated based on assuming that failure occurs if the arch can not sustain a chosen threshold load. By varying the threshold, the cumulative distribution of the failure load can be constructed as shown in Fig. 1 for an example problem.

Random Field Discretization

The application of the reliability theory to structural problems involving continuous random fields requires the use of a discretization technique in order to transform the continuous problem into a component reliability problem. The proposed approach is based on the use of mean, variance and autocorrelation coefficient functions for one-dimensional random fields to characterize the uncertainty in the geometric and the material imperfections.

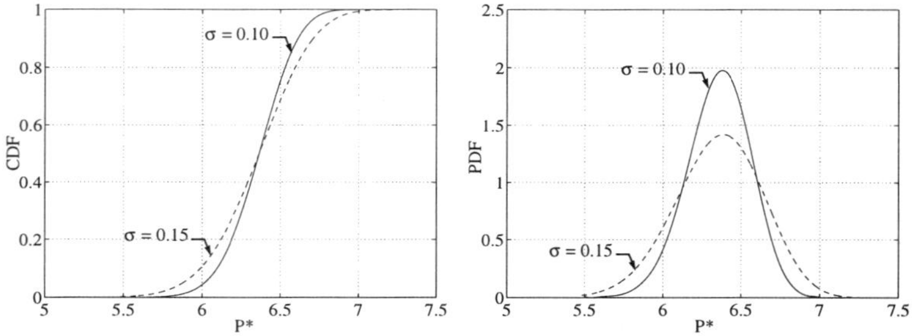

Figure 1. Cumulative distribution and probability density functions for the symmetric buckling load of the arch under uniformly distributed loading for different standard deviations of the shape imperfection.

Four methods of random field discretizations are considered within the scope of this study. The first two are the midpoint (MP) method (Der Kiureghian and Ke, 1988) and the local averaging (LA) method (Vanmarcke and Grigoriu, 1983). These methods discretize the field as a stepwise function with discontinuity along the element boundaries. A relatively fine mesh is required for accurate representation of the field.

A series expansion method based on the Karhunen-Loeve (KL) theorem was used by Spanos and Ghanem (1989). In this method the field is expanded in terms of a set of independent standard normal variates and the eigenvalues and eigenfunctions of the correlation function.

The last method is the series expansion (SE) method given by Zhang and Ellingwood (1994). Here the field is represented in terms of a vector of correlated random variables with zero means, a vector of constant coefficients, and any complete set of orthogonal deterministic basis functions. Solving for the constant coefficients and the covariances between the random variables provides a complete discrete representation of the random field. The SE method does not require solution of a complicated integral equation as in the KL method. The different methods are compared in terms of convergence rate in Fig. 2 for two typical problems.

The SE method allows for the use of an orthogonal set of functions which satisfy the boundary conditions of the problem, usually a set of trigonometric functions. This eliminates the need for a special treatment of the discretization technique in order to take into account the non-stationarity of the autocorrelation function representing the geometric imperfections near the ends.

A novel technique, for representing non-stationarity near the ends, postulates that the curvature at any point in the arch is stationary and then by integrating twice, the configuration of the arch is given in a non-stationary form as

$$y = \rho(x, x') + C_1 x + C_2 \tag{3}$$

where C_1 and C_2 are random coefficients computed for each realization to satisfy

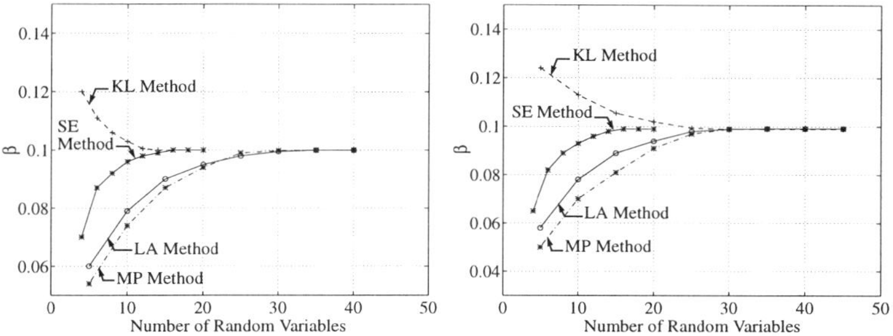

Figure 2. Rate of convergence of the reliability index for different discretization methods for cases of a triangular and a sinusoidal autocorrelation functions.

the boundary conditions. This approach is used with the MP, LA and KL methods. Non-stationarity effects are found to be small in general.

Conclusions

Results, such as shown in Fig. 1, indicate that uncertainty in the random imperfections may have a significant effect on the probabilistic limit stability of shallow elastic arches. It can be concluded that the KL method proves to be the most efficient method for random field discretization for cases where exact formulas are available for evaluating the eigenfunctions of the correlation kernel. The SE method is more efficient in cases where the KL method requires lengthy numerical analysis to evaluate the spectral decomposition of the field. SORM proves to be in very good agreement with MCS results and FORM can be considered as a quick and fair first approximation for the failure probability.

Appendix I. References

Der Kiureghian, A. and Ke, J.-B. (1988). "The stochastic finite element method in structural reliability." *Probabilistic Engrg. Mech.*, 3(2), 83-91.

Gjelsvik, A. and Bodner, S. R. (1962). "Energy criterion and snap buckling of arches." *J. Engrg. Mech. Div.*, ASCE, 88(5), 87-134.

Liu, P.-L., Lin, H.-Z. and Der Kiureghian, A. (1989). CALREL user manual. Dept. of Civil Engineering, University of California, Berkeley.

Madsen, H. O., Krenk, S. and Lind, N. C. (1986). Methods of structural safety. Prentice-Hall, Inc., Englewood Cliffs, N. J.

Spanos, P. D. and Ghanem, R. (1989). "Stochastic finite element expansion for random media." *J. Engrg. Mech.*, ASCE, 115(5), 1035-1053.

Vanmarcke, E. H. and Grigoriu, M. (1983). "Stochastic finite element analysis of simple beams." *J. Engrg. Mech.*, ASCE, 109(5), 1203-1214.

Zhang, J. and Ellingwood, B. (1994). "Orthogonal series expansions of random fields in reliability analysis." *J. Engrg. Mech.*, ASCE, 120(12), 2660-2677.

STABILITY OF BOUNDED-NOISE EXCITED SYSTEM

Q. C. Li[1]

ABSTRACT

This paper discusses the asymptotic sample stability for a modified Mathieu-Hill equation in which the periodic excitation is replaced by a random noise of bounded amplitude. It is found that the bandwidth of the excitation has a significant impact on the stability boundary, either positive or negative, depending on the location of the central excitation frequency.

INTRODUCTION

Consider the equation of motion: $\ddot{x} + 2\zeta\ \Omega\dot{x} + \Omega^2 x\,[1 + \varepsilon \sin\Theta(t)] = 0$ (1)

where x is the displacement, an overdot represents one differentiation with respect to time, ζ and Ω are the damping coefficient and the circular natural frequency of the system, respectively. In the parametric excitation term $\varepsilon\sin\Theta(t)$, ε is the amplitude and $\Theta(t) = \nu\, t + \sigma B(t) + \chi$, in which $B(t)$ is a unit Wiener process, χ is a random initial phase, ν and σ represent the peak frequency location and noise bandwidth, respectively. Eq.(1) reduces to the well-known Mathieu-Hill equation if $\sigma = 0$. The excitation is a random noise of bounded total energy when $\sigma>0$; it is a narrow-band noise if $\sigma \ll \nu$, and a wide-band noise if $\sigma \gg \nu$. By adjusting the values of σ and ν, its spectral density can be made to match closely some real spectral densities such as wind turbulence (Lin and Li, 1993).

The term "exact tuning" is referred to $\nu = 2\Omega$, i.e., the central frequency of the excitation is twice as the natural frequency of the system. A detuning parameter is consequently defined as $\Delta=2\Omega-\nu$.

Now transform the displacement and the velocity into two new variables amplitude $A(t)$ and phase $\Phi\,(t)$ by letting $x = A\cos\Phi$ and $\dot{x} = -\Omega A\sin\Phi$. Eq. (1) is replaced by two first order differential equations:

$$\dot{A} = \Omega[-\zeta(1-\cos 2\Phi) + \frac{1}{2}\varepsilon\sin 2\Phi\sin\Theta]A \tag{2}$$

[1] Assistant Professor, Department of Mathematics and Center for Applied Stochastics Research, Florida Atlantic University, 777 Glades Road, Boca Raton, Florida 33431-0991

$$\dot{\Phi} = \Omega[1 - \zeta \sin 2\Phi + \frac{1}{2}\varepsilon(1 + \cos 2\Phi)\sin\Theta] \tag{3}$$

ANALYTICAL SOLUTIONS

The following two cases are amenable to analytical solutions.

NARROW-BAND NEARLY-TUNED EXCITATION Using the deterministic averaging procedure of Bogoliubov and Mitropolsky (1961) and letting $\Psi=\Phi-\Theta/2$, Eqs. (2-3) are converted to the following stochastic differential equations in the Itô (1951) sense:

$$dA = (-\zeta + \frac{1}{4}\varepsilon \cos 2\Psi)\Omega A dt \qquad d\Psi = (\frac{\Delta}{2} - \frac{1}{4}\varepsilon\, \Omega \sin 2\Psi)dt - \frac{\sigma}{2}dB(t) \tag{4}$$

It is noted that the Ψ equation is decoupled from the A equation, therefore the probability density of Ψ, denoted by $p(\psi)$, can be solved independently from the corresponding Fokker-Planck equation. The asymptotic behavior of the dynamical system is determined by the Lyapunov exponent, and the system is asymptotically stable if the Lyapunov exponent is negative. At the stability boundary, the Lyapunov exponent λ_{LE} is zero, namely $\lambda_{LE} = E[-\zeta + (\varepsilon/4)\cos 2\Psi] = 0$. The critical excitation amplitude ratio is then obtained as $\varepsilon_c/\varepsilon_0 = 1\Big/\int_0^{2\pi} \cos 2\psi\, p(\psi) d\psi$, where $\varepsilon_0=4\zeta$ is the critical excitation amplitude when $\Delta=0$ and $\sigma = 0$. The same stability boundary was obtained by Dimentburg (1991) using the stochastic averaging method.

WIDE-BAND EXCITATION When the excitation is wide-banded, its correlation time is much smaller than the relaxation time of the dynamical system; therefore, the solution processes may be approximated by Markov processes for which stochastic averaging procedure (Stratonovich 1967) can be applied to obtain the Itô equation:

$$dA = [-\zeta + \frac{3\pi}{8}\Omega\, \Phi_{FF}(2\Omega)]\Omega A\, dt + \frac{\pi}{4}\Omega^2 \Phi_{FF}(2\Omega) A^2 dB(t) \tag{5}$$

where Φ_{FF} is the spectral density function of the excitation $F(t) = \varepsilon \sin\Theta(t)$. It is known that A=0 is a "trap" type singular boundary (Zhang and Kozin 1990). The stability condition is

$$\frac{\varepsilon_c}{\varepsilon_0} = 1\Bigg/\left[\Omega\,\sigma\sqrt{\frac{2\zeta}{4(2\Omega-\nu)^2+\sigma^4} + \frac{2\zeta}{4(2\Omega+\nu)^2+\sigma^4}}\right] \tag{6}$$

COMPUTER SIMULATIONS

Analytical solutions for other parameter ranges are not obtainable at the present time. Thus, use of computer simulation is in order. The simulation scheme is based on a method developed by Wedig (1989), in which the maximum Lyapunov exponent is calculated from the simulated time history of the phase variable. In simulating the phase differential equation (3), the fourth order Runge-Kutta method, which provides higher accuracy and faster convergence than the Euler method, is used. Müller's method is employed for finding the critical excitation amplitude, namely the excitation amplitude that makes the Lyapunov exponent be zero.

The simulation results are shown in Figs. 1-3. The case of perfect periodic excitation, i.e., σ=0, in Fig. 1 agrees with the well-known Strutt diagram. It is seen that a slight increase of σ, the bandwidth parameter of the random noise, results in a stabilizing effects in the neighborhoods of exact tuning ($\nu/(2\Omega)\approx1$). At a detuned frequency, such as $\nu/(2\Omega)$ =1.5, the noise destabilizes the system if σ<3. This destabilizing trend can also be seen in Fig. 2 in the detuned case of Δ=-3 (same as $\nu/(2\Omega)$=1.5). It reverses to a stabilizing trend as σ is further increased beyond σ=3. The theoretical predictions for the cases of exact tuning (Δ=0) and nearly tuned (Δ=0.1) are plotted in Fig. 2 along with the simulated results. It is seen that the theoretical predictions, shown by lines, agree with the simulation results, shown by scattered points, very well.

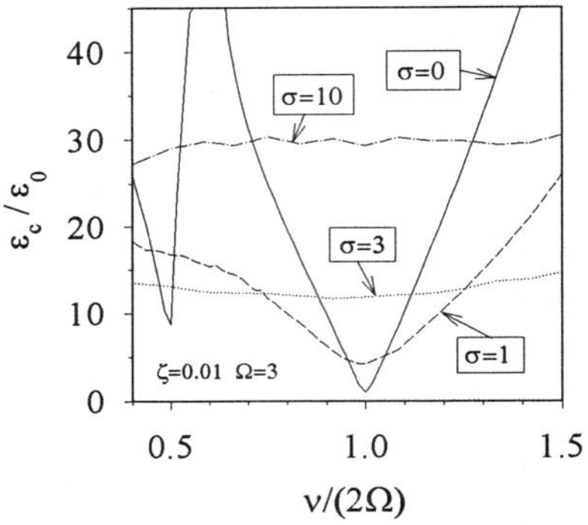

Fig. 1 Stability boundary by simulation: critical excitation amplitude ratio vs. non-dimensionalized peak location

The wide-band case is shown in Fig. 3. It can be seen from both the theoretical and simulation results that the critical excitation amplitude increases as σ increases and the peak location has little effect on the stability (also seen in Fig. 1 for the curve σ=10). It is noted that the theoretically predicted critical excitation amplitude is higher than that obtained by simulation.

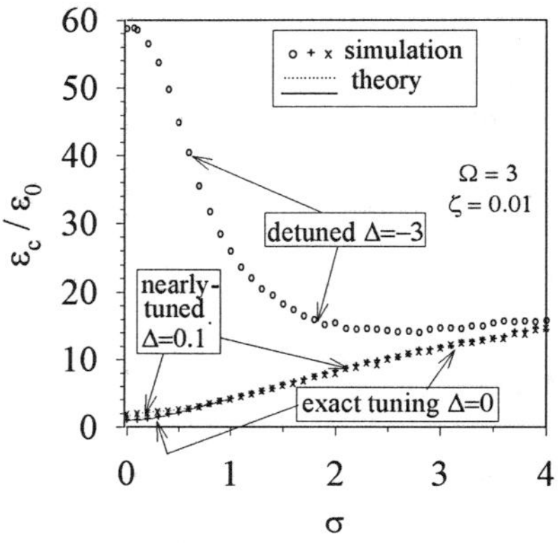

Fig. 2 Stability boundary: critical excitation amplitude ratio vs. bandwidth parameter

CONCLUDING REMARKS

The stability boundary of a modified Mathieu equation under randomly perturbed parametric excitation is investigated. Analytical results for the cases of wide-band excitation and nearly-tuned narrow-band excitation are obtained by

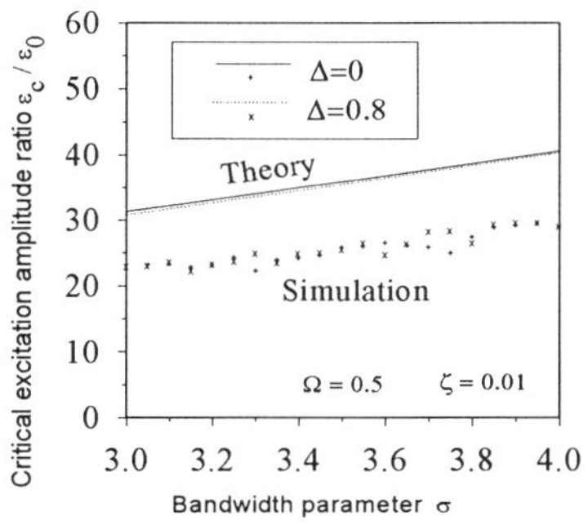

Fig. 3 Stability boundary for wide-band noise

using averaging techniques. In addition, computer simulation is performed to verify the accuracy of the analytical results, and to obtain supplemental results in the regions in which analytical solutions are presently not available. The results show that the noise component stabilizes the system for both the cases of wide-band excitation and nearly-tuned narrow-band excitation, in the sense that an increase in σ results in an increase in the critical excitation amplitude. This suggests that the noise acts as a damper in such cases. However, the noise can play an opposite role as well. In the case of detuned narrow-band excitation, the noise destabilizes the system.

REFERENCES

Bogoliubov, N. N., and Mitropolsky, Y. A. (1961). *Asymptotic Methods in the Theory of Non-Linear Oscillations*. Gordon & Breach, New York, NY.

Dimentburg, M. F. (1991). "A stochastic model of parametric excitation of a straight pipe due to slug flow of a two-phase fluid." *Proc., the Fifth Intl. Symp. on Flow-Induced Vibrations*, Brighton, UK.

Itô, K. (1951). "On a formula concerning stochastic differentials." *Nagoya Math. J.*, Japan, Vol. 3, pp. 55-65.

Lin, Y. K. and Li, Q. C. (1993). "New stochastic theory for bridge stability in turbulent flow." *J. Engrg. Mech.*, ASCE, 119, 113-117.

Stratonovich, R. L. (1967) *Topics in the Theory of Random Noise I & II*, Translated by Silverman, R. L., Gorden and Breach, New York, NY.

Wedig, W.V. (1989). "Analysis and simulation of nonlinear stochastic system." *Nonlinear Dynamics in Engineering Systems*, W. Schiehlen ed., Springer-Verlag, Berlin, Germany, 337-344.

Zhang, Z. Y. and Kozin, F. (1990). "On almost sure sample stability of nonlinear stochastic dynamic systems" *Proc., IEEE Intl. Conf. on System Engrg.*, Pittsburgh, PA.

Application of Numerical Limit Analyses for Shallow Foundations on Clay

Andrew J. Whittle[1] & Boonchai Ukritchon[1]

Abstract

Robust and efficient numerical methods have recently been developed to solve rigorous upper and lower bound theorems of plastic collapse for planar problems in cohesive-frictional soils. The analyses are formulated and solved by a linear programming method with finite element discretization. This paper illustrates the application of the numerical limit analyses for computing the undrained bearing capacity of eccentrically loaded footings on non-homogeneous clay layers.

Introduction

Limit analyses, based on the lower and upper bound theorems of classical plasticity, provide a powerful framework for analyzing stability problems in soil mechanics (e.g. Chen, 1975). However, it is often difficult to apply limit analyses for practical problems which involve non-homogeneous ground conditions, complex geometries and loading conditions. Instead, most stability calculations in geotechnical practice are based on approximate limit equilibrium methods. A generalized numerical method for solving the lower bound theorem, using finite element discretization and linear programming, was first proposed by Lysmer (1970). A major limitation of this original formulation was the large computational time required to solve the large system of constraint equations, arising primarily from the linearization of the yield function. Sloan (1988) developed a more robust lower bound formulation for plane strain problems, which solves the linear programming problem efficiently using an active set algorithm. More recently, Sloan and Kleeman (1995) have applied the same algorithm to solve the dual upper bound linear programming problem. Ukritchon et al. (1996) have applied and extended the lower bound (Assadi & Sloan, 1988) and upper bound (Sloan & Kleeman, 1995) formulations in order to develop a generalized failure criterion for the undrained

[1]Associate Professor & Graduate Research Fellow, Department of Civil & Environmental Engineering, Massachusetts Institute of Technology, Cambridge, MA02139.

stability of footings on clay subject to inclined and eccentric loading. The analyses consider inhomogeneous clay layers where the undrained shear strength varies linearly with depth. However, the same methodology can readily be extended to generalized soil profiles.

Limit Analyses by Finite Elements and Linear Programming

The lower bound analysis discretizes the soil mass into 3-noded triangular elements ('extension' elements are introduced to model infinite boundaries of a half-space) and assumes a linear variation of the unknown stresses (σ_x, σ_y, τ_{xy}) within each element. In contrast to conventional finite element methods, each node is unique to a particular element such that stress discontinuities can occur along shared edges between element. The unknown nodel stresses are subject to constraints which ensure that the stress field is statically admissible at all points. This is achieved by imposing i) the static equilibrium equations at all points inside each element and along each stress discontinuity; ii) prescribed surface (normal and/or shear) traction boundary conditions; and iii) the yield function, $F(\sigma_x, \sigma_y, \tau_{xy})$ must not be violated at any point (i.e., $F \leq 0$). For a cohesive clay layer, the non-linear Tresca criterion is transformed $F([\sigma_x-\sigma_y]/2, \tau_{xy})$ to form a circular locus, which is then replaced by a series of linear inequalities in the form of an n-sided, interior polygon (the examples in this paper all use n = 24). The lower bound collapse load is found by maximizing the force $\mathbf{Q} = \mathbf{C}^T\sigma$ applied over some specified part of the boundary (where σ is the global vector of nodal stresses), subject to the constraints that the stress field is statically admissible. Sloan (1988) solves this linear programming problem (Table 1) in the canonical form used by the active set algorithm:

Minimize $-\mathbf{C}^T\sigma$ Subject to: $\begin{cases} \mathbf{A}_1\sigma = \mathbf{B}_1 \\ \mathbf{A}_2\sigma \leq \mathbf{B}_2 \end{cases}$	Minimize $\left\{\mathbf{C}_1^T\dot{\lambda} + \mathbf{C}_2^T\mathbf{u}\right\}$ Subject to: $\begin{cases} \mathbf{A}_1\dot{\lambda} + \mathbf{A}_2\mathbf{u} + \mathbf{A}_3\mathbf{u}_t^+ + \mathbf{A}_3\mathbf{u}_t^- = 0 \\ \mathbf{u}_t^+, \mathbf{u}_t^-, \quad \dot{\lambda} \geq 0 \end{cases}$ $\mathbf{u}_t^+$, $\mathbf{u}_t^-$ tangential interface velocity components
Lower Bound	Upper Bound

Table 1. Linear Programming Forms for Numerical Limit Analyses

The upper bound analysis assumes a linear variation of the velocity field (v_x, v_y) within each triangular element, with velocity discontinuities permitted along edges of adjacent elements. The conditions for a kinematically admissible velocity field are imposed through a set of linear constraints which impose: i) compatibility between velocities and strains; ii) the associated flow rule is based on the linearized form of the Tresca failure criterion, where the plastic strain rates , $\dot{\varepsilon}_{ij}^{p} = \dot{\lambda}\, \partial F / \partial \sigma_{ij}$ (where $\dot{\lambda}$ is the plastic multiplier); and iii) prescribed velocity boundary conditions. The plastic strains are constant within each element (for linear variation of velocities). Sloan and Kleeman (1995) also describe an algorithm to ensure that the associated flow rule is also satisfied along all velocity discontinuities where the relative tangential velocity is not known a priori.

The principle of virtual work links the kinematically admissible velocity field with the applied external load, by equating the rate of work done by the external loads with the power dissipated in the plastically deformed region (within the elements and along the velocity discontinuities). The upper bound collapse load can then be obtained by minimizing the power dissipation for the kinematically admissible velocity field as shown in Table 1 (**u** and $\dot{\lambda}$ are the global vectors of nodel point velocities and plastic multiplier rates).

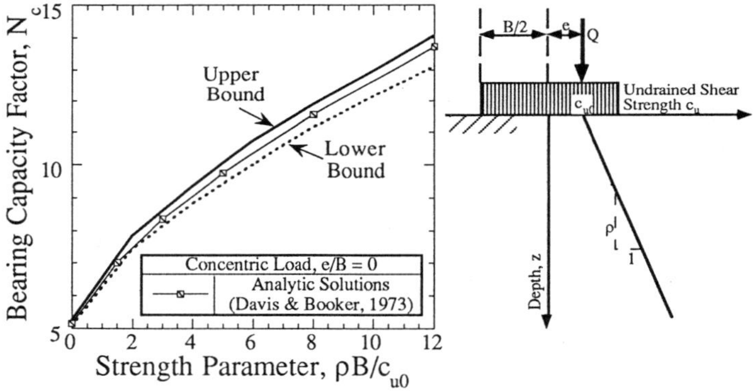

Figure 1. Capacity of Concentrically Loaded Footings on Non-Homogeneous Clay

Rigid Surface Footing with Eccentric Loading

Figure 1a shows a rigid surface footing with an eccentric vertical load Q, on a clay layer whose undrained shear strength increases linearly with depth. The lower bound analysis of this problem solves Max.$\{Q\} = \{\int \sigma_n dx\}$, and assumes that the shear traction, $|\tau| \leq c_{u0}$, and normal traction $\sigma_n \geq 0$, at the soil-footing interface. The upper bound analysis allows separation to occur at the soil-footing interface. Assuming a Tresca yield function with a tension cut-off, then the normal velocity jump at points along the interface, $\Delta u_n \leq 0$. The external work done by the applied load is defined in terms of the rigid body motions of the footing which moves in the direction of the applied vertical load Q through a unit virtual displacement.

Figure 1 show that the numerical solutions do provide lower and upper bounds on the analytical solution for the bearing capacity of concentrically loaded footings (Davis & Booker, 1973). In all cases, the numerical limit analyses bound the exact collapse load within ±3 - 5%. Typical computation (CPU) times for each analysis range from 10 - 60mins (on a Dec Alpha 3000-300x workstation). Figure 2 shows the numerical upper and lower bounds for eccentrically loaded footings with e/B = 0 - 0.35. There are no analytic solutions for these conditions, although conventional practice assumes that the collapse load, $Q_u = [N_c\ c_u]\ B'$, where $B' = [B-2e]$ is the effective width of the footing. The figure shows that this assumption is valid for the homogeneous clay layer ($\rho B/c_{u0} = 0$), where it matches the lower bound numerical solution. However, for non-homogeneous clay layers, there is a much more rapid loss of bearing capacity for eccentrically loaded footings on non-homogeneous clay layers.

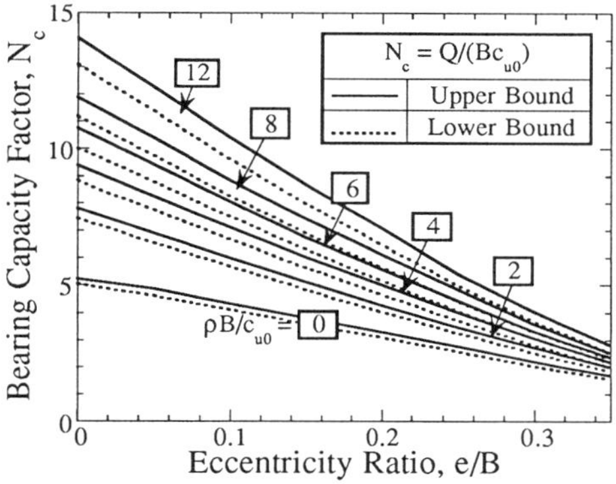

Figure 2. Effect of Load Eccentricity on Bearing Capacity of Footings on Clay

References

Assadi, A. & Sloan, S.W. (1990) "Undrained stability of shallow square tunnel," *ASCE J. of Geotechnical Engineering*, 117(8), 1152-11173.

Davis, E.H. & Booker, J.R. (1973) "The effect of increasing strength with depth on the bearing capacity off clays," *Géotechnique*, 23(4), 551-563.

Lysmer, J. (1970) "Limit analysis of plane problems in soil mechanics," *ASCE, J. of Soil Mechanics and Foundations Division*, 96(4), 1311-1334.

Sloan, S.W. (1988) "Lower bound limit analysis using finite elements and linear programming," *Intl. J. for Numerical and Analytical Methods in Geomechanics*, 12, 61-77.

Sloan, S.W. & Kleeman, P.W. (1995) "Upper bound limit analysis using discontinuous velocity fields," *Computer Methods in Applied Mechanics and Engineering*, 127, 293-314.

Ukritchon, B., Whittle, A.J., & Sloan, S.W. (1996) "The undrained stability of footings on clay with inclined and eccentric loading," submitted for publication.

An Analysis of Strong-Discontinuities in Inelastic Solids with Applications to the Finite Element Simulation of Strain Localization Problems

F. ARMERO* & K. GARIKIPATI♯

Abstract

This contribution summarizes recent advances obtained by the authors in the analysis and numerical simulation of strain localization in inelastic solids. The localized patterns of the deformation are considered in the limit case of strong discontinuities (discontinuous displacements), and analyzed in the framework of distribution theory. Furthermore, these results have led to the formulation of enhanced finite element methods that resolve accurately these discontinuous solutions. In contrast with the strong mesh dependence observed for standard finite element techniques when solving strain localization problems, the proposed methods are mesh-size and mesh-alignment independent, without the introduction of any ad-hoc numerical parameters (e.g. characteristic length) or other regularizations.

1. Introduction

The final stages of the deformation of solids are often characterized by highly localized patterns of strain. The formation of shear bands in metals and soils, the cracking of concrete and the failure surfaces in rocks (faulting) are few representative examples. In all cases, the overall response of the solid is dominated by the presence of the localization bands, leading to highly inhomogeneous states of deformation, with a softening overall response of the solid. A strong concentration of the disspative mechanisms associated to the failure is observed in these small regions of intensive straining. The actual thickness of these bands varies considerably depending on the physical mechanisms involved. We can observe shear bands of the order of microns in metals (ASARO [1983]), of order of millimeters in overconsolidated sands (VARDOULAKIS [1978]), and well-defined failure surfaces dominating over distributed micro-cracks in concrete (READ et al [1984]). In general, the small scale characteristic of the localization bands is to be contrasted with the typical scales in practical engineering applications. The presence of multiple scales in the problem is then apparent.

The above physical considerations result in important consequences for the numerical simulation of failure of solids. We first observe that a classical rate independent continuum model does not define a characteristic length. Therefore, the modeling of such localized patterns of deformation leads necessarily to the consideration of surfaces of discontinuity of the displacement field, the limit

* Assistant Professor of Civil Engineering, UC Berkeley.

♯ Postdoctoral Fellow, Division of Applied Mechanics, Stanford University.

case of zero thickness. Numerical solutions based on standard finite element methods do possess a characteristic length, namely the element diameter, and therefore, the solution is completely influenced by the mesh size.

The approach proposed herein studies the phenomenon of strain localization in the limit case of *strong discontinuities* (discontinuities in the displacements). An analysis of classical plasticity models in the context of distribution theory characterizes completely this type of discontinuous solutions maintaining the classical continuum framework, and leading to a *localized softening mechanism along the discontinuity.* The main results are summarized in Section 2 below. The introduction of the identified localization mode as an *enhancement* of the element strain field has led to the formulation of the new finite element method described briefly in Section 3, which leads to mesh insensitive solutions.

2. Analysis of Strong-Discontinuities

In contrast with the discontinuous strains associated to the so-called *weak discontinuities, strong discontinuities* exhibit a displacement field of the form

$$\boldsymbol{u}(\boldsymbol{x}) = \bar{\boldsymbol{u}}(\boldsymbol{x}) + [\![\boldsymbol{u}]\!]\, H_\Gamma(\boldsymbol{x}) \,, \tag{2.1}$$

with a jump discontinuity $[\![\boldsymbol{u}]\!]$ across a surface Γ, with $H_\Gamma(\boldsymbol{x})$ denoting the Heaviside function on Γ. The strains associated to (2.1) are singular distributions on Γ (namely, the Dirac delta δ_Γ). The analyses presented in SIMO, OLIVER & ARMERO [1992] for the infinitesimal case, and ARMERO & GARIKIPATI [1996] for the finite strain range have characterized this type of solutions in a classical continuum model of associated plasticity with strain-softening. The main results can be summarized in the following three points:

1. Discontinuous solutions of the form (2.1) are consistent with a localized plastic flow on Γ. The softening response is given by a softening modulus $\mathcal{H}$ of the form $\mathcal{H} = \bar{\mathcal{H}}\delta_\Gamma$, i.e, the strain-softening is localized on Γ.
2. The orientation of the discontinuity surface Γ is characterized by the unit normal $\boldsymbol{n}$ satisfying the *localization condition*

$$\boldsymbol{Q}^{ep}\boldsymbol{m} = \boldsymbol{0} \,, \quad \text{with} \quad \boldsymbol{Q}^{ep} = \boldsymbol{n}\cdot\boldsymbol{C}^{ep}\boldsymbol{n} = \boldsymbol{n}\cdot\left[\boldsymbol{C} - \frac{\boldsymbol{C}\boldsymbol{N}_\phi \otimes \boldsymbol{C}\boldsymbol{N}_\phi}{\boldsymbol{N}_\phi : \boldsymbol{C}\boldsymbol{N}_\phi}\right]\boldsymbol{n} \,, \tag{2.2}$$

   where $\boldsymbol{m} = [\![\dot{\boldsymbol{u}}]\!]/\|[\![\dot{\boldsymbol{u}}]\!]\|$ and $\boldsymbol{N}_\phi = \partial_{\boldsymbol{\sigma}}\phi$ is the normal to the yield surface $\phi \leq 0$, $\boldsymbol{\sigma}$ denoting the stress tensor. Here, $\boldsymbol{C}$ is the elastic tangent tensor, $\boldsymbol{C}^{ep}$ the *perfectly plastic tangent tensor,* and $\boldsymbol{Q}^{ep}$ the so-called *acoustic tensor.*
3. Furthermore, the evolution of $\zeta := \|[\![\dot{\boldsymbol{u}}]\!]\|$ is given by

$$\zeta = \frac{1}{\Xi\bar{\mathcal{H}}}\left[\boldsymbol{N}_\phi : \dot{\boldsymbol{\sigma}}\right]_\Gamma \quad \text{where} \quad \Xi := \left[\frac{\boldsymbol{C}\boldsymbol{N}_\phi : (\boldsymbol{n}\otimes\boldsymbol{m})}{\boldsymbol{N}_\phi : \boldsymbol{C}\boldsymbol{N}_\phi}\right]_\Gamma . \tag{2.3}$$

   Hence, the softening modulus $\bar{\mathcal{H}}$ determines the rate of change of the discontinuity jump $\|[\![\boldsymbol{u}]\!]\|$. Equation $(2.3)_1$ leads to an evolution equation for the resolved stress $r_\Gamma = \boldsymbol{m}\cdot\boldsymbol{\sigma}\boldsymbol{n}$ in terms of ζ.

Infinitesimal strains are assumed in (2.2) and (2.3). See ARMERO & GARIKIPATI [1996] for the equivalent expressions including nonlinear geometric effects.

3. A Finite Element Method for Localization Problems

A new finite element method has been proposed in ARMERO & GARIKIPATI [1995, 1996] that incorporates the distributional character of the strains at the element level. The method falls within the class of enhanced strain methods as originally proposed in SIMO & RIFAI [1990] for the infinitesimal case, and SIMO & ARMERO [1992], and SIMO, ARMERO & TAYLOR [1993] for finite deformation problems. The method consists of a local enhancement of the *displacement gradient* $\boldsymbol{H}$ of the form

$$\boldsymbol{H} = \underbrace{\text{Grad}\bar{\boldsymbol{u}}^h}_{Conforming} + \underbrace{\left[-\frac{1}{h^{(i)}} (\boldsymbol{\alpha}_e \otimes \boldsymbol{n}^{(i)}) + (\boldsymbol{\alpha}_e \otimes \boldsymbol{n})\, \delta_\Gamma \right]}_{Enhanced}, \tag{3.1}$$

where $\bar{\boldsymbol{u}}^h$ denotes a standard isoparametric interpolation of the displacement field based on triangles, and $\boldsymbol{\alpha}_e$ are element based parameters approximating the discontinuity jump of the displacements. The enhanced part of $\boldsymbol{H}$ in (3.1) can be obtained as the gradient of the discontinuous interpolation function depicted in Figure 3.1.a, where the quantities $\boldsymbol{n}^{(i)}$ and $h^{(i)}$ are defined. The final finite element equations have the form

$$\left.\begin{aligned} \boldsymbol{R} &:= \boldsymbol{f}^{ext} - \mathop{\mathbf{A}}_{e=1}^{n_{el}} \int_{\Omega_e} \boldsymbol{B}_e^T \boldsymbol{\sigma} d\Omega = 0\,, \\ b_e &:= -\int_{\Omega_{e,loc}} \boldsymbol{G}_e^T \boldsymbol{\sigma} d\Omega - r_\Gamma = 0. \end{aligned}\right\} \tag{3.2}$$

where the $\boldsymbol{B}_e$ denotes the standard strain operator associated to the interpolation of the regular displacements $\bar{\boldsymbol{u}}^h$, and $\boldsymbol{G}_e = -(1/A_e)\ (\boldsymbol{m} \otimes \boldsymbol{n})^s$ with $A_e =$ area of element Ω_e. The resolved stress r_Γ is given in terms of the jump $\boldsymbol{\alpha}_e$ by (2.3). The numerical implementation of (3.2) involves the consistent linearization of (3.2), including geometric terms in the finite strain range, and involves the static condensation of the localization parameters $\boldsymbol{\alpha}_e$ at the element level. These parameters are introduced when condition (2.2) is satisfied. Any standard triangular element (e.g. mixed triangles) can be used as a basis for the enhancement (3.1). Extensions to quadrilateral elements are easily devised.

Figure 3.1.b depicts the solution obtained for the representative problem of slope stability. A finite strain J_2-flow theory model is considered, based on the multiplicative decomposition of the deformation gradient $\boldsymbol{F} = \boldsymbol{F}^e \boldsymbol{F}^p$, with a localized strain softening law along the discontinuity (see ARMERO & GARIKIPATI [1996]). The simulation is performed by imposing a downwards displacement at the center of the top surface of the rigid block. Quadratic 7-node triangles are considered. It is to be noted that the orientation of the discontinuity is determined as it propagates through the localization condition, and no predetermined direction is fixed a-priori. The plot in Figure 3.1.c depicts the displacement obtained by two different discretizations, showing the mesh-size independence of the numerical solutions. We observe the same rate of dissipation in the two cases as indicated by the equal slopes of the softening branches. Furthermore, the proposed method is able to capture the localized deformation, involving in this case a circular slip-line, in general *unstructured meshes*.

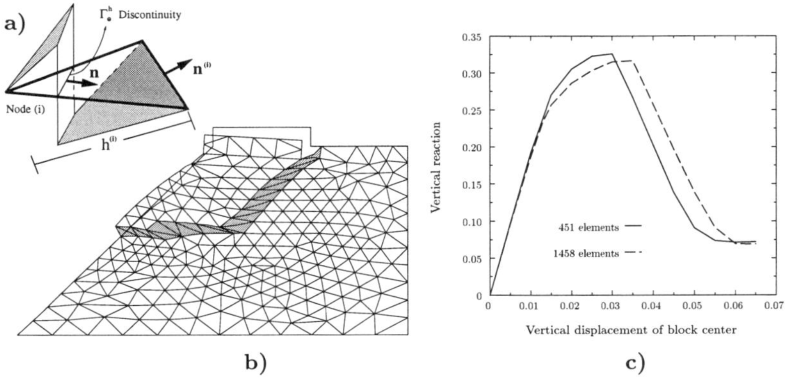

FIGURE 3.1. Slope stability. **a)** Discontinuous interpolation function. **b)** Solution obtained with the proposed finite element method (451 elements). **c)** Load vs. imposed displacement at top center of rigid block.

4. Concluding Remarks

An analysis of strong-discontinuities in continuum plasticity models has shown the consistency of these models with a localized dissipative mechanism. Based on these results, a new enhanced finite element method has been proposed for the numerical simulation of localization problems, leading to mesh-size independent solutions without the need of numerical regularization parameters.

References

ARMERO, F. & GARIKIPATI, K. [1995] "Recent Advances in the Analysis and Numerical Simulation of Strain Localization in Inelastic Solids," *Proc. Comp. Plas. IV,* ed. by D.R.J. Owen et al, CIMNE, Barcelona.

ARMERO, F. & GARIKIPATI, K. [1996] "An Analysis of Strong-Discontinuities in Multiplicative Finite Strain Plasticity and their Relation with the Numerical Simulation of Strain Localization in Solids", *Int. J. Sol. Struct.,* in press.

ASARO [1983] "Micromechanics of Crystals and Polycrystals", *Advances in Applied Mechanics,* **23**, 1-115.

READ, H.E. & HEGEMIER, G.A. [1984] "Strain Softening of Rock, Soil and Concrete – A Review Article", *Mechanics of Materials,* **3**, 271-294.

SIMO, J.C. & ARMERO, F. [1992] "Geometrically Non-Linear Enhanced Strain Mixed Methods and the Method of Incompatible Modes", *Int. J. Num. Meth. Eng.,* **33**, 1413-1449.

SIMO, J.C.; ARMERO, F. & TAYLOR, R.L. [1993] "Improved Versions of Assumed Enhanced Strain Tri-Linear Elements for 3D Finite Deformation Problems", *Comp. Meth. App. Mech. Eng.,* **110**, 359-386.

SIMO, J.C.; OLIVER, J. & ARMERO, F. [1993] "An Analysis of Strong Discontinuities Induced by Softening Solutions in Rate-Independent Solids", *J. Comp. Mech.,* **12**, 277-296.

SIMO, J.C. & RIFAI, M.S. [1990] "A Class of Mixed Assumed Strain Methods and the Method of Incompatible Modes", *Int. J. Num. Meth. Eng.,* **29**, 1595-1638.

VARDOULAKIS, I.; GOLDSCHEIDER, M. & GUDEHUS, G. [1978] "Formation of Shear Bands in Sand Bodies", *Int. J. Num. Meth. Geom.,* **2**, 99-128.

Implicit Integration Procedures and Consistent Tangent Operators for Bounding Surface Plasticity Models

P. Rahulkumar[1] and S. Saigal[2]

Abstract

Bounding surface type plasticity models used to predict the behavior of soils under monotonic and cyclic loading conditions are characterized by complex pressure sensitive criterion, dependence of yield criterion on the third stress invariant, softening/hardening response, nonlinear elasticity. Here the development of efficient procedures to treat these various complexities through the use of algorithms based on implicit schemes for the integration of stresses and plastic variables and consistent tangent moduli will be addressed.

Introduction

The bounding surface type of plasticity models (Dafalias & Herrmann, Morz, Norris & Zienkiewicz) although quite impressive in their analytical development for predicting the behavior of soils have relatively few numerical implementations (Herrmann et. al.) available in literature today. The various analytical complexities in these models have forced the use of explicit stress integration procedures. The use of explicit stress integration schemes lead to conditionally stable procedures which are vulnerable to numerical breakdown for even small strain increments. Recently Borja & Lee (1990) and Borja (1991) have developed implicit stress integration procedures for the relatively simple modified Cam-Clay model. It was demonstrated that the stress integration algorithm is capable of handling the nonlinear pressure dependent elastic moduli which results in a nonobvious return path and is amenable to exact linearization for the development of the consistent tangent moduli.

In this paper the need for the development of accurate stress integration algorithms for bounding surface type of plasticity models will be addressed. The stress integration algorithms based on stable implicit schemes for such models that account for nonlinear elasticity and plasticity based on the treatment given by Borja (1991) and dependence of yield criterion on third stress invariant will be developed.

1

[1] Development Engineer, ANSYS, Inc., 201 Johnson Rd., Houston, PA 15342

[2] Associate Professor, Department of Civil and Environmental Engineering, Carnegie Mellon University, Pittsburgh, PA 15217.

Constitutive Equations

The constitutive equations for the radial mapping version of the associative bounding surface plasticity model (Dafalias & Herrmann) under plastic isotropic conditions assumes an additive decomposition of the total strain rate into an elastic and plastic part, given as $\dot{\varepsilon}_{ij} = \dot{\varepsilon}^e_{ij} + \dot{\varepsilon}^p_{ij}$. The elastic rate equation is given as, $\dot{\sigma}_{ij} = C_{ijkl}\dot{\varepsilon}^e_{kl}$, where C_{ijkl} is the elasticity tensor given as

$$C_{ijkl} = K\delta_{ij}\delta_{kl} + \mu\left(\delta_{ik}\delta_{jl} + \delta_{il}\delta_{jk} - \frac{2}{3}\delta_{ij}\delta_{kl}\right)$$

where K is the bulk modulus dependent on the hydrostatic pressure, p; μ is the shear modulus. The bulk modulus is given as $K = (1 + e/\kappa)p$, where e is the void ratio and κ is the swelling/unloading line slope. The shear modulus is given as $\mu = 3K(1-2\nu)/2(1+\nu)$ for a given Poisson's ratio, ν. It is advantageous to separate the stress into a spherical part, p, and deviatoric part s_{ij}, as $p = (1/3)\sigma_{ii}$, $s_{ij} = \sigma_{ij} - p\delta_{ij}$. The state functions that describe the plastic state under plastic isotropy depend on the three stress invariants, I, J_2, and J_3 defined as

$$I = \sigma_{ii}, \quad J_2 = \frac{1}{2}s_{ij}s_{ij}, \quad J_3 = \frac{1}{3}s_{ij}s_{jk}s_{ki} \tag{1}$$

The plastic strain rate is given as $\dot{\varepsilon}^p_{ij} = \langle L\rangle L_{ij}$, where $L = (1/K_p)L_{ij}\dot{\sigma}_{ij}$. L_{ij} is the loading direction; L is the loading index, the Macaulay bracket is defined as $\langle x\rangle = x$ if $x > 0$, else $\langle x\rangle = 0$; K_p is the hardening modulus which is function of the distance of the current stress state from the corresponding image stress state on the bounding surface. The image stress state is given by the image stress invariants, $\bar{I}$, $\bar{J}_2$, $\bar{J}_3$, which are given by the radial mapping rule as, $\bar{I} = b(I - CI_0) + CI_0$, $\bar{J}_2 = b^2 J_2$, and $\bar{J}_3 = b^3 J_3$, where b is a scalar valued parameter, C is constant and I_0 is scalar valued hardening parameter . The bounding surface is given in terms of image stress invariants and hardening parameter as $F(\bar{I}, \bar{J}_2, \bar{J}_3, I_0) = 0$. The variation of the hardening parameter in the rate form is given as

$$\dot{I}_0 = \frac{1+e}{\lambda - \kappa} I_0 \dot{\varepsilon}^p_{ii} \tag{2}$$

Integration of Discrete Form of Rate Equations in Stress Invariant Space

The plastic rate equation discretization by using the unconditionally stable backward Euler scheme for the incremental plastic strains from time t_n to t_{n+1} is given as

$$\Delta\varepsilon^p_{ij} = L\left[\frac{\partial F}{\partial \bar{I}}\delta_{ij} + \frac{\partial F}{\partial \bar{J}_2}\bar{s}_{ij} + \frac{\partial F}{\partial \bar{J}_3}\left(\bar{s}_{ij}\bar{s}_{kl} - \frac{2}{3}\bar{J}_2\delta_{ij}\right)\right]_{n+1} \tag{3}$$

Following the treatment for the integration of nonlinear elasticity given in Borja (1991), the integration of the elastic stress rate equation results in

$$\sigma_{ij(n+1)} = \sigma_{ij(n)} + \int_{t_n}^{t_{n+1}} K\hat{C}^e_{ijkl} d\varepsilon^e_{kl} \approx \sigma_{ij(n)} + \overline{K}\hat{C}^e_{ijkl}\left(\Delta\varepsilon_{kl(n+1)} - \Delta\varepsilon^p_{kl(n+1)}\right) \quad (4)$$

where $\hat{C}^e_{ijkl} = \left[\delta_{ij}\delta_{kl}\right] + 2\eta\left(\frac{1}{2}\left[\delta_{ik}\delta_{kl} + \delta_{il}\delta_{jk}\right] - \frac{1}{3}\delta_{ij}\delta_{kl}\right)$, $\eta = 3(1-2\nu)/2(1+\nu)$. Using equations (3) and (4), it is required to compute the stress invariants I_{n+1}, $J_{2(n+1)}$, $J_{3(n+1)}$, the hardening parameter $I_{0(n+1)}$, and the scalar parameter b_{n+1} used in the radial mapping rule to obtain the image stress.

The first equation for the computation of these unknowns is obtained by applying the trace operator to equation (4) resulting in

$$I_{n+1} = \mathrm{tr}\left(\sigma_n + \overline{K}\hat{C}^e_{ijkl}\left(\Delta\varepsilon_{kl} - \Delta\varepsilon^p_{kl}\right)\right) = f_1\left(I_{n+1}, J_{2(n+1)}, J_{3(n+1)}, I_{0(n+1)}, b_{n+1}\right) \quad (5)$$

The second equation is obtained as

$$J_{2(n+1)} = \frac{1}{2} s_{ij(n+1)} s_{ij(n+1)} = f_2\left(I_{n+1}, J_{2(n+1)}, J_{3(n+1)}, I_{0(n+1)}, b_{n+1}\right) \quad (6)$$

where the deviatoric stress, $s_{ij(n+1)}$, is given as

$$s_{ij(n+1)} = s_{ij(n)} + 2\overline{\mu}\left(\Delta\gamma_{ij(n+1)} - \Delta\gamma^p_{ij(n+1)}\right) \quad (7)$$

$\Delta\gamma_{ij}$ is the total deviatoric strain increment, and $\Delta\gamma^p_{ij}$ is the deviatoric plastic strain increment. The third equation is obtained by the use of Caley-Hamilton theorem given as, $s_{ik}s_{kl}s_{lj} = J_2 s_{ij} + J_3\delta_{ij}$, resulting in

$$J_{3(n+1)} = f_3\left(I_{n+1}, J_{2(n+1)}, J_{3(n+1)}, I_{0(n+1)}, b_{n+1}\right) \quad (8)$$

The fourth equation is obtained by integrating the hardening law given by equation (2) over the finite time increment and is given as

$$I_{0(n+1)} = I_{0(n)} \exp\left[\frac{1+e}{\lambda-\kappa}\Delta\varepsilon^p_{ii}\right] = f_4\left(I_{n+1}, J_{2(n+1)}, J_{3(n+1)}, I_{0(n+1)}, b_{n+1}\right) \quad (9)$$

The fifth equation is obtained by using the fact that the image stress given by the image stress invariants must always lie on the bounding surface and is given as

$$F\left(I_{n+1}, J_{2(n+1)}, J_{3(n+1)}, I_{0(n+1)}, b_{n+1}\right) = 0 \quad (10)$$

Equations (5), (6), (8), (9), (10) are the five nonlinear equations in implicit form which need to be solved for the three stress invariants, hardening parameter, and the scalar parameter. This is done by the use Newton-Raphson iterative scheme by casting the five equations in the form $\{R(x)\} = 0$, where the $\{x\}$ is the vector of unknowns at time t_{n+1}. The Newton-Raphson iterative scheme is setup as

$$\left[\frac{\partial\{R\}}{\partial\{x\}}\right]\left\{\Delta x^k\right\} = -\left\{R\left(x^k\right)\right\},\ \left\{x^{k+1}\right\} = \left\{x^k\right\} + \left\{\Delta x^k\right\} \qquad (11)$$

Once the above Newton-Raphson procedure converges within a specified tolerance, the stress components, $\sigma_{ij(n+1)}$, at time t_{n+1} are to be recovered from the three stress invariants, I_{n+1}, $J_{2(n+1)}$, $J_{3(n+1)}$.

The consistent material tangent modulus is required in the formulation of the global tangent stiffness matrix used in the Newton-Raphson iterative solution of the global finite element equations. The stress integration function is given by equation (4) and the material tangent modulus consistent with this stress function is obtained as its strain derivative as

$$D_{ijkl} = \frac{\partial\sigma_{ij(n+1)}}{\partial\varepsilon_{ij(n+1)}} = \delta_{ij}\frac{\partial p_{n+1}}{\partial\varepsilon_{kl(n+1)}} + \frac{\partial s_{ij(n+1)}}{\partial\varepsilon_{kl(n+1)}} \qquad (12)$$

The spherical and deviatoric part of the stress components obtained from equation (4) will be used to evaluate the strain derivatives in equation (12).

The accuracy and robustness of the above algorithm in terms of the rate of convergence of the global Newton-Raphson iterations and stress update will be presented and discussed.

References

1) Borja, Ronaldo, I., and Lee, Seung, R., "Cam-Clay Plasticity, Part I; Implicit Integration of Elasto-Plastic Constitutive Relations," Comp. Meth. in App. Mech. and Engrg., Vol. 78, 1990, pp. 49-72.
2) Borja, Ronaldo, I., "Cam-Clay plasticity, Part II: Implicit Integration of Constitutive Equations Based on a Nonlinear Elastic Stress Predictor," Comp. Meth. in App. Mech. and Engrg., Vol. 88, 1991, pp. 225-240.
3) Dafalias, Y.F., and Herrmann, L.R., "Bounding Surface Plasticity II: Application to Isotropic Cohesive Soils," J. of Engrg. Mech., ASCE, Vol. 112, No. 12, 1986, pp. 1263-1291.
4) Herrmann, L.R., Kaliakin, V., Shen, C.K., Mish, K.D., Zhu, Z.Y., "Numerical Implementation of Plasticity Model for Cohesive Soils," J. of Engrg. Mech., ASCE, Vol. 113, No. 4, 1987, pp. 500-519.
5) Morz, Z., Norris, V.A., Zienkiewicz, O.C., "Application of an Anisotropic Hardening Model in the Analysis of Elasto-Plastic Deformations of Soils," Geotechnique, Vol. 29, No. 1, 1979, pp. 1-34.

Improving Robustness of Algorithms to Implement HiSS models in FEM

G.W. Wathugala[1] and S. Pal[2]

ABSTRACT

A constitutive model has only a limited value if it is not implemented in a computer procedure. Many algorithms have been developed for different constitutive models including the Hierarchical Single Surface (HiSS) models. It is important that these algorithms be (a) accurate, (b) efficient, and (c) robust. In this paper, several methods have been presented to improve the robustness of algorithms to implement HiSS models in finite element programs. These algorithms are based on the elastic predictor plastic corrector method and the plastic predictor plastic corrector method.

INTRODUCTION

In this paper, several methods have been presented to improve the robustness of algorithms to implement HiSS models in finite element programs. These algorithms are based on the elastic predictor plastic corrector method and the plastic predictor plastic corrector method. An algorithm may be very stable and robust under laboratory stress and strain paths but could blowup in a boundary value problem due to the complex and irregular nature of the strain paths in boundary value problems. Therefore, it is very important that we explicitly treat all the possible conditions where our algorithm fails. Some of of the cases investigated in this paper are (a) $J_1^I < 0$, (b) $J_1^C < 0$, (c) $F(\sigma_{ij}^I, \alpha_{ps}^I) < 0$, (d) $F(\sigma_{ij}^C, \alpha_{ps}^C) < 0$, and (e) F(full increment)> 0 but F(subincrement)< 0. Here, J_1 is the first invariant of the stress tensor. Superscripts I and C denote the intermediate (predictor) and converged (after the correction) points respectively. F is the yield function and α_{ps} is the hardening function.

HIERARCHICAL SINGLE SURFACE (HiSS) MODELS

The hierarchical modelling approach allows for progressive development of models of higher grades corresponding to different levels of complexities (Desai et al., 1986). The same yield surface is used for all the HiSS models and, in general, can be defined as

$$F \equiv F(\sigma_{ij}, \alpha_{ps}) = 0 \tag{1}$$

where α_{ps} is the hardening function and can be a function of various internal variables related to the plastic deformations. In general, it is defined as $\alpha_{ps} \equiv f_\alpha(\xi, \xi_D, \xi_V)$ where ξ,

[1]Assistant Professor, Dept. of Civil and Env. Eng., Louisiana State University, Baton Rouge, LA 70803.
[2]Graduate Student, Dept. of Civil and Env. Eng., Louisiana State University, Baton Rouge, LA 70803.

ξ_D, and ξ_V are trajectories of total, deviatoric and volumetric plastic strains, respectively. The potential function is defined as a deviation of the loading function. In general, it is defined as $Q \equiv Q(\sigma_{ij}, \alpha_Q)$ where α_Q is defined as a function of α_{ps} and some other history parameters such as induced anisotropy.

STRAIN TO STRESS ALGORITHMS

The basic problem in calculating stress increment, $d\sigma_{ij}$ corresponding to a given strain increment, $d\varepsilon_{kl}$ which causes virgin loading, may be expressed as follows Given: σ^O_{ij}, α^O_{ps}, ξ^O_i, $d\varepsilon_{kl}$ and $F(\sigma^O_{ij}, \alpha^O_{ps}) = 0$. Find: σ^C_{ij} and α^C_{ps} so that

$$F(\sigma^C_{ij}, \alpha^C_{ps}) = 0 \tag{2}$$

Where superscripts O and C refer to initial and final converged quantities respectively. Most of the iterative schemes for solution of the above system starts with a trial decomposition for the incremental strain tensor. The elastic predictor plastic corrector method uses $d\varepsilon^p_{kl} = 0$ as the trial solution, whereas the plastic predictor plastic corrector method decomposes $d\varepsilon_{kl}$ based on the initial elasto-plastic constitutive tensor, C^{ep}_{ijkl}. Let us assume that these trial solutions moves stress to an intermediate state I and a single correction moves it to the final converged solution at C. Now, Eq.(2) can be expressed as

$$F(\sigma^I_{ij} + d\sigma^{IC}_{ij}, f_\alpha(\xi^I_i + d\xi^{IC}_i)) = 0 \tag{3}$$

Taylor's series expansion of Eq. (3) around the point I and the flow rule in plasticity $d\varepsilon^{p,IC}_{kl} = \lambda^{IC} n^Q_{kl}$ give the expression for λ^{IC} (plastic correction) as

$$\lambda^{IC} = \frac{-F(\sigma^I_{ij}, f_\alpha(\xi^I_i))}{\left\{ \left[\frac{\partial f}{\partial \sigma_{ij}}\right]_I C^e_{ijkl} n^Q_{kl} + \left[\frac{\partial F}{\partial \alpha_{ps}} \frac{\partial f_\alpha}{\partial \xi_i}\right]_I f_i(n^Q_{kl}) \right\}} \tag{4}$$

Here n^Q_{kl} is the direction of plastic strain. Since higher order terms in Eq.(4) have been neglected, the solution obtained here would not satisfy the yield function given by Eq.(2). The quantities obtained here actually refer to an intermediate state closer to the final converged solution. Substituting the solution obtained here as the intermediate state in Eq.(4), the procedure is repeated until λ^{IC} or $F(\sigma^I_{ij}, \alpha^I_{ps})$ is less than a prescribed tolerance.

PROCEDURES TO IMPROVE THE ROBUSTNESS

There are some situations where the algorithm described above does not work. Special procedures to avoid these situations are presented here. Even though some of these cases are rare, it is important that they be dealt explicitly to assure the reliability and robustness of the algorithm.

Case I: $J^I_1 < 0$

Yield function, F and the potential function, Q in the HiSS models have non-integer exponents of J_1. Therefore when J^I_1 is negative, the terms $F(\sigma^I_{ij}, \alpha^I_{ps})$, n^Q_{kl} and $(\partial F/\partial \alpha_{ps})$ in

Eq.(4) can not be evaluated. This problem can be avoided by using smaller subincrements that assure $J_1^I < 0$. This is done by scaling down the strain increments by a factor $k = J_1^O/(J_1^O - J_1^I)$. This situation normally occurs at the dilation region.

Case II: $J_1^C < 0$

This is somewhat rare situation which might occur during large strain increments due to numerical inaccuracies. Even though the use of subincrements for large strain increments avoids this situation for many problems, a simple procedure is introduced to avoid blowing up in those rare cases. Since this error is due to over correction of an intermediate stress, the corrector is scaled down until J_1^C is positive by $\sigma_{ij}^C = \sigma_{ij}^C(J_1^I/(J_1^I - J_1^C))$. Generally this stress point will not be on the yield surface and the iteration process has to be continued until the convergence is reached.

Case III: $F(\sigma_{ij}^I, \alpha_{ps}^I) < 0$

In plastic predictor plastic corrector method, we have come across $F(\sigma_{ij}^I, \alpha_{ps}^I) < 0$ in certain situations. It is required that $F(\sigma_{ij}^I, \alpha_{ps}^I) > 0$ for the algorithm to work. This problem is overcome by modifying the predicted stress increment $d\sigma_{ij}^{OI}$ by a factor k which is determined by solving $F(\sigma_{ij}^O + kd\sigma_{ij}^{OI}, \alpha_{ps}^I) = 0$ for k, using the Newton-Raphson procedure. Now, the stress increment from O is given by $k \times d\sigma_{ij}^{OI}$. A detail procedure is given in Wathugala (1990).

Case IV: $F(\sigma_{ij}^C, \alpha_{ps}^C) < 0$

Even though this is rare, it is also possible to get $F(\sigma_{ij}^C, \alpha_{ps}^C) < 0$ in some situations. Since $F(\sigma_{ij}^C, \alpha_{ps}^C)$ is the intermediate value for the next iteration, this problem has to be avoided to ensure a robust algorithm. This situation arises due to over correction and the problem is solved by scaling down the stress correction, $d\sigma_{ij}^{IC}$ by a factor k which is determined by solving $F(\sigma_{ij}^I + kd\sigma_{ij}^{IC}, \alpha_{ps}^C) = 0$ for k, $(0 < k < 1)$ using the Newton-Raphson procedure. Now, the new stress correction is given by $k \times d\sigma_{ij}^{IC}$. A detail description of the procedure is given in Wathugala (1990).

Case V: F**(full increment)**> 0 **but** F**(subincrement)**< 0

During the subincrementation, the elastic prediction of the stress follows a nonlinear path when shear stress components are present. Therefore, one could have a situation when F(full increment)> 0 but F(subincrement)< 0 (as shown in Fig.1) which could lead to numerical instability. Here, we need to find point D from which virgin loading takes place. Since point A is also on the yield surface, the Newton-Raphson procedure mentioned earlier does not gaurantee that it will find point D as a solution instead of point A. This problem can be fixed as follows.

Before proceeding for the subincrementation, first the quantity $d\sigma_{ij}^e n_{ij}^F$ is calculated. If it is less than zero that means unloading takes place at the beginning which could be followed by reloading and virgin loading. Now, the neutral loading point C is found as the stress point with maximum α_r. Here α_r is defined as $\alpha_r = R^{-1}(\sigma_{ij})$, where $R \equiv R(\sigma_{ij}, \alpha_r) = 0$ is the reference function which is algebraically similar to F. Higher α_r values refer to smaller reference surfaces (Wathugala, 1990). When point C is known, we can limit our Newton-

Raphson search to stress points after point C and find the point D where virgin loading starts. Now, AD is considered as nonvirgin loading and DB is considered as virgin loading.

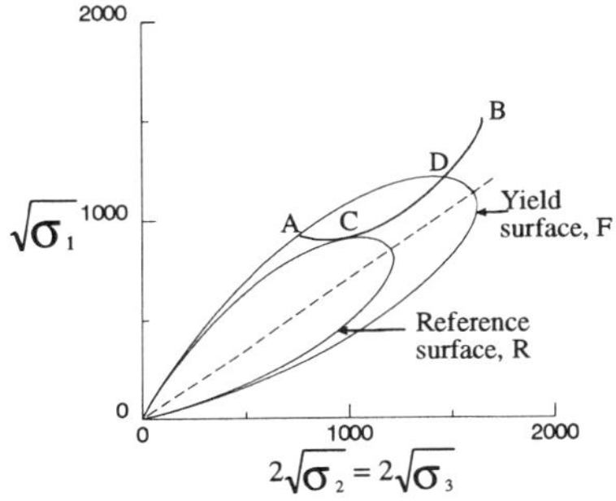

Figure 1: Stress path in subincrementation

CONCLUSION

Several procedures to improve the robustness of algorithms to implement Hierarchical Single Surface (HiSS) models have been presented. The basic algorithms used to implement HiSS models are based on elastic predictor plastic corrector method and plastic predictor and plastic corrector method. The robustness of an algorithm is very important because a gauss integration point in an element in a finite element mesh for a complex boundary value problem experiences very complex strain path which is not common in a laboratory test. Individual treatment of the cases (a) $J_1^I < 0$, (b) $J_1^C < 0$, (c) $F(\sigma_{ij}^I, \alpha_{ps}^I) < 0$, (d) $F(\sigma_{ij}^C, \alpha_{ps}^C) < 0$, and (e) F(full increment)> 0 but F(subincrement)< 0, where the present algorithms fail, have been made. Here, J_1 is the first invariant of the stress tensor. Superscripts I and C denote the intermediate (predictor) and converged (after the correction) points respectively. Considerations of all the possibilities, even the very rare ones have made the algorithm very robust and reliable.

ACKNOWLEDGEMENT

Partial financial support from Louisiana Transportation Research Center, Louisiana Board of Regents (Grant Nos. LEQSF(1994-97)-RD-A-08 and LEQSF(1994-96)-ENH-TR-28) and National Science Foundation (Grant No. CMS-9410482) are greatfully acknowledged.

APPENDIX

References

Desai, C.S., Somasundaram, S., and Frantziskonis, G.N., (1986), "A Hierarchical Approach for Constitutive Modelling of Geologic Materials," *Int. J. Num. Anal. Meth. Geomech.*, Vol. 10, pp. 225-257.

Wathugala, G.W.,(1990), "Finite Element Dynamic Analysis of Nonlinear Porous Media with Applications to Piles in Saturated Clays," *Ph.D. Dissertation,* University of Arizona, Tucson, Arizona, 475 pages.

CRITICAL STATE MODEL AT FINITE STRAINS

Ronaldo I. Borja[1] and Claudio Tamagnini[2]

Abstract

The performance of a numerical model which casts a finite deformation critical state theory within the framework of finite element (FE) analysis is assessed from the results of numerical simulations. The finite deformation theory is formulated within the context of multiplicative plasticity on a hyperelastic-plastic modified Cam-clay model with a bilogarithmic compressibility law.

Introduction

A finite deformation version of the modified Cam-clay model of Roscoe and Burland (1968) was recently proposed (Borja and Tamagnini 1996; Borja et al. 1996) employing a mathematical approach based on a multiplicative decomposition of the deformation gradient. The product decomposition approach gives rise to an additive split of natural (logarithmic) strains, and allows the conventional return mapping scheme of the infinitesimal plasticity theory to be carried over to the finite deformation regime without modification (Simo 1992). Central to the formulation is the assumed existence of a stored energy function and the validity of a bilogarithmic compressibility law in the infinitesimal and finite deformation regimes (Houlsby 1985; Butterfield 1979). Results of numerical simulations are presented describing the performance of a nonlinear FE model which uses the proposed constitutive theory.

Constitutive theory

Roscoe and Burland (1968) proposed a two-invariant plasticity model, called modified Cam-clay, to describe the load-deformation behavior of normally consolidated clays under 'triaxial' stress conditions. The model captures many essential features of a soil's drained deformation behavior in the region of small strains, such as pressure sensitivity, coupled volumetric-deviatoric behavior, and plastic hardening with volumetric compaction. The model has since been re-modified,

[1] Assoc. Prof., Dept. of Civ. Engrg., Stanford Univ., Stanford, CA 94305.
[2] Post-Doctoral Student, Dept. of Civ. Engrg., Stanford Univ., Stanford, CA.

enriched, and even extended to three-dimensional states of stress for use with general purpose analysis codes. Fairly recently, the modified Cam-clay model was cast within the framework of finite deformation elastoplasticity based on a multiplicative decomposition of the deformation gradient (Borja and Tamagnini 1996; Borja et al. 1996). This latter version of the model is the subject of study in the present paper.

Extension of the constitutive model to the finite deformation regime requires two modifications to the infinitesimal version of the model. The first involves the assumption that there exists a stored energy function describing the hyperelastic component of the model. The pressure-dependent nature of the elastic bulk and shear moduli observed in many laboratory tests requires that this stored energy function be exponential in the elastic volumetric strains, and results in a coupled volumetric-deviatoric elastic behavior (Houlsby 1985). The second modification involves replacing the linear void ratio-logarithm of volumetric stress compressibility law of the old model with a bilogarithmic compressibility law. The bilogarithmic approximation agrees better with results of many laboratory tests over a wider range of pressures than does the original hardening law (Butterfield 1979). Furthermore, it is compatible with the product decomposition scheme in the region of finite strains, and hence results in significant numerical advantages. Complete details of the finite deformation Cam-clay model are presented by Borja et al. (1996) and Borja and Tamagnini (1996).

Numerical simulations

Figure 1 shows a FE mesh in which a 16 m-width strip load of 100 kPa is applied gradually (drained analysis) on the surface of a 20 m-thick normally consolidated clay layer. The domain is discretized into 160 bilinear square elements with a side length of 2 m, for a total of 187 displacement nodes and 320 degrees of freedom. The material parameters for the clay layer are as follows: swelling-recompression indices $\hat{\kappa} = 0.05$, $\widetilde{\kappa} = \hat{\kappa}/(1+\hat{\kappa}) = 0.048$; virgin compression indices $\hat{\lambda} = 0.20$, $\widetilde{\lambda} = \hat{\lambda}/(1+\hat{\lambda}) = 0.167$; elastic shear modulus $\mu^e = 10,000$ kPa (assumed constant for this example); reference preconsolidation pressure $p_0 = -10$ kPa.

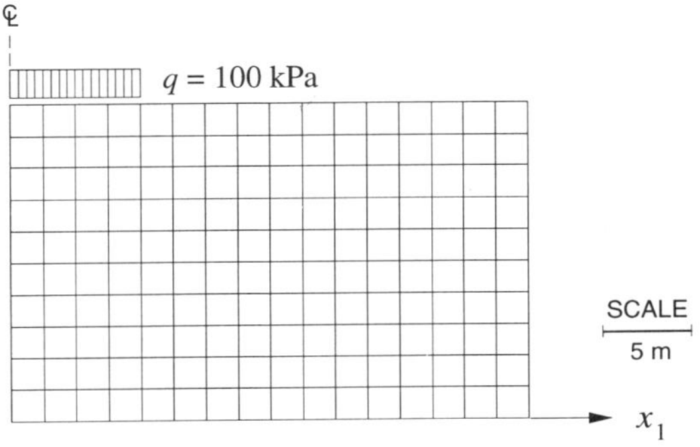

Figure 1. Finite element mesh for the strip load problem.

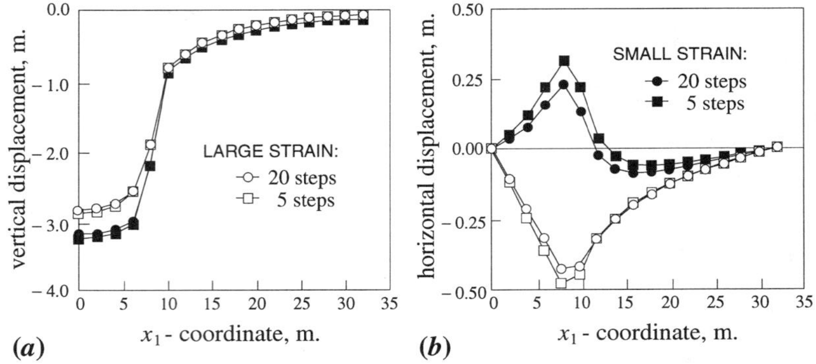

Figure 2. Displacement profiles: (a) vertical; (b) horizontal.

Figures 2(a) and 2(b) show the profiles of final vertical and horizontal displacements on the ground surface, respectively. As expected, the small strain analysis results in larger vertical displacements than the large strain analysis due to the use of natural (logarithmic) strains in the latter analysis. Opposite trends are manifested in the horizontal ground displacement profiles, which predicted generally outward lateral movement for small strain analysis and consistently inward movement for large strain analysis. The results shown in these figures indicate that the two analyses could produce responses that differ not only quantitatively but also qualitatively.

The accuracy of the algorithm is also depicted in Figs. 2(a) and 2(b). Comparing the solution in which the total load is applied in 5 increments with that in which the load is applied in 20 increments, the relative error (defined as the difference between the predicted displacements divided by the displacement predicted by the 20-increment solution) is seen to be relatively small, and remains generally below 3%. The large strain solution exhibits smaller relative error than the small strain solution.

Figure 3 compares the global convergence profiles of the small-strain and finite-strain analyses corresponding to the 5-increment solution. Convergence is achieved in 10 iterations or less, and exhibits a quadratic rate in the limit of small error. Interestingly, the large strain solution tends to converge about one or two iterations faster than the small strain solution even though the degree of nonlinearity of the former solution is higher. Consistent linearization appears to have made the contribution of additional nonlinearity relatively insignificant.

Closure

A finite deformation Cam-clay model is implement into a fully nonlinear FE program. The introduction of geometric nonlinearities into the solution alters the

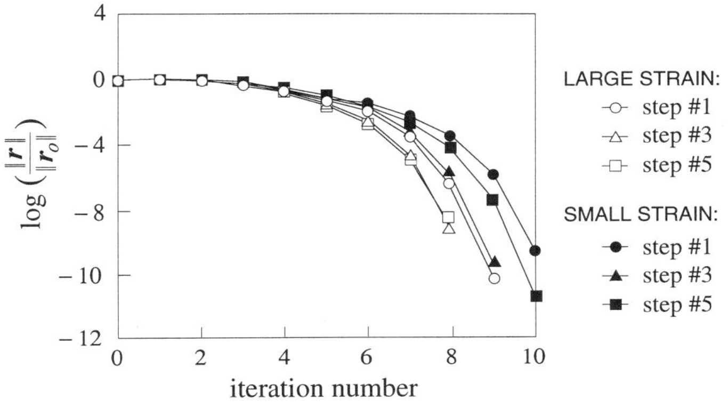

Figure 3. Convergence profile for the 5-increment solution.

predicted response of the structure at large strains. The additional nonlinearity due to large deformation appears to have no significant effect on the efficiency of the computer code. The accuracy of the finite deformation algorithm as well as the rate of convergence of the iteration compare well with those of the small strain algorithm provided that the external loads are applied in sufficiently small increments.

References

Borja, R.I. and Tamagnini, C. (1996). Cam-clay plasticity, Part III: Extension of the infinitesimal model to include finite strains, Comput. Methods Appl. Mech. Engrg., in review.

Borja, R.I., Tamagnini, C., and Amorosi, A. (1996). Hyperelastic-plastic Cam-clay model, J. Geotech. Engrg., ASCE, in review.

Butterfield, R. (1979). A natural compression law for soils, Géotechnique, 29, 469–480.

Houlsby, G.T. (1985). The use of a variable shear modulus in elastic-plastic models for clays, Computers and Geotechnics, 1, 3–13.

Roscoe, K.H. and Burland, J.B. (1968). On the generalized stress-strain behavior of 'wet' clay, in: J. Heyman and F.A. Leckie, eds., Engineering Plasticity, Cambridge Univ. Press, 535–609.

Simo, J.C. (1992). Algorithms for static and dynamic multiplicative plasticity that preserve the classical return mapping schemes of the infinitesimal theory, Comput. Methods Appl. Mech. Engrg., 99, 61–112.

Structural Model Updating Using Expanded Modeshapes

James L. Beck[1], M., ASCE and Michael W. Vanik[2]

Abstract

A model updating procedure is being developed as part of a global structural health monitoring project which utilizes Bayesian probabilistic inference and the concept of expanded modeshapes. The method is described and results of tests using a small-scale laboratory model are discussed.

Introduction

A methodology is presented for structural model updating using limited modal information. The model is identified through an iterative procedure which involves solving a small number of linear systems of equations during each iteration. The method has two key advantages: there is no need to specify which experimental and model modes correspond, and the eigenparameters of the model do not need to be found. This makes automation of the process easier and reduces computational cost.

Theory

Model Definition: The method utilizes a class $\mathcal{M}_{N_d}$ of N_d degree-of-freedom finite-element models of the form $M\ddot{\underline{x}} + K(\underline{\theta})\underline{x} = \underline{f}(t)$ where $K(\underline{\theta}) = K_0 + \sum_{i=1}^{N_\theta} \theta_i K_i$, $\underline{\theta} \in \Theta \subseteq \mathbb{R}^{N_\theta}$, and M is assumed known. The K_i are the uncertain stiffness contributions from elements or substructures which are to be updated while K_0 represents known contributions. The structural data, $\mathcal{D}$, consists of N_m experimentally determined modal frequencies, $\hat{\omega}_r$, and corresponding incomplete modeshapes, $\hat{\underline{\psi}}_r \in \mathbb{R}^{N_o}$, where N_o is the number of observed degrees of freedom. The goal is to find the model(s), $\hat{\underline{\theta}}$, which are the most plausible based on $\mathcal{D}$, that is, which "best" explain these data.

[1]Associate Professor, Division of Engineering and Applied Science, California Institute of Technology, MS 104-44, Pasadena, CA 91125

[2]Graduate Student, California Institute of Technology, MS 104-44

Probabilistic Analysis: In order to make inferences about the relative plausibilities of various models based on the data, the problem is cast in a Bayesian probabilistic framework. This approach facilitates understanding of how to handle multiple "best" models and how to update these models when new data becomes available (Beck and Katafygiotis 1992; Beck 1996). When analyzed in this framework, the updated probability distribution conditional upon $\mathcal{D}$ and $\mathcal{M}_{N_d}$ of the model parameters, $\underline{\theta}$, and the *expanded modeshapes*, $\underline{\Phi} = [\underline{\phi}_1^T, \ldots, \underline{\phi}_{N_m}^T]^T$, described later, is given by

$$p(\underline{\theta}, \underline{\Phi} \,|\, \mathcal{D}, \mathcal{M}_{N_d}) = k(N_o, N_m, \mathcal{D}) J(\underline{\theta}, \underline{\Phi})^{\frac{-N_o N_m}{2}} \tag{1}$$

where

$$J(\underline{\theta}, \underline{\Phi}) = \sum_{r=1}^{N_m} J_r(\underline{\theta}, \underline{\phi}_r) = \sum_{r=1}^{N_m} \left[\frac{1}{\epsilon_{\omega_r}^2 \hat{\omega}_r^4} \|(K(\underline{\theta}) - \hat{\omega}_r^2 M)\underline{\phi}_r\|_{M^{-1}}^2 + \frac{1}{\epsilon_{\psi_r}^2 \|\underline{\hat{\psi}}_r\|_2^2} \left(\|\underline{\hat{\psi}}_r\|_2^2 \|\Gamma \underline{\phi}_r\|_2^2 - \langle \underline{\hat{\psi}}_r, \Gamma \underline{\phi}_r \rangle_2^2 \right) \right] \Big/ \|\underline{\phi}_r\|_M^2 \tag{2}$$

The measure of fit, $J(\underline{\theta}, \underline{\Phi})$, is a weighted sum of modal equation and modeshape component errors. The parameters ϵ_{ω_r} and ϵ_{ψ_r} measure the relative confidence in the observed modal components. The matrix Γ is an observation matrix which selects the components of a vector corresponding to those measured. The vectors $\underline{\phi}_r$ are *expanded modeshapes* which must be found along with the structural model parameters, $\underline{\theta}$, in order for the equation error term to be used. Expressing an equation error using only the incomplete measured modeshapes, $\underline{\hat{\psi}}_r$, would require employing Guyan reduction or a similar technique to reduce the model. Doing so would destroy the structure of the equations which is exploited later. The modeshape expansion technique used here is similar to that used by Levine-West, Kissil, and Milman (1994).

The second term in (2) is a measure of the error between corresponding measured and expanded modeshape components. This form arises from finding the optimal scalings on the $\underline{\phi}_r$ which minimize the modeshape component error for given $\underline{\phi}_r$ and $\underline{\hat{\psi}}_r$. It measures the discrepancy in angle between the two vectors. Since the scaling can now be arbitrarily assigned, a choice of $\|\underline{\hat{\psi}}_r\|_2^2 = 1$ (Euclidean norm) can be made, while $\|\underline{\phi}_r\|_M^2 = 1$ (mass-weighted norm).

Given the probability density function (PDF) in (1), the probability that the parameters of interest, $\underline{\theta}$, are in a set $\Theta_s \subseteq \Theta$ corresponding to "damage" involves integration over the set Θ_s and over the entire range of $\underline{\Phi}$. Papadimitriou, Beck, and Katafygiotis (1995) present an asymptotic method for approximating integrals of this type. It requires evaluation of most probable model(s), $\underline{\hat{\theta}}$, which maximize the updated PDF in (1). This is equivalent to finding the minimum of $J(\underline{\theta}, \underline{\Phi})$. A novel method to evaluate this minimum is as follows.

Updating Algorithm: Minimizing J is equivalent to minimizing a modified measure of fit, $H = \sum_r \left[\|\underline{\phi}_r\|_M^2 J_r + \lambda_r(\|\underline{\phi}_r\|_M^2 - 1)\right]$, which leads to

$$(3) \quad \underline{0} = \frac{\partial H}{\partial \underline{\theta}} = A^T \tilde{M}^{-1} A\,\underline{\theta} - A^T \underline{\Phi} \quad \text{and} \quad \underline{0} = \frac{\partial H}{\partial \underline{\phi}_r} = \left[\tilde{K}_r - \lambda_r M\right]\underline{\phi}_r,$$

where λ_r is a Lagrange multiplier for the constraint $\|\underline{\phi}_r\|_M^2 = 1$,

$$A = A(K_{1,\ldots,N_\theta}, \mathcal{D}, \underline{\Phi}) \in \mathbb{R}^{N_d N_m \times N_\theta}, \quad \tilde{K}_r = \tilde{K}_r(K(\underline{\theta}), M, \mathcal{D}, \Gamma) \in \mathbb{R}^{N_d \times N_d}$$

and $\tilde{M}^{-1}$ is a block diagonal matrix of the appropriate dimension with M^{-1} as each block of the diagonal. It is clear that setting the derivative with respect to $\underline{\theta}$ to zero while holding the $\underline{\phi}_r$ fixed gives a linear system of equations for $\underline{\theta}$. Doing the same for each of the $\underline{\phi}_r$ derivatives gives N_m eigenvalue problems for $\hat{\underline{\phi}}_r$, the resulting normalized eigenvector of $\tilde{K}_r$. It is easily shown that the value of J_r at the optimal expanded modeshape is $\hat{\lambda}_r$. Hence, minimizing J involves finding the smallest eigenvalue and corresponding normalized optimal expanded modeshape, $\hat{\underline{\phi}}_r$, for each of the N_m eigenvalue problems.

This iterative solution procedure can be summed up as a) Select an initial model, $\underline{\theta}_o$, b) Solve N_m eigenvalue problems $\partial H/\partial \underline{\phi}_r = \underline{0}$ for the smallest eigenvalues, $\hat{\lambda}_r$, and their associated eigenvectors, $\hat{\underline{\phi}}_r$, c) Using the expanded modeshapes, solve $\partial H/\partial \underline{\theta} = \underline{0}$ for $\underline{\theta}_1$, d) Repeat the procedure until a set convergence criterion is achieved.

Results

Validation Testing: Experimental testing was performed on an aluminum structure capable of being configured with one, two, or three stories. A story was modelled as two rigid plates connected by four beam elements, one at every plate corner. Each plate had three DOF (degrees-of-freedom): one "torsional", α_i, and two horizontal, y_i and z_i. Results are presented for the 3-story structure. In this case, θ_1, θ_3, and θ_5 scaled the y-stiffnesses for each story while θ_2, θ_4, and θ_6 scaled the z-stiffnesses. The original finite-element model was updated using all 9 modal frequencies and modeshapes. The elemental stiffnesses were then re-calibrated so that the θ_i's were unity for the undamaged structure. One column was then removed from the first story and the model was updated using full or partial modal information from this damaged structure.
Case I: Updating using the full 9 modes (frequencies and modeshapes at 9 DOF) gave a most probable model $\hat{\underline{\theta}} = (0.74\ 0.73\ 0.72\ 0.94\ 0.86\ 0.96)$.
Case II: Updating using the 9 modal frequencies gave $\hat{\underline{\theta}} = (0.50\ 0.67\ 0.87\ 0.74\ 1.46\ 1.57)$.
Case III: Updating using the 5 dominant modes with all the modeshape components at the first floor and the roof gave $\hat{\underline{\theta}} = (0.62\ 0.73\ 0.63\ 0.84\ 0.45\ 1.0)$.

It was expected that the updated model would have a θ_1 and θ_2 reduced to about 0.75, reflecting the loss of one of the 4 beam elements. For case I, this loss was observed. In cases II and III, the values were lower. This is a combined effect of having less modal data and model error. The model error is most readily demonstrated by the fact that modeshapes for the system are fully three dimensional, while the model possesses symmetries which always produce pure (uncoupled) translational and rotational modes. Notice also that the other parameters changed rather that remaining at their undamaged values of unity. This is thought to be a result of two effects. First, removal of a leg in the first story changes the boundary condition above that leg, reducing the effective stiffness of the second story. The model error also effects the ability to identify these parameters.

Conclusion

A model updating method which utilizes modal data from a structure to update a structural model has been presented. The selected measure of fit possesses partial derivatives which are linear in the derivative parameters. This enables finding the most probable models by iteratively solving linear systems given by the zeros of the derivatives. This simplicity is made possible through utilization of expanded modeshapes and a comparison of measured and expanded modeshapes which is independent of their normalization. The experimental testing showed that model error necessitates a need for careful interpretation of the updated model. Future testing will explore the use of more detailed structural models. Work is also underway that will address the problem of finding and interpreting multiple most probable models which occur with more complex models, but incomplete modal data.

References

Beck, J. L. (1996). System identification methods applied to measured seismic response. In *Proceedings Eleventh World Conference on Earthquake Engineering*, Acapulco, Mexico.

Beck, J. L. and L. S. Katafygiotis (1992). Probabilistic system identification and health monitoring of structures. In *Proceedings Tenth World Conference on Earthquake Engineering*, Madrid, Spain.

Levine-West, M., A. Kissil, and M. Milman (1994). Evaluation of mode shape expansion techniques on the micro-precision interferometer truss. In *Proceedings Twelfth International Modal Analysis Conference*, Honolulu, Hawaii.

Papadimitriou, C., J. L. Beck, and L. S. Katafygiotis (1995). Asymptotic expansions for reliabilities and moments of uncertain dynamic systems. EERL Report No. 95-05, California Institute of Technology, Pasadena.

MODAL FILTER BASED ON-LINE MONITORING OF UNCERTAIN STRUCTURAL SYSTEMS

E.A. Johnson[1] (Student Member, ASCE), L.A. Bergman[2] (Member, ASCE), P.G. Voulgaris,[3] and L.C. Freudinger[4]

ABSTRACT

Identification of a structural system can be performed in a number of ways. The modal filter method has been used to develop single degree-of-freedom transfer functions from multimodal data for various linear time-invariant structures and has also been used in the implementation of modal control strategies. It may be desirable to apply these methods to systems whose behavior changes over time. Such engineering structures are numerous; in the aerospace field, aircraft are a classic example, where structural system identification performed in flight may yield vastly different results from similar tests on the ground due to aerodynamic coupling and loading. Some civil engineering structures, *e.g.*, suspension bridges, may display the same type of behavior, though generally to a lesser extent. On-line monitoring of these time-varying structures, with special regard to structural integrity, poses a difficult problem, especially since the number of sensors in on-line monitoring may be limited. Various aspects of this problem will be discussed, particularly from the standpoint of monitoring critical modes of the structure.

INTRODUCTION

Numerous methods have been developed in recent decades to handle the task of experimental system identification; some excellent books on the subject have been written (*e.g.*, Ljung, 1987) and computer programs such as MATLAB® have facilitated the easy access by the user. Structural system identification has come to be recognized as a particular subset of this field. Recent papers have examined several of

1. Research Assistant, Dept. of Aeronautical & Astronautical Engrg., Univ. of Illinois at Urbana-Champaign, Urbana, IL 61801.

2. Professor, Dept. of Aeronautical & Astronautical Engrg., Univ. of Illinois at Urbana-Champaign, Urbana, IL 61801.

3. Assistant Professor, Dept. of Aeronautical & Astronautical Engrg., Univ. of Illinois at Urbana-Champaign, Urbana, IL 61801.

4. Project Engineer, NASA Dryden Flight Research Facility, Edwards, CA 93523.

these methods and concluded that some of them operate effectively on civil engineering structures (Ghanem and Shinozuka, 1995; Shinozuka and Ghanem, 1995).

Off-line, or batch, system identification performs well for systems that are time-invariant. There are, however, numerous examples of time-varying systems. For example, in the aerospace field, the structural parameters of an aircraft may exhibit time-varying characteristics due to aeroelastic effects that alter as the aircraft environment (*e.g.*, air properties, weather, etc.) and trajectory and angle of attack change, or due to damage to the aircraft. Civil engineering structures may also display such time-dependence; for example, suspension bridges and tall buildings display aeroelastic effects similar to those of aircraft, though generally to a significantly lesser extent. On-line, or adaptive, system identification is one method of treating such a problem with time-dependent systems (see, *e.g.*, Ljung and Söderström, 1983).

Monitoring critical modes of a structural system is required in certain applications, especially when the stability or integrity of the structure must be observed. This is particularly true when the structure is actively controlled and the controller must accommodate the significant changes that may occur in such time-varying systems. One method of determining modal response is to use a *spatial modal filter.* Meirovitch and his colleagues used spatial eigenfunctions or interpolation functions to determine, off-line, a modal filter for certain one dimensional distributed-parameter systems for the purpose of independent modal state control (Meirovitch and Baruh, 1982, 1985; Öz and Meirovitch, 1983; Meirovitch and Öz, 1984; Meirovitch and Ghosh, 1987). Shelley *et al.* developed another formulation for the spatial modal filter based on experimental frequency response functions (Shelley, 1991; Freudinger, 1991) and have utilized their method for active modal structural control of a highway bridge (Shelley *et al.*, 1994, 1995).

In its simplest form, Shelley's reciprocal modal filter method, given transfer functions experimentally measured at various frequencies from one control input to all sensor outputs, and given the location of a complex pair of poles of the system, determines the coefficients of a linear combination of those transfer functions (essentially a transformation to modal coordinates) that results in a minimal difference (in a least-squares sense) from a single-degree-of-freedom transfer function with the given complex pole pair. This has a strong physical basis if the system is classically damped, where the ideal modal filter is simply a function of the eigenvectors of the system.

PROPOSED METHOD FOR ON-THE-FLY MONITORING

There are two, often conflicting, goals in on-line monitoring: the accuracy of estimated system behavior and the pace at which the estimates can be determined. Active control of the system may require both goals; quick reaction to the detection of damage or instability certainly requires the latter. In order to provide a continuous and rapid monitoring of system integrity, this paper proposes a two-stage monitoring system to address both goals. A diagram of this system can be seen in Figure 1.

The "outer" on-line monitoring loop is the system identification loop, performed as fast as is possible with the available computational resources, but generally not providing real-time updates of the system parameters. Various methods could be used, including adaptive or recursive identification. A high-precision baseline, or laboratory, test is useful in establishing basic system parameters and initial information for the second loop.

Modal response monitoring is the task of the inner loop, which uses the most recent estimate of the system parameters from the outer loop to compute modal response. This loop must be designed to function as rapidly as possible for monitoring system integrity in real-time. Further simplicity in this loop may rely upon the fact that often a small number of modes of the system are considered critical and, thus, require on-line monitoring. For example, in an aircraft structure, aeroelastic flutter in one or more critical modes may warrant surveillance, and monitoring the damping in those modes can provide a warning against the onset of such instabilities.

Several other real-world complexities must also be included. First, consider that in on-line monitoring, one often is limited to the use of a small number of sensors compared to what might be available in a baseline test. Furthermore, the types of sensors may be different. For example, in baseline ground tests of aircraft vibration, the state-of-the-art is to mount hundreds of accelerometers on the exterior of the aircraft, whereas monitoring a plane in flight requires that the sensors, predominantly strain gages, be permanently installed to the interior to the structure. The usual method of dealing with a limited array of sensors is to use a Kalman filter in the inner loop to estimate the unsensed data. A further advantage of the Kalman filter is its property to filter out unknown sensor and process noise.

The determination of reciprocal modal filter characteristics may be done, following Shelley, in the frequency domain. But since the system identification is performed on-line, it is more efficient to simply compute the modal filter from the identified system parameters. For example, if the system can be cast in a continuous-time state space model, a transformation into modal state space is possible using the eigenfunctions of the state matrix (assuming that the system is classically damped).

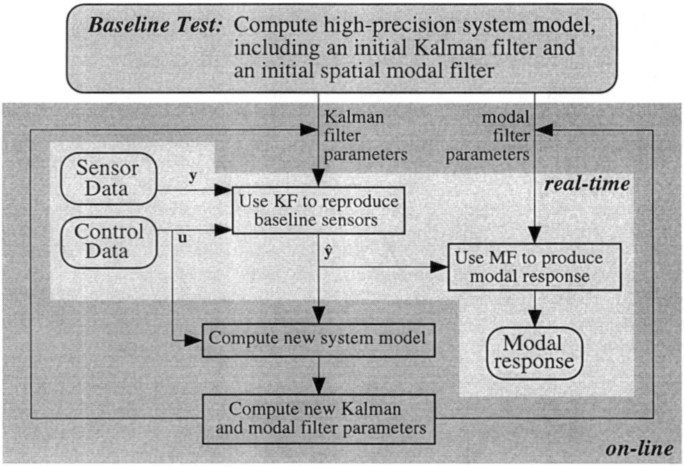

Figure 1: Two-stage monitoring system.

CONCLUSIONS

A strategy for the on-line identification of structures with parameters that vary fairly rapidly with time has been discussed. It involves a two stage process, the first of which is on-line system identification, updating from a baseline model. The second stage, which is performed in real-time, permits continuous monitoring of the stability of the system, especially regarding pathological behavior such as flutter, through the use of a spatial modal filter. Numerical and experimental studies of this two-stage monitoring are in progress, especially as related to the integration of the identification and filtering procedures.

ACKNOWLEDGMENTS

The authors would like to acknowledge the partial support of this research by NASA-Dryden under grant NASA NAG 2-4001.

REFERENCES

Ghanem, R. and Shinozuka, M. (1995). "Structural System Identification, I: Theory." *J. Engineering Mechanics*, ASCE, **121**(2), 255-264.

Freudinger, L.C. (1991). "Analysis of Structural Response Data Using Discrete Modal Filters." NASA CR-179448.

Juang, J-N., Phan, M., Horta, L.G., and Longman, R.W. (1993). "Identification of Observer/Kalman Filter Markov Parameters: Theory and Experiments." *J. Guidance, Control, and Dynamics*, **16**(2), 320-329.

Ljung, L. (1987). *System Identification: Theory for the User.* Prentice-Hall, Englewood Cliffs, NJ.

Ljung, L. and Söderström, T. (1983). *Theory and Practice of Recursive Identification.* MIT Press, Cambridge, MA.

Meirovitch, L., and Baruh, H. (1982). "Control of Self-Adjoint Distributed-Parameter Systems." *J. Guidance, Control, and Dynamics*, **5**(1), 60-66.

Meirovitch, L. and Baruh, H. (1985). "The Implementation of Modal Filters for Control of Structures." *J. Guidance, Control, and Dynamics*, **8**(6), 707-716.

Meirovitch, L. and Ghosh, D. (1987). "Control of Flutter in Bridges." *J. Engineering Mechanics*, ASCE, **113**(5), 720-736.

Meirovitch, L. and Öz, H. (1984). "Digital Stochastic Control of Distributed-Parameter Systems." *J. Optimization Theory and Applications*, **43**(2), 307-325.

Öz, H. and Meirovitch, L. (1983). "Stochastic Independent Modal-Space Control of Distributed-Parameter Systems." *J. Optimization Theory and Applications*, **40**(1), 121-154.

Shinozuka, M. and Ghanem, R. (1995). "Structural System Identification, II: Experimental Verification." *J. Engineering Mechanics*, ASCE, **121**(2), 265-273.

Shelley, S.J. (1991). "Investigation of Discrete Modal Filters for Structural Dynamic Applications." Ph.D. dissertation, Dept. of Mechanical, Industrial, and Nuclear Engineering, Univ. of Cincinnati, Cincinnati, Ohio.

Shelley, S.J., Aktan, A.E., Brown, D.L., and Allemang, R.J. (1994). "University of Cincinnati Experience with Implementing Active Structural Vibration Control." *First World Conf. on Structural Control*, Los Angeles, CA, August 3-5, 1994. Proceedings, FP4-62-70.

Shelley, S.J., Freudinger, L.C., and Allemang, R.J. "Development of an On-line Modal State Monitor." *11th International Modal Analysis Conference* (IMAC), Kissimmee, FL, Feb. 1-4, 1993. Proceedings, 606-612.

Shelley, S.J., Lee, K.L., Aksel, T., and Aktan, A.E. (1995). "Active-Control and Forced-Vibration Studies on Highway Bridge." *J. Structural Engineering*, **121**(9), 1306-1312.

Advances in System Identification Using Output Measurements

N. P. Jones[1], M. ASCE, J. H. Ellis[2] M. ASCE, and K. Pan[1].

Introduction

Identification of system characteristics is often accomplished by using both input and output data (e.g., in the case of earthquakes). For large structures i.e., buildings and bridges, only output measurements are usually possible under ambient conditions. Under such circumstances, the system parameters have typically been estimated under the assumption that the input excitation is Gaussian white. Recent efforts by the authors have led to the development of a Kalman filter algorithm implemented in the frequency domain which was shown to successfully avoid many of the problems typically encountered in this "unknown input" excitation case.

In this paper, the extended Kalman filter algorithm is applied to a SDOF system excited by a Kanai-Tajimi input spectrum. The algorithm is shown to be successful in estimating the input spectrum both parametrically and non-parametrically. Comparison to the results presented in (Conte et al. 1995) is given to compare the accuracy and reliability of the proposed methods.

SDOF System under Base Excitation

The equation of motion of a SDOF system under base excitation can be written

$$\ddot{x}(t) + 2\zeta_s\omega_s\dot{x}(t) + \omega_s^2 x(t) = -\ddot{z}(t) \qquad (1)$$

where ω_s and ζ_s are the natural frequency and damping ratio of the SDOF system, respectively. $x(t)$ is the relative displacement, and $\ddot{z}(t)$ is the absolute ground acceleration.

In the power-spectral density domain, Eqn.(1) becomes

$$\mathbf{S}_{\ddot{\mathbf{x}}} = \omega^4 \mathbf{H}(\omega)\mathbf{S}_{\ddot{\mathbf{z}}}\mathbf{H}(\omega)^* \qquad (2)$$

[1]Dept. Civ. Engrg. Johns Hopkins Univ., Baltimore, MD 21218
[2]Dept. Geog. & Env. Engrg., & Civ. Engrg., Johns Hopkins Univ., Baltimore, MD 21218

where $\mathbf{H}(\omega)$ is the system transfer function

$$\mathbf{H}(\omega) = \frac{1}{\omega_s^2 - \omega^2 + 2\omega_s\omega\zeta_s j} \tag{3}$$

Extended Kalman Filter in Frequency Domain

Although the Kalman filter was originally formulated in the time domain, extensions have been made to the other domains such as the frequency and spatial domains. In a frequency-domain Kalman filter application, state variables, the system dynamic matrix and observation matrix are all functions of the *frequency* variable instead of time. Details of the formulation can be found in Shi et al. (1995) and Shi (1995).

In this application, the ground acceleration can be identified by either parametric or non-parametric (i.e., point by point) methods; the latter is the more general case in practice where the form of the excitation is truly unknown.

With a parametric input PSD, the state can be expressed quantitatively as:

$$\text{s} = \{\text{natural frequencies; damping ratios; input PSD parameters}\}^T \tag{4}$$

With a non-parametric input PSD, the input PSD itself can be expressed as a component of the state variable:

$$\mathbf{s} = \{\text{natural frequencies; damping ratios; input PSD}\}^T \tag{5}$$

The input PSD is identified at each step, at the same time as the frequency, and damping ratio estimates are updated. Since the estimate of $S_f(\omega)$ at the current frequency step is dependent only on its estimate at the previous step, the system dynamics equation (describing the evolution of the state vector $\mathbf{s}(\omega)$) is therefore represented as

$$\frac{d\mathbf{s}(\omega)}{d\omega} = \mathbf{F}_D\mathbf{s} + \mathbf{w}(\omega) = \mathbf{0} + \mathbf{w}(\omega) \qquad \mathbf{s} \sim N(\mathbf{s}(\omega_0), \mathbf{P}(\omega_0)) \tag{6}$$

where $\mathbf{w}(\omega)$ is assumed to be Gaussian process noise, with zero mean and covariance matrix $\mathbf{Q}(\omega)$.

In the matrix $\mathbf{Q}(\omega)$, the element associated with the state variable $S_f(\omega)$ represents the uncertainty in modeling the input PSD function as constant in the frequency domain, which is generally assigned a relatively large value considering the fact that the input PSD is not necessarily Gaussian white; the entries associated with system parameters represent the uncertainty in modeling the vibratory system as a classically-damped SDOF system, which should assume a small (if nonzero) value considering that an SDOF system is a very good representation of the real vibratory system. Note as well that the choice of replacing $\mathbf{F}_D$ with a zero matrix reflects the lack of knowledge of the true input PSD function (a common situation). If, however, the form of the input PSD function were somehow known (e.g., a Kanai-Tajimi model), then $\mathbf{F}_D$ could be suitably reformulated.

Simulation of Ground Motion

A Kanai-Tajimi random process model is assumed for the ground acceleration and used for simulation purposes:

$$\mathbf{S}_{\ddot{\mathbf{z}}}(\omega) = \frac{\phi_0(\omega_g^4 + 4\omega_g^2\omega^2\zeta_g^2)}{(\omega_g^2 - \omega^2)^2 + 4\omega_g^2\omega^2\zeta_g^2} \tag{7}$$

The Kanai-Tajimi earthquake model is derived from a simplified representation of the dynamic behavior of the uppermost layer between the ground surface and the nearest bedrock. The parameters ω_g and ζ_g represent the predominant ground frequency and damping ratio, respectively, while ϕ_0 denotes the intensity of the white noise exciting the soil layer at bedrock level.

Numerical Example

A SDOF system with ω_s = 12.0rad/sec and ζ_s = 0.05 was simulated under input parameters ϕ_0 = 1.0, ω_g = 15.708rad/sec and ζ_g = 0.60. The simulation was performed under cut-off frequency ω_c = 40.0rad/sec, 262144 time points with interval Δ_t = 0.005sec and 8344 frequency points with interval Δ_ω = 0.0047937rad/sec. The simulated input PSD,transfer function and output PSD are shown in Figure 1. The system was then identified using three approaches: (1) Non-parametric identification without knowledge of the input PSD; (2) Non-parametric identification with knowledge of the input PSD; and (3) Parametric identification with assumed knowledge of the *form* of the input PSD. The results are shown in Table 1 and the input identification in Case 1 is shown in Figure 2, after backward filtering and averaging were used to obtain an improved estimate of the input PSD.

Table 1. Identification of SDOF System and Excitation

Cases	Iter.	ω_s	ζ_s	ϕ_0	ω_g	ζ_g
Initial		7.8540	0.1200	0.7000	10.4720	0.4000
1	1st	12.3116	0.0430	-	-	-
	3rd	11.9904	0.0482	-	-	-
	5th	11.9835	0.0480	-	-	-
2	1st	12.0030	0.0493	-	-	-
	3rd	11.9970	0.0499	-	-	-
	5th	11.9967	0.0499	-	-	-
3	1st	13.7559	0.1301	1.3228	13.2031	0.6222
	10th	12.0037	0.0505	1.0074	15.6327	0.6010
	20th	11.9965	0.0500	0.9989	15.7069	0.6000
Theoretical		12.0000	0.0500	1.0000	15.7080	0.6000
Exact ACF *		15.7080	0.6000	1.0000	12.0000	0.0500
Estimated ACF *,**		15.4949	0.4169	0.8407	12.0046	0.0577

* Continuous time domain method (Conte, et al., 1995)

** Using estimated Auto-Correlation Function (ACF)

Conclusions

It is observed that the extended Kalman filter in the frequency domain gives very good results even when the input excitation is unknown. Also, the input excitation can be simultaneously and reliably identified with system characteristics from the output response only.

References

Conte, J.P. and Krishnan, S. (1995) "Modal Identification Method for Structures Subjected to Unmeasured Random Excitation." In em Proc. 10th Engineering Mechanics Specialty Conference, ASCE, Boulder, CO, 1995.

Shi, T. "Identification of System Characteristics and Random Excitation from Output Measurements." Dissertation submitted in partial fulfilment of the degree of Doctor of Philosophy, The Johns Hopkins University, March, 1995.

Shi, T., Jones, N.P., and Ellis, J.H. (1995) "Estimation of dynamic structural parameters and loading using output measurements." In em Proc. 10th Engineering Mechanics Specialty Conference, ASCE, Boulder, CO.

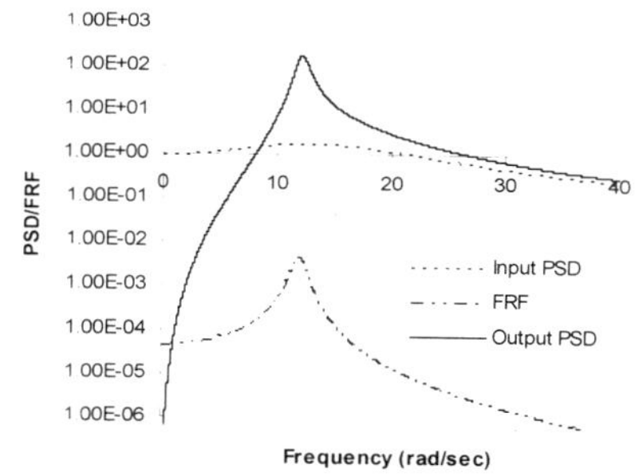

Figure 1: Input and Output PSD, Frequency Response Function

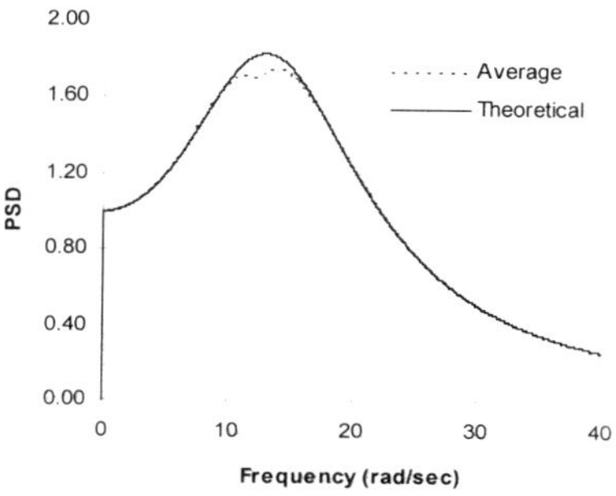

Figure 2: Comparison of Actual and Estimated Input Spectra

Nonlinear Identification of Semi-Active Control Devices

B.F. Spencer Jr.,[1] M. ASCE, S.J. Dyke,[1] Stud. M. ASCE
M.K. Sain[2] M. ASCE and J.D. Carlson[3]

Abstract

Semi-active control devices have received significant attention in recent years because they offer the adaptability of active control devices without requiring large, external power sources. To take advantage of the unique features of these devices for structural response reduction, control-oriented models first must be determined. Model development and identification are challenging tasks due to inherent device nonlinearities. This paper examines the identification problem for a magnetorheological (MR) damper, a semi-active control device that is particularly promising for civil engineering applications. Based on a recently developed model of the MR damper, an output-error method is proposed to solve the model parameter identification problem.

Introduction

Passive and active control systems represent the two ends of the spectrum in the use of supplemental damping strategies for response reduction in civil engineering structures subjected to strong earthquakes and severe winds. Semi-active control systems appear to combine the best features of both approaches, offering the reliability of passive devices, yet maintaining the versatility and adaptability of fully active systems.

Because of their mechanical simplicity, controllable fluid dampers have attracted considerable attention as a class of semi-active devices for structural response reduction. The essential characteristic of controllable fluids is their ability to reversibly change from a free-flowing, linear viscous fluid to a semi-solid with a controllable yield strength in milliseconds when exposed to an electric or magnetic field. Two fluids that are viable contenders for development of controllable dampers are: (i) electrorheological (ER) fluids and (ii) magnetorheological (MR) fluids. ER fluid dampers have been considered by a number of authors for civil engineering applications. Although the use of MR dampers for this purpose has more recent origins, the high yield strength, low viscosity, environmental robustness and stable hysteretic behavior of modern MR fluids make associated devices quite promising for structural response reduction (Spencer, *et al.*, 1996; Dyke, *et al.*, 1996; *http://www/nd.edu/~quake/*).

1. Dept. of Civil Engrg. and Geo. Sci., Univ. of Notre Dame, Notre Dame, IN 46556.
2. Dept. of Elect. Engrg., Univ. of Notre Dame, Notre Dame, IN 46556.
3. Lord Corp., Mech. Prod. Div., Thomas Lord Res. Center, Cary, NC 27511.

To develop effective strategies for use of MR dampers in dynamic hazard mitigation, a control-oriented model must be developed that can portray the salient features of the device. This paper discusses a nonlinear model recently proposed for this purpose and presents an output error method for model parameter identification.

Magnetorheological Dampers

A prototype MR damper, shown schematically in Fig. 1, has been obtained from the Lord Corporation for evaluation and testing (Carlson, 1994; Carlson and Weiss, 1994); Carlson, *et al.*, 1995; Spencer, *et al.*, 1996; see also *http://www. rheonetic.com/mr-fluid/*). The damper is 21.5 cm long in its extended position, and the main cylinder is 3.8 cm in diameter. The main cylinder houses the piston, the magnetic circuit, an accumulator and 50 ml of MR fluid, and the damper has a ±2.5 cm stroke. The magnetic field produced in the device is generated by a small electromagnet in the piston head. The current for the electromagnet is supplied by a linear current driver which generates a current that is proportional to the applied voltage. The peak power required is less than 10 watts, which would allow the damper to be operated continuously for more than an hour on a small camera battery. Forces of up to 3000 N can be generated with the device. The damper has fast response time, typically reaching rheological equilibrium in less than 10 msec after applying the magnetic field.

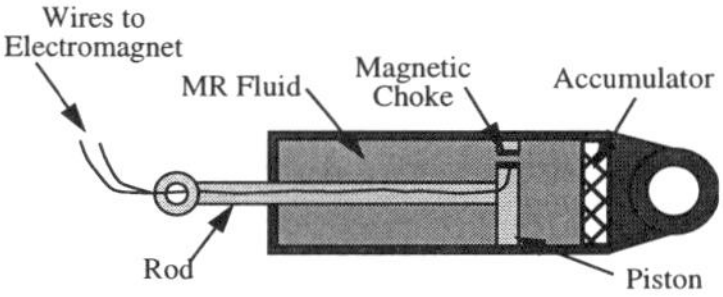

Fig. 1. Schematic of MR Damper.

The response of the MR damper due to a 2.5 Hz sinusoid with an amplitude of 1.5 cm is shown in Fig. 2 for four constant voltage levels, 0 V, 0.75 V, 1.5 V, and 2.25 V, being applied to the current driver for the device. These voltages correspond to 0 A, 0.25 A, 0.5 A and 0.75 A, respectively. Saturation of the MR effect begins in the tested device when the applied voltage is 2.25 V (0.75 A). Thus, attention is restricted to voltages between 0 and 2.25 V. The forces generated are shown as a function of time in Fig. 2a, the force-displacement loops are shown in Fig. 2b, and the force-velocity loops are shown in Fig. 2c. Note that the force-displacement loops in Fig. 1b progress along a clockwise path with increasing time, whereas the force-velocity loops in Fig. 2c progress along a counter-clockwise path with increasing time.

The simple mechanical idealizations of the MR damper depicted in Fig. 3 has been shown to accurately predict the behavior of the prototype MR damper over a broad range of inputs (Spencer, *et al.*,

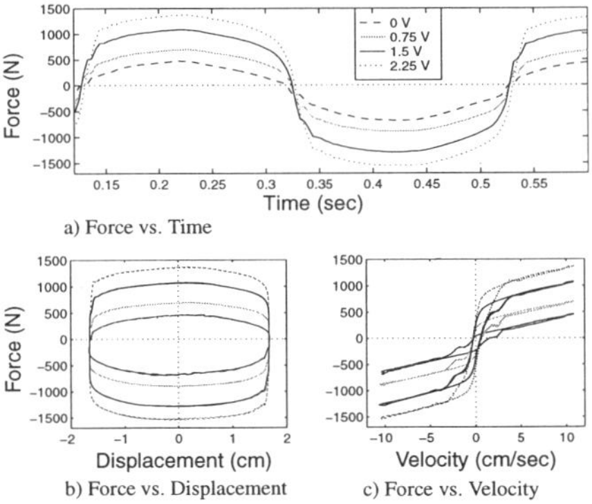

Fig. 1. Experimentally Measured Force for 2.5 Hz Sinusoidal Excitation with an Amplitude of 1.5 cm.

1996). The equations governing the force f predicted by this model are given by

$$f = c_1\dot{y} + k_1(x - x_0) \tag{1}$$

$$\dot{z} = -\gamma|\dot{x} - \dot{y}|z|z|^{n-1} - \beta(\dot{x} - \dot{y})|z|^n + A(\dot{x} - \dot{y}) \tag{2}$$

$$\dot{y} = \frac{1}{c_0 + c_1}\{\alpha z + c_0\dot{x} + k_0(x - y)\} \tag{3}$$

In this model, z is an the evolutionary variable that accounts for the history dependence of the response.

A general model which is valid for fluctuating magnetic fields can be formulated by considering the dependence of the model parameters on the voltage applied to the current driver. Spencer, *et al.* (1996) have suggested this dependence is of the form

$$\alpha = \alpha_a + \alpha_b u, \quad c_1 = c_{1a} + c_{1b}u, \quad c_0 = c_{0a} + c_{0b}u \tag{4}$$

where u is given as the output of a first-order filter given by

$$\dot{u} = -\eta(u - v) \tag{5}$$

and v is the commanded voltage sent to the current driver. Eq. (5) is necessary to model the dynamics involved in reaching rheological equilibrium.

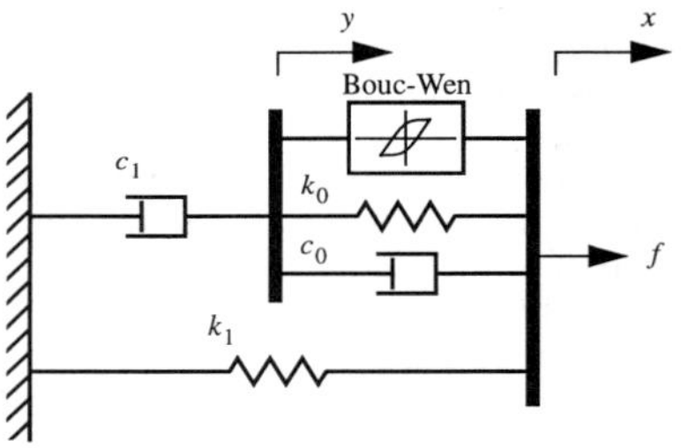

Fig. 2. Simple Mechanical Model of the MR Damper.

Parameter Identification

The basic problem is to estimate the parameters of a nonlinear two-input (*i.e.,* the applied voltage u, and the displacement x), single-output (*i.e.*, the force f) model of the damper. Because all of the states of the model are not measurable, the available methods for model parameter identification are limited. A least-squares output-error method can be employed in conjunction with a constrained nonlinear optimization procedure to estimate the 14 model parameters in Eqs. (1–5). In this example, the objective is to minimize the root mean square of the output error over n measured responses, *i.e.,*

$$E_{\text{rms}} = \left\{\sum_{i=1}^{n} \frac{1}{T_i}\int_0^{T_i}\{f_{ei}(t) - f_i(t)\}^2 dt\right\}^{1/2} \tag{6}$$

where $f_{ei}(t)$, $f_i(t)$ and T_i are the measured force, the force predicted by the model and the length of the measured record, respectively, for the *i*th experiment considered.

A set of parameters was determined to fit the generalized model in Eqs. (1–5) to the experimental data in three tests (*i.e.*, $n = 3$): (i) step voltage response of the damper under a constant piston velocity, (ii) constant voltage response under random piston displacement, and (iii) random voltage response under random piston displacement.

The optimization was performed using the sequential quadratic programming algorithm available in MATLAB (1994). The resulting parameters are: c_{0a} = 20.2 N·sec/cm, c_{0b} = 2.68 N·sec/cm/V, k_0 = 15 N/cm, c_{1a} = 350 N·sec/cm, c_{1b} = 70.8N·sec/cm/V, k_1 = 5.37 N/cm, x_0 = 18.4 cm, α_a = 44.9, α_b = 638 V^{-1}, γ = 39.3 cm^{-2}, β = 39.3 cm^{-2}, A = 47.2, n = 2, η = 251 sec^{-1}.

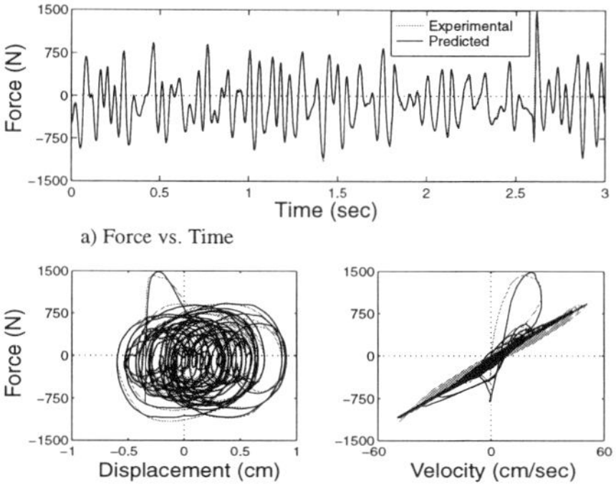

a) Force vs. Time

b) Force vs. Displacement

c) Force vs. Velocity

Fig. 1. Comparison of the Model Results and the Experimental Data for the Random Voltage, Random Displacement Test.

Fig. 1 shows a comparison between the experimental and predicted force generated by the MR damper. The applied displacement was a 0-20 Hz broadband Gaussian excitation with an rms value of 0.34 cm and a peak of 1.1 cm. The voltage to the current driver was a 0-20 Hz lognormal random excitation with a mean of 0.052 V, an rms of 0.11 V and a maximum value of 2.4 V. As can be seen, the comparison is excellent. The interested reader is directed to Spencer, *et al.* (1996) for more information.

Acknowledgment

This research is supported in part by National Science Foundation Grant Nos. CMS 93–01584 and CMS 95–00301. In addition, the authors from Notre Dame would like to express their appreciation to the Lord Corporation of Cary, North Carolina for providing the prototype magnetorheological damper. In particular, the encouragement and advice of Mr. Thomas Loftus is gratefully acknowledged.

References

Carlson, J.D. (1994). "The Promise of Controllable Fluids." *Proc. of Actuator 94* (H. Borgmann and K. Lenz, Eds.), AXON Technologie Consult GmbH, pp. 266–270.

Carlson, J.D. and Weiss, K.D. (1994). "A Growing Attraction to Magnetic Fluids," *Mach. Des.*, August, pp. 61–64.

Carlson, J.D., Catanzarite, D.M. and St. Clair, K.A. (1995). "Commercial Magneto-Rheological Fluid Devices." *Proc. 5th Int. Conf. on ER Fluids, MR Fluids and Associated Tech.*, Univ. Sheffield, UK.

Dyke, S.J., Spencer Jr., B.F., Sain, M.K. and Carlson, J.D. (1996). "Seismic Response Reduction Using Magnetorheological Dampers." *Proc. IFAC,* San Francisco, California.

MATLAB (1994). The Math Works, Inc. Natick, Massachusetts.

Spencer Jr., B.F., Dyke, S.J., Sain, M.K. and Carlson, J.D. (1996). "Phenomenological Model of a Magnetorheological Damper." *J. of Engrg. Mech., ASCE* (in press).

Identification of Nonlinear Systems under Random Excitation

B.A. Zeldin[1] and P.D. Spanos[2]

Abstract. The paper introduces a method of identification of non-linear systems under random excitation. The non-linearity is accounted for by a combination of linear subsystems and known zero-memory non-linear transformations; a linear non-anticipative multi-input single-output (MISO) system is developed. This MISO system can be readily represented by using an Exogeneous Auto-Regressive Moving-Average (ARMAX) discrete-time model; a two-stage method for identification of the developed ARMAX model is adopted. A numerical example involving the identification of unknown parameters of a randomly excited Duffing oscillator elucidates the usefulness of the proposed method.

Key words: non-linear system identification, ARMAX models, Duffing oscillator,

1. Introduction

A short survey of methods of identification of linear systems can be found in [5]. Further, several approaches, such as those presented in Refs. [1,2,4,6], have been proposed to determine parameters of non-linear systems subjected to random and/or deterministic excitation.

This paper proposes a non-linear identification method which is based on the interpretation of the measurements as the inputs and output of an associated linear system lending itself to parametric ARMAX identification. Specifically, the non-linear restoring force is represented as a set of known zero-memory non-linear transformations that are combined with linear subsystems; it reduces a single-input-single-output nonlinear system to a linear MISO system. This method is particularly efficient if preliminary studies of the non-linear system may be used to speculate with regards to the mathematical form of the restoring force. The identification of the equivalent linear MISO system is performed by using a parametric procedure; the linear subsystems are represented by using an ARMAX model. Note that parametric methods of system identification have obvious advantages over the non-parametric frequency domain methods which are based on windowed periodograms [1]; the transfer functions associated with dynamic structural systems usually have sharp peaks which can be smeared out by improperly selected windows.

2. Non-linear System Modeling and Simplifications

Random vibrations of nonlinear systems can be described by the differential equation

$$m\ddot{x} + g(\dot{x}, x) = f(t) \ , \tag{1}$$

where $x(t)$ is the system response, g is the nonlinear restoring force, and $f(t)$ is the random excitation; the dot signifies differentiation with respect to time. This paper considers systems with nonlinearities expressible as a superposition of zero-memory non-linear transformations and linear subsystems. That is,

[1] Research Associate, Mech.Eng., Rice University, P.O.Box 1892, Houston, Texas 77251.
[2] L.B.Ryon Chair in Eng., Rice University, P.O.Box 1892, Houston, Texas 77251.

$$\sum_{k=0}^{l-1} P_k\left(\frac{d}{dt}\right)x_k(t) = f(t), \tag{2}$$

where $P_k(s)$ are polynomials, and $x_k(t) = g_k(\dot{x}, x)$ are non-linear zero-memory transformations of the response with $x_0(t) = x(t)$. Note that by using concepts of modern theory of linear *behavioral* systems [9], Eq. (2) can be interpreted as a linear AR time-invariant system. Indeed, the vector $\bar{w} = (x_0, \ldots x_{l-1}, f)^T$ can be treated as the vector of external variables describing the linear system associated with Eq. (2). Then, Eq. (2) can be rewritten as

$$R(d/dt)\bar{w}(t) = 0 , \tag{3}$$

where $R(s) = [P_0(s), \ldots P_{l-1}(s), -1]$ is the polynomial matrix containing just a single row. Note that the rank of the matrix $R(s)$ is always equal to one for all values of the complex variable s, since the last component of this matrix is equal to -1. Clearly, the system of Eq. (3) is *controllable* and it allows steering its external variables to any desirable *future*.

It is known that the external variable of any AR system described by Eq. (3) can be partitioned into the input $\bar{u}(t)$ and output $y(t)$ variables [9]. The element $x_i(t)$ of the vector $\bar{w}(t)$ associated with the highest degree polynomial function $P_i(s)$ can be taken to be the output variable $y(t)$ yielding the, so-called, *non-anticipative* input-output system [9]. Without loss of generality it can be assumed that $y(t) = x_0(t)$. The following transfer function

$$y = G(s)\bar{u} = P_0^{-1}(s)[-P_1(s), \ldots -P_{l-1}(s), 1](x_1, \ldots x_l, f)^T \tag{4}$$

completely characterizes the linear system of Eq. (3). Note that the transfer function $G(s)$ is a *proper* rational matrix [9].

Eq. (4) provides a representation of a non-linear system as a linear non-anticipative MISO system. Note the difference in the representation of the equivalent linear MISO system of Eq. (4) and one proposed in Ref. [1]. Here, the random excitation $f(t)$ is treated as one of the components of the input vector, and the element $x_i(t)$ associated with the highest degree polynomial function $P_i(s)$ is taken to be the output of the equivalent linear system.

3. Two Stage ARMAX Identification of an Equivalent MISO System

An input/output non-anticipative linear system always has the input/state/output (ISO) representation [9]

$$\dot{\bar{q}} = A\bar{q} + B\bar{u} \ , \quad y = C\bar{q} + D\bar{u}, \tag{5}$$

where the vector $\bar{q}$ represents the state variables of the system. The ISO representation of Eq. (5) can be properly discretized and condensed to the standard discrete-time ARMA model associated with Eq. (4). That is,

$$\sum_{l=0}^{p} a(l)y(t-l) = \sum_{k=1}^{l}\sum_{l=0}^{q} b_k(l)u_k(t-l) \ , \tag{6}$$

where $y(l)$ and $u_k(l)$ are sampled values of the corresponding continuous functions, and $a(l)$ and $b_k(l)$ are the AR and MA parameters, respectively, with $a(0) = 1$. The transfer functions

$$G_k(z) = B_k(z)/A(z) = \sum_{l=0}^{q} b_k(l)z^{-l} / \sum_{l=0}^{p} a(l)z^{-l} \tag{7}$$

can be viewed as an approximation of the continuous time transfer functions $G_k(i\omega) = P_0^{-1}(i\omega)P_k(i\omega)$; here, $z = e^{i\omega\Delta t}$, and Δt is the sampling interval.

Eq. (6) provides a discrete-time model for the non-linear system of Eq. (2). However, it does not account for some additional disturbances resulting from non-measured forces acting on the system of Eq. (1) and measurement noises. These disturbances can be properly modeled by introducing an exogeneous input into Eq. (6). This yields the ARMAX model

$$\sum_{l=0}^{p} a(l)z(t-l) = \sum_{k=1}^{l}\sum_{l=0}^{q} b_k(l)u_k(t-l) + \sum_{l=0}^{q_e} c(l)e(t-l), \tag{8}$$

where $e(k)$ is a white noise process. That is, $E[e(k)e(l)] = 2\pi\delta_{kl}/\Delta t$, where δ_{kl} is the Kronecker delta.

ARX models are derivable form ARMAX models by setting q_e equal to 0. An ARX model can be readily identified by minimizing the power of the exogeneous disturbance

$$E[e^2(t)] = E[(\sum_{l=0}^{p} a(l)z(t-l) - \sum_{k=1}^{l}\sum_{l=0}^{q} b_k(l)u_k(t-l))^2]. \tag{9}$$

This criterion yields the system of linear equations

$$\sum_{l=0}^{p} a(l) R_{zz}(l-j) - \sum_{k=1}^{l} \sum_{l=0}^{q_k} b_k(l) R_{zu_k}(j-l) = 0$$
$$\sum_{l=0}^{p} a(l) R_{zu_m}(l-j) - \sum_{k=1}^{l} \sum_{l=0}^{q_k} b_k(l) R_{u_m u_k}(j-l) = 0 \qquad (10)$$

for the ARMA parameters of equation (8). The symbols $R_{(\circ)}(l)$ denote the corresponding correlation functions; for the implementation of the method they must be estimated from the data.

Unfortunately, this "direct" ARX model yields a biased estimate of the ARMAX parameters. As an alternative, this paper utilizes a two-stage linear identification procedure similar to one used in Ref. [3] for ARMA identification of structural system, and in Ref. [7] for parametric modeling of stochastic processes. First, the ARMAX system of Eq. (8) is approximated by a long ARX (p^*, q^*) model where $p^* > p$, $q^* > q$. The long ARX model is determined by solving the system of linear equations (10). Then, ARMAX parameters are estimated by reducing this long ARX model and removing the extraneous degrees of freedom. The following error criterion

$$\int_{-2\pi/\Delta t}^{2\pi/\Delta t} |G_{ARX}(z) - G_{ARMAX}(z)|^2 d\omega \qquad (11)$$

is used in this regard. An efficient iterative procedure to solve this non-linear minimization problem was developed in Ref. [8]. The physical structural parameters associated with the non-linear dynamic system of Eq. (1) can be readily found from the parameters of the ARMA portion of the derived ARMAX model.

4. A Numerical Example

A randomly excited Duffing oscillator

$$m\ddot{x} + c\dot{x} + k(1 + \varepsilon x^2)x = f(t) \qquad (12)$$

is considered to illustrate the concepts of the method. The random excitation $f(t)$ is modeled as white noise with unit spectral density. A simulation study was performed by selecting the values $m = 1$, $c = 0.2$, $k = 1$, and $\varepsilon = 0.25$. The response of the Duffing oscillator was evaluated by numerical integration of Eq. (12); the obtained records were sampled with time step $\Delta t = 0.25$, and additional noise was added to simulate the measurement noise corresponding to *signal to noise ratio* equal to 40dB; that is, the standard deviation of the added discrete white noise was equal to 10% of the standard deviation of the corresponding signal.

Following section 2, auxiliary inputs $x_0(t) = x(t)$, $x_1(t) = x^3(t)$ are taken with corresponding polynomial matrices $P_0(s) = ms^2 + cs + k$ and $P_1(s) = \varepsilon k$. The equivalent linear non-anticipative MISO system has the variable x_0 as its output and the variables $x_1(t)$ and $f(t)$ as its inputs. The transfer function of this linear system is the matrix containing two elements. That is,

$$\begin{bmatrix} G_1(s) & G_2(s) \end{bmatrix} = \begin{bmatrix} -\dfrac{\varepsilon k}{ms^2 + cs + k} & \dfrac{1}{ms^2 + cs + k} \end{bmatrix}. \qquad (13)$$

The results of the ARX modeling of the transfer functions $G_1(s)$ and $G_2(s)$ are shown in Figures 1 and 2, respectively. Clearly, the "direct" ARX(3,3) model yields a biased estimate of the system transfer functions, whereas the "long" ARX(9,9) model provides a quite accurate approximation. The ARX(9,9) model for the transfer function $G_2(s)$ is then reduced to the ARMA(3,3) model by using the error criterion of Eq. (11) and the numerical algorithm of Ref. [8]; related results are shown in Figure 3.

The natural frequency and the damping of the linear system associated with the transfer function $G_2(s)$ are found by relating the poles of the continuous-time and corresponding discrete-time ARMA(3,3) models [3]. Specifically,

$$\omega_n = \sqrt{k/m} = \sqrt{\delta^2 + \lambda^2}/\Delta t \ , \quad \xi = c/(2\sqrt{km}) = \delta/(\omega_n \Delta t), \qquad (14)$$

where $\delta = -ln(|\mu|)$, $\lambda = arg(\mu)$ with μ denoting the root of the denominator of the rational function $G_2(s)$ associated with the ARMA(3,3) model. The mass m and the nonlinearity coefficient ε are estimated by equating the total power associated with the continuous and discrete models of the transfer functions $G_1(s)$ and $G_2(s)$. The results of these computations are $\hat{m} = 1.034$, $\hat{k} = 1.041$, $\hat{c} = 0.211$, and $\hat{\varepsilon} = 0.239$.

5. Concluding Remarks

The presented results show that the proposed method merits additional attention regarding reliability and computational efficiency aspects. Also, its applicability to multi-degree-of-freedom systems may be pursued.

Acknowledgment

The financial support of this work by the grant DE-FG03-95ER14500 from the Department of Energy is gratefully acknowledged.

References

1. Bendat, J.S., Palo, P.A., and Coppolina, R.N., (1992). "A General Identification Technique for Nonlinear differential Equations of Motion", *Probabilistic Engineering Mechanics*, 7, 43-61.
2. Billings, S.A., Tsang, K.M., and Tomlinson, G.R., (1983). "Application of the NARMAX methods to Nonlinear Frequency Response Estimation", *Proceedings of the 6th International Modal Analysis Conference*, 1-6.
3. Gersch, W., Taoka, G.T., and Liu, R., (1976). "Structural System Parameter Estimation by Two-Stage Least-Squares Method", *ASCE Journal of Engineering Mechanics*, 102, 883-899.
4. Gifford, S.J., and Tomlinson, G.R., (1989). "Recent Advances in the Application of Functional Series to Nonlinear Structures", *Journal of Sound and Vibration*, 135, 289-317.
5. Kozin, F., and Natke, H.G., (1986). "System Identification Techniques", *Structural Safety*, 3, 269-316
6. Masri, S.F., Sassi, H., and Caughey, T.K., (1982). "Nonparametric Identification of Nearly Arbitrary Nonlinear Systems", *ASME Journal of Applied Mechanics*, 49, 619-628.
7. Mignolet, M.P., and Spanos, P.D., (1987). "Recursive Simulation of Stationary Multi-Variate Random Processes - Part I", *Journal of Applied Mechanics*, 109, 674-680.
8. Spanos, P.D., and Zeldin, B.A., (1996) "Efficient Iterative ARMA Approximation of Multi-Variate Random Processes for Structural Dynamics Applications", *International Journal of Structural Dynamics and Earthquake Engineering* (to appear).
9. Willems, J.C., (1991). "Paradigms and Puzzles in the Theory of Dynamical Systems", *IEEE Transactions on Automatic Control*, 36(2), 259-294.

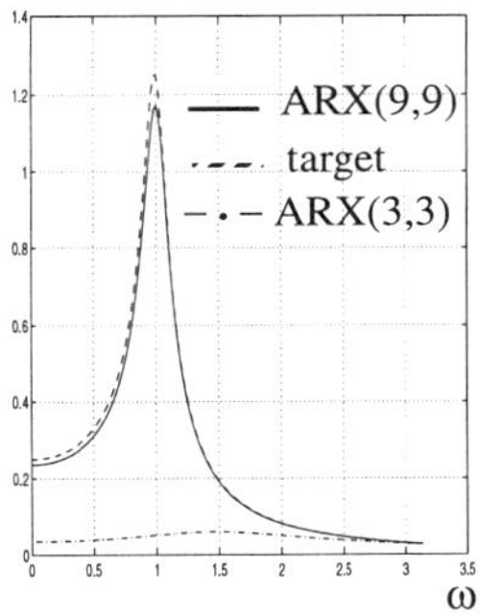

Figure 1.
Transfer function $|G_1(i\omega)|$

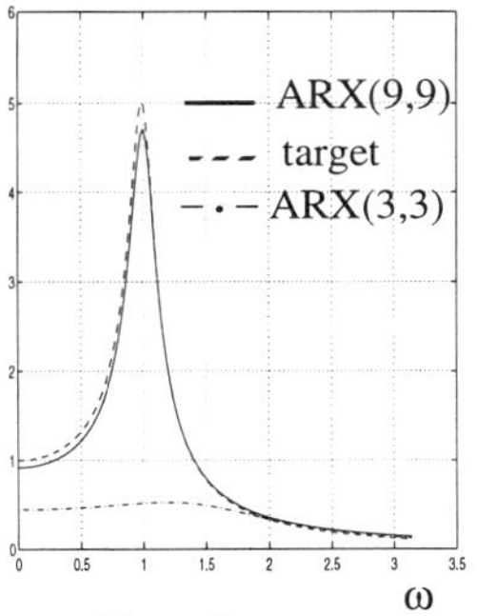

Figure 2.
Transfer function $|G_2(i\omega)|$

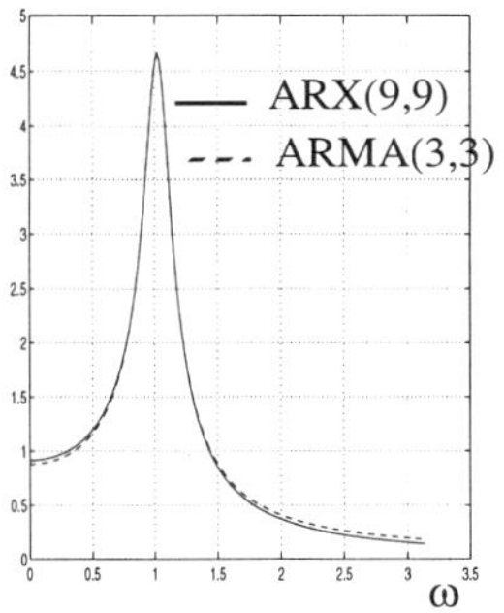

Figure 3.
Transfer function $|G_2(i\omega)|$

Soil-Structure Interaction for Base-Isolated Buildings

Maria I. Todorovska[1]

Abstract:

In earthquake resistant design of base-isolated buildings, the dynamic soil-structure interaction, the wave passage and the contribution of the higher modes of vibration are commonly ignored. This paper considers possible consequences of such simplifications. It is concluded that fixed-base models excited by horizontal shaking only may not be adequate in determining design loads for such buildings.

Introduction:

In controlled conditions, during experiments, the performance of models of base-isolated buildings has been found to be satisfactory. However, these experiments consider oversimplified excitation (horizontal driving motion at the base). In the near-field of a strong earthquake, large vertical acceleration and rotational components of ground motion can be expected. So far, base-isolated buildings have not been tested under such conditions. Moreover, with few exceptions (e.g., Wolf and Obernhuber 1981; Wolf et al. 1983; Todorovska and Trifunac 1993) numerical analyses also neglect the effects of soft soil and wave passage. This paper presents results of behavior for a simple structural model, base-isolated shear-beam, but on flexible soil and excited by more realistic excitation (plane SV-waves under oblique incidence).

Model

The soil-structure interaction model is two-dimensional and linear (Fig. 1). The soil is represented by a homogeneous half-space, the foundation is semi-circular, and assumed to be rigid, and the building is represented by a shear-beam. Expansion in cylindrical wave functions is used to represent the scattered and the radiated field (Todorovska 1993a). The base-isolators are modeled as massless rigid bars connected by springs and dashpots to the foundation and by pins to a thin rigid platform supporting the shear-beam. Base-isolated shear-beam model and equivalent SDOF oscillator building models were analyzed in more detail, by Todorovska and Trifunac (1993). They also present exact equations of motion, including both geometric and material non-linearity of the oscillator and the isolators, and coupling of translations

[1] Res. Assist. Prof., Univ. Southern California, Dept. of Civil Engng, Los Angeles, California 90089-2531

and rotations with vertical motions. Soil structure effects for the same models but without isolators were studied by Todorovska (1993b), Todorovska and Trifunac (1992) and Todorovska (1992).

Results

For a three storey building model (height H_{sb}=15 m, width W_{sb}=15 m, fundamental natural period $T_1 = 0.3$ sec, damping ratio $\varsigma = 0.02$), response was calculated for the case when only dynamic soil-structure interaction is considered, and when the wave passage is also considered. These results are then compared with those of a fixed-base model excited by horizontal translation. The foundation has same width as the shear-beam ($W_{sb} = 2a$), and the ratio of the densities of the building, ρ_b, of the foundation, ρ_f, and of the soil, ρ_s, is $\rho_b/\rho_s = \rho_f/\rho_s = 0.2$. The isolators have height $h_i = 0.4$ m, damping ratio is $\varsigma_i = 0.08$, and the mass of the platform is such that $m_p/m_b = 0.25$. The ratio $\varepsilon_i = \omega_i/\omega_1$, where $\omega_1 = 2\pi/T_1$ and ω_i is the fixed-base frequency of the model if the shear-beam is rigid, is used to express the stiffness of the isolators. Value $\varepsilon_i = 0.58$ is considered as typical for the linear analysis in this paper. Fig. 2 shows results for horizontal driving motion and soil-structure interaction effects ignored, Fig. 3–for dynamic soil-structure interaction considered but wave passage ignored, and Fig. 4–for both dynamic soil-structure interaction and wave passage considered (plane SV-wave, incidence at 30° off vertical). In all figures, $\eta = \frac{2a}{\beta T} = \frac{\omega a}{\pi\beta}$ is dimensionless frequency and $\beta = 250$ m/sec is soil shear-wave velocity. u_b^{rel} and $\alpha_i h_i$ are as indicated in Fig.1, $f_{x,O_p}^{(ss)}$ is shear force acting on both isolators, and $M_O^{(ss)}$ is overturning moment of the base-isolated structure about point O.

Discussion and Conclusions

The results indicate that the dynamic soil-structure interaction results in further shift of the first peak of the transfer-function towards longer periods (softening of the system), and in further reduction of its amplitude. The higher modes are not much affected. Due to wave passage, the foundation is driven by rocking, in addition to translations, the amount of which depends on type of incident waves and on dimensionless frequency η. The amplitude of rocking motion increases with frequency up to a limit when the short wavelength excitation starts to be filtered out. The higher frequency modes are more affected by the rocking and may be significantly amplified, especially the second mode. Excluding their contribution to the response may underestimate the design loads. It is concluded that the wave passage, and the dynamic soil-structure interaction must be considered in detailed analyses of base-isolated buildings.

References:

Todorovska, M.I. (1992). Effect of the depth of the embedment on the system response during building-soil interaction, *Soil Dyn. & Earthqu. Eng.*, **11**(2), 111–123.

Todorovska, M.I. (1993a). In-plane foundation-soil interaction for embedded circular foundations, *Soil Dyn. & Earthqu. Eng.*, **12**(5), 283–297.

Todorovska, M.I. (1993b). Effects of the wave passage and the embedment depth during building-soil interaction, *Soil Dyn. & Earthqu. Eng.*, **12**(6), 343–355.

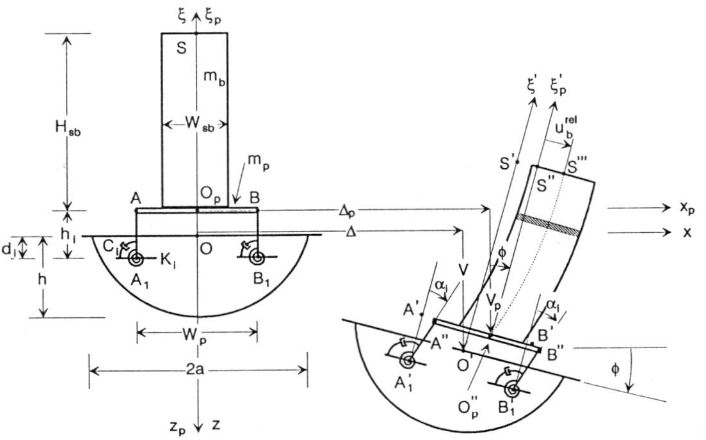

Fig. 1

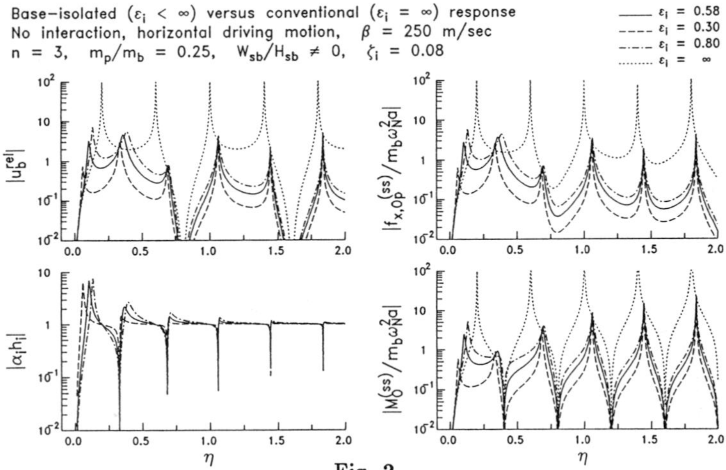

Fig. 2

Todorovska, M.I., & M.D. Trifunac (1992). The system damping, the system frequency and the system response peak amplitudes during in-plane building-soil interaction, *Earthqu. Eng. & Struct. Dyn.*, **21**(2), 127–144.

Todorovska, M.I., & M.D. Trifunac (1993). The effects of the wave passage on the response of base-isolated buildings on rigid embedded foundations, Rep. No. 93-10, Dept. Civil Eng., Univ. of Southern California, Los Angeles, California.

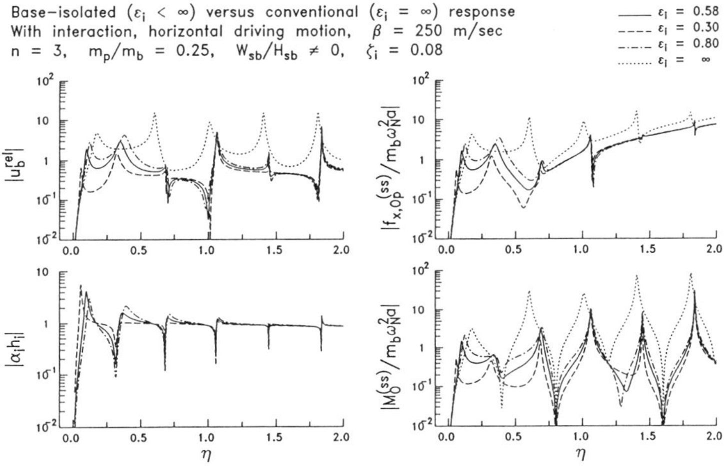

Fig. 3

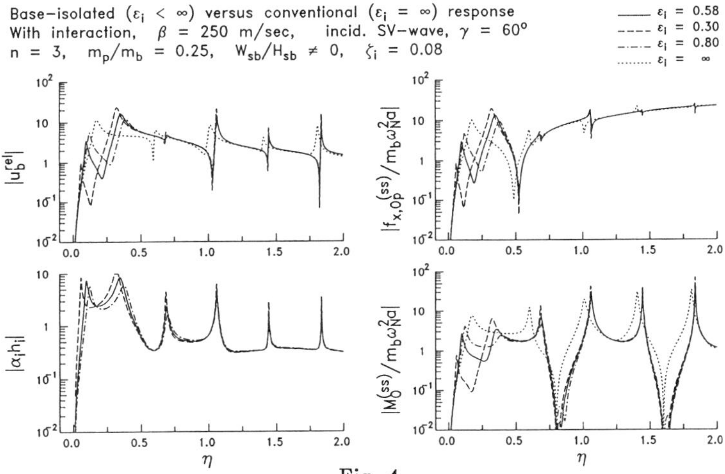

Fig. 4

Wolf, J.P. & P. Obernhuber (1981). Effects of horizontally propagating waves on the response of structures with a soft first story, *Earthqu. Eng. & Struc. Dyn.*, **9**, 1–21.

Wolf, J.P., P. Obernhuber & B. Webber (1983). Response of a nuclear power plant on aseismic bearings to horizontally propagating waves. *Earthqu. Eng. & Struc. Dyn.*, **11**, 483–499.

CONSISTENT INFINITESIMAL FINITE-ELEMENT CELL METHOD - A BOUNDARY FINITE-ELEMENT PROCEDURE

CHONGMIN SONG[1] and JOHN P. WOLF[1]

Abstract

The consistent infinitesimal finite-element cell method, a boundary-element method based on finite elements, models unbounded and bounded media, combining the advantages of the boundary-element and finite-element methods. For a bounded medium the static-stiffness and mass matrices with respect to the nodes on the boundary are constructed.

Unbounded Medium

Inspired by the cloning algorithm pioneered by Dasgupta (1982), the consistent infinitesimal finite-element cell method to analyze unbounded media for hyperbolic, parabolic and elliptic partial differential equations has been developed. This method is described in great detail in the book by Wolf and Song (1996). Wave propagation analyzed in the time domain is also addressed in Song and Wolf (1996a) with a summary presented at this conference (Song and Wolf, 1996b).

The consistent infinitesimal finite-element cell method, which is based solely on the finite-element method, captures the radiation condition without using a fundamental solution as is necessary in the boundary-element method. Only the structure-medium interface is discretized leading to a reduction of the spatial dimension by one. This method converges to the exact solution in the finite-element sense on the structure-medium interface and is exact in the radial direction towards infinity. The consistent infinitesimal finite-element cell method is thus a *boundary finite-element method* combining the advantages of the boundary-element method and the finite-element method.

Bounded Medium

The consistent infinitesimal finite-element cell method can also analyze a bounded medium. Again, similarity and finite-element assemblage are used in the derivation (Figure 1) with the limit of the infinitesimal cell width performed analytically. The concept is outlined in Song and Wolf (1996b).

[1]Engineer, Institute of Hydraulics and Energy, Department of Civil Engineering, Swiss Federal Institute of Technology Lausanne, CH-1015 Lausanne, Switzerland

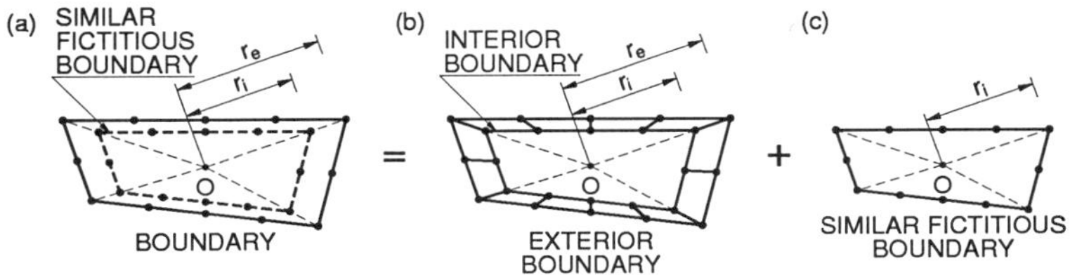

Figure 1. Concept of Consistent Infinitesimal Finite-Element Cell Method for a Bounded Medium with Infinitesimal Cell Width Leading to Finite-Element Discretization of Boundary Only

The consistent infinitesimal finite-element cell equation in the frequency domain for a bounded medium with the unknown dynamic-stiffness matrix $[S^b(\omega)]$ equals

$$(-[S^b(\omega)] + [E^1])[E^0]^{-1}(-[S^b(\omega)] + [E^1]^T) + (s-2)[S^b(\omega)] + \omega\frac{\mathrm{d}[S^b(\omega)]}{\mathrm{d}\omega} - [E^2] + \omega^2[M^0] = 0 \quad (1)$$

with the spatial dimension s (= 2 or = 3) and the coefficient matrices $[E^0]$, $[E^1]$, $[E^2]$ and $[M^0]$ calculated similarly as the static-stiffness and mass matrices of a finite element on the structure-medium interface (Wolf and Song, 1996).

Static-Stiffness and Mass Matrices

For use in a practical procedure to perform a dynamic analysis of a bounded medium, the static-stiffness matrix $[K^b]$ and mass matrix $[M^b]$ are determined. Substituting the low-frequency expansion

$$[S^b(\omega)] = [K^b] - \omega^2[M^b] \quad (2)$$

in (1) yields the eigenvalue problem with eigenvalues $\lceil\lambda\rfloor$ and eigenvectors $[\Phi]$

$$\begin{bmatrix} [E^0]^{-1}[E^1]^T - \frac{s-2}{2}[I] & -[E^0]^{-1} \\ -[E^2] + [E^1][E^0]^{-1}[E^1]^T & -[E^1][E^0]^{-1} + \frac{s-2}{2}[I] \end{bmatrix} \begin{bmatrix} [\Phi_{11}] & [\Phi_{12}] \\ [\Phi_{21}] & [\Phi_{22}] \end{bmatrix} = \begin{bmatrix} [\Phi_{11}] & [\Phi_{12}] \\ [\Phi_{21}] & [\Phi_{22}] \end{bmatrix} \begin{bmatrix} \lceil\lambda\rfloor & \\ & -\lceil\lambda\rfloor \end{bmatrix} \quad (3)$$

The static-stiffness matrix follows as

$$[K^b] = [\Phi_{21}][\Phi_{11}]^{-1} \quad (4)$$

and the mass matrix is calculated solving the linear equation in $[M^b]$

$$\left((-[K^b] + [E^1])[E^0]^{-1} - \frac{s}{2}[I]\right)[M^b] + [M^b]\left([E^0]^{-1}(-[K^b] + [E^1]^T) - \frac{s}{2}[I]\right) + [M^0] = 0 \quad (5)$$

The analytical expression for the static displacements at an interior point equals

$$\{u\} = [\Phi_{11}] \left[\left(\frac{r}{r_0}\right)^{-\lambda_i - \dfrac{s-2}{2}} \right] [\Phi_{11}]^{-1}\{u_0\} \tag{6}$$

where $\{u_0\}$ are the displacements on the boundary with the characteristic length r_0.

Benchmark Examples

Unbounded media are calculated with the consistent infinitesimal finite-element cell method in Song and Wolf (1996b) and in Wolf and Song (1996). The first example of a bounded medium consists of a cantilever (Figure 2) with one end fixed of length $l = 24m$, height $h = 6m$, Poisson's ratio =0.2 and $\sqrt{E/\rho} = 100m/s$ (Young's modulus E, mass density ρ). 1 3-node element is used over the height and n_l elements over the length l. $[K^b]$ and $[M^b]$ of order $8(n_l + 1) \times 8(n_l + 1)$ follow from (4) and (5). After introducing the boundary condition at the fixed end, the natural frequencies ω_i are calculated. The periods $T_i = 2\pi/\omega_i$ are plotted in Figure 2, where the solution using a very fine finite-element mesh is also specified (dotted line). The consistent infinitesimal finite-element cell method yields higher accuracy than a boundary-element solution (Nardini and Brebbia, 1984).

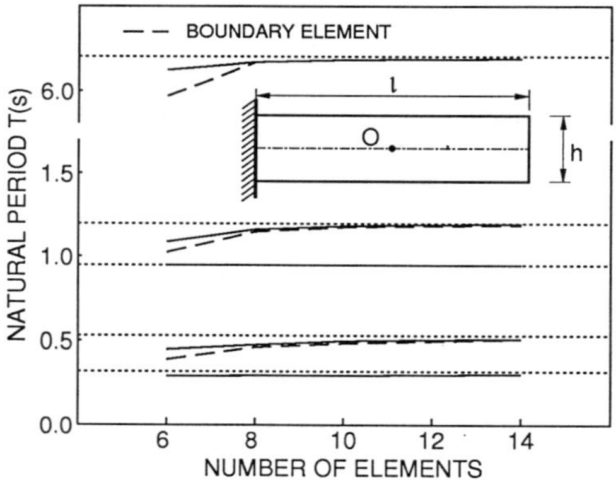

Figure 2. Natural Periods of Cantilever in Plane Stress

The second example addresses an edge-cracked plate (Figure 3). The similarity center O is chosen at the tip of the crack. The straight crack faces are traction free. As the latter pass through the similarity center, no discretization is necessary, in contrast to both the boundary-element method and the finite-element method. This not only reduces the number of degrees of freedom, but

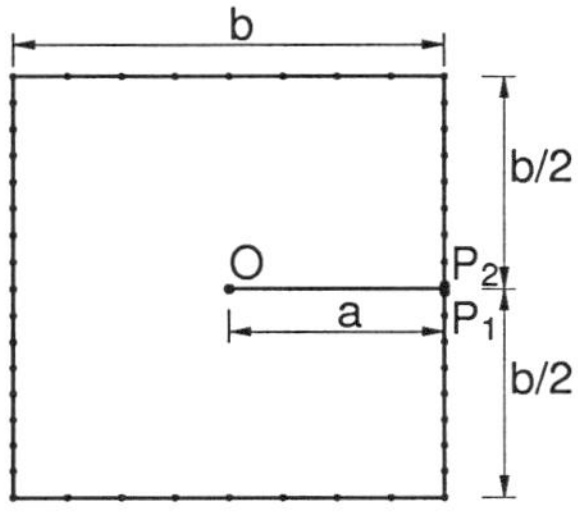

Figure 3. Edge-Cracked Plate

also eliminates the error due to the approximation introduced through the discretization of the crack faces. At the intersection point of each crack face with the side of the plate, a node is chosen. The two nodes P_1 and P_2 with the same coordinates are independent. The discretization of the sides with 24 3-node line elements is shown in the figure. The displacement in the interior of the plate is described in (6) with r measured from the crack tip. For $s = 2$, the spatial variation of the displacements in the radial direction is represented by $r^{-\lambda_i}$ with the i-th eigenvalue λ_i determined from (3). It is thus instructive to calculate the first few λ_i. The eigenvalues 0, 0 and -1 correspond to the 2 translational and the rotational rigid-body motions. For $a = 0.5b$ the first 5 of the remaining eigenvalues are -0.49998, -0.50001, -1.00000, -1.49998, -1.50004. The values are very close to those of the eigenvalues of the theory of elasticity -0.5, -0.5, -1, -1.5, -1.5. The first 2 eigenvalues result in the singularities of the stresses at the crack tip ($\sim r^{-0.5}$) and the third to constant stresses. No error from a practical point of view is introduced in the radial direction permitting a rigorous representation of the stress singularities around the crack tip.

Appendix References

Dasgupta, G. (1982). "A finite element formulation for unbounded homogeneous continua." *Journal of Applied Mechanics*, ASME, Vol. 49, 136–140.

Nardini, D. and Brebbia, C. A. (1984). "Boundary integral formulation of mass matrices for dynamic analysis." *Topics in Boundary Element Research*, Vol. 2, Chapter 7, Ed. Brebbia, C.A., Springer, Berlin, 191-208.

Song, Ch. and Wolf, J. P. (1996a). "Consistent infinitesimal finite-element cell method: three-dimensional vector wave equation." *International Journal for Numerical Methods in Engineering*, (in press).

Song, Ch. and Wolf, J. P. (1996b). "Consistent infinitesimal finite-element cell method for an anisotropic unbounded medium." *Proceedings of the 11th Engineering Mechanics Conference*, ASCE, Fort Lauderdale, FL, May 19-22.

Wolf, J. P. and Song, Ch. (1996). *Finite-Element Modelling of Unbounded Media*, John Wiley & Sons Ltd.

Dynamic Through-the Soil Interaction of Adjacent Surface or Buried Structures

D.C. Rizos[1],Assoc. Member ASCE, and D.L. Karabalis[2], Assoc. Member ASCE

Abstract

The dynamic through-the-soil interaction of adjacent 3-D surface or buried structures is studied using a new advanced direct time domain Boundary Element formulation. The proposed formulation is based on the B-spline fundamental solutions corresponding to a homogeneous, isotropic, and linear elastic full space. The solution is accomplished in a time marching fashion, using higher order spatial and temporal discretizations.

Introduction

The prediction of the transient response of elastic systems of infinite or semi-infinite extent subjected to time dependent loading and boundary conditions has been the subject of numerous analytical and numerical methodologies. Among them is the direct time domain Boundary Element Method (BEM), which is based on integral equation formulations, developed in the beginning of the century. The variations of the method reported in the literature (Karabalis and Beskos 1984, Ahmad and Banerjee 1988) are due to different expressions of the fundamental solutions of the associated integral equations. Extensive reviews on the development and application of the BEM in elastodynamic problems can be found in various sources, e.g. (Manolis and Beskos 1988).

In the present work, a general 3-D direct time domain BEM proposed by (Rizos and Karabalis 1994) is used to study the interaction of adjacent surface or buried structures in infinite and semi-infinite domains. The method is based on the family of the "B-spline fundamental solutions" developed originally by (Rizos 1993), and allows for virtually any order of time variation of the field variables, without excessive computational effort. Higher order isoparametric elements are used for

[1] *Department of Civil Engineering, University of South Carolina, Columbia SC, 29208*
[2]*Department of Civil Engineering, University of Patras, GR-26500,* Patras, Greece

spatial discretization along with a numerical treatment of singularities. Consequently, the proposed methodology is proven accurate and efficient for all numerical applications attempted so far.

Formulation

The proposed BEM formulation is based on the assumption that the response, $R[\underline{x},t|f(\underline{x},t)]$, of the elastodynamic system due to an arbitrary transient excitation, $f(\underline{x},t)$, can be expressed as a linear combination of characteristic responses, $R[\underline{x},t-t_i|B_0^k(t)]$, due to typical time dependent forcing functions. An attractive selection for such a typical external function is the family of the B-spline piecewise smooth polynomials $B_0^k(t)$ of various degrees k (Shoenberg 1967). In view of the proposed scheme, the actual system response due to an arbitrary transient excitation can be computed as,

$$R[\underline{x},t|f(\underline{x},t)] \approx \sum_{i=1}^{j} f(\underline{x},\tau_i) R[\underline{x},t-t_i|B_0^k(t)], \quad t \in [t_1,t_j] \tag{1}$$

The validity of this assumption is justified by the linearity of the elastodynamic system, and does not introduce approximations of any kind. In order to evaluate the characteristic response of the system, the fundamental solutions, U_{ij}^B and T_{ij}^B, appearing in the associated Boundary Integral Equation (BIE) are developed on the basis of a concentrated body force of B-spline time variation (Rizos 1993, Rizos and Karabalis 1994). The BEM requires discretization of the domain and field variables in both space and time. In this work quadratic isoparametric elements have been implemented for the spatial discretization. The order of the temporal discretization is the one dictated by the adopted B-spline functions. Cubic B-spline functions are implemented in the present applications. Following well established procedures, the BIE can be cast in a matrix form and solved in a time marching fashion as indicated by the following equation,

$$\begin{aligned} c_{ij}^p\{u_i^p(t_N)\} = & \sum_{n=1}^{N}\left[U_{ij}\left[\left(\underline{x}^q,t_n;\underline{x}^p|B_0^k\right)\right]\right]\left\{t_{(\underline{n})i}^q(\tau_{N-n})\right\} - \\ & \sum_{n=1}^{N}\left[T_{ij}\left[\left(\underline{x}^q,t_n;\underline{x}^p|B_0^k\right)\right]\right]\left\{u_i^q(\tau_{N-n})\right\} \end{aligned} \tag{2}$$

where p and q denote points on the boundary, U_{ij} and T_{ij} are the B-spline fundamental solutions integrated over the boundary of the domain, $u_i^q(\tau_{N-n})$ and $t_{(\underline{n})i}^q(\tau_{N-n})$ are the vectors of nodal displacements and tractions, respectively, and t_n and τ_{N-n} are the time instances dictated by the temporal discretization and the associated B-spline.

Numerical applications

Two examples are considered in this section involving the through the soil interaction of embedded and surface bodies. In the first example two cavities of radius α=212 are embedded in an elastic full space with their centers located at points (-318,0,0) and (318,0,0), respectively. The soil characteristics are: Lame' constants $\mu=\lambda=3.5972 \times 10^6$, and mass density $\rho=2.5 \times 10^{-4}$, in consistent units. The source cavity is subjected first to internal pressure of B-spline time variation and the characteristic response is calculated as indicated by Eq. 1. Figure 1 (a)-(c) shows the characteristic deformed configurations of the system at representative times. Subsequently, a constant internal pressure P=1000 is applied in the source cavity and the response in the X direction of points (-106,0,0) and (106,0,0) on the source and receiver cavity, respectively, is computed by Eq. 1. Figure 2 shows this response along with the analytic solution for a single cavity subjected to the same excitation.

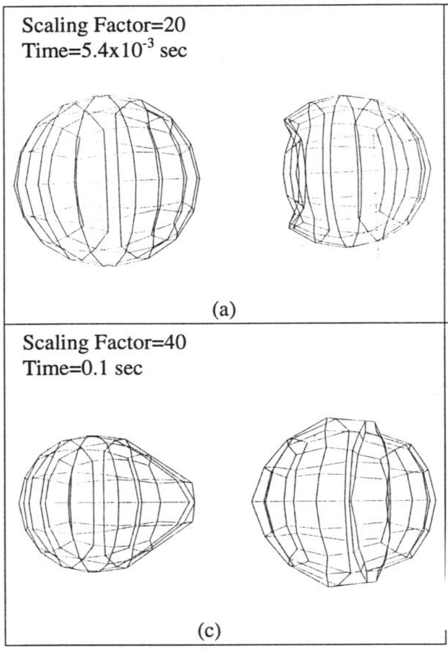

Figure 1(a)-(c) Deformed configurations

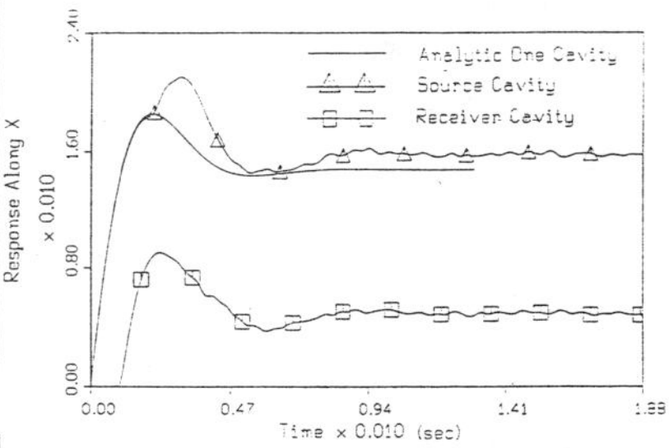

Figure 2. Response of points S and R

The second example pertains to the interaction of a 60x60 rigid surface footing resting on the half space with an embedded cavity of radius α=30. The center of the cavity is located at depth d=45. The soil properties are: Lame' constants $\lambda=6.63 \times 10^6$, $\mu=3.35 \times 10^6$, and mass density $\rho=2.82 \times 10^{-4}$, in consistent units. The system is subjected to vertical load on the center of the footing. Figure 3 a-b shows the deformed configurations of the system at selected times due to B-spline time varying vertical force applied on the footing. Figure 4 shows the response of the center of the footing and the north pole of the cavity, due to a vertical load of triangular time variation applied at the center of the foundation.

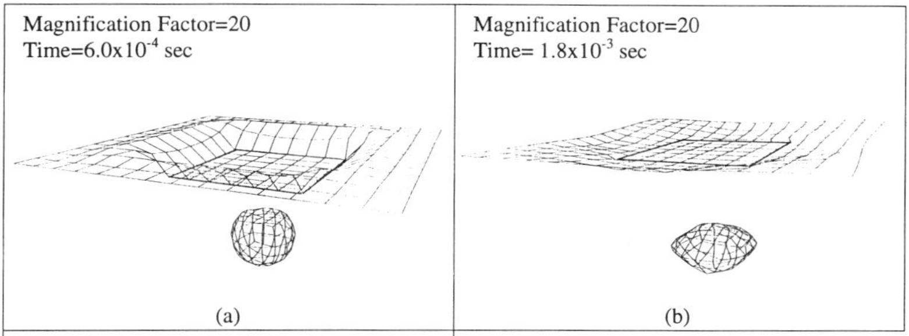

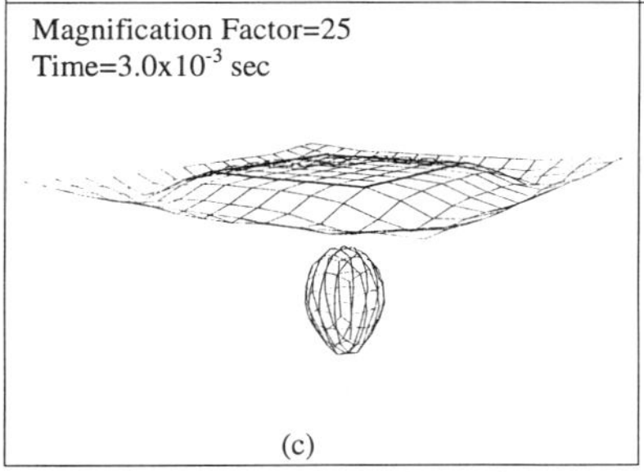

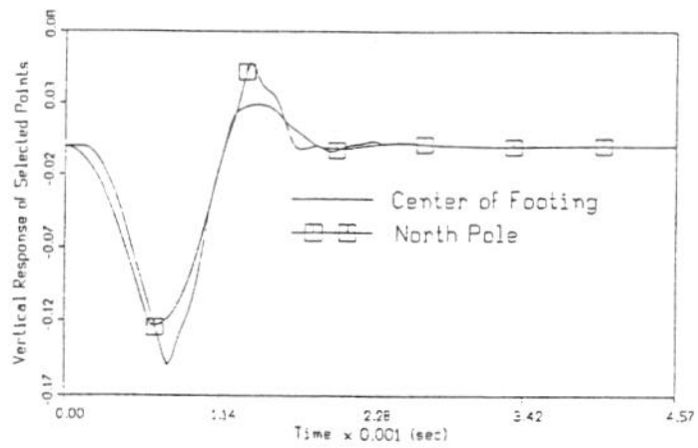

Figure 3. Deformed configurations

Figure 4. Response due to triangular load

<u>References</u>

Ahmad, S. and Banerjee, P.K. (1988). "Time-Domain Transient Elastodynamic Analysis of 3-D Solids by BEM", *International Journal of Numerical Methods in Engineering*, Vol. 26, pp. 1709-1728.

Karabalis, D.L. and Beskos, D.E., (1984). "Dynamic Response of 3-D Rigid Surface Foundations by Time Domain Boundary Element Method", *Earthquake Engineering and Structural Dynamics*, Vol. 12, pp. 73-93.

Manolis, G.D. and Beskos, D.E. (1988). *Boundary Element Methods in Elastodynamics*, Unwin Hyman, London.

Rizos, D.C. (1993). *An Advanced Time Domain Boundary Element Method For General 3-D Elastodynamic Problems*, Ph.D. Dissertation, Univ. of South Carolina, Columbia, South Carolina.

Rizos, D.C. and Karabalis, D.L. (1994). "An Advanced Direct Time Domain BEM Formulation for 3-D Elastodynamic Problems", *Computational Mechanics*, Vol. 15, No. 3, pp. 249-269.

Schoenberg, I.J. (1967). "On Spline Functions", pp. 255-291, in: Shisha, O. (Ed.), *Inequalities*, ,Academic Press, New York.

Dynamic Sub-Structure Method in Time Domain Using Analytical Representation of Dynamic Stiffness of Soil

Nagayuki Yoshida[1]

Abstract

The impulse function of dynamic stiffness of soil and its convolution integral are calculated to evaluate the dynamic interaction forces between soil and structures in time domain. This paper formulates three kinds of representation for the dynamic interaction forces and the related transform rules with causality. The impulse function can be analytically represented by using a discrete Hilbert transform and the causal inverse fast Fourier transform together with the complex exponential algorithm.

Introduction

The impulse function of the dynamic stiffness of soil is used in the time domain analysis of soil-structure interaction (Wolf and Obernhuber, 1985; Wolf and Darbre, 1986). The dynamic stiffness can be decomposed into the regular and the singular parts in the three types (Yoshida *et al*, 1986; Hayashi and Katukura, 1990). If the real part of the dynamic stiffness could not be obtained, the real part of the regular part is approximately calculated from its imaginary part in frequency domain in order to satisfy the causality by using a discrete Hilbert transform (DHT) (Yoshida *et al*, 1986; Simon and Tomlinson, 1984). Applying the causal inverse fast Fourier transform (CIFFT) (Hayashi and Katukura, 1990) to this real part, the impulse function of the dynamic stiffness of soil is obtained as a large amount of data on the time axis. The impulse data is replaced to a series of analytical functions approximately by using the complex exponential algorithm (CEA) (Brown *et al*, 1979). This paper presents only the formulation.

Dynamic Sub-Structure Method in Time Domain

The equations of motion for the inner field V_I containing the structures can be discretized by the numerical method such as the finite element method as follows:

[1]Department of Structural Engineering, Hiroshima University
1-4-1, Kagamiyama, Higashi-Hiroshima 739, Japan

$$\begin{bmatrix}[M_{SS}] & [M_{SB}] \\ [M_{BS}] & [M_{BB}]\end{bmatrix}\begin{Bmatrix}\{\ddot{u}_S(t)\} \\ \{\ddot{u}_B(t)\}\end{Bmatrix}+\begin{bmatrix}[C_{SS}] & [C_{SB}] \\ [C_{BS}] & [C_{BB}]\end{bmatrix}\begin{Bmatrix}\{\dot{u}_S(t)\} \\ \{\dot{u}_B(t)\}\end{Bmatrix}+\begin{bmatrix}[K_{SS}] & [K_{SB}] \\ [K_{BS}] & [K_{BB}]\end{bmatrix}\begin{Bmatrix}\{u_S(t)\} \\ \{u_B(t)\}\end{Bmatrix}$$

$$=\begin{Bmatrix}\{0\} \\ -\{R(t)\}\end{Bmatrix} \quad (1)$$

in which $\{u\}$ and $\{R\}$ denote the displacement vector and the dynamic interaction force vector in time domain on the pseudo boundary surface S_B respectively (Figure 1). $\{R(t)\}$ is represented by the inverse Fourier transform or the convolution integral using the dynamic stiffness matrix $[S]$ obtained by the boundary integral equation method or the transmitting boundary representation as follows:

$$\{R(t)\}=\frac{1}{2\pi}\int_{-\infty}^{\infty}[S(\omega)]\{u(\omega)\}e^{i\omega t}d\omega$$

$$=\int_0^t[S(t-\tau)]\{u(\tau)\}d\tau \quad (2)$$

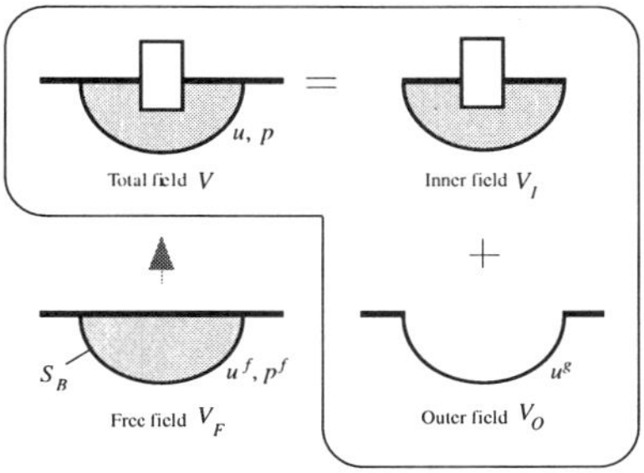

Figure 1. Substructure model

in which $\{u\}$ denotes the relative displacements from the free motion $\{u_B^g\}$ in the outer field V_O

$$\{u(\omega)\}=\{u_B(\omega)\}-\{u_B^g(\omega)\}. \quad (3)$$

The free surface condition on S_B in the outer field V_O is given by

$$\{p_B^g(\omega)\}=\{p_B^f(\omega)\}+[S(\omega)]\left(\{u_B^g(\omega)\}-\{u_B^f(\omega)\}\right)=0 \quad (4)$$

or

$$\{u_B^g(\omega)\}=\{u_B^f(\omega)\}-[S(\omega)]^{-1}\{p_B^f(\omega)\} \quad (5)$$

in which $\{p\}$ is the surface traction and the super suffix f denotes the free field with no structures.

Table 1. Regular and singular parts of $[S(\omega)]$

	$l=1$	$l=2$	$l=3$
$[k^n]$	$[k(\infty)]$	$[k(0)]$	$[k(0)]$
$[c^n]$	$[c(\infty)]$	$[c(\infty)]$	$[c(0)]$
$[k^r(\omega)]$	$[k(\omega)]-[k(\infty)]$	$[c(\omega)]-[c(\infty)]$	$\frac{[k(0)]-[k(\omega)]}{\omega^2}$
$[c^r(\omega)]$	$[c(\omega)]-[c(\infty)]$	$\frac{[k(0)]-[k(\omega)]}{\omega^2}$	$\frac{[c(0)]-[c(\omega)]}{\omega^2}$

$[S(\omega)]$ can be decomposed into the regular and the singular parts in the three following ways (Table 1):

$$[S(\omega)]=[S_l^n]+[S_l^r(\omega)](i\omega)^{l-1} \quad : l=1,2,3 \quad (6)$$

where

$[S(\omega)]=[k(\omega)]+i\omega[c(\omega)]$,

$[S_l^n]=[k^n]+\mathrm{i}\omega[c^n]$: singular part, $[S_l^r(\omega)]=[k^r(\omega)]+\mathrm{i}\omega[c^r(\omega)]$: regular part. Substituting equation (2), (3) and (6) in equations (1) leads to the fundamental equations in the dynamic sub-structure method:

$$\begin{bmatrix}[M_{SS}] & [M_{SB}]\\ [M_{BS}] & [M_{BB}]\end{bmatrix}\begin{Bmatrix}\{\ddot{u}_S(t)\}\\ \{\ddot{u}_B(t)\}\end{Bmatrix}+\begin{bmatrix}[C_{SS}] & [C_{SB}]\\ [C_{BS}] & [C_{BB}]+[c^n]\end{bmatrix}\begin{Bmatrix}\{\dot{u}_S(t)\}\\ \{\dot{u}_B(t)\}\end{Bmatrix}$$

$$+\begin{bmatrix}[K_{SS}] & [K_{SB}]\\ [K_{BS}] & [K_{BB}]+[k^n]\end{bmatrix}\begin{Bmatrix}\{u_S(t)\}\\ \{u_B(t)\}\end{Bmatrix}+\begin{Bmatrix}\{0\}\\ \int_0^t[S_l^r(t-\tau)]d^{l-1}\{u_B(\tau)\}/d\tau^{l-1}\cdot d\tau\end{Bmatrix}$$

$$=\begin{Bmatrix}\{0\}\\ [c^n]\{\dot{u}_B^g(t)\}+[k^n]\{u_B^g(t)\}+\int_0^t[S_l^r(t-\tau)]d^{l-1}\{u_B^g(\tau)\}/d\tau^{l-1}\cdot d\tau\end{Bmatrix}. \tag{7}$$

Causality

If the impulse function of the regular part is real and satisfies the causality $\left([S_l^r(t<0)]=0\right)$, both $[k^r(\omega)]$ and $[c^r(\omega)]$ satisfy the Hilbert transform

$$[k^r(\omega)]=-\frac{1}{\pi}\mathrm{P}\int_{-\infty}^{\infty}\frac{z[c^r(z)]}{z-\omega}dz=-\frac{2}{\pi}\mathrm{P}\int_0^{\infty}\frac{z^2[c^r(z)]}{z^2-\omega^2}dz \tag{8}$$

$$[c^r(\omega)]=\frac{1}{\pi\omega}\mathrm{P}\int_{-\infty}^{\infty}\frac{[k^r(z)]}{z-\omega}dz=\frac{2}{\pi}\mathrm{P}\int_0^{\infty}\frac{[k^r(z)]}{z^2-\omega^2}dz, \tag{9}$$

where $\mathrm{P}\int$ denotes the principal value of the integral. Furthermore, $[S_l^r(t)]$ can be represented only from $[k^r(\omega)]$ or $[c^r(\omega)]$ by the causal inverse Fourier transform

$$[S_l^r(t)]=\frac{1}{2\pi}\int_{-\infty}^{\infty}[S_l^r(\omega)]e^{\mathrm{i}\omega t}d\omega=4U(t)\cdot\mathrm{Re}\left[\frac{1}{2\pi}\int_0^{\infty}[k^r(\omega)]e^{\mathrm{i}\omega t}d\omega\right]$$

$$=4U(t)\cdot\mathrm{Re}\left[\frac{1}{2\pi}\int_0^{\infty}\mathrm{i}\omega[c^r(\omega)]e^{\mathrm{i}\omega t}d\omega\right] \tag{10}$$

where $U(t)=1\ (t>0),\ 0\ (t<0)$ denotes the step function. Equations (9) are evaluated discretely (DHT), and Equations (10) can be calculated by applying the fast Fourier transform algorithm (CIFFT).

Complex Exponential Algorithm (CEA)

$S_l^r(\omega)$ can be approximately expressed in a partial-fraction expansion

$$S_l^r(\omega)=\sum_{k=1}^{N_1}\left(\frac{D_k}{\mathrm{i}\omega-\lambda_k}+\frac{D_k^*}{\mathrm{i}\omega-\lambda_k^*}\right)+\sum_{k=1}^{N_2}\frac{C_k}{\mathrm{i}\omega+\gamma_k} \tag{11}$$

where γ_k is real and λ_k^* is the conjugate of a complex number $\lambda_k=-\beta_k+\mathrm{i}\omega_k$. Applying the inverse Fourier transform results in

$$S_l^r(t_m)=2\mathrm{Re}\left\{\sum_{k=1}^{N_1}D_kX_k^m\right\}+\sum_{k=1}^{N_2}C_kY_k^m$$

$$=\sum_{k=1}^{N_1}\left(A_k\,e^{-\beta_k m\Delta T}\cos\omega_k m\Delta T+B_k\,e^{-\beta_k m\Delta T}\sin\omega_k m\Delta T\right)+\sum_{k=1}^{N_2}C_k\,e^{-\gamma_k m\Delta T} \tag{12}$$

with $X_k = e^{(-\beta_k + i\omega_k)\Delta T}$, $Y_k = e^{-\gamma_k \Delta T}$ and $A_k = 2\mathrm{Re}[D_k]$, $B_k = -2\mathrm{Im}[D_k]$.
The polynomial of $n(=2N_1 + N_2)$th order with the roots X_k, X_k^*, Y_k is defined as

$$\alpha_n X^n + \alpha_{n-1} X^{n-1} + \cdots + \alpha_1 X + \alpha_0 = \prod_{k=1}^{N_1}(X - X_k)(X - X_k^*)\prod_{k=1}^{N_2}(X - Y_k) = 0 \quad (13)$$

where $\alpha_n = 1$. Multiplying the both sides of equation (12) by α_m and adding it from $m = 0$ to n lead to

$$S_l^r(t_n) + \sum_{m=0}^{n-1} S_l^r(t_m)\cdot\alpha_m = 2\mathrm{Re}\left\{\sum_{k=1}^{N_1} D_k\left(\sum_{m=0}^{n}\alpha_m X_k^m\right)\right\} + \sum_{k=1}^{N_2} C_k\left(\sum_{m=0}^{n}\alpha_m Y_k^m\right) = 0. \quad (14)$$

To determine coefficients α_m, $p(>>n)$ linear equations for n unknowns are constructed by using p sets of n discrete points each starting from one delayed point. They can be solved by a generalized inverse technique based on the singular value decomposition. X_k, Y_k and A_k, B_k, C_k are obtained from equation (13) and (12) respectively.

Concluding Remarks

The CIFFT to obtain the impulse function of the dynamic stiffness of soil has been presented. This method becomes more effective to be combined with the DHT when the impulse function has to be calculated from the imaginary part of the dynamic stiffness. The obtained data in time history can be expressed by the linear combination of stably dumped harmonic functions by using the CEA method. This method can reduce the storage requirement compared with the ordinarily used FFT.

Reference

Brown, D. L., R. J. Allemang, R. Zimmerman and M. Mergeay (1979). Parameter estimation techniques for modal analysis, *SEA Paper*, No.790221.

Hayashi, Y. and H. Katukura (1990). Effective time-domain soil-structure interaction analysis based on FFT algorithm with causality condition, *Earthquake Engineering and Structural Dynamics*, 19, 693-708.

Simon, M. and G. R. Tomlinson (1984). Use of the Hilbert transform in modal analysis of linear and non-linear structures, *Journal of Sound and Vibration*, 96, No.4, 421-436.

Wolf, J. P. and P. Obernhuber (1985). Non-linear soil-structure-interaction analysis using Green's function of soil in the time domain, *Earthquake Engineering and Structural Dynamics*, 13, 213-223.

Wolf, J. P. and G. R. Darbre (1986). Non-linear soil-structure-interaction analysis based on the boundary-element method in time domain with application to embedded foundation, *Earthquake Engineering and Structural Dynamics*, 14, 83-101.

Yoshida, N., Y. Fujitani and D. Fujii (1986). Dynamic soil-foundation interaction analysis based on interrelations of fundamental physical quantities, *Journal of Structural and Construction Engineering*, 367, 103-110 (*in Japanese*).

Assessments of Lateral and Torsional Structural Responses Induced by Incoherent Ground Motions

G. D. Hahn[1] and X. Liu[2]

Abstract

Reported assessments of dynamic responses of structures to incoherent ground motions are relevant to the practical definition of related base shear reductions, accidental eccentricities, and structural design for torsional effects induced by earthquakes.

Introduction

Effects of kinematic soil-structure interaction are studied in the present paper. Such effects are defined as the differences in the responses of a structure to foundation input motions obtained under two different sets of conditions: (1) Specifying uniformly distributed (in space) free-field ground motions, leading to a foundation input motion (FIM) with a translational component that is the same as the free-field ground motion at any point in the soil; and (2) accounting for the spatial variation of the free-field ground motions, leading to a FIM that contains both translational and "torsional" components. The FIM is that experienced by the foundation if it and the superimposed structure are assumed to be massless, and the free-field ground motion is that experienced at a point in the soil if no structure were present.

Spatial variations of ground motions are contributed by the so-called "wave passage" and "incoherence" effects (Veletsos and Prasad 1989, Hahn and Liu 1994). However, observations have been made (Luco and Wong 1986, Veletsos and Prasad 1989) that suggest the wave passage actions can be approximately taken as equivalent incoherence actions. Based on such observations, the present paper assesses structural responses to vertically propagating, incoherent seismic motions. These motions are further assumed to be horizontally polarized, resulting from the propagation of shear waves in the soil.

The main influences of the incoherence effects are to reduce the lateral component of the FIM and the associated lateral structural response and base shear, and to induce a "torsional" component of FIM that leads to torsion in the structure (Veletsos and Prasad 1989, Hahn and Liu 1994). These torsional actions in the structure have been interpreted in terms of "accidental eccentricities" (Hahn and Liu). There is a need, however, for simple methods to estimate such shear reductions and accidental eccentricities. Assessments presented in the present paper intend to be responsive to this need. These assessments further address technical issues involved in the design of structures for the torsional effects induced by earthquakes.

Formulation of Problem

Results to be presented have been obtained for a particular case of the one-story structure studied by Hahn and Liu (1994). The structure consists of a rigid circular slab of radius R

[1]Associate Professor of Civil Engineering, Vanderbilt University, P.O. Box 124-B, Nashville, TN 37235
[2]Engineer, Hart Consultant Group, Santa Monica, California

supported by elastic columns that are in turn supported by a rigid circular mat foundation also of radius R. The structure is assumed to be symmetric; its centers of geometry, mass, and stiffness coincide. The ground motions act laterally in a conveniently defined x-direction. The spatial variation of such ground motions is described by using an incoherence function employed in previous studies (Veletsos and Prasad 1989, Hahn and Liu 1994). The associated temporal variation of the ground excitation is indirectly defined in the frequency (ω) domain in terms of a power spectral density (PSD) function, $S_{\ddot{g}}(\omega)$, numerically obtained from an appropriate analysis of the first seven seconds of the acceleration record for the 1940 El Centro earthquake. More details about the structure and the excitation can be found in Liu (1995). Response quantities of interest are the relative lateral response of the structure (in the x-direction) and the associated torsional response.

Let $\ddot{x}_b(t)$ and $\ddot{\theta}_b(t)$ be the lateral and torsional components of FIM accelerations, respectively; and let $\ddot{y}_b(t) = R\ddot{\theta}_b(t)$. The PSD functions for $\ddot{x}_b(t)$ and $\ddot{y}_b(t)$, $S_{\ddot{x}}(\omega)$ and $S_{\ddot{y}}(\omega)$, respectively, are defined by $S_{\ddot{x}}(\omega) = |T_{xg}|^2 S_{\ddot{g}}(\omega)$ and $S_{\ddot{y}}(\omega) = |T_{yg}|^2 S_{\ddot{g}}(\omega)$. The quantities $|T_{xg}|^2$ and $|T_{yg}|^2$, the squares of the absolute values of appropriate transfer functions, are given by (Veletsos and Prasad 1989):

$$|T_{xg}|^2 = |T_{xg}(\omega)|^2 = (1/b_o^2)\{1 - \exp(-2b_o^2)\,[I_0(2b_o^2) + I_1(2b_o^2)]\,\} \tag{1}$$

$$|T_{yg}|^2 = |T_{yg}(\omega)|^2 = (1/b_o^2)\{1 - \exp(-2b_o^2)\,[I_0(2b_o^2) + 2I_1(2b_o^2) + I_2(2b_o^2)]\,\} \tag{2}$$

where I_0, I_1, and I_2 are modified Bessel functions of the first kind of the order indicated by the subscript; $b_o = \tilde{\tau}\omega$, and $\tilde{\tau} = \gamma R/v_s$. The dimensionless quantity γ, with values in the approximate range of 0 – 0.5, is the incoherence parameter; v_s is the velocity of propagation of shear waves in the soil.

Let $X(t) = X(t,\tilde{\tau})$ and $\Theta(t) = \Theta(t,\tilde{\tau})$ be the relative slab-foundation lateral response measured at the center of the structure and the corresponding relative torsional response, respectively. Because the structure is symmetric, these relative responses are uncoupled. Also, let $Y(t) = R\,\Theta(t)$. The response $X(t)$ is induced by $\ddot{x}_b(t)$ only; $Y(t)$ is due to $\ddot{y}_b(t)$ only. Consequently, the PSD functions for $X(t)$ and $Y(t)$, $S_X(\omega)$ and $S_Y(\omega)$, respectively, are given by $S_X(\omega) = |T_X|^2 S_{\ddot{x}}(\omega)$ and $S_Y(\omega) = |T_Y|^2 S_{\ddot{y}}(\omega)$, where T_X and T_Y are appropriate response transfer functions. T_X is involved in the definition of the steady-state response of a linear oscillator with natural frequency ω_X, the lateral natural frequency of vibration of the structure, to a complex-valued harmonic excitation. Similarly, T_Θ results from the consideration of an oscillator with a natural frequency ω_Θ, the torsional natural frequency of the structure. In terms of $S_{\ddot{g}}(\omega)$, the response PSD functions are given by

$$S_X(\omega) = |T_X|^2 |T_{xg}|^2 S_{\ddot{g}}(\omega) \tag{3}$$

$$S_Y(\omega) = |T_Y|^2 |T_{yg}|^2 S_{\ddot{g}}(\omega) \tag{4}$$

Mean peak responses can be obtained from the PSD functions $S_X(\omega)$ and $S_Y(\omega)$ (e.g., see Der Kiureghian 1980). For $\gamma = 0$, $b_o = \tilde{\tau} = 0$, $|T_{xg}| = 1$, $|T_{yg}| = S_Y(\omega) = 0$, the ground motions are uniformly distributed in space, and there are no torsional effects in the structure.

Lateral Response Reduction

Figure 1 reports mean peak values of $X(t)$, $X_m(\tilde{\tau})$, in the form of normalized pseudo velocity spectra, with $\dot{x}_g$ = mean peak free-field ground velocity. The plots are for a damping ratio of 0.02 and show reductions in lateral response with increasing values of the effective transit time, $\tilde{\tau}$, particularly for intermediate to high frequency systems. The mean peak response, $X_m(\tilde{\tau})$, depends on the square root of the integral of $S_X(\omega)$ and other factors. To obtain simple approximations for the response reduction $\psi = X_m(\tilde{\tau} > 0)/X_m(\tilde{\tau} = 0)$, it is noted that $|T_X|^2$ in (3) is a function with a sharp peak at $\omega = \omega_X$. A critical examination of this problem would reveal that when ω_X falls within the frequencies of the excitation, or when ω_X is smaller than such frequencies, $\psi \approx |T_{xg}|$. When ω_X is greater than the excitation frequencies, then ψ can be taken as $|T_{xg}|_o$ = square root of the average of $|T_{xg}|^2$ over $0 \leq \omega \leq \omega_{me}$. The quantity ω_{me} is an approximation for the maximum frequency present in the excitation. Since $|T_{xg}|$ is a decreasing function of ω, taking ω_{me} smaller than the actual maximum frequency of the excitation leads to conservative results. For the El Centro earthquake considered, ω_{me} is taken as $2\pi(7.5\,\text{Hz})$ in rad/sec. The final approximation is $\psi = |T_{xg}|$ or $|T_{xg}|_o$, whichever is larger. Figure 2 presents results of this simple approximation for a damping ratio of 0.02 compared with corresponding highly accurate numerical values; the approximation is seen to give relatively good results. The mean peak base shear is proportional to X_m.

Accidental Eccentricity

Mean peak values of $Y(t, \tilde{\tau})$, $Y_m(\tilde{\tau})$, are reported in Fig. 3 for a damping ratio of 0.02. The lateral displacement at an important point on the periphery of the structure is $X_P(t, \tilde{\tau}) = X(t, \tilde{\tau}) + Y(t, \tilde{\tau})$. Because $X(t, \tilde{\tau})$ and $Y(t, \tilde{\tau})$ are uncorrelated (Veletsos and Prasad 1989, Hahn and Liu 1994), an approximation for the mean peak value of $X_P(t, \tilde{\tau})$ can be obtained as $X_{Pm}(\tilde{\tau}) = \sqrt{X_m(\tilde{\tau})^2 + Y_m(\tilde{\tau})^2}$. The mean peak base shear for the structure is $V_m = K_X X_m(\tilde{\tau})$, where K_X = lateral stiffness of structure. The accidental eccentricity is defined as the distance from the center of the structure at which V_m must be statically applied to induce by itself the displacement $X_{Pm}(\tilde{\tau})$ at the periphery. The normalized accidental eccentricity, β, is the accidental eccentricity divided by the diameter $D = 2R$. It is given by $\beta = 0.25\,\varepsilon[(1 + X_{Pm}^2/X_m^2)^{1/2} - 1]$, which can be approximately expressed as (Liu 1995)

$$\beta = 0.25\,\varepsilon\,\{[1 + (1/\varepsilon^2)\,(|\bar{T}_{yg}|^2/|\bar{T}_{xg}|^2)\,\rho_X^2(\omega_\Theta, \zeta_\Theta)/\rho_X^2(\omega_X, \zeta_X)]^{1/2} - 1\} \qquad (5)$$

where $\varepsilon = \omega_\Theta^2/\omega_X^2$; ζ_X and ζ_Θ = damping ratios for lateral and torsional responses, respectively; and $\rho_X(\omega_X, \zeta_X) = \omega_X^2 X_m/\ddot{x}_g = \omega_X^2 X_m(\tilde{\tau}, \zeta_x)/\ddot{x}_g$ = normalized pseudo acceleration, with $\ddot{x}_g$ = mean peak free-field ground acceleration; $|\bar{T}_{xg}|^2 = |T_{xg}(\omega_X)|^2$ if $\omega_X \leq \omega_{me}$; $|\bar{T}_{xg}|^2 = |T_{xg}|_o^2$ if $\omega_X \geq 2\omega_{me}$; for $\omega_{me} \leq \omega_X \leq 2\omega_{me}$, $|\bar{T}_{xg}|^2$ is calculated by a linear interpolation between its previously defined values at $\omega_X = \omega_{me}$ and $2\omega_{me}$. $|\bar{T}_{yg}|^2$ is defined in a manner similar to $|\bar{T}_{xg}|^2$, by replacing $|T_{xg}(\omega)|$ with $|T_{yg}(\omega)|$ and ω_X with ω_Θ. Figure 4 shows approximate values of β from (5) compared with corresponding values obtained by accurate numerical calculations. These data are for $\zeta_X = \zeta_\Theta = 0.02$, and it is seen in Fig. 4 that the approximation is appropriate.

Practical Implications and Conclusion

Two approaches need to be identified to assess relevant practical implications. Approach 1: calculate the peak displacement at the periphery of the structure and the force in a resisting structural member located at the periphery by using the code-defined (UBC 1994) base shear (i.e., unreduced base shear that does not account for incoherence) in combination with the code-specified 5% normalized eccentricity. Approach 2: repeat the calculations identified in Approach 1 by using the incoherence-reduced base shear in combination with the accidental eccentricity defined in the present paper. It has been found in a number of calculations (Liu 1995) that Approach 1 usually gives much more conservative results than Approach 2 for torsionally stiff systems (i.e., for $\varepsilon > 1$); its results tend to be less conservative than but not particularly different from those of Approach 2 for torsionally flexible systems (i.e., for $\varepsilon < 1$). It then appears that for torsionally flexible systems, the codified approach, based on the results of the present study, would lead to just slightly unconservative designs of structures against the torsional effects induced by earthquakes. This appearance could be misleading. Examinations in progress at the present time reveal that the torsional responses shown in Fig. 3 are associated, under some conditions of practical importance, with a torsional rotation which could itself induce relatively high shearing stresses in reinforced concrete columns. A careful assessment would further reveal that the results of such examinations are basically consistent with results of analyses based on experimental data (De la Llera and Chopra 1994). The effects of such a rotation itself are not explicitly considered in codified provisions. This problem deserves further attention.

References

De la Llera, J. C., and Chopra, A. K. (1994)."Evaluation of code accidental-torsion provisions from building records." *J. Struct. Engrg.*, ASCE, 120(2), 597-616.

Der Kiureghian, A. (1980). "Structural response to stationary excitation." *J. Engrg. Mech.*, ASCE, 106(6), 1195-1213.

Hahn, G. D., and Liu, X. (1994). "Torsional response of unsymmetric buildings to incoherent ground motions." *J. Struct. Engrg.*, ASCE, 120(4), 1158-1181.

Liu, X. (1995). "Seismic torsional response of structures." Ph.D. Thesis, Dept. of Civ. and Environmental Engrg., Vanderbilt University.

Luco, J. E., and Wong, H. L. (1986). "Response of a rigid foundation to a spatially random ground motion." *J. Earthquake Engrg. and Struct. Dynamics*, 14, 891-908.

UBC (1994). International Conference of Building Officials. Uniform Building Code (UBC), Whittier, California.

Veletsos, A. S., and Prasad, A. M. (1989). "Seismic interaction of structures and soils: Stochastic approach." *J. Struct. Engrg.*, ASCE, 115(4), 935-956.

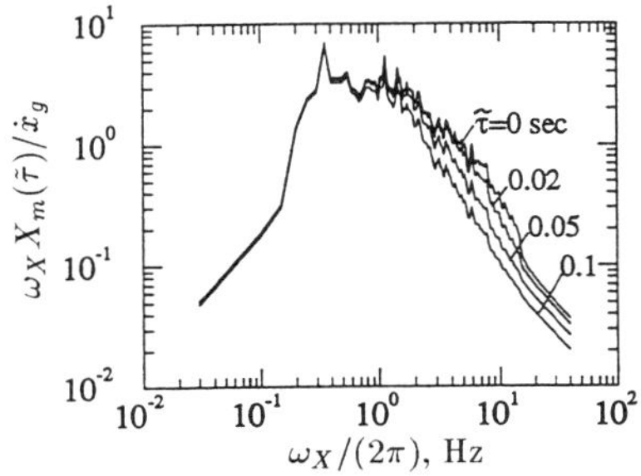

Fig. 1. Mean Peak Lateral Responses

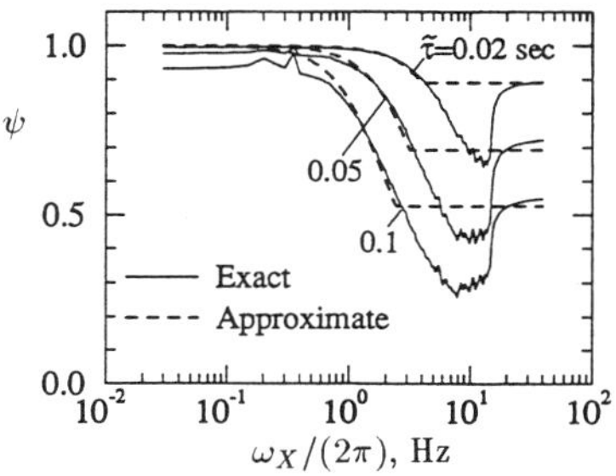

Fig. 2. Lateral Response Reductions

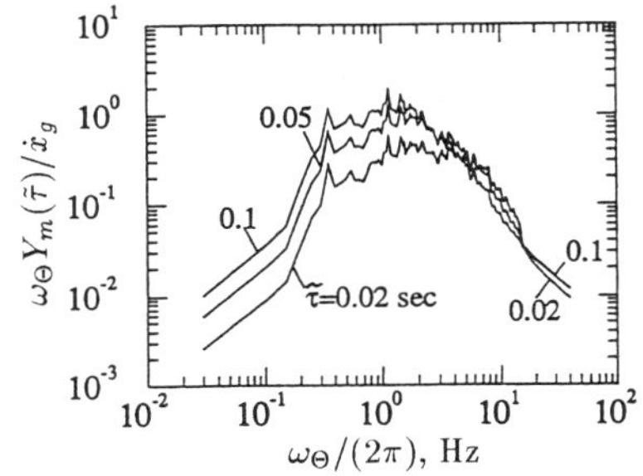

Fig. 3. Measures of Mean Peak Torsional Responses

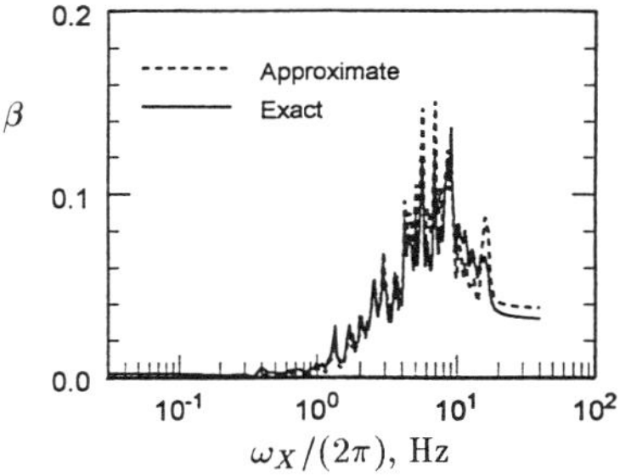

Fig. 4. Values of β for $\varepsilon = 0.75$ and $\tilde{\tau} = 0.05$ sec.

Minimally Invasive Endoluminal Vascular Grafts

L. Pinchuk*, J.P. Dereume, H. Kontges, N.Frid, Y.P. Kato, B.A. Weber, J.B. Martin, I.J. Khan, R. Alcime, G.J. Wilson and D.C. MacGregor

Abstract

Endoluminal grafts now form the basis for minimally invasive vascular surgery and are fraught with mechanical challenges relating to ease of introduction, prevention of migration, adequacy of dilation, long-term fatigue and effectiveness in the body. The Corvita Endoluminal Graft, presented in overview below, is an advanced design which may serve to replace conventional vascular surgery as the procedure of choice for repairing diseased or otherwise damaged arteries in the future.

Introduction

Endoluminal vascular grafts, often called "endovascular grafts" or simply "stent-grafts", are combinations of vascular stents and vascular grafts. A stent, in simplicity, is a generally cylindrical metallic expandable mesh used to prevent a vessel from collapsing. A vascular graft is a replacement vessel or conduit used to bypass diseased segments in the vascular system. Thus, an endoluminal graft is a vascular graft that is attached to a metallic mesh that can be inserted into an artery from a remote site and essentially reline a diseased segment of an artery. In essence, these novel combination devices form the basis for minimally invasive vascular surgery.

The requirements for an endoluminal graft are: 1) it must be capable of being compressed into a catheter delivery system and delivered to the site of deployment where it expands or is expanded into place; 2) it must have sufficient hoop strength to dilate the vessel, prevent closure of the vessel by muscle spasm and stop leakage around the device; 3) once expanded into place, it must not migrate; 4) it must be both biocompatible and hemocompatible; and 5) it must be biodurable and not catastrophically degrade, corrode or fracture in the body.

----

* Vice President, Product Development, Corvita Corporation., 8210 N.W. 27 St., Miami, Florida. 33122

Endoluminal grafts have been used in the vascular system to: 1) repair defects such as holes or dissections in arteries due to trauma, fistulas, iatrogenic injury, etc.; 2) bridge aneurysms and pseudoaneurysms; 3) bridge defects such as occlusions, stenoses, etc; and 4) to prevent strictures due to vessel spasm, tumor encroachment, etc.

The first published report of an endoluminal graft to treat an arterial lesion was by Becker in 1991[1]. A year later, Parodi presented his early work on bridging aortic aneurysms and on repairing fistulas in peripheral arteries[2]. The Parodi device consisted of two balloon-expandable stents between which was sutured a Dacron® vascular graft. The initial success of the Parodi device inspired many investigators around the world to develop smaller, stronger and more "user friendly" versions of this device. This paper describes the Corvita Endoluminal Graft, conceived by Dr. Jean-Pierre Dereume from the Erasmus University Hospital in Brussels, Belgium, which addresses many of the shortcomings of the initial endoluminal graft designs.

The Corvita Endoluminal Graft

The Corvita Endoluminal Graft (CEG) consists of a self-expanding braided wire stent integrally bound to a non-woven porous polycarbonate urethane vascular graft liner throughout its entire length. The stent component is comprised of a diamond pattern braid comprised of a plurality of highly corrosion-resistant cobalt-chromium-nickel super alloy wires with a Young's modulus in the order of 200,000 MNm^{-2} and a rigidity modulus of approximately 82,000 MNm^{-2}.

The vascular graft component[3] is comprised of hundreds of layers of elastomeric polycarbonate urethane fibers[4], 10 to 20 microns in diameter, spun in a braiding pattern similar in angle to the metallic stent to which it is attached. The polycarbonate urethane fibers are bound to each other at cross-over points to form a non-woven scaffold. The pores of the liner are sized to allow tissue ingrowth, yet heal with time to prevent blood leakage and pressure transfer through the vessel wall. The braid or pick angle formed by both the wire stent and the polycarbonate urethane fibers is approximately 45° as measured relative to the longitudinal axis of the device. The 45° braid angle allows for an expansion ratio from fully collapsed to fully expanded of approximately 1:5 with a change in length of less than 30 percent. The integral design also allows the user to easily cut the device to the proper length at the time of surgery. Figure 1 is an illustration of the CEG being deployed from a catheter and bridging an aneurysm in a blood vessel.

The wire stent component of the CEG provides the hoop stress necessary to self-expand in the vasculature and to dilate a stenotic vessel. The equations governing the outward pressure of the stent on the arterial wall are reviewed by Jedwab and Clerc[5], and are a function of the wire diameter, the number of wires, the stent

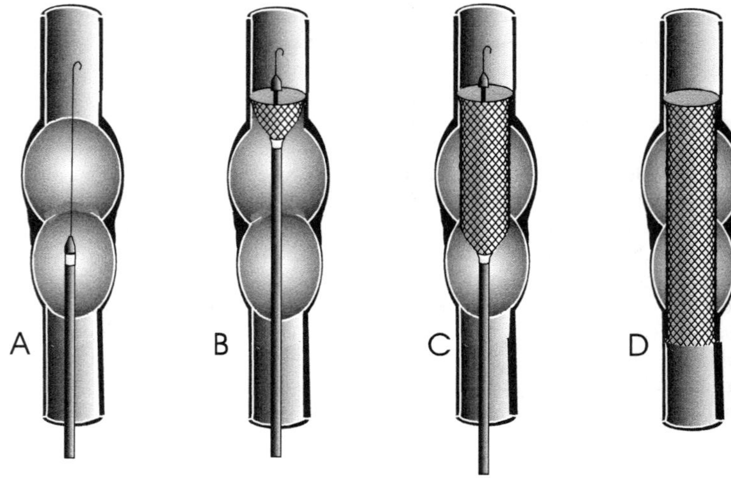

Figure 1: *A) A guidewire, followed by a catheter introducer containing the compressed CEG, is maneuvered to the damaged vessel, an aneurysm in this case; B) the CEG self-expands as the sheath is withdrawn; C) the sheath is further withdrawn to expose more of the CEG and bridge the aneurysm; D) the CEG is fully deployed and the catheter removed.*

diameter, both the Young's and rigidity moduli of the wire and the deformation of the stent (change in pick angle) relative to its resting position.

It was found empirically for small diameter endoluminal grafts of diameters of 6 mm to 12 mm that braids consisting of 24 to 32 wires (0.10 to 0.17 mm diameter) demonstrate sufficient hoop stress to bypass stenotic segments of arteries and to repair fistulas. Larger diameter stents of 25 mm to 40 mm, required to bridge aneurysms and prevent leaks in the aorta, require 0.24 mm diameter wires with 96 wires per device. Although the outward force of the stent component prevents initial migration of the device in the vasculature, the actual method of securing the CEG in place in flowing blood in a pulsating tube is by means of the outwardly flared wire ends of the device lodging in the arterial wall.

Although initial indications from many explants suggest that the stent, when ingrown with tissue, demonstrates little to no flexion, the device nevertheless has been subjected to a battery of mechanical testing. The more severe tests include: 1) individual wire fatigue testing (in excess of 100 million cycles at 550 MPa stress levels) with extrapolations by finite element analysis to worse case *in vivo* performance conditions; 2) accelerated flex fatigue cycling in a saline environment within a latex balloon cycled to the extremes of physiological pressure; and 3) exaggerated and accelerated radial and longitudinal cyclic testing at 5% length change conducted in a dry environment to investigate wire wear properties.

The polycarbonate urethane graft component has been tested in animals and humans for approximately five years in a vascular graft embodiment with the microfibers showing some biodegradation but essentially maintaining the scaffolding structure of the liner intact. The polycarbonate urethane has also undergone a battery of specific testing, including mutagenicity, one-year chronic toxicity and two-year carcinogenicity testing.

To date, hundreds of these devices have been implanted in animals to evaluate migration potential, healing and effectiveness. Approximately one hundred CEGs have been implanted in patients in Europe, South America and the United States. The results, although preliminary (less than three years in duration), are very encouraging, with a very low failure rate due mainly to difficulties in placement. The use of these minimally invasive endoluminal grafts can change the way vascular surgery is performed in the future as they provide less trauma, reduced morbidity, faster healing time, fewer blood transfusions, less anesthetic, reduced hospitalization and lower cost than conventional vascular surgery.

References

1) G.J. Becker, *et al.*, "Percutaneous Placement of a Balloon-Expandable Intraluminal Graft for Life-Threatening Subclavian Arterial Hemorrhage". *J. Vasc. Intervent. Radiol.*, Vol. 2, pp 225-229(1991).

2) J.C. Parodi and H.D. Barone, "Transluminal Treatment of Abdominal Aortic Aneurysms and Peripheral Arteriovenous Fistulas". Presented at the *19th Annual Symposium on Current Problems and New Technologies in Vascular Surgery.* NY, NY, Nov. 1992.

3) G. J. Wilson *et al.*, "A Compliant Corethane/Dacron Composite Vascular Prosthesis: Comparison with 4 mm ePTFE Grafts in a Canine Model". *ASAIO Trans.*, Vol. 39, pp M526-531(1993).

4) L. Pinchuk, "A Review of the Biostability and Carcinogenicity of Polyurethanes in Medicine and the New Generation of 'Biostable' Polyurethanes". *J. Biomater. Sci., Polymer. Edn.,* Vol. 6, No. 3, pp 225-267(1994).

5) M.R. Jedwab and C. Clerc, "A Study of the Geometrical and Mechanical Properties of a Self-Expanding Metallic Stent - Theory and Experiment". *J. of Applied Biomaterials,* Vol. 4, pp 77-85(1993).

The Effect of Diameter Mismatch Upon Hemodynamics in the Distal Anastomoses of Vascular Bypass Grafts

Robert S. Keynton[1], Mary M. Evancho[2], Rick L. Sims[2] and Stanley E. Rittgers[3]

Abstract

It is the purpose of this study to evaluate the effect of graft caliber upon hemodynamics at the distal anastomosis of vascular bypass grafts in an animal model. Tapered e-PTFE synthetic grafts were placed in an end-to-side fashion in the common carotid arteries of 16 adult mongrel dogs with graft-to-artery diameter ratios (DR) of either 1.0 or 1.5. Following graft implantation, three point arterial velocities were measured simultaneously at 0.35 mm radial increments from the vessel wall within the anastomotic region by a new *in vivo* triple crystal ultrasonic transducer, and wall shear rates were computed. The results from this study showed that the graft-to-artery diameter ratio does significantly affect the hemodynamic flow patterns in the distal anastomoses of arterial bypass grafts.

Introduction

Several investigators [Binns, 1989; Rittgers, 1978] have suggested that hemodynamics play a key role in the development of IH. Moreover, geometric factors have been found to be of primary importance in determining graft patency and local hemodynamics[Friedman, 1983]. Currently, the graft-to-artery diameter ratio at the distal anastomosis of bypass grafts varies widely in surgical practice. In the case of the commonly used reversed saphenous vein, the anastomotic DR is much greater than 1.0 due to the juxtaposition of the large proximal vein with the smaller distal artery. Likewise, in the case of a synthetic graft, the DR may be greater than 1.0 due to the fixed dimension of a straight prosthesis to a smaller artery. On the other hand, the graft-to-artery diameter ratio is a variable which can

---

[1]Department of Biomedical Engineering, Louisiana Tech University, P.O. Box 10348, Ruston, Louisiana 71272-0046.

[2]Falor Center for Vascular Studies, Summa Health Systems, 525 East Market St., Akron, Ohio 44309.

[3]Department of Biomedical Engineering, The University of Akron, 301 Sidney L. Olson Research Center, Akron, Ohio 44325-0302.

be readily controlled at the time of surgery. Therefore, it is the purpose of this study to determine the effects of changing the graft-to-artery diameter ratio upon the near wall hemodynamics within the distal anastomosis of vascular bypass grafts in an acute animal model.

Materials and Methods

Sixteen animals were placed in dorsal recumbency, draped with sterile towels and the carotid arteries were isolated bilaterally via a 20 cm midline neck incision. All dogs were cared for in compliance with the "*Principles of Laboratory Animal Care*" and the "*Guide for the Care and Use of Laboratory Animals*". Six centimeter lengths of tapered (4-7 mm I.D.) e-PTFE synthetic grafts (W.L. Gore and Associates, Inc., Flagstaff, AZ) were placed in each of the common carotid arteries of adult, mongrel dogs weighing between 25 and 30 kg. The distal anastomoses consisted of a graft-to-artery diameter ratio, **DR**, of either 1.0 or 1.5 with end-to-side proximal and distal anastomoses of 30°.

Using a new *in vivo* triple crystal pulse ultrasonic Doppler wall shear rate transducer [Keynton, 1995], simultaneous velocity measurements were made at radial increments of 0.35, 0.70 and 1.05 mm, respectively, from the artery wall in the plane of the anastomosis. Each crystal was interfaced with pulse ultrasonic Doppler velocity detectors (Crystal Biotech, Hopkinton, MA) which operated at a transmitting frequency of 20-MHz and a pulse repetition frequency of 62.5 kHz. The analog zero-crossing counter outputs from all three crystals together with ECG timing information were simultaneously recorded on a 4 channel digital oscilloscope at 100 Hz sampling rate, 512 data points, for 4 individual cycles and stored on disk.

Immediately following implantation, flow velocities were measured at 6 axial locations along the near-wall and 9 axial locations along the far-wall from the toe of the graft at the distal anastomosis in 0.5 D_a increments. The velocities were recorded with the transducer in two orientations (forward and reverse facing) in order to obtain two velocity components which were used to compute a resultant velocity vector at each measurement site. A two way ANOVA was used to determine the effects of DR and axial position upon the localized shear rates.

Results

The native artery average mean wall shear rate for all animals was found to be 692.6 $\pm$ 286.1 s^{-1} (24.5 $\pm$ 10.1 dynes/cm^2). The complete original and corresponding cubic spline regression velocity waveforms at each radial position per axial location are shown in Figure 1 for the far-walls of a typical animal. In addition, Figure 3 directly demonstrates how well the cycle mean velocities from the cubic spline regressions compare to the original data at each radial position. In comparing the vector plots of the two diameter ratios, the flow appeared to return to a more fully developed state further downstream for DR = 1.5 than for the DR = 1.0 case (Fig.2). Along the far-wall, the magnitudes of the velocities in the 1.0 case typically increased monotonically over a shorter distance downstream of the toe and appeared to be lower than compared to the 1.5 case. Additionally, it was noticed that for the DR=1.0 case the stagnation zone was located further downstream than the DR=1.5 case. Along the near-wall, the velocities were found to increase monotonically for both diameter ratios and were observed to be higher in the DR=1.5 case than in the DR=1.0 case.

Similarly, the far-wall shear rate data demonstrates that the flow exiting the graft is divided into two streams, one flowing in the upstream direction and the other in the downstream direction (Fig. 4). The flow moving in the downstream direction is greater than the upstream as evidenced by the magnitudes of the shear rates measured upstream being lower than the values measured downstream. In general, it was noticed that the shear rates on both the near- and far-walls tended to return to a more constant level within 2-3 diameters downstream of the toe. In evaluating the effects of axial position upon the mean and peak wall shear rates, the location along the length of the artery did have a significant effect ($p<0.0001$) with the differences lying strictly in the upstream section of the distal anastomosis. The absolute values of the normalized mean and peak shear rates were found to be significantly different for both the axial position ($p<0.0001$) and diameter ratio ($p<0.0049$ and $p<0.001$, respectively) with the higher shear rates occuring in the DR=1.0 case. The near-wall shear rate data showed no significant differences between groups due to diameter ratio or axial position.

Conclusions

The *in vivo* hemodynamic patterns observed in the present study do correspond well with *in vitro* [Keynton, 1991] and numerical [Fei, 1994] model studies of vascular bypass graft distal anastomoses. In particular, the observance of points of stagnation, recirculation regions upstream of the stagnation zones, and elevated wall shear rates along the far-wall of the artery. Finally, the results from this study show that the graft-to-artery diameter ratio does significantly affect the hemodynamic flow patterns in the distal anastomoses of arterial bypass grafts.

References

1. Binns, R.L., Ku, D.N., et al.: J Vasc Surg 10:326-327, 1989.
2. Fei, D.Y., Thomas, J.D., and Rittgers, S.E.: J Biomech Engr 116:355-360, 1994.
3. Friedman, M.H., Deters, O.J., and Mark, F.F.: Athero 46:225- 231, 1983.
4. Keynton, R.S., Shu, M.C.S., and Rittgers, S.E.: J Biomech Engr 113:458-463, 1991.
5. Keynton, R.S., Nemer, R.E., et al.: IEEE Trans Biomed Engr 42:5:433-441, 1995.
6. Rittgers, S.E., Karayannacos, P.E., et al.: Circ Res 42:792-801, 1978.

This work was supported by grants from the Summa Health System Foundation, NIH/NIA grant #1-R03-AG103098-1, NCRR grant #1-F1-GM15566-01, and W.L. Gore and Associates, Inc.

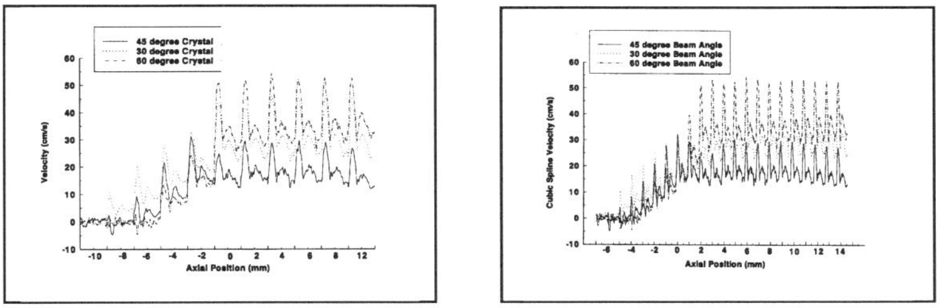

(a) (b)

Figure 1. *In vivo* velocity waveforms at different axial locations along the far-wall for DR=1.0: a) original data; and, b) cubic spline interpolation.

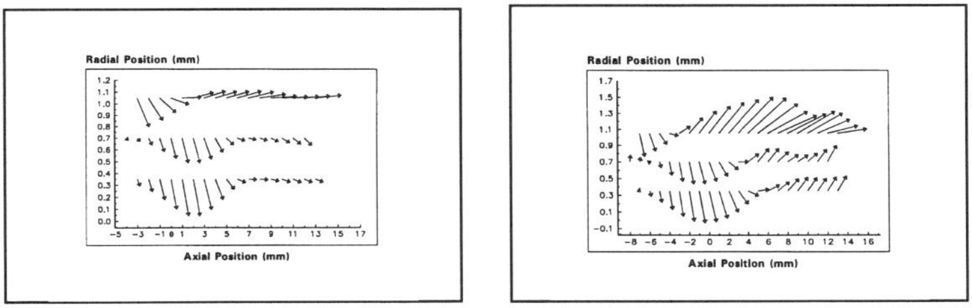

(a) (b)

Figure 2. Velocity vector plot of a vascular bypass graft distal anastomosis along the far-wall for: a) DR=1.0; and, b) DR=1.5.

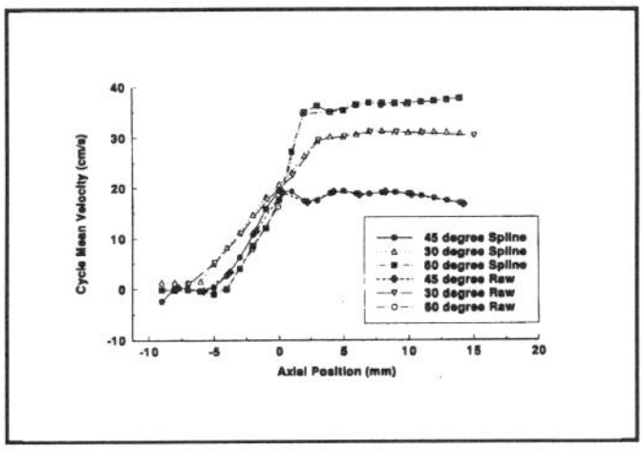

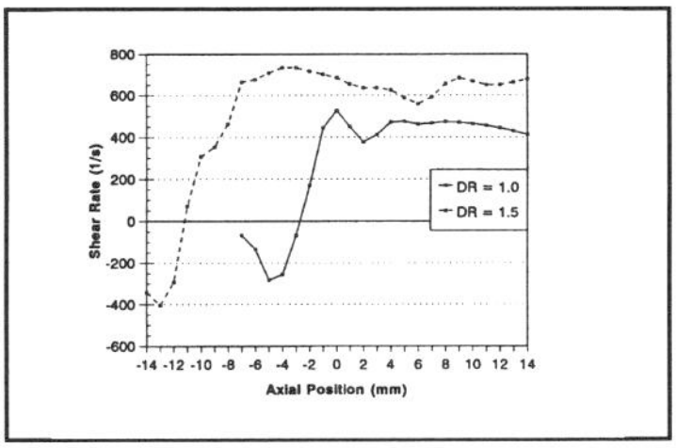

Figure 3. Direct comparison between the cycle mean velocities of the raw and cubic spline data shown in Fig. 1.

Figure 4. Cycle mean shear rates along the far-wall of the artery for both diameter ratio cases.

Biaxial Mechanical Behavior of Bovine Pericardium as a Bioprosthetic Material

Michael S. Sacks[1]

Abstract

Chemically treated bovine pericardium (BP) is used in many current bioprosthetic heart valves (BHV) designs. Although successful in the first five years post-implantation, BHVs suffer from long-term degradation and failure due to fatigue and calcification. A first step toward a sound engineering basis for improved BHV designs is the development of a realistic constitutive model for chemically treated BP. With the use of an exponential-type strain energy function W, the quasi-static mechanical behavior of BP was accurately modeled. The particular form of W used also allowed for the separation of axial and coupling components of the tissue stress, better revealing the characteristics of the mechanical coupling behavior.

Introduction

Chemically treated bovine pericardium (BP) is used in many current bioprosthetic heart valves (BHV) designs. Although successful in the first five years post-implantation, BHV suffer from long-term mechanical fatigue and calcification, resulting in degradation and failure. A further difficulty lies BP's intrinsic structural and mechanical variability, which confounds the traditional design process. As a first step toward developing a more sound engineering basis for improved BHV designs, we have developed a constitutive model using an exponential-type strain energy function, W, that models BP quasi-static biaxial mechanical behavior. A small angle light scattering (SALS)-based sorting/scanning procedure was used to pre-select BP specimens with a regular and relatively homogeneous collagen fiber architecture (Sacks, 1994), The particular form of W used allows for the decomposition of the mechanical response along each axis into an axial component representing the stress due to axial strain, and a coupling component representing the response due to the cross-axis strain.

----

[1]Assistant Professor, Department of Biomedical Engineering University of Miami, Coral Gables, FL 33124-0621 email: msacks@coeds.eng.miam.edu

Methods

Whole anterior sections of the BP sac were placed flat in a glass well filled with a hyper-osmotic storage solution. Vector plots of each anterior section were generated, and a 25 mm x 25 mm test specimen that had a uniform collagen fiber architecture was selected for biaxial testing. Each biaxial test specimen was fixed in 0.625% glutaraldehyde in a stress-free state. The test specimens were then mounted to the biaxial tester by continuous strands of 000 silk suture looped five times through the specimen along each of the four sides, and aligned such that the preferred collagen fibers are parallel to the x_1 stretch axis (Figure 1-a). All testing was performed with the specimens completely immersed in phosphate buffered normal saline at room temperature, pH 7.4. Tissue deformations were measured by monitoring the real-time movements of four small beads forming a ~4 mm square region in the specimen center. The test protocol began with preconditioning the specimen under equi-biaxial strain, i.e. where the ratio of the axial strains E_{11}:E_{22} = 1:1, where E_{ij} are the Green's strains. Next, five consecutive tests were performed in which the E_{11}:E_{22} ratio was kept constant, using ratios of 3:1, 2:1, 1:1, 1:2, and 1:3 (Figure 1-b). Each test consisted of 15 contiguous cycles using a 30 second period, with the data from the last three cycles of considered as representative.

To characterize the anisotropic mechanical behavior, we assumed that chemically treated BP can be treated as a hyperelastic material (Fung, 1965). Thus, the stresses can be computed from $S_{ij} = \partial W/ \partial E_{ij}$, where W is the strain energy function, and S_{11} and S_{22} are the Kirchoff stresses along the preferred and cross-preferred collagen fiber directions, respectively. For canine pericardium, Choi and Vito (1994) proposed the following strain energy function:

$$W = b_0\left[\exp\left(\frac{1}{2}b_1E_{11}^2\right)+\exp\left(\frac{1}{2}b_2E_{22}^2\right)+\exp\left(b_3E_{11}E_{22}\right)-3\right] \quad (1)$$

Where b_i are the material constants. Using $S_{ij} = \partial W/ \partial E_{ij}$, the b_i are determined by simultaneously fitting data from all five protocols for each specimen to:

$$\begin{aligned} S_{11} &= b_0\left[b_1\,E_{11}\exp(\frac{1}{2}b_1E_{11}^2)+ b_3\,E_{22}\exp(b_3E_{11}E_{22})\right] \\ S_{22} &= b_0\left[b_2\,E_{22}\exp(\frac{1}{2}b_2E_{22}^2)+ b_3\,E_{11}\exp(b_3E_{11}E_{22})\right] \end{aligned} \quad (2)$$

An advantage of this form of W is that the second exponent term in (2), which represents the axial coupling, can be algebraically separated from the first exponent term so that the total stress can be divided into the axial and coupling contributions.

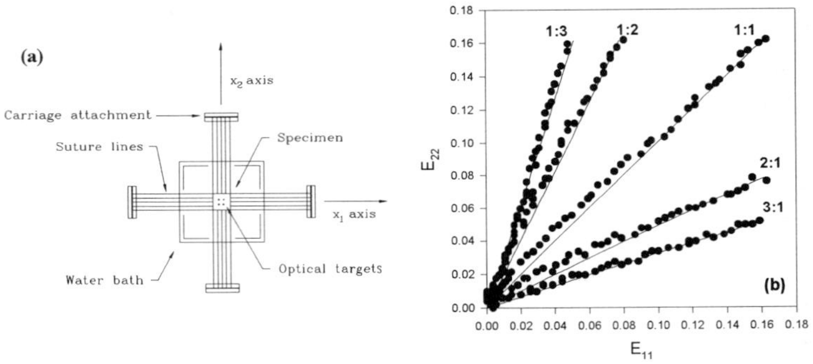

Figure 1 - (a) An illustration showing the attachment of a BP test specimen to the biaxial mechanical test device, and (b) five E_{11}:E_{22} strain protocols used to characterize the BP anisotropic biaxial mechanical behavior. In this study, E_{11} and E_{22} are the strains along the preferred fiber and cross preferred directions, respectively.

Results and Discussion

Representative biaxial mechanical behavior of glutaraldehyde treated BP for all stretch protocols are shown in Figure 2. Equation (2) fit biaxial data from all nine BP specimens data well, with a mean r^2 of 0.97 (Table 1). S_{11} was consistently larger in magnitude than S_{22}, demonstrating mechanical anisotropy. The axial component of the total stress was dominant (Figure 3-a). However, the axial component increases with strain for S_{11} but decreases for S_{22} relative to the total stress, indicating different mechanical coupling behavior for each axis (Figure 3-b). The basis for this complex coupling behavior is likely due to tissue microstructure. The nature of this interaction is currently under investigation in our laboratory.

Table 1 - Results of 9 glutaraldehyde treated BP specimens fitted to equation (2).

	b_0	b_1	b_2	b_3	r^2
mean	1.873	106.92	71.39	45.52	0.97
sem	0.49	8.86	19.20	8.27	0.01

Acknowledgments The support of the Whitaker Foundation is greatly appreciated.

References

Choi, HW, and Vito, RP, J. Biomech. Eng., 1990, **112**, 153-159.
Fung, Y.C. Foundations of Solid Mechanics, Prentice-Hall, 1965.
Sacks, MS, Chuong, C, More, R. ASAIO Journal, 1994; **40**(3):M632-7

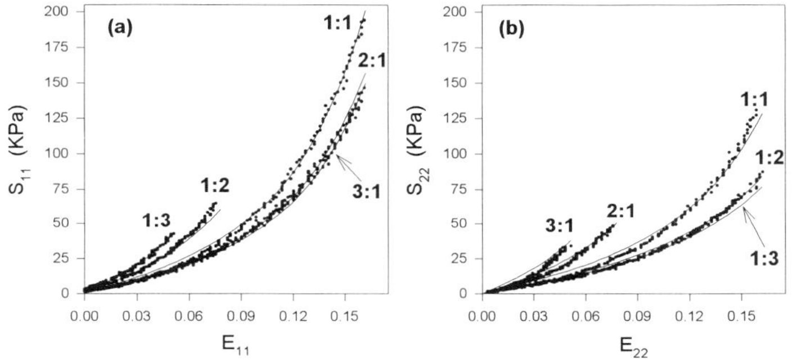

Figure 2 - Representative biaxial mechanical behavior of glutaraldehyde treated BP for all 5 E_{11}:E_{22} strain history protocols (indicated beside each curve). S_{11} is larger in magnitude than S_{22}, demonstrating BP mechanical anisotropy.

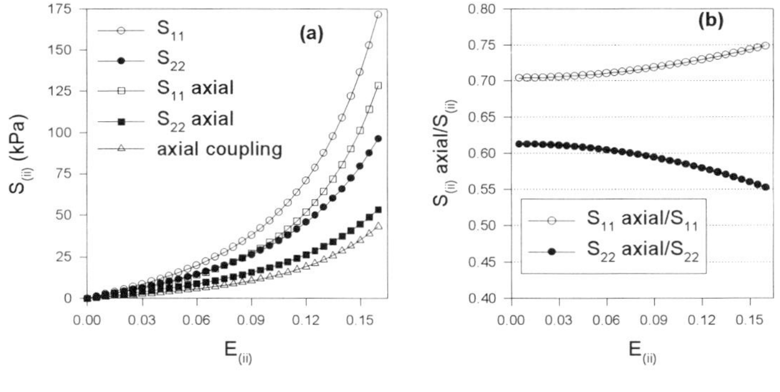

Figure 3 - (a) Total, axial, and coupling components of the Kirchoff stresses for equi-biaxial strain (E_{11}:E_{22}=1:1), computed using equation (3) and the mean material constants listed in Table 1. For both axes, the axial component is dominant. However, (b) demonstrates that the axial component increases with strain for S_{11} but decreases for S_{22} relative to the total stress, indicating different mechanical coupling behavior for each axis.

Pulsatile Flow Circuitry for Dynamic Study of a Volume Plethysmograph

Alex A. Yu[1] Jeff Raines[2] Ned H.C. Hwang[1]

Abstract

A pulsatile flow loop has been established to simulate arterial flow at the level of the human thigh. We are using an artificial artery with mechanical characters similar to human artery. The subcutaneous tissue is simulated with a commercially available artificial fat. The hemodynamic and volume plethysmographic results show pulsatile and mean pressure and flow values similar to values found in the human circulation.

Introduction

Pulsations in pressure and flow are generated with each heart contraction and produce pulse wave propagation in arteries. These waveforms are a function of ventricular contractile performance and the geometric and viscoelastic properties of the arterial tree. These properties are known to change considerably with age and pathological conditions. There are several different approaches in studying the pressure-flow conditions (1). 1) transmission line or wave propagation models; 2) reduced models, such as the Windkessel , which are the special cases of propagation models; 3) hemodynamic flow loops that simulate the circulatory condition of the arterial system in question. The hemodynamic model is more straight forward. Further, all important parameters can be measured and analyzed. It is also the experimental basis for establishing the transmission line or lumped element theory.

In this study, we focused on establishing a pulsatile flow arterial circulation at the level of the human thigh. The pressure and flow conditions with respect to the arterial compliance are of major interest.

----

1 Department of Biomedical Engineering, University of Miami, Coral Gables, FL 33124

2 Raines, University of Miami School of Medicine, Miami, FL 33101

The results indicate this circulatory flow loop can be used to study normal subjects as well as patients with different arterial compliance associated with various physiological condition including atherosclerosis. Our ultimate goal was to simulate realistic human pressure-flow conditions at the thigh level for the purpose of investigating arterial pressure-volume changes as measured by a specific volume plethysmograph.

Background

A brief review of the anatomical structure of the thigh is indicated (2). The external iliac artery passes out of the pelvic cavity beneath the inguinal ligament and becomes the femoral artery. It is close to the surface and not only can be palpated but is an important pressure point. Several vessels arise from the femoral to serve the thigh region. The largest of these, the deep femoral artery (profunda femoris), passes posteriorly to serve the hamstring muscles. The lateral and medial femoral circumflex arteries encircle the proximal end of the femur and serve muscles in this region. The femoral artery becomes the popliteal artery as it passes above the posterior aspect of the knee. The popliteal artery divides into peroneal artery and the anterior tibial and posterior tibial arteries.

The volume plethysmograph for the indirect measurement of arterial compliance requires a material that simulates human thigh tissue and transmits volume changes from the surrounded vascular bed. In this paper, we call this material artificial fat.

The arterial flow rate to supply the human thigh varies as a function of the flow state. At rest the flow rate is between 100 and 200 mL/min; with exercise this value can easily reach > 1000 mL/min.

System and Method

The Hemodynamic Bench-top Model (HBM) was designed to simulate the hemodynamic conditions found in human lower extremities. Pulsatile flow in the HBM is generated by a linear actuator (Aura System, El Segundo, CA) which delivers a continuous force of 10 pounds and a stroke of 2 inches. The Electronic Control Unit (ECU) is used to provide the position control and position feedback which is needed to precisely reproduce pulsatile flow conditions on a beat-to-beat basis. Two 17 mm mechanical heart valves (Omni Science and St. Jude) are used in the flow loop system to control flow direction.

A physiological pressure transducer (Ohmeda Medical Device Division Inc., Oxnard, CA) and an electromagnetic flow meter (Zepeda Instruments, Seattle, Washington) are used to measure instantaneous pressure and flow rate in the HBM loop, respectively.

The structure of the human thigh is simulated by wrapping sheets of artificial fat (Muller Sports Medicine Inc., Prairle Du Sac, WI) around 6.4 mm penrose drain tubing (Davol Inc. Cranston, RI) whose compliance is similar to that of the human artery at the level under investigation. The volume plethysmograph is the Vasogram 2 System currently under development for the early detection of coronary artery disease (Vasocor Inc., Miami, FL).

A personal computer (Gateway 2000 P4D-66) and data acquisition card (Microstar Laboratories, Bellevue, WA) acquire, monitor and store the instantaneous pressure and flow signals. The PC also controls the movement of the linear actuator to obtain physiological pressure and flow simulation. The HBM flow loop is shown in Fig.1. A software package written in C language is used to interface all the inputs and outputs and was developed in the author's Laboratory.

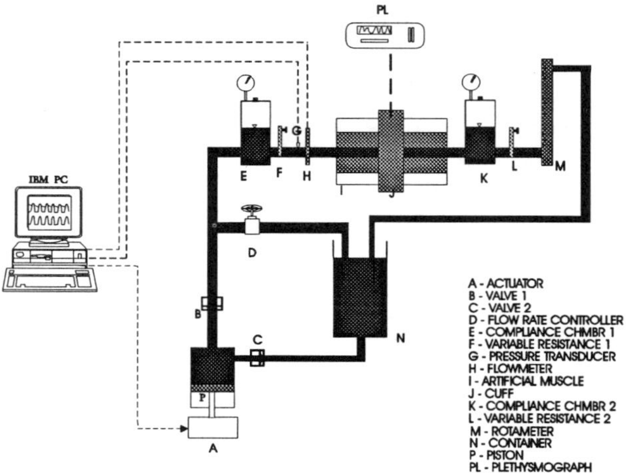

Fig.1 Diagram of the Simulation System

The system has two adjustable impedance units which can be individually adjusted to control pressure and flow rate. Two compliance chambers represent the lumped arterial compliance of vessels upstream and downstream of the human thigh. By adjusting the impedance and compliance in the loop, we can change the corresponding pressure and flow rate. A bypass valve is also installed to provide additional flow rate control when needed. The computer controlled linear actuator is activated by a specially designed waveform to simulate the physiological flow conditions in the human thigh. The fluid used in the experiment is distilled water; the experiments are performed at room temperature (20° $\pm$ 2° C).

Results

The pressure and the flow rate curves for our Control Case are shown in Fig. 2. Raw data from the Vasogram 2 which can be used to determine local arterial compliance is also shown in this figure. These three curves represent tracings found in healthy human subjects and suggests our pulsatile flow loop can be used to simulate the human lower extremity circulatory system in various physiological conditions, including decreased arterial compliance found in atherosclerosis.

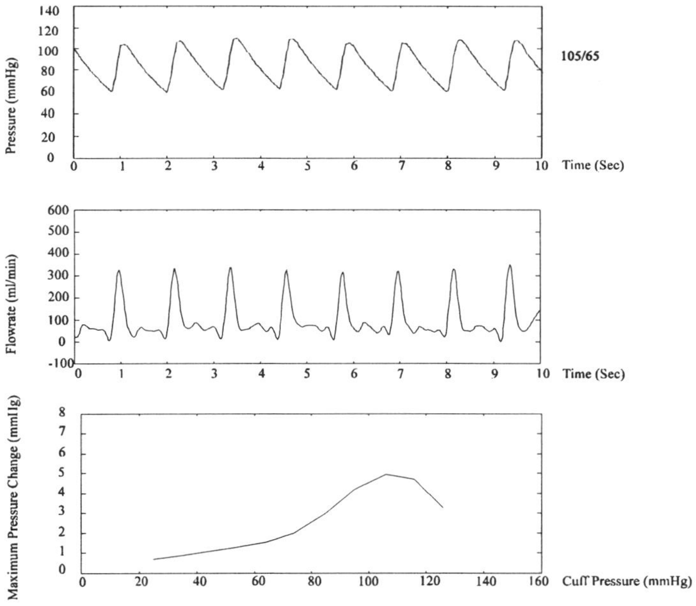

Fig. 2 Pressure, flowrate and cuff pressure change at 50 bpm.

Reference

1. David S. Berger, John K. J. Li, Warren K. Laskey, and Abraham Noordergraaf. Repeated reflection of wave in the systemic arterial system. Am. J. Physiol. 264, 1993.
2. Kent M. Van De Graaff. Human Anatomy. 3rd edition. 1992.
3. W.W. Nichols, M.F.Orouke. Mcdonald's Blood Flow in Arteries. Lea & Feligh Publisher, 1990

A Proposed Poroelastic Model for Hydraulic Fracturing Breakdown Pressure in Porous Rocks

Xiaofeng Huang†

Abstract

A general poro-elastic model for hydraulic fracturing breakdown pressure is derived by applying the true effective stress into the solution of the minimum tangential total stress at the borehole wall during hydraulic fracturing, under the condition that the porous material is fully saturated prior to the hydraulic fracturing test. The true value of the effective stress is found to be the difference between the total stress and the fluid pressure, which is the sum of two parts of fluid pressure. One part is equal to the pre-existing pore fluid pressure, the other is the pressure difference between injection fluid pressure and the pre-existing pore fluid pressure multiplied by the Biot's coefficient α. The new model matches the experimental results of breakdown pressure on Indiana limestone (Huang, 1989) the best. Haimson and Fairhurst, and Hubbert and Willis' models predict the lower and upper bounds, respectively.

Introduction

The elastic model for breakdown pressure (P_c) was derived by Hubbert and Willis (1957) as:

$$P_c = 3S_h - S_H + T - P_o \tag{1}$$

where: S_h and S_H = least and largest horizontal principal stresses, respectively;
T = tensile strength of rock, and
P_o = initial pore pressure.

Equation (1) is based on the assumption that no fluid flow from the borehole into the surrounding rock occurs. However, in practice, most rocks are permeable so that borehole injection fluid flows into the surrounding rock. In this situation, equation (1) does not model what is occurring; therefore, another model is needed. The first poro-elastic model for the breakdown equation, which considers the effect of fluid flow, was derived by Haimson and Fairhurst (1967) as:

$$P_c - P_o = \frac{3S_h - S_H + T - 2P_o}{K} \tag{2}$$

where: $K = 2 - \alpha\dfrac{1-2\nu}{1-\nu}$ and $\alpha = 1 - \dfrac{C_r}{C_b}$ is Biot's coefficient,

† Merit Engineering, Inc., 1200 Harris Avenue, Suite #203, Bellingham, WA 98225, U.S.A. Telephone: (360)738-6083, Fax: (360)738-1499

ν = Poisson's ratio, and
C_r and C_b = compressibility of rock solid and rock bulk, respectively.

Equation (2) was derived by using Terzaghi's effective stress law which is:

$$\sigma = S - P_c \qquad (3)$$

where: σ = effective principal stress, and
S = total principal stress.

The H&F (Haimson and Fairhurst) model has an obvious problem as being questioned by Schmitt and Zoback (1988). In the limiting case of impermeable rock, which implies $\alpha = 0$, the model does not converge to the elastic model (H&W model) as expected. The parameter K becomes K=2, but it should be K=1 if the model is to agree with equation (1). Schmitt and Zoback (6) rederived the poro-elastic model by applying Nur and Byerlee's exact effective stress law which is:

$$\sigma = S - \alpha P_c \qquad (4)$$

to the same solution of the minimum tangential total stress at the borehole wall during hydraulic fracturing used by Haimson and Fairhurst (1), and obtained

$$P_c = \frac{3S_h - S_H + T + \alpha(K-2)P_o}{K + \alpha - 1} \qquad (5)$$

In the case of highly porous rock, which implies that $\alpha = 1$, equation (5) is the same as equation (2). In the limiting case of impermeable rock, which implies $\alpha = 0$, equation (5) and equation (2) are different. In this situation, however, they both cannot merge to equation (1) which was proved correct for the no fluid flow case.

Considering the procedures from which equations (2) and (5) were derived, it is apparent that the difference between equations (2) and (5) was only induced by using different effective stress laws, equations (3) and (4). Therefore, to clarify the exact poro-elastic model for breakdown pressure, it is necessary to re-examine the effective stress law first.

Effective Stress

The concept of effective stress was first introduced for soil consolidation by Terzaghi (1923). The effective stress was defined as the stress which constitutes the stress-strain relation (Terzaghi and Peck, 1948). The effective stress expressed by Terzaghi as equation (3) is based on experimental grounds only. Nur and Byerlee (1971), by using the constitutive equation derived by Biot (1941), derived the so-called "exact effective stress" as equation (4). Their experimental results on rocks showed that the stress-strain relation is controlled by equation (4). They explained the reason for the validity of Terzaghi's effective stress, equation (3), is that when $C_r \ll C_b$, $\alpha \approx 1$, then equation (4) reduces to equation (3). Although Nur and Byerlee did not conduct experiments on rock for strength, they limited their effective stress law to elastic stress-strain relation only by stating that their results are not applicable to inelastic processes, such as fracture, or elastic process other than strain. This is perhaps an attempt not to raise controversy because most previous tests on soils showed Terzaghi's effective stress law (equation 3) holds for failure.

From the above review, it is apparent that the expression for effective stress is controversial. Neither Terzaghi's concept of effective stress nor Nur Byerlee's exact effective stress can satisfactorily explain the phenomena of effective stress. A method which can better explain the concept of effective stress is definitely needed.

Effective Stress From Energy Conservation Point of View

In geomechanics, the energy conservation method, i.e., Bernoulli's equation, is commonly used when the moving fluid pressure in the porous media needs to be known. In a case like hydraulic fracturing, the relationship between injection fluid pressure (P_i) and fluid pressure inside the rock, i.e., pore pressure (p), can be qualitatively expressed as:

$$P_i = p + P_v \quad (6)$$

where: $P_v = \frac{\gamma v^2}{2g}$

g = acceleration of gravity, and

γ = unit weight of the fluid.

Equation (6) shows that the larger the velocity of fluid flow, the smaller the pore pressure becomes for a constant borehole fluid pressure. If the fluid does not move, the pore pressure is equal to the borehole fluid pressure. From equation (6), the effective stress for the general case, when fluid motion can occur, can be written as:

$$\sigma = S - p = S - (P_i - P_v) \quad (7)$$

In a fully saturated medium, such as the type analyzed by Biot (1941), suppose the initial pore pressure is P_o. The injection of fluid into borehole causes a fluid pressure change $\Delta P = P_i - P_o$, which in turn causes fluid flow. In this case, the fluid pressure in the surrounding medium is the sum of two parts. One part is the pressure from the moving fluid, which can be found from equation (6) as $\Delta p = (P_i - P_o) - (P_i - P_o)_v$. The other part is from the original pore fluid pressure P_o. Thus,

$$p = P_o + \left[(P_i - P_o) - (P_i - P_o)_v\right] \quad (8)$$

where: $0 \le \left[(P_i - P_o) - (P_i - P_o)_v\right] \le (P_i - P_o)$

The effect of the incremental fluid pressure ΔP, which causes fluid flow, on the state of stress during fluid flow was solved by Biot (1941). Nur and Byerlee (1971) perhaps were the first to use Biot's consolidation theory to explain effective stress. However, they did not include one important point which is the basis for Biot's theory held. Biot (1941) mentioned that the fluid pressure in his derivation is the incremental fluid pressure. Therefore, equation (4) given by Nur and Byerlee should be explicitly given as

$$\sigma = S - \alpha(\Delta P) = S - \alpha(P_i - P_o) \quad (0 \le \alpha \le 1) \quad (9)$$

It is apparent that Biot's theory is only dealing with a fluid pressure difference. It is not applicable to the static fluid situation. And it is necessary to mention that an important basic requirement for both Biot's theory and Terzaghi's theory to be valid is the assumption that the porous medium is fully saturated. This assumption is essential. For dry or partially saturated porous medium, the problem is controversial.

Application of Effective Stress to Hydraulic Fracturing

The total pore pressure in hydraulic fracturing test when fracture initiates can be obtained by combining equations (8) and (9):

$$p = P_o + \alpha(P_c - P_o) \quad (10)$$

Applying equation (10) into the solution of minimum tangential effective stress at the borehole wall, Huang's model is obtained (Huang, 1995):

$$P_c - P_o = \frac{3S_h - S_H + T - 2P_o}{1 + \alpha \frac{\nu}{1-\nu}} \quad (11)$$

The model merges to Hubbert and Willis' model in the limiting case of $\alpha=0$ and is identical to the Haimson and Fairhurst's model when $\alpha=1$. The experimental results of hydraulic fracturing breakdown pressures with pre-existing pore pressures (Huang, 1989) agree well with Huang's model. While the other models show discrepancies with the experimental results and the discrepancies become greater as effective confining pressure increases. Hubbert and Willis' model predicts the upper bound and Haimson and Fairhurst's the lower bound (Huang, 1996).

Conclusions

The poroelastic model for hydraulic fracturing breakdown pressure was derived by using the total minimum tangential stress at the borehole wall analyzed by Haimson and Fairhurst (1967) with the "true" pore pressure demonstrated in this paper. The pore pressure component α (P_c-P_o) induced by fluid flow is the same as that analyzed by Nur and Byerlee (1971). This pore pressure component is implied in Biot's (1941) constitutive equation of consolidation (Huang, 1996). Therefore, the use of this component to obtain effective stress in hydraulic fracture initiation is consistent with the application of the total stress component due to fluid flow from Biot's equation in derivation of hydraulic fracturing breakdown equation.

The experimental results of breakdown pressure from Indiana limestone with the existence of initial pore pressure showed good agreement with Huang's model, especially when the effective confining pressures are greater than 14 MPa (Huang, 1996). This indicates that Huang's model may be superior when hydraulic fracturing stress measurement is conducted at great depth for this type of rock.

References

Biot, M.A. General theory of three-dimensional consolidation, *J. of Appl. Physics*, **12**, 155-165 (1941).

Haimson, B.C. and Fairhurst, C., Initiation and extension of hydraulic fractures in rock, *Soc. Petr. Engr. J.*, 310-318 (1967).

Huang, X., The effect of Pore Pressure on Hydraulic Fracturing, *M.S. Thesis*, University of Wisconsin-Madison (1989).

Huang, X., Effective stress and poro-elastic model for hydraulic fracturing stress measurements, *Trans Am. Geoph. Union, EOS*, 76, pp. , 1995.

Huang, X. A general poroelastic model for hydraulic fracturing,*in preparation,* 1996.

Hubbert, M.K. and Willis, D.G, Measurement of hydraulic fracturing, *Transactions AIME*, **210**, pp.153 (1957).

Nur A. and J.D. Byerlee, An exact effective stress law for elastic deformation of rock with fluids, *J. Geoph. Res.*, **76**, 6414-19 (1971).

Schmitt, D.R. and M.D. Zoback, The effect of an exact effective stress law on the hydrofracture breakdown equations, *Procs. of the 2nd Int'l Workshops on Hydraulic Fracturing Stress Measurements*, **1**, pp. 431 (1988).

Terzaghi, K. Van, Die Berechnung der Durchlassigkeitsziffer des Tones aus dem Verlauf der hydrodynamischen Spannungserscheinungen, *Sber. Akad. Wiss. Wien,* 132, 105, (1923).

Terzaghi, K. Van, and R.B. Peck, Soil Mechanics in Engineering Practice, *John Willey & Sons, Inc.,* New York, Chapman & Hall, Ltd., London, (1948).

Cracks in poroelastic materials with a highly anisotropic fluid response

By C. Atkinson and R.V. Craster

Department of Mathematics, Imperial College of Science, Technology and Medicine, London, SW7 2BZ, U.K.

Department of Theoretical Mechanics, University of Nottingham, Nottingham, NG7 2RD, U.K.

A porous elastic material with an isotropic elastic response, but highly anisotropic permeability (the vertical permeability is zero, so fluid flow is only in horizontal directions) is considered. It is shown that, in the special cases when the crack faces are impermeable and are orientated to be parallel to the flow direction, the highly anisotropic results are very good approximations to those found for a material with isotropic permeability. Many rocks or soils have a layered structure so the results are also of interest in their own right.

1. Introduction

In the majority of solutions employing Biot's theory of poroelasticity, Biot (1941), the porous material is taken to be isotropic in both the fluid and solid responses, although Biot generalised his theory to encompass anisotropy, Biot (1956). For most geophysical materials an assumption of isotropy is rather crude, in particular many geophysical materials such as shales are laid down in a highly stratified manner so have a layered structure, Swan *et al* (1989), and can exhibit strongly anisotropic behaviour in both their fluid and solid responses. These materials could be considered to be transversely isotropic such materials have permeabilities which are anisotropic and for some layered materials the permeability along the bedding plane may be several times greater than that perpendicular to the bedding plane. To approach such problems, and to consider the effect of anisotropic permeability, we consider a specialisation of the usual isotropic problems where the solid elastic matrix is still isotropic, but the fluid response is allowed to have a particularly simple anisotropy. This anisotropic fluid response is easy to introduce to the usual isotropic poroelastic equations, using the reformulation of Biot's equations by Rice and Cleary (1976) the stresses are

$$\sigma_{ij} = 2G\epsilon_{ij} + \frac{2G\nu}{(1-2\nu)}\delta_{ij}\epsilon_{kk} - \alpha p\delta_{ij}, \tag{1.1}$$

with ϵ_{ij} as the strain tensor. The pore pressure, p, satisfies the following relationship between the variation of fluid volume content, ζ, and the dilatation, $e = \epsilon_{kk}$, $p = Q\zeta - \alpha Q\epsilon_{kk}$. The constants and their physical interpretation are in the standard notation. The governing equations can be written as: the equilibrium equation for the stresses $\sigma_{ij,j} = 0$. A one dimensional Darcy's law which relates mass flux to the gradient of pore pressure in the x direction, $q_1 = -\rho_0\kappa_1 p_{,1}$. This equation differs from the isotropic permeability case where $q_i = -\rho_0\kappa p_{,i}$. Finally there is a mass conservation equation for the pore fluid $\partial m/\partial t = -q_{1,1}$, with m the mass of fluid per unit volume and ρ_0 as the reference density. In the isotropic case $\partial m/\partial t = -q_{i,i}$. The governing equations are written as an elastic Navier equation with a coupling term for the pore pressure and as a generalised diffusion equation for the pore pressure with a coupling term for the dilatation, e, i.e.,

$$G\nabla^2 u_i + \frac{G}{(1-2\nu)}e_{,i} - \alpha p_{,i} = 0, \quad \frac{\partial p}{\partial t} - Q\kappa_1 \frac{\partial^2 p}{\partial x^2} = -\alpha Q \frac{\partial e}{\partial t}. \tag{1.2}$$

The highly anisotropic case is a specialised version of the isotropic permeability case where flow is only allowed in the horizontal direction, but note that the pore fluid pressure is still fully coupled to the elastic dilatation and vice-versa. The advantages of considering the highly anisotropic case are that in some special cases further insight is gained into the physical interactions occuring, and that the results are often substantially simpler.

In a series of recent papers Atkinson and Craster have considered time dependent fracture in a poroelastic material; the material is isotropic in both its fluid and solid responses. Various semi-infinite crack problems have been solved exactly, and more recently some finite crack problems have been treated using an asymptotic method. Much of this work is outlined in Craster and Atkinson (1995). Most of the analysis was aimed at solving problems for which the crack faces would have a different pore pressure boundary condition from that found by symmetry conditions upon the fracture plane ahead of the crack tip, this will allow more complex fluid-solid interactions to be considered. The highly anisotropic versions of these problems, at least in the rather extreme case considered here, do not exist. Another important class of problems which are also treated, although somewhat more briefly, are cases when the pore pressure boundary condition upon the crack faces and the fracture plane ahead of the crack tip are the same. It is with those cases where the boundary condition is that this fracture plane is impermeable that there is a similarity with a highly anisotropic permeability material. The assumption that flow occurs only horizontally implies that a comparison with impermeable cracks in a material with isotropic permeability, (which from symmetry have an impermeable pore pressure boundary condition ahead of the crack tip) might be worthwhile. In the isotropic case the effectively impermeable half plane will ensure that, at least local to the crack tip, much of the fluid diffusion will be horizontal. The specific problem to be considered is a semi-infinite crack lying on $y = 0$ $x < 0$, the material ahead of the crack is assumed to be undamaged. In the highly anisotropic case the crack is assumed to lie in a plane parallel to the direction of fluid flow. The boundary conditions upon $y = 0$ are

$$\sigma_{yy} = -\tau_0 e^{\frac{x}{a}} H(t) \quad \text{for} \quad x < 0, \quad u_y = 0 \quad \text{for} \quad x > 0 \tag{1.3}$$

$$\sigma_{xy} = 0 \quad \text{for all } x \text{ and } \frac{\partial p}{\partial y} = 0 \quad \text{for all } x. \tag{1.4}$$

The boundary conditions for $x > 0$ are deduced from symmetry, and for analytical convenience the internal stress loading is taken to be spatially exponentially decaying; more elaborate loadings can be deduced by superposition. The near crack tip stress fields are given by the usual elastic crack tip for a tensile crack solution, where the stress intensity factor K_1 is a function of time. For the situations considered here the pore pressure gradient is not singular at the crack tip.

In the case of purely horizontal fluid flow one can extract the exact solution for the mode 1 stress intensity factor in real time as

$$K_1(t') = \tau a^{\frac{1}{2}} 2^{\frac{1}{2}} \left(1 + \left(\left(\frac{1}{N_+(0)} - 1\right) e^{\frac{t'}{N_+^2(0)}} \operatorname{erfc}\left(\frac{\sqrt{t'}}{N_+(0)}\right)\right)\right) H(t') \tag{1.5}$$

and $t' = ct/a^2$. This solution is shown in figure 1 as $K_1(t')/\tau a^{\frac{1}{2}} 2^{\frac{1}{2}}$ versus t', for $\nu = 0.1$, $\nu_u = 0.5$; the limiting small time and large time behaviour can be extracted directly from formula (1.5). They are that

$$\text{as } t' \to 0 \ K_1(t') \sim \tau a^{\frac{1}{2}} 2^{\frac{1}{2}} \frac{(1-\nu_u)^{\frac{1}{2}}}{(1-\nu)^{\frac{1}{2}}}, \quad \text{and as } t' \to \infty \ K_1(t') \sim \tau a^{\frac{1}{2}} 2^{\frac{1}{2}}. \tag{1.6}$$

These limiting cases can be deduced from physical arguments, for large times the material becomes effectively drained and the stress intensity factor is that of an elastic material with drained Poisson ratio. For small times the crack tip behaves as an inclusion with drained Poisson ratio embedded within an undrained elastic material. This can be made rigorous using an invariant integral.

For an isotropic permeability in those special cases when the fracture plane is impermeable (or permeable) for all x these limiting results are independent of the anisotropy of the permeabilty. The issue when the pore pressure boundary condition is different upon the crack faces from that on the fracture plane ahead of the crack is considerably more complicated, see Atkinson and Craster (1992).

The dotted line in figure 1, for the same material parameters and with $\kappa = \kappa_1$, is the stress intensity factor for an identically loaded impermeable crack under tension in an isotropic elastic material with an isotropic permeablity. The fully coupled system of equations is somewhat harder to analyse; the solution is found after a Laplace transformation in time, due to the complexity of the analysis the inversion has to be performed numerically. The Laplace transformed (s is the transform parameter) stress intensity factor is, from Craster and Atkinson (1992), that

$$K_1(s) = \frac{\tau a^{\frac{1}{2}} 2^{\frac{1}{2}}}{s\overline{N}_+(\frac{i}{a_1})} \tag{1.7}$$

where $a_1 = a(s/c)^{\frac{1}{2}}$; it is inverted numerically to obtain the real time results. The final results are remarkably close; similar accuracy occurs for other values of ν, ν_u. The isotropic permeability crack problem has (from symmetry) the boundary condition that $\partial p/\partial y = 0$ on $y = 0$ for $-\infty < x < \infty$. The closeness of the two solutions suggests that, at least, local to the crack tip much of the fluid motion is purely in the x direction (due to the impermeable boundary condition) and this leads to the similarity. Initially this is a somewhat surprising result as, in the isotropic permeability case one would expect some fluid transport in the vertical direction due to the loading and subsequent changes in the elastic stresses. These in turn would effect the stresses near the crack tip.

It is worth noting that in the cases for an isotropic permeability material, when the crack is permeable the comparison with these highly anisotropic results is not good.

2. Conclusion

It is shown that for a rather special poroelastic problem an assumption that the material allows only horizontal flow leads to highly accurate approximations to those results found if the material permeability is isotropic. The highly anisotropic results are considerably simpler than the equivalent isotopic results; since many engineering/ geomechanical applications may just require a quick assessment of the order of the pore fluid effect this approximation is of value. Note the solutions are good global approximations and are not asymptotic. Due to the layered structure of many rocks and soils these results are not without value in their own right. In particular they suggest that for those problems considered here the specific value of the vertical permeability, κ_2, is not important, (provided it is less than or equal to the horizontal permeability) although further detailed analysis is required to fully quantify this conjecture.

References

Atkinson C. and Craster R.V. 1992, "The application of invariant integrals in elastic diffusive solids", *Phil. Trans. Roy. Soc.*, **A339**, 231-262

Biot M.A. 1941, "General theory of three-dimensional consolidation", *J. Appl. Phys.*, **12**, 155-164

Biot M.A. 1956, "Theory of deformation of a porous viscoelastic anisotropic solid", *J. Appl. Phys*, **27**, 459-467

Craster R.V. and Atkinson C. 1992, "Shear cracks in thermoelastic and poroelastic media", *J.Mech. Phys. Solids*, **40**, 4, 887-924

Craster R.V. and Atkinson C. 1995, "Theoretical aspects of fracture in porous elastic media", in Mechanics of Poroelastic Media (Ed. A.P.S. Selvadurai), Kluwer Academic Press

Rice J.R. and Cleary M.P. 1976, "Some basic stress-diffusion solutions for fluid saturated elastic porous media with compressible constituents," *Rev. Geophys.Space Phys.*, **14**, 227-241

Swan G., Cook J., Bruce S. and Meehan R. 1989, "Strain rate effects in Kimmeridge Bay Shale", *Int. J. Rock Mech. Min. Sci. & Geomech. Abstr.*, **26**, 135-149

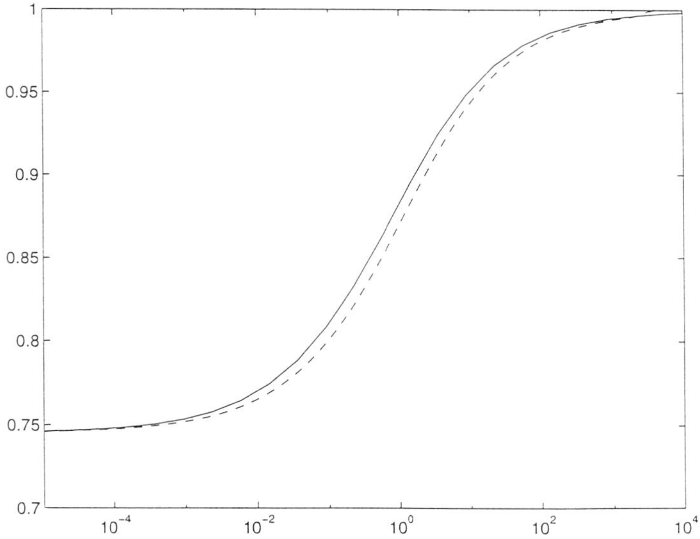

EQUIVALENT STRENGTH OF POROUS FRACTURED ROCK

William G. Pariseau[1]

Abstract

Numerous discontinuities, "joints", in natural rock masses suggest an equivalent material properties approach for averaging such important geologic features up to a level of detail that can be accommodated by the computer. A non-representative volume element (NRVE) formulation shows that the form of the elastic-plastic constitutive equations is the same for homogeneous material elements and multiple-material elements, subsequently homogenized. Thus, $\{<d\sigma>\}=([C^*]-[C_p^*])\{<d\epsilon>\}$ where σ is effective stress and $[C^*]$ and $[C_p^*]$ are equivalent elastic moduli and plastic "correction" matrices, respectively. However, there is no equivalent plastic potential Y^* or failure criterion F^*.

Introduction

Equivalent material properties relate averages of variables that appear in the constitutive equations adopted for the problem of interest. The averaging scale is large compared with a relevant microstructural scale but small compared with the scale of the region of interest. For example, in application of Hooke's law to a tunnel in massive sandstone, the microstructural scale would be characterized by grain size (millimeter scale). The elastic moduli would be measured on drill core samples containing many grains (centimeter scale), while tunnel diameter (meter scale) would be much larger than laboratory sample size. In wet ground, under saturated conditions, Biots's well-known extension of Hooke's law to poroelastic behavior applies. In differential form this is

$$\begin{Bmatrix} \{<d\sigma>\} \\ <d\zeta> \end{Bmatrix} = \begin{bmatrix} [C^*] & \{c^*\} \\ \{c^*\}^t & -c \end{bmatrix} \begin{Bmatrix} \{<d\epsilon>\} \\ <d\pi> \end{Bmatrix} \tag{1}$$

----

[1]Professor, Department of Mining Engineering, 315 WBB, University of Utah Salt Lake City, UT 84112-1183

where tension is positive, $\{\cdot\}$ is a 6x1 matrix, $[\cdot]$ is a 6x6 matrix, $<\cdot>$ is an overall average; σ, π, ϵ, ζ, C^*, c^* and c are effective stress, fluid stress, strain, moisture content, equivalent elastic moduli, coupling constants and hydrostatic constant, respectively. Usually, the angle brackets and star superscript are not used, while the physical meanings of stress, strain, moduli and so forth are understood intuitively. In this regard, *effective* stress determines strength and is associated with the strain contribution to the total (equilibrium) stress in (1).

In natural rock masses where joints are too numerous to be treated individually, an equivalent properties approach becomes desirable. The relevant "microscale" is joint spacing. Although spacing generally varies within and between joint sets, the meter scale is typical. A representative volume element (RVE) for properties measurements would need to contain many joints and thus would have to be tens of meters on edge and weigh many tonnes. Obviously such an RVE is an impractical size for testing. In a numerical analysis of a jointed rock mass, no one element could be considered representative. A non-representative volume element (NRVE) approach to equivalent rock mass properties is therefore needed.

The NRVE approach is an analytical approximation that treats joints as thin layers of material different from intact rock between joints (Pariseau, 1988, 1993). The approach builds on results from composite mechanics (Hill, 1963, 1985; Willis, 1981) within a finite element context that also provides a measure of true equivalent properties for comparison. Huet (1982) presents a systematic discussion of general conditions necessary for equivalent properties. In essence, volume averages of products of deviation terms must vanish in applicable balance laws, so an average of a product pair is then equal to the product of the two averages. Balance laws and constitutive equations for averages then have the same form as point-wise balance laws and constitutive equations.

NRVE Approach

Under the assumption of small, additive elastic and plastic strain increments and with the adoption of a plastic potential Y and failure criterion F, the elastic-plastic effective stress strain relations for a homogeneous material are

$$\{d\sigma'\}_\alpha = ([C^\alpha]-[C_p^\alpha])\{d\epsilon\}_\alpha \tag{2}$$

where $[C_p^\alpha]=[C]\{Y_\sigma\}(\{F_\sigma\}^t[C]\{Y_\sigma\}-\{F_\epsilon\}^t\{Y_\sigma\})^{-1}\{F_\sigma\}^t[C]$ and subscripts σ and ϵ imply differentiation with respect to stress and plastic strain. Averaging (2) over an element of homogeneous material results in the same form. However, the constitutive variables are now understood to be *local* averages over the α-material volume, indicated by the label α. This is the same intuitive understanding associated with application of (1) to tunneling in sandstone.

When (2) is averaged over a heterogeneous element of volume V containing several materials, the result is

$$\{<d\sigma'>\} = (\sum_{\alpha} f_\alpha([C^\alpha]-[C_p^\alpha])[A]^\alpha])\{<d\epsilon>\} \tag{3}$$

where f_α is the volume fraction (V_α/V) of the α-material and a strain influence matrix $[A^\alpha]$ is introduced that relates the local (α) average strain increment to the *overall* element average, that is, $\{d\epsilon\}_\alpha = [A^\alpha]\{<d\epsilon>\}$. The two sums in (3) are the equivalent elastic properties and plastic "correction" matrices. Thus, $[C^*] = \sum f_\alpha[C^\alpha][A^\alpha]$ and $[C_p^*] = \sum f_\alpha[C_p^\alpha][A^\alpha]$ which show that equivalent properties are closely linked to strain influence tensors.

Although the homogenized, equivalent elastic-plastic effective stress strain relationship (3) for porous, fractured rock is similar to that for homogeneous media (2), there are no equivalent plastic potentials Y^* and failure criteria F^* that when differentiated lead directly to (3), as in the homogeneous case. There are no simple equivalent strengths. For example, there is no equivalent un-confined compressive strength C_o^* that can be calculated in terms of joint and intact rock compressive strength, C_j and C_r, respectively. There is no analogy between metal dislocations and rock joints.

When a single joint is embedded in a homogeneous or previously homogenized volume, the element becomes a two-material composite $(\alpha=1,2)$. The strain influence matrices then may be obtained from the equivalent properties of the composite, the properties of the joint and the intact rock between. For example, $[A^1] = (1/f_1)([C^1]-[C^2])^{-1}([C^*]-[C^2])$. Conversely, if the influence matrices are known, then the composite properties can be determined. Unfortunately, the relationship does not indicate how either may be found in the first place. The NRVE approach overcomes this obstacle.

Figure 1 shows a comparison of NRVE and true results for a generic element containing five joints loaded by an applied shear stress, $\tau_{yz}=1.38$ MPa. Associated flow rules are used with "parabolic" yield conditions: $J_2+AI_1=B$ where J_2 is the second invariant of deviatoric stress and I_1 is the first invariant of stress. Each joint occupies about 1% of the total volume. Intact rock Young's modulus is 16.6 GPa, Poisson's ratio is 0.2, unconfined compressive strength is 33 MPa which corresponds to a 0.2% strain to failure., tensile and shear strength are 3.31 MPa and 6.03 MPa, respectively. Joint Young's modulus is 1/100th intact rock modulus; joint strengths are 1/10th rock strength. All joints have the same properties, although this is not necessary. Joints 2 and 5 fail during the loading process. The equivalent response is subsequently nonlinear seen in the departure from linearity in the stress strain curves. In conclusion, the strong similarity in trends between NRVE and true responses indicates the NRVE elastic-plastic approach is worthy of further investigation.

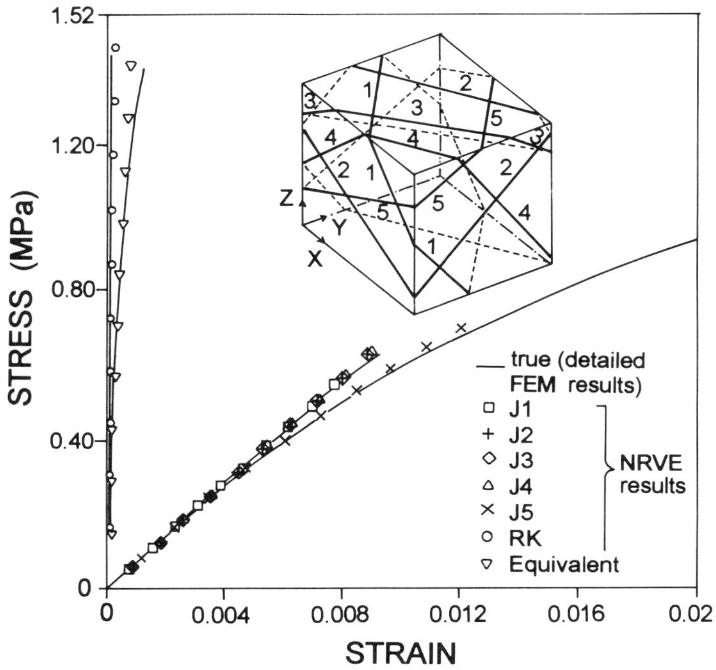

Fig. 1. Comparison of NRVE and true equivalent elastic-plastic response.

References

Hill, R. 1963. Elastic Properties of Reinforced Solids: Some Theoretical Principles. *J. Mech. Phys. Solids.* **11**, 357-372.

Hill, R. 1985. On the Micro-to-Macro Transition in Constitutive Analyses of Elastoplastic Response at Finite Strain. *Math. Proc. Cam. Phil. Soc.* **98**, 579-590.

Huet, C. 1982. Universal Conditions for Assimilation f a Heterogeneous Material to an Effective Continuum. *Mechanics Research Communications.* **9**, 165-170.

Pariseau, W.G. 1988. On the Concept of Rock Mass Plasticity. *Proc.* 29th U.S. Symp. Rock Mech., Balkema, 291-302.

Pariseau, W.G. 1993. Equivalent Properties of a Jointed Biot Material. *Int. J. Rock Mech. Min. Sci. & Geomech. Abstr.* **30**, 1151-1157.

Willis, J.R. 1981. Variational and Related Methods for the Overall Properties of Composites. *Advances in Applied Mechanics* (C-S Yih, ed.) **21**, pp 1-78.

Coupled processes of gas hydrates dissociation and fluid filtration in saturated porous media

Alexey Maksimov[1]

Abstract

A mathematical model of dynamics of interconnected processes of heat transfer, water and gas two-phase filtration, and salt diffusion in water phase with simultaneous gas hydrate dissociation in porous media is constructed by methods of continuum mechanics and thermodynamics of phase transitions.

Introduction

Modern interest [*Kvenvolden (1988), Istomin, et al (1992), and many others*] in gas hydrates in natural conditions is connected with two main aspects: firstly, it is expected by number of geological evaluations that amounts of natural gas comparable with gas deposits capacity are kept in deposits of gas hydrates; secondly, decomposition of gas hydrates in reservoirs lead to some undesirable technological consequences, such as sharp growth of pressure near wells in low-permeable or impermeable local areas.

For finding of optimal technological solutions, concerned with the development of gas hydrates deposits, or with designing trouble-free well operations, an analysis of mechanisms of processes of heat and mass transfer in accordance with phase transiton of gas hydrates is needed.

Gas hydrate dissociation/formation in reservoirs is connected with change of defining thermodynamic parameters (pressure, temperature, concentration of soluble admixture) and is accompanied by transfer of heat, gas, water and admixtures dissolved in it. A mathematical model describing interconnected processes of heat and mass transfer with phase transition in porous media can be formulated by geterogeneous media mechanics methods. Following this approach it was shown in [*Bondarev, et al (1990)*] that heating of gas hydrates, coexisting with gas and water under thermodynamic equilibrium conditions lead to extended dissociation region formation instead of traditional conception of sharp front formation. In [*Maksimov (1992)*] it was confirmed the sufficient effect of water phase transport on the common gas hydrate dissociation dynamics. In this work the model is developed with additional account of soluble admixture.

[1]Oil & Gas Research Institute, Leninsky prosp. 65, Moscow, 117917, Russia

Model formulation

Main system of equations of conservation laws has following form:

$$m\frac{\partial}{\partial t}(\nu\epsilon\rho_h + (1-\nu)s\rho_w) + \text{ div } \rho_w\vec{v}_w = 0;$$

$$m\frac{\partial}{\partial t}(\nu(1-\epsilon)\rho_h + (1-\nu)(1-s)\rho_g) + \text{ div } \rho_g\vec{v}_g = 0;$$

$$m\frac{\partial}{\partial t}((1-\nu)sc\rho_w) + \text{ div } c\rho_w\vec{v}_w = mD \text{ div } \rho_w(1-\nu)s \text{ grad } c;$$

$$\vec{v}_w = -\frac{k}{\mu_w} f_w(\nu, s) \text{ grad } P; \qquad \vec{v}_g = -\frac{k}{\mu_g} f_g(\nu, s) \text{ grad } P;$$

$$(\rho C)_e\frac{\partial T}{\partial t} - mq\rho_h\frac{\partial \nu}{\partial t} - m(1-\nu)(1-s)\frac{\partial P}{\partial t} +$$
$$+(\rho_w\vec{v}_w C_w + \rho_g\vec{v}_g C_g) \text{ grad } T = \text{ div } \lambda_e \text{ grad } T, \tag{1}$$

where $\lambda_e = (1-m)\lambda_s + m\nu\lambda_h + m(1-\nu)s\lambda_w + m(1-\nu)(1-s)\lambda_g;$

$$(\rho C)_e = (1-m)\rho_s C_s + m\nu\rho_h C_h + m(1-\nu)s\rho_w C_w + m(1-\nu)(1-s)\rho_g C_g;$$

$$q = \epsilon h_w + (1-\epsilon)h_g - h_h;$$

$$\rho_s = const; \quad \rho_h = const; \quad \rho_w = \rho_w^{(0)}(1 + (P - P^{(0)})/K_w); \quad \rho_g = P/RT$$

Condition of thermodynamic equilibrium of mixture [*Istomin, et al (1992)*]:

$$F(T, P, c) \equiv -\ln P/P^{(0)} + \alpha_1/T + \alpha_2 + \alpha_3 c + \alpha_4 c^2 + \alpha_5 \ln(1 + \alpha_6 c) = 0 \tag{2}$$

Let at the fixed boundary $x = 0$ of porous medium $0 < x < \infty$ are supporting values of temperature T, salt concentration c, and pressure P, corresponding to the phase diagram area of gas and salt solution mixture existance and gas hydrate absence. Then naturally to expect formation in the process of gas hydrate decomposition of two areas, divided by the free moving boundary $X(t)$. Area $(X(t) < x < \infty)$ represents extended region of phase transition, for which are written equations (1)-(2). In the area $(0 < x < X(t))$ gas hydrate is absent, equations for this area easy to get from the system (1), if place $\nu \equiv 0$. Two areas are divided by $X(t)$, at which continuity of pressure, temperature and salt concentration and breakups of ν and s (hydrate and water saturations) are expected.

$$x = X(t): \qquad T_- = T_+ = T_*; \quad P_- = P_+ = P_*; \quad c_- = c_+ = c_*;$$

$$m(s_-\rho_w - \nu_+\epsilon\rho_h - (1-\nu)s_+\rho_w)\dot{X} - \rho_w(v_{w-}^n - v_{w+}^n) = 0;$$

$$m((1-s_-)\rho_g - \nu_+(1-\epsilon)\rho_h - (1-\nu_+)(1-s_+)\rho_g)\dot{X} - \rho_g(v_{g-}^n - v_{g+}^n) = 0;$$

$$c_*\nu_+\epsilon\rho_h\dot{X} + \rho_w D(s_-(\text{grad } c)_-^n - s_+(1-\nu_+)(\text{grad } c)_+^n) = 0;$$

$$m\nu_+\rho_h q\dot{X} = \lambda_{e_+}(\text{grad } T)^n_+ - \lambda_{e_-}(\text{grad } T)^n_-; \quad F(T_*, P_*, c_*) = 0 \tag{3}$$

Initial conditions and conditions at the permeable fixed boundary:

$$t = 0: \quad T = T_0, P = P_0, c = c_0, \nu = \nu_0, s = s_0, X = 0, F(T_0, P_0, c_0) = 0;$$

$$x = 0: \quad T = T^0, \quad P = P^0, \quad c = c^0 \tag{4}$$

Self-similar solution

Let in conditions (4) $T_0, P_0, c_0, \nu_0, s_0, T^0$, and P^0 – constants. Then problem (1)-(4) allows one-dimensional self-similar solution:

$$T = T(\xi), P = P(\xi), c = c(\xi), \nu = \nu(\xi), s = s(\xi), X = \gamma t^{1/2}, \xi = x t^{-1/2}$$

On Figure 1 non-dimensional functions T/T_0 (1), P/P_0 (2), c/c_0 (3), ν (4) and s (5) versus non-dimensional self-similar variable $\zeta = \xi D^{-1/2}$ are presented in two different scales. Here β denotes a non-dimensional value of γ. The results are obtained numerically under typical values of parameters [*Istomin, et al (1992)*], porosity $m = 0,2$, and permeability $k = 10^{-15}$ m^2. Values of parameters for the gas and gas hydrate correspond to the methane and its hydrate, and parameters of phase equilibrium correspond $NaCl$ as a soluble admixture.

The distributions correspond to following initial values and values at the fixed permeable boundary: $P_0 = 10$ MPa, $c_0 = 0,01, T_0 = 285,9$ K, $\nu_0 = 0,45$, $s_0 = 0,35$, P^0= 6,14 MPa, c^0= $0,01, T^0 = 303,4$ K.

One may see large scale extended region of phase transition formation. It also accompanied by narrow diffusional boundary layer near the moving boundary.

Evaluation of porous pressure growth in closed system

Significant growth of pressure may result the hydrate decomposition near impermable fixed boundary (for instance, well casing), if saturated by hydrate rock is limited by the impermeable border or hydrate reservoir is practically impermeable. In this case an estimation of pressure growth as a result of hydrate dissociation can be obtained for completely saturated by hydrate reservoir. Then $c \equiv 0$ and in the limited area processes of redistribution of water and gas occur practically instant in contrast with the scale of front motion. In such case it's sufficient to consider the conditions at the moving boundary, having place flows equal zero:

$$\epsilon\rho_h + (1 - \epsilon)\rho_g = \rho_w, \text{ where} \tag{5}$$

$$\rho_w = \rho_w^{(0)}(1 + (P - P^{(0)})/K_w), \quad \rho_g = P_*/RT_*, \quad \ln P_*/P^{(0)} = \alpha_1/T_* + \alpha_2$$

Estimations of maximum pressure may be obtained from transcendental equations system (5). Under typical values of parameters it turned out to be approximately 50 MPa. Obviously, such high pressure may cause hydraulic fracturing or even collapse of casing.

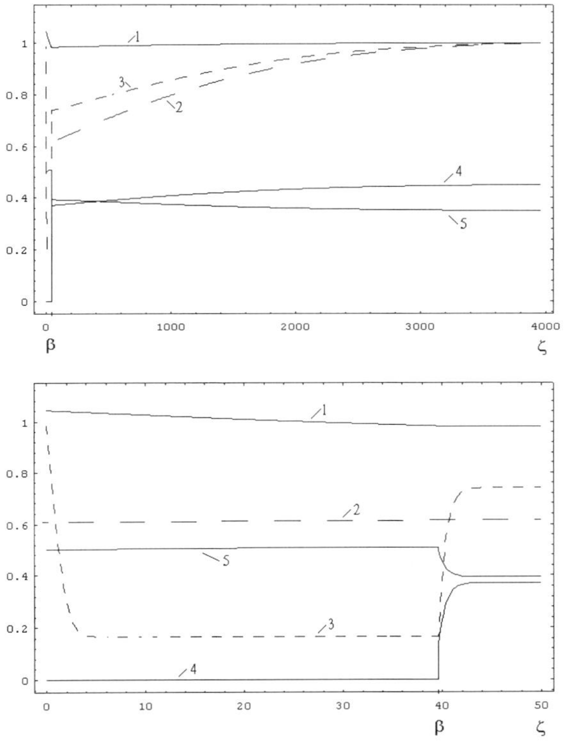

Figure 1

References

Bondarev E.A., Kapitonova T.A., Maksimov A.M., and Tsypkin G.G., (1990), *Mathematical simulation of hydrate formation and dissociation in the systems of natural gas production and transportation*, Modelling, Simulation & Control, C, AMSE Press, **21**, *N* 1, p. 53-63.

Istomin V.A., and Yakushev V.S., (1992), *Gas hydrates in natural conditions*, Nedra, Moscow, 236p. (in Russian).

Kvenvolden, K.A., (1988), *Methane hydrate – a major reservoir of carbon in the shallow geosphere?*, Chemical Geology, **71**, p. 41-51.

Maksimov A.M., (1992), *Mathematical model of volumetric gas hydrates dissociation in porous media: account of water phase mobility*, Inzhenerno-phyzichesky zhurnal, **62**, *N* 1, p. 76-81 (in Russian).

Structural Sensing with Fiber Optic Systems

Raymond M. Measures[1]

Abstract

Fiber optic Bragg gratings represent the most versatile structural monitoring devices available today. This new sensing technology is opening the door to a range of monitoring not practical, or even possible, with conventional strain sensing technology. An overview of the prospects will be presented and illustrated.

Introduction

Bragg gratings are formed in optical fibers by a flash of intense UV light. As such they are very simple, rugged and compact and they should become inexpensive, when used in large numbers. Optical fiber Bragg gratings constitute one of the most versatile structural sensing device ever created for they can be used as short or long gauge strain sensors, they can be used to measure temperature, or strain and temperature simultaneously. A number of these sensors can be strung along a length of optical fiber making quasi-distributed sensing possible, and more recently we have demonstrated that truly distributed sensing is possible within a single grating with a spatial resolution of better than 1mm. These sensors can be used in a self-diagnostic mode to warn, for example, if their bond to a host structure has started to fail. [LeBlanc et al 1996] They can also be used to track very slow changes in strain, such as the relaxation of prestressing tendons within a bridge girder, while simultaneously monitoring fast strain changes from transient events like the passage of a vehicle over the bridge [Measures et al 1995]. In principle, their frequency response extends from 0 to the MHz range. They can be used for acoustic emission detection and optoacoustic probing and could monitor the cure of concrete or composite material. The small diameter (~125μm) and linear geometry of optical fibers makes it quite practical to embed then within composite material. They have been installed in prepreg layups, and included in filament wound structures, or pultruded. Because the gratings are simply a modified segment of an optical fiber, this form of sensor is immune to electromagnetic interference, are nonconducting, and corrosion-free. Also, improvements in fabrication of the gratings could lead to sensors with the same high strength and fatigue resistance as fused silica.

----

[1]Dr. R.M. Measures, University of Toronto Institute for Aerospace Studies,
4925 Dufferin St., Downsview, Ont. M3H 5T6 Canada

Technology Overview

A fiber optic Bragg grating is a periodic variation in the index of refraction along a small segment of the light guiding core of an optical fiber. This grating behaves as a tiny mirror that selectively reflects only a very narrow band of light centred about the Bragg wavelength. What makes this device so attractive as a structural sensor is the linear variation of this Bragg wavelength with applied strain or temperature:

$$\lambda_B\ (\varepsilon,T) = \lambda_B\ (\varepsilon_o, T_o) + S_\varepsilon\ (\varepsilon - \varepsilon_o) + S_T\ (T - T_o)$$

where $\lambda_B\ (\varepsilon,T)$ and $\lambda_B\ (\varepsilon_o, T_o)$ represent the Bragg wavelengths at the state of interest (ε,T) and the reference state (ε_o, T_o) respectively. ε and ε_o and T and T_o are the corresponding states of strains and temperature. S_e and S_T are the "strain-wavelength" and "temperature-wavelength" sensitivities of the Bragg grating. If there is no change of temperature then we can write

$$\Delta\lambda/\lambda_o = G_F\ (\varepsilon - \varepsilon_o)$$

where $\Delta\lambda \equiv \lambda_B\ (\varepsilon, T) - \lambda_B\ (\varepsilon_o, T_o)$, $\lambda_o \equiv \lambda_B\ (\varepsilon_o, T_o)$,
and G_F is the gauge factor, typically about 0.8 in value when $\lambda_o = 1550$ nm.

The measurement of strain is effectively the same as the measurement of the centre wavelength of the back reflected light from the sensing grating situated at some point along the optical fiber. A simple system for making this measurement has been developed, [Measures et al 1992] and an engineered practical system is commercially available from ElectroPhotonics Corp. Where both the strain and the temperature vary, then some form of temperature compensation is required. Recently, it has been shown that two Bragg gratings operating at two very different wavelengths can be collocated and used to measure both strain and temperature simultaneously [Xu et al. 1995].

If the strain applied to a grating is nonuniform, then the reflection spectrum degenerates from a narrow peak to a more complex form. We have shown that as long as the strain profile along the grating is monotonic, the spectrum can be analysed to provide the strain profile [Huang et al 1996]. The grating can thus be thought of as a "distributed" strain sensor and we have demonstrated that the strain profiles measured in this manner are reliable [LeBlanc et al 1996]. We have also shown that any debonding of the sensor from the host (a common mode of failure for strain sensors in general) can be detected from changes in the spectral form of the signal [LeBlanc et al 1996].

Fiber Optic Structural Sensing Measurements

A fiber optic Bragg strain sensor array was installed in the Beddington Trail bridge in Calgary (Alberta, Canada) to monitor the use of three types of prestressing tendons [Maaskant et al 1996] conventional steel strand; Tokyo Rope Carbon Fiber Composite Cable (CFCC) (Tokyo Rope Manufacturing Co. Ltd., Japan); and Leadline Rod (Mitsubishi Chemical, Japan). The three lane two span skew bridge consists of 26 precast prestressed girders, 6 of which contain the CFRP prestressing strands. The fiber optic grating sensors were attached to prestressing tendons of each type which were subsequently embedded in the concrete girders. The motivation for the use of fiber optic strain sensors was the desire to monitor the long term characteristics of the tendon/concrete bond and the relaxation behaviour of the new CFRP materials. Monitoring of traffic usage was also explored with this strain sensor array. This information is useful in assessing bridge designs, comparing actual loading to the loads used in the design and specified in bridge design codes, and may prove to be helpful for bridge diagnostics and assessments with regard to bridge maintenance or retirement when structural damage has been sustained. The fiber optic grating sensors used in the Calgary bridge were surface bonded onto the prestressing tendons. Since the tendons were to be prestressed to levels of approximately 8000 με, long term durability was ensured by bonding the sensors after the application of prestress. The fiber optic sensor array consists of a total of 18 strain and temperature sensors mounted on prestressing tendons in various locations.

The sensor measurements are designated by the tendon type (SS: steel strand; TR: Tokyo Rope CFCC; LL: Mitsubishi Chemical Leadline Rod), girder identification and sensor location are identified. **Figure 1** provides an indication of the initial relaxation in the strain on a number of the sensors and the subsequent relaxation that occurred over the 18 months that the bridge was in service. The measurements of September, 1993 were taken immediately following post-tensioning of the girders. These measurements represent the stress relaxation in the tendons from the combined effects of destressing, concrete shrinkage and creep, the dead loading of the bridge deck, and the post-tensioning applied across the two spans. Subsequent measurements began approximately one month after the bridge was opened with the last recorded measurement occurring in April, 1995, 18 months later.

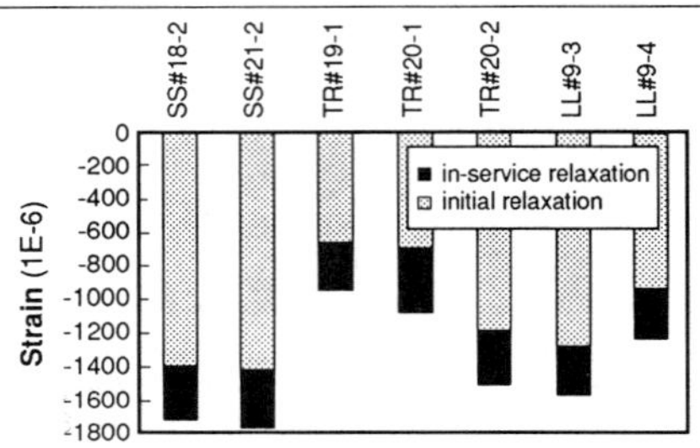

Fig. 1 Strain relaxation measured by fiber optic sensors for graphite fiber composite and steel prestressing tendons in Beddington Trail Bridge over 18 months.

The measurements presented in **Figure 1** have been corrected for thermal apparent strain arising from both the sensor itself as well as the thermal expansion mismatch between sensor and host, and hence represent the stress-induced component of the strain. It must be recognized, however, that these corrected strains reflect the prevailing temperature via the overall thermal behavior of the structure. Since it is known that the prestressing materials possess widely different thermal expansivities, an attempt was made to normalize the data with respect to temperature so that a more meaningful comparison can be made regarding permanent tendon relaxation. The steel reinforcement is thermally well matched with the concrete, and hence the girders are assumed to possess an overall expansivity equal to that of the concrete. The measurements were normalized to a temperature of 20°C using the relative thermal expansion of the concrete and the prestressing tendon. **Figure 2** provides the strain history monitored by the sensors.

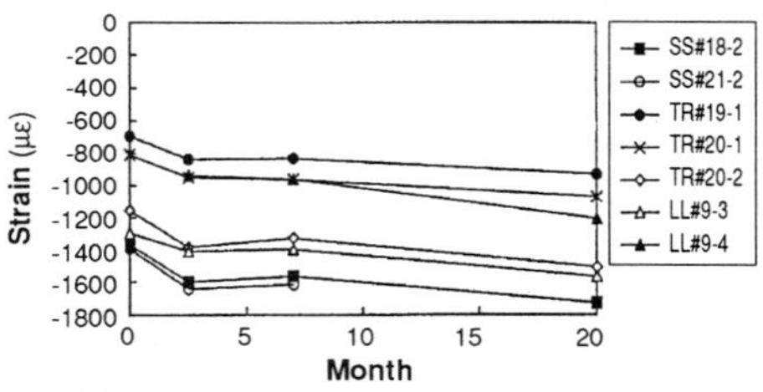

Fig. 2 Strain history for both steel and graphite fiber composite prestressing tendons in the Beddington Trail Bridge.

The total expected loss of prestress in the steel tendons has been estimated as 325 MPa or 1600με. The superior relaxation characteristics of CFRP tendons led to expectations of approximately one half of this value. Indeed the measurements

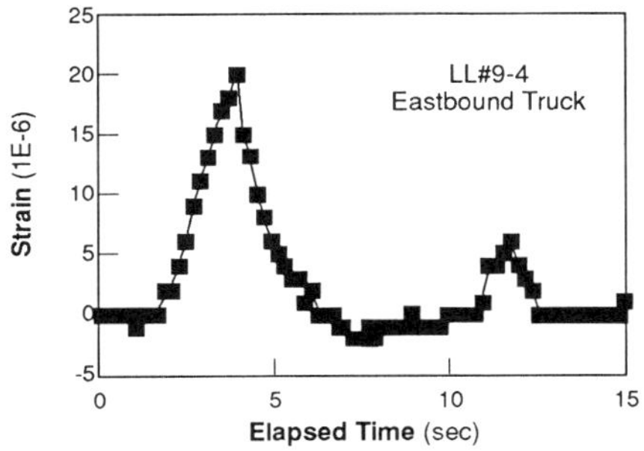

Fig. 3 Strain response to vehicle traffic by integrated fiber optic sensors in the Beddington Trail Bridge.

show less relaxation for the CFRP tendons, but in locations 2 and 3 in particular, the measurement value exceeds the expectations to some extent. Part of this discrepancy may arise from the measurement location due to the combined effects of post-tensioning and dead load effects.

Figure 3 provides an example of dynamic strain monitoring of traffic loads by the same sensors. These measurements were made with a commercially available Fiber Optic Grating Strain Indicator (ElectroPhotonics Corp., Toronto, Canada) which possesses a strain resolution of 1 me over a range of 8,000 to 10,000 $\mu\varepsilon$. A 25 ton single axle truck was passed over the bridge on the eastbound and westbound lanes at a slow speed (10 to 30 kph) for these measurements. This resulted in consistent peak strains of 20 to 25 $\mu\varepsilon$ for truck passes directly over the sensor location. Note the response to a second smaller truck following the 25 ton truck.

Concluding Remarks

Fiber optic structural sensing based on Bragg gratings represents a powerful new technology that could greatly strengthen the arsenal of the construction engineer.

Acknowledgements

Support for this research was provided by the Natural Sciences and Engineering Research Council of Canada, Ontario Laser and Lightwave Research Centre, Institute for Space and Terrestrial Science and City of Calgary. Special thanks to all the members of the Fiber Optic Smart Structures Laboratory at University of Toronto, Institute for Aerospace Studies and to Ken Hill of the Canadian Research Centre, Sami Rizkalla of the University of Manitoba, C. Wade and A. Guha-Thakurta of the City of Calgary and G. Tadros of SCI Inc.

References

Meltz G., Morey W.W. and Glenn W.H. (1989), "Formation of Bragg gratings in optical fibers by a transverse holographic method, *Optic Letters,* Vol. 14, No. 15.

LeBlanc M., Huang S., Ohn M., Measures, R.M., Guemes A. and Othonos, A. (1995), *Distributed strain measurement using a fibre Bragg grating based on reflection spectrum analysis,* in preparation.

Measures R.M., Alavie A.T., Maaskant R., Ohn M., Karr S. and Huang S.Y., "A structurally integrated Bragg grating laser sensing system for a carbon fiber prestressed concrete highway bridge", *Smart Materials and Structures,* Vol. 4., (1995) pp.20-30.

Measures R.M., Melle, S. and Liu, K., (1992) "Wavelength demodulated Bragg grating fiber optic sensing systems for addressing smart structure critical issues", *Smart Mater. Struct. 1,* pp.36-44.

Xu, M.G., Archambault, J.L., Reekie L. and Dakin P. (1994), "Discrimination between strain and temperature effects using dual wavelength fibre grating sensors", *Electronics Letters,* Vol. 30, No. 13, pp. 1085-1087.

Huang S., Ohn M. and Measures R.m. (1996) "Phase-based Bragg Intragrating distributed strain sensor", *Applied Optics,* (accepted for publication).

LAUNCH VIBRATION ISOLATION SYSTEM

Eugene R. Fosness[1], Rory R. Ninneman[1] , Paul S. Wilke[2] and Conor D. Johnson[3]

Abstract

One of the most severe environment that a satellite will be subjected to during its lifetime will occur during launch. This paper summarizes the results and status of research efforts in the area of satellite isolation from the launch environment. The objective of this effort was to reduce the launch-induced dynamic acceleration of the satellite by insertion of an isolator. Isolation issues involving the use of passive elements and launch vehicle system-level requirements will be discussed.

Introduction

The vibrations that occur in a spacecraft during launch are both structure-borne and acoustic in nature. The focus of this paper is to provide information on current Air Force Phillips Laboratory (PL) programs that involve technologies encompassing passive launch vibration isolation. Small Business Innovative Research (SBIR) efforts show extreme promise due to the teaming arrangement developed by the small business (CSA Engineering) to include a Launch Vehicle (LV) manufacturer (McDonnell Douglas Aerospace), spacecraft component manufacturer (Honeywell Satellite Systems), and a satellite manufacturer (Space Systems/Loral). The result of these SBIR efforts will be a rigorous design methodology for whole-spacecraft isolation and a prototype that will have undergone basic functional testing.

Motivation/Payoff

Spacecraft adapters or Payload Attach Fittings (PAF) are used to provide an interface between the LV and payload. Typical adapters are stiff and rigid, which provide an efficient transmission path for dynamic and inertia loads. The traditional

1. Research Structural Engineer, Air Force Phillips Laboratory (VTSS), Albuquerque, NM 87117-5776.
2. Senior Engineer, CSA Engneering, Inc., Palo Alto, CA 94303-3843
3. Principal Engineer, CSA Engineering, Inc., Palo Alto, CA 94303-3843

approach to spacecraft design against launch vibration has been through structural stiffening or component isolation. However, this approach is not only costly, time consuming and adds weight, but can lead to other liabilities once the spacecraft is in orbit. Current payload adapters do not provide isolation from launch loads except on a case-by-case basis. Implementing an isolation system into the adapter is the logical place for a payload isolator. However, whole spacecraft isolation is a substantial change in the dynamic properties of the combined system and is bound to have side effects which must be addressed. A spacecraft isolation system is envisioned to replace the traditional spacecraft adapters used to physically attach satellites to LV as shown in Figure 1. Note that this figure shows a conceptual design only; details of the actual hardware design are proprietary under SBIR rights. The implementation of this technology will result in: 1) greater survivability at launch; 2) a reduction of launch vehicle loads; 3) a minimization of dynamic-related spacecraft failures; 4) a reduction of cost, size, weight of some spacecraft; 5) a lowering of certain test requirements; 6) the allowance for "tuning" of the isolator instead of spacecraft requalification; 7) a reduction of the number of analysis load cycles.

Phase I SBIR Effort

In 1993, the PL awarded CSA Engineering a Phase I SBIR (Maly, 1994) to determine the feasibility of a payload isolation system. A system analysis was performed to insure that interactions between the LV and the dynamics of the isolator and spacecraft were taken into account during the design process. CSA's approach was to perform coupled dynamic analysis to determine the isolators specifications.

A generic medium LV model, a 6000 lb. pseudo-satellite model, and transient loads for the liftoff flight event were obtained from MDA. The LV model had 234 degrees of freedom (DOF) and was in the form of mass and stiffness matrixes and modal DOF representations. The PAF (Figure 1) was modeled with beam, bar, spring, and viscous elements in NASTRAN. The transient loads consisted of 5 pad reaction forces, main engine thrust, two lateral loads due to the gimbaling, and the solid booster thrust load. The pseudo-satellite model consisted of an elastic representation of a satellite with tuned modes with the correct c.g. The primary structure of the satellite model had frequencies that were similar to those of past spacecraft launched on the LV (12 Hz lateral and 35 Hz axial when constrained at the PAF interface). A number of cases were analyzed in which the PAF model was modified to simulate an axial isolator of a given break frequency. The baseline case consisted of the unmodified PAF (67 Hz axial isolator). Based on the analysis results, the following assessments were made for the system under study. Note that these are all based on a large satellite and do not necessarily apply to smaller satellites: 1) High axial dynamic loads occur at low frequencies (<15 Hz); 2) Isolators with break frequencies <8 Hz are too soft and consume to much rattlespace; 3) Isolation modes are in the middle of system modes; 4) High-g steady state loads on a soft isolator cause excess clearance reduction and have unrealistic stroke requirements; 5) There are no quasi-static loads in the lateral direction; 6) Clearance reduction of a <20 Hz axial isolator was too great; 7) Cannot isolate against the high axial steady-state loads; 8) Need hybrid active/passive system to isolate axial disturbances.

The analysis indicates that incorporating axial isolation into the PAF introduces axial compliance, which tends to allow low-frequency lateral rocking of the spacecraft. Axial passive isolation was not feasible because of geometric restrictions such as excessive large rattlespace requirements and static sag due to the large quasi-static loads. The axial constraint required that an isolation system be designed between system frequencies, which also makes it hard to meet the isolation criteria. To avoid dynamic coupling, the isolation system must be designed with axial frequencies above 35 Hz and lateral frequencies above 15 Hz, which again limits the level of passive isolation to be achieved. In order to accommodate the static sag, the isolator must be designed for accelerations that exceed 6 g's. Based on the results of the analysis, it was believed that low frequency passive axial isolation was not feasible under current LV constraints for a large satellite on a medium LV. This is mainly applicable for large satellites with small footprints and a high CG. Smaller satellites are more amenable to axial passive isolation. However, these analyses also indicate that lateral isolation is both feasible and useful to protect secondary spacecraft structures. It was found that the lateral response of the LV in the 25-35 Hz range would excite the spacecraft secondary structural modes resulting in possible damage.

Phase II SBIR Effort

In 1994, CSA was awarded a Phase II SBIR to design an isolator that provides lateral isolation without excessive lateral rocking. The goal of the Phase II SBIR was to develop a passive lateral isolator that would replace an existing payload attach fitting (PAF) for a medium launch vehicle. The isolator would be designed to reduce the structure-borne dynamic vibrations by a factor of 2 RMS and satisfy all requirements imposed on current PAFs. A launch loading condition, known to cause the most severe lateral loads, was identified for use in design and analysis trades studies for an isolating PAF. This loading condition, common to most launch vehicles, is typically caused by a main-engine resonance which couples with an axial mode of the LV. As fuel is depleted, the axial mode tunes to the engine resonance. This excited axial mode couples laterally to the spacecraft as an almost purely sinusoidal lateral excitation to the spacecraft base. This loading condition can therefore be modeled as a complex frequency response analysis at a single frequency.

For the detailed design of the isolator, MDA provided a high-quality system model that consisted of the generic but realistic LV, PAF, and satellite. The LV model was in Craig-Bampton substructure format and consisted of 139 modal DOF and 12 physical boundary DOF. This model represents the LV configuration and dynamics during the appropriate flight stage and was not intended for all flight events. The spacecraft consisted of an 18 ft. tall, 6500 lb. structure that had a 1st bending mode at ~10 Hz. The spacecraft model, with its bus structure plus several substructured components, totalled 20,200 DOF. Analyses were performed to investigate the performance of an isolator using transient loads applied to the LV. The transient loads were derived from MDA's vast amount of measured flight data. Several trade studies were performed on parameters to optimize the isolator's performance. A unique isolating PAF was designed that allows lateral isolation but retains axial stiffness. Based on the analysis results, the isolator would have a lateral break frequency of 15 Hz. The performance of the isolator within the PAF is shown in Figure 2. For the frequency range of interest for the spacecraft (20-40 Hz), the analysis indicates that a factor of 7 reduction in lateral vibration is obtainable using a passive isolator design. For the integration of an isolator to replace the existing PAF, there was only a 16 lb. weight penalty (increases from 185 to 201 lbs.)

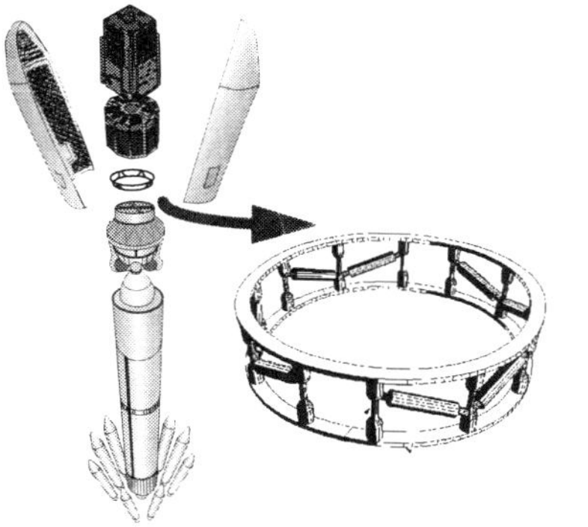

Figure 1. Spacecraft with isolator

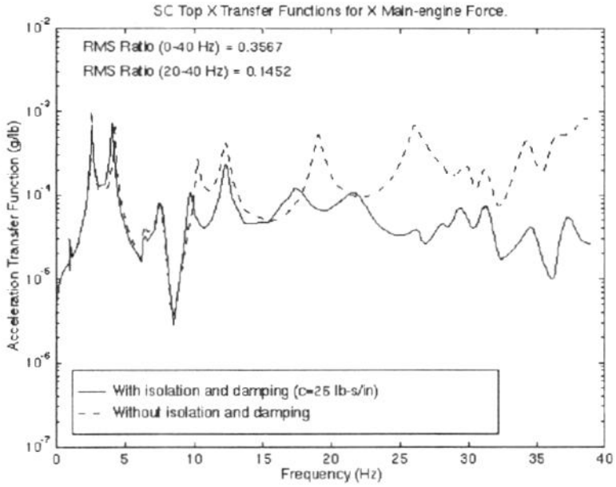

Figure 2. Predicted Transfer Function for X Main-engine Force

Conclusions

In the Phase I SBIR, it was determined through system level dynamic analysis that lateral vibration isolation was the most applicable for the system under study. In the Phase II, a lateral isolating PAF was designed, built, and tested to replace an existing PAF. This isolating PAF reduced the structure-borne lateral vibrations by at least a factor of two, and satisfied all PAF requirements. Also, MDA expects for most cases the weight savings due to lower satellite design loads will outweigh the isolation system weight penalty. It is expected that the successful implementation of this program will have an impact on all future military and commercial satellites programs by decreasing the launch loads for which satellite systems are designed.

REFERENCES

1. Maly, J., "Missile Payload Vibration Isolation", PL-TR-94-1052, Phillips Laboratory, Kirtland Air Force Base, NM, Feb., 1994. (SBIR restricted distribution)

Seismic Active Control by a Heuristic-Based Algorithm

Yu Tang[1]

Abstract

A heuristic-based algorithm for seismic active control is generalized to permit consideration of the effects of control-structure interaction (CSI) and actuator dynamics. The control force is computed at one time step ahead before being applied to the structure. Therefore, the proposed control algorithm is free from the problem of time delay. A numerical example is presented to show the effectiveness of the proposed control algorithm. Also, two indices are introduced in the paper to assess the effectiveness and efficiency of control laws.

Introduction

This work is a sequel to Tang (1995) in which a heuristic-based control law was proposed, and the results presented showed that the proposed control law can reduce the structural response by one order of magnitude and has the ability to reduce the peak that occurs during the first few cycles of the response time history. However, in Tang (1995), the effects of control-structure interaction (CSI) and actuator dynamics were not considered. The importance of accounting for the effects of CSI and actuator dynamics in seismic active control has been demonstrated in Dyke et al. (1995a). Also, a recent paper by Dyke et al. (1995b) has indicated that these effects do not result in a pure delay in the system and cannot be accounted for by the time-delay compensation.

Structural System Considered

The structural system considered in this paper is

[1]Engineer, Reactor Engineering Division, Argonne National Laboratory, 9700 S. Cass Avenue, Argonne, IL 60439.

basically the same single-degree-of-freedom (SDOF) system presented in Fig. 10 in Dyke et al. (1995a) except that different values of mass and damping coefficient of the structure are used. The structural properties of the SDOF system are: Mass, m: 67.33 N-sec^2/cm; Stiffness, k: 14.06 kN/cm; Damping Coefficient, c: 38.09 N-sec/cm; Mass of the stiff frame and the actuator, m_0: 0.073 N-sec^2/cm; and Stiffness of the four tendons, k_0: 14,879 N/cm.

Equations of Motion

The equations of motion solved in this paper that govern the response of the system are given by

$$m\ddot{x} + c\dot{x} + \left(k + k_0\cos^2\theta\right)x + k_0\cos\theta\, y = -m\ddot{x}_g \tag{1}$$

for the SDOF structure, and

$$m_0\ddot{y} + k_0 y + k_0\cos\theta\, x = -m_0\ddot{x}_g + F \tag{2}$$

for the stiff frame that is used to connect the actuator to the four pretensioned tendons. The notations used in Eqs. (1) and (2) are x = displacement of the building relative to the ground, y = displacement of the actuator, θ = tendon inclination angle relative to base (θ = 36° in this study), and F = force applied by the hydraulic actuator to the rigid frame. F is related to y by the actuator dynamics given by (Dyke et al. 1995a)

$$\dot{F} = \alpha_1(u-y) - \alpha_2 F - \alpha_3\dot{y} \tag{3}$$

in which u = control command, and α_1, α_2, and α_3 = the actuator dependent coefficients. In this study the values of α_i, i=1,2,3 used are those corresponding to a real actuator used in Dyke et al. (1995a) that were determined experimentally. The force developed in the tendons during vibration consists of two parts: $k_0\cos\theta x$ and $k_0 y$. $k_0\cos\theta x$ is a passive force, whereas $k_0 y$ is an active force. The horizontal component of $k_0 y$, $k_0 y\cos\theta$, is the effective control force acting on the structure. Notation F_a is introduced to denote this effective control force, i.e., $F_a = k_0 y\cos\theta$. In Tang (1995), F_a was regarded as an individual control force acting on the structure, and only Eq. (1) was considered in the analysis.

The control algorithm implemented in this paper is an extension of that presented in Tang (1995). Due to space limitations, only the concept is presented in this paper. The control command, u, needed for $t=t_i+ t$ is computed at $t=t_i$, and it is determined such that the structural velocity at $t=t_i+ t$ is zero assuming that $\ddot{x}_g(t_i+ t)=0$.

In this study a control performance index (CPI) is introduced to assess the control performance. CPI is defined by the equation

$$CPI = \frac{p_0 - p_c}{p_0} \tag{4}$$

where p_0 = absolute value of maximum peak of interested response quantity under the condition of zeroed control, and p_c = absolute value of maximum peak of interested response quantity with control. Also, an energy efficiency index (EEI) is introduced to assess the efficiency of the energy consumption of control. EEI is defined by the equation

$$EEI = \text{maximum value of } F_a \text{ / maximum value of } F \tag{5}$$

Numerical Results

The base excitation considered is the first 6.29 sec of the El Centro 1940 N-S accelerogram (scaled to 25% of its magnitude). In the analysis an additional 3-second free vibration response is also computed. The sampling rate is assumed to be 100 Hz. The following strategies are considered in the numerical example. "Zeroed Control" represents the case in which the active bracing is attached, but the control command, u(t), is set equal to zero; "Ideal Control" corresponds to the case in which there are no limitations on the control; and "Limited Control" is the case in which the control is constrained by the cylinder velocity of 63.5 cm/sec, and the maximum value of u(t), u_{max}, is set equal to 0.635 cm. For comparison purposes, the time history of the relative displacement of the SDOF system for the three cases considered are shown in Fig. 1. It can be seen from this figure that with the proposed control law, not only is the system response remarkably reduced, but the first few peaks of time history for the zeroed control condition are also significantly reduced. The value of CPI for x(t) for the case of ideal control is 0.9; however, the value of EEI is about zero due to the excessively large actuator force needed. Therefore, the "ideal control" is not feasible for actual implementation. This unrealistic control, however, would not have been found if the effects of CSI and actuator dynamics were not included in this study. For the case of limited control, the values of CPI for x(t) and EEI are 0.68 and 0.15, respectively, and the corresponding values for the zeroed control case are 0 and 0.19, respectively.

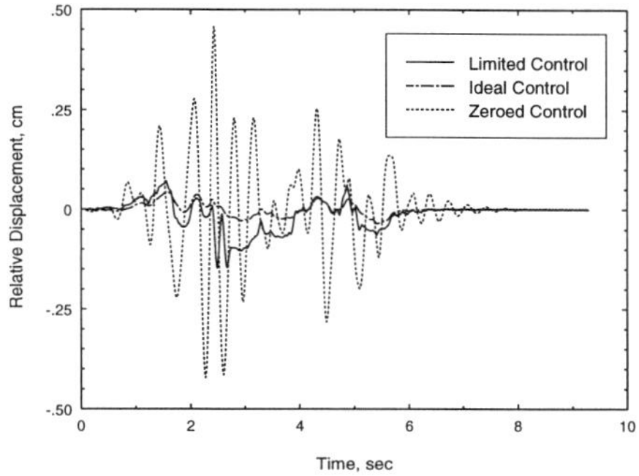

Figure 1. Comparison of Time Histories of Relative Displacement

Concluding Remarks

A preliminary study of seismic active control considering the effects of control-structure interaction and actuator's dynamics is presented. The control law implemented is a heuristic-based control algorithm. The response of a SDOF system is studied. The results obtained indicate that the effectiveness of the proposed control law depends on the capacity of the actuator; also, without considering the capacity limitation of the actuator used, the control results may not be realistic.

Acknowledgments

The author wishes to thank Dr. B. F. Spencer, Jr. and Ms. S. J. Dyrk of the University of Notre Dame for providing the numerical data of the parameters in the equation of the actuator dynamics. This work was performed under the auspices of the U.S. Department of Energy, Contract No. W-31-109-Eng-38.

References

Dyke, S. J., Spencer Jr., B. F., Quast, P., and Sain, M. K. (1995a). "Role of control-structure interaction in protective system design," *J. Engrg. Mech.*, ASCE, 121(2), 322-338.

Dyke, S. J., Spencer Jr., B. F., Quast, P., Sain, M. K., Kaspari Jr., and Soong, T. T. (1995b). "Acceleration feedback control of MDOF structures," to appear in *J. Engrg. Mech.*, ASCE.

Tang, Y. (1995). "A heuristic-based algorithm for active control," to appear in *J. Engrg. Mech.*, ASCE.

Micromechanics Based Design of Optical Fiber Crack Sensor

Christopher K.Y. Leung[1], Associate Member ASCE, and Neill Elvin[2]

Abstract

A novel optical fiber sensor to detect and monitor flexural cracks in concrete structures is currently under development at MIT and Brown University. For the sensor to be useful in practical applications, its sensitivity, range and loss have to be optimized. The paper presents a model which combines micromechanics with electromagnetic wave theory to predict the light signal loss as a function of crack opening. The model will provide guidelines for the choice of design parameters in the optimization of sensor performance.

Introduction

The condition of concrete structures can be assessed through the monitoring of crack openings. For example, opening of cracks beyond about 0.2mm may lead to steel reinforcement corrosion problems, while excessive crack opening (up to a few mm's) is a sign of significant structural damage. The monitoring of crack opening with optical fiber sensors has been demonstrated by several investigators [1-3]. However, available sensors either require a-priori knowledge of the crack location [1, 2], or is of the integrated type [3] which cannot distinguish between a single widely opened crack and a large number of smaller cracks. For reliable crack monitoring in real structures, a new approach needs to be developed.

A Novel Crack Detection and Monitoring Concept

Researchers at MIT and Brown University are currently developing a novel sensor for the detection and monitoring of cracks in a concrete structure. The principle of the sensor is illustrated in Fig.1, which shows a 'zig-zag' sensor at the bottom of a bridge deck. When an optical fiber intersects a crack at an angle other than 90 degrees, it has to bend to stay continuous as the crack opens. Bending will introduce loss in both the forward light power and the backscattered power. If the backscattered power is measured as a function of time (with Optical Time Domain Reflectometry or OTDR), the power signal will decrease slowly with time (due to scattering/absorption) and drop suddenly at a particular point (where concrete cracking leads to fiber bending), after which it will decrease slowly again (Fig.1b).

[1] Assistant Professor, Dept. of Civil & Environmental Engineering, MIT, Cambridge, MA 02139

[2] Doctoral Candidate, Dept. of Civil & Environmental Engineering, MIT, Cambridge, MA 02139

From the point of sudden drop, the crack location along the fiber can be deduced. From the magnitude of the drop, the crack opening can be calculated if a 'calibration' relation is available. If there are several cracks along the fiber, the backscattered signal will show a number of sudden drops. A number of cracks can hence be located and monitored with a single fiber. The proposed technique does not require prior knowledge of the crack locations, which is a significant advancement over existing crack monitoring techniques. However, the crack directions need to be known. An ideal application of the technique is in the monitoring of flexural cracks at the bottom of concrete bridge decks, where cracks may form at arbitrary locations along the deck, but essentially perpendicular to the spanning direction. To sense cracks effectively, several sensors should be employed. With a single fiber, if a crack intersects the 'zig-zag' fiber at a location where the fiber direction is changing, results will be difficult to interpret. The crack therefore needs to be picked up by another fiber with its turning point at other sections. For new structures, the sensor can be cast in the structure. For existing structures, the fiber needs to be first incorporated into a polymeric sheet which will then be glued to the structure surface.

Preliminary experimental results have shown that crack opening well below 0.05mm can be detected with the above concept. The successful development of the sensor for practical applications, however, relies on the optimization of its sensitivity (i.e., loss vs crack opening), range (which is limited by fiber breakage) and loss (which determines the number of cracks one can detect before the signal drops below the noise threshold). This paper presents a theoretical framework for the modeling of light signal loss as a function of crack opening. The analysis involves two major steps (i) micromechanical modeling of fiber bending as the crack opens, and (ii) modeling of electromagnetic wave propagation in a deformed waveguide.

Micromechanical Modeling of Fiber Bending

As crack opening induces bending on the optical fiber, antisymmetry leads to the formation of an inflection point in the middle of the crack. In the analysis, the fiber is first cut at the inflection point to give two parallel but separate segments. A displacement is then applied to each of the cut ends to put the fiber back together. The deformation along the fiber corresponding to the applied displacement is obtained through a three-dimensional finite element analysis. In the analysis, 8-node solid elements are employed to model the fiber, the matrix as well as the polymeric coating around the fiber. The boundary of the matrix is placed at 5 fiber diameters from the fiber center. This is found to be sufficient to simulate a matrix of infinite extent.

With the finite element analysis, fiber displacement is only obtained at a given number of points. To obtain an accurate estimate of the second derivative of the displacement, which governs the signal loss, a reliable interpolation technique is required. It has been shown in [4] that a fiber in a cementitious matrix can be modeled as a beam on an elastic foundation, with the foundation stiffness increasing with distance from the crack face. Here, we assume the foundation stiffness to be a linear function of distance from the crack face. A fourth order differential equation can then be set up and solved numerically. In really, the stiffness will reach a constant rather than increasing monotonically. However, since the reaction forces drop rapidly with distance from the crack face, the assumptions of a continuously increasing stiffness and one that reaches a plateau give no noticeable difference in the results. By varying β, the proportional constant between the foundation stiffness and the distance from the crack face, the best fit to the finite element result can be obtained

(Fig.2). From the best fit curve, the curvature along the fiber is calculated and the signal loss can then be calculated.

Modeling of Signal Loss in a Curved Optical Fiber

Bending loss along an optical fiber with arbitrary curvature can be obtained with the beam propagating method [5]. For a time harmonic E field, one can separate the time variable from the wave equation to obtain the Helmholtz equation in space variables. Instead of solving the Helmholtz equation in curvilinear coordinates, the curved fiber is first transformed into an equivalent straight fiber with a modified refractive index profile [6]. The Helmholtz equation is then solved with the finite difference method. As an initial condition, the E field corresponding to the propagation mode in a straight optical fiber is imposed at a point far away from the crack (where no bending loss has yet occurred). The analysis is carried out along the fiber until it becomes straight again at the other side of the crack. The ratio of power at the input and output sections allows the calculation of dB loss as a result of fiber bending.

Results

Following the computational process described above, the dB loss in the fiber as a function of crack opening is obtained for several fiber inclination angles. The results are shown in Fig.3. The trend of the theoretical prediction is identical to our preliminary experimental results. Quantitative agreement is yet to be achieved because: (i) as a first attempt, the wave propagation analysis is carried out assuming the fiber to be a slab rather than a cylindrical waveguide (to reduce computation time), and (ii) the stiffness of the polymeric coating is assumed rather than directly measured. Improvement of the theoretical model as well as further experimental verification is currently being carried out.

Conclusions

A model which combines micromechanics with electromagnetic wave theory is developed to derive the relation between signal loss and crack opening for a novel optical fiber sensor. Results from the model compares well with preliminary experimental results. While full verification of the model is yet to be carried out, its potential as a tool for sensor design is illustrated.

References

1. Ansari, F. and Navalurkar, R.K. (1993). "Kinematics of Crack Formation in Cementitious Composites by Fiber Optics", ASCE J. Eng. Mech., Vol.119, No.5, 1048-1061.
2. Voss, K.F. and Wanser, K.H. (1994). “Fiber Sensors for Monitoring Structural Strain and Cracks”, Proceedings, 2nd European Conference on Smart Structures and Materials,Glasglow, 144-147.
3. Wolff, R. and Miesseler, H. (1992). "Monitoring of Prestressed Concrete Structures with Optical Fiber Sensors", Proceedings, 1st European Conference on Smart Structures and Materials, Glasgow, 23-29.
4. Leung, C.K.Y. & Li, V.C. (1992). "Effects of Fiber Inclination on Crack Bridging Stresses in Brittle Matrix Composites", J. Mech. Phys. Solids, Vol.40, No.6, 1333-1362.
5. Yamauchi, J., Ikegaya, M. and Nakano, H. (1992). "Analysis of Bent Step-Index Optical Fibres by the Beam Propagation Method", IEE Proceedings-J,

Vol.139, No.3, 201-207.
6. Heiblum, M. and Harris, J.H. (1975). "Analysis of Curved Optical Waveguides by Conformal Transformation", IEEE J. of Quantum Electronics, Vol. QE-11, No.2, 75-83.

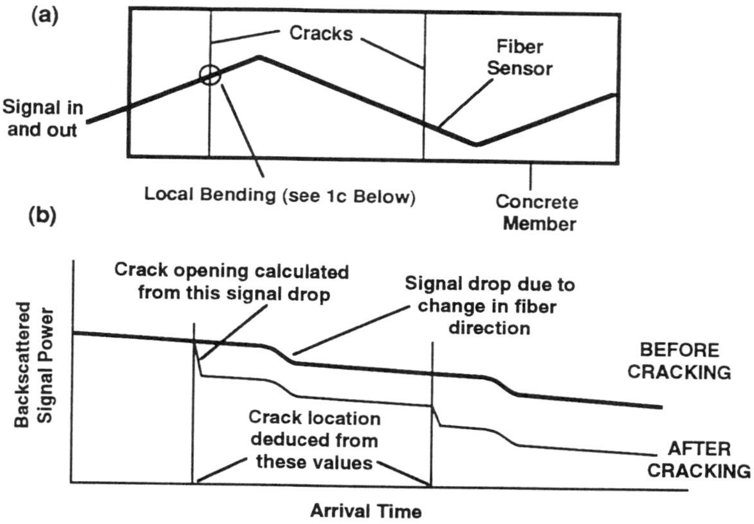

Fig.1 The Novel Optical Fiber Crack Sensor

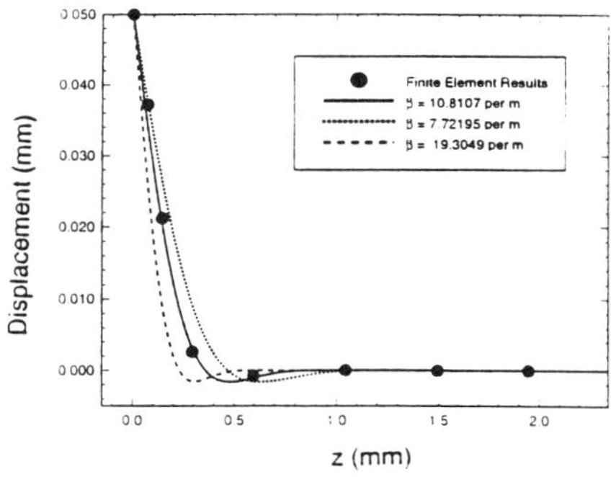

Fig.2
Finite Element Results and Fitting
with Beam on Elastic Foundation Model

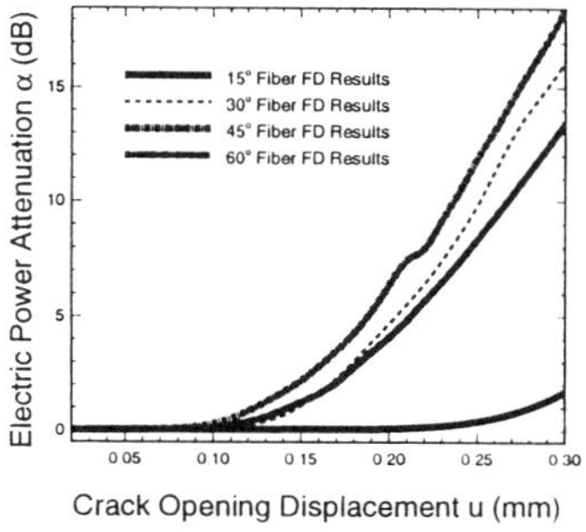

Fig.3
Model Prediction of
Signal Loss vs Crack Opening

Interference Assemblies

R. R. Little [1]

G. R. Frederick [2]

B. K. Park [3]

Abstract

This paper contains the development of a method for determining the contact pressure distribution along the mating surfaces of two members assembled together with an initial interference fit. This method utilizes superposition, Fourier series expansion techniques and the finite element method to determine the desired contact pressure distribution. The general approach to the solution of this problem entails assuming a Fourier series representation of the contact pressure distribution expressed in terms of unknown Fourier coefficients. Superposition and finite element methods are used to determine the displacement vector at common nodal points along the mating surfaces of the two members. The correct distribution of contact pressure results when these nodal displacements satisfy the initial interference between the two members. Enforcing this displacement requirement yields a system of equations from which the unknown Fourier pressure coefficients can be determined. Once the correct contact pressure distribution is known, the resulting stress distributions and deformations in the inner and outer members can be calculated. This method is illustrated and numerical results obtained for the case where an outer member having an irregular shaped boundary is shrunk onto a solid circular shaft representing the inner member.

[1]Associate Professor of Mechanical Engineering, University of Toledo
[2]Professor of Civil Engineering, University of Nevada, Las Vegas
[3]Technical Specialist, Saginaw Division - General Motors Corporation

Digital computer programs were written to generate and solve the governing equations for the desired Fourier series coefficients, the resulting contact pressure distribution along the mating surface of the two members, and the resulting system of finite element equations in terms of the nodal displacements of both the inner and outer members using the methods previously indicated. Pre- and post-processing phases of the solution were included in the program development which also utilized bandwidth reduction methods in order to minimize the storage requirements. Numerical results were obtained considering outer members that had circular internal boundaries and square, octagonal, and sixteen-sided polygonal external boundaries. Although all cases investigated have symmetric conditions with respect to the geometry of the outside member, and thus the contact pressure distribution about 45 degree lines, this condition is not necessary. The systematic procedure developed in order to determine the contact pressure distribution along mating surfaces in interference fit assemblies, together with the resulting stress distributions in the two members, can be a valuable tool for the design engineer who must deal with interference fitted assemblies.

A New Method for Solving Large Deformation Problems

Xiangjun Qiu [1]

Abstract

Most commercial softwares use the standard continuum mechanics method to treat large deformation elasticity problems, in which the material's natural configuration is chosen as the reference configuration. In this paper a new method is developed, in which a material non-natural configuration is selected as the reference configuration. It is shown that a good choice of reference configuration may significantly reduce the complexity of the large deformation elasticity problems and sometimes the nonlinear problems may be reduced to linear problems. Two examples are given in the paper, which demonstrate the merits of the new method.

Introduction

In the frame of continuum mechanics, any possible deformed configuration of a body is allowed to be chosen as the reference configuration for the body. However, in FEM for solids it is natural and popular to use the natural configuration, which is defined as the configuration corresponding to unconstrained and unloaded state of the body, as the initial reference configuration. In case of large deformation, the CPU time needed to solve the nonlinear problem directly depends on the degree of difference between the current configuration and the initial reference configuration of the body. Therefore, it is logical to think about using a non-natural reference configuration in order to efficiently solve large deformation problems. In this paper, an attempt is made at solving certain large

[1]Manager, Applied Mechanics, Conveyor Dynamics Inc., 1111 West Holly Street, Bellingham, WA 98225

deformation problems with the help of choosing appropriate non-natural reference configurations. It is shown that a good choice of reference configuration may simplify the solution procedure significantly. Two examples are given in the paper. The method developed in the paper has been proved very useful in solving large deformation of thin plates, shells and mambranes.

Theory

We use $\mathbf{R}_0$, $\mathbf{R}_f$, and $\mathbf{r}$ to denote position vectors emanating from the same origin O and terminating at a particle of a body in the natural configuration, reference configuration, and the current configuration, respectively (see Fig.1). The actual motion carries particles from the natural configuration to the current configuration. Let us consider a large deformation motion of a body and imagine that the process of the motion could be detoured in such a way that the body would first move from the natural configuration to the reference configuration and then from the reference configuration to the current configuration. As long as the material behavior follows elastic or hyperelastic constitutive law, such a detouring will not change the final solution at the current configuration of the body. The following steps will lead to the solution for an elastic body undergoing large deformation.

1. Choose an appropriate reference configuration.
2. Use the constitutive law of the material to map the deformation gradient $\partial\mathbf{R}_f/\partial\mathbf{R}_0$ to the stress at the reference configuration. Such stress is called initial reference stress.
3. Find the body force and boundary force, under which the body would be kept in equilibrium in the shape of reference configuration with the initial reference stress. Such a set of body force and boundary force is called initial reference equilibrium load (IREL). The IREL can be determined through the equilibrium equations:

$$\mathbf{f}_0 = -\mathrm{Div}(\sigma_0)$$

$$\mathbf{t}_0 = \sigma_0 . \mathbf{n}$$

   where σ_0 is the initial reference stress tensor of Cauchy type, Div denotes the divergence operator corresponding to the reference configuration, $\mathbf{n}$ is a unit normal to the surface of the reference configuration, $\mathbf{f}_0$ is the body force of IREL, and $\mathbf{t}_0$ is the boundary force of IREL.
4. Apply the actual load and boundary condition to the reference configuration.
5. Release IREL by adding $-\mathbf{f}_0$ and $-\mathbf{t}_0$ to the actual body force and boundary force.

6. Solve the elasticity problem with the above boundary condition and load.
7. Verify if the deformation in the solution is small enough. If it is, the solution is final. If it is not, we can either improve the solution iteratively by using nonlinear methods or go back to Step 1 with selecting a better reference configuration.

The examples in the next section will illustrate the above solution procedure.

Examples And Conclusions

The first example deals with the large deformation of two rubber bands, which are connected end-to-end (see Fig.2). The two rubber bands have the same cross section areas S_0 and lengths $0.5L_0$ before elongation and are subjected to a total elongation of $L/L_0=4$ (see Fig.2). Assuming both rubber follow Mooney-Rivlin constitutive law and neglecting the stress concentration around the connecting area, we can write the following 1-D form constitutive equations

$$F_i=2(\lambda_i-\lambda_i^{-2})(A_i+\lambda_i^{-1}B_i)S_0 \quad (i=1,2) \quad (3.1)$$

where the subscripts 1 and 2 denote the two rubber materials, F_i are the uniaxial forces in cross section areas, λ_i are the elongation ratios ($\lambda_i=L_i/0.5L_0$), A_i and B_i are Mooney-Rivlin constants (in general A_1 and B_1 are different from A_2 and B_2). The reference configuration lengths are selected by the following ad hoc formulae

$$L_{fi}=L(A_i+B_i)^{-1}/[(A_1+B_1)^{-1}+(A_2+B_2)^{-1}] \quad (i=1,2) \quad (3.2)$$

The initial reference forces can be calculated by

$$F_{fi}=2(\lambda_{fi}-\lambda_{fi}^{-2})(A_i+\lambda_{fi}^{-1}B_i)S_0 \quad (i=1,2) \quad (3.3)$$

where $\lambda_{fi}=L_{fi}/0.5L_0$. In this case the body force of IREL is zero and the boundary force of IREL is shown in Fig.2b. Applying the actual boundary condition to the reference configuration and releasing the IREL, we have a system shown in Fig.2c, in which $F_f=F_{f1}-F_{f2}$, to solve. Assuming the deformation from the reference configuration to the current configuration is small, we can linearize the Eq.(3.1)

$$\Delta F_i=k_i\,\varepsilon_i \quad (i=1,2) \quad (3.4)$$

where $k_i=2[(\lambda_{fi}+\lambda_{fi}^{-2})(A_i+\lambda_{fi}^{-1}B_i)-(\lambda_{fi}-\lambda_{fi}^{-2})\lambda_{fi}^{-1}B_i]S_0$. Solving the linear system of Fig.2c and superposing the result to the initial reference forces, we finally obtain the solution for the uniaxial force in rubbers

$$F=(k_1+k_1)^{-1}(k_2F_{f1}+k_1F_{f2}) \quad (3.5)$$

The numerical results of Eq.(3.5) for 4 cases are shown in Table 1. It is seen that comparing with the exact solution, the Eq.(3.5) has less than 6% error and is obtained without solving any nonlinear equations.

The second example deals with small strain but large bending deformation of a flat rectangular thin sheet with a width of πD. The sheet is rolled until one of its edge meets the other, then the two edges are connected by a hinge, and finally the sheet is inserted into a slot with an openning of D (see Fig.3). Select a circular cylinder with diameter of D as the reference configuration and note that the sheet has a flat natural configuration. Following the above mentioned steps, we can finally reach a closed form solution, which agree with experiment measurements very well (see in Fig.3). We neglect the details for the solution procedure because of space limitation.

Table 1.

Case#	A_1	B_1	A_2	B_2	λ_{f1}	λ_{f2}	λ_{*1}	λ_{*2}	F/S_0	F_*/S_0	*Err%*
1	1.0	0.5	2.0	1.0	4.0	2.0	3.985	2.015	8.812	8.829	-0.19
2	2.0	1.0	6.0	3.0	4.5	1.5	4.387	1.613	18.763	19.316	-2.86
3	1.0	1.0	3.0	0.0	3.6	2.4	4.067	1.933	10.382	9.985	3.98
4	2.0	1.5	5.0	2.5	4.5	1.5	4.277	1.723	16.932	17.878	-5.29

*Note: the subscript * denotes the exact solution and* $Err=(F-F_*)/F_*$. *The units for* A_i, B_i *and* F/S_0 *are MPa.*

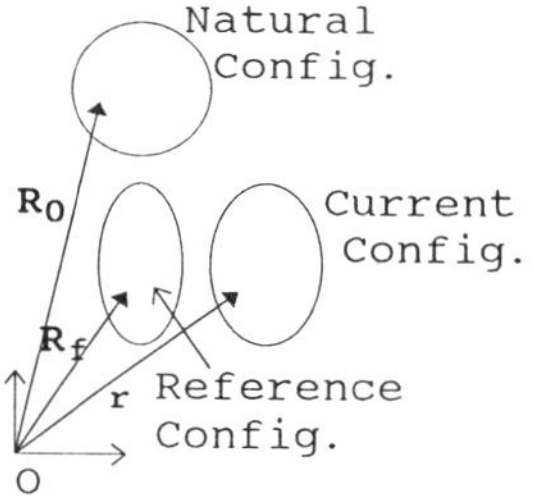

Fig.1 Body Motion

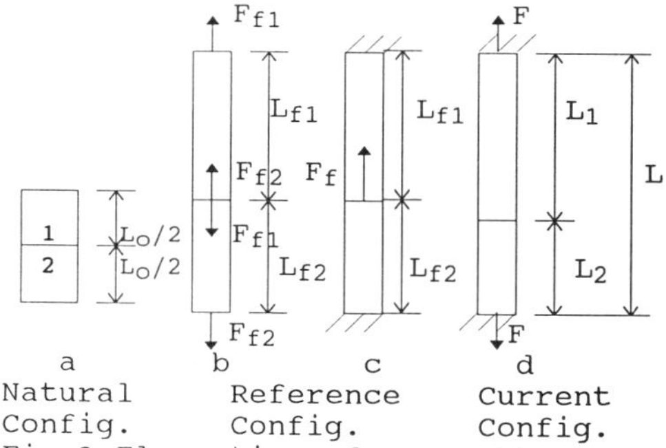

Fig.2 Elongation of Two Rubber Bands

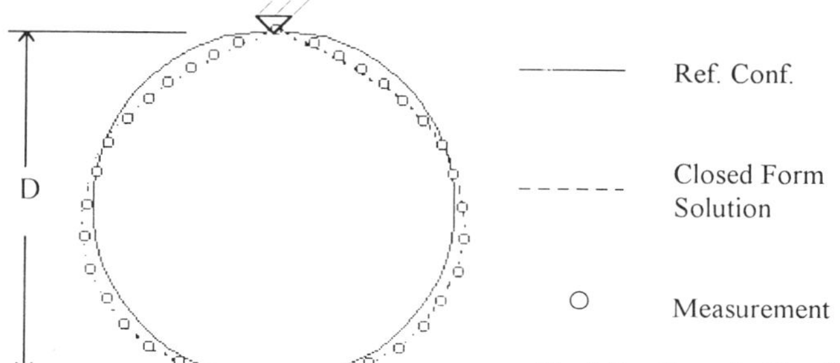

Fig. 3 Bending of a Flat Sheet

Viscoelastic Modeling of Paper

Edmond P. Saliklis[1]

Abstract

In this study, a viscoelastic model was used to predict the constitutive behavior of paper subjected to compressive stress. The same model was then used to solve for the response of paper subjected to creep loading. Results show that the model can account for both types of loading.

Introduction

To fully understand the effect that rate of loading has on the constitutive response of an engineering material, experimental measurements must be correlated to a mechanistic model. In this study, a viscoelastic model was used to predict the constitutive behavior of paper subjected to compressive stress. A suitable one-dimensional model for studying viscoelastic materials incorporates instantaneous elasticity, retarded elasticity, and plastic flow.

The model, shown in Figure 1, has been used by other researchers to describe a material's viscoelastic behavior (Alfrey 1948, Tobolsky and Eyring 1943). It has been applied to paper mechanics problems (Mason 1948) and used to study the creep response of wood structures (Fridley et al. 1992, Pelvris and Triantafillou 1995).

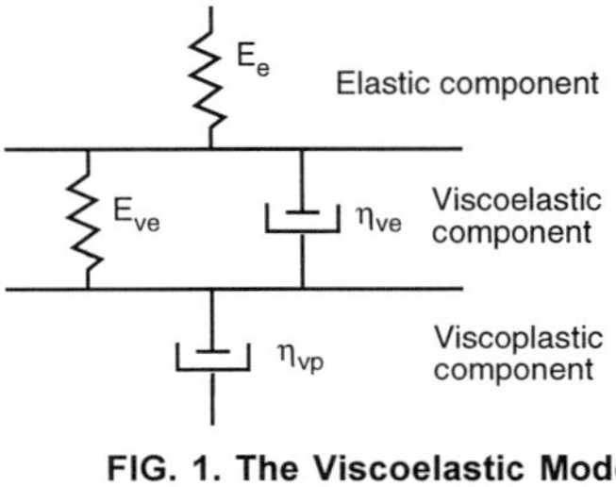

FIG. 1. The Viscoelastic Model

[1]Edmond P. Saliklis, Research Engineer, USDA Forest Service, Forest Products Laboratory, One Gifford Pinchot Drive, Madison WI 53705–2398

Method

The study reported herein applied the model (Fig. 1) to predict the response of paper subjected to various loading histories. The fundamental differential equation of motion of this system was solved for monotonically increasing load and creep load. The elastic component was modified to allow for nonlinear behavior. The modification was such that the modulus of elasticity E no longer remained constant and decreased proportionately to the increase in stress. In other words, the elastic spring stiffness became a function of the stress in the element.

$$E \equiv E_e(1 - ckt) \qquad (1)$$

where k = rate of applied stress and t = time; therefore, kt = stress and c is the constant creating nonlinearity in the spring. Solving the equation of motion for monotonically increasing loads resulted in the expression:

$$\varepsilon(t) = \frac{kt}{E_e(1 - ckt)} + kD\tau\left(-1 + \frac{t}{\tau} + e^{-t/\tau}\right) + \frac{kt^2}{2\eta_{vp}} \qquad (2)$$

where E_e is the modulus of the elastic spring; D is $1/E_{ve}$, τ is the time constant relating E_{ve} to η_{ve}; and η_{vp} is the viscoplastic viscosity. This expression was able to model the constitutive behavior of paper loaded at three markedly different rates: 9.86×10^{-5} MPa/s, 9.86×10^{-3} MPa/s, and 9.86×10^{-1} MPa/s. Calibration was performed by curve-fitting four parameters, (c, D, τ, η) to the slowest rate of loading, i.e., 9.86×10^{-5} MPa/s. Curve fitting was accomplished with a commercially available package (Jandel Scientific 1994). The remaining parameter, E_e, was found through fitting the data of a standard uniaxial test (Urbanik 1982). This same viscoelastic model was solved for creep loading, where the stress was previously represented by kt, it is now a constant σ_0. The solution for this case is the following:

$$\varepsilon(t) = \sigma_0\left(\frac{1}{E_e} + D(1 - e^{-t/\tau}) + \frac{t}{\eta_{vp}}\right) \qquad (3)$$

In this solution, the constants have the same value as in Equation (2). Thus, having calibrated a mechanistic response for a monotonically increasing load (rate of load, ROL), you can automatically predict the response to creep loading, (duration of load, DOL). This results in an intrinsic coupling of ROL response to DOL response.

An alternative way to arrive at creep response is an isobaric response method. Here, the stress versus strain response of various rates of monotonically increasing load is plotted or tabulated such that for different constant stresses (isobars), the strain at a particular time can be collected (Fig. 2). Then, it is easy to fit a curve through the points (Fig. 2) (A, t_1), (A′, t_2), and (A″, t_3). Repeating this process for the B and C sets of points results in a generated three-dimensional map relating ROL response to DOL response (Fig. 3).

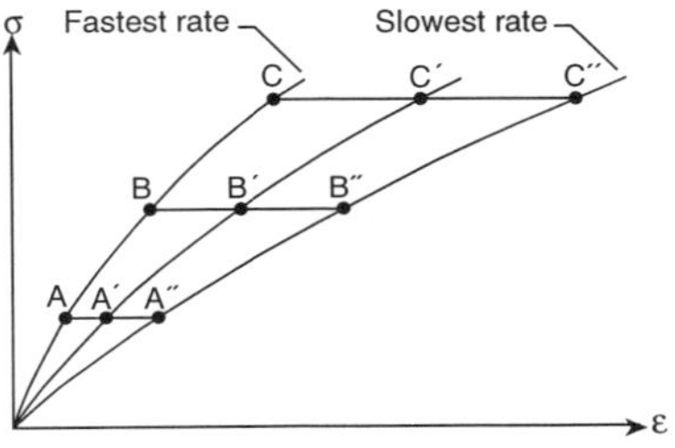

FIG. 2. Rate of Load Response Variations

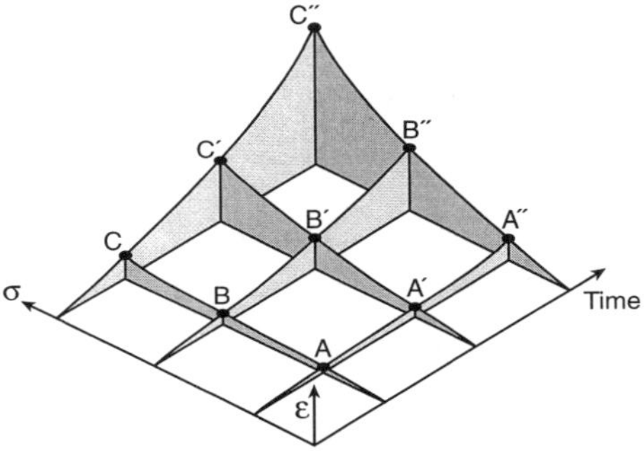

FIG. 3. Generation of Isobaric Lines

A virgin paper (205 g/m^2 basis weight) was experimentally investigated by Gunderson et al. (1988) for ROL response in the machine direction (MD). The paper was tested at two levels of environmental relative humidity: 50% and 90%. Six replications were performed for each test. Newly acquired DOL data had three replications of each test. The viscoelastic model was able to account for both ROL and DOL. The DOL data are shown with both methods of prediction, i.e. the viscoelastic model prediction and the isobaric response prediction (Figs. 4 to 7).

Concluding Remarks

In this study, a viscoelastic model was applied to paper compressed at various rates of load. The same model was then used to solve for the response of paper subjected to creep loading. Results show that the model can account for both types of loading. In addition, a predictive isobaric response method was described and applied to the data.

Appendix. References

Alfrey, T. (1948). *Mechanical behavior of high polymers.* Interscience Publishers, Inc., New York, N.Y.

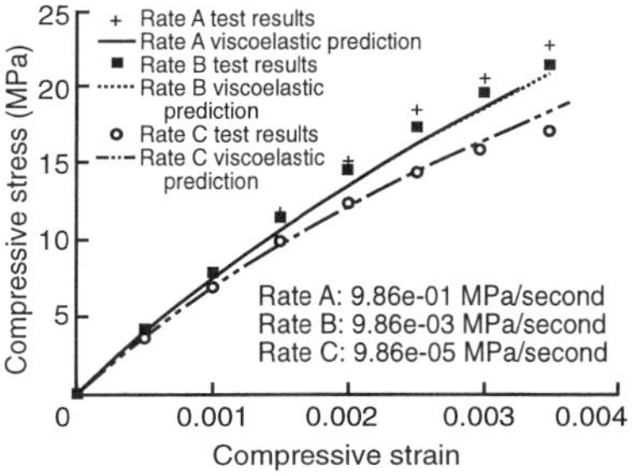

FIG. 4. Analysis of 6756 Paper, Machine Direction at 50% Relative Humidity

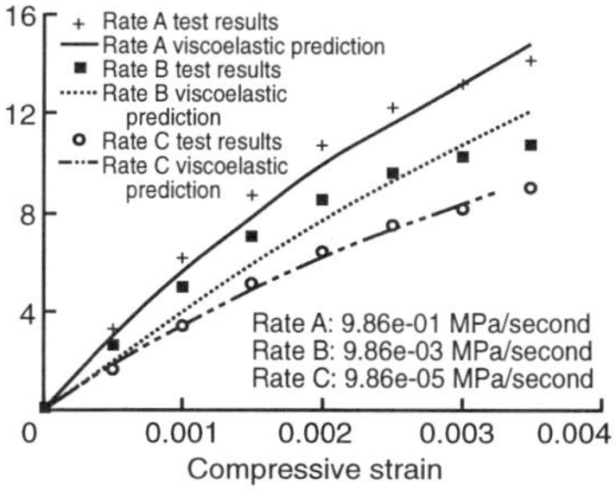

FIG. 6. Analysis of 6756 Paper, Machine Direction at 90% Relative Humidity

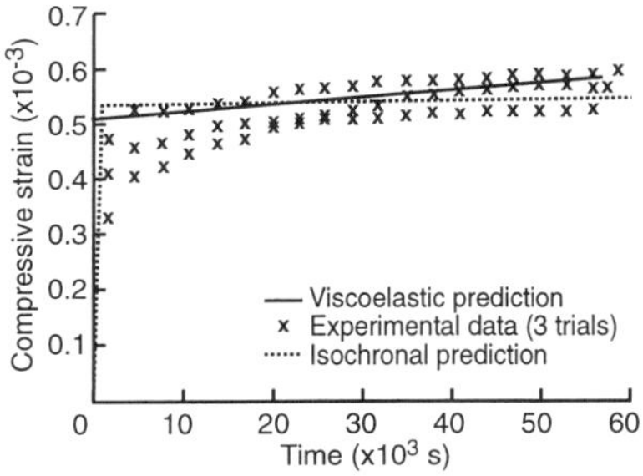

FIG. 5. Analysis of 6756 Paper, Machine Direction at 50% Relative Humidity

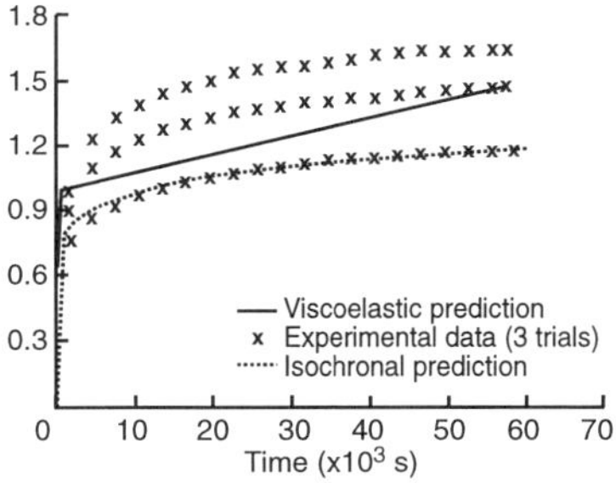

FIG. 7. Analysis of 6756 Paper, Machine Direction at 90% Relative Humidity

Fridley, K.J., Tang, R.C., and Soltis, L.A. (1992). "Creep behavior model for structural lumber." *J. of Struct. Engrg.*, 118(8), 2261–2277.

Gunderson, D.E., Considine, J.M., and Scott, C.T. (1988). "The compressive load-strain curve of paperboard: rate of load and humidity effects." *J. of Pulp and Paper Sci.*, 14(2), J37–J41.

Jandel Scientific. (1994). *Table curve 2D, automated curve fitting software, user's manual.* Jandel Scientific, San Rafael, CA.

Mason, S.G., (1948). "The rheology of paper," *Pulp and Paper Magazine of Canada*, 49, Pt. 1, 207–214.

Plevris, N. and Triantafillou, T.C. (1995). "Creep behavior of FRP-reinforced wood members." *J. Structural Engrg.*, 121(2), 174–186.

Tobolsky, A.V., and Eyring, H. (1943). "Mechanical properties of polymeric materials." *J. Chemical Physics*, (11),125.

Urbanik, T., (1982). "Method analyzes analogue plots of paperboard stress–strain data," *Tappi J.* 65(4), 104–133.

THE VISCOELASTIC-LARGE DEFORMATION RESPONSE OF THE TAYLOR IMPACT CYLINDER

H.L. Schreyer[1], M. ASCE, and D. Sulsky[2]

Abstract

A formulation for viscoelasticity is given in which evolution equations similar to those of viscoplasticity are used. The result is a convenient algorithm that is implemented in the material point method, a new numerical algorithm for analyzing solid bodies subject to large deformations. The approach is demonstrated with a Taylor impact problem in which the large deformations of both the impact and recovery phases are exhibited.

Introduction

The integrals that involve relaxation functions in the conventional formulation of viscoelastic constitutive equations have been a severe problem for the development of efficient numerical algorithms. A rate form of viscoelasticity similar to those of plasticity would be useful in that one efficient algorithm could then be used for a wide range of constitutive equations. The starting point for the current development is an evolution equation for a viscoelastic backstrain similar to that of plastic strain rate based on a viscoplastic model. If the evolution equations are linearized, the result is the standard viscoelastic model.

To illustrate the application of such a model, a simulation of the Taylor impact problem for a viscoelastic cylinder is performed. The material point method, which is a variation of a recent development at Los Alamos National Laboratory of a particle-in-cell method, is used for the simulation. An impact velocity is chosen that causes the cylinder to deform to approximately one-third of its initial height. As a consequence of the viscoelastic constitutive equation, the cylinder then recovers to its original configuration.

Viscoelasticity

The concept of a backstress is common in metals. In a general sense, plastic strain can be considered a "back strain." In a similar vein, suppose that a dominant aspect of viscoelastic behavior is a back strain, $\mathbf{e}^b$. With the assumption of linear

----

[1]Professor, Dept. of Mechanical Engineering; [2]Professor, Dept. of Mathematics and Statistics; University of New Mexico, Albuquerque, NM 87131

elasticity, suppose the constitutive equation is

$$\boldsymbol{\sigma} = \mathbf{E} : (\mathbf{e} - \mathbf{e}^b) \tag{1}$$

in which $\boldsymbol{\sigma}$ denotes the stress, $\mathbf{e}$, the total strain, and $\mathbf{E}$ the elaticity tensor. Let $\boldsymbol{\sigma}^b$ denote a back stress. If the back strain and stress rates, $\dot{\mathbf{e}}^b$, are assumed to be proportional to the difference between $\boldsymbol{\sigma}$ and a back stress $\boldsymbol{\sigma}^b$, then the evolution equations are taken to be

$$\dot{\mathbf{e}}^b = \frac{\eta^e}{2G}[\boldsymbol{\sigma} - \boldsymbol{\sigma}^b] \qquad \dot{\boldsymbol{\sigma}}^b = \frac{\eta^\sigma}{2G}[\boldsymbol{\sigma} - \boldsymbol{\sigma}^b] \tag{2}$$

in which η^e and η^σ are material parameters with the dimension of inverse time, and G is the shear modulus. It can be shown that for uniaxial stress (1) and (2) reduce to the standard linear viscoelastic model.

A simplification of the theory is to assume that (2) holds only for the deviatoric parts of the tensors in these equations. Then the spherical parts of stress and strain are governed by elasticity. Here, the linear relation

$$P = -Be_v \tag{3}$$

has been assumed in which B is the bulk modulus, and e_v, the volumetric strain. A nonlinear volumetric elastic relation is easily acommodated and a nonlinear viscoelastic constitutive equation is achieved by replacing η^e and η^σ with terms proportional to dimensionless invariants of $\boldsymbol{\sigma} - \boldsymbol{\sigma}^b$, raised to a power. Only linear constitutive equations were used for the numerical simulation

Numerical Algorithm

A variation of the particle-in-cell method (FLIP) developed at Los Alamos National Laboratory by Brackbill and Ruppel (1986) and Brackbill et al. (1988) has been used as the base for an approach in which material points are tracked (material point method). History dependent constitutive equations are applied only at these material points so the error associated with the mapping of history variables is avoided. This class of numerical algorithms utilizes two meshes, one a material or Lagrangian mesh defined over the region of the body under consideration, and the second a spatial or Eulerian mesh defined over the computational domain. The advantages of both Eulerian and Lagrangian methods are achieved by using the appropriate frame for each aspect of the computation, with a mapping performed between the two meshes at each step in the loading process. The numerical dissipation due to convection normally associated with an Eulerian method is avoided by using a Lagrangian step; the mesh distortion seen in large-deformation problems with a Lagrangian method is avoided by using a user-controlled spatial mesh. As a consequence, problems which have caused difficulties in the past with conventional approaches are now handled fairly routinely. An example is the Taylor problem consisting of a cylindrical rod axially impacting a rigid surface.

The initial formulation of the material point method has been limited to problems of plane stress or strain (Sulsky et al., 1994; Sulsky et al., 1995) as a testbed for the numerical algorithm. Recently, the formulation has been extended to three-dimensional axisymmetric problems which permits the anaysis of the Taylor problem and the upsetting of cylindrical billets (Sulsky and Schreyer, 1996).

Example Problem

The Taylor problem in which the cylinder consists of a viscoelastic material was chosen as a model problem for two reasons. First, if viscoelasticity is cast into the form of evolutionary equations similar to those of plasticity, the numerical algorithm is much more efficient than algorithms based on hereditary integrals. Second, the problem serves to demonstrate that the material point method can be used to simulate the large deformations associated with both the compressive phase of the impact and the recovery phase that occurs with viscoelasticity.

For the purpose of this demonstration, artificial parameters have been selected. The cylinder is taken to be of length 2.35 cm and radius 0.39 cm with an initial velocity of 263 m/s. Material parameters consist of the elastic parameters which are Young's modulus (39.1 Gpa) and Poisson's ratio (0.3), from which the bulk and shear moduli are obtained, and the viscoelastic parameters of $\eta^e = 5.0/s$ and $\eta^\sigma = 0.01/s$. The mass density is 2700 kg/m^3. The spatial grid consists of a 16 x 34 mesh with the cylinder represented by 4 material points per element.

Figure 1a shows the cylinder in its initial configuration just before impact at time t = 0. The spatial grid is shown as are the material points describing the domain of the cylinder. Subsequent frames in the figure show the cylinder at sequential times during the impact and rebound phase. Notice that at the time of maximum deformation shown in Fig. 1d, the cylinder has compressed to approximately 1/3 of its original length with no evidence of a numerical breakdown. The last frame, Fig. 1f, shows the cylinder at a much later time when the cylinder has recovered to approximately 7/8 of its original length.

Conclusions

The material point modification to the latest version of the particle-in-cell method provides a natural procedure for handling the large deformations of solids. With the formulation of viscoelasticity in terms of evolution equations, the example problem shows that the method can provide technically significant solutions in a few minutes on a work station. Other important applications including metal rolling, upsetting and metal forming are currently being investigated.

Acknowledgment

The support of Los Alamos National Laboratory for part of this work is greatly appreciated.

References

Brackbill, J.U., and Ruppel, H.M., 1986, "FLIP: A Method for Adaptively Zoned, Particle-in-cell Calculations in Two Dimensions," J. Comput. Phys., **65**, 314-343.

Brackbill, J.U., Kothe, D.B., and Ruppel, H.M., 1988, "FLIP: A Low-dissipation, Particle-in-cell Method for Fluid Flow," Comput. Phys. Comm.., **48**, 25-38.

Sulsky, D., Chen, Z., and Schreyer, H.L., 1994, "A Particle Method for History-Dependent Materials," Computer Methods in Applied Mechs. and Engineering, **118**, 179-196.

Sulsky, D., Zhou, S.-J., and Schreyer, H.L., 1995, "Application of a Particle-in-Cell Method to Solid Mechanics," Computer Physics Communications, **87(1&2)**, 236-252.

Sulsky, D., and Schreyer, H.L., "Axisymmetric form of the Material Point Method with Applications to Upsetting and Taylor Impact Problems," Submitted for publication.

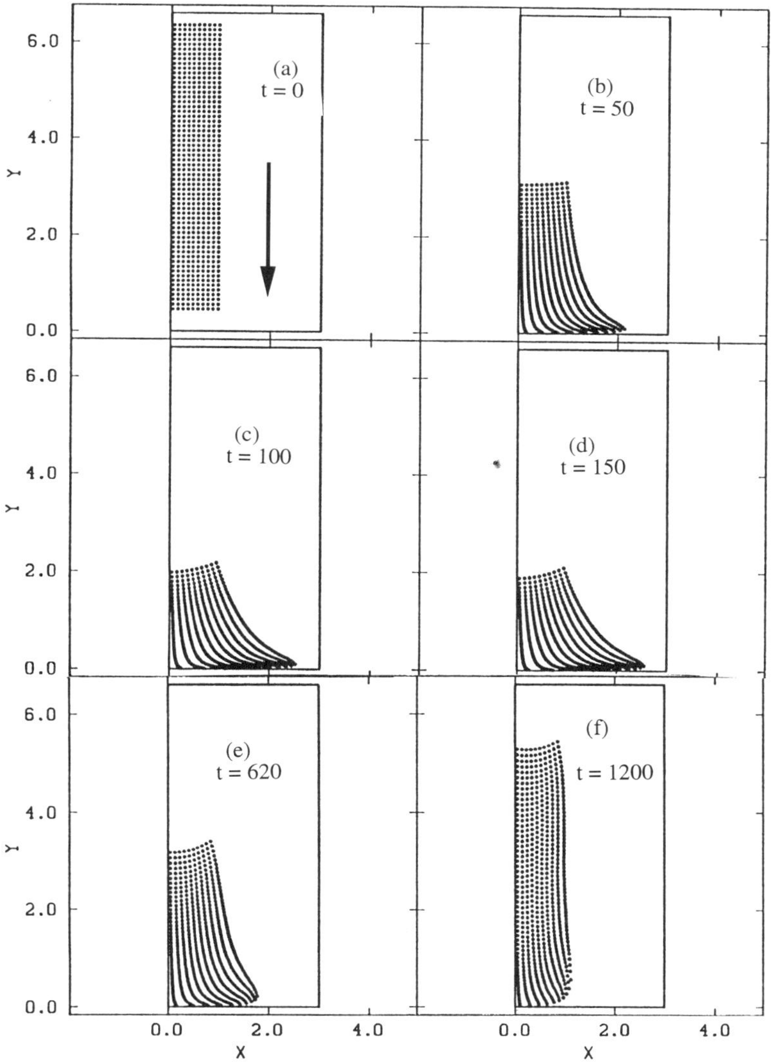

Figure 1. Configurations of the cylinder at various times.

EFFECTIVE MOMENT OF INERTIA OF ELASTO-PLASTIC BEAMS

Barry T. Rosson[1], Member ASCE
Ronald K. Faller[2], Student Member ASCE

Abstract

An effective moment of inertia is developed for a rectangular, prismatic elasto-plastic beam with elastic, linear-hardening material behavior. The particular solution for a beam with elastic, perfectly plastic material behavior is also presented with applications for predicting beam deflections. Formulas are given for the direct application of I_{eff} for predicting deflections in closed-form of elasto-plastic beams with concentrated and distributed loads.

Introduction

Inelastic analysis procedures are necessary for predicting the behavior of structural members stressed above the yield limit of the material and for determining displacements in both the elastic and plastic regions of the member. Almost all of the current inelastic procedures have addressed the reduced rigidity of the member by making use of a reduced modulus of elasticity, commonly referred to as the tangent modulus of elasticity, E_t.

Fertis, Taneja, and Lee (1991) and Fertis and Taneja (1991) showed that elastic and inelastic analysis of members with both constant and variable rigidity could be approximated using equivalent linear systems and applying known methods of linear elementary mechanics. For inelastic members, an expression was developed for a reduced modulus of elasticity, which was later solved for by using substitution of variables and trial-and-error procedures.

This paper describes a new procedure that reduces the rigidity of a beam stressed beyond the elastic limit of the material by using an effective moment of inertia in the elasto-plastic regions. Closed-form equations are given for

[1] Assist. Professor, Civil Engr. Dept., Univ. of Nebraska, Lincoln, NE 68588.
[2] Ph.D. Candidate, Civil Engr. Dept., Univ. of Nebraska, Lincoln, NE 68588.

determining the effective moment of inertia for rectangular, prismatic beams with elastic, linear-hardening material behavior. The particular condition of a beam with elastic, perfectly plastic material behavior is further developed for predicting beam deflections in closed-form.

Effective Moment of Inertia

The second-order differential equation that is routinely used for the determination of the curvature and deflection of elastic beams can be used for the same purpose for elasto-plastic beams provided the proper reduction in rigidity is made for moments beyond the specified yield moment. This can be done by modifying the modulus of elasticity with the use of a tangent modulus, E_t, or by using a reduced moment of inertia, which in this paper will be referred to as the effective moment of inertia, I_{eff}. Equation (1) shows the moment-curvature relationship with the effective moment of inertia for elasto-plastic moments, M_{ep}, above the yield moment, M_y.

$$\frac{d^2y}{dx^2} = \frac{M_{ep}}{EI_{eff}} \tag{1}$$

It is assumed that the elasto-plastic beam has small deflections, the shear stresses are small compared with the longitudinal stresses, the shear deformations are negligible, and that plane sections remain plane after bending.

It can be shown that a cubic equation in I_{eff}/I can be found (Faller & Rosson 1996) in terms of M_{ep}/M_y and the linear strain-hardening coefficient α as

$$\left[\frac{1}{2\left(\frac{M_{ep}}{M_y}\right)^3}(1-\alpha)\right]\left(\frac{I_{eff}}{I}\right)^3 - \left[\frac{3}{2\left(\frac{M_{ep}}{M_y}\right)}(1-\alpha) - 1\right]\left(\frac{I_{eff}}{I}\right) - \alpha = 0 \tag{2}$$

Plots of this equation show that for any $\alpha > 0$, and as M_{ep}/M_y approaches infinity, the ratio of I_{eff}/I approaches α. Equation (2) can be shown to have two solutions that have real roots and are given by Equations (3) and (4). Equation (4) is used to find I_{eff} when $\alpha < 1 - (2/3)M_{ep}/M_y$ for $1 < M_{ep}/M_y < 1.5$; for all other conditions of α and M_{ep}/M_y Equation (3) is used to calculate I_{eff}.

$$I_{eff} = \left[\sqrt[3]{-\frac{b}{2}+\sqrt{\frac{b^2}{4}+\frac{a^3}{27}}} - \sqrt[3]{\frac{b}{2}+\sqrt{\frac{b^2}{4}+\frac{a^3}{27}}}\right]I \tag{3}$$

$$I_{eff} = \left[\sqrt[3]{-\frac{b}{2} + \sqrt{\frac{b^2}{4} + \frac{a^3}{27}}} + \sqrt[3]{-\frac{b}{2} - \sqrt{\frac{b^2}{4} + \frac{a^3}{27}}} \right] I \tag{4}$$

where a and b are given as

$$a = \frac{2}{(1-\alpha)} \left(\frac{M_{ep}}{M_y} \right)^3 - 3 \left(\frac{M_{ep}}{M_y} \right)^2 \tag{5}$$

$$b = -\frac{2\alpha}{(1-\alpha)} \left(\frac{M_{ep}}{M_y} \right)^3 \tag{6}$$

A particular solution is found for $\alpha = 0$ in Equation (2), which represents an elastic, perfectly plastic material behavior. Equation (7) is the solution for this case and, because of its relative simplicity, it is particularly well suited to solutions in closed-form of elasto-perfectly plastic beam flexure.

$$I_{eff} = \left[\frac{M_{ep}}{M_y} \sqrt{3 - 2\frac{M_{ep}}{M_y}} \right] I \tag{7}$$

This equation can be conveniently used in Equation (1) for beam flexure in the elasto-plastic region. Area and centroid formulas of M_{ep}/EI_{eff} are presented in Table 1 for use with customary methods of analysis for predicting deflections of elasto-plastic beams with concentrated and distributed loads.

References

Fertis, D.G., Taneja, R., and Lee, C.T. (1991), "Equivalent Systems for Inelastic Analysis of Non-Prismatic Members," *Computers and Structures*, Volume 38, No. 1, pp 31-39.

Fertis, D.G., and Taneja, R. (1991), "Equivalent Systems for Inelastic Analysis of Prismatic and Nonprismatic Members," *Journal of Structural Engineering*, Volume 117, No. 2, pp 473-488.

Faller, R.K., and Rosson, B.T. (1996), "Effective Moment of Inertia for Elastoplastic Beams," submitted to *Structural Engineering and Mechanics*, 1996.

Moment	$\frac{M}{EI_{eff}}$	Area A_{ep}	Centroid $\bar{x}_{ep}$
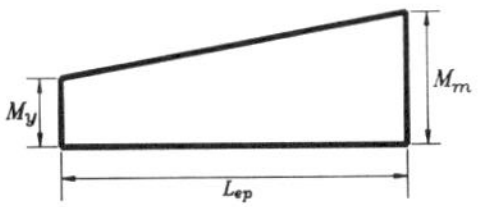 **Linear Moment Region**	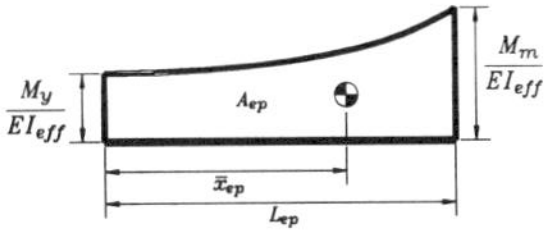	$A_{ep} = \frac{L_{ep} M_y^2 (1-K_0)}{(M_m - M_y) EI}$	$\bar{x}_{ep} = \frac{L_{ep}(M_y - K_0 M_m)}{3(1-K_0)(M_m - M_y)}$
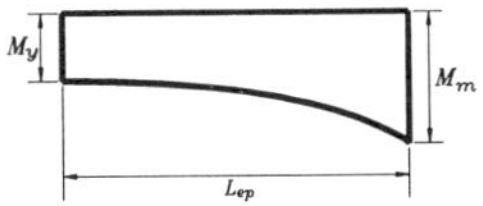 **Quadratic Moment Region: Case I**	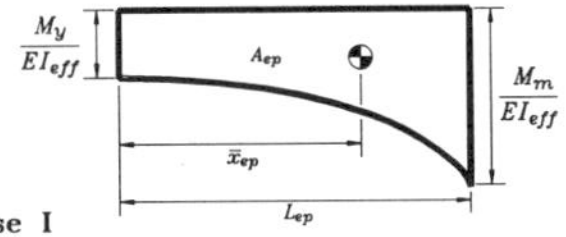	$A_{ep} = \frac{\sqrt{2}}{2} \frac{M_y(K_2 - K_3)}{K_1 EI}$	$\bar{x}_{ep} = \frac{\sqrt{2}}{4} \frac{L_{ep}^2[2K_1 M_y(K_0 - 1) + \sqrt{2} V_n (K_2 - K_3)]}{(M_y - M_m + V_n L_{ep})(K_2 - K_3)}$
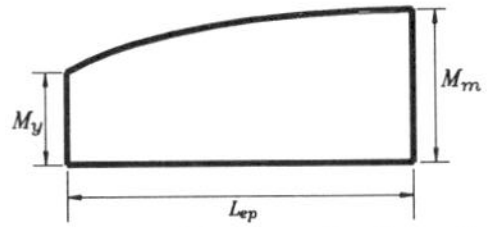 **Quadratic Moment Region: Case II**	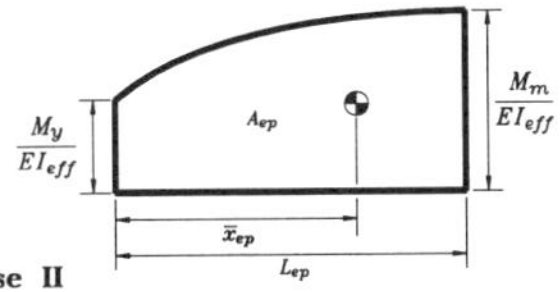	$A_{ep} = \frac{\sqrt{2}}{4} \frac{M_y K_6}{K_5 EI}$	$\bar{x}_{ep} = \frac{\sqrt{2}}{2} \frac{L_{ep}[-2L_{ep} M_y K_5 (1-K_0) + \sqrt{2} K_6 (M_m - M_y)]}{K_6 (M_m - M_y)}$

Constants:

$$K_0 = \sqrt{3 - 2\frac{M_m}{M_y}}$$

$$K_1 = \frac{1}{L_{ep}} \sqrt{1 - \frac{M_m}{M_y} + \frac{V_n L_{ep}}{M_y}}$$

$$K_2 = \ln[K_4(M_y - M_m) + \sqrt{2} K_1 V_n L_{ep}^2 + 2K_0(M_y - M_m + V_n L_{ep})]$$

$$K_3 = \ln[2(M_y - M_m + V_n L_{ep}) - \sqrt{2} K_1 V_n L_{ep}^2]$$

$$K_4 = 2\sqrt{2} K_1 L_{ep}$$

$$K_5 = \frac{1}{L_{ep}} \sqrt{\frac{M_m}{M_y} - 1}$$

$$K_6 = K_8 - \ln(M_y) - 2K_7$$

$$K_7 = \ln(1 - \sqrt{2} K_5 L_{ep})$$

$$K_8 = \ln(3M_y - 2M_m)$$

Table 1. Area and Centroid Formulas for Linear and Quadratic Moments in the Elasto-Plastic Region

Mechanism Study of Landslides

Dagang Zhang[1] and Mostafa A. Foda[2]

Abstract

A model using an elastic sliding mass riding on a basal granular shear flow layer is established to study the mechanism of large landslides. The balance and the stability of this structure are investigated by using a complex wave frequency approach. The results show that once the sliding velocity reaches certain critical value, a Kelvin-Helmholtz type of instability wave can be generated at the interfaces of the quasi-static regimes and the rapid granular flow regime. The generated basal pressure waves will force favorable phase shifts in overburden pressure and sliding velocity, leading to a net reduction in the total power loss to friction. Theoretical analysis and numerical results show that the sliding with small depth of sliding mass can produce stronger instability wave than the sliding with larger one.

Introduction

One important observation from large volume of field data on natural landslides, avalanches, etc., confirms that the primary mode of motion is that of *sliding* rather than *flowing*, especially in many of the large-scale events. Most vigorous shearing is confined to an active zone of thin basal layer resulting in a nearly uniform velocity distribution over most of the depth away from the base (see, for example, Foda 1994, Melosh 1986). This field observation indicates that these large-scale landslides can be described by a slowly deforming mass riding on a thin layer of granular flows. Various hypotheses and models have been proposed based on this primary sliding mode, such as different kinds of lubrication mechanisms, basal layer fluidization mechanisms and newly proposed pressure wave mechanisms, etc. Despite so many investigations, no satisfactory explanations have yet been given for the unexpectedly long traveling distance of landslides. These hypotheses are incomplete either because of the lack of a generating or triggering mechanism, or due to the conflict between adopted assumptions and experimental results.

Recently, Foda (1994) showed the existence of a Kelvin-Helmholtz type of

---

[1] Han-Padron Associates, LLP, 1400 Broadfield Blvd., Houston, TX 77084, U.S.A.
[2] Environmental Resources Engineering, Department of Civil Engineering, University of California, Berkeley, CA 94720, U.S.A.

instability wave at the interface of an infinite elastic mass moving at an uniform sliding velocity U and an underlying thin basal layer of high shear granular flow. The instability study of this system showed that above a threshold sliding velocity U, instability wave will grow exponentially, and this will lead to a net reduction in the basal drag on the sliding mass. The problem was idealized by assuming an infinite depth of sliding mass and rigid base boundary. Obviously, real slides can not be represented by this model when the depth of slides becomes shallow. Kobayashi (1994) investigated Foda's theory and showed that there are two types of waves excited along the basal layer. The most interesting second type is guided along the basal layer, as predicted in Foda's theory, and is shown to cause a "sonic boom" phenomenon when slides move above certain critical velocities, thus contributing to the long run-out of large landslides.

This study uses the complex wave frequency approach and examines in some detail the dynamics of landslides by extending the Kelvin-Helmholtz type of internal wave concept to general case of landslides: a finite sliding mass with deformable basal boundary. The main ideas of this paper are summarized as in the following. The basal shear zone behaves like rapid granular flow under large overburden pressure; and these high kinetic energy of fluctuating granular particles supplies the energy for generating internal pressure wave as well as heat loss. This propagating pressure wave can reduce the energy losses in landslides, thus being equivalent to a lower drag coefficient for the sliding. The kinetic energy of granular flow depends on the shear flow rate. Only when the mean flow velocity reaches certain critical values, can the basal layer granular material behave like rapid granular flow and obtain enough kinetic energy, thus generating strong enough pressure waves.

Formulation and Solution of Landslides Problems

The schema of a slide on a deformable base with finite depth is shown in Figure 1. Both the sliding mass and the ground are considered as elastic media. The elastic sliding mass has a depth of d and an uniform sliding velocity of 2U. Between these two elastic medium, there is a thin basal layer of thickness h, where a high-shear granular flow is concentrated. Because of the presence of fluctuating pressure, the sliding becomes nonuniform along the interfaces. Most of the sliding becomes simply focused at isolated regimes when the overburden pressure is minimum, and hence is moving, on the average, under lower frictional resistance than that for the uniform sliding case (no instability waves). The instability waves propagate in sliding direction with same phase, but with different amplitudes: $\eta_1=a_1e^{i(\lambda x-\omega t)}$ at the interface of sliding mass and basal layer, and $\eta_2=a_2e^{i(\lambda x-\omega t)}$ at the interface of basal layer and underlying ground. The pressures corresponding to the perturbations are $(p_{yy})_1=p_{01}e^{i(\lambda x-\omega t)}$ and $(p_{yy})=p_{02}e^{i(\lambda x-\omega t)}$, respectively.

We take wave frequency ω to be complex and leave wave number λ to be real. Within the quasi-static sliding mass and the elastic ground, we assume that the elastodynamic equation and the Hooke's law are still applicable. For the basal zone, we impose the assumption of shallowness of this shear thin layer. By solving the governing equations of the basal layer, we get $p_{01}=p_{02}$, which suggests that if the

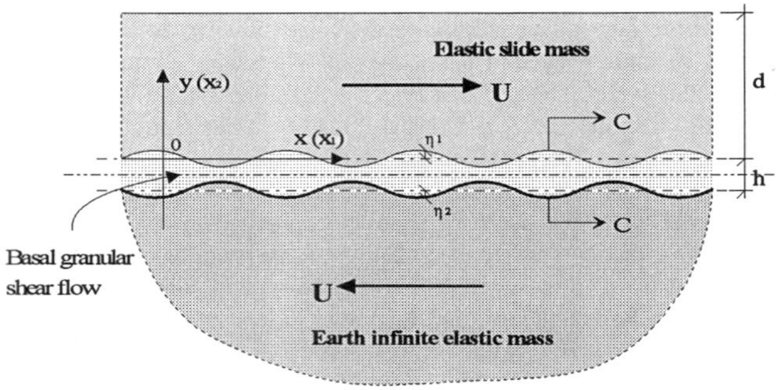

Figure 1. Definition sketch for the model of sliding on a deforming base boundary. The generated pressure waves force both sides of granular boundaries to deform.

shear basal layer is shallow, the perturbation pressure will keep constant throughout the depth. The solutions for this landslides problem are solved by combining the governing equations in three fields (sliding mass, basal layer, and elastic ground) and applying the corresponding boundary conditions (at the interfaces of basal layer and far fields). Here we give the simplified dispersion relation for a small depth of sliding mass situation:

$$2(1-\nu)C^4 + iK(1-2\nu)VC^3 - h\kappa\lambda C^2 + h\kappa\lambda V^2 = 0 \qquad (1)$$

where C is the nondimensional instability wave speed, the wave speed scaled by shear wave velocity: $C=\omega/(\lambda c_s)$. C is complex because the wave frequency ω is complex. K is the Coulomb friction coefficient for granular material; and ν is Poisson's ratio. For the case of incompressible material (ν=0.5), we can carry out the explicit solution for this simplified dispersion relation:

$$C=\pm\sqrt{\frac{h\kappa\lambda}{2}}\left(1 \pm \sqrt{1-\frac{4V^2}{h\kappa\lambda}}\right)^{\frac{1}{2}} \qquad (2)$$

An instability wave solution will be expected when nondimensional sliding velocity (U scaled by c_s) $V> \frac{(h\kappa\lambda)^{\frac{1}{2}}}{2} = V_c$. We call V_c the critical sliding velocity, which is proportional to the square-root of the basal layer depth. For very thin basal layer, the perturbation wave at the interface becomes unstable when the sliding velocity exceeds the smaller critical velocity, while for larger basal layer depth a larger sliding velocity is needed to generate this instability wave. For a fixed basal layer depth, the longer the instability wave length, the smaller the critical sliding velocity. The solution for κ=0.4 is plotted in Figure 2 for the unstable ranges. As the range of instability increases with increasing sliding velocity, so does the growth rate. Figure 3a shows the relation between phase speed and sliding velocity in the unstable region for certain growth rates. A slower wave is always expected for this structure to reach unstable condition, as shown in the figure. The same is true for the basal layer depth and sliding velocity, as shown in Figure 3b.

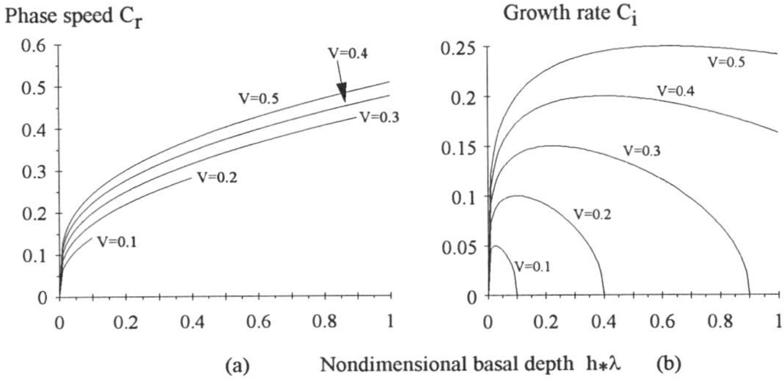

Figure 2. Nondimensional wave phase speed and growth rate for a range of sliding velocities and nondimensional basal layer depth, with $\kappa = 0.4$.

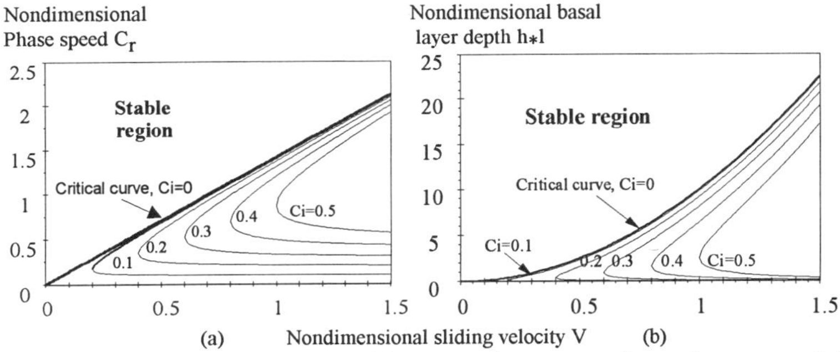

Figure 3. Unstable regions of instability waves for a range of growth rates, $\kappa = 0.4$.

The depth effects of the sliding mass on the characteristics of instability waves are also studied. The results indicate that a much slower wave speed is expected as the thickness of sliding mass decreases. The slower wave will induce more instability on the system with smaller sliding mass, especially for the cases with the conditions approaching those of debris flows. The comparison of results between rigid base and deformable base shows that the existing perturbation wave at the interface between the basal layer and the underlying ground also reduces the power loss due to friction, thus producing a lower drag resistance. The instability waves at both interfaces favor the reduction of drag resistance, so that a longer sliding distance may be reached.

References

[1] Foda, M. A. 1994 Landslides riding on basal pressure waves. *Continuum Mech. Thermodyn.* vol **6**, 61-79.

[2] Kobayashi, Y. 1994 Effect of basal guided waves on landslides. *Pageoph*, vol. **142**, No. 2.

[3] Melosh, H. J. 1986 The physics of very large landslides. *Acta Mechanica* **64**.

Scaling-Up Of Small-Scale Granular Sediment Transport Laws

J. Raghuraman and P. K. Haff[1]

Abstract

Practical applications of laboratory sediment transport studies include the attempt to predict erosion of soil on hillslopes subject to rainfall-induced runoff. Most scientific studies of the mechanics of sediment transport have been performed at laboratory-scale, or in limited reaches of streams where bed and flow conditions are relatively well-known. Here we examine the application of basic sediment transport laws to prediction of large-scale sediment fluxes on hillslopes where topographic details of the slope are unknown at scales below the cell-size of the simulation. In particular, we examine the situation in which small channels or rills on the hillslope are not resolved. Recent model studies of hillslope erosion have invoked constitutive laws that are functions of powers of slope and of surface discharge. These transport laws are suggested by the results of small-scale studies of sediment transport that have been carried out mainly in flumes. On large-scale natural surfaces, however, rills are often present. If the rills are too small to be explicitly modeled, then one may consider averaging the flow over the rills. This is generally not a satisfactory strategy because of the nonlinear nature of sediment transport.

Small scale studies of transport of particulate sediment by running water indicate that the sediment flux ($kg\, m^{-1} s^{-1}$) scales with a power of the bed shear stress. The Einstein-Brown formula [Vanoni, 1975] predicts that sediment flux is proportional to the cube of the bed-shear stress. For steady flow, shear stress is proportional to depth of water, so that, all else being equal, deeper flows will be (much) more effective at transporting sediment than shallower flows. Under natural conditions on a hillslope, irregularities in the surface will cause local variations in flow depth and

[1]Department of Geology and Center for Hydrologic Science, Duke University, Durham, NC 27708-0230.

hence shear stress. If the location and nature of these variations is known explicitly, then one may perform a calculation in which surface irregularity is included in the description of the bed, and the sediment transport rate and its divergence (which determines the magnitude of erosion and deposition) can be computed "exactly" using a transport model such as that of Einstein-Brown. The results of this calculation can then be compared with the results of a calculation in which the channel structure is not resolved.

A constitutive law that has been used recently by a number of workers relates sediment transport to powers of surface slope S and water discharge q, i. e., $q_s = kq^nS^m$. Here q and S are averaged quantities defined over a cell. In practical applications, the cell can be of sufficient size to contain (unresolved) rills or other irregularities that locally affect the flow and hence the transport rate. We compare the sediment transport rates computed from these effective, large-scale rules, with the results of an Einstein-Brown calculation in which rill structure is preserved. For simplicity, and to illustrate the effects of unresolved rills on sediment transport, we study the case of a single rill on a hillslope. The rill runs straight down the slope. Infiltration rates are assumed to be constant, and rainfall intensity uniform over the area of interest. The rills may be overtopped by the flow, in which case Manning's equation is used to estimate the allocation of water between the channel and the interfluve. Sediment transport is then calculated locally using the Einstein-Brown equation.

The sediment transport rate computed on the basis of the average shear stress in the cell first decreases as actual channel width increases from zero, and then increases with further increase in channel width, exhibiting a minimum at an intermediate value of channel width. The "exact" sediment transport rate, on the other hand, at first increases as channel width increases from zero, then decreases as channel width increases further. Thus the transport rate computed on the basis of correct stress computation in a resolved channel shows a maximum at intermediate channel widths.

Despite this discrepancy, parameters appearing in the empirical power-law large-scale transport rules can be varied to obtain a calibration of the rule to actual local sediment transport rates. Good fits are obtained as long as the fit is applied to a slope with the same channel width as that for which the fit was obtained. However, in general one wants to apply the empirical equation to an entire hillslope, and channel geometry will tend to change with position on the slope. One approach is to obtain fits for a variety of channel widths, and then to employ average values of the locally calibrated power-law parameters. For common channel geometries, however, this results in overprediction of the actual transport at small channel sizes (compared to plot width), and underprediction at intermediate to large channel sizes. Mispredictions by a factor of 2 or 3 seem common. More importantly, since channel sizes can change systematically with position on the slope, generally becoming larger in the downslope direction, use of power laws with average fit parameters will tend to result in spatially biased, systematic errors of prediction of sediment transport (and hence erosion and deposition).

We have studied the performance of a power-law rule on a hillslope that is convex-up in its upper portion, and concave-up in its lower portion. The hillslope is broken into a number of cells, and the sediment transport rate is computed for each cell. Although average values of k, n and m in the power law rule $q_s = kq^n S^m$ lead to systematic discrepancies between the fitted and the exact transport calculation across the hillslope, there is, in fact, for certain channel geometries, a choice of parameters for which the fit is very good everywhere on the hillslope, even though channel width, water discharge, and slope are changing continuously.

If the channel width (and depth) increase in the downslope direction as power laws of the distance from the divide, then a single choice of parameters in the power law can provide a good fit to the exact sediment transport for every point on the hillslope, independent of slope. However, for channel geometries that change character in any other way as a function of hillslope position, the fit deteriorates. The conclusion of this study is that empirical power-law sediment transport rules defined in terms of a single set of parameter values may generate systematic errors in their prediction of erosion and deposition rates. These errors will be correlated with position on the hillslope, and will be a function of hillslope rill geometry. The projected evolution of landforms based upon use of such models will reflect these systematic errors. The long term integrated effect of these errors on detailed geomorphic prediction or reconstruction of land surface configurations remains to be studied.

This work was supported by the Army Research Office, Engineering and Environmental Sciences Division, Grant No. DAAH04-94-G-0067.

Jamming of the Flow of Granular Materials

Yi Sun[1] and Oleg Vinogradov[2]

Abstract

The flow of a solid particles-fluid mixture is numerically simulated in order to determine the critical parameters of the system leading to a jam. Two parameters are varied: the volumetric density of solid particles and the velocity of fluid flow. The results presented are based on the mathematical models of granular materials treated as multibody systems with variable topology. The fluid flow is considered to be potential. It is shown that jamming strongly depends on the volumetric density of particles. The results of numerical experiments are in qualitative agreement with physical experiments of flow of spherical particles in a channel.

Introduction

Particles carried by a fluid in a channel interact with each other and with the boundaries. This interaction results in a loss of kinetic energy and thus reductions in particles velocities. If the rate of particles supply remains the same then the fraction of particles in a given volume, where the speed reduction takes place, will be increasing. If the inflow exceeds outflow then the net balance will grow until the particles density reaches its maximum which will correspond to a complete stagnation or a flow jam.

The laboratory experiments have shown that the flow of spherical beads in a straight channel (Ettema 1989) and blocks in a sinuous channel (Urroz and Ettema 1994) leads to a jam depending on the initial volume density of particles, their relative size, and flow velocity. For the range of parameters used in the experiments on beads (Ettema 1989), the volume density of beads appears to be the most important one.

---

[1]Graduate student, The University of Calgary, Calgary, Alberta, Canada T2N 1N4

[2]Professor, The University of Calgary, Calgary, Alberta, Canada T2N 1N4

This parameter is also important in applications where material transport and its volumetric discharge is the main objective.

It is clear that any attempt to analyze the phenomenon of jam numerically must take into account the variability of particles density. At the moment a discrete model of the system of particles seems to be the only alternative to study the non-steady flows of fluid-particle mixtures. In order to capture the essence of the process, namely, the losses due to interactions and the variability of particles density a discrete model must allow the clustering of particles. A cluster, being a system of interacting particles, is by itself a variable entity due to the variable number of particles it may comprise and the changing inter-connectivity between the particles, i.e. due to the variability of cluster topology. A numerical model of a cluster thus poses computational difficulties since it results in a variable system of coupled nonlinear differential-algebraic equations, i.e. any change in a cluster topology requires a new set of the governing equations and their solution. The numerical efficiency becomes an important factor in simulating the flow.

In this paper a discrete system comprises isolated particles and clusters. A cluster is treated as a multibody system and a Newton-Euler approach is used to formulate its equations of motion. A fluid flow is considered to be potential. An iterative method with a variable time-step and error control is developed to maintain the desired accuracy. The simulation is done for a plane problem and the results can be interpreted as conditions for the jam initiation. A comparison with the tests results obtained by Ettema (1989) shows a qualitative agreement.

Physical Model of a System

A particle is modelled by a rigid disk, the radius of which is chosen from within some range with equal probability. A subsystem of disks is generated first with a predetermined surface density within the area in the channel which is called the generation area (Fig.1). This area is not a part of the simulation space, and can be viewed as a feeder. The disks from the generation area are "moved" with constant speed into a buffer area. When a disk crosses the boundary of the buffer area it acquires the velocity of the corresponding potential fluid. A buffer area is needed to allow a sufficient subsystem of disks to maintain a potential flow distribution of velocities before the mixing due to interactions with the boundaries starts. From the buffer area the disks enter the"real" pipe with corrugated boundaries. New disks are generated in the process of simulation, while the disks leaving the control area are discarded.

The interactions between the disks and disks and walls are treated as dynamic (collisions) or quasi-static depending on the relative velocity between the two bodies. A collision is considered to be central. The corresponding energy loss is reflected in the coefficient of restitution. It is assumed that the flow is moderate enough so that a collision of a disk with a cluster does not cause the break up of the cluster. It does, however, change the velocities of disks belonging to a cluster. Thus a collision produces two effects: loss of energy and instantaneous change of velocities without the change of positions.

A quasi-static interaction results in a disk joining a cluster (or a disk) without the above instantaneous changes in the state of the system, but producing the "long-time" effect by altering the system and thus the governing equations.

The potential flow theory is used here in a sense described in (Wierzba and Vinogradov, 1991), namely, to determine the drag forces acting on disks. It is assumed that a disk forming a cluster with an upstream disk along the stream is not experiencing its own drag force (fully shielded). On the other hand, an isolated downstream disk would have a velocity of the corresponding potential flow.

Two events lead to the loss of energy, namely, the collisions mentioned earlier and sliding between the disks and disks and walls. The motion of disks between any two consecutive topological changes is considered to be continuous since it is governed by the same set of equations. The sliding takes place during this continuous motion. The internal forces are not treated as constants during the time step Δt but as variables thus taking into account that when disks slide and change their relative positions, as it is supposed to be in an underconstrained system or in a system with one-sided constraints, the constraint (internal) forces in the system also change.

An iterative method is used to solve the coupled system of equations. Among other advantages, the iterative method allows to split the system into one, defining the constraint forces for the given positions and velocities, and another, defining the positions and velocities for the given constraint forces. Both, forces and state coordinates, are updated in the process until a predetermined accuracy is achieved.

Numerical Results

The numerical simulations were carried out for the following channel and disk geometries (Fig.1): D = 0.4m, d = (0.05 - 0.055)m, L_G = 1.25D, L_B = 0.5D, L_C = 3.75D, e = 0.125D, b = 0.05D, r = 0.1D. The coefficient of friction was μ = 0.3, and the coefficient of restitution e_r = 0.7. The maximum velocity at the centre of the channel was v_m = 1.5v, where v is the mean velocity corresponding to a given discharge and it varied in numerical experiments from v=0.25m/sec to v=1.0m/sec.

The simulations were carried out on the IBM-RS6000 computer. There was no limit on the number of disks, those out of control area were discarded and new ones were generated as long as the simulation was running. The simulation run was stopped when the buffer area was full, i.e. when the discharge was zero, which was the sign of a jam in the pipe. For a typical run with a number of disks about 300 the CPU time was about 3min, while the real time about 12min.

The results of numerical experiments are plotted in Fig.1 in relative surface density - Froude number ($Fr = v/(gD)^{0.5}$) coordinates. It is seen that for the relative size of the diks d/D = 0.1375 and relative size of the wall roughness b/D = 0.05 the boundary between the no-jam and certain-jam areas is almost a vertical line (or transition zone). Although these results could not be checked quantitatively against the experimental data, since in (Ettema 1989) the initiation of a jam is associated with the volume discharge of the spheres, the experimental data indicate that for a relatively large particle size (in experiments it was d/D = 0.02) the no-jam-certain-jam boundary is almost a vertical line. Similar results are obtained for relatively large blocks in a curved channel (Urroz and Ettema 1994). It is also of interest to note that the time to the beginning of the onset of jamming depends on the initial flow velocity, namely, it is almost in inverse proportion to the latter.

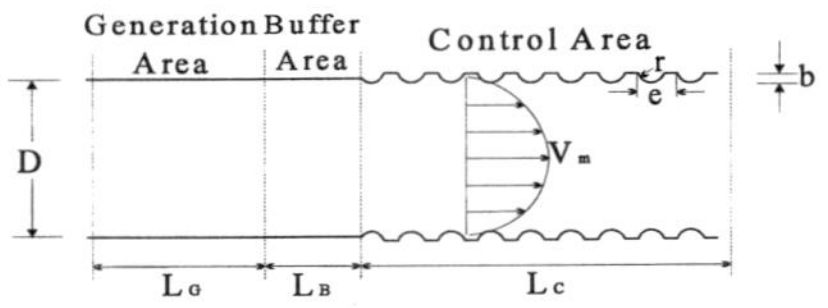

Fig.1 A schematic diagram of simulation area

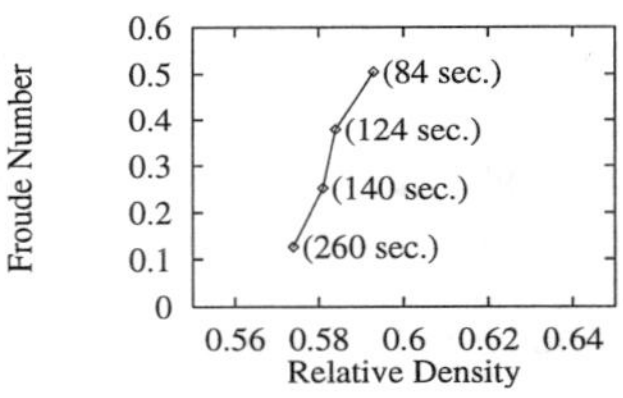

Fig.2 Relative surface densities of disks resulting in a flow jam, the numbers in brackets indicate the time to jam initiation

Acknowledgment

The financial assistance provided by the Natural Sciences and Engineering Council of Canada is gratefully acknowledged.

References

Ettema, R. , 1989, "Jam Initiation in Unobstructed Channels", *Proceedings of 23d Congress of IAHR*, Ottawa, pp.19-29.

Urroz, G.E. and Ettema, R., 1994, "Small-Scale Experiments on Ice-Jam Initiation in Curved Channel", *Can. J. of Civ. Eng.*, 21, pp.719-727.

Wierzba, P., and Vinogradov, O.G., 1991, "Simulation of Topologically Variable Multibody System in Plane Motion", *Proceedings of 1991 European Simulation Multiconference Modelling & Simulation*, Society for Computer Simulation, San Diego, CA, pp.935-940.

Insight into the Inelastic Response to a Fully Nonstationary Earthquake Ground Motion Model

Bor-Feng Peng[1] and Joel P. Conte[2]

Abstract

The statistics of various inelastic response parameters indicative of structural damage are examined for constant ductility design. A versatile fully nonstationary earthquake ground motion model, accounting for time-varying intensity and frequency content typically exhibited by actual earthquake ground motions, is used for the study. Four different hysteretic SDOF models are considered, namely the bilinear hysteretic model, the Clough's stiffness degrading model, the slip model and the Bouc-Wen model. A comprehensive parametric study is carried out to investigate the importance and relative effects of the type of hysteretic behavior, the initial structural period, T_0, the yield strength, R_y, and the strain hardening ratio, α, on the statistics of various inelastic response spectra.

Introduction

In earthquake resistant design, an important issue is the determination of the yield strength, R_y, or yield strength coefficient, η (= R_y expressed as a propotion of the weight of the structure), required to limit the displacement ductility demand imposed by the ground motion to a specified value. The maximum displacement ductility is defined as the ratio of the maximum displacement response to the yield displacement, U_y. Although the maximum ductility response is an important damage index, it is well recognized that a structure can be damaged or weakened by a combination of stress reversals (cumulative type damage) and high stress excursion (first-excursion type damage). Thus, numerous damage indices have been proposed and defined as functions of structural response and ground motion parameters. A comprehensive review of the concepts and procedures used in deterministic seismic damage prediction was given by Powell and Allahabadi (1988). Park and Ang (1985) proposed a damage index defined as a linear function of the maximum deformation and the effect of repeated cyclic loading. They validated their damage index with experimental data on reinforced concrete members and calibrated it against a significant amount of observed seismic damage. More recently a state-of-the-art review of seis-

[1] Graduate student, Dept. of Civil Engrg, Rice University, P.O. Box 1892, Houston, TX 77251
[2] Assistant professor, Dept. of Civil Engrg, Rice University, P.O. Box 1892, Houston, TX 77251

mic damage indices for concrete structures was presented by Williams and Sexsmith (1995). Since inelastic structural response can be very sensitive to earthquake ground motion time histories, which are inherently random, the probabilistic approach is used here. The statistics of some basic damage response parameters are analyzed for a stochastic earthquake ground motion process calibrated against an actual earthquake accelerogram. Ensemble-average inelastic response spectra conditioned on a constant ductility response are obtained through Monte Carlo simulation from a set of 100 artificial earthquake ground motions. The sensitivities of these average response spectra to the type of hysteretic behavior, the target constant ductility level, and the strain-hardening ratio are investigated.

Stochastic Earthquake Ground Motion Model

The nonstationary stochastic ground motion model used in this study is defined as a sum of zero-mean, independent, uniformly modulated Gaussian processes:

$$\ddot{U}_g(t) = \sum_{k=1}^{p} X_k(t) = \sum_{k=1}^{p} A_k(t)S_k(t) \tag{1}$$

Each uniformly modulated process consists of the product of a deterministic time modulating function, $A_k(t)$, and a stationary, unit variance, Gaussian process, $S_k(t)$. Thus, this earthquake model is a sigma-oscillatory Gaussian process. The modified gamma function is used as time modulating function, i.e.,

$$A_k(t) = \alpha_k(t-\zeta_k)^{\beta_k}e^{-\gamma_k(t-\zeta_k)}H(t-\zeta_k) \tag{2}$$

in which α_k and γ_k are positive constants, β_k is a positive integer, and ζ_k represents the "arrival time" of the k-th sub-process, $X_k(t)$; $H(t)$ denotes the unit step function. The k-th zero-mean stationary Gaussian process, $S_k(t)$, is characterized by its autocorrelation function

$$R_{S_kS_k}(\tau) = e^{-\nu_k\tau}\cos(\eta_k\tau) \tag{3}$$

or power spectral density function

$$\Phi_{S_kS_k}(\omega) = \frac{\nu_k}{2\pi}[1/(\nu_k^2+(\omega+\eta_k)^2)+1/(\nu_k^2+(\omega-\eta_k)^2)] \tag{4}$$

in which ν_k and η_k are the two free parameters representing the frequency bandwidth and predominant (or central) frequency of the process $S_k(t)$, respectively. The time variation of the mean square ground acceleration is given by

$$E[|\ddot{U}_g(t)|^2] = \int_{-\infty}^{\infty}\sum_{k=1}^{p}|A_k(t)|^2\Phi_{S_kS_k}(\omega)d\omega = \sum_{k=1}^{p}|A_k(t)|^2 \tag{5}$$

and the evolutionary power spectral density function of $\ddot{U}_g(t)$ takes the form

$$\Phi_{\ddot{U}_g\ddot{U}_g}(t,\omega) = \sum_{k=1}^{p}|A_k(t)|^2\Phi_{S_kS_k}(\omega) \tag{6}$$

Note that the ground acceleration process, $\ddot{U}_g(t)$, is not separable, although its component processes are individually separable (i.e., uniformly modulated). Each component process captures the complex time-frequency distribution of the earthquake ground acceleration in a local time-frequency region. The model identification and parameter estimation procedures used to calibrate this earthquake model against any target earthquake accelerogram can be found in (Conte and Peng 1996).

Statistics of Inelastic Response Parameters for Constant Ductility Design

Results are presented for the stochastic ground motion model calibrated against the N-S component of the Imperial Valley earthquake of May 18, 1940, recorded at El Centro site. In a first phase, for each artificial ground excitation, the yield strength, R_y, required for a displacement ductility demand of $\mu_d = 6$ is determined iteratively for each SDOF structural model considered. The average strength coefficients and corresponding strength reduction factors (defined as $R_y(\mu_d = 1)/R_y(\mu_d = 6)$) are shown in Figs. 1(a) and (b). In the second phase, the statistics of various inelastic response parameters are determined under the condition of a fixed displacement ductility response $\mu_d = 6$ (for each realization!). The response parameters presented here include the hysteretic energy dissipated E_H, the energy ratio E_H/E_I in which E_I denotes the earthquake input energy, the cumulative displacement ductility μ_{acc}, and the maximum normalized plastic deformation range $(\Delta U_{p,i}/U_y)_{max}$. All the energy-related response quantities are normalized by $R_y U_y$. The ensemble-average (or estimated mean denoted by $\hat{E}[...]$) of the response spectra corresponding to these response parameters are shown in Fig. 1 for an initial damping ratio of 5%, a zero strain-hardening ratio, α, and for the various hysteretic models. Based on these results and those obtained by simulating the N-S component of the San Fernando earthquake of February 9, 1971, recorded at Orion Blvd. site, the following conclusions can be drawn.

(1) As shown in Fig. 1(a), the strength coefficient response spectrum $\hat{E}[\eta]$ for the slip model is larger than those for the other hysteretic models, especially in the low period range (below one sec). This difference is also reflected in Fig. 1(b) which displays the lowest strength reduction spectrum $\hat{E}[R]$ for the slip model. It is also worth noting that under constant ductility design, the slip model yields much lower inelastic response spectra in Figs. 1(c) to (f) than the other hysteretic models, which is at least in part due to the higher yield strength (and yield displacement) required by the slip model to limit the ductility.

(2) The ensemble-average E_H response spectrum, see Fig. 1(c), for the Clough's stiffness degrading model is significantly larger than those for the other hysteretic models in the low period range from 0.1 to 0.4 sec. On the other hand, for an initial structural period exceeding 0.5 sec, the Bouc-Wen hysteretic model with model parameters chosen in this study dominates the other hysteretic models. The same trends appear in the spectrum $\hat{E}[E_H/E_I]$ shown in Fig. 1(d).

(3) In constant ductility design, the ensemble-average inelastic response spectra are highly dependent on the initial natural period of the structure in the period range less than about one sec. The damage related structural response parameters in this low period range are generally higher than in the initial period range above one sec. In the low period range, the ensemble-average constant-ductility damage spectra are very sensitive to the hysteretic model used. These trends can be observed clearly in Figs 1(c) to (f).

(4) In constant ductility design, it is traditionally believed that the influence of the hysteretic model on the strength reduction factor or the yield strength required can almost be neglected. However, the structural engineer should select carefully the hysteretic model which best represents for the properties of the structural system and materials under consideration, since other non-ductility based response parameters indicative of damage are sensitive to the hysteretic model.

(5) Although not shown here, it was also found that the strain-hardening ratio α does not have a substantial influence on the required yield strength for a specified displacement ductility. However, its effects on the other ensemble-average damage related response spectra cannot be underestimated, especially in the low period range.

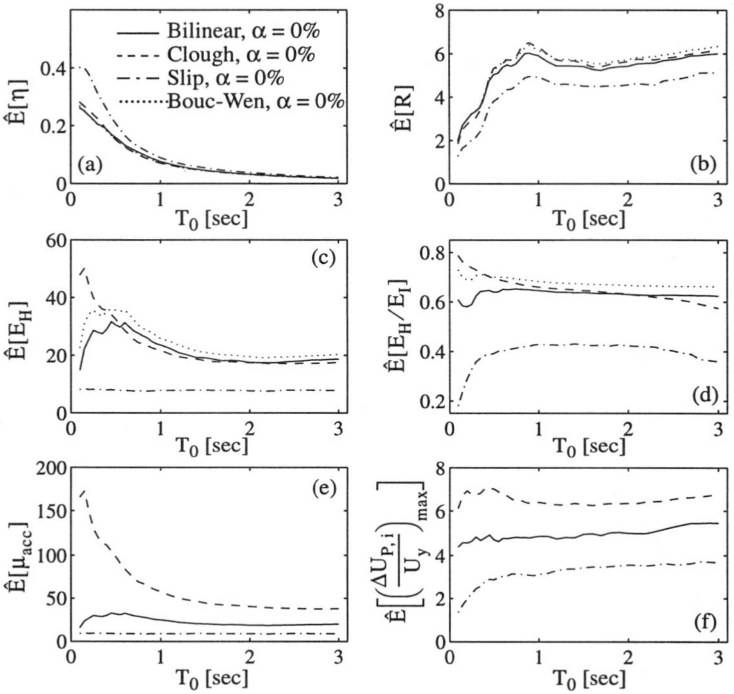

Fig. 1. Ensemble-Average Response Spectra ($\mu_d = 6$, $\xi_0 = 5\%$) for El Centro 1940 Earthquake Model

Conclusions

Ensemble-average response spectra of various damage related structural response parameters were investigated under the assumption of constant ductility design. A newly developed fully nonstationary stochastic earthquake ground motion model was applied to simulate the ensemble of input ground motion time histories. The influence and relative importance of the hysteretic model and structural parameters on the ensemble-average inelastic spectra were analyzed.

References

Conte, J. P. and Peng, B.-F. (1996). A fully nonstationary analytical earthquake ground motion model. Accepted for publication in *J. eng. mech.*, ASCE.

Powell, G. H. and Allahabadi, R. (1988). Seismic damage prediction by deterministic methods: concepts and procedures. *Earthquake eng. struct. dyn.*, 16, pp. 719-734.

Park, Y.-J. and Ang, A. H.-S. (1985). Mechanistic seismic damage model for reinforced concrete. *J. struct. eng.*, ASCE, Vol. 111, pp. 722-739.

Williams, M. S. and Sexsmith, R. G. (1995). Seismic damage indices for concrete structures: a state-of-the-art review. *Earthquake spectra,* Vol. 11, pp. 319-349.

A METHOD FOR EXTRAPOLATION OF EXTREME VALUE DATA

Arvid Naess[1], Member ASCE

Abstract

This paper presents a method for extrapolation of extreme value data for estimating high return period design values. It is assumed that the data represent statistically independent samples from an extreme value probability distribution, and that the underlying phenomenon giving rise to the extreme values can be modelled as a stochastic process. A specific feature of the proposed method is a transformation of the observed extreme value data to obtain better fit to a Gumbel-type extreme value distribution.

Introduction

The focus of this paper is on the problem of extrapolation of a limited set of extreme value data to establish a design level with a return period of typically 50 or 100 years. Usually it can be decided that the extreme value distribution converges to a Gumbel or Type-I extreme value distribution, Castillo, 1988. The general procedure is then to plot the data on a Gumbel probability paper, and carry out a linear regression on the data in order to fit a straight line, which would correspond to perfectly Gumbel distributed data. Usually, a confidence band on the straight line fit is also provided. Within this approximation, an extrapolation can then be carried out to reach the desired return period. Based on work related to the extrapolation of extreme wind data (Cook, 1985), a general procedure for plotting extreme value data for efficient extrapolation is proposed. The idea behind the method is to transform the data to obtain faster convergence to the Gumbel distribution. It will be shown that for a wide class of practical engineering cases, the required transformation can be derived from the underlying probability distribution of the process under study.

Transformation of extreme value data

Let $X(t)$ denote a stochastic process with a long-term PDF (probability density function) $f_X(x)$ and a corresponding CPD (cumulative probability distribution) $F_X(x)$. It is asssumed that

$$f_X(x) = a\exp\{-\alpha(x)\} \tag{1}$$

where a is a positive constant and $\alpha(x)$ is a well-behaved function.

The focus of this paper is the largest value that $X(t)$ assumes during some specified period of time T, e.g. one year. Let $\hat{X}(T) = sup\,\{X(t);\ 0 \le t \le T\}$ denote this value. The goal is therefore to determine the statistical distribution (CDF) $F_{\hat{X}(T)}(\cdot)$ of $\hat{X}(T)$. It turns out that this extreme value distribution is intimately related to another function of the process $X(t)$, viz. the so-called mean level-upcrossing rate, which we denote by $\nu_X^+(x)$. $\nu_X^+(x)$ is equal to the mean number of times per time unit that the process $X(t)$ crosses the level x from below. For the processes of engineering relevance to be studied here, it is the case that

[1]Faculty of Civil Engineering, The Norwegian University of Science and Technology, Trondheim, N-7034 Norway

$$\nu_X^+(x) = \nu \cdot \exp\{-\alpha(x) + \delta(x)\} \tag{2}$$

where $|\delta(x)|$ is of much slower increase than $\alpha(x)$ as $x \to \infty$.

As is well known, even in the case of a Gaussian process, no closed-form expression is known for the exact extreme value distribution. However, by adopting the assumption that upcrossings of high levels by $X(t)$ are statistically independent, the extreme value distribution of $X(t)$ is given as follows

$$\begin{aligned} F_{\hat{X}(T)}(x) &= Prob\{\hat{X}(T) \leq x)\} = \exp\{-\nu_X^+(x)T\} \\ &= \exp\{-\exp\{-\alpha(x) + \delta(x) + \ln(\nu T)\}\} \end{aligned} \tag{3}$$

On the basis of the general theory of extreme value distributions (Castillo, 1988; Leadbetter et al., 1983), it can be shown that $F_{\hat{X}(T)}(x)$ can be expected to approach a Gumbel distribution when T increases to large values. This can be expressed as follows

$$F_{\hat{X}(T)}(x) \to \exp\{-\exp\{-a_T(x-b_T)\}\}, \quad T \to \infty \tag{4}$$

for suitable choice of the parameters $a_T > 0$ and b_T. The problem with this asymptotic result is that no information on the rate of convergence is available. In specific cases the convergence of $F_{\hat{X}(T)}(x)$ toward a Gumbel distribution turns out to be very slow. In spite of this, the asymptotic result of equation (4) is used extensively in practical extreme value analyses to justify the application of a Gumbel distribution for extrapolating to long return period design values from a limited number of extreme value observations.

It will now be shown that equations (1) and (3) can be used to develop a general method which will provide a better theoretical basis for extrapolation of extreme value data. For this purpose a new stochastic process $Y(t)$ is introduced as follows: $Y(t) = \beta[X(t)]$, where $\beta(\cdot)$ is a strictly increasing function such that $\beta(x) = \alpha(x)$ for $x \geq x_o$, for some x_o. The extreme value distribution $F_{\hat{Y}(T)}(y)$ of $Y(t)$ is then given by the expression

$$F_{\hat{Y}(T)}(y) = \exp\{-\exp\{-\alpha[\beta^{-1}(y)] + \delta[\beta^{-1}(y)] + \ln(\nu T)\}\} \tag{5}$$

Hence, for $y \geq y_o = \alpha(x_o)$, it is obtained that

$$F_{\hat{Y}(T)}(y) = \exp\{-\exp\{-y + \delta[\beta^{-1}(y)] + \ln(\nu T)\}\} \tag{6}$$

It may be generally assumed that $\delta[\beta^{-1}(y)]/y$ will be negligible for relevant extreme values. This leads to the conclusion that the transformed process $Y(t)$ has an extreme value distribution that is very close to the Gumbel distribution for all values of T of practical interest. This is in contrast to the original process $X(t)$ whose extreme value distribution converges only slowly toward a Gumbel distribution if $\alpha(x)$ deviates significantly from a linear function.

The theory developed above can now be implemented in a practical procedure in the following manner. Given a finite, independent sample of observed extreme value data $\{\hat{x}_1, ..., \hat{x}_n\}$ of the process $X(t)$, which is assumed to satisfy the conditions required by the theoretical developments of the previous section. That is, it is assumed that the data have been sampled from the distribution $F_{\hat{X}(T)}(\cdot)$, and that $\hat{x}_i \geq x_o$, $i = 1, ..., n$. It is further assumed that the underlying distribution is estimated so that the function $\alpha(x)$ can be determined.

The transformed extreme value sample $\{\hat{z}_1, ..., \hat{z}_n\}$, where $\hat{z}_i = \alpha(\hat{x}_i)$, $i = 1, ..., n$, can therefore be expected to follow very closely the Gumbel distribution given by equation (7). By plotting the transformed extreme value data on Gumbel probability paper using a suitable plotting position formula (Castillo, 1988), a straight line can then be fitted to the resulting data points by invoking for instance a maximum likelihood method or a least squares method.

It is realized that the prerequisite for a successful application of the proposed method in practice is that $\alpha(x)$ can be stably predicted. This basically means that the 'tail'-behaviour of $\alpha(x)$ is determined to a large extent by the main body of data.

It should also be noted that the above development in fact shows that improved extrapolation accuracy can be expected even if $\alpha(\cdot)$ is only approximately determined.

Central questions of statistical extreme value estimation like confidence bands, bias of estimates, etc. will not be discussed in this paper. The main reason for this is that for such questions there will be no significant difference between the conventional and the transformation method.

Numerical examples

The performance of the transformation method will be illustrated by application to two numerical examples. The extreme value analyses have all been carried out by using the software package EXTPAR (Mathiesen, 1993). The fitting of the straight line to the plotted data were in all cases based on the maximum likelihood method.

Example 1 - Gaussian Process

Let $X(t)$ denote a stationary Gaussian $N(0,1)$-process satisfying the adopted assumptions. The corresponding extreme value distribution is accordingly

$$F_{\hat{X}(T)}(x) = \exp\{-\exp\{-\frac{x^2}{2} + \ln(\nu^+T)\}\} \qquad (7)$$

where ν^+ denotes the mean zero upcrossing rate. It is seen that $\alpha(x)=x^2/2$. Denoting the return period in multiples of T by R, the level ξ_R with a return period R is derived from equation (7): $\xi_R = \{2\ln[\nu^+T/(-\ln(1-1/R))]\}^{1/2}$

Five random, independent samples of 20 values in each sample are produced from this extreme value distribution for three different values of the effective duration, viz. $\nu^+T = 100$, 500 and 1000. For return periods of 10, 50 and 100, we have in Table 1 listed the the deviation in per cent from the exact value given by ξ_R of the corresponding mean estimates obtained from the 5 samples by using the two extrapolation methods discussed in this paper, viz. the conventional method and the transformation method.

It is seen that both extrapolation methods overpredict the exact values. However, the transformation method produce accurate predictions for all three effective durations. While there is a significant deviation between corresponding predictions given by the two extrapolation methods for the shortest effective duration, it is much less pronounced for the longer effective durations. This indicates that the effective duration is an important parameter that largely determines the accuracy that can be expected from the conventional method. An inspection of equation (7) reveals that this is directly related to the accuracy of approximating the expression $x^2 - 2\ln(\nu^+T)$ by a linear function in the vicinity of $x = [2\ln(\nu^+T)]^{1/2}$, which is indeed clearly dependant on how large ν^+T is.

Table 1 - Deviation from exact values (%)

Effective duration	Extrapol. method	Return period R		
		10	50	100
100	Transform.	1.0	1.2	1.2
	Convent.	5.2	9.3	11.1
500	Transform.	0.8	1.0	1.0
	Convent.	0.5	2.0	3.1
1000	Transform.	0.7	0.9	1.0
	Convent.	1.3	3.0	3.7

Example 2 - Wind Data

The prediction of long return period design values for wind speed has been the subject of numerous research efforts over many years. In spite of this, there does not seem to be a general concencus on the type of extreme value distribution that would best serve this purpose. In fact, all the three standard extreme value distributions (Frechet, Gumbel, reverse Weibull) have been suggested. The wind loading provisions of the American National Standard ANSI A58.1-1972 were based on the Frechet-distribution for locations not subjected to mature hurricane winds. Later studies revealed that a Gumbel-distribution may be more appropriate (Simiu et al., 1978). However, recent research based on peak-over-threshold (POT) methods indicate that even a reverse Weibull distribution may be used to represent extreme wind speeds (Simiu and Heckert, 1994). However, by applying the POT method to transformed data, it appears that the Gumbel distribution is just as good a choice as the reverse Weibull. This is of course in agreement with the theoretical developments in this paper which tells us that one should not use the wind speed data themselves, but rather the transformed data to test whether a Gumbel distribution is appropriate or not.

The long-term distribution of the wind speed in temperate latitudes, and in particular in Norway, will be assumed to be fairly well described by a Rayleigh distribution. This is substantiated by more than 50 years of wind speed data for U.K. sites revealing that a two-parameter Weibull distribution decribes the data very well with shape parameter ranging from 1.8 to 2.2 (Cook, 1985). Hence a Rayleigh distribution, which corresponds to a shape parameter of 2.0, is an appropriate representative distribution of wind speed data to use for the transformation of extreme values. That is, for the wind speed data we shall use $\alpha(x) = x^2$.

Table 2 summarizes the results for the 100-year return period values for four air field sites in the northern part of Norway. The length of the data records available were as follows. Kirkenes: 24 years, Bardufoss: 29 years, Andøya: 31 years, Bodø: 23 years. The table reveals that there are significant differences between long return period values predicted by the two alternative methods also for extreme wind speed analysis.

Table 2 - 100-year wind speed (m/s)

	Kirkenes	Bardufoss	Andøya	Bodø
Conv. meth.	46.2	37.6	57.6	46.8
Trans. meth.	43.7	36.3	52.9	43.3

Conclusions

A method for transformation of extreme value data for optimal plotting and extrapolation is described. The example studies presented in the paper clearly support the conclusion that this method, which is based on knowledge of the underlying probability distribution of the process under study, performs generally better than the conventional method, which is based on assuming that the untreated extreme value data follow a Gumbel distribution.

Acknowledgements

The wind data were provided by The Norwegian Meteorological Institute. The author also gratefully acknowledges the support of The Norwegian Building Standardization Council, as well as the assistance of Mr. L.E. Storm and Mr. H. Storli with the numerical work.

References

Castillo, E. (1988). *Extreme Value Theory in Engineering*. Academic Press Inc., San Diego.

Cook, N.J. (1985). *The designer's guide to wind loading of building structures*, Part 1. Butterworths, London.

Simiu, E., Filliben, J.J. and Biétry, J. (1978). "Sampling Errors in the Estimation of Extreme Wind Speeds". *Journal of the Structural Division*, ASCE, **104**, 491-501.

Simiu, E. and Heckert, N.A. (1994). *Extreme Wind Distribution Tails: A 'Peaks Over Threshold' Approach*. NIST Building Science Series 174, National Institute of Standards and Technology, Gaithersburg, MD 20899, USA.

Stochastic Prediction and Control of Uncertain Nonlinear Dynamical Systems

R. Ghanem[1], Member, ASCE

M. Grigoriu[2], Member, ASCE

Abstract

A novel procedure is developed for the control of nonlinear dynamical systems with uncertain parameters. The algorithm is based on the concept of sliding mode control. Traditionally, sliding mode control has evolved around deterministic procedures with the variability and uncertainty in system parameters accounted for through the adaptivity of the control algorithm. This approach in effect results in a deterministic extrapolation of the state and parameters of the system given a single realization up to the present. In this paper, an approach is presented that features a stochastic extrapolation of the dynamics and the state of the system. This has the benefit of allowing the observed realization up to the present to be contaminated by noise. It also provides a consistent treatment of stochastic predictive control. The proposed development relies on the assumption that the system is subjected to a filtered gaussian white noise. The evolution in time of the probabilistic structure of the state, driven by the white noise and the control force, can then be evaluated.

Nonlinear Stochastic Control

Consider the Duffing oscillator subjected to a filtered white noise acceleration $y(t)$ and a control force $u(t)$. The motion of the oscillator is governed by the following equation,

$$\ddot{x} + 2\zeta\omega\dot{x} + \omega^2 x(1 + \epsilon x^2) = y(t) + \frac{1}{m}u(t) \tag{1}$$

$$dy(t) = -\alpha y(t)dt + \sigma_y\sqrt{2\alpha}dB(t)\ , \qquad \alpha > 0\ , \tag{2}$$

where $B(t)$ denotes a Brownian motion with variance dt, ω and ζ denote dynamical properties of the oscillator, ϵ is a small parameter, and α denotes a spectral parameter of the non-white input acceleration process $y(t)$, while σ_y^2 denotes is variance. Using the following notation,

[1]Johns Hopkins University, Baltimore, MD 21218.

[2]Cornell University, Ithaca, NY 14853.

$$x_1 = x, \qquad x_2 = \dot{x} \ , \qquad x_3 = y \ , \tag{3}$$

the equation of motion is recast as the following first order differential equation,

$$d\begin{bmatrix} x_1 \\ x_2 \\ x_3 \end{bmatrix} = \begin{bmatrix} x_2 \\ -\omega^2 x_1(1+\epsilon x_1^2) - 2\zeta\omega x_2 + x_3 \\ -\alpha x_3 \end{bmatrix} dt + \begin{bmatrix} 0 \\ 0 \\ \sigma_y\sqrt{2\alpha} \end{bmatrix} dB(t) + \begin{bmatrix} 0 \\ \frac{1}{m} \\ 0 \end{bmatrix} u(t)dt \ . \tag{4}$$

This last equation is rewritten as,

$$d\mathbf{X} = [\mathbf{a}(\mathbf{X}) + \mathbf{c}u(t)]\, dt + \mathbf{b}dB(t) \ . \tag{5}$$

The objective of the control force is to track a specified motion $x_d(t)$ in the state space. The tracking error is then given by,

$$e(t) = x_1(t) - x_d(t) \ . \tag{6}$$

Let a combined error be defined as,

$$s(t) = e + \lambda\dot{e} \ , \tag{7}$$

where λ denotes some design parameter. It can then be shown (Slotine, 1991; Ghanem and Bujakov, 1994) that tracking along the sliding surface $s(t) = 0$ trajectory in state space can be asymptotically achieved with the following choice of control force,

$$u(t) = m\ddot{x}_r(t) - ks(t) + a_1 x(t)(1 + \epsilon x^2(t)) + a_2\dot{x}(t) \ , \tag{8}$$

where,

$$\ddot{x}_r(t) = \ddot{x}_d(t) - \lambda\dot{e}(t) \ , \tag{9}$$

k is some positive constant, and the coefficients a_1 and a_2 denote the quantities ω^2 and $2\zeta\omega$, respectively. Note that $\ddot{x}_r(t)$ is obtained by modifying $\ddot{x}_d(t)$ according to the tracking errors. In the presence of uncertainties regarding the dynamical parameters ω and ζ, the above control algorithm has to be modified as follows,

$$u(t) = \hat{m}\ddot{x}_r(t) - ks(t) + \hat{a}_1 x(t)(1 + \epsilon x^2(t)) + \hat{a}_2\dot{x}(t) \ , \tag{10}$$

where,

$$\begin{aligned} \dot{\hat{m}} &= -\gamma s\ddot{x}_r \\ \dot{\hat{a}}_1 &= -\gamma s x(t)(1 + \epsilon x^2(t)) \\ \dot{\hat{a}}_2 &= -\gamma s\dot{x}(t) \ . \end{aligned} \tag{11}$$

The adaptation law given by equation (11) ensures asymptotic tracking of the control law. The sliding mode control problem defined and solved by the above equations is performed in continuous time. Its discrete time implementation consists of evaluating the right hand side of equation (8) at the last available time step, and thus involves delays. The efficiency of the control can be enhanced through a judicious one-step prediction, of the various quantities on the right hand side of equations (10) and (11). In the following, stochastic prediction will be performed by assuming locally gaussian behavior of the state process conditioned on its past. Thus, given the state and control force at a given time step t, the value of the state will be replaced by its predicted value prior to evaluating the updated coefficients and the control force to be applied the next time step, $t+1$. Thus,

$$u(t+\Delta t) = m\ddot{x}_r^*(t+\Delta t) - ks^*(t+\Delta t) + a_1^* x^*(t+\Delta t)(1+\epsilon x^{*2}(t+\Delta t)) + a_2^* \dot{x}^*(t+\Delta t)\,, \tag{12}$$

where the starred quantities refer predicted values.

Applying an Euler scheme to equation (4), and taking the right hand side in this last equation at the previous time step, $\mathbf{X}(t+\Delta t)|_{\mathbf{X}(t)=\mathbf{x}}$ becomes a gaussian process, $\mathcal{N}(\mathbf{m_X}, \mathbf{K_X})$, where,

$$\mathbf{m_X} = E[\mathbf{X}(t+\Delta t)|_{\mathbf{X}(t)=\mathbf{x}}] = \mathbf{x} + [\mathbf{a}(\mathbf{x}) + \mathbf{c}u(t)]\,dt \tag{13}$$

and

$$\mathbf{K_X} = COV[\mathbf{X}(t+\Delta t)|_{\mathbf{X}(t)=\mathbf{x}}] = \mathbf{b}\mathbf{b}^T\,. \tag{14}$$

The results from implementing the above procedure are shown in Figures (1) and (2). These figures are correspond to a value of $\epsilon = 0.1$, $\zeta = 0.05$, $\omega_0 = 20$, $m = 1$, $b = 1$, $\lambda = 10$, $k = 64$. Moreover, the filtered white noise process is characterized by a standard deviation $\sigma_y = 10$ and a correlation coefficient, $\alpha = 0.12$. Figure (1) shows the results associated with the controlled motion, the uncontrolled motion, and the partially controlled motion, where the input motion is assumed constant over the previous time interval, and stochastic prediction is not implemented. It is observed that prediction of the input motion, as described in this paper, improves the behavior of the controlled system. Figure (2) shows the convergence of the estimate for the parameter ω_0, starting with an initial guess of 15 (25% error).

References

Ghanem, R., and Bujakov, M., "Adaptive control of nonlinear dynamical systems with uncertainties," *First World Conference on Structural Control,* Los Angeles, 3-5 August 1994.

Slotine, J.-J., and Li, W., *Applied Nonlinear Control,* Prentice-Hall, 1991.

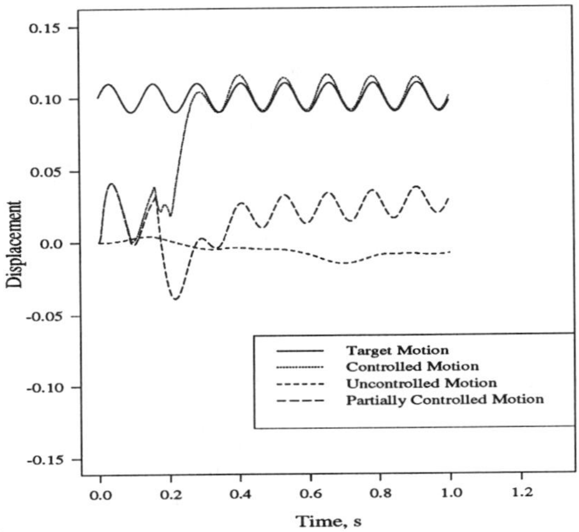

Figure 1: Controlled, Uncontrolled, and Partially Controlled Motions; $\sigma_y = 10$, $\alpha = 0.12$.

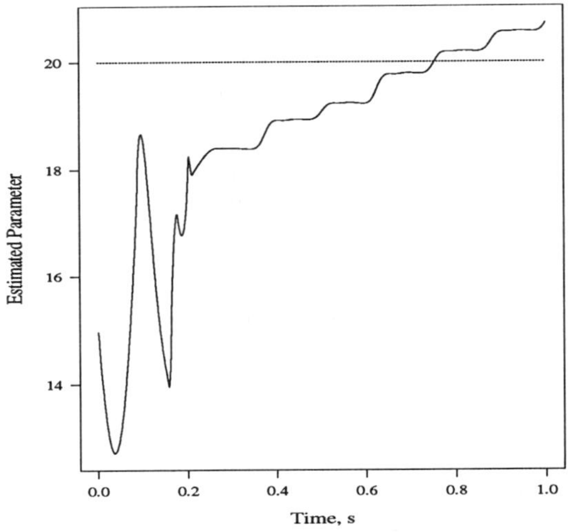

Figure 2: Time Evolution of the Estimated Uncertain ω_0.

A UNIFIED DESCRIPTION OF SOIL BEHAVIOR

by Juan M. Pestana[1], Associate Member, ASCE

ABSTRACT:

This paper describes the essential features of a unified effective stress model for sands and clays valid over a wide range of confining stresses and densities. The model introduces a modification to the conventional incrementally linearized theory of rate independent, elasto-plasticity in order to explicitly incorporate the current void ratio as an additional state variable. This enables the model to describe characteristic transitions from dilative to contractive shear response for sands as the confining pressure increases (non-normalizable behavior). The model predicts "normalizable" behavior as the stress state approaches the LCC regime, thus describing the behavior of normally consolidated clays as a particular case. Model evaluation is accomplished through comparisons with measured soil behavior at the element level.

Introduction

The description of soil behavior through constitutive equations is a difficult task attempting to extend predictive capabilities, while controlling model complexity and input parameters. One of the most important limitations of (most) existing elasto-plastic models for sands is that their input parameters are density and confining pressure dependent. As an example, stress-strain-strength properties of a given sand at two different void ratios are characterized as two separate materials with different sets of input parameters. In contrast, most models developed for clays are based on the critical state soil mechanics framework and exhibit normalizable behavior which is independent of current density. Furthermore, most sands are typically deposited (formed at nearly zero effective confining stress) within a narrow range of densities, while clays can be formed at a wide range of void ratios depending on depositional conditions.

These two very simple observations impose important restrictions on the derivation of constitutive equations applicable for both sands and clays over a wide range of stresses and void ratios. The proposed model, referred to as MIT-S1, addresses this problem by including the (current) void ratio as a state parameter (in addition to the current stress state, σ'). This paper describes the essential features of the proposed model which is capable of describing the behavior of sands, highly overconsolidated clays and silts (Pestana, 1994). Model performance is evaluated here through comparisons between model predictions and data from laboratory triaxial tests on a freshly deposited sand while model evaluation for clays can be found elsewhere (Pestana and Whittle, 1994).

Compression Behavior

Constitutive models for clays and sands have generally been developed independently due to the perceived differences, among other things, of compressibility in the practical range of confining stresses (Figure 1). Pestana and Whittle (1995b) presented a new compression model for cohesionless soils describing the recoverable and irrecoverable strain components during hydrostatic and one-dimensional compression over a wide

[1] Assistant Professor, University of California, Berkeley, CA

range of densities and confining pressures. Sand specimens compressed from different initial formation densities approach a unique response at high stress levels, referred to as the Limiting Compression Curve (LCC). The behavior in the LCC regime is described by a linear relationship in a loge-logσ' space while the transitional regime is controlled primarily by the formation density (Fig. 1). Pestana (1994) shows that the consolidation behavior of normally consolidated clays can also be described, for modeling purposes, by a linear relationship in a loge-logσ' space, over a wide stress range.

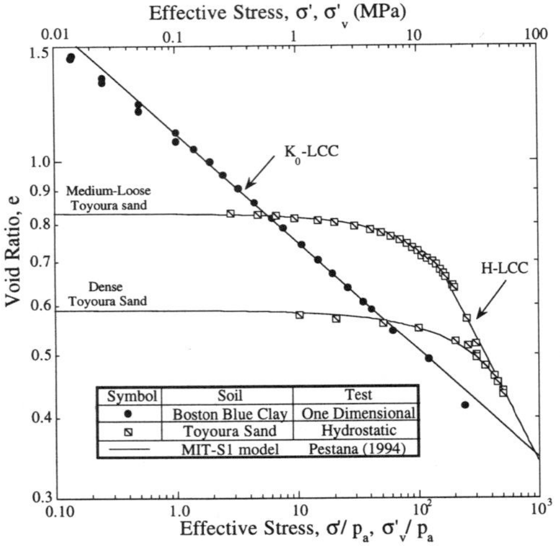

Figure 1: Compression Curves for Boston Blue Clay and Toyoura Sand

Figure 1 shows isotropic compression curves for a medium-loose and a dense specimen of Toyoura sand, as well as one dimensional compression of Boston Blue Clay over a wide range of stresses (0.01 MPa - 50 MPa). Toyoura sand is a fine, poorly graded quartzitic sand, well documented in the Japanese soil mechanics literature, while Boston Blue Clay is a low plasticity clay extensively used in research at MIT. The MIT-S1 model accurately describes the observed behavior over a wide range of stress and void ratios for both soils.

A detailed evaluation and comparisons of model predictions with observed compression behavior of sands and clays was presented by Pestana (1994) and Pestana and Whittle (1995b).

Model Formulation

The new model, referred to as MIT-S1, is based on the incrementally linearized theory of rate independent, elasto-plasticity (e.g., Prévost, 1978) and uses the same structure as a previously reported prototype for clays (MIT-E3; Whittle and Kavvadas, 1994). The model introduces significant changes in the geometry of the bounding surface and hardening laws, while a new framework of compression behavior unifies the modeling of clays and sands (Pestana and Whittle, 1994; Pestana and Whittle, 1995a)

Most sand models assume that conditions describing yield behavior and peak friction angle are a function of current density (i.e., void ratio) but do not provide for expressions describing the variation of input parameters as density changes. As a result, a sand having two different density states (e.g., dense and loose) will be described as two different materials. The MIT-S1 model explicitly introduces the current void ratio as an additional state variable to describe the behavior of sands in the transitional regime. Although this approach has been extensively used in clay models such as those based on the Critical State Soil Mechanics Framework (e.g., Schofield and Wroth, 1968; Wood, 1990), these formulations assume that the shape of the yield function is independent of current density, thus predicting normalizable behavior (i.e., with respect to previous stress history)

The new model formulation incorporates a single yield surface whose geometry resembles a distorted 'teardrop' or lemniscate with an aperture (shape) function controlling the peak friction angle and dilation rates as a function of formation density. The yield function, f, is explicitly written in terms of the current void ratio:

$$f\ (\sigma', e, \alpha', \mathbf{b}) = 0 \tag{1}$$

where σ' and e are the state variables defining the current stress tensor and void ratio, α' and $\mathbf{b}$ are the internal variables (function of stress history) defining the size and orientation (i.e., anisotropy tensor) of the bounding surface. The model formulation assumes that the hardening of the size of the yield function, α', is given by:

$$\frac{\dot{\alpha}'}{\alpha'} = F_1(\dot{\varepsilon}^p) + F_2(\dot{\mathbf{e}}^p)\,F_3(\delta_p); \quad \text{where } \delta_p = 1 - \alpha'/\alpha'_p \tag{2}$$

where F_1 and F_2 are generic functions of the plastic volumetric strain, ε^p and plastic shear strain tensor, $\mathbf{e}^p$, respectively, and F_3 is a generic function of the normalized distance, δ_p, of the current bounding surface to the Limiting Compression Curve at the same void ratio, α'_p. The functional form in equation 2 states that hardening in the size of the bounding surface for sands is decoupled from changes in void ratio (or plastic volumetric strains) in the transitional regime ($\delta_p \neq 0$). This assumption is similar to those used by other models for sands (e.g., Jefferies, 1993; Nova and Wood, 1979) However, in contrast to previous formulations, the change in size of the surface is neither related to critical (or steady) state conditions (e.g., Nor-Sand Model, Jefferies, 1993, Crouch et al., 1994), nor unbounded (e.g., Lade, 1977; Nova and Wood, 1979), but is controlled by the Limiting Compression Curve, LCC. For soils in the LCC regime ($\delta_p = 0$), the hardening function becomes only a function of volumetric plastic strains and resembles the one used in the critical state soil mechanics framework, thus allowing the representation of normally consolidated clays as a particular case. Details of the formulation are presented by Pestana (1994) and are beyond the scope of this paper.

Figure 2 shows the observed behavior of three specimens of Toyoura sand with void ratios, e_0= 0.735, 0.833, and 0.905 isotropically consolidated to the same confining stress of 1 MPa. As can be seen, the behavior is not normalizable as a function of stress history alone and ranges from contractive (positive pore pressure) with strain softening after peak for high void ratios, to dilative with strain hardening characteristics for lower void ratios and is well described by the model.

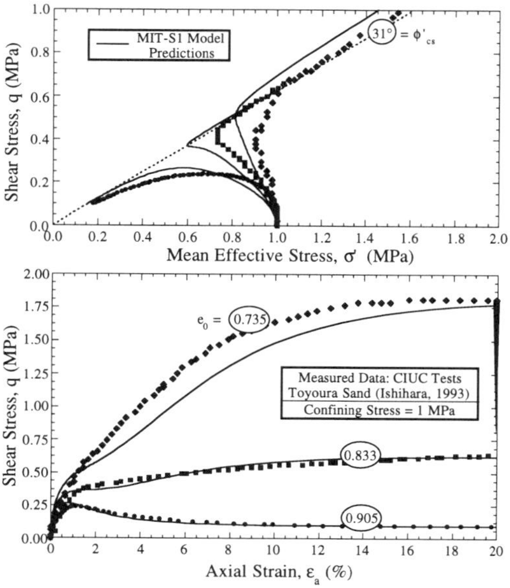

Figure 2: Predicted Effects of Formation Void Ratio on the Undrained Behavior of Toyoura Sand

Figure 3 shows the undrained behavior of medium-loose (e_0 = 0.833) Toyoura sand specimens at four different confining stresses, σ'_c = 0.1, 1, 2, and 3 MPa. The observed behavior ranges from strongly dilative (i.e., negative pore pressures) for tests at low confining pressures (σ'_c= 0.1 MPa) to contractive behavior at high stresses (σ'_c= 2-3 MPa) and is accurately described by the model. Stress-strain curves predicted by MIT-S1 are qualitatively in good agreement with measurements at all stress levels. In particular, the model predicts that all samples having the same pre-shear void ratio will have the same undrained strength at large strains.

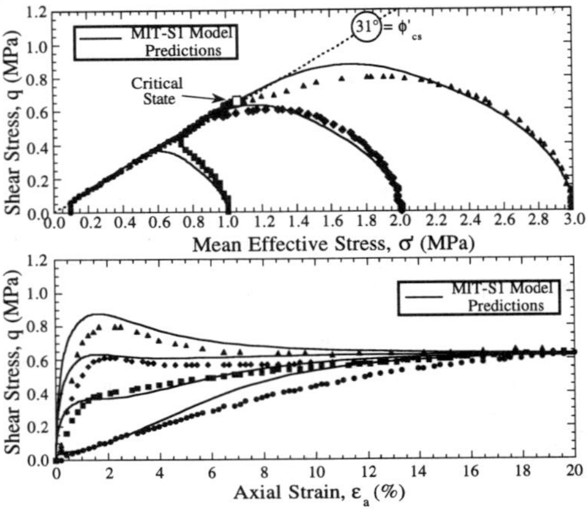

Figure 3: Effects of Confining Stress on the Measured Undrained Behavior of Toyoura Sand

Concluding Remarks

This paper described the essential features of a new effective stress soil model, MIT-S1, which unifies the description of observed sand and clay behavior over a large range of confining stresses and densities. The model explicitly incorporates the current void ratio as a state variable, in addition to the stress state, σ', to describe non-normalizable behavior, ranging from contractive to dilative behavior as a function of current stress level and density in the transitional regime ($\delta_p \neq 0$). Behavior in the LCC regime is independent of initial formation density and exhibits normalized stress-strain-strength characteristics, thus allowing the behavior of clays to be described as a particular case.

Acknowledgments

Support for this research was partially provided by INTEVEP, S.A. The comments and suggestions of Prof. A.J. Whittle and C.C. Ladd are gratefully acknowledged.

References

Crouch, R.S., Wolf, J.P and Dafalias, Y.F. (1994) "Unified critical state bounding surface plasticity model for soil," ASCE J. of Engineering Mechanics, 120(11), 2251-2270.

Ishihara, K. (1993) "Liquefaction and flow failure during earthquakes," Géotechnique, 43 (3), 351-415.

Jefferies, M.G. (1993) "Nor-Sand: a simple critical state model for sand," Géotechnique, 43 (1), 91-103

Lade, P.V. (1977) "Elasto-Plastic Stress-Strain Theory for Cohesionless Soil with Curved Yield Surfaces," Int'l. Journal of Solids and Structures, 13, 1019-1035.

Nova, R. and Wood, D.M. (1979)"A constitutive model for sand in triaxial compression," Intl. J. for Numerical & Analytical Methods in Geomechanics, 3, 255-274.

Pestana, J.M. (1994) A unified constitutive model for clays and sands, Sc.D. Thesis, MIT, Cambridge, MA

Pestana, J.M. & Whittle, A.J. (1994) "Model prediction of anisotropic clay behavior due to consolidation stress history", 8th Int. Conf. Computer Meth. & Adv. in Geomech. (IACMAG '94)

Pestana, J.M. and Whittle, A.J. (1995a) "Predicted effects of confining stress and density on shear behavior of sand," Proc. 4th Int. Conf. on Comp. Plasticity, Barcelona, Vol. II, 2319-2330.

Pestana, J.M. and Whittle, A.J. (1995b) Pestana, J.M. and Whittle, A.J. (1995b) "Compression model for cohesionless soils," Géotechnique, 45 (4), 611-631.

Prévost, J.H. (1978) "Plasticity theory for soil stress-strain behaviour," ASCE, J. Engrg. Mech. Div., 104 (EM5), 1177-1194.

Schofield, A.N. & Wroth, C.P. (1968) Critical state soil mechanics, McGraw-Hill Book Co., London.

Whittle, A.J. and Kavvadas, M. (1994) "MIT-E3: A constitutive model for overconsolidated clays", ASCE, J. of Geotech. Engrg., 120(l), 173-198.

Wood, D.M. (1990) Soil behaviour and critical state soil mechanics. Cambridge University Press.

Formulation and Implementation of Improved Zero-Thickness Interface Elements

V. N. Kaliakin[1] , M. ASCE

Abstract

Realistic numerical simulation of soil-structure interaction requires a robust idealization of the interfaces between soil and the structural elements. This is commonly realized through the use of suitable interface elements possessing zero thickness. Recent work in such simulations has led to the identification of kinematic inconsistencies of traditionally used zero-thickness interface elements, and to the development of improved interface elements. In this paper, these kinematic inconsistencies are reviewed and an overview of the improved family of zero thickness interface elements is given, along with the requisite implementation schemes.

Introduction

Requisite to realistic numerical simulation of the interaction between soils and structures is a robust idealization of the interfaces between the two constituents. This is typically realized through the use of suitable interface elements possessing zero thickness. Recent work in such simulations has led to the identification of kinematic inconsistencies of traditionally used zero-thickness interface elements. An improved linear interface element was subsequently developed and its predictive capabilities ascertained (Kaliakin and Li 1995). More recently, this work was extended to quadratic elements (Kaliakin and Li 1996). Past analyses of quadratic zero-thickness interface elements suggest that such elements are plagued by spurious tangential force oscillations that are attributed to the type of numerical integration scheme employed Gens et al., 1989; Schellekens and De Borst, 1993). As shown by Kaliakin and Li (1995), such force oscillations are due not to the integration scheme employed, but instead to fundamental kinematic deficiencies associated with the element. Once these deficiencies are suitably removed, an improved element is obtained that exhibits superior performance regardless of the integration scheme employed.

This paper reviews the kinematic inconsistencies associated with existing zero-thickness interface elements. An overview of the improved family of such elements is next be given, along with the requisite implementation schemes.

Illustration of a Fundamental Kinematic Deficiency

When using quadratic continuum elements in two-dimensional analyses, nodal compatibility requires the use of quadratic (six-node) isoparametric zero-thickness interface elements. For brevity in the subsequent discussion, the standard quadratic interface element shall be referred to as the "S6N" element. This element possess six

[1]Assistant Professor, Department of Civil and Environmental Engineering, University of Delaware, Newark, DE 19716.

nodes and twelve displacement degrees of freedom (Fig. 1). At each nodal location along the interface a pair of nodes is placed at the same initial geometric location; the thickness of the element is thus initially zero.

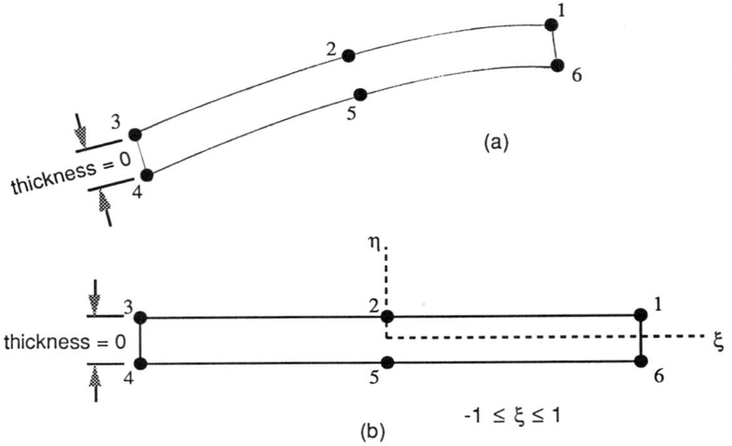

Figure 1. Standard Six-Node Zero-Thickness Interface Element in (a) Global Coordinate; and, (b) Natural Coordinates

The vector of absolute nodal displacement degrees of freedom with respect to the local coordinate system $(x' - y')$ is given by

$$\{u'\} = \{u'_{1x} \quad u'_{1y} \quad u'_{2x} \quad u'_{2y} \quad u'_{3x} \quad u'_{3y} \quad u'_{4x} \quad u'_{4y} \quad u'_{5x} \quad u'_{5y} \quad u'_{6x} \quad u'_{6y}\}^T \tag{1}$$

where the superscript T denotes the operation of matrix transposition. Associated with the above displacement components is a corresponding vector of nodal forces $\{f^{e'}\}$. The vector of relative displacements is defined by

$$\{w\} = \begin{Bmatrix} w_t \\ w_n \end{Bmatrix} = [N^*]\{u'\} \tag{2}$$

where w_t and w_n represent tangential and normal relative displacements, respectively, along the interface. The matrix relating relative and absolute displacements is given by

$$[N^*] = \begin{bmatrix} N_1 & 0 & N_2 & 0 & N_3 & 0 & -N_3 & 0 & -N_2 & 0 & -N_1 & 0 \\ 0 & N_1 & 0 & N_2 & 0 & N_3 & 0 & -N_3 & 0 & -N_2 & 0 & -N_1 \end{bmatrix} \tag{3}$$

where N_1, N_2 and N_3 represent standard quadratic interpolation functions associated with the ξ-coordinate system (Fig. 1); note that $N_4 = N_3$, $N_5 = N_2$, and $N_6 = N_1$.

The relative displacements are related to forces acting along the interface through the following general constitutive relation

$$\begin{Bmatrix}\sigma_t\\ \sigma_n\end{Bmatrix}=\begin{bmatrix}d_{11} & d_{12}\\ d_{12} & d_{22}\end{bmatrix}\begin{Bmatrix}w_t\\ w_n\end{Bmatrix}=[D]\{w\} \tag{4}$$

where σ_t and σ_n represent the tangential and normal forces per unit length, respectively, acting along the interface. The constitutive parameters d_{11} and d_{22} represent the tangential and normal stiffness per unit length, respectively, along the interface. The presence of the coupling parameters d_{12} allows for the simulation of dilatational phenomena.

Employing a standard minimization of the total potential energy, the element equations are obtained in terms of the absolute displacements and associated forces. The element stiffness matrix is computed from

$$\left[K^{e'}\right]=\int_{x'}[N^*]^T[D][N^*]dx'=\int_{-1}^{1}[N^*]^T[D][N^*]|J|d\xi \tag{5}$$

The Jacobian determinant is given by

$$|J|=\frac{\partial x'}{\partial \xi}=\frac{1}{2}(x_1'-x_3')+\xi(x_1'-2x_2'+x_3') \tag{6}$$

Finally, the element equations must be transformed from $(x'-y')$ coordinates to system coordinates $(x\text{–}y)$ in the usual manner.

If the mid-side nodes are equidistant from the corner nodes, and if the length of the element is denoted by L, then equation (6) gives $|J|=L/2$. In this case the element stiffness matrix can be integrated explicitly. For simplicity, the explicitly integrated form of the element stiffness matrix shall now be used to illustrate a fundamental deficiency associated with the S6N element. Consider a single element example in which the local element coordinates are aligned with the system coordinates and nodes 4 to 6 are restrained. Accounting for the homogeneous displacement specifications at nodes 4 to 6, and assuming $d_{12}=0$, the element equations reduce to

$$\frac{L}{30}\begin{bmatrix}4d_{11} & 0 & 2d_{11} & 0 & -d_{11} & 0\\ 0 & 4d_{22} & 0 & 2d_{22} & 0 & -d_{22}\\ 2d_{11} & 0 & 16d_{11} & 0 & 2d_{11} & 0\\ 0 & 2d_{22} & 0 & 16d_{22} & 0 & 2d_{22}\\ -d_{11} & 0 & 2d_{11} & 0 & 4d_{11} & 0\\ 0 & -d_{22} & 0 & 2d_{22} & 0 & 4d_{22}\end{bmatrix}\begin{Bmatrix}u_{1x}'\\ u_{1y}'\\ u_{2x}'\\ u_{2y}'\\ u_{3x}'\\ u_{3y}'\end{Bmatrix}=\begin{Bmatrix}f_{1x}'\\ f_{1y}'\\ f_{2x}'\\ f_{2y}'\\ f_{3x}'\\ f_{3y}'\end{Bmatrix} \tag{7}$$

For the special case of $f_{1x}'=P, f_{2x}'=f_{3x}'=0$ and $f_{2y}'=-4Q=4f_{1y}'=4f_{3y}'$, equations (7) give

$$\{u'\}=\begin{Bmatrix}\frac{9P}{Ld_{11}} & \frac{-6Q}{Ld_{22}} & \frac{-3P}{2Ld_{11}} & \frac{-6Q}{Ld_{22}} & \frac{3P}{Ld_{11}} & \frac{-6Q}{Ld_{22}}\end{Bmatrix}^T \tag{8}$$

Prior to the mobilization of the frictional resistance along the interface, the quantities u'_{1x}, u'_{2x} and u'_{3x} represent the "pre-slip" tangential displacements. Even if, through the choice of suitably large values for d_{11}, these displacements are minimized, the fact remains that a kinematic inconsistency is associated with the element. That is, the mid-side node displaces in a direction opposite to that associated with both the node at which the load P is applied, and the other unconstrained corner node. As shown by Kaliakin and Li (1995, 1996), in multi-element meshes this seemingly simple inconsistency leads to force oscillations under simulated sliding conditions.

Development of an Improved Element

Consider a "macro-element" consisting of two S6N elements. Following assembly of the matrices for the two sub-elements, the degrees of freedom associated with the four mid-side nodes are "condensed" from the equations. For the single element example considered above, the element equations give

$$\{u'\} = \left\{ \frac{17P}{Ld_{11}} \quad \frac{-30P}{Ld_{22}} \quad \frac{3P}{Ld_{11}} \quad \frac{-42P}{Ld_{22}} \quad \frac{P}{Ld_{11}} \quad \frac{-30P}{Ld_{22}} \right\}^T \tag{9}$$

Contrary to the tangential response of the S6N element (equation 8) where the mid-side node displaced in a direction opposite to applied load, the tangential displacements associated with the LK2 element are seen to be consistent. As such, the kinematic deficiency noted for the S6N element has been removed from the formulation.

Finally, it is important to point out that the algorithm used in implementing the improved element differs somewhat from that typically associated with zero-thickness interface elements. Past experience with the improved element indicates that this algorithm leads to a robust implementation (Kaliakin and Li 1995, 1996).

Conclusion

Investigation of the standard quadratic zero-thickness interface (S6N) element has demonstrated a fundamental kinematic deficiency inherent in its response. When used to simulate sliding and/or pull-out behavior, this element exhibits oscillations in the tangential force per unit length. In past investigations, such oscillations were attributed to the particular numerical integration scheme used in computing the element arrays. However, as shown herein, the aforementioned force oscillations can be removed not by altering the numerical integration scheme used, but by removing the kinematic deficiency that plagues the element.

References

Gens, A., Carol, I. and Alonso, E. E. (1989), "An interface element formulation for the analysis of soil-reinforcement interaction," *Comp. and Geotech.*, **7** 133–151.

Kaliakin, V. N. and Li, J. (1995), "Insight Into Deficiencies Associated with Commonly Used Zero-Thickness Interface Elements." *Comp. and Geotech.* **17**(2), 225-252.

Kaliakin, V. N. and Li, J. (1996), "A Kinematically Consistent Quadratic Zero-Thickness Interface Element," submitted.

Schellekens, J. C. J. and De Borst, D. (1993), "On the numerical integration of interface elements," *Int. J. Num. Meth. Eng.*, **36** 36–43.

Generalized Plane Strain Finite Element Analysis : Geomechanical Applications

V. N. Kaliakin[1] , M. ASCE, L. Cui [2], A. H-D Cheng[3] , M. ASCE

Abstract

Many problems commonly analyzed in mining and petroleum engineering applications are three-dimensional in nature. In an effort to simplify the inherently intensive solution effort associated with their solution, it is often possible to idealize these problems under conditions of generalized plane strain, thus simplifying the formulation by rendering most physical quantities functions of only two spatial coordinates. Since the material properties may still be anisotropic, and the far-field stresses can retain their three-dimensional nature, the solution will not, however, be entirely two-dimensional. In this paper, the assumptions underlying generalized plane strain analyses are reviewed. An overview of a generalized plane strain finite element formulation for an anisotropic poroelastic material, its numerical implementation and subsequent performance are likewise discussed.

Introduction

The need for analyzing borehole and/or tunnel problems commonly arises in the fields of rock and petroleum engineering. Such problems are complicated by the anisotropic nature of the geomaterials, the pores of which are typically filled with fluid, and by the three-dimensional geometry. Past analyses of borehole/tunnel problems have typically involved certain simplifying assumptions to render the problems more tractable. For example, in view of the associated special geometry, material idealizations and loadings, these three-dimensional problems have been analyzed assuming conditions of generalized plane strain (Amadei, 1983).

One shortcoming of the above solutions is the fact that the material was assumed to be elastic, and no account was made for the presence of pore fluid. In practice, many geomaterials are known to be fluid saturated. When subjected to external load, or when a fluid flow is driven through the material, a coupled hydraulic-mechanical response is known to take place. This coupling requires a modification of the traditional theories of elasticity and flow through porous media. The theory commonly used to account for this coupling is poroelasticity (Detournay and Cheng, 1993). Poroelastic solutions differ from their counterparts in elasticity in not only magnitude, but also due to the fact that the solution is, due to the presence of the flow, time dependent.

[1]Assistant Professor, Department of Civil and Environmental Engineering, University of Delaware, Newark, DE 19716.

[2] Rock Mechanics Institute, The University of Oklahoma, Norman, OK 73019.

[3] Professor, Department of Civil and Environmental Engineering, University of Delaware, Newark, DE 19716.

Several generalized plane strain problems in poroelastic materials have recently been solved analytically (Cui, 1995; Abousleiman et al. 1996). However, the inclusion of material anisotropy, the presence of pore fluid, and the possibility of non-circular geometries (e.g., tunnels) precludes the analysis of realistic borehole/tunnel problems by analytical means. Instead, numerical techniques must be employed. For example, three-dimensional finite element analyses of a borehole have recently been performed by Cui et al. (1996). Although these analyses successfully simulated the behavior of a borehole, they were computationally intensive. An alternative to many three-dimensional finite element analyses of borehole/tunnel problems is a generalized plane strain analysis. Since only a two-dimensional geometry must now be analyzed, a significant decrease in computational effort is realized. This paper reviews the assumptions underlying generalized plane strain idealizations. An overview of the finite element formulation for an anisotropic poroelastic material is next presented, and its numerical implementation and subsequent performance discussed.

The Generalized Plane Strain Problem

Consider a homogeneous elastic body having general rectilinear anisotropy and bounded by a cylindrical surface. The length of the body is assumed infinite. Its cross-section is arbitrary, either finite or infinite, and either simply-or multiply connected. The origin of the coordinate system is placed at an arbitrary point within the cross-section. The *z*-axis is taken parallel to the generators, and the *x* and *y*-axes are placed in accordance with the shape of the particular cross-section (Fig. 1). The body is assumed to be loaded by surface and body forces acting in the planes of the cross-section; i.e., normal to the generators, and not varying along its length. Clearly, for the body described above, the stresses and deformations of all cross-sections are only functions of the *x* and *y*-coordinates. In an isotropic or anisotropic body at each point of which there exists a plane of elastic symmetry normal to the generators, the cross-sections remain plane. The deformation is then said to be "plane strain". If, however, in the more general case none of the planes of elastic symmetry are parallel to the *x-y* plane, and particularly in the general case of anisotropy, the deformation is no longer plane; the cross-sections will now warp, but all identically. Such deformation is commonly referred to as "generalized plane strain" (Lekhnitskii,1981).

Based upon the above discussion, and since the material is assumed to be poroelastic, it follows that the strain and stress fields shall be functions of only *x*, *y* and time *t*. Also, from Darcy's law it is easy to show that since the pore pressure is a function of *x*, *y* and *t*, that the flow of pore fluid shall likewise be *z*-independent. Although the strain and stress fields are independent of z, unlike the case of plane strain, quantities such as ε_{zz} and the engineering shearing strains γ_{xz} and γ_{yz} do not, in general, vanish. From the compatibility equations for linear systems, it follows that under generalized plane strain conditions ε_{zz} must be a linear function of *x* and *y*; i.e.,

$$\varepsilon_{zz} = A(t)x + B(t)y + C(t) \tag{1}$$

where A, B and C are arbitrary functions of time. In the present development, ε_{zz} is assumed to be a function of time only, thus $A = B = 0$.

Basic Equations of Poroelasticity

The governing equations of poroelasticity used to account for the coupling between mechanical and pore fluid response are unchanged from those commonly appearing in the literature (e.g., Detournay and Cheng, 1993). As such, these equations are not repeated here.

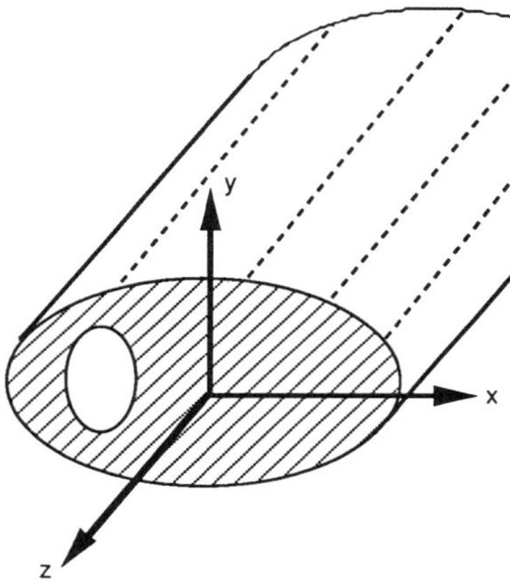

Figure 1. Body Associated with a Generalized Plane Strain Idealization

Finite Element Formulation, Implementation and Performance

The finite element formulation of the generalized plane strain problem is quite straightforward. The weak form of the Galerkin weighted residual statement of the governing equations is obtained in a manner very similar to that for a standard three-dimensional poroelasticity formulation (Cui et al., 1996), and results in a two-field problem with displacements and pore pressure as primary dependent variables. The only difference is associated with the strain-displacement transformation matrix. For a typical element, this matrix is written in terms of sub-matrices as

$$\left[B^u\right]=\left[B_1^u \quad B_2^u \quad \cdots \quad \cdots \quad B_{nen}^u\right] \tag{2}$$

where each sub-matrix has the following general form

$$\left[B_i^u\right]=\begin{bmatrix} \frac{\partial N_i^u}{\partial x} & 0 & 0 & \frac{\partial N_i^u}{\partial y} & 0 & 0 \\ 0 & \frac{\partial N_i^u}{\partial y} & 0 & \frac{\partial N_i^u}{\partial x} & 0 & 0 \\ 0 & 0 & \frac{\partial N_i^u}{\partial z} & 0 & \frac{\partial N_i^u}{\partial x} & \frac{\partial N_i^u}{\partial y} \end{bmatrix}^T \quad ; \quad i=1,2,\cdots,nen \tag{3}$$

In the above equation, N_i^u represents an interpolation function associated with the displacement field approximation, *nen* equals the number of element nodes related to this approximation, and a superposed "T" denotes the process of array transposition.

The numerical implementation of the generalized plane strain element is also quite straightforward. Besides the usual "overhead" associated with the addition of an element to an existing program, the element routines are very similar to those for a three-dimensional formulation. However, the calculation of strains and stresses differs slightly from such a formulation in that $\left[B_i^u\right]$ is now given by equation (3).

The final aspect of the generalized plane strain element to be addressed is its performance. To this end, a borehole problem in an anisotropic material, with plane of material symmetry inclined with respect to the axis of the borehole, was analyzed. No analytical solution exists for such a problem. A three-dimensional idealization of the problem involved 1,344 8-node hexahedral elements (with linear pore pressure idealization). A generalized plane strain idealization required only 240 4-node quadrilateral elements (with linear pore pressure idealization). An excellent agreement between results of the two analyses was obtained (Cui et al., 1996). The most compelling aspect of the generalized plane strain analysis is the savings in computational effort. Whereas the three-dimensional analysis required approximately 20 hours of CPU time on a Sun SPARC II workstation, the generalized plane strain analysis required only about 6 minutes.

Conclusion

Generalized plane strain analyses represent an efficient approach for analyzing certain three-dimensional problems in geomechanics. The formulation and implementation of a generalized plane strain element for poroelastic media is quite straightforward. Compared to a full three-dimensional analysis, the performance of such an element is quite good, with a substantial savings in computational effort.

References

Abousleiman, Y., Cheng, A. H-D., Jiang, C. and Roegiers, J.-C. (1996), "Poroviscoelastic Analysis of Borehole and Cylinder Problems," *ACTA Mech.*, in press.

Amadei, B. (1983), *Rock Anisotropy and the Theory of Stress Measurements*, Springer-Verlag.

Cui, L. (1995), Poroelasticity with Application to Rock Mechanics, PhD Dissertation, Department of Civil Engineering, University of Delaware.

Cui, L., Abousleiman, Y., Cheng, A. H-D., Kaliakin, V. N., and Roegiers, J.-C. (1996), "Finite Element Analysis of Anisotropic Poroelasticity: A Generalized Mandel's Problem and an Inclined Borehole Problem," *Int. J. Num. Anal. Meth. Geom.*, in press.

Cui, L., Kaliakin, V. N., Cheng, A. H-D. and Abousleiman, Y. (1996), "Poroelastic Analysis of Generalized Plane Strain Problems," submitted.

Detournay, E. and Cheng, A. H-D. (1993), "Fundamentals of Poroelasticity", Chap. 5 in *Comprehensive Rock Engineering: Principles, Practice & Projects, Vol. II, Analysis and Design Method*, C. Fairhurst, ed., Pergamon Press, 113-171.

Lekhnitskii, S. G. (1981), *Theory of Elasticity of an Anisotropic Body*, Mir Publishers, Moscow.

On a Conceptual Model for Turbulent Skin Friction

Chao Si and Manhar Dhanak *

Abstract

A model for turbulent skin friction, proposed by Orlandi and Jimenez, involving consideration of quasi-streamwise vortices in the cross-stream plane, is utilized to study the effect on the skin friction of oscillating the surface beneath the boundary layer in the spanwise direction. It is shown that the interaction between the streamwise vortex and the Stokes layer on the surface leads to a significant reduction in skin friction and turbulence production, consistent with predictions based on experiments and direct numerical simulations.

Introduction

There is now significant experimental and numerical evidence that coherent structures in the inner layer of a turbulent boundary layer play an essential role in the generation of Reynolds Stress. They act to maintain the convection of momentum normal to the wall, resulting in the formation of long streamwise streaks, separated by more intermittent ridges (see for example, Kim et al, 1987, Spalart, 1988, and Jimenez and Moin, 1991). These coherent structures are typically in the form of long quasi-streamwise vortices which act to redistribute the longitudinal velocity field into alternating high and low speed regions. Orlandi and Jimenez (1994) suggest a cross-plane model for the interaction between these flow structures and the rigid surface; they show that the model predicts the characteristics of turbulent skin friction fairly well. Several recent experimental and direct numerical simulation studies (Bradshaw and Potikos, 1985, Driver et al, 1987 , Moin et al., 1990 and Jung et al., 1992) also show that imposition of an oscillatory spanwise pressure gradient or spanwise oscillation of the wall beneath the boundary layer temporarily inhibits production of turbulence in the flow, leading to transient reductions in all turbulent quantities including the Reynolds shear stress and the turbulent kinetic energy. It seems evident that these changes are related to the modification of the coherent flow structures. In this paper, we utilize the Orlandi-Jimenez model and numerically study the evolution of the coherent structures and their interaction with the surface beneath, when the latter is subjected to spanwise oscillation. Hence, we determine the mechanism of turbulence suppression and skin friction reduction associated with the oscillatory motion.

Formulation

We consider the flow in a Cartesian co-ordinate system $Oxyz$ with Ox along the streamwise direction so that $y - z$ is the cross-plane of interest(figure

*Florida Atlantic University Boca Raton, Florida FL 33431

1). All of the flow variables are non-dimensionalized with respect to the wall units. Then the x component of vorticity ω satisfies

$$(\frac{D}{Dt} - \nabla^2)\omega = \omega(q_1\phi') \tag{1}$$

where $D/Dt = \partial/\partial t + (v - \phi)\partial/\partial y + w\partial/\partial z$ and $\nabla^2 = \partial^2/\partial y^2 + \partial^2/\partial z^2$. The deformation of the flow field by a Hiemenz straining flow with straining rate a is represented by q_1 (with the streamwise component of velocity given by $u = q(y) + xq_1(y, z, t)$), which satisfies,

$$(\frac{D}{Dt} - \nabla^2)q_1 = -q_1(2\phi' + q_1) - v\phi'' - C_2 \tag{2}$$

where, in view of the boundary condition at the far boundary, C_2 is a constant, to be determined from initial conditions. The cross-stream components of velocity are obtained by solving a Poisson equation. The flow is considered to be periodic in the z-direction with a wavelength Z. At the upper boundary of the computational domain the flow is assumed to be undisturbed and along the wall no-slip boundary condition is enforced. Equations (1) and (2) are solved using a third-order Runge-Kutta scheme in time.

Results

We choose $a = 0.2$, with the initial streamwise velocity profile, $q(y)$, close to that of a Couette flow as shown in figure 1. The choice of the initial profile is unimportant since it does not affect the results qualitatively. The presence of a single coherent structure or a pair of structures, which are imposed initially on the uniform field, causes the average streamwise skin friction to increase due to the convection of the momentum normal to the wall. Figure 2 (a),(b) show the vorticity contours in the cross-stream plane at two times during the evolution of a vortex pair from two finite vortex sheets in the absence of any oscillation; as shown by Orlandi and Jimenez (1994), the straining flow acts to attenuate the convection of the vortices away from the wall, thereby enhancing the interacrion with the wall. In order to examine the effect of spanwise oscillation of the wall on the flow, a spanwise velocity $C_z\cos(2\pi(t + \chi T)/T)$ is added to the base flow, where T and χ are respectively the period and phase of the oscillation. For $C_z = 30$, the corresponding evolution of the vortex pair from the initial state shown in figure 2(a) to a subsequent state at $t^+ = 0.88$ is shown in figure 2(c). The Stokes layer acts to destroy the apparent symmetrical evolution, shown in figure 2(b), in the absence of oscillation. The corresponding contours of streamwise velocity in the cross-plane are shown in figure 3. Comparisons between the skin friction coefficient, c_f, in the presence of oscillation and its value, c_{fo}, in the absence of oscillation, are made in two different ranges of the oscillation period T. For $1 < T < 10$, the evolution is followed for $t^+ = 20$. As can be seen from figure 4, for values of T in this range, spanwise oscillation produces a reduction in skin friction, the maximum reduction increasing with increase in the oscillation period and the vortex Reynolds number, Γ, and c_f/c_{fo} can reduce to 90% as shown. For $50 > T > 400$, computational limitations imply that we cannot follow the evolution long enough to cover several oscillation periods when a quasi-steady state may be reached. Besides, the coherent structures decay through viscous diffusion and do not persist for much longer than $t^+ = 20$. Therefore, for such values of T, initial structures are introduced in a well-established Stokes layer during different phases of the oscillation; this represents the appearance of a coherent streamwise vortex in an oscillatory boundary layer during such a phase. The skin friction is approximated by averaging over these phases. From figure 4, we note that, c_f/c_{f0}, achieves its minimum value for $T^+ = 150$ in the case of vortex pairs and for $T^+ = 80$ in the case of single structures. The oscillation also acts to attenuate the other attributes of turbulent flow, such as the Reynolds stress and the Reynolds stress production term in the energy equation, as shown in figure 5 and 6, consistent with the results of direct numerical simulation due to Jung et al. (1990).

Discussion

The mechanism of suppression or attenuation of turbulence through imposed spanwise wall oscillations can be explained using the present model as follows. As shown above, the dominant feature of the inner layer is the formation and maintenance of long streamwise streaks with momentum burst in the middle and down-wash at each side. When a periodic spanwise cross flow is established, the streaks or the coherent structures are deformed in a way which promotes the interaction of the latter with the rigid surface beneath, leading to a rapid annihilation of the coherent structures and a consequent attenuation of turbulence production. Further, the spanwise oscillations induce spanwise mixing of momentum associated with the 'burst' regions with that associated with the 'sweep' regions, resulting in a reduction in the rate of momentum convection normal to the wall, which in turn has a direct impact on the Reynolds stress in the streamwise direction.

Acknowledgement

This work is supported by the Office of Naval Research under grant N00014-94-1-0453 and by the National Science Foundation under grant BCS-9211 847.

References

Bradshaw, P and N. Pontikos, J. Fluid Mech. 159,105, (1985).
Jimenez, J. and P. Moin, J. Fluid Mech. 225, 221, (1991).
Jung, W., N. Mangiavacchi, and R. Akhavan, Phys. Fluids A4, 1605, (1992).
Kim, J, P. Moin, and R. Moser, J. Fluid Mech. 177, 133, (1987).
Moin, P., T. Shih, D. Driver, and N. Mansour, Phys. Fluids A2, 1846, (1990).
Orlandi, P. and J. Jimenez, Phys. Fluids. 6, 634-641 (1994).
Spalart, P. R., J. Fluid Mech. 187, 61, (1988).

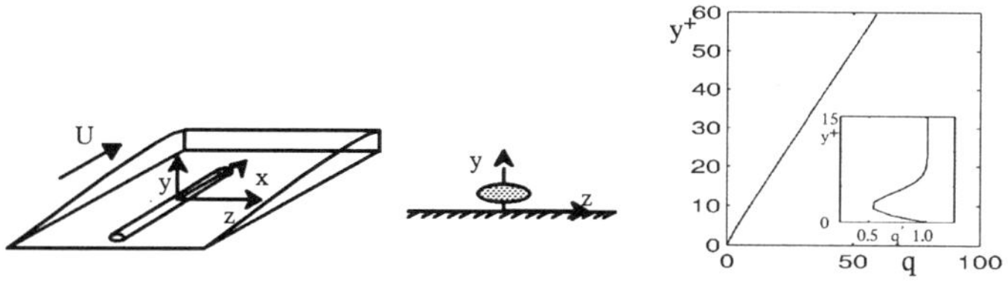

Figure 1. Schematics of the model flow and Initial streamwise velocity $q(y^+)$. The inset shows $q'(y^+)$ in the vicinity of $y^+=0$.

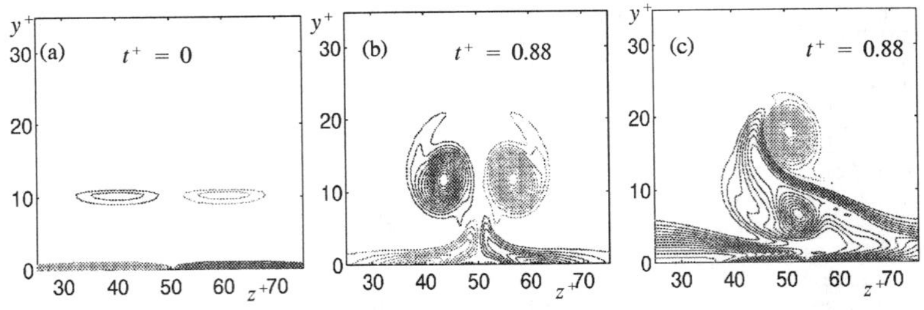

Figure 2. Evolution of a pair of streamwise vortices of opposite strength, at (a) $t^+ = 0$, and at $t^+ = 0.88$ in (b) the absence of any oscillation, and (c) the presence of spanwise surface oscillations with $C_z = 30$. The figures show vorticity contours. $\Gamma = 500$, $a = 0.2$.

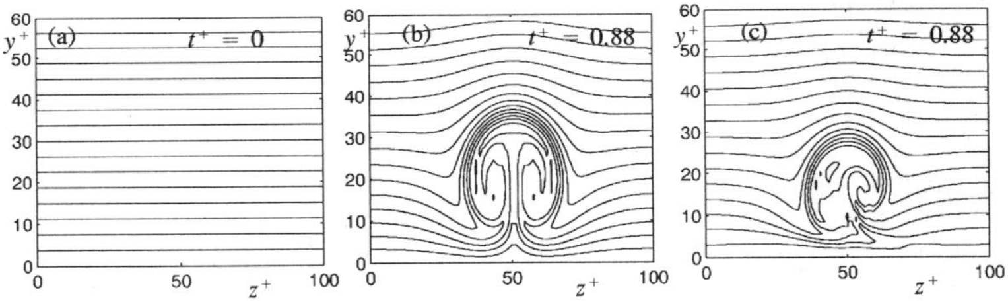

Figure 3. Streamwise velocity contours corresponding to the cases shown in figure 2, showing the influence of spanwise oscillation of the surface beneath the boundary layer on the streamwise velocity streaks. 3(a) in the absence of and 3(b) in the presence of wall oscillation.

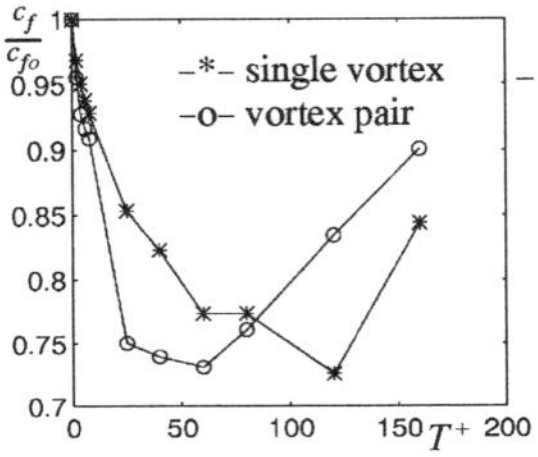

Figure 4. Skin friction reduction due to wall oscillation, as a function of period T^+. c_f is averaged over various phases of the oscillation.

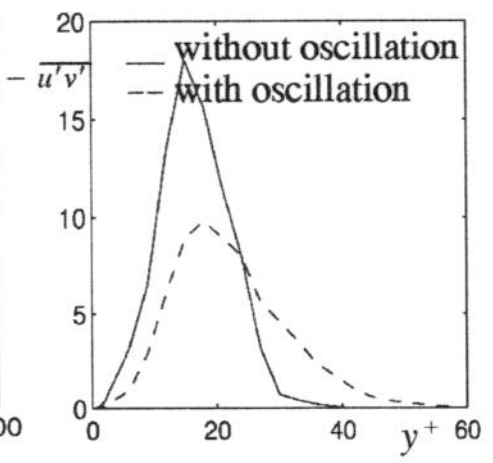

Figure 5. Effect of oscillation on the distribution of the Reynolds stress.

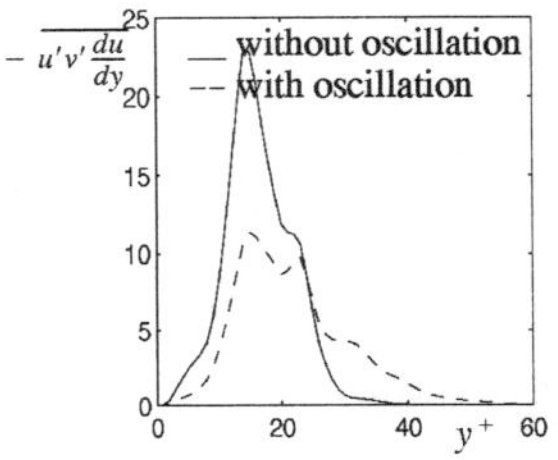

Figure 6. Effect of the oscillation on the distribution of the Reynolds stress production term in the energy equation.

Discussion of Some Grid-Independence Issues in the Context of $k - \epsilon$ and $k - \omega$ Models of Turbulence

Nabil Elkouh, Simone Sebben, and B. Rabi Baliga*

Abstract

Grid-independent numerical solutions obtained with high- and low-Reynolds number $k-\epsilon$ and $k-\omega$ models of turbulence are presented and discussed. Attention is limited to simple fully-developed turbulent flow of a constant-property Newtonian fluid in a parallel-plate channel. Reynolds number values in the range 10,000 to 500,000 are investigated. For each case considered, numerical solutions are obtained with at least six different uniform or nonuniform grids, as appropriate, in conformity with strict convergence criteria. An extrapolation scheme is proposed and used to estimate the grid-independent solutions.

Introduction

The importance of undertaking grid-independence studies in computer simulations of turbulent flows came into sharp focus at the 1980-81 AFOSR-HTTM-Stanford Conference on Complex Turbulent Flows (Wilcox 1993). However, there is still a great need for benchmark numerical solutions of turbulent flow problems, with various available turbulence models. This paper represents an effort to fulfil a small part of this need. Attention in this work is limited to simple fully-developed turbulent flow of a constant-property Newtonian fluid in a parallel-plate channel. Reynolds number values in the range 10,000 to 500,000 are investigated. The objective is to present and discuss grid-independent numerical solutions obtained with high- and low-Reynolds number $k-\epsilon$ and $k-\omega$ models of turbulence, proposed by Launder and Spalding, Jones and Launder, and Wilcox (Wilcox 1993). These two-equation eddy-viscosity based models and their variants are heavily used in engineering design calculations.

*Department of Mechanical Engineering, McGill University, 817 Sherbrooke Street West, Montreal, Quebec, H3A 2K6, Canada.

Mathematical Model and Numerical Method

The Reynolds-averaged x-momentum equation in the fully-developed regime is the following:

$$0 = -\frac{\partial \bar{P}}{\partial x} + \frac{\partial}{\partial y}\left(\mu_{eff}\frac{\partial \bar{u}}{\partial y}\right) \text{ ; where } \mu_{eff} = \mu + \mu_t \quad (1)$$

Here, μ is the dynamic viscosity of the fluid; μ_t is the turbulent or eddy viscosity; x is the streamwise coordinate; y is the cross-stream coordinate; and the origin of this Cartesian coordinate system is located on the top surface of the bottom plate of the channel.

Detailed discussions of the high- and low-Reynolds number $k-\epsilon$ and $k-\omega$ models of turbulence used in this work, and appropriate implementation strategies, are available in the work Wilcox (1993). Due to space restrictions, these discussions will not be repeated in this paper. The governing equations were solved using a finite-volume method (Patankar 1980). A sequential iterative method, based on repeated applications of the tridiagonal matrix algorithm, was used to solve the discretized equations: Convergence was assumed when the sum of the nodal values of the absolute, normalized residues, for each dependent variable, was less than 10^{-7}.

In all cases, $\beta^* = (-d\bar{P}/dx)D_H/(\rho(\nu/D_H)^2)$, the nondimensional pressure gradient, was specified and the Reynolds number, $Re = \rho\bar{u}_{av}D_H/\mu$, and the Darcy friction factor, $ff = (-d\bar{P}/dx)D_H/(0.5\rho\bar{u}_{av}^2)$, were calculated. Here, $D_H = 4H$ is the hydraulic diameter of the channel. Other results presented later in this paper include the following nondimensionalized, centerline ($y/H = 1$) values: $k^* = k/u_\tau^2$; $\epsilon^* = \epsilon H/u_\tau^3$; and $\omega^* = \omega H/u_\tau$, where u_τ is the friction velocity (Wilcox 1993). Results yielded by the experimental correlation of Jones (1976), Re_{Jones} and ff_{Jones}, are also reported.

Extrapolation Procedure

Following the Richardson procedure (Wilcox 1993), any numerical result of interest, say Φ, was assumed to be an analytical function of a grid-size parameter, h, and expressed in a Taylor series expansion about $\mathsf{h} = 0$:

$$\Phi_i = \Phi_{GI} + C_1\mathsf{h}_i + C_2\mathsf{h}_i^2 + C_3\mathsf{h}_i^3 + \ldots \quad (2)$$

where the subscript 'GI' implies extrapolated, or grid-independent, value, obtained in the limit of $\mathsf{h} \to 0$. The series in Eq. (2) was truncated at $C_3\mathsf{h}_i^3$. Extrapolated Φ_{GI} values were then obtained with solutions from at least three successively finer *sets* of four grids, for example, Grids 1-4, Grids 2-5, and Grids 3-6. Grid-independent solutions were assumed only when the three extrapolated values agreed to at least six significant figures.

Results and Discussion

With the high-Reynolds number $k-\epsilon$ model, uniform grids were used in the core region in between the near-wall control volumes. For the most challenging of the cases considered, $Y^+ = 30$ and $\beta^* = 2000\text{x}10^6$, to obtain grid-independent extrapolated results, 2501, 2601, 2701, and 2801 nodes, wall-to-wall, were required. Such results for $\beta^*\text{x}10^{-6} = 2$, 100, and 2000 are given below in Table 1. Re and ff were obtained with $\bar{u}_{av}$ values that were calculated assuming the nodal values of $\bar{u}$ prevailed over their respective control volumes in the core region, and using an exact integration of a two-layer wall function ($u^+ = y^+$ for $0 < y^+ \leq 11.63$; and the log-law for $11.63 < y^+ < Y^+$). Re_P and ff_P were calculated assuming the nodal values of $\bar{u}$ prevailed over their respective control volumes across the whole channel, as is typically done in multidimensional finite-volume computer codes. The Re_P and ff_P values compare better with the corresponding values ofRe_{Jones} and ff_{Jones}, respectively, than the Re and ff values, suggesting that the less accurate prevailing-assumption integration may have influenced the fine-tuning of the high-Reynolds number $k-\epsilon$ model.

$\beta^* = 2\text{x}10^6$; $Re_{Jones} = 10899.8$; $ff_{Jones}\text{x}10^2 = 3.36687$						
Y^+	Re	Re_P	$k^*_{y/H=1}$	$\epsilon^*_{y/H=1}$	$ff\text{x}10^2$	$ff_P\text{x}10^2$
30	10160.2	10555.9	.850103	.700413	3.87488	3.58977
60	9774.26	10445.9	1.50553	1.47671	4.18690	3.66580
90	No meaningful solutions can be obtained:					
120	Wall-normal distance of near wall node exceeds H.					
$\beta^* = 100\text{x}10^6$; $Re_{Jones} = 101100$; $ff_{Jones}\text{x}10^2 = 1.95672$						
Y^+	Re	Re_P	$k^*_{y/H=1}$	$\epsilon^*_{y/H=1}$	$ff\text{x}10^2$	$ff_P\text{x}10^2$
30	101793	102189	.911056	.709791	1.93015	1.91523
60	100649	101320	.906712	.708461	1.97431	1.94822
90	99469.3	100417	.901405	.707240	2.02140	1.98343
120	98342.3	99565.6	.896411	.707094	2.06800	2.01749
$\beta^* = 2000\text{x}10^6$; $Re_{Jones} = 534417$; $ff_{Jones}\text{x}10^2 = 1.40055$						
Y^+	Re	Re_P	$k^*_{y/H=1}$	$\epsilon^*_{y/H=1}$	$ff\text{x}10^2$	$ff_P\text{x}10^2$
30	537737	538133	.923000	.716835	1.38331	1.38128
60	537047	537719	.922586	.716686	1.38687	1.38341
90	535882	536829	.922054	.716491	1.39291	1.38800
120	534618	535841	.921385	.716249	1.39950	1.39312

Table 1: Results obtained with the high-Reynolds number $k-\epsilon$ model.

The absolute percentage differences $(100\text{x}(\Phi_{Y^+=30} - \Phi_{Y^+=120})/\Phi_{Y^+=120})$ in the values of Re, $k^*_{y/H=1}$, $\epsilon^*_{y/H=1}$, and ff are the following, respectively: 0.58%, 0.18%, 0.08%, and 1.2% for $\beta^* = 2000\text{x}10^6$; 3.5%, 1.6%, 0.38%, and 6.7% for $\beta^* = 100\text{x}10^6$; and 4.0%, 44%, 53%, and 7.5% for $\beta^* = 2\text{x}10^6$, where the results for $Y+ = 60$, rather than $Y^+ = 120$, have been used. These results show that the influence of Y^+ can be substantial, and simulations based on the wall-function treatment must be treated with a healthy dose of skepticism.

With the low-Reynolds number $k-\epsilon$ and $k-\omega$ models, first, for each specified value of β^*, a relatively coarse, geometrically expanding, nonuniform grid, symmetric about $y/H = 1.0$, was constructed, with $y^+ < 1.0$ for the first near-wall node. In subsequent grid-refinement studies, the number of coarse-grid control volumes ($NCGCV$) and the geometric expansion factor ($GFAC$) were maintained constant; and each control volume, or interval, of the coarse grid was subdivided into an integral number of control volumes ($NCVCGI$). With this pattern of nonuniform grids, the proposed extrapolation scheme can be used assuming that $h_i = 1/NCVCGI_i$. Computational experiments with $\beta^* = 2000\text{x}10^6$, the most challenging of the cases considered, showed that to obtain extrapolated results that are invariant to six significant figures, grids of $NCGCV = 45$, $GFAC = 1.4$, and $NCVCGI = 11, 13, 15$, and 17, wall-to-wall, are needed. Such results are presented Tables 2 and 3.

$\beta^*\text{x}10^{-6}$	Re	Re_{Jones}	$ff\text{x}10^2$	$k^*_{y/H=1}$	$\epsilon^*_{y/H=1}$
2	11015.0	10899.8	3.29681	.964646	.746450
100	101415	101100	1.94458	.953422	.733708
500	531567	534417	1.41561	.933322	.722789

Table 2: Results obtained with the low-Reynolds number $k-\epsilon$ model.

$\beta^*\text{x}10^{-6}$	Re	Re_{Jones}	$ff\text{x}10^2$	$k^*_{y/H=1}$	$\omega^*_{y/H=1}$
2	10422.7	10899.8	3.68211	.710224	.673491
100	100340	101100	1.98647	.786566	.643581
500	533214	534417	1.40688	.797759	.639358

Table 3: Results obtained with the low-Reynolds number $k-\omega$ model.

Acknowledgements

We are grateful to the Natural Sciences and Engineering Research Council (NSERC) of Canada, for awarding a fellowship to Nabil Elkouh and a Research Grant to Professor B. Rabi Baliga. Financial support from the Brazilian government (CNPq scholarship to Simone Sebben) and from Hydro Québec (McGill Major fellowship to Nabil Elkouh) are also gratefully acknowledged. Support of Professor Baliga's research activities by the Centre de Recherche en Calcul Appliqué (CERCA) in Montreal is deeply appreciated.

References

Jones, O.C. (1976). "An Improvement in the Calculation of Turbulent Friction Factors in Rectangular Ducts." *ASME J. Fluids Eng.*, 98, 173-181.

Patankar, S.V. (1980). *Numerical Heat Transfer and Fluid Flow.* McGraw-Hill.

Wilcox, D.C. (1993). *Turbulence Modelling for CFD.* DCW Industries, Inc.

A Secondary Flow Correction for
Depth-Averaged Flow Calculations

John Finnie[1],
Barbara Donnell, Joe Letter, and Robert S. Bernard[2]

Abstract

A secondary flow (or "bendway") correction has been added to the RMA2 computer program of the U.S. Army Corps of Engineers. The modified program solves a transport equation for streamwise vorticity and converts it to accelerations due to secondary currents. These additional accelerations result in improved predictions of depth-averaged velocity. In particular, their effect is to reduce depth-averaged velocities on the inside of curves and increase them on the outside of curves. Results of the bendway correction are compared to data from an experiment conducted at the Riprap Test Facility at the Waterways Experiment Station.

Introduction

The RMA2 computer program is a depth-averaged hydrodynamic, finite element computer program used by the US Army Corps of Engineers to model river and estuary flows. It computes horizontal velocities and water surface elevations for subcritical, free-surface flow in 2D flow fields. The bendway correction of Bernard and Schneider (1992) has been found to greatly improve the prediction of depth-averaged velocities in curved sections of rivers and estuaries, and will be applied to RMA2.

Theory

---

[1]Department of Civil Engineering, University of Idaho, Moscow, ID, USA 83844-1022, (208) 885-7559

[2]Hydraulics Laboratory, Waterways Experiment Station, Corps of Engineers, 3909 Halls Ferry Road, Vicksburg, MS, USA 39180-6199

The equations governing depth-averaged, free surface flows are

$$h \frac{\partial u}{\partial t} + hu \frac{\partial u}{\partial x} + hv \frac{\partial u}{\partial y} - \frac{h}{\rho}\left(\epsilon_{xx} \frac{\partial^2 u}{\partial x^2} + \epsilon_{xy} \frac{\partial^2 u}{\partial y^2} \right) + hg \frac{\partial a}{\partial x} + hg \frac{\partial h}{\partial x} - \tau_x + s_x = 0 \tag{1}$$

$$h \frac{\partial v}{\partial t} + hu \frac{\partial v}{\partial x} + hv \frac{\partial v}{\partial y} - \frac{h}{\rho}\left(\epsilon_{yy} \frac{\partial^2 v}{\partial x^2} + \epsilon_{xy} \frac{\partial^2 v}{\partial y^2} \right) + hg \frac{\partial a}{\partial x} + hg \frac{\partial h}{\partial y} - \tau_y + s_y = 0 \tag{2}$$

$$\frac{\partial h}{\partial t} + h \left(\frac{\partial u}{\partial x} + \frac{\partial v}{\partial y} \right) + u \frac{\partial h}{\partial x} + u \frac{\partial h}{\partial y} = 0 \tag{3}$$

where h is the local depth, u and v are the depth-averaged velocities in the x and y directions, respectively, t is time, ρ is mass density, ε is eddy viscosity, g is the acceleration of gravity, a is the bottom elevation, S_X and S_Y are the stresses that result from the non-uniformity of u and v in the vertical direction, and τ_X and τ_Y are external tractions (bottom friction, wind, and coriolis effects).

In order to evaluate S_X and S_Y, the shear stress due to secondary currents must be calculated. This stress is determined from the following equation (Johannesson and Parker, 1989), $\tau_S = \rho\, h\, |\underline{u}|\, \Omega \sqrt{C_f}$, where C_f is a coefficient of friction, $|\underline{u}|$ is the magnitude of the vector $\underline{u}$ and $\Omega = C_2\, \omega_S / 12$, and the subscript s indicates the streamwise direction. C_2 is an experimentally determined coefficient, and

$$\omega_s = \frac{1}{2} \left(\frac{\partial w}{\partial y} - \frac{\partial v}{\partial z} \right) \tag{4}$$

The value of Ω is calculated from the non-linear, second order transport equation

$$\frac{\partial \Omega}{\partial t} + u \frac{\partial \Omega}{\partial x} + v \frac{\partial \Omega}{\partial y} = \frac{A_s \sqrt{C_f}\, |\underline{u}|^2}{Rh\ (1 + 9\, h^2 / R^2)} - D_s \sqrt{C_f}\, \Omega \frac{|\underline{u}|}{h} + \frac{1}{h} \nabla \cdot (\nu\, h\, \nabla \Omega) \tag{5}$$

where R is the local radius of curvature and $A_S = 5.0$ and $D_S = 0.5$. These two constants have been evaluated from experimental data and include the coefficients C_1 and C_2. The local radius of curvature can be calculated by

$$R = \frac{|\underline{u}|^3}{uv(v_y - u_x) + u^2 v_x - v^2 u_y} \tag{6}$$

where the subscripts x and y indicate derivatives.

Boundary and initial conditions are needed for equation (5). At the inlet, Ω is assumed to be zero. At the sidewalls and outlets, the gradient of Ω is zero. Since the secondary flow disappears on the sidewalls, the production of Ω is reduced by one-half in elements next to the sidewalls.

Once the value of Ω is known, τ_S is evaluated. The components of $\underline{S}$ (the streamwise acceleration caused by this shear stress) are given in the following equation.

$$\underline{S} \approx \rho^{-1} \frac{\underline{u}}{|\underline{u}|} \left[\underline{n} \cdot \nabla (h\tau_s) + \frac{2h\tau_s}{R} \right] \tag{7}$$

where $\underline{n}$ is the unit vector normal to the vector $\underline{u}$. The normal derivative of the depth-averaged secondary shear stress is calculated as

$$\frac{\partial (h\tau_s)}{\partial R} = \underline{n} \cdot \nabla (h\tau_s) = \frac{v(h\tau_s)_x - u(h\tau_s)_y}{|\underline{u}|} \tag{8}$$

The secondary stress is positive in the direction of motion. The sign of R is positive for counterclockwise motion and negative for clockwise motion. The normal derivative is positive when $h\tau_S$ increases in the positive radial direction.

Results

The modified RMA2 computer program was tested against an experiment conducted at the Riprap Test Facility of the Waterways Experiment Station (WES) by Maynord (1987). The Riprap Test Facility is an outdoor channel approximately 900 feet long, and in plan view looks like the capital letter "B". The channel is 3.63 M (11.9 feet) wide with 2H:1V side slopes, Manning's n = .026, and bottom slope is .00216. The flow for this experiment was 4.25 cms (150 cfs). The channel is lined with stone riprap varying in diameter from 13 to 64mm (0.5 to 2.5 inches).

Comparisons between the experiment, and the calculations of the original RMA2, and RMA2 with the bendway correction are shown in Figure 1. For these calculations, the eddy viscosity was set to 29 N-sec/m^2 (0.6 lb-sec/ft^2).

In order to test the effect of eddy viscosity on the predictions, eddy viscosity was varied from 10 to 86 N-sec/m^2 (0.2 to 1.8 lb-sec/ft^2) and the calculation repeated for each new value of eddy viscosity. Figure 2 shows the results of this sensitivity analysis at a typical station.

Acknowledgements

This work was supported by the Engineer Waterways Experiment Station under the auspices of the U.S. Army Research Office Scientific Services Program administered by Battelle. The views, opinions, and/or findings contained in this report are those of the authors and should not be construed as official Department of the Army position, policy, or decision, unless so designated by other documentation.

References

Bernard, R.S, and Schneider, M.L. (1992). "Depth-Averaged Numerical Modeling for Curved Channels," Technical Report HL-92-9, U.S. Army Engineer Waterways Experiment Station, Vicksburg, MS.

Johannesson, H., and Parker, G. (1989). "Velocity Redistribution in Meandering Rivers," Journal of Hydraulic Engineering, Vol. 115, No. 8, pp 1019-1039.

Maynord, S.T. (1987) Unpublished Test Data, US Army Engineer Waterways Experiment Station, Vicksburg, MS.

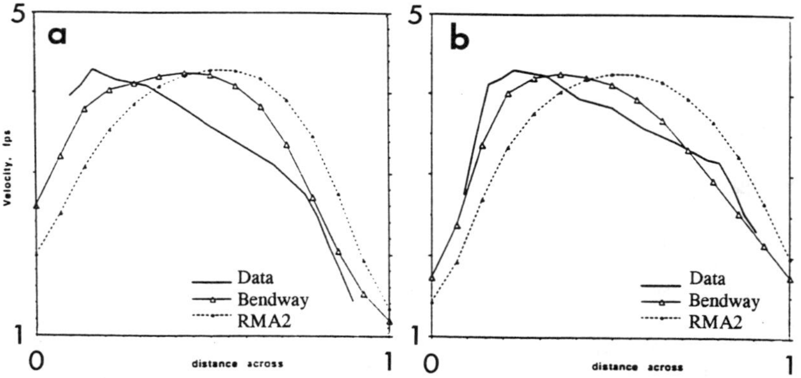

Figure 1 Comparison of Experimental and Computed Velocity, with and without the Bendway Correction Vorticity Equation: a) Station 2+81, b) Station 3+31

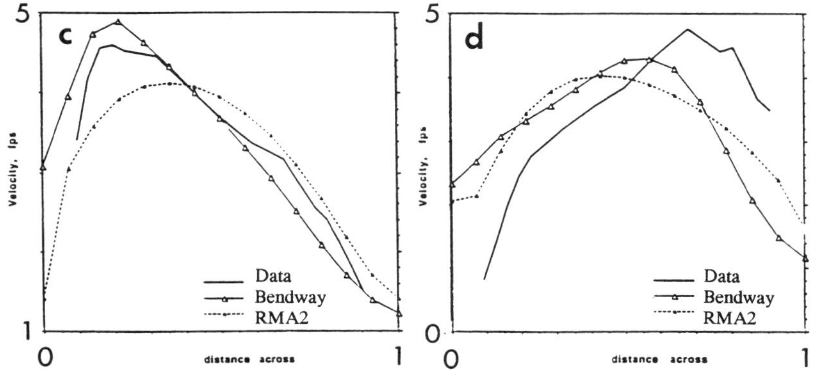

Figure 1 Comparison of Experimental and Computed Velocity, with and without the Bendway Correction Vorticity Equation: c) Station 4+97, d) Station 5+78

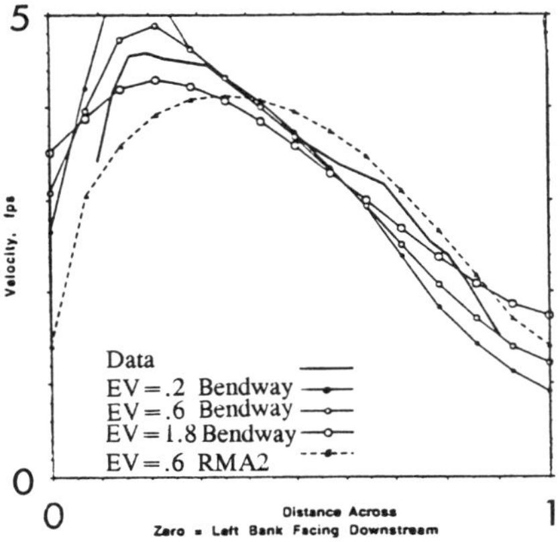

Figure 2 Effect of Eddy Viscosity on Computed Velocities, Station 4+97

CONSISTENT INFINITESIMAL FINITE-ELEMENT CELL METHOD FOR AN ANISOTROPIC UNBOUNDED MEDIUM

CHONGMIN SONG[1] and JOHN P. WOLF[1]

Abstract

The consistent infinitesimal finite-element cell method is applied to anisotropic unbounded media where the fundamental solution used in the boundary-element procedure is very complicated.

Concept

To analyze dynamic medium-structure interaction by the substructure method, the interaction force-motion relationship in the degrees of freedom on the structure-medium interface of the unbounded medium is required which is described in the frequency domain by the *dynamic-stiffness matrix* $[S^{\infty}(\omega)]$ and in the time domain by the *acceleration unit-impulse response matrix* $[M^{\infty}(t)]$. $[S^{\infty}(\omega)]$ or $[M^{\infty}(t)]$ is calculated with the *consistent infinitesimal finite-element cell method.* As only the structure-medium interface is discretized and the formulation is based solely on finite elements, the method is a *boundary finite-element procedure.*

The consistent infinitesimal finite-element cell method is described in great detail in the book by Wolf and Song (1996), where partial differential equations of the hyperbolic type (e.g. wave propagation in time domain), the parabolic type (e.g. diffusion in time domain) and of the elliptic type (statics and e.g. wave propagation in frequency domain) are addressed. The application to bounded media is also examined. This paper describes only the concept and demonstrates the high versatilities and accuracy in modeling anisotropic unbounded media. For the derivations and a literature survey the reader is referred to Wolf and Song (1996).

The consistent infinitesimal finite-element cell method is based on assemblage of similar unbounded media and a finite-element cell for which the limit of the infinitesimal cell width is performed analytically (Figure 1), resulting in the discretization of the structure-medium interface only.

[1]Engineer, Institute of Hydraulics and Energy, Department of Civil Engineering, Swiss Federal Institute of Technology Lausanne, CH-1015 Lausanne, Switzerland

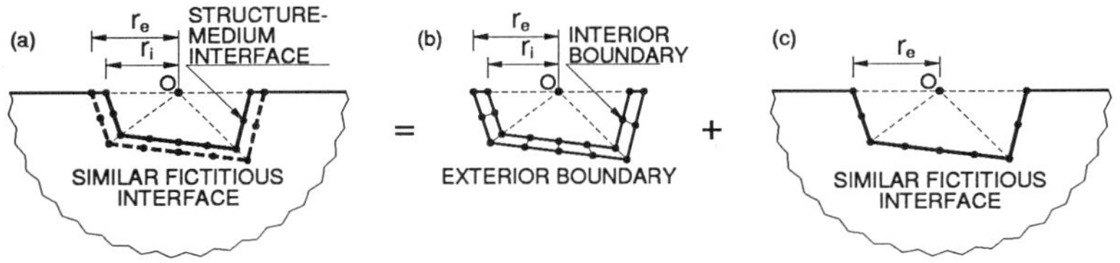

Figure 1. Concept of Consistent Infinitesimal Finite-Element Cell Method with Infinitesimal Cell Width Leading to Finite-Element Discretization of Structure-Medium Interface Only

Characteristic Features

The consistent infinitesimal finite-element cell method is a *stand-alone finite-element formulation* capable of capturing the radiation condition at infinity without using analytical solutions. Only the coefficient matrices of the surface finite elements are calculated, which are then used in standard matrix operations to obtain the unit-impulse response matrix of the unbounded medium. The method can also calculate problems for which the fundamental solution (which is necessary in the boundary-element method) does not exist in closed form as for certain anisotropic materials without increasing the computational effort.

A *reduction of the spatial dimension by one* in the discretization results as in the boundary-element method. When a boundary extending from the structure-medium interface to infinity (such as a half-space with a free surface) exists, this novel method automatically incorporates this boundary condition in contrast to the boundary-element method. Material inhomogeneities which satisfy similarity can also be processed without any additional effort. As no other approximation than that of the finite-element method is introduced, the consistent infinitesimal finite-element cell method *converges to the exact solution in the finite-element sense in the circumferential directions.* Furthermore, as the limit of the infinitesimal cell width is performed analytically, the method is *exact in the radial direction.* The consistent infinitesimal finite-element cell method is thus a *boundary finite-element method* combining the advantages of the boundary-element method and the finite-element method.

The consistent-boundary method (Kausel, Roësset and Waas, 1975) to analyze a horizontally layered medium is a special case of the consistent infinitesimal finite-element cell method where the characteristic lengths of the interior and exterior boundaries are the same.

Examples

Only a summary demonstrating the accuracy and versatility of the consistent infinitesimal finite-element cell method is presented. For details Wolf and Song (1996) should be consulted.

First, a rectangle ($e = b$) embedded in a transversely isotropic half-plane with 5 material constants presented in Figure 2 is calculated with 24 3-node line elements on the structure-medium interface. For a rigid interface the rocking coefficient $M^{\infty}(t)$ is determined which is then transformed to the dynamic-stiffness

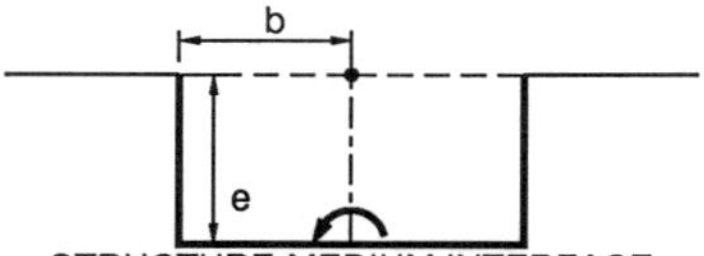

Figure 2. Strip Foundation with Rectangular Cross Section Embedded in Half-Plane

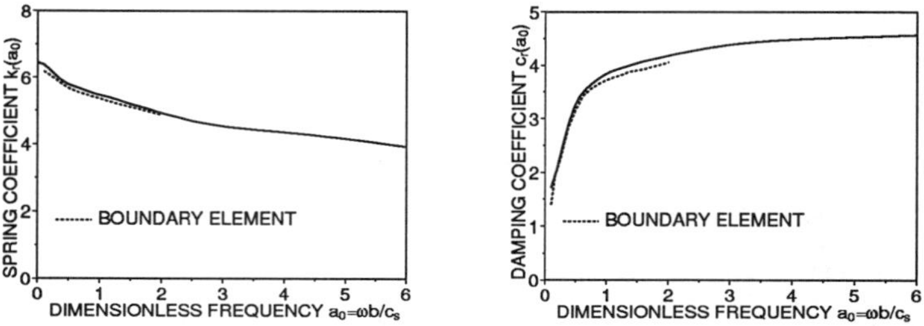

Figure 3. Rocking Dynamic-Stiffness Coefficient of Rigid Strip Foundation

coefficient $S^{\infty}(\omega)$ in the frequency domain. The latter is decomposed into the spring coefficient $k(a_0)$ and the damping coefficient $c(a_0)$ with the dimensionless frequency a_0. Excellent agreement (Figure 3) with the boundary-element solution of Wang and Rajapakse (1991) results.

Second, a prism ($e = 2/3b$) embedded in a transversely isotropic half-space shown in Figure 4a is analyzed with the finite-element mesh of the structure-medium interface of Figure 4b. For a rigid interface, the horizontal and rocking $M^{\infty}(t)$ are plotted in Figure 5. Hardly any deviation from the results of an extended mesh analysis exists.

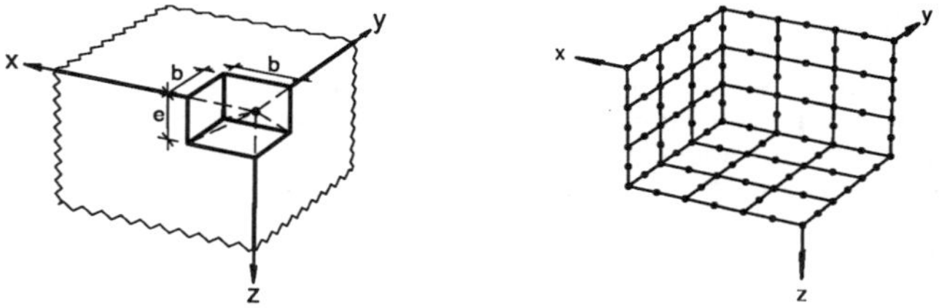

(a) Geometry with Similarity Center (b) Finite-Element Discretization of Structure-Medium Interface

Figure 4. One Quarter of Square Prism Embedded in Half-Space

Third, a railroad tunnel intersecting a fault (zone C) with 2 different trans-

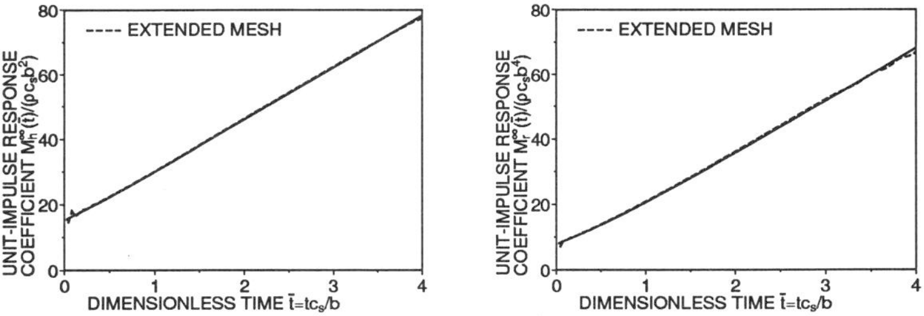

(a) Horizontal (b) Rocking

Figure 5. Acceleration Unit-Impulse Response Coefficients of Rigid Prism Embedded in Half-Space

versely isotropic zones A and B shown in Figure 6a is analyzed for plane strain with 31 3-node line elements on the structure-medium interface. The static displacements of the wall relative to the center of the track for the applied loads are plotted in Figure 6b. Excellent agreement with the results of an extended mesh analysis is obtained.

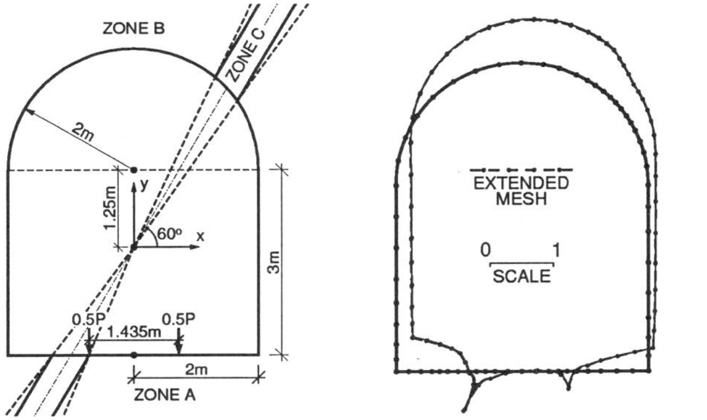

(a) Geometry with Similarity (b) Deformed Tunnel Wall

Figure 6. Tunnel in Inhomogeneous Unbounded Rock

Appendix References

Kausel, E., Roësset, J. M. and Waas, G. (1975). "Dynamic analysis of footings on layered media." *Journal of Engineering Mechanics*, ASCE, Vol.101, 679–693.

Wang, Y. and Rajapakse, R. K. N. D. (1991). "Dynamics of rigid strip foundations embedded in orthotropic elastic soils." *Earthquake Engineering and Structural Dynamics*, Vol.20, 927–947.

Wolf, J. P. and Song, Ch. (1996). *Finite-Element Modelling of Unbounded Media*, John Wiley & Sons Ltd.

DYNAMIC RESPONSE OF FLEXIBLE RETAINING WALLS

A. H. Younan,[1] Associate Member, ASCE, and A. S. Veletsos,[2] Member ASCE

Abstract

A study of the effects of wall flexibility on the response to horizontal ground shaking of cantilever walls retaining a uniform viscoelastic stratum of constant thickness and infinite extent in the horizontal direction is made. The response quantities examined include the total dynamic wall force and its line of application.

Introduction

The recent earthquake in Kobe, Japan provided further confirmation of the fact that the dynamic response of retaining walls is far from being well understood. The objective of the present study is to assess the effects of wall flexibility on the response of such systems considering the wall to respond as a cantilever beam and the soil as a linear viscoelastic material. The effects of both low-frequency, quasi-static excitations and of an actual earthquake ground motion are examined.

As far as it can be determined, the only previous study of flexible retaining walls is the one reported by Sun and Lin (1995), who employed an extension of the authors' analysis for rigid walls (Veletsos and Younan, 1994a). However, the limited numerical data presented are believed to be in error. In particular, in their expressions for the pressures induced on rigid walls, the factor ε^3 should appear without the exponent.

System Considered

The system examined consists of a semiinfinite, uniform layer of viscoelastic material of height H that is free at its upper surface, is bonded to a rigid base, and is retained along one of its vertical boundaries by a prismatic wall of the same height. The bases of both the stratum and the wall are presumed to experience a space-invariant horizontal motion, the acceleration of which at any time t is $\ddot{x}_g(t)$, and its maximum value is $\ddot{X}_g$. Material damping for the medium is considered to be of the constant hysteretic type.

The properties of the soil layer are defined by its mass density ρ_s, shear modulus of elasticity G_s, Poisson's ratio ν_s, and the material damping factor δ_s, which is considered to be frequency-independent and the same for both shearing and axial deformations. The wall is modeled as a flexural cantilever beam rigidly attached to the base.

[1] Post-Doctoral Fellow, Dept. of Civil Engrg., Rice Univ., Houston, Texas 77005-1892

[2] Brown & Root Prof., Dept. of Civil Engrg., Rice Univ., Houston, Texas

Its thickness is denoted by t_w, and its physical properties are defined by Young's modulus of elasticity E_w, Poisson's ratio ν_w, and the material damping factor δ_w. Its mass is considered to be negligible compared to the participating soil mass.

Method of Analysis

Fundamental to the analysis employed is the assumption that, under the horizontal excitation considered, no vertical normal stresses develop anywhere in the medium, i.e. $\sigma_y = 0$. It is further assumed that the horizontal variations of the vertical displacement of the medium are negligible, so that the horizontal shearing stresses τ_{xy} can be expressed as $\tau_{xy} = G_s^*(\partial u/\partial y)$, where u is the horizontal displacement of an arbitrary point of the medium relative to the moving base, $G_s^* = G_s(1 + i\delta_s)$ is the complex-valued shear modulus, and $i = \sqrt{-1}$. The reliability of these assumptions has been confirmed by comparing the solutions for rigid walls (Veletsos and Younan, 1994b) with those obtained by Wood's 'exact' solution (1973).

The steps involved in the analysis are as follows: First, the horizontal displacements of the medium relative to the moving base are expressed as a linear combination of the natural modes of vibration of the stratum when it is considered to act as an unconstrained, vertical cantilever shear-beam. Similarly, the relative displacements of the wall are expressed as a linear combination of its own natural modes, i.e. those of a clamped-free beam. These expressions satisfy the boundary conditions of zero displacement at the rigid base, and zero stresses and forces at the top surface. Next, the differential equation governing the motion of the medium in the horizontal direction is solved subject to the condition that the displacements of the stratum and wall are the same at their interface. For the purpose of satisfying this boundary condition, each natural mode of vibration of the beam is expanded in terms of the corresponding modes of the stratum. The result of this step are medium displacements and wall pressures that are functions of the as yet undetermined wall displacements. Finally, the wall displacements are evaluated by satisfying the dynamic equilibrium of the forces acting on the beam using Lagrange's equations of motion. The analysis is first implemented for harmonic excitations. The response to arbitrary transient ground motions is then evaluated by Fourier transform techniques.

Presentation of Results

Consideration is first given to the responses obtained for excitations the dominant frequencies of which are small compared to the fundamental natural frequency of the soil-wall system. Such excitations and the resulting effects will be referred to as 'static', a term that should not be confused with that normally used to represent the effects of gravity forces.

Static Effects: The 'static' value of the total force or base shear per unit of wall length, $(Q_b)_{st}$, is plotted in Fig. 1, along with the distance h from the base to the line of action of the resultant wall force. Referred to as the effective height, h represents the distance by which $(Q_b)_{st}$ must be multiplied to obtain the static value of the overturning base moment per unit of wall length. The results are normalized with respect to $\rho \ddot{X}_g H^2$ and H, respectively, and are plotted against the dimensionless measure of the relative flexibility of the wall and retained medium $d_w = G_s H^3/D_w$, where $D_w = E_w t_w^3/12(1-\nu_w^2)$. The range of d_w values considered is representative of that encountered in practice. Poisson's ratio for the retained material in these solutions is considered to be $\nu_s = 1/3$.

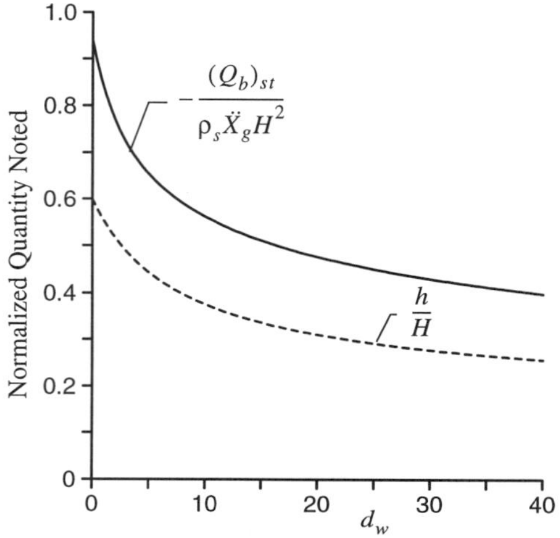

Fig. 1 Effects of Wall Flexibility on Static Values of Base Shear and Effective Height; $\nu_s = 1/3$.

It is observed that, by reducing the horizontal extensional stiffness of the retained material relative to its shearing stiffness, the flexibility of the wall decreases the magnitudes of the resulting pressures on and forces in the wall. An increase in wall flexibility also reduces the effective height h, with the result that the reduction in the overturning base moment is generally much greater than that for the corresponding shear. These reductions are substantial even for rather small values of d_w. For example, $(Q_b)_{st}$ is reduced from $0.941\rho\ddot{X}_g H^2$ for a rigid wall to $0.561\rho\ddot{X}_g H^2$ for a system with $d_w = 10$. The corresponding reduction in effective height is from $0.6H$ to $0.375H$, and in base moment is from $0.565\rho\ddot{X}_g H^3$ to $0.210\rho\ddot{X}_g H^3$.

Dynamic Effects: Figure 2 shows normalized values of the maximum dynamic base shear per unit of length, $|(Q_b)_{max}|$, induced in the walls of systems excited by the first 6.3 sec of the N-S component of the 1940 El Centro earthquake ground motion record, for which $\ddot{X}_g = 0.312$g. The results are plotted for four values of the relative flexibility factor d_w as a function of T_1, the fundamental period of the stratum when it is considered to act as an unconstrained, cantilever shear-beam. The latter period is defined by $T_1 = 4H/v_s$, where $v_s = \sqrt{G_s/\rho_s}$ is the shear-wave velocity of the medium. As before, the wall in these solutions is presumed to be massless, $\nu_s = 1/3$, and the damping factors for the soil and wall are $\delta_s = 0.1$ and $\delta_w = 0.04$ (or 2 percent of critical damping).

The plots in Fig. 2 are similar to, but by no means the same as, the response spectra for similarly excited, viscously damped single-degree-of-freedom systems. Specifically, for low-natural-period, stiff strata, the maximum base shears are equal to their static values. With increasing flexibility of the stratum, the base shears increase, and after attaining nearly horizontal plateaus, they reach values that may be substantially lower than the static values. In all cases, the forces for the flexible wall are lower than for the corresponding rigid wall. Although not shown, the effective height h is

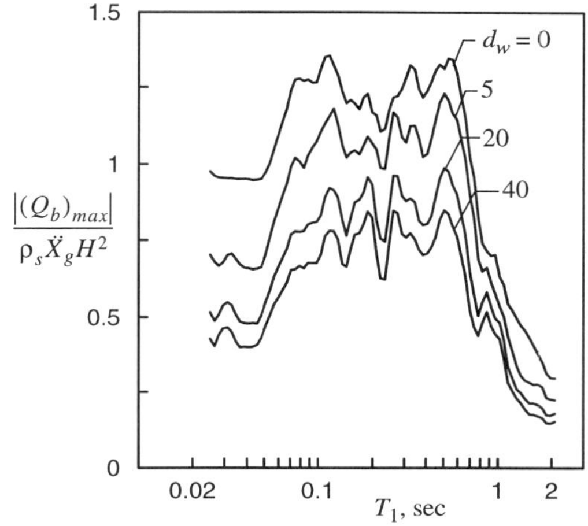

Fig. 2 Maximum Values of Dynamic Base Shear in Walls of Systems Excited by El Centro Record; $\nu_s = 1/3$, $\delta_s = 0.1$ and $\delta_w = 0.04$.

relatively insensitive to variations in T_1, and may, for all practical purposes, be assigned the values applicable to the statically excited systems.

In the practically important period range of $T_1 = 0.1$ to 0.5 sec, in which the base shear values are nearly constant, the average amplification factor (defined as the ratio of the dynamic to the corresponding 'static' values) range from 1.43 for a rigid wall to 1.85 for a wall with $d_w = 5$, and to 2.04 and 2.12 for walls with values of d_w equal to 20 and 40, respectively.

Conclusions

The dynamic forces induced by ground shaking in flexible walls retaining a viscoelastic stratum may be significantly lower than those for non-deflecting rigid walls. It is hoped that the information presented will lead not only to a more rational interpretation of the results of test programs but also to a greater appreciation of the value of elastic solutions for the problem examined.

Acknowledgment

This study was carried out under Project 568821 with Brookhaven National Laboratory, Upton, New York. This support is acknowledged gratefully.

Appendix I: References

Sun, K. and Lin, G. (1995). "Dynamic response of soil pressure on retaining wall." *Proc. Third Int. Conf. on Recent Advances in Geotech. Earthquake Engrg. and Soil Dyn.*, University of Missouri, Rolla, Mo.,Vol. I, 347-350.

Veletsos, A. S. and Younan, A. H. (1994a). "Dynamic soil pressures on rigid retaining walls," *Earthquake Engrg. & Struct. Dyn.*, 23(3), 275-301.

Veletsos, A. S. and Younan, A. H. (1994b). "Dynamic modeling and response of soil-wall systems." *J. Geotech. Engrg.,* ASCE, 120(12), 2155-2179.

Wood, J. H. (1973). "Earthquake-induced soil pressures on structures." *Report EERL 73-05*, Earthquake Engrg. Research Lab., Cal Tech, Pasadena, CA.

Dynamic Interaction Between Embedded Foundations
by the Substructure Deletion Method

Raimondo Betti[1], Euclides de Mesquita Neto[2] and Edivaldo Romanini[3]

Abstract

This paper presents a detailed analysis of the effects of cross-interaction through the soil for embedded foundations using the Substructure Deletion method. First, a Boundary Element formulation of the substructure deletion method to determine the impedance matrix for multiple embedded foundations of arbitrary shape is presented. Then, the two-dimensional and three-dimensional cases of two identical embedded foundations are analyzed, considering the effects of the relative distance between the foundations on the value of the impedance functions.

Introduction

In the calculation of the dynamic response of closely spaced foundations, it is extremely important to consider not only the interaction effects between the individual foundation and the surrounding soil but also the cross--interaction through the soil between adjacent foundations. Many studies have been done in order to investigate the response of both superficial and embedded foundations in uniform or layered half--space, subjected to external excitations, using various techniques applied in the analysis of a single foundation. Unfortunately, as for the single foundation case, theoretical and computational limitations have constrained the analysis only to simple geometries for the foundations. Using the Substructure Deletion Method (SDM) as presented in this study, the analysis of the dynamic response of multiple embedded foundations of arbitrary shape can be easily performed, keeping the computation efforts within reasonable limits.

---

[1]Asst.Prof., Dept. Civil Engr. & Engr. Mech., Columbia University, New York, NY 10027, USA

[2]Professor, DMC/FEM/UNICAMP - C.P. 6122 - CEP 13083-070 - Campinas, SP, Brazil

[3]DEX/CEUL/UFMS

Impedance Matrix for Multiple Foundations

The analysis of multiple embedded foundations, bonded to the soil, can be easily performed by applying the SDM. The basic consideration of the SDM (Betti (1994), Mesquita (1995)) is that, if the volumes V_i^b (i = 1,2,.... N_f), with N_f being the total number of foundations) are inserted into the cavities occupied by the foundations, and the continuity conditions are imposed along the embedded surfaces S_i^e (i=1,2,... N_f), the flat unexcavated half-space is recreated (Fig. 1). Hence the solution of the radiation problem in the case of embedded foundations can be obtained from the solution of the unexcavated half-space (exterior problem) and from the analysis of the various bounded domains V_i^b (interior problems). In this study, each bounded excavated domain V_i^b is analyzed using a Direct version of the Boundary Element Method (DBEM) with constant elements. For the generic i-th removed bounded domain, the force-displacements relation can be expressed as:

$$\begin{Bmatrix} \mathbf{u}_h^i \\ \mathbf{u}_e^i \end{Bmatrix} = \begin{bmatrix} C_{hh}^i & C_{he}^i \\ C_{eh}^i & C_{ee}^i \end{bmatrix} \begin{Bmatrix} \mathbf{t}_h^i \\ \mathbf{t}_e^i \end{Bmatrix}$$

where $\mathbf{u}_h^i$ and $\mathbf{t}_h^i$ are the displacement and traction vectors, respectively, for points on the horizontal surface S_i^h , while $\mathbf{u}_e^i$ and $\mathbf{t}_e^i$ are associated with points on the embedded surface S_i^e . The matrix $[C^i]$ represents the compliance matrix for the i-th soil inclusion.

The compatibility conditions for the displacement and the equilibrium conditions must be satisfied over the embedded surface $\hat{S}_t^e$, defined as:

$$\hat{S}_t^e = \hat{S}_1^e \cup \hat{S}_2^e \cup \ldots \cup \hat{S}_{N_f}^e$$

The boundary conditions on each embedded surface $\hat{S}_i^e$ are represented by:

$$\hat{\mathbf{u}}_e^i = \mathbf{u}_e^i$$
$$\hat{\mathbf{t}}_e^i + \mathbf{t}_e^i = 0$$

where now $\hat{\mathbf{u}}_e^i$ and $\hat{\mathbf{t}}_e^i$ are the displacement and traction vectors, respectively, for points on $\hat{S}_i^e$ in the excavated unbounded soil. The final compliance matrix $[\hat{C}]$, which provides the required force-displacement relationship for the excavated infinite domain, is given by:

$$[\hat{C}] = -[C_{eh}]\,[\,[C_{hh}^f] - [C_{hh}]\,]^{-1}[C_{he}] - [C_{ee}]$$

where the half-space matrix $[C_{hh}^f]$ is a full matrix and is obtained from the analysis of the flat surfaces S_i^h (i =1,2,...., N_f). The matrices $[C_{hh}]$, $[C_{he}]$, $[C_{eh}]$, $[C_{ee}]$ represent diagonal matrices, whose general expression is given by:

$$[C_{lk}] = \begin{bmatrix} [C_{lk}^1] & & & \\ & [C_{lk}^2] & & \\ & & \ddots & \\ & & & [C_{lk}^{N_f}] \end{bmatrix} \qquad (l,k=e,h)$$

By taking the inverse of the compliance matrix, it is possible to obtain the impedance matrix $[\hat{K}]$ for the multiple foundation system. For the case of two adjacent foundations, the compliance matrix $[\hat{C}]$ has the following expression:

$$[\hat{C}] = -\begin{bmatrix} [C_{eh}^1] & [0] \\ [0] & [C_{eh}^2] \end{bmatrix} \left[\begin{bmatrix} C_{hh}^{f11}] & [C_{hh}^{f12}] \\ [C_{hh}^{f21}] & [C_{hh}^{f22}] \end{bmatrix} - \begin{bmatrix} [C_{hh}^1] & [0] \\ [0] & [C_{hh}^2] \end{bmatrix} \right]^{-1} \begin{bmatrix} [C_{he}^1] & [0] \\ [0] & [C_{he}^2] \end{bmatrix} - \begin{bmatrix} [C_{ee}^1] & [0] \\ [0] & [C_{ee}^2] \end{bmatrix}$$

Two Rigid, Embedded Foundations Case

The three-dimensional analysis of two rigid identical foundations, at distance l, embedded in a homogeneous half--space, has been investigated in order to point out the effects of the cross-interaction through the soil. Each foundation has a square base, of dimension 2a X 2a, and it is embedded in the soil at a depth h. The distance l between the two foundations varies in order to emphasize the effects of the cross--interaction. The surrounding soil is represented by a viscous--elastic, homogeneous half--space. The investigation has been conducted for different values of the embedment ratio h/a (0.333, 0.5 and 1) for a value of the dimensionless frequency $a_0 = 1.0$. In the two-dimensional case, the rigid--body impedance matrix is represented by a 6 X 6 matrix, with 12 independent impedance functions, while, in the three-dimensional case, such a matrix becomes a 12 X 12 matrix.

In fig. 2 the contributions to the frequency-dependent horizontal impedance function of the first foundation from the motion of the second foundation for a_0=1 and h/a=1 are expressed as functions of the dimensionless distance l/a. The dimensionless distance l/a is defined as the distance between the foundations, normalized with respect to the characteristic dimension of the base of the foundation. From the analysis of the results, it is possible to conclude that the cross-interaction effects should be considered in the analysis of the soil--structure interaction effects on the response of multiple embedded foundations when the distance between adjacent foundations is less or equal to 40 times the characteristic length of the foundation. It is interesting to point out that, at small distances, displacement in the vertical direction contributes to the horizontal force on the foundation; this effect was not present in the single foundation analysis. For higher values of l/a, the interaction through the soil can be neglected and the multiple foundation system can be analyzed as a group of single foundations. For bridges with massive foundations and anchorage blocks, the cross-interaction effects become extremely important, especially in the analysis of the lateral spans, where the dimensionless distance l/a is in the range of 10-15.

Bibliography

1. Betti, R. and Abdel-Ghaffar, A.M. (1994):"Analysis of Embedded Foundations by the Substructure Deletion Method", *J. Engrg. Mech.*, ASCE, 120(6), 1283-1303.
2. Mesquita, E., Romanini, E., and de Pontes, B.R. (1995):"A Boundary Element Implementation of the Substructure Deletion Method", Proc. BEM17, Computational Mechanics Publications, 375-386.

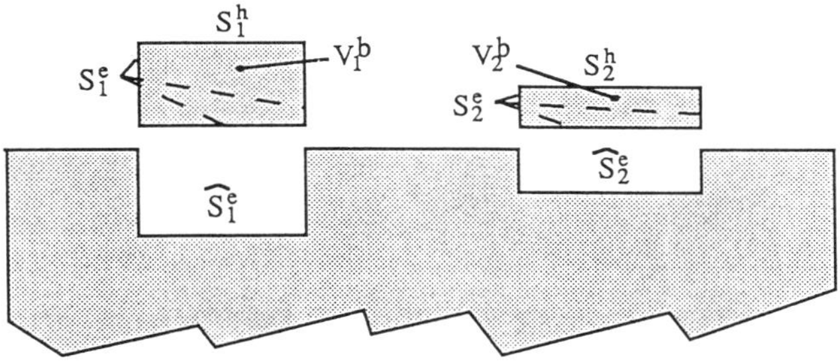

Figure 1

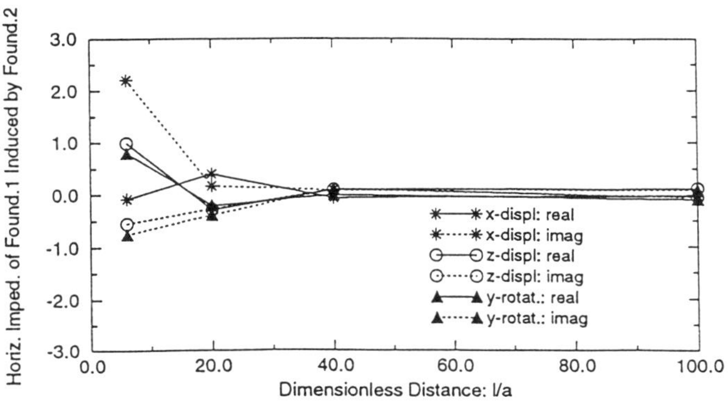

Figure 2

Experimental Study Of Steady And Pulsatile Flows In Models Of Abdominal Aortic Aneurysms

Robert. A. Peattie[1], Tiffany. J. Riehle[1], Matthew. L. Parsons[1], Brian. P. Giles[1] and Edward. I. Bluth[2]

Introduction

An aneurysm is a permanent, balloon-like, blood filled dilation of an arterial vessel, occurring most commonly in the abdominal aorta. The growth and eventual rupture of aneurysmal lesions is dependent on mechanical interaction between the aneurysm wall and the motion of the blood within it, as flowing blood exerts both normal and shear stresses on the wall. Failure of the aneurysm occurs when these stresses exceed the strength of the wall.

Rupture of AAAs accounts for 15,000 deaths per year in the U.S., ranking as the 13th leading cause of death. It has been estimated that as much as 2-3% of the adult population may eventually develop AAAs. Present clinical practice is to regard all AAAs with diameter greater than 5cm as being at high risk of rupture, and to manage them by resection. However, it is not likely that aneurysmal diameter alone is a sufficient criterion for unambiguously establishing risk of rupture. As a first step towards understanding the mechanical stresses that the flow imparts onto the aneurysm walls, and thus to improve assessment of risk of rupture, we have therefore studied flow patterns and the resulting wall stresses in simplified models of AAAs.

Methods

Simulated fusiform AAAs were cast in clear elastomeric blocks. Seven such models, each having a nondilated inlet tube inner diameter, d, of approximately 1.30cm, and a bulge length, L, of $4d$, were fabricated. The maximum bulge diameters, D, were distributed from 1.73 to 4.48cm. As a result, the ratio of the bulge diameter to the nondilated entrance tube diameter of the models, D/d, varied from 1.4 to 3.3. The models thus simulated *in vivo* aortic aneurysms of diameters from 3.3 to 7cm.

The model aneurysms were incorporated into a gravitationally driven flow loop that permitted either steady or pulsatile flow. Governing values of the Reynolds number, Re, were calculated as $Re = 4Q/\pi d\nu$ with Q the volume flow rate and ν the fluid viscosity. Assuming Q to vary from 36mL/sec at rest to several times this value under moderately active conditions in the human aorta, the kinematic viscosity of blood to be 0.033cm^2/sec and the inner diameter of the abdominal aorta as 2.0cm, Re for mean abdominal aorticflow *in vivo* varies from approximately 700 at rest to approximately 3000 under active conditions. Matching steady flows were produced in the experiment.

Flow patterns were visualized with a color flow ultrasonography. Quantitative measurements of the instantaneous local flow velocities were then made by Laser Doppler

1 Dept. of Biomedical Engineering, Tulane University, New Orleans LA 70118
2 Dept. of Radiology, Alton Ochsner Medical Foundation, New Orleans LA 70121

Velocimetry (LDV), using a dual-beam LDV system (ThermoSystems Inc.). For all LDV experiments, a solution of approximately 2kg of sodium thiocyanate in 3L of distilled water was used as working fluid. This solution was titrated *in situ* to maximize the match between its index of refraction and that of the model. Data for eight axial velocity profiles were collected in each model. Shear stresses at the model walls were evaluated by fitting a polynomial curve to the data points closest to the wall, and numerically differentiating the polynomial.

To measure the wall pressure, a series of 5 equally spaced 3mm diameter holes was drilled through the model wall and connected to pressure transducers. The taps were numbered sequentially from the most proximal, 1, to the most distal, 5. Station 1 was positioned 1.2cm proximal to the entrance of the model bulge, station 3 was positioned at the bulge midpoint and station 5 was positioned 1.2cm distal to the exit of the bulge. At each station, the pressure was recorded digitally for 20 seconds. Four such records were averaged, and the resulting mean pressure and sample standard deviation were calculated.

Results

As a representative example of the velocity fields developed in the models, figure 1 illustrates the flow in model 5 ($D/d = 2.75$, simulating a 5.5cm *in vivo* aneurysm). Figure 1a shows the axial velocity field at $Re = 500$. The inlet profile is found to be parabolic in shape, and this parabolic shape is maintained throughout the core without spreading. Slight retrograde velocities, indicating a slowly recirculating vortex, are found within the bulge region. The most rapid retrograde velocities are seen in the distal half of the model, and are about 7% of the peak forward velocity at this axial location.

Figure 1b shows the mean velocity field at $Re = 2600$. At this higher Re, the inlet profile is much more flattened than at $Re = 500$. The blunt core flow spreads radially outwardly as it travels through the aneurysm, until it is reconstricted by the exit tube. As a result, steep velocity gradients are created at the wall of the aneurysm exit. As in the laminar case, a mean recirculation is observed outside the core flow. The greatest retrograde velocity in this case is about 13% of the peak forward velocity.

Critical conditions for transition to turbulence are given as a function of the model aneurysm diameter ratio, D/d, in figure 2. The lower curve shows the Reynolds number at which initial bursts of intermittent turbulence first appeared, while the upper curve displays the Reynolds number at which continuous turbulence became fully developed. It is clear from figure 3 that the flow in the larger models tended to become turbulent at lower Reynolds numbers than the flow in the smaller ones. In the smallest model ($D/d = 1.41$) initial transition occurred at about $Re = 1900$. Conversely, flow in the largest model

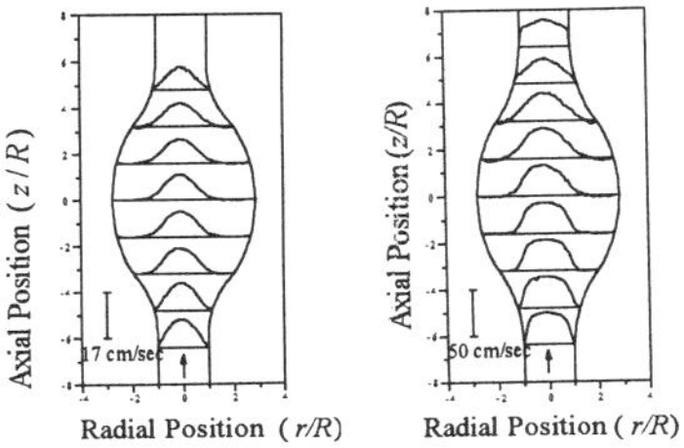

(a) (b)

Figure 1. Velocity profiles in model 5 ($D/d = 2.75$).

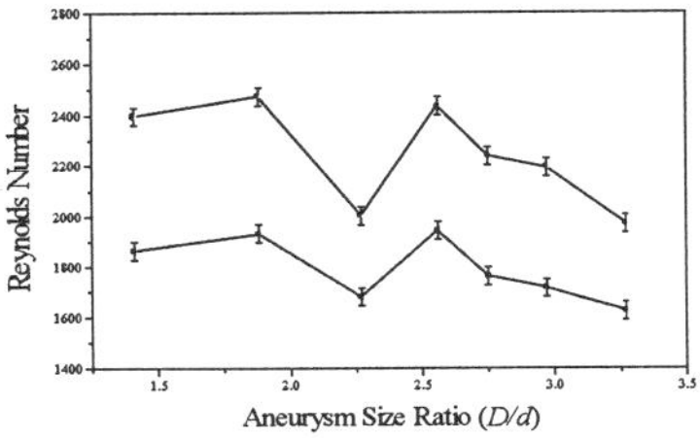

Figure 2. Critical conditions for transition to turbulence.

($D/d = 3.27$) started to become turbulent at $Re = 1600$, and nearly became fully turbulent by $Re = 2000$.

There was a clear maximum in the pressure at station 3, the midpoint of the bulge, at both laminar and turbulent flow rates, for all the models. In the proximal half of the model the pressure increased slightly over the bulge entrance, from station 1 to 2, and then more rapidly along the bulge expansion, from station 2 to 3. In the distal half, the pressure then decreased rapidly from station 3 to 4, returning at station 4 to approximately the same level as station 2. Finally, over the bulge exit, for flow at $Re = 800$ the pressure fell again, while for flow at $Re = 2900$ it rose slightly. The magnitude of the proximal pressure increase grew with the size of the model up to $D/d = 2.75$, then was approximately constant for the two largest models.

Figure 3 depicts the wall shear stresses in model 5 ($D/d = 2.75$), as a function of position along the bulge, for both laminar and turbulent flows. In this figure the shear stress has been normalized by the corresponding value of the wall shear for flow in a straight tube at the same value of Re. When the flow was laminar, shear stresses in the bulge had very small magnitude, a consequence of the slow recirculating velocities in the bulge region. In addition, shear stresses at the entrance and exit of the bulge both had the same magnitude as for flow in a straight tube.

For turbulent flows, both the mean shear stress and the maximum value of the instantaneous stress are given in figure 3. When the flow was turbulent, the shear stress at the bulge exit had approximately twice its value at the bulge entrance. Three-dimensional velocities associated with turbulent velocity eddies caused the axial velocity field to spread radially into the bulge area as flow convected through the model. At the exit of the bulge, the flow was then forced to reconstrict, leading to rapid shear layers at the exit wall of the bulge. The result was a doubling of the wall shear stress from its value in a straight tube. Further, when the flow was turbulent, the instantaneous shear stress at the bulge exit wall was far greater than the mean stress, a result of intrinsic fluctuations in the velocity. In model 5, the peak instantaneous stress exceeded the mean by a factor of 5.5 (40.3 *vs.* 7.2 dynes/cm^2). In the various models, this factor varied between 3 and 20.

Discussion

The velocity results show that there is a core of fluid that passes through the center of the model, surrounded laterally by a recirculating vortex in the bulge. The sign changes in the wall shear from positive to negative, then back to positive, indicate changes in the direction of the shear stress that the flow exerts on the wall. At the exit of the bulge there is a region at which shear stress reversal tends to pull the model walls apart.

When the flow was laminar, the largest models experienced the greatest pressure at the bulge midpoint. This is an important point, as it suggests that the largest aneurysms *in vivo* are exposed to the greatest pressure forces for two separate reasons. The Law of Laplace (wall stress = pressure x bulge radius) shows that the wall stress magnitude increases as the bulge diameter increases, even if the pressure in the fluid remains constant as the bulge diameter increases. But as the bulge diameter increases, the pressure at the bulge midpoint increases as well. Hence there are two independent mechanisms acting to increase the wall normal stress in the largest models: (i)

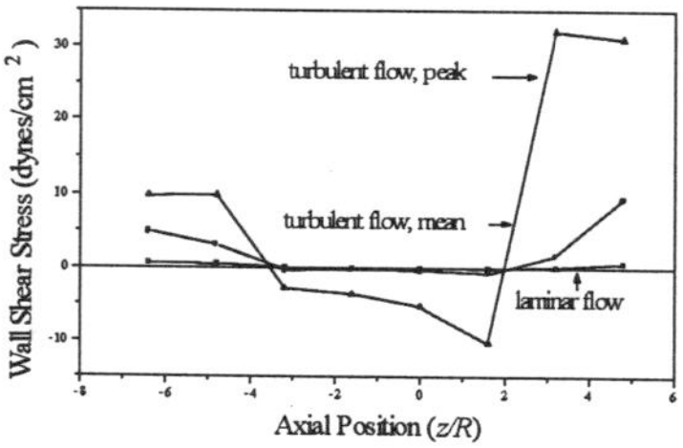

Figure 3. Wall shear stress distribution, model 5.

Laplace's Law and (ii) growth in the maximum wall pressure as the bulge diameter increases. The combination of these two mechanisms may be one reason for the increased risk of rupture associated with aneurysm size reported by other authors.

In addition to pressure effects, figure 3 shows that peak instantaneous stress levels also increased with bulge diameter. Increased shear stress may be a third source of increased risk for larger aneurysms *in vivo*.

The shear stress data may be rescaled for the human abdominal aorta by applying physiologic parameters. Taking $\rho = 1.3$ g/cm^3, $\nu = 0.033$ cm^2/sec and $d = 2.0$ cm, the mean wall stress in an undilated aorta at $Re = 2600$ will be approximately 15 dynes/cm^2. An aneurysmal dilation in this vessel with a diameter ratio of approximately 2:1 (that is, a 4 - 4.5cm aneurysm) will be exposed to a mean shear stress of about twice that at its exit, about 30 dynes/cm^2. Peak shear stresses at the exit will then be approximately 100 dynes/cm^2.

To summarize, then, the following conclusions can be drawn from our studies:

- Larger models become turbulent at lower flow rates than smaller models.
- As fluid flows through the models, turbulence builds at a much more rapid rate than it would in undilated tubes. In addition, the rate of growth of turbulence intensity in the distal half of the models increases with the bulge diameter. In the largest models, turbulence intensities can reach 40%.
- A maximum pressure is found at the midpoint of the bulge in all models. This maximum pressure increases with model size, so that the largest models experience the greatest wall pressure.
- When the flow is turbulent, the mean wall shear stress at the distal end of the model is twice as great as that in an undilated tube. As a result of velocity fluctuations, the maximum instantaneous wall shear stress can exceed the mean stress by an order of magnitude. The largest models also are exposed to the greatest peak wall shear stress.

Shear stresses at the model exit are found to approach critical values for lesion formation reported by Fry[1]. We therefore hypothesize that over long times, accumulated bursts of shear stress eventually produce_lesions of the aneurysmal endothelium, initiating a clotting cascade and leading to thrombus formation. Thrombus deposition then has the following beneficial effects:

- It limits endothelial lesion development, protects the underlying tissue from mechanical damage and prevents hemorrhage.
- It reduces the effective hydraulic diameter of the aneurysm, thus increasing the critical *Re* for turbulence development and reducing the likelihood of turbulence formation.
- When turbulence does occur, the reduced lumenal diameter in the presence of mural thrombus causes a reduction in the turbulence intensity. The net effect is to further reduce the risk of rupture.

Acknowledgments

The authors acknowledge the support of the Whitaker Foundation and the Tulane University Committee on Research.

References

1. Fry, D.L. Acute vascular endothelial changes associated with increased blood velocity gradients. *Circ. Res.* **22**:165-197, 1968

Validation of A 3-D Numerical Model of LV Ejection

Tong Ding
Richard T. Schoephoerster

Abstract

The single most common cause of death in western culture is ischemic heart disease, which results from insufficient coronary artery blood flow. A numerical left ventricular model was developed to detect the abnormal wall motion by analyzing the velocity field inside of the left ventricle. The aim of this study is to validate the numerical model by comparing its results with the experiment results. There was good qualitative agreement between the experimental and numerical results.

Introduction

A characteristic of ischemic heart disease is the abnormal left ventricular wall motion during systole. Besides many methods to quantify and localize ischemic regions of the myocardium, some studies have focused their attention to the flow dynamics inside the LV. Ray et al.(1977) showed through experiments that abnormal wall motion changes the flow parameters.

Schoephoerster(1993) introduced a 2-D numerical model of LV systolic flow dynamics. This model provides a numerical simulation of flow in a contracting cavity with a single outlet and arbitrary user-supplied wall motion. Two LV assessment parameters were also introduced :1)local pressure gradient near the LV wall, and 2) the central ejection region(CER), defined as the region of flow domain in which the obtained velocity field vectors are aligned +/- 5 degree from the LV long axis. The study shows that the local wall pressure gradient may not be suitable for localization of coronary occlusion but for presence of disease only. The CER shifted toward the region of LV wall which exhibits the abnormal motion, and the CER coefficient decreased with increasing severity of abnormal wall motion. Most recently, this model has been expanded to three dimensions (Schoephoerster et al. 1996). The purpose of this study is to validate the three

----

Mechanical Engineering Dept., Florida International University, Miami, FL 33199 dimensional numerical model by measurements of the velocity vector field using Digital particle image Velocimetry (DPIV).

Experimental Methods

DPIV is a non-intrusive technique that permits the mapping of instantaneous 2-D flow-fields. Particles in the flow field are illuminated with a laser light sheet and act as tracers which are captured by a video imaging camera. Two sequential digital images are subsampled with the use of a variable size interrogation window. The cross correlation of the 2 image samples is computed with a Fourier analysis technique resulting in displacement vectors (Willert et al. 1991).

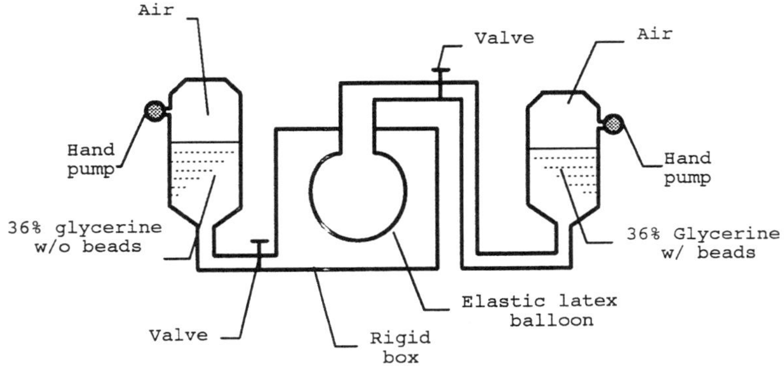

Figure 1. Experimental setup

In the experiments, a transparent latex balloon was used to simulate the LV. The fluid was 36% glycerine by volume in saline to match both density and viscosity of the human blood. Velocity measurements at two perpendicular cross sections were taken -- the long axis plane and the short axis plane. Both normal and ischemic cases were simulated. In order to simulate the reduced wall motion, transparent silicon glue was put on the balloon to increase the rigidity of the wall so as to reduce the contractile capability. For each case, pictures of eight parallel planes were used to rebuild the 3-D geometry. The numerical models were based on these geometries and also the boundary conditions were calculated from these conditions.

Numerical Methods

The flow is assumed to be viscous homogeneous laminar

Newtonian flow, with gravity effects ignored. So the governing equations for the flow will then be the Navier-Stokes equations for incompressible, unsteady flow with constant properties. Numerical modeling was conducted using FIDAP (Fluid Dynamics International, Evanston, IL) CFD package. The FIDAP program utilizes a finite element procedure, using the Galerkin form of the weighted residuals method to solve the Navier-Stokes equations.

Results and Discussion

Figure 2 compares the experimental and numerical results of the velocity field within the normal LV. There is good qualatative agreement between the experimental and numerical results, where the flow is symmetric with respect to the long axis of the LV. Figure 3 compares numerical and experimental results for the case of abnormal LV wall motion. The region of abnormal wall motion is shown on the graph. Again, there is good qualatative agreement between each, in which the flow is generally directed towards the simulated ischemic region. The experimental results validate the adequacy of the numerical model to accurately predict left ventricular velocity vector fields based on computed wall motion.

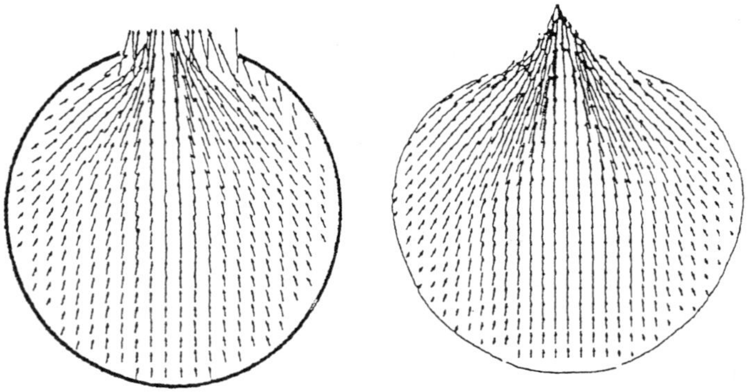

(a) (b)

Figure 2. The normal LV (a) numerical result (b)experimental result

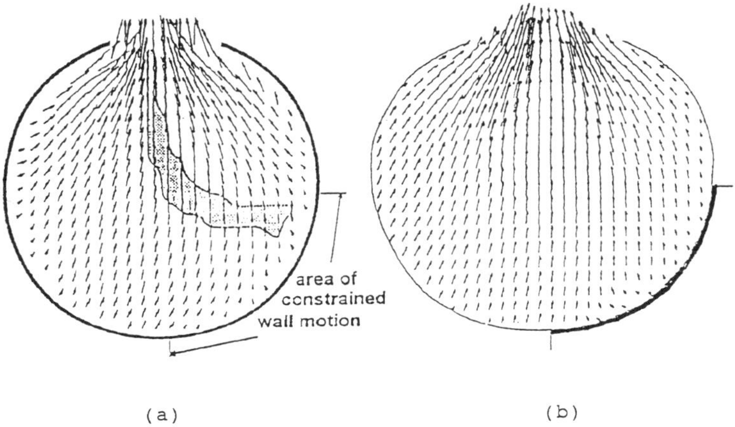

Figure 2. The abnormal LV (a) numerical result (b)experimental result

References

Gonzalez,E., Schoephoerster,R.T., "A simulation of Three-Dimensional Systolic Flow Dynamics in a Spherical Ventricle: Effects of Abnormal Wall Motion", Annuals of Biomedical Engineering, Vol. 24, pp. 48-57, 1996

Ray,G., Liu,Y.K., Chadran,K.B., and Weiting,D.W., "Pressure Gradient Changes in Experimental Acute Myocardial Infarction Dogs", Proc. ASME Biomechanics Symp., Vol2, pp 134-138, 1977.

Schoephoerster,R.T., Silva,C.L., and Ray,G., "Finite Analytic Model for Left Ventricular Systolic Flow Dynamics", Journal of Engineering Mechanics, Vol. 119, No. 4, April 1993, pp. 733-747.

Schoephoerster,R.T., Silva,C.L., and Ray,G., "Evaluation of Left Ventricular Function Based on Simulated Flow Dynamics Computed from Regional Wall Motion", Journal of Biomechanics, Vol. 27, No. 2, Feb. 1994, pp. 125-136.

Willert,C.E., Gharib,M., "Digital Particle Image Velocimetry", Experiments in Fluids, Vol. 10, No. 4, 1991.

Computation of Velocity Fields of Intravenous Balloon Pumping

Huaqiang Li[1], Tin-Kan Hung[2], Chiuping Chang[3] and Pat Sawzik[4]

Intravenous oxygenation as an alternative therapy has been under development at several institutions in the U.S. An intravenous membrane oxygenator was invented by Dr. B. G. Hattler at the University of Pittsburgh (Hattler et al. 1994). The oxygenator is characterized by a bundle of hollow fibers, supplying oxygen to the inferior vena cava with a balloon in the core region. The incorporation of the balloon pumping is to disrupt diffusional boundary layer around the fibers, enhancing intravenous gas exchange through convective mixing of blood around hollow fiber membranes. As demonstrated in the animal experiments, the device is placed in the inferior vena cava, right atrium and terminal superior vena cava (Fig. 1).

A computational analysis is employed to investigate the velocity field of the venous return with the balloon pumping. The complexity of the three-dimensional flow of blood past the oxygenator makes it impossible to quantify the velocity field. A simplified axisymmetric flow without the presence of the fibers could provide some insight into the pumping mechanism useful for experimental model studies and the prototype development.

The flow of blood can be treated as Newtonian, incompressible and laminar. The temporal variation of the flow field associated with the balloon motion is mapped onto a fixed domain. The transformation converts the radial axis r to R (Hung 1981) by

$$R = [\, r - b(z,t) \,] / [\, a - b(z,t) \,]$$

where "a" is the radius of the vena cava, and $b(z,t)$ the instantaneous radius of the balloon. For simplicity, the latter are assumed to be circular in cross section, and it is a function of the longitudinal axis, z, and time, t. This transformation maps the centerline, the balloon and catheter surface to R=0, and the vessel wall to R=1 in the computational domain, making it convenient to solve the moving boundary value problem by computer.

----

[1] Research Assistant; [3] Postdoctoral Fellow.

[2] Professor, Department of Civil and Environmental Engineering, University of Pittsburgh, Pittsburgh, PA 15261.

[4] Manager, Walter s Scientific Instrument Labs, Pittsburgh, PA.

The Navier-Stokes equations and the equation of conservation of mass are transformed to the fixed coordinate system: R and Z=z/a (Hung 1981). They are expressed in the integral form over a mesh (Li 1995). The SIMPLER algorithm (Semi-Implicit Method for Pressure Linked Equations Revised) developed by Patankar (1981) is employed to express these equations in a finite difference form with the up-wind scheme.

A fully developed laminar flow in the annular tube is prescribed as the upstream boundary condition. The flow rate is characterized by the Reynolds number: Re = VD/υ (where V is the mean velocity in a circular vena cava of diameter D, and υ the kinematic viscosity of the blood). The reference pressure at this inlet can be treated as a constant in the computation while the pressures at the downstream end are part of the computational solutions. The vena cava wall motion is relatively small in comparison with the balloon inflation and deflation. It is, therefore, considered to be stationary. The same condition is assumed on the catheter wall. The balloon wall motion is prescribed by a sinusoidal function shown in Fig. 2.

A two-chamber balloon pumping is considered in this study. The geometry of the fully inflated balloon is shown in Fig. 3. As shown in Figs. 4, 5 and 6, the prescribed motions of these two chambers are the same. The balloon motion generates the radial velocity component of the venous flow, altering the longitudinal velocity profiles. Fig. 4 shows the velocity profiles along the vena cava at the time between the end of balloon deflation and the onset of balloon inflation. The velocity of the balloon is momentarily zero. The radial balloon velocity reaches the peak at $\omega T = 3\pi/4$, the longitudinal velocities at this instant increase in the downstream direction (see Fig. 5). They are indicated by the increase in the Reynolds number along the balloon. The velocity profiles associated with the onset of balloon deflation are shown in Fig. 5. At this instant, the velocity of the balloon also vanishes. During the balloon deflation phase, the Reynolds number decreases with the wall flux along the balloon. Because of the momentum associated with the inflation and deflation of the balloon, no boundary layer separation occurs in the trailing edge of the balloon.

The study was supported by an NIH grant RO1-HL33934 to TKH and a grant DAMD-17-94 of the U.S. Army Medical Research & Development Command to Dr. B.G. Hattler of the Department of Surgery of the University of Pittsburgh.

REFERENCES

Li, H. (1995). "A Computational Fluid Dynamics Analysis of Axisymmetric Balloon Pumping in Vena Cava." Ph.D. Dissertation, University of Pittsburgh.

Hung, T.K. (1981). "Forcing Functions in Navier-Stokes Equations." Journal of Engineering Mechanics Division, ASCE, Vol. 107, No. EM3, 643-648.

Patankar, S.V.(1981)."A Calculation Procedure for Two-Dimensional Elliptic Situations." Numerical Heat Transfer, Vol.4, No.4, 409-425.

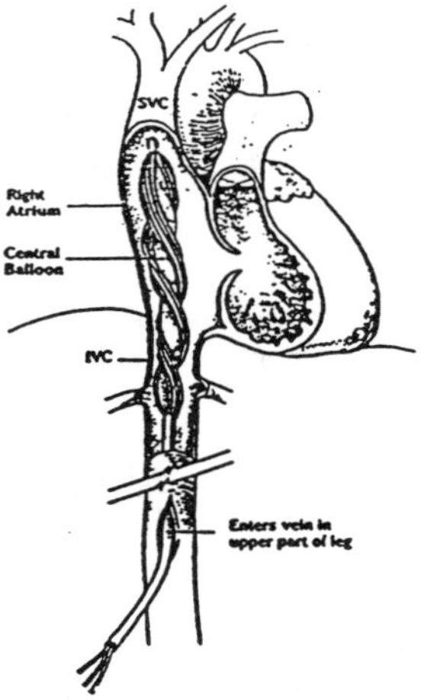

Figure 1 Intravenous Membrane Oxygenation

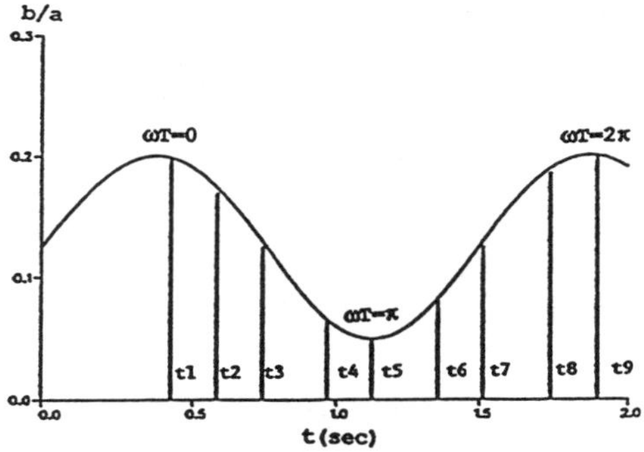

Figure 2 Balloon Wall Motion

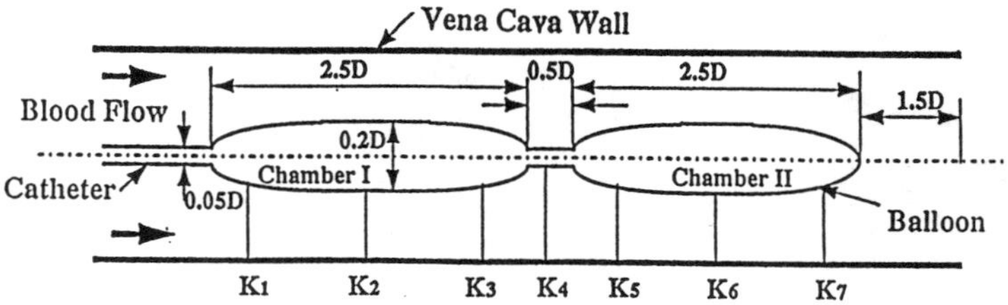

Figure 3 Schematic of the Model

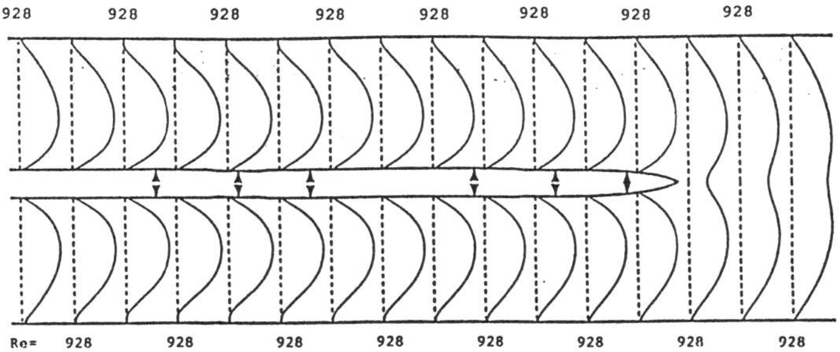

Figure 4 Longitudinal Velocity Profiles (40 bpm) at $\omega T=\pi$

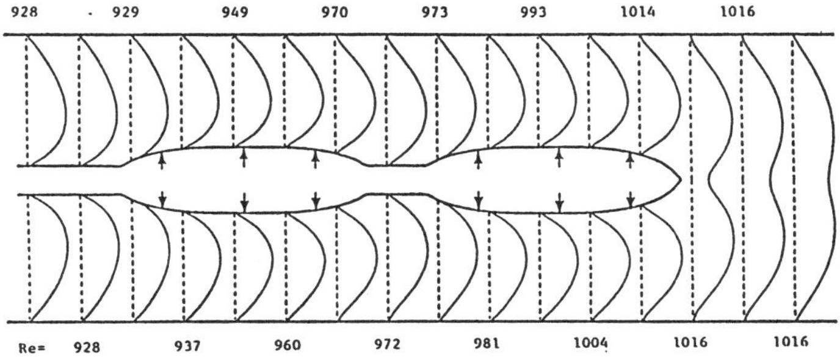

Figure 5 Longitudinal Velocity Profiles (40 bpm) at $\omega T=3\pi/2$

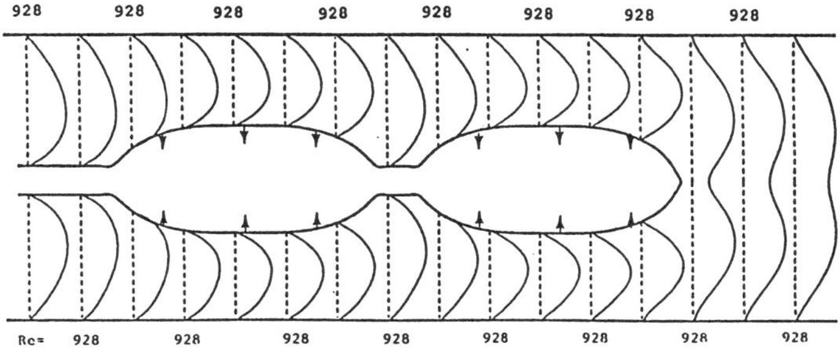

Figure 6 Longitudinal Velocity Profiles (40 bpm) at $\omega T=2\pi$

Simulation of Pulsatile Flow Past a St. Jude Valve

L. Niu, D. Bluestein, and R.T. Schoephoerster

Abstract

Numerical solutions of the time-varying, turbulent flow through a St. Jude bileaflet valve in the aortic position have been obtained using the commercially available FIDAP package. To ease computational complexity, these initial simulations were performed in two dimensions with rigidly fixed leaflets. A numerical simulation such as this allows the quantitative examination of the cumulative effects of elevated shear stress as a blood platelet passes through the valve region, and this can be used to indicate the likelihood of stress-induced platelet activation due to flow produced by the artificial valve.

Introduction

Thromboembolism and anticoagulant-related hemorrhage still represent the major causes of valve-related morbidity and mortality in patients with implanted heart valve prostheses and left ventricular assist devices. Dewanjee (1990) has summarized results of a series of systematic studies on the imaging of platelet deposition and quantitation of platelet thrombus in a variety of cardiovascular devices, including heart valve prostheses, synthetic vascular grafts, oxygenators, and hemodialyzers. Also, since local disturbances in the flow of blood contribute significantly to the initiation of platelet deposition and aggregation, many studies have been performed to determine the effects of valve geometry on the dynamics of the flow downstream of the valves (Schoephoerster and Chandran, 1991). We have recently measured platelet deposition in an axisymmetric model of arterial stenosis(Schoephoerster et al., 1993) and correlated these results with numerical simulations of the flow (Bluestein et al., 1996). Platelet deposition was increased in areas with enhanced convective transport of platelets due to velocity components perpendicular to the vessel wall, and decreased in regions where high local wall shear stresses allowed embolization of the adhered platelets. Flow through mechanical heart valves produces phenomena similar in nature to those described, and a detailed analysis of the local fluid mechanics in the close vicinity of the valve may provide information as to the likelihood of platelet activation and aggregation in and around the valve structure. We present here some initial results of flow through a St. Jude valve in the aortic position.

Mechanical Engineering Dept., Florida International University, Miami, FL 33199

Methods

The geometry for the problem at hand is the left ventricular outflow tract and the aortic root, including the sinuses (Figure 1). Due to the complex geometry and to save computational time, these initial simulations were performed in two dimensions, with rigid valve leaflets, and thus only the systolic portion of the cardiac cycle was simulated. The orientation and dimensions of the 2-D projections of the leaflets of the bileaflet St. Jude valve were provided by St. Jude Medical, Inc. (St. Paul, MN). The geometry for the sewing ring was ignored. The computational domain includes an inlet length of 10 diameters upstream of the valve to allow development of the flow parameters, and a likewise exit length of 10 diameters for the same reason. The time-averaged continuity and Navier-Stokes equations were solved using FIDAP (Fluid Dynamics International, Evanston, IL). The governing equations were closed in terms of the turbulent flow parameters using the standard two equation k-ε turbulence model. Specialized wall elements were used for near wall modeling, in which the variation of the turbulent momentum diffusivity is modeled using van Driest's mixing length approach. The inlet boundary condition was a parabolic flow profile, with a flow rate varying in time and modeled after Guyton (1991). A zero stress condition was applied at the outlet, and a no-slip condition applied at the boundary walls. Several cycles were run at a cardiac output of 3.11 liters/min and cycle rate of 50 beats/min until the solution at each time step (30 total during systole) was identical to the previous time step.

Results

Results are shown for the computed velocity vector field and the velocity profile downstream of the leaflets at early systole (acceleration phase, Figure 1) and late systole (deceleration phase, Figure 2). Early in systole, vortices are formed and are confined within the sinus cavities, and the valve leaflets produce only a small wake region immediately after the leaflets. This results in a bi-peaked velocity profile, with each peak corresponding to the two major outflow sections of the valve. Later in systole, the wake from the leaflets extends far downstream, resulting in a tri-peaked velocity profile. The sinus recirculation region also extends further downstream, and this is displayed in the velocity profile as reversed flow near the wall. In addition, a second recirculation region forms downstream of the sinus region due to the deceleration of flow. In both cases, the flow region between the valve leaflets provides a region of high fluid shear stress. These flow conditions are similar to those observed in our previous work with a constricted tube, in which platelet aggregation was enhanced downstream of the constriction due to upstream platelet activation in the high stress throat region followed by areas of recirculation where activated platelets may aggregate.

References

Bluestein, D., et al. (1996) Ann. Biomed. Engg., in press.
Dewanjee, M.K. (1990) Platelets and Atherosclerosis, Ed.: Kessler, Springer-Verlag, Berlin, pp.71-86.
Guyton, A.C. (1991) Textbook of Medical Physiology. Saunders, Philadelphia, Eighth Edition, p. 102.
Schoephoerster, R.T., and Chandran, K.B. (1991) J. Biomechanics, 24:549-562.
Schoephoerster, et al. (1993) Arterioscler. Thromb., 13:1806-1813.

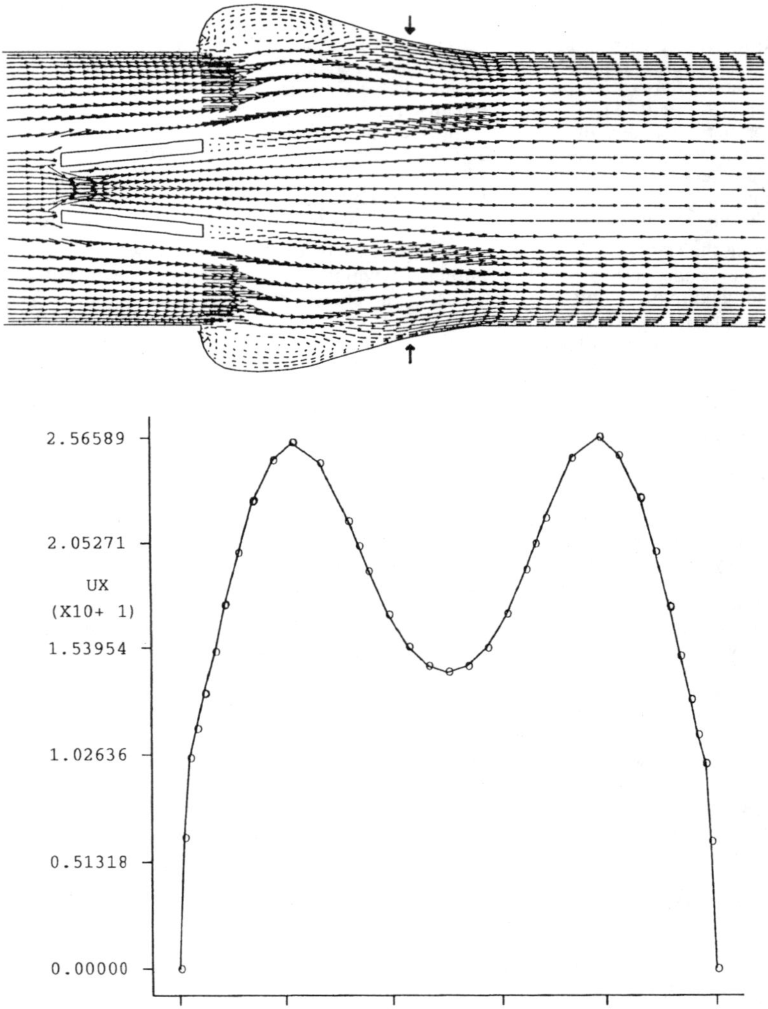

Figure 1. Velocity vector field early in systole, during the acceleration phase, and a corresponding velocity profile (in cm/s) at the cross section shown by the arrows.

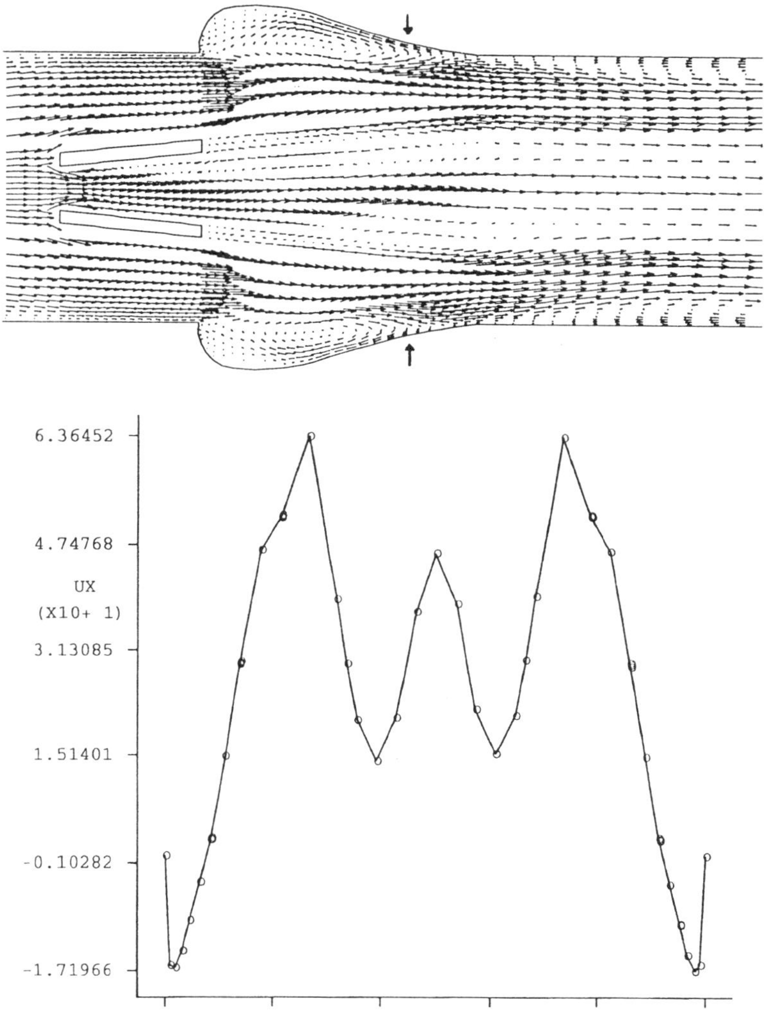

Figure 2. Velocity vector field late in systole, during the deceleration phase, and a corresponding velocity profile (in cm/s) at the cross section shown by the arrows.

Computational Modeling of Fluid Dynamics in Aortopulmonary Shunts: Comparison to in Vitro Studies

Kevin K. Whitehead[1], Theresa A. Tacy, Edward G. Cape

The use of a modified Blalock-Taussig aortopulmonary shunt in infants with insufficient pulmonary blood flow is a palliative procedure used to sustain the patient until a more curative procedure can be performed. Accurate knowledge of the pressure gradient (PG) across a modified Blalock-Taussig shunt (BTS) is critical in surgical planning for the staged palliative approach, but the fluid dynamics across these shunts have not been adequately described. While it is common to estimate the aortopulmonary pressure gradient (PG) using the modified Bernoulli equation (MBE), $\Delta P = 4v^2$, the accuracy of this technique is limited due to viscous energy losses in the BTS which are unaccounted for in the MBE. The study tests an equation derived from the macroscopic energy balance and based on established empirical data for turbulent fluid contraction into a tube and flow through conduits. **Methods:** We studied the relationship between PG predicted by MBE and actual aortopulmonary PG using an in vitro pulsatile flow model with Gore-Tex BTS connecting systemic and pulmonary circuits. Using variable compliance and resistance, 5 peak PG's (20-100 mmHg) were generated. Effects of BTS length (L) and diameter (D) were investigated by studying all combinations of L = 3, 6, 9 cm and D = 4, 5, 6 mm. The peak velocity (V) of the blood analog fluid within each BTS was measured by continuous wave Doppler for each condition, and the MBE was tested as a predictor of the PG across the BTS. Furthermore, the newly proposed energy equation was tested and compared to the MBE as a predictor of the PG. A computational model using finite element analysis implemented on a Cray C90 is being used to understand scatter in the resulting in vitro data. **Results**: The slope of the regression line (y=0.80x, r=0.97) reflected a tendency for underestimation of the PG by MBE and was significantly different from unity (p=0.01). However, the error was not limited to underestimation. At higher Reynolds numbers, the MBE overestimated the PG, suggesting that when inertial forces predominate, pressure recovery becomes an important factor in the error of Doppler predictions. The average error was 21% for all BTS and flow conditions. Using the newly derived energy equation, the mean error decreased to 5%, and the slope of the regression line relating Doppler predicted to actual PG (y=1.04x, r=0.97) was not significantly different from the line of unity (p>0.5). Also presented will be ongoing investigations using computer modeling to assess the extent of pressure recovery effects in the PA and determine the extent to which the flow in the BTS develops. **Conclusions:** The improvement provided by the energy equation over the MBE supports the hypothesis that the underestimation of the true pressure gradient has a viscous etiology. Application of fluid dynamic theory may yield a truly accurate predictor of PG across BTS which is critical in surgical decision making.

[1]Cardiac Dynamics Laboratory, Children's Hospital of Pittsburgh, University of Pittsburgh, 3705 5th Avenue, Pittsburgh, PA 15213

System Identification Using Earthquake Acceleration Records

A.-W. Elgamal,[1] Member, ASCE, M. Zeghal[2], Associate Member, ASCE, H. T. Tang[3], and J. C. Stepp[4]

Abstract

Free-field downhole-array earthquake records are employed to identify seismic site response properties. Correlation and spectral analyses are used to evaluate shear wave propagation characteristics, variation of shear wave velocity with depth, and site resonant frequencies and modal configurations. The actual seismic shear stress-strain histories are directly evaluated from downhole accelerations using a newly developed identification technique.

Introduction

The Lotung (Taiwan) site was equipped in 1985 with an extensive array of downhole (DHA and DHB, Fig. 1a) and surface accelerometers (Tang 1987) by the Electric Power Research Institute (EPRI) and the Taiwan Power Company (TPC). Eighteen earthquakes (referred to as LSST 1-LSST 18) were recorded during the period 1985–1986.

In this study, simple identification procedures were used to: (1) evaluate shear wave propagation characteristics, (2) identify site resonant frequencies and modal configurations, and (3) estimate the associated seismic shear stress-strain histories. A detailed description of the conducted studies is given in Elgamal *et al.* (1995), Zeghal and Elgamal (1993), and Zeghal *et al.* (1995). This paper outlines the employed analysis techniques, and summarizes a number of significant findings.

Identification of Low-Strain Shear Wave Velocity Profile

Correlation analyses were employed to evaluate average shear wave velocity between Lotung free field DHB stations, and investigate shear wave vertical propagation characteristics (Elgamal *et al.* 1995). The cross-correlation function between downhole acceleration histories $a_i(t)$ and $a_j(t)$ recorded at stations i and j may be expressed as:

$$c_{a_i\, a_j}(t = m\Delta t) = \frac{\frac{1}{N-m}\sum_{n=1}^{N-m} a_i(n\Delta t)a_j((n+m)\Delta t)}{\sqrt{\frac{1}{N}\sum_{n=1}^{N} a_i^2(n\Delta t)}\sqrt{\frac{1}{N}\sum_{n=1}^{N} a_j^2(n\Delta t)}},\quad m = 0, 1, 2, \cdots, N \qquad (1)$$

in which Δt is time step of digital data, $t = m\Delta t$ is time delay, and $(N\Delta t)$ is acceleration record length. This cross-correlation function reaches a major peak at a time delay $t =$

---

[1]Associate Professor, Civil Eng. Dept., Rensselaer Poly. Inst., Troy, NY 12180.
[2]Research Associate, Civil Eng. Dept., Rensselaer Poly. Inst., Troy, NY 12180.
[3]Program Manager, Electric Power Research Institute, Palo Alto,CA 94304.
[4]Earthquake Hazards Solutions, 6607 Cuesta Trail, Austin, TX 78730.

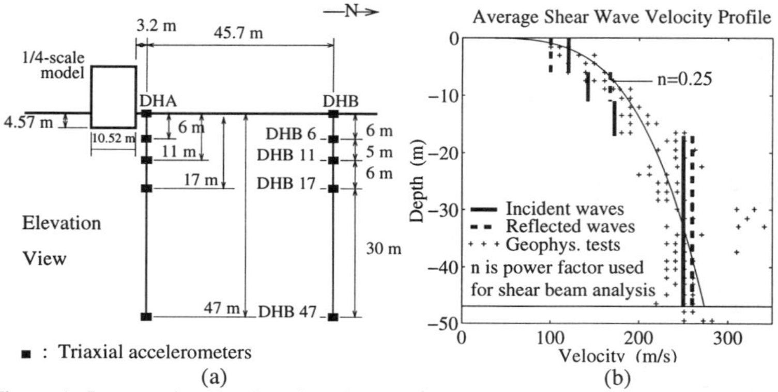

Figure 1: Lotung site: (a) downhole instrumentation; (b) average shear wave velocity profile estimated from the LSST records.

t_d, where t_d is time for seismic waves to travel from station i to station j. Thus, the apparent velocity v_a of wave propagation between stations i and j may be estimated as:

$$v_a = d/t_d \tag{2}$$

in which d is the known distance between stations i and j. Within the instrumented zone at Lotung, the angle of seismic wave incidence with respect to the vertical direction was found to vary in the range of 5° to 15°. Thus, actual shear wave velocity v is practically equal to v_a ($0.96v_a \leq v \leq v_a$). The overall average shear wave velocity profile, evaluated from the LSST records (Fig. 1b), fell within a data band of geophysical measurements, as documented by EPRI.

In view of wave reflection at the free ground surface, the DHB array acceleration records include incident and reflected components. Thus, the cross-correlation function of a station i underlying a station j may display four major peaks. For instance, the cross-correlation between LSST 10 NS acceleration records at 6 m and 11 m depths (Fig. 2a), displayed four peaks at -0.135 sec, -0.035 sec, 0.035 sec and 0.135 sec. These peaks correspond to correlations between the 11 m and 6 m depth incident and reflected components, i.e., R_{11} and I_6, R_{11} and R_6, I_{11} and I_6, and I_{11} and R_6 (Fig. 2b). Time lag at such correlation peaks may be employed to evaluate, shear wave velocities between each two DHB downhole stations and between these station and the ground surface (Zeghal and Elgamal 1993, Elgamal *et al.* 1995).

Identification of Resonant Site Characteristics

Using downhole accelerations $a_i(t)$ and $a_j(t)$, the cross-spectrum energy function $s_{a_i\,a_j}(f)$ may be estimated as follows:

$$s_{a_i\,a_j}(f) = A_i^*(f)A_j(f) \tag{3}$$

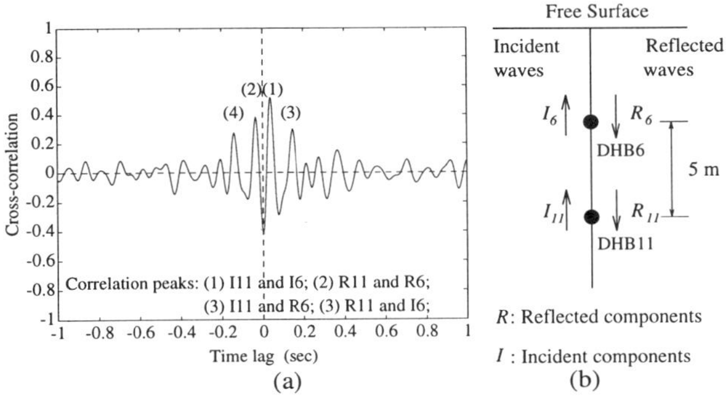

Figure 2: (a) Cross-correlation between 11 m and 6 m depth NS accelerations during LSST 10; (b) Schematic description of wave reflection at free ground surface.

in which f is frequency, and $A_i^*(f)$ is complex conjugate of the finite Fourier transform $A_i(f)$ of $a_i(t)$. The amplitude of $s_{a_i\, a_j}(f)$ peaks at either a resonant or an input frequency (an excitation spectrum peak). In general, such peaks represent a resonance if the corresponding phase-angle approaches 0° or 180°. A coherence function approaching 1.0 at a given resonant frequency ensures quality and accuracy of estimates.

The relative displacement of the kth natural mode ϕ_k (of frequency f_k), at the ith sensor location may be approximated by:

$$\phi_k(z_i) = \pm\sqrt{s_{a_i\, a_i}(f_k)} \tag{4}$$

where $s_{a_i\, a_i}(f_k)$ is a_i auto-spectral density function amplitude at the kth resonant frequency f_k, and z_i is depth coordinate of a_i. Fig. 3a shows the identified 4th spectral modal configuration, along with the corresponding mode shape of a calibrated shear beam. The shear beam and spectral estimates were found to be in general agreement over a wide range of resonances.

Shear Stress-Strain Histories

Using a shear beam model to describe seismic lateral site response, with linear interpolation between downhole accelerations, the discrete counterpart of shear stresses at level z_i reduces to (Zeghal *et al.* 1995):

$$\tau_i(t) = \tau_{i-1}(t) + \rho\frac{\ddot{u}_{i-1} + \ddot{u}_i}{2}\Delta z_{i-1}, \quad i = 2, 3, \cdots \tag{5}$$

in which subscript i refers to level z_i (of the ith accelerometer), $\tau_i(t) = \tau(z_i, t)$, $\tau_1 = \tau(0, t) = 0$ at the stress-free ground surface, $\ddot{u}_i = \ddot{u}(z_i, t)$, and Δz_i is spacing interval between accelerometers. This stress estimate (Eq. 5) is second order accurate. The corresponding second-order accurate shear strain may be expressed as:

$$\gamma_i(t) = \frac{1}{\Delta z_{i-1} + \Delta z_i}\left((u_{i+1} - u_i)\frac{\Delta z_{i-1}}{\Delta z_i} + (u_i - u_{i-1})\frac{\Delta z_i}{\Delta z_{i-1}}\right), \quad i = 2, 3, \cdots \tag{6}$$

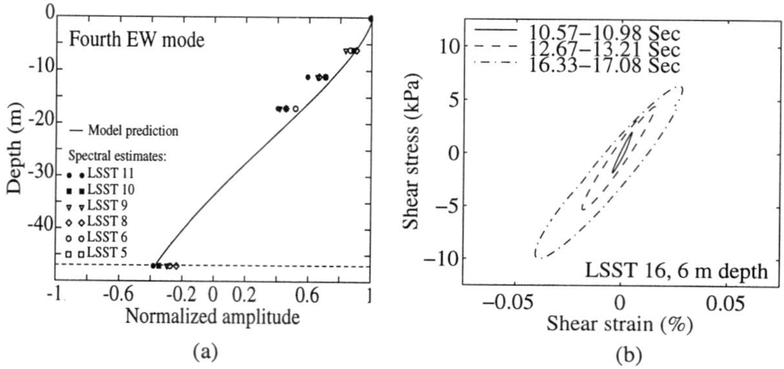

Figure 3: Lotung site dynamic response: (a) 4th modal configuration, (b) selected cycles of LSST 16 stress-strain history at 6 m depth.

in which $u_i = u(z_i, t)$ is absolute displacement (evaluated through double integration of the recorded acceleration history $\ddot{u}(z_i, t)$). Fig. 3b depicts selected cycles of LSST 16 shear stress-strain histories at 6 m depth, and shows stiffness reduction due to the increase in shear strain amplitude.

CONCLUSIONS

The conducted studies showed that downhole acceleration records of free field site response and simple identification tools may be utilized to evaluate: (1) low-strain shear wave velocity profile, (2) resonant frequencies and modal configurations, and (3) downhole seismic shear stress-strain histories.

ACKNOWLEDGMENTS

The research reported herein was supported by grants from EPRI (grant No. RP1444-18), USGS (grant No. 1434-94-G-2397), and NSF (grant No. MSS-9057388). This support is gratefully acknowledged.

REFERENCES

Elgamal, A.-W., Zeghal, M., Tang, H. T., and Stepp, J. C. (1995). "Lotung Downhole Array. I: Evaluation of Site Dynamic Properties," *Journal of Geotechnical Engineering*, ASCE, Vol. 121, No. 4, 350-362.

Tang, H. T. (1987). "Large-Scale Soil Structure Interaction," *Report No. NP-5513-SR*, Electric Power Research Institute, Palo Alto, California.

Zeghal, M. and Elgamal, A.-W., (1993). "Lotung Site: Downhole Seismic Data Analysis," *Report,* Department of Civil Engineering Rensselaer Polytechnic Institute (to appear as Electric Power Research Institute Report TR-106093).

Zeghal, M., Elgamal, A.-W., Tang, H. T. and Stepp J. C. (1995). "Lotung Downhole Array. II: Evaluation of Soil Nonlinear Properties," *Journal of Geotechnical Engineering*, ASCE, Vol 121, No. 4, 363-378.

Sectional Analysis for Nonlinear System Identification of Concrete Structures

Jie Wang[1], Manoj B. Chopra[2] and Sashi K. Kunnath[3]

ABSTRACT

This paper presents a systematic methodology for identifying force-deformation parameters required by nonlinear programs for the analysis of concrete structures. It describes a sectional analysis technique based upon a fiber-element model for characterizing the cyclic moment-curvature hysteresis of reinforced concrete sections. The cyclic, degrading constitutive behavior of concrete and steel are taken into account and the analysis includes the effects of confinement, strain-hardening of steel and the cracking of concrete. The control parameters to define the hysteresis loops are identified graphically in this paper, however, the automation of this procedure is essential. A typical structural frame is analyzed to illustrate the utility of this method.

INTRODUCTION

Practical analysis of large reinforced concrete (RC) structural systems is accomplished using member by member discretization in which all nonlinear action is concentrated in potential plastic hinge zones. The inelastic behavior at these sections is characterized by cyclic, degrading force-deformation hysteresis loops. Since inelastic action in reinforced concrete tends to spread along the length of the member (spread plasticity), it is generally preferable to use flexibility-based formulations in developing the element stiffness matrices (Kunnath et al.[1]). Such flexibility-based methods rely on accurate prescription of the moment-curvature characteristics at critical cross-sections. A sectional analysis technique is described in this paper to derive the cyclic moment-curvature behavior of typical RC sections. The proposed procedure may be incorporated as part of a system identification routine that provides direct input into a nonlinear computer program such as IDARC or DRAIN-2DX. This routine must also include a procedure for automatic identification of the control parameters that define the degrading behavior of the hysteresis loops. In the present paper, the control parameters are calibrated graphically, however, work is underway to automate this task.

---

1 Graduate Assistant, Dept. of Civil and Environ. Engg, University of Central Florida, Orlando FL 32816

2 Assistant Professor, Dept. of Civil and Environ. Engg, University of Central Florida, Orlando FL 32816

3 Associate Professor, Dept. of Civil and Environ. Engg, University of Central Florida, Orlando FL 32816

HYSTERETIC MODELING OF STEEL AND CONCRETE

The general relationship recommended by Richard and Abbott[2] is used here to model the monotonic loading behavior of steel and its cyclic properties when the reversal happens within the strain hardened branch. For the case that reversal happens in the yield plateau range, the model developed by Mander et al.[3] is used herein to calculate the loading and unloading of the stress-strain curve. The different branches specifying the behavior of steel are briefly described in Fig. 1.

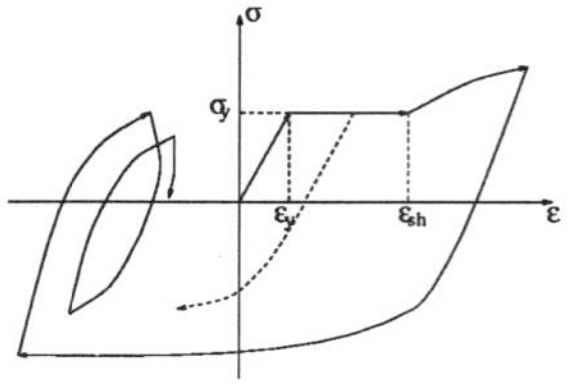

Fig. 1 Cyclic stress-strain curve of steel

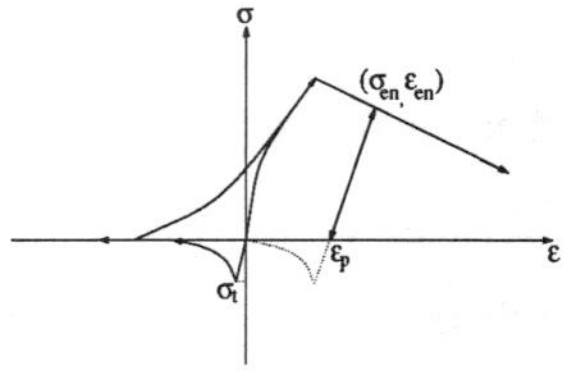

Fig. 2 Cyclic stress-strain curve of concrete

The model used for concrete under monotonic loading is composed of compression and tension portions in which the compression portion is a simplified version of the one used by Darwin [4]. The rule for cyclic loading in the compression portion is characterized by two points, which are the reversal point from the envelop curve determined by the strain ϵ_{en}, and the point on the strain axis, determined by the plastic strain ϵ_p. The plastic strain, ϵ_p is given by the following equation proposed by Karsan[5].

$$\frac{\epsilon_p}{\epsilon_0} = 0.145\left(\frac{\epsilon_{en}}{\epsilon_0}\right)^2 + 0.13\left(\frac{\epsilon_{en}}{\epsilon_0}\right), \qquad if\ \epsilon_p>0,\ then\ \epsilon_p=0 \quad (1)$$

The behavior of concrete under cyclic loading is shown in Fig. 2.

MOMENT-CURVATURE ANALYSIS FOR UNIAXIAL BENDING

A computer program is developed to obtain the moment-curvature behavior of typical reinforced concrete sections, with or without axial loads, under reversed cyclic loading. The moment-curvature analysis is based on a "Fiber-Element" model, in which the cross-section is divided into a number of layers or strips. The governing equations of the fiber-element model is explained in detail in Kunnath et al.[1]. The strain in any fiber is given by $\epsilon=\epsilon_0+\phi y$, where ϵ_0 is the strain at the centroid of the cross section and ϕ is the curvature. For the case of a known constant external force F, and a given value of curvature ϕ, ϵ is calculated using the following Newton iteration scheme:

$$\begin{cases} (\varepsilon_0)|_n^{(i+1)} = (\varepsilon_0)|_n^{(i)} - \dfrac{\partial G}{\partial \varepsilon_0}\Big|_n^{(i)} G(\varepsilon_0^{(i)}) \\ (\varepsilon_0)|_n^{(0)} = (\varepsilon_0)|_{n-1} \end{cases} \quad (2)$$

Thus, M can be obtained by substituting ϕ and the calculated ϵ into the governing equations. For the case where F and M are specified, ϵ and ϕ can be calculated by a two-dimensional Newton iteration scheme as follows:

$$\begin{Bmatrix}\varepsilon_0\\ \phi\end{Bmatrix}_n^{(i+1)} = \begin{Bmatrix}\varepsilon_0\\ \phi\end{Bmatrix}_n^{(i)} - (J_n^{(i)})^{-1}\begin{Bmatrix}G_1(\varepsilon_0,\phi)\\ G_2(\varepsilon_0,\phi)\end{Bmatrix}_n^{(i)}$$

$$\begin{Bmatrix}\varepsilon_0\\ \phi\end{Bmatrix}_n^{(0)} = \begin{Bmatrix}\varepsilon_0\\ \phi\end{Bmatrix}_{n-1} \qquad \text{where } J_n^{(i)} = \begin{bmatrix}\dfrac{\partial G_1}{\partial \varepsilon_0} & \dfrac{\partial G_1}{\partial \phi}\\ \dfrac{\partial G_2}{\partial \varepsilon_0} & \dfrac{\partial G_2}{\partial \phi}\end{bmatrix}_n^{(i)} \tag{3}$$

The above procedure is used to generate the cyclic moment-curvature behavior of a selected cross-section. The curvature increments are typically applied until yielding of the section takes place. The loads are then reversed and at least two additional cycles up to two times and four times the yield curvature are applied. This provides adequate data to calibrate all branches of the hysteresis loops.

CASE STUDY

A three-story frame model described in Ref. [1], as shown in Fig. 3, is selected for analysis. The size and reinforcement of each member are presented in the adjoining table.

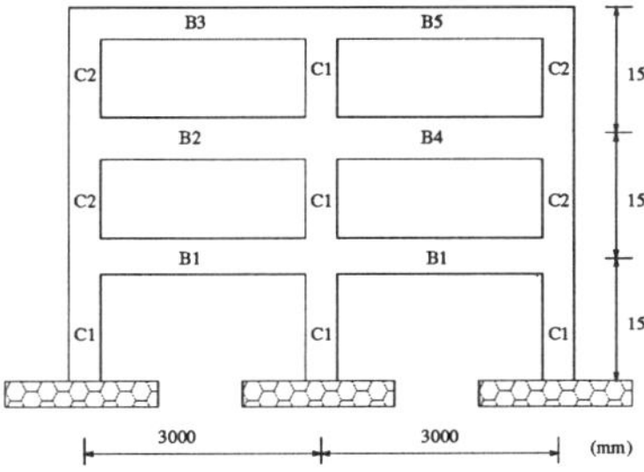

Fig.3 Three-story building

SECTION	NO. OF BARS & BAR DIA	HOOPS
COLUMN C1	4#14	12mm@75mm
COLUMN C2	4#12	8mm@75mm
BEAM B1	TOP: 2#16; BOT: 2#16	6mm@75mm
BEAM B2	TOP: 2#16; BOT: 2#16, 1#10	6mm@75mm
BEAM B3	TOP: 2#12; BOT: 2#14	6mm@75mm
BEAM B4	TOP: 2#18; BOT: 2#16	6mm@75mm
BEAM B5	TOP: 2#14; BOT: 2#14	6mm@75mm

The moment-curvature diagram for each beam and column is obtained using the developed computer program. Two typical diagrams are presented below. One is for column C1 subjected to an external compressive axial load, the other is for beam B1. The generated moment-curvature loops and corresponding system parameters are presented in Figure 4.

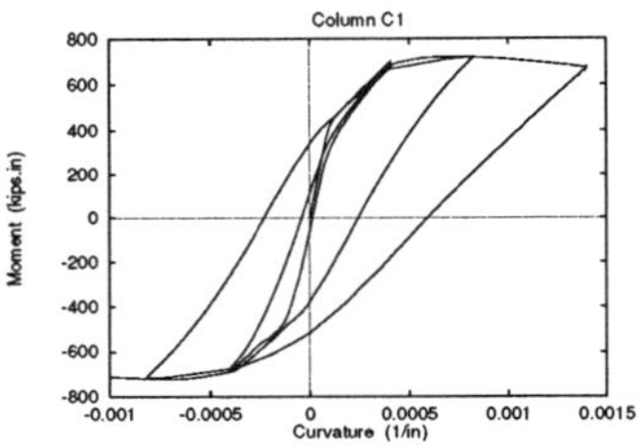

System Parameters

M_{cr}= 400 kip·in
ϕ_{cr}= 10^{-4} 1/in
M_y= 700 kips·in
ϕ_y= $4*10^{-4}$ 1/in
$(EI)_3$= 0
α= 0.94
β= 0.08 1/kips
ϕ_u= 1.36×10^{-3} 1/in

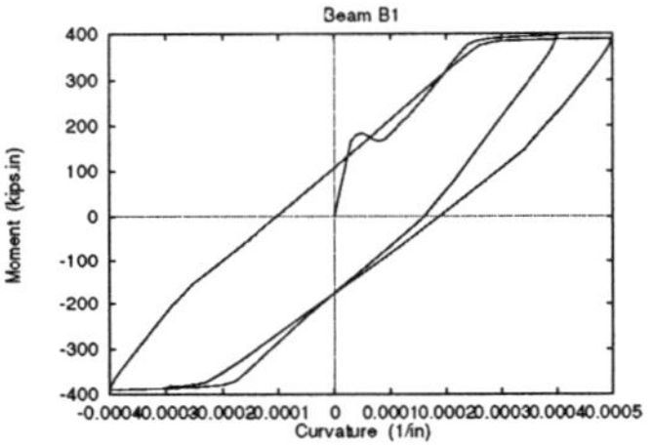

System Parameters

M_{cr}= 130 kip·in
ϕ_{cr}= 2.9×10^{-5} 1/in
M_y= 360 kip·in
ϕ_y= 2.4×10^{-4} 1/in
$(EI)_3$= 10^5 kip·in^2
α= 1.05
β= 0.33 1/kips
ϕ_u= 6.04×10^{-3} 1/in

Fig.4. Simulated Hysteresis Loops and Identified Parameters

These parameters can be input into any of the well-known computer programs for inelastic damage analysis such as IDARC[1] or DRAIN-2DX to determine the response of the entire structure under general loading conditions.

REFERENCES

1. Kunnath, S. K.et al., "IDARC version 3.0: A program for the inelastic damage analysis of reinforced concrete structures", Technical report NCEER-92-0022

2. Richard, R. M. and Robert, B. J., "Versatile elastic-plastic stress-strain formula", Technical notes, Journal of the Engineering Mechanics Division, ASCE, Vol. 101, No. EM4, August 1975, pp. 511-515.

3. Mander, J. B. et al., "Seismic energy based fatigue damage analysis of bridge columns: Part I- Evaluation of seismic capacity", technical report NCEER-94-0006.

4. Darwin, D., "Inelastic model for cyclic biaxial loading of reinforced concrete", thesis represented to the University of Illinois at Urbana-Champaign in partial fulfillment of the requirement for the degree of Doctor of Philosophy, 1974.

5. Karsan, D. C., "Behavior of plain concrete under variable load histories", thesis represented to Rice University in partial fulfillment of the requirement for the degree of Doctor of Philosophy, 1968.

DEFLECTION OF BEAMS WITH INTEGRAL ELASTIC SUPPORTS

Karl K. Stevens[1]

Abstract

Maximum deflections (displacements normal to the center-line) of cantilever and fixed-fixed beams with integral elastic supports are determined using the finite-element program IMAGES©. Factors by which deflection calculations based on the assumption of rigid supports can be corrected to account for the effects of deformations in the supports are presented for select cases where the elastic modulus of the support is greater than, equal to, or less than that of the beam. Small elastic displacements and plane-stress conditions are assumed; the effects of fillets at the interface between the beam and sujport are not considered.

Introduction

All "fixed-end" beams actually terminate in supports that are less than perfectly rigid. The resulting shear force and bending moment transmitted to the support results in localized deformations that can contribute significantly to the overall beam deflections (O'Donnell, 1960). Here, we consider beams with integral, but dissimilar, elastic supports. This work is a modification and extension of that of Cook (1967) and Ramakrishna (1994).

A key question is how best to represent the effects of deformations in the support on overall beam deflections. Cook (1967) curve-fit deflection values obtained using finite element methods to the parabolic deflection shape predicted by simple beam theory. The location at which the slope of this curve becomes zero was then determined and used to establish the length of an equivalent cantilever with rigid supports. A similar approach was used by Ramakrishna (1994). Beam deflections obtained from finite element calculations were substituted into the deflection equations obtained from simple beam theory, resulting in a set of linear, over-determined, algebraic equations

[1] Associate Dean, College of Engineering, Florida Atlantic University, 777 Glades Road, Boca Raton, FL, 33431.

with the beam deflection and slope at the support as unknowns. These equations were solved in a least-squares sense using the matrix pseudo-inverse.

Further work by this author has verified that the approaches used by Cook (1967) and Ramakrishna (1994) both tend to give erratic results, strongly dependent upon the number and location of the deflection "data" points used. Equations involved in the computations are badly conditioned. A sketch of the deflected shape of the beam, along with simple geometric considerations, readily reveals why this may be so. Cook (1967) reported that his results were consistent, possibly because his finite element results may have been of very high precision. Here, we simply report factors by which results calculated for beams with perfectly-fixed ends can be corrected to account for the effects of deformations in the supports. These factors can be obtained consistently and reliably, using realistic finite element methods.

Finite Element Model

The cantilever configuration considered is shown in Figure 1. By enforcing the symmetry condition of zero longitudinal displacement at the free (left) end, this model also represents a fixed-fixed beam of length $2L$ subjected to a loading of $2P$. A perfect bond between the beam and support is assumed. For the case where the beam and support have the same elastic moudulus, this configuration represents a beam machined integral with its supports.

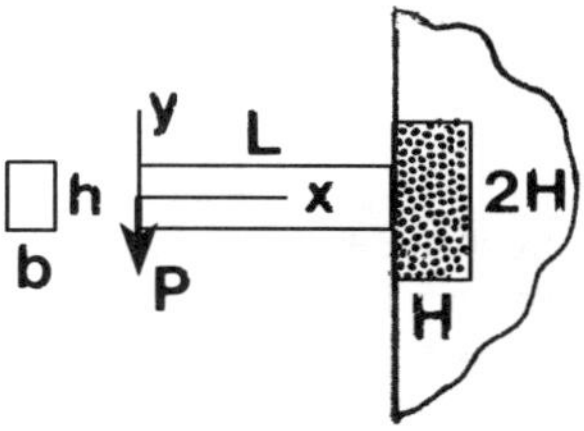

Figure 1. Beam Configuration

With the further assumptions of small elastic displacements, plane-stress conditions, and equal values of Poisson's ratio for the beam and support, dimensional analysis provides the following relationship for the beam deflection, y:

$$y/y_0 = f(L/h, E_s/E_b, \upsilon) \quad (1)$$

Here, y_0 is the maximum theoretical bending deflection of an identical beam with rigid supports, L/h is the beam slenderness ratio, E_s/E_b is the ratio of the elastic modulus of the support to that of the beam, and ν is Poisson's ratio.

Rectangular plane-stress elements were used to calculate the system strain energy, U, and the displacements at all nodes. The mesh size was adjusted to give accurate results with reasonable computational effort. "Check" solutions for beams with rigid supports gave deflection results that differed from theoretical values by less than 1%.

All results presented in this paper are based on the parameter values $\nu = 0.3$, $b=1$, and $L/h=5$, with $E_s/E_b = 0.1$, 1, and 10. For this value of the slenderness ratio, shear effects are small, but not negligible. This is particularly true for the fixed-fixed case, for which bending effects are smaller than for the cantilever. Shear effects are discussed in more detail in the *Results* section.

The semi-infinite support was modeled as a rectangular region (shown shaded in Figure 1) sufficiently large that displacements at all nodes on its outer boundary could be considered zero. These nodes were completly restrained in the finite element calculations. An appropriate boundary for the support region was determined by increasing the size of the region in increments until the incremental change in system strain energy (proportional to the beam tip displacement) was less than 1%.

Effect of the support region size on the strain energy of a cantilever system with $E_s/E_b = 1$ is shown in Figures 2 and 3. The region approaches the support conditions of a semi-infinite plane, as defined in the preceding paragraph, at a value of H/h ≈3.

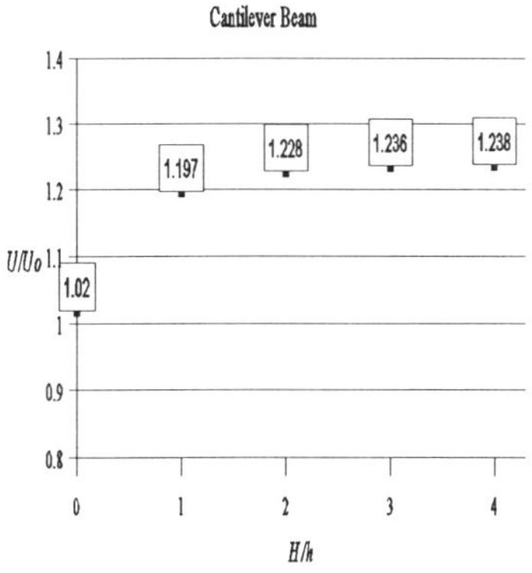

Figure 2. Effect of Support Region Size on System Strain Energy (Es/Eb=1;L/h=5;ν=0.3;b=1)

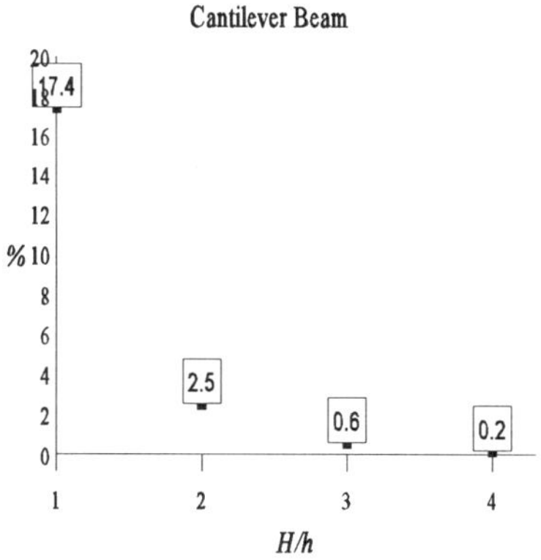

Figure 3. Incremental Change in System Strain Energy (Es/Eb=1;L/h=5;ν=0.3;b=1)

Minimum acceptable values of *H/h* for all configurations considered are summarized in Table 1. As expected, the required values of *H/h* decrease as the relative stiffness of the support increases.

Table 1
Minimum Required Support Size, *H/h*
$L/h=5$; $b=1$; $\nu=0.3$ (Fig. 1)

E_s/E_b	0.1	1	10
Cantilever	4	3	2
Fixed-Fixed	4	3	2

Results

Table 2 provides factors, k, by which the maximum deflections can be obtained using the relationship

$$y_{max} = k y_0 \quad (2)$$

where y_0 is the theoretical maximum bending deflection for identical beams with fixed supports. It is important to note that these correction factors include the beam deflection due to shear. The relative effects of shear increase as the stiffness of the support increases, because of the smaller overall deflection. For $E_s/E_b = 10$, shear contributed about 3% to the maximum deflection and system strain energy for the cantilever considered; the corresponding value for the fixed-fixed configuration was 12%.

Table 2
Deflection Correction Factors, k (Eq.2)
L/h=5; H/h=4; b=1; ν=0.3 (Fig. 1)

E_s/E_b	0.1	1	10
Cantilever	3.108	1.239	1.042
Fixed-Fixed	2.508	1.327	1.131

Direct comparisons with the results of Cook (1967) and O'Donnell (1960) are not possible. Their results do not include shear deflections in the beam and their beam and support confiturations are not exactly the same as considered here. However, for the case of a cantilever with $E_s/E_b = 1$, approximate comparisons can be made. By computing the shear deflection from equations found in many strength of materials texts and adding it to results obtained from Cook (1967) and O'Donnell (1960), we get k=1.29 (Cook) and k=1.30 (O'Donnell). The comparable value from Table 2 is k=1.24, for a maximum difference of about 5%. This further confirms the reliability of the results presented here.

References

Cook, R.D., 1967, "Deflections of a Series of Cantilevers Due to Elasticity of Support", *J. of Applied Mechanics*, Vol. 34, No. 3, Transactions ASME, pp. 760-761.

O'Donnell, W.J., 1960, "The Additional Deflection of a Cantilever due to Elasticity of Support", *J. of Applied Mechanics,* Vol. 27, No. 3, Transactions ASME, pp. 461-464.

Ramakrishna, S.B., 1994, *Identification of Support Conditions on a Beam Using Finite Element Analysis,* M.Sc. Thesis, Florida Atlantic University, Boca Raton, FL.

Comparisons Between Laboratory Measured and FWD Backcalculated Resilient Moduli

Anand J. Puppala, A.M., ASCE[1], Steven L. Cumbaa[2] and William H. Temple[3]

Abstract: Several factors influence the backcalculation analysis methods performed on FWD measured data. This paper compares the backcalculated resilient moduli of four programs, Modulus, Modcomp3, Bousdef and Peddl with the laboratory measured moduli. Comparisons are also evaluated with respect to the pavement configuration aspects.

Introduction

Flexible pavement design, evaluation and overlay designs require material characterizations of individual pavement layers. The fundamental property required in the flexible pavement material characterization is known as the resilient modulus. Currently, there are two approaches to determine the resilient modulus of pavement layers. The first approach uses repeated load triaxial test or a dynamic natured testing method on the subgrade core samples. The other approach is to conduct non-destructive testing using the falling weight deflectometer (FWD) device on pavements and then use the backcalculation procedures to analyze the FWD data to determine pavement layer moduli. The second approach is considered less expensive, since it does not require expensive sampling and laboratory testing. Also, the second approach is less time consuming than the first approach. Because of the advantages, more transportation agencies are using the falling weight deflectometer test method in their pavement evaluation studies. Several computer programs with different analytical models are available to determine the moduli of pavement layers. A list of these programs is summarized in Roberts et al. 1993 and Uzan, 1994. Often, these solutions provide different moduli for the same pavement layers under similar conditions. This discrepancy is attributed to several factors (Uzan, 1994). The type of analysis, material stiffness variations and accuracy of deflection measuring sensors are a few of the factors that can lead to erroneous interpretations.

[1]Research Associate, [2]Management Systems Research Administrator, [3]Associate Director of Research, Louisiana Transportation Research Center, Baton Rouge, LA 70808

This paper attempts to compare the backcalculated moduli with laboratory measured resilient moduli. Four types of backcalculation programs are used in this comparison. The influence of pavement configuration aspects on the comparisons is also evaluated.

Backcalculation Programs

Several types of backcalculation procedures and their approaches are detailedly documented by Uzan (1994) and Roberts et al. (1993). The final analyses of these procedures depend on several factors such as, the type of analysis (static or dynamic), the type of material characterization (linear or non-linear behavior) and input variables such as the layer thickness and the bed rock depth (Roberts et al. 1993; Uzan, 1994). Four programs are used in the present analysis. These are PEDD1, Bousdef, MODULUS and MODCOMP3. Details on these programs can be found elsewhere (ASTM Proceedings and in Roberts et al. 1993).

Pavement Configurations

The pavement sections are selected from the Louisiana Experimental Base Project, a project study conducted by the Louisiana Transportation Research Center to evaluate several factors associated with the AASHTO flexible pavement design procedures (Hadley, 1983). The test sections are located on US 71-167 in central Louisiana in Rapides Parish which accommodates a moderate volume of mixed traffic. The subgrade material is a fine grained soil ranging from silty clay loam to heavy clay. The test pavements include eighteen sections, fourteen experimental and four control sections (Hadley, 1983). Among these sections, four sections are selected since they represent two types of base materials, black base and soil cement and two different surface thicknesses, 8.5 and 6.5 in. of hot mix asphalt surface. Sections T-1 and T-5, as well as T-2 and T-10, are identical except for the variations in asphalt thicknesses and minor differences in base layer thicknesses. The modular ratio defined as the ratio of elastic moduli of top asphalt and base layers is about 2 for T-1 and T-5 sections and around 1 for the T-2 and T-10 sections. Table 1 provides details of various test sections considered in the present analysis.

Analysis

Laboratory properties such as the resilient moduli of asphalt, base, and subgrades of various test sections are available and they are used to compare with backcalculated moduli. These properties were obtained from the tests conducted by the materials research laboratory at Louisiana Tech and the Asphalt Institute in Lexington, Kentucky (Hadley, 1983). The properties are presented in Table 1.

The fwd test data on these sections was obtained from the field tests conducted with a KUAB type falling weight deflectometer device at an operating temperature of approximately 62° F. Test data obtained with a 9000 lb load was only considered in this analysis. These data were used in the backcalculation programs to interpret the moduli properties. A comparison between the laboratory measured moduli (Table 1) and the backcalculated moduli (Table 2) provides the following observations. All backcalculated programs consistently predicted higher moduli than the laboratory measured moduli. Peddi1 provided reasonable predictions for base layers, however, it overpredicted the moduli of asphalt layers. All four programs predicted subgrade moduli closer to one another and greater than the laboratory moduli values. All four programs have not followed a consistent trend with regard to providing reliable predictions of moduli of each pavement section. In certain sections, certain programs provided better predictions and for other remaining sections, the same programs provided poorer predictions. This observation is in agreement with findings of other investigations (ASTM Proceedings).

Table 1: Details of Pavement Configurations (HMAC - Hot Mix Asphalt Concrete; BB - Black Base; SC -Soil Cement)

Section	**Pavement Material Type**	**Measured Moduli in ksi**
T-1	8.5" HMAC, 8" BB, Subgrade	500 (HMAC), 249 (BB),6.7 (Subgrade)
T-2	8.5" HMAC, 9" SC, Subgrade	500 (HMAC), 557 (SC), 6.7 (Subgrade)
T-5	6.5" HMAC, 7.5" BB, Subgrade	500 (HMAC), 241 (BB), 7.5 (Subgrade)
T-10	6.5" HMAC, 12" SC, Subgrade	500 (HMAC), 554 (SC), 10.0 (Subgrade)

The modular ratio variations appear to influence Bousdef and Modulus program predictions, particularly in the case of base layers. Higher modular ratio values (T-1 and T-5) have resulted in higher backcalculated base layer moduli. In the case of asphalt layers, all four programs provided higher moduli values for T-2 section than for T-10. This indicates that asphalt layer thickness also influences the backcalculations. Overall, no consistent or conclusive trend or pattern with respect to modular ratio is observed when all four program results are grouped together. This may be attributed to the small amount of data used in this analysis and the variations in the backcalculation algorithms used in the programs. This study signifies the need for further studies to understand these aspects.

Summary

This paper compared the predictions of four backcalculation programs with the measured moduli properties of four pavement sections. Backcalculated moduli of pavement layers are significantly higher than laboratory measured properties. Also noted in this analysis is the influence of modular ratio and the asphalt layer thickness on the backcalculations. However, the number of sections analyzed in this paper are not significant enough to draw valid conclusions.

Table 2: Backcalculated Moduli in ksi of Four Programs (1 ksi = 7 MPa)

Section	PEDD1	BOUSDEF	MODULUS	MODCOMP3
T-1	1101 643 17.6	540 1109 17.6	531 6423 15.8	451 6170 14.2
T-2	1030 462 17.7	757 685 18.4	710 420 11.1	1050 188 19.4
T-5	1200 346 16.1	272 1434 15.7	568 3271 13.3	282 428 16.9
T-10	463 565 20.6	256 645 22.3	267 516 12.3	284 4880 18.0

References

1. Hadley, W. O. (1983) "Experimental Base Project Reports 78 - 1 and 78 - 2," Louisiana Tech University, Ruston, Louisiana.
2. Proceedings, Nondestructive Testing of Pavements and Backcalculation of Moduli (Second Volume), ASTM STP 1198, H.L. Von Quintus, A. Bush, III, and G. Y. Baladi, Eds., ASTM, Philadelphia, 1994.
3. Roberts, F.L., Wedgeworth, J.B. and Pumphrey Jr., N. D., "Analysis and Evaluation of Methods for Backcalculation of Mr Values," FHWA/LA-9-263, Louisiana Transportation Research Center, Baton Rouge, 1993, 91 pages.
4. Uzan, J., "Advanced Backcalculation Techniques," Nondestructive Testing of Pavements and Backcalculation of Moduli (Volume II), ASTM STP 1198, H.L. Von Quintus, A. Bush, III, and G. Y. Baladi, Eds., ASTM, Philadelphia, 1994.

Simultaneous Measurement of Strain and Temperature Using Fiber Grating Sensors

Faramarz Farahi[1]

Abstract

An electronic phase demodulation scheme is presented which allows simple and practical method for measuring temperature and strain applied to a fiber Bragg grating sensor. The scheme uses a spectral scanning technique which produces two periodic waves, with their phases simultaneously affected by temperature and strain. However, the effects of temperature and strain on the phase of this periodic signals are not identical. This leads to simultaneous recovery of two measurands, strain and temperature, applied to the sensing element.

Introduction

Recently, a considerable research effort has been focused on the development of "sensing structures," that is, structures that have the capability to provide real-time information on their health and integrity. Fiber Bragg grating (FBG) sensors (Meltz et al, 1989) appear to be ideal for these application, for example, for strain measurement due to their immunity to EMI, their size and format which make them suitable to be easily embedded in the structure to provide absolute strain measurement which is inherently encoded in the wavelength of the signal returned by the device (Measures, 1992). However, for a practical system, the measurement of a small Bragg wavelength shift with conventional spectrometers or monochromators is not a good solution owing to their bulk-optical nature, size and lack of ruggedness and, in some cases, limited resolution. Several techniques have been developed for demodulation of these sensors. Among all these schemes, the filter based technique which uses a second FBG for the interrogation of signal from the sensor FBG has been the most attractive techniques for demodulation of grating bases sensing systems (Davis and Kersey, 1995). It is important to emphasize that in the systems utilizing these tech-

---

1. Professor of Physics, Physics Department, University of North Carolina, Charlotte, NC 28223

niques stability of the optical power and the system loss are essential. Additionally, all these type of sensors have a drawback since they are not specific. For example, a fiber strain sensor shows serious sensitivity to temperature, therefore, strain measurement can only be performed in a well controlled environment (Farahi, 1993). This paper presents an optical configuration which uses a matched FBG grating to detect the signal in reflective mode. In addition, it will also be shown that this technique can be easily implemented to provide simultaneous measurement of strain and temperature. The presented signal recovery technique can produce two periodic output signals with their phases proportional to the temperature and strain applied to FBG sensor. The major advantage of this scheme is its immunity to optical power fluctuations in the system. A system using this signal recovery is inherently independent to variation in the power of the optical source and the loss in the system.

The configuration used to demonstrate this signal recovery scheme is shown in Fig.1. A superluminescent source was used (λ_{peak} = 826.7 nm, $\Delta\lambda$ = 20.2 nm) to illuminate the system. The spectral characteristics of the sensing grating, FBG1, and processing grating, FBG2, were measured at room temperature and with no axial strain applied to the fiber grating, they were; λ_B = 835.4 nm and $\delta\lambda$ = 0.2 nm. Light reflected from the sensing grating was sent to the processing grating using a 3 dB coupler, C1. Light after being reflected again from the processing grating was detected by a photodiode using coupler C2.

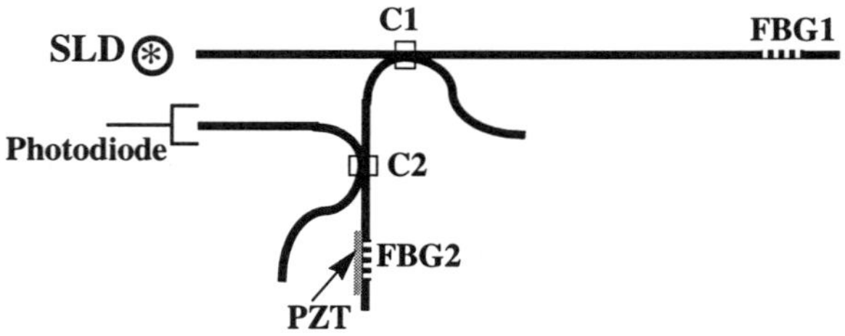

Figure 1: Schematic of the FBG sensor with dynamic tuning of a matched grating.

A sawtooth strain signal with a large amplitude at frequency of 3 Hz was applied to this grating to periodically match and mismatch the processing grating with respect to the sensing grating. The photodiode output is a sinusoidal signal with its phase linearly dependent on the applied strain to the sensor (FBG1). This phase can be measured when it is compared relative to a reference signal with a stable phase. In this experiment a lock-in amplifier was used for this purpose. Fig. 2 presents some of the results obtained and demonstrates that this signal recovery scheme is immune to fluctuation of optical power in the system. The following steps were taken to obtain these results. The injection current applied to SLD was kept constant to achieve stable optical power. The phase shift was recorded over a period of approximately 20 seconds for different values of applied axial strain to FBG1. Midway through the experiment, the strain applied to the sensing fiber was kept constant and the injection current to SLD was varied over a range from 120 mA to 135 mA, that corresponds to ~ 35% change in the optical power. As it can be observed form Fig. 2, the measured electronic phase shift is unaffected by variation in optical power. Results of Fig. 3 were obtained for quasi static applied strain over a wide range. It is evident from this plot that the electronic phase shift of output signal is linearly dependent on the applied strain to the sensing fiber. The DC noise level was measured and it was correspond to a $0.19^{o}/\sqrt{Hz}$. Considering the slope of this plot, $0.24^{o}/\mu\varepsilon$, and the conversion factor of the lock-in amplifier, $18^{o}/V$, the static strain sensitivity of $0.8\ \mu\varepsilon/\sqrt{Hz}$.

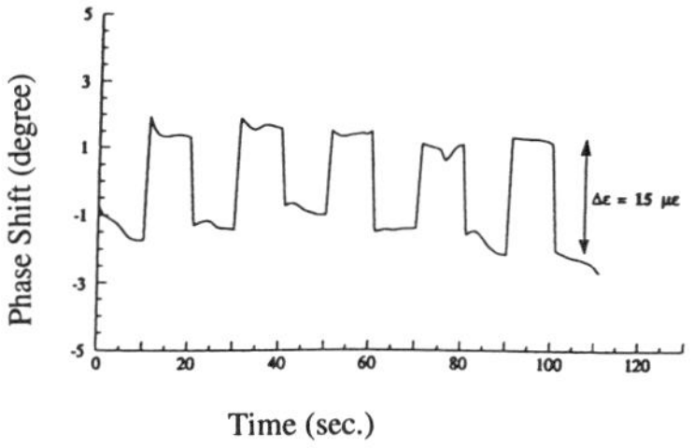

Figure 2:
Phase shift for different values of strain and different injection current to SLD (different optical power).

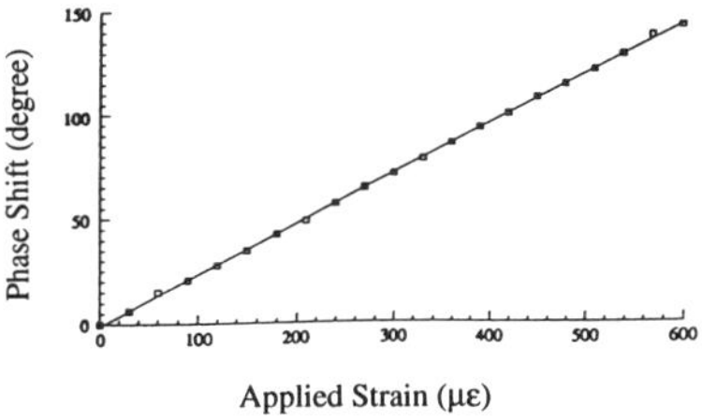

Figure 3:
The measured phase shift as a function of applied strain to the sensing fiber.

Simultaneous recovery of strain and temperature can be achieved by using two superimposed gratings with spatial periods Λ_1 and Λ_2 to form a sensing element (Fig. 4). At the receiving end, two gratings with the same spatial periods Λ_1 and Λ_2, but at two different positions, are attached to two piezoelectric transducers. These transducers can be modulated at frequencies f_1 and f_2 to produce two signals similar to those obtained when only one grating was used. The phase of each signal is now a linear function of temperature and strain applied to the sensing element, i.e.

$$\Delta\phi_1 = \alpha_T^1 \Delta T + \beta_\varepsilon^1 \Delta\varepsilon$$
$$\Delta\phi_2 = \alpha_T^2 \Delta T + \beta_\varepsilon^2 \Delta\varepsilon \quad \text{(EQ 1)}$$

where $\alpha_T{}^i$ and $\beta_T{}^i$ are thermal and strain coefficients of the sensor for grating of spatial period Λ_i. This set of equations can be used to simultaneously determine two unknowns, ΔT and $\Delta\varepsilon$. These two equations are independent since, in general, coefficients α_T^1, α_T^2, β_ε^1 and β_ε^2 are unique and different from each other.

In conclusion, the feasibility of a new signal recovery scheme for fiber Bragg grating sensor has been experimentally demonstrated. This scheme uses a processing fiber Bragg grating similar to that of sensor which is periodically scanned to produce an output signal. It has been shown that the phase change of this signal is linearly related to strain applied to the sensor grating. The simplicity, ruggedness, small size, and immunity to optical power fluctuations are the main features of the sensing system utilizing this demodulation scheme. Another major advantage of this signal recovery scheme is its capability to be readily used for simultaneous measurement of strain and temperature.

Acknowledgments

This work is in part supported by NSF grant DMI-9413966.

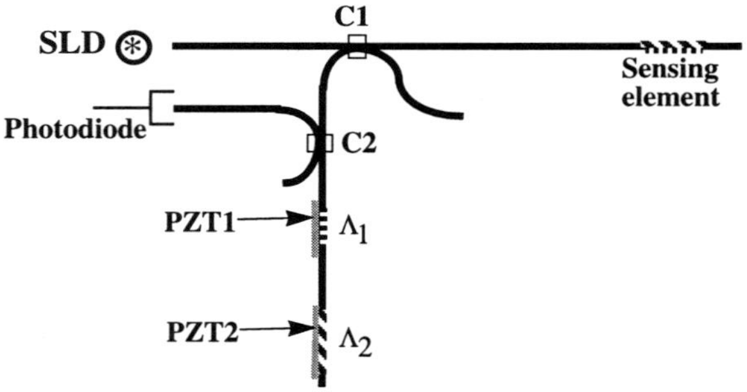

Figure 4: Schematic of the FBG sensor for temperature and strain measurement.

References

Davis, M.A., and Kersey, A.D. (1995), Matched-filter interrogation technique for fiber Bragg grating array, *Electron. Lett.*, **31**, 822.

Farahi, F. (1993), Strain and temperature measurement using optical fiber, *Application of Fiber Optic Sensors in Eng. Mechanics*, By: F. Ansari, Published by ASCE, 236.

Measures, R.M. (1992), Smart composite structures with embedded sensors, *Composites Eng.*, **2**, 597.

Meltz, G., Morey, W.W., and Glenn, W.H. (1989), Formation of Bragg gratings in optical fiber by a transverse holographic method, *Opt. Lett.*, **14**, 823.

Fiber sensors for damage detection on large structures and for assessment of deformation behavior of cementitious materials

W. R. Habel, D. Hofmann, B. Hillemeier, F. Basedau[1]

Abstract

Optical fibers adhered to steel and concrete surfaces for on-line crack detection and localization of the damaged area as well as fiber optical microstrain sensors for deformation measurements of concrete elements and mortar samples at early ages are presented. Examples of application of these sensor types point out their capability in comparison with conventional sensors or methods. For the first time, deformations of mortar in the first hours after casting could be measured by directly embedded glass tubes.

Introduction

The use of fiber optical sensing technique provides new opportunities in monitoring and evaluation of structure components and building materials. On the one hand, long optical fibers adhered on structure surfaces or embedded in hydraulically curing building materials, are able to detect crack formation including a remote localization of the crack point along the fiber (section A.). On the other hand, strain sensitive fiber optical sensor arrangements allow one to record very small changes of deformation or to measure vibration parameters of structure components (section B.).

Section A.: Crack monitoring on large structures

Old steel structures which have reached the end of their service life have to be evaluated. Cracks form, as often as not one does not know the area in which they have formed. Single cracks found in prestressed concrete structures, on the other hand, have to be monitored in order to obtain information about their condition.

A cheap way of obtaining early information about a deterioration in structure quality is the attachment of a fiber slope to the component's surface. However, optical fibers - polymeric or glass fibers with a polymeric coating - normally possesses a high tensile strength. When cracks appear in the matrix material the fiber commonly slips and cannot

---

[1] All with the Institut für Erhaltung und Modernisierung von Bauwerken (IEMB) Berlin
(Institute for Rehabilitation & Maintenance of Structures)
c/o TU of Berlin, FB 9, Straße des 17. Juni 135, 10623 Berlin
Correspondence to be addressed to Prof. Dr. Hillemeier

indicate the crack activity. For this reason, the tensile strength must be decreased by etching the fiber surface or by partially removing the coating.

Application examples:

In order to detect cracks in an old Berlin steel bridges made of puddle iron, a special sensitizing technique and an attachment method usable on site were used to make fiber slopes fit for long-term monitoring of steel structures. Fatigue tests on real structure parts in a testing hall of the Federal Institute for Material Research and Testing (BAM) Berlin, until formation of fatigue cracks, enabled the detection of the crack formation already during period of plastic deformation round the crack tip by recording the optical intensity signal of the slope. Fig. 1 clearly shows the changes in signal when the crack forms. The first order of the signal, from the crack formation until its opening (center curve in fig 2.), could be used as a trigger signal for e. g. traffic affection. In laboratory tests a threshold of response of about 10 ... 30 μm crack width could be achieved for such fiber slopes.

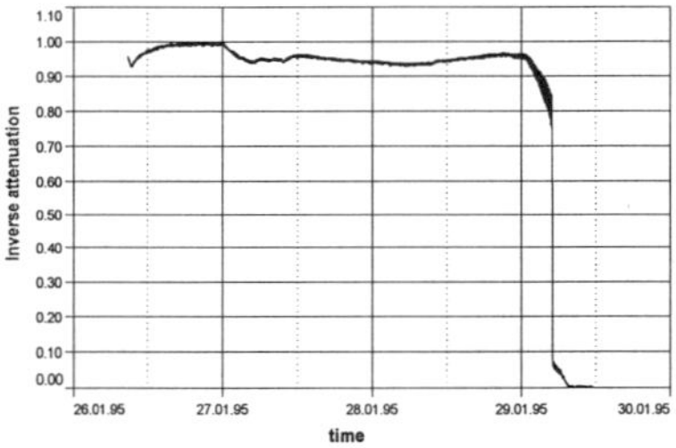

Fig. 1. Slope intensity signal when a crack appears during the fatigue test

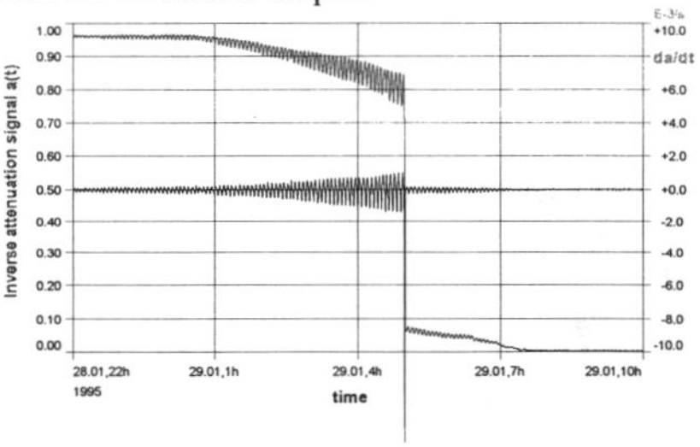

Fig. 2. Detail of fig.1; center curve shows the first order of the slope intensity

The same method was used to prepare fiber slopes for application on concrete surfaces inside a box girder of a prestressed concrete bridge. The attachment method for concrete surfaces differs from that for steel. The adjustment of the crack detection threshold for a fiber to be adhered to concrete is more difficult; in laboratory, a threshold range of 50 μm up to 250 μm - depending on the adhesive and the fiber treatment - was achieved for crack formation detection. Such a fiber slope is being tested at a prestressed concrete bridge in Berlin in order to monitor the propagation of existing cracks.

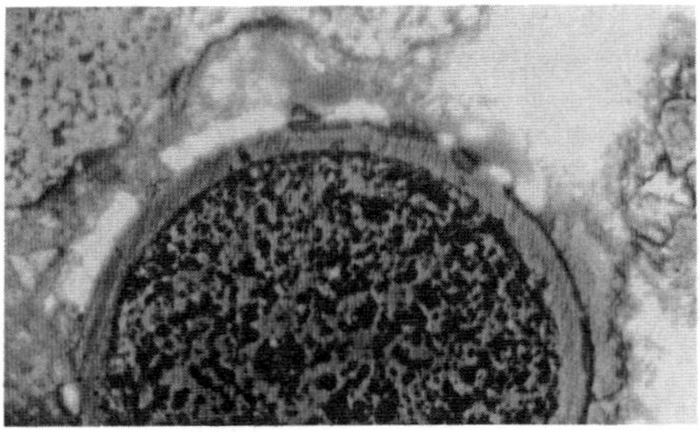

Fig. 3. Coating impairment due to contact with mortar

The use of sensitized fiber slopes for the detection of single cracks or crack fields inside tank walls, which are possibly formed during the curing process due to constraint, is a very important task. However, the sensitizing procedure of the optical fibers has to differ considerably from the application on surfaces. We took advantage of the fact of a self-sensitizing process (due to contact to a cementitious material with a special admixture). First tests in laboratory confirmed that untreated (non-sensitized) optical fibers survived the casting and compaction process, but the polymeric coatings were influenced by chemical attack during the hydration reaction. One of the coatings tested was influenced on its surface in such a way as to be used as sensitized fiber (fig. 3). This weakening of the quite hard and relatively resistant coating material is the basis for our self-sensitizing method. Very small cracks in the coating material could also be observed in the micro section of samples.

Section B.: Micro deformation measurements at early ages

The majority of the measuring tasks on structures concerning their static and dynamic behavior can be solved by using resistive strain gages, displacement meters or vibration sensors. In some cases, e. g. in power plants, in the vicinity of motors or high speed railroad lines, high electromagnetic interference leads to distortion with the data recording. On the other hand, very small deformations are to be measured in not completely cured concrete, which means, sensors must not stiffen the measuring object. All types of conventional sensors make the measuring object stiffer than it would be without sensor; the deformation behavior of the measuring object is disturbed. This causes problems, e. g. when mortar deformations are to be measured in the interphase between rheology and solid state.

Fiber optical micro strain sensors of Fabry-Pérot type, including its leading cables, enable the solution of these measuring tasks , because they are extremely small, and it is even possible to design an arrangement which does not stiffen a soft measuring object.

The Fabry-Pérot sensor type is comprehensively described by Murphy et al. (1992) and Habel et al. (1996). It consists of a hollow glass tube, about 15 mm long, which can be embedded into building material. Bare (uncoated) optical fibers are inserted into the opposite sides of the tube, in such a way as that their ends form a gap inside the tube. Before fixing the fibers to the ends of the tube by fusion, a specific distance s_o between both fiber end faces has to be adjusted (about 25 μm). The measured signal results from the superposition of coherent light waves reflected in the two fiber end faces and changes when the tube is deformed. Due to this superposition fringes are formed, and on the output of the sensor (one optical fiber of the sensitive element) the intensity signal of the interfering light can be recorded.

A typical resolution of this sensor arrangement in combination with the equipment is 0.06 με; at maximum 0.047 με. The sensitivity is strongly affected by the way of attachment or embedment.

Application examples:

The effectiveness of a combined heating and cooling procedure during construction of large concrete tanks in a sewage treatment works was to be evaluated. For this purpose a number of conventional strain sensors (resistive strain gages) embedded in a rather massive mortar prism (stiffness !) were installed in the reinforcement cage of the tank wall. Because of great demands on resolution (< 1 με), additional fiber optical microstrain sensors were installed. Interferometric sensors of Fabry-Pérot type were used again.

In order to minimize the affection of the measuring zone by the sensor, these sensor tubes (∅ = 0.25 mm) were equipped with thin metal strips in both ends and covered with a thick silicon rubber layer afterwards. These strips protruded from this rather soft material and could clinge to the surrounding mortar during casting. For this arrangement the load transfer from the measuring object to the sensitive element showed a satisfactory linearity, however a low sensitivity. This method certainly reduces stress concentrations which are caused by shrinkage of the concrete at stiff mortar prisms containing conventional sensors. Such disturbances produce an error of measurement, which is especially significant when very small deformations are measured.

In order to take full advantage of the resolution of interferometric fiber sensors, a direct bonding of the sensor tube to the mortar, as far as possible, was aspired. For this reason, the design of an appropriate coating is essential.

A number of tests were carried out in order to evaluate the behavior of the coating material when are in direct contact with cementitious materials (high alkalinity in the fresh state of mortar/concrete). Two specially modified polymeric coatings and one inorganic one showed satisfactory results. Fig. 4 shows the interface region of an inorganically coated sensor between the sensitive element (cylindrical tube) and the surrounding mortar. The bonding behavior was good and provides a linear load transfer over the deformation range.

Two kinds of sensor designs for a direct bonding to mortar were used and investigated in laboratory:

- a stiff sensor tube for recording strain deformations of cured mortar bodies, which was to be compressed as well as influenced by moisture and temperature;
- a stress-free sensing arrangement on the basis of a Fabry-Pérot sensor specially developed for this purpose, which allowed us to record deformations of mortars at early ages which have only just developed a very low strength.

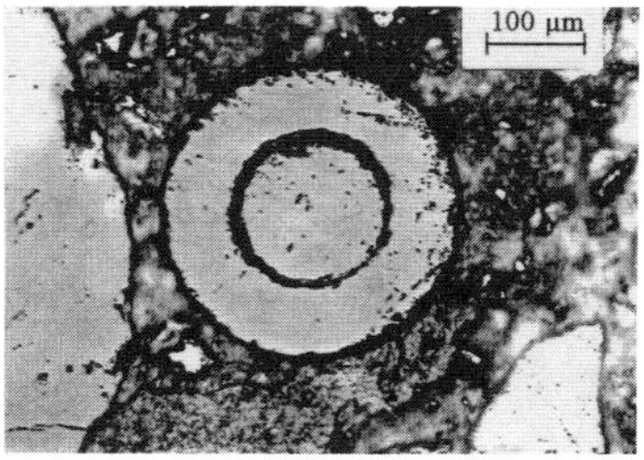

Fig. 4. Micro section of a carbon-coated fiber sensor embedded in cement mortar (grain size: 0/4 mm, age: 8 months)

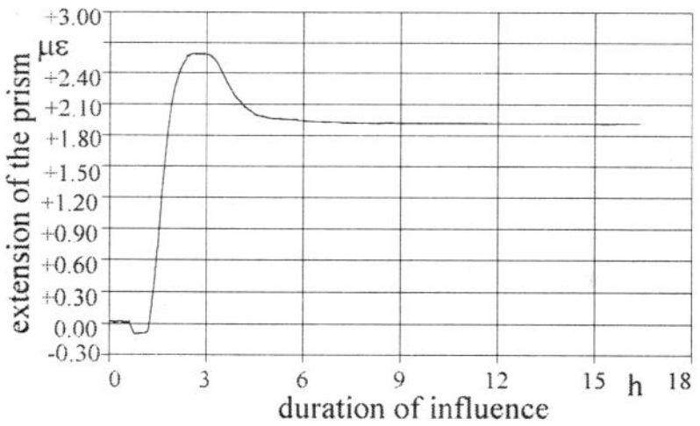

Fig. 5. Length changes of an hydrating injection mortar prism (Sensor: non-reactive fiber sensor)

It has to be pointed out that a stiff sensor with a hard inorganic coating showed good results in hardened mortars, but it is not suitable for recording deformations at an early state of curing. In this case we used the non-reactive sensor arrangement, coated with a special polymeric material, and applied a special embedment method. Fig. 5 shows the measured deformation of an injection mortar with swelling behavior in the first hours after casting.

References

Habel, W.R. et al. 1996: Deformation measurements of mortars at early ages and of large concrete components on site by means of embedded fiber optic microstrain sensors. To be published in Cement and Concrete Composites. 1996

Murphy, K.A. et al. 1992: Extrinsic Fabry-Pérot optical fiber sensor. 8th Opt. Fib. Sens. Conf., Jan. ´92 Monterey, CA. (USA). Conf. Proc. 193-196.

Embedded Fiber Optic Displacement Sensor for Concrete Elements

Xi Chen[1], Farhad Ansari[2], and Hong Ding[1]

Abstract

Development of a fiber optic sensor for embeddment in concrete and other types of cementitious composites is described. The sensor transduction mechanism is based on the measurement of the mean absolute intensity variation of the speckle pattern generated at the output end of the optical fiber. Calibration results on 2.5 cm gauge fiber sensors indicate a resolution of 1.2 microstrain per unit of optical signal (MSIV), and a dynamic range of 4000 microstrain. Fiber optic sensors were employed for measurement of the crack tip opening displacements (CTOD) in notched concrete beams, and results are compared with crack opening displacements measured at the crack mouth (COD).

Introduction

Over the past few years, a number of embedded optical fiber sensors have been developed for measurement of strains and deformations in a variety of materials (1978, 1988). One of the earliest applications of optical fiber sensors to concrete was developed for the determination of entrained air void system in fresh concrete (1990, 91). Other embedded fiber sensors have also been developed for applications to concrete structures (1992). Optical fiber sensors can be categorized into different types, depending on the application and requirements in terms of sensitivity and dynamic range. Review of technical literature is beyond the scope of the present work. It suffices to indicate that in general, optical fiber sensors can be classified under three categories, namely, phase, intensity, and spectrometric types.

----

[1] Research assistant, Smart Sensors and NDT Laboratory, New Jersey Institute of Technology (NJIT), Newark, NJ 07102

[2] Professor and Director, Smart Sensors and NDT Laboratory, New Jersey Institute of Technology (NJIT)

The present work involves development of a new fiber optic sensor for measurement of internal strains and deformations due to cracks in cementitious composites. Strains and deformations are measured based on mean intensity variation of speckle patterns due to the mode redistribution within a multimode optical fiber. The development and calibration of this fiber optic sensor for embeddment within cementitious composites is described. To examine the applicability of the sensor for measurement of displacements associated with cracks, optical fibers were embedded in single-edge notched three point bend concrete specimens. One parameter of significant importance in determining fracture properties is crack tip opening displacement (CTOD) (1988,89). For instance, crack extension or fracture may be assumed to occur as soon as the crack opening displacement exceeds a critical value. To date, it has not been possible to measure CTOD, and instead, researchers have resorted to measurement of crack openings at the mouth of the crack by means of clip gauge extensometers, or Linear Variable Differential Transformers (LVDT). Therefore, the crack mouth opening displacement (COD) has been employed in the technical literature in lieu of CTOD. The new optical fiber sensor allows for direct measurement of CTOD.

Sensor Methodology

The transduction mechanism for the sensor is based on correlation between the applied strain to the fiber, and the light intensity variation of speckle patterns due to mode redistribution within a multimode fiber. A frame grabber image digitizer is employed for acquisition of speckle patterns from the output end of the fiber. Image data is then numerically evaluated according to the formulations described in the following

The undisturbed (unstrained) speckle field of observation plane at any position *(x,y)*, assuming that all the fiber modes (total number of *M*) are equally excited, can be considered as the superposition of all modes. The complex phasor of the *m*th mode at point Q in status S_o can be expressed as (1994):

$$e_{om}(x,y) = a_{om}(x,y)\exp[j\phi_{om}(x,y)] \tag{1}$$

where $a_{om}(x,y)$ and $\phi_{om}(x,y)$ are the amplitude and phase respectively, and light intensity field at point Q is:

$$E_o(x,y) = \sum_{m=1}^{M} e_{om}(x,y) = \sum_{m=1}^{M} a_{om}(x,y)\exp[j\phi_{om}(x,y)] \tag{2}$$

When the fiber status change as a result of perturbation, the speckle field emerging from the fiber will also be changed. The light intensity variation caused by fiber status change between status S_o and the new fiber status S at point Q can be written as:

$$\Delta I(x,y) = |E_o(x,y)|^2 - |E(x,y)|^2 \tag{3}$$

where $E_o(x,y)$ and $E(x,y)$ are the initial and the new light fields at point Q. For simplicity, we drop the *(x,y)* coordinate in the following derivation. By integrating over the speckle field of the observation plane, the Mean-absolute Speckle Intensity Variation (MSIV) can be evaluated as:

$$MSIV = \iint |\Delta I| dxdy \tag{4}$$

Assuming the fiber is perturbed by a strain $\varepsilon = \Delta L / L$, and L is the length of the probing section (gauge length), we have

$$MSIV \approx \varepsilon \iint L \left| \sum_{m=1}^{M-1} \sum_{n=m+1}^{M} 2a_{om}a_{on}(\frac{2\pi}{\lambda})\xi\eta(\frac{1}{\cos\theta_M} - 1)\frac{n-m}{M}\sin\varphi_{nm} \right| dxdy \tag{5}$$

where $\varphi_{mn} = \phi_m - \phi_n$, ($\phi_m$ and ϕ_n are the phases for *m*th and *n*th modes respectively), θ_M is the incident angle of the *M*th modal light field with respect to the fiber axis, η the refractive index of the fiber, ξ is the strain optics correction factor, which has typical value of 0.78, λ is the wavelength of the laser source.

From Eq.(5), we know that the MSIV will be linearly proportional to the fiber perturbation, e.g. the strain ε.

Experimental Investigation

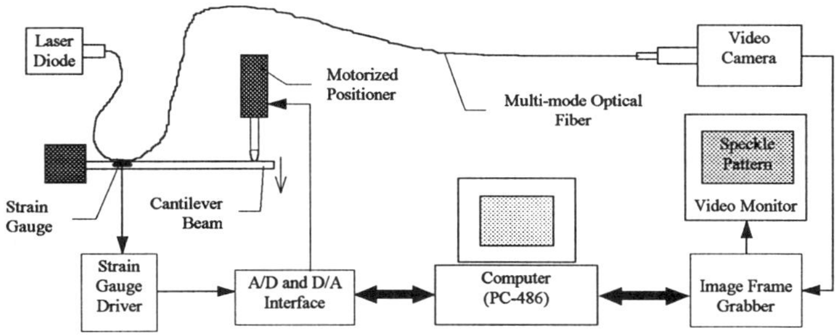

Fig. 1 Calibration test setup

Correlation between the displacements and corresponding signal modulations were accomplished through calibration. The configuration shown in Fig.1 corresponds to the calibration setup by which the mean absolute speckle intensity variation (MSIV) is related to CTOD values. As shown in Fig.1, the multimode fiber is adhered to one side of a plexi-glass cantilever beam. A strain gauge is affixed on the opposing surface of the beam with respect to the fiber, and the beam is loaded via a motorized micrometer positioner. A visible laser diode with a wavelength of 670 nanometer is used as input light source for the fiber, and a video camera captures the speckle pattern at the output end of the fiber. An image capture board is used for digitization and storage of the speckle pattern for subsequent numerical operations by a microcomputer. A video monitor is also employed for real-time display of the speckled image from within the fiber. The entire calibration process, including application of displacement at the tip of the cantilever, and acquisition of data is performed by a 486 processor. The adhered length of the fiber (2.5 cm) comprises the gauge length of the fiber. Typical calibration results include variation in mean absolute speckle intensity (MSIV) as a function of strain, and as shown in Fig.2, this relationship is very linear (R^2=0.98).Correlation between deformations and MSIV is simply computed from integration of strain over the 2.5 cm gauge length. This procedure is not as accurate (especially for large gauge lengths) as a direct correlation between MSIV and displacement. Elastic analysis of deformation on the surface of the cantilever at the location of the sensor indicated six percent computational error. To obtain optimum correlation results, it was decided to perform the calibrations on-line with the actual experiments. In this way, variations in the calibration constants due to differences in optical fiber characteristics would be totally eliminated. The on-line calibration cantilever beam is shown along with the experimental three-point-bend beam setup in Fig.3.

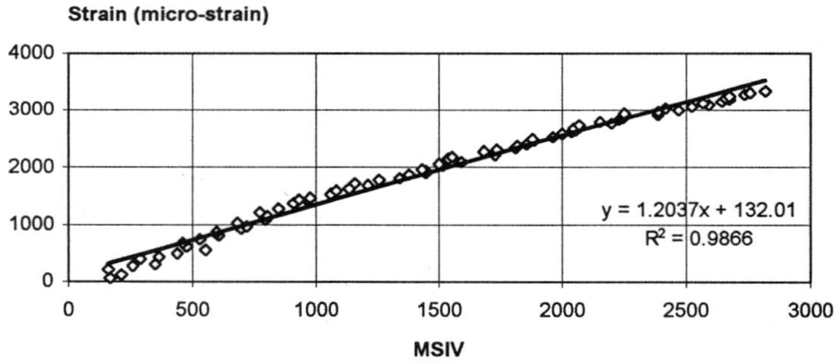

Fig. 2 Variation of MSIV as a function of strain

All specimens were 10cm×10cm×86cm, and tested on a span length of 62 cm under three-point bending. The mix proportions by weight of the concrete used in this study were designed to be 1:1.1:2.0:0.35:0.10, corresponding to cement:sand:aggregate: water: silica fume by weight. Type I Portland cement conforming to ASTM C 150, ASTM No. 2 grade river sand passing through sieve No. 8, and maximum coarse aggregate size passing 9.5 mm and retained on No. 4 sieve were used. Specimens were cast in plexiglass molds, and all the specimen had a center edge notch of 5 cm . The 2.5cm gauge optical fiber sensor was placed at the tip of the notch (approximately at 2mm above the notch tip) by the aid of two thin bars of 0.5 mm diameter each. Specimens were cured in a curing room for about three weeks before testing. A number of companion cylinders were also cast. The average compressive strength of the concrete was 44 MPa. As shown in Fig. 3, the three point bend tests were performed in a closed loop testing system under constant rate of Crack Mouth Opening Displacement (COD) in order for the crack to grow in a controlled manner. COD was measured by an LVDT attachment.

Results, Discussions, and Conclusions

The resolution and dynamic range of the speckle based optical fiber sensor can be obtained from the calibration test results (Fig.2). As our experiments indicate, the resolution, and the dynamic range are functions of the gauge length. This will be discussed in future publications as our experiments are still at a stage of infancy. For the 2.5 cm gauge length reported here, and as shown in Fig.2, the sensor's strain

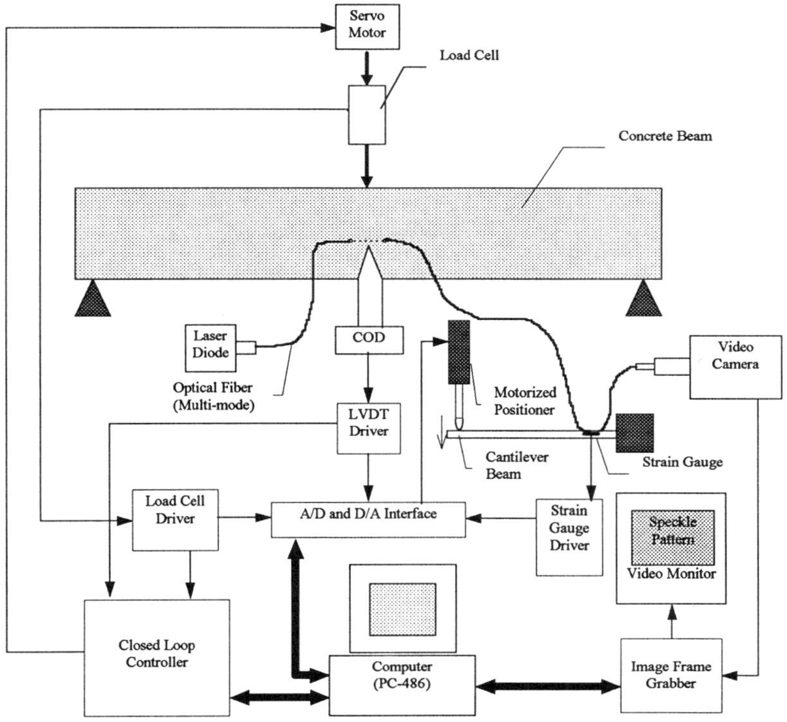

Fig. 3 Three point bend test setup

resolving power, and the dynamic range are 1.2 microstrain per MSIV, and 4000 microstrain respectively. The dynamic range of the sensor is sufficient for applications in conjunction with cementitious composites

Results given in Fig.4 depicts measured COD, CTOD, and load data in a typical three point bend test against time. Comparison of CTOD, and COD results given in Fig. 4 indicate the microcrack detection capability and resolution power of the fiber

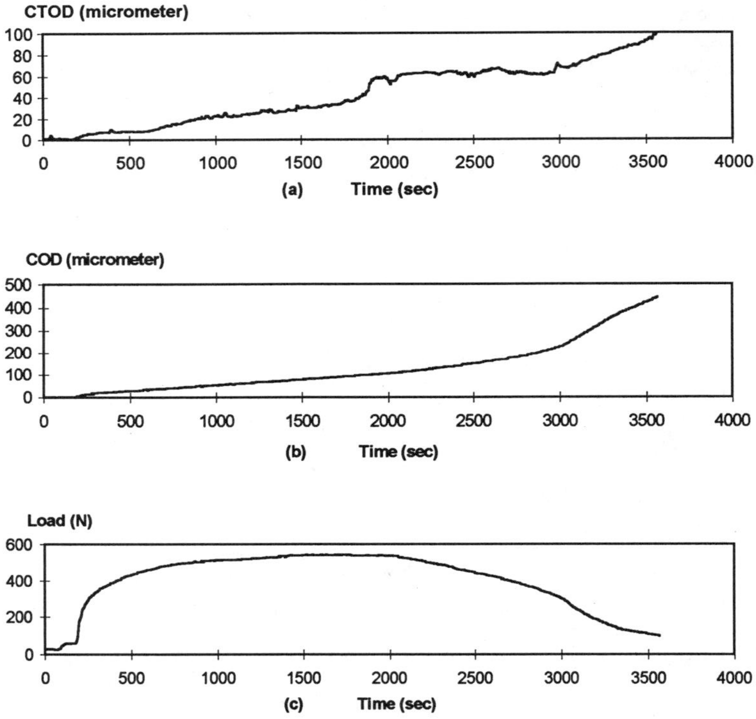

Fig.4 Experimental data from a typical three point bend test

optic sensor in terms of displacements. Change in the slope of CTOD versus time data in Fig.4(a) signify opening of microcracks within the material. COD and CTOD values are further compared in the load deformation diagram given in Fig.5. CTOD results indicate more detailed information about the microcracking process within the damaged zone in concrete. The signal to noise ratio for the speckle based sensor is rather high (S/N = 250), and the noisy appearance of the CTOD curve may be misleading. This curve is indicative of the actual opening and closing of the microcracks in the damaged zone. minute crack closing displacements shown in Fig.5 are attributed to the control signal associated with the closed-loop displacement rate parameter. COD values are much larger than CTOD results mainly due to its distance being further away (by 5 cm) from the center of curvature of the deflected beam.

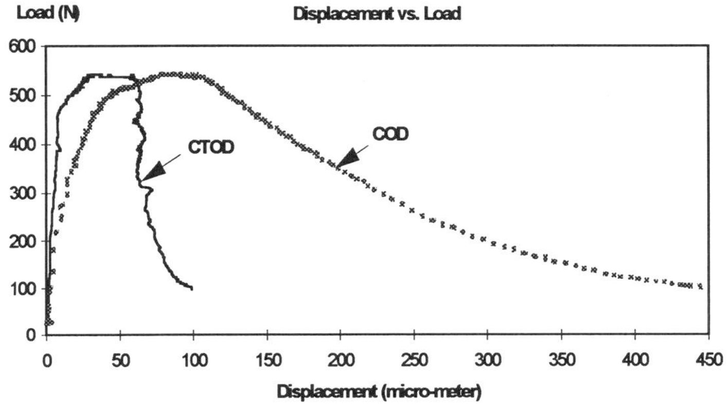

Fig.5 Load-deformation diagram for a typical three-point-bend test

Preliminary results from the experiments reported here are promising. The sensor developed in the present study is very low cost as it employs a typical multimode telecommunication fiber. The fiber does not require any modifications, and it can be used as is. The optical fiber is 250 micron in diameter, and its embeddment in concrete does not perturb the matrix.

Acknowledgment

This study was made possible by a grant from the National Science Foundation, grant number CMS-9402671, to the New Jersey Institute of Technology.

Appendix I-References

Ansari, F. (1989), "Mechanism of Microcrack Formation in Concrete," ACI, Mat. Jnl, 86 (5), 459-464.

Ansari, F. (1990), "A New Method for Assessment of Air Voids in Plastic Concrete," Cement and Concrete Research, 20 (6), 901-910.

Ansari, F., Chen, Q.Y. (1991), "Fiber-Optic Refractive Index Sensor for Use with Concrete," Applied Optics, 30 (25), 236-239.

Ansari, F. (1992), "Real-time Condition Monitoring of Concrete Structures by Embedded Optical Fibers", ASCE-EMD, SP, Ansari, F., Sture, S., Eds, 49-59.

Butter, C.D., Hocker,G.B. (1978), "Fiber Optic Strain Sensor," Applied Optics, 17, 2867-73.

Claus, R.O., Bennett, K.D., May R.G. (1988), "Optical Fiber Methods for the NDE of Smart Skins and Structures", Proc. Fall Conf., SEM, 19-24.

Pan, K., Uang, C. (1994), "Multimode Fiber Sensing by Using Mean-absolute Speckle-intensity Variation", Applied Optics, 33 (10), 2095-2098.

Looping Behavior and Strength of Prestressed Arches

Zefang Xu[1], Student Member, ASCE, and Amir Mirmiran[2], Member, ASCE

Abstract

Buckling of prestressed elastica arches and effects of loading conditions on the stability and strength of arches have been studied. A prestressed arch is made by elastically buckling an initially flat thin strut into a curved shape, and then, attaching it to the supports. A modified Riks-Wempner method is incorporated in a corotational finite element formulation to effectively trace the snap-through, snap-back, and looping paths of arches.

Introduction

With the advent of stressed-arch frames for large clear-span structures in Australia and the U.S. (Clarke and Hancock 1995), stability of arches has received renewed attention in recent years. Compared with a rigid arch that is rolled to produce a purely stress-free system, prestrains and prestresses exist at each section of a prestressed elastica arch. The induced stresses are functions of the geometric shape of the arch (i.e., elastica curve). A prestressed arch returns to its original flat configuration, if released from its supports. Two widely different methods; the discrete element, and the nonlinear finite element, have usually been applied to the large displacement analysis of rigid arches. However, these methods are not readily applicable for the stability analysis of prestressed elastica arches, since elastica arches are formed through a sequential fabrication process. Therefore, an algorithm is required to follow the loading history of the arch during its assembly. The few studies on the stability of prestressed elastica arches all indicate that they are generally less stable than their equivalent rigid arches. However, these studies were limited to the initial segment of the post-buckling path, since they used an augmentation technique for traversing the limit points. The present study utilizes a

[1] Doctoral Student, Dept. of Civ. & Envir. Engrg., Univ. of Central Florida, Orlando, FL 32816.
[2] Asst. Prof., Dept. of Civ. & Envir. Engrg., Univ. of Central Florida, Orlando, FL 32816.

modified Riks-Wempner method that is embedded in a corotational finite element formulation to effectively trace the snap-through, snap-back, and looping paths of an arch in its entirety. The model is augmented with a special algorithm that simulates the sequential fabrication process of elastica arches. A computer program is developed based on the mathematical model, and its accuracy is confirmed by comparing the results for several benchmark problems with the available analytical solutions. Effects of location and inclination angle of concentrated loads have been investigated. These parameters not only affect the magnitude but also the shape of the response. Results indicate that a prestressed arch is least stable when a point load acts within the range of 10 to 25 percent of the span length from the crown. The exact range depends on the inclination angle of the load.

Method of Analysis

A corotational displacement-based formulation of a two-dimensional, two-node straight beam element is developed to model the prestressed arch at all stages of construction. The model incorporates large displacements, large rotations, and material nonlinearities. A local convected coordinate system is attached to the beam element in such a way that it moves and rotates with the element, but does not deform with it. As such, one can easily separate the large rigid body displacements from the infinitesimal deformational displacements of the element, and discard the former as they do not affect the strains. In order to traverse the limit points along the equilibrium path, the model is augmented with the modified Riks-Wempner method (Tsai and Palazotto 1990). A special numerical algorithm is devised to trace the stresses and strains at each section of the prestressed arch throughout its fabrication process. The model considers two load stages. In the first stage of construction, a flat strut is subjected to an axial compressive load with a magnitude larger than its Euler buckling load, and a moment equal to 5% of the axial load to initiate the buckling process. During this stage, the rise-to-span ratio of the arch is continuously monitored until it reaches the desired value, upon which the buckling load is removed and the arch is attached to its supports (pinned or clamped). The elastica arch thus formed is then subjected to the specified loads, which is considered as the second load stage. A Fortran-77 computer program has been developed to perform the above procedures. The program can be used for any structural frame, rigid or prestressed, that could be modeled with two-dimensional straight beam elements. The specifics of the program and its underlying mathematical models are described in detail elsewhere (Xu 1995).

Results

1. *Looping Behavior of a Prestressed Elastica Arch*

A prestressed elastica arch with a prismatic rectangular cross section and pinned supports, subjected to a concentrated load at its crown is shown Figure 1.

Huddleston (1970) and Mirmiran and Amde (1995) have discussed the stability of prestressed arches, and calculated the starting portion of the post-buckling path of the arch. The former used a shooting method to solve a set of nonlinear boundary value problem, while the latter used a displacement finite element with an augmentation scheme. The present method, however, can continue to trace the entire looping paths up to point *c* for the symmetric mode and point *d* for the asymmetric mode (see Figure 1). In the figure, the normalized load parameter PL^2/EI is plotted against the normalized deflection v/L, where P = applied load at the crown, L = length of arch centerline, E = modulus of elasticity of the arch material, I = moment of inertia of the arch section, and v = vertical deflection at the crown. Table 1 shows a comparison of the bifurcation and limit loads at points *a* and *b* for the present method as well as the shooting method by Huddleston (1970). Close agreement is noted. Several other benchmark problems have been analyzed (Xu 1995), and it is concluded that the program can be reliably used to study the complete buckling response of arches and frames.

Table 1. Bifurcation and Limit Loads for Centrally Loaded Prestressed Elastica Arch

	P_aL^2/EI	P_bL^2/EI
Shooting method (Huddleston 1970)	34.2	48.3
Present method	33.96	51.34
Difference (%)	0.70	6.29

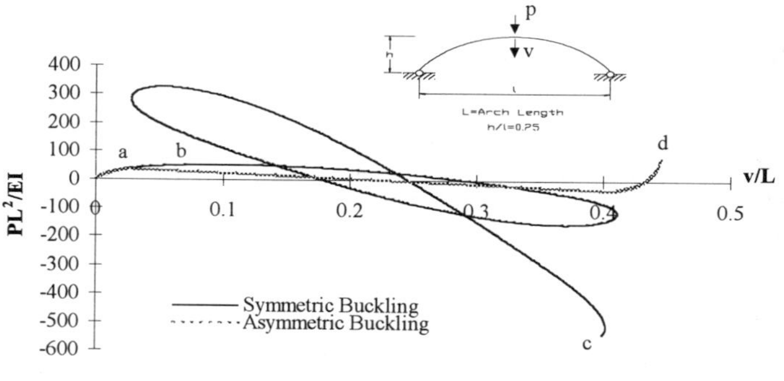

Figure 1. Looping Paths for Prestressed Elastica Arch under Point Loading

2. *Effect of Loading Conditions*

A prestressed elastica arch with the same characteristics as the one in Figure 1 is used for the parametric study. The location and the inclination angle of concentrated loads have been chosen as parameters. The load parameters for the first buckling points are plotted against the location of the applied load, i.e., a/l, in Figure 2. The figure includes three curves for different inclination angle of the load, i.e., 0, 30, and 45 degrees, respectively. At each inclination angle, there is a range within which the bifurcation load of the arch is not affected by the location of the load. Moreover, within this range, the arch is least stable. As the inclination angle increases, the width of this range decreases. It is also clear from Figure 2 that for a point load at the crown, the load's inclination angle slightly reduces the stability of the arch, as the bifurcation load for 45° and 30° inclination angles are about 3% and 1% lower than that of a vertical load, respectively.

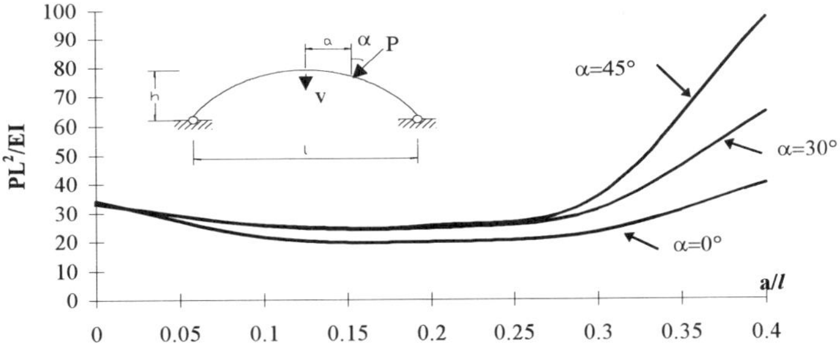

Figure 2. Critical Load versus a/*l* Ratio for Point Loads

References

Clarke, M. J., and Hancock, G. J. (1995). "Design of top cord of stressed-arch frames." *J. Struct. Engrg.*, ASCE, 121(2), 201-213.

Huddleston, J. V. (1970). "Behavior of a steep prestressed arch made from a buckled strut." *J. Applied Mech.*, ASME, 37(4), 984-994.

Mirmiran, A., and Amde, A. M. (1995). "Effects of fabrication process on prestressed composite arches." *J. Struct. Engrg.*, ASCE, 121(1), 124-131.

Tsai, C. T., and Palazotto, A. N. (1990). "A modified Riks approach to composite shell snapping using a high-order shear deformation theory." *Comput. Struct.*, 35(3), 221-226.

Xu, Z. (1995). "Buckling and postbuckling of prestressed arches using corotational finite element." M. S. thesis, Univ. of Central Florida, Orlando, Fla.

Collapse Analysis of Steel Frame Structures Under Earthquake Loading

Scott C. Martin[1] and Roberto Villaverde,[2] Member, ASCE

Abstract

Current code provisions for aseismic design rely on the capacity of the structure to resist deformations beyond its elastic limit. As a result, a severe ground motion may induce large lateral displacements and, hence, a partial or total collapse by lateral instability. An analytical tool is thus needed to check a structure's safety margin against collapse, and to identify the structural components that may be in need of reinforcement. This paper describes a proposed algorithm to perform this type of analysis for three-dimensional steel frame structures, and presents the results of a numerical test carried out to illustrate its application and its effectiveness. The algorithm is formulated on the basis of a step-by-step finite element analysis and utilizes nonlinear large-deformation three-dimensional shell elements to account for material and geometric nonlinearities and spread of plasticity. It detects a partial or total collapse when a zero or negative eigenvalue is found, and operates on the updated stiffness matrix of the structure at the time the zero or negative eigenvalue is found to identify the buckled regions that lead to its collapse

Introduction

As is well known, the foremost objective of seismic codes is to insure structures will resist severe shaking without collapse. However, the damage and many collapses in Mexico City during the Michoacan earthquake of September 19, 1985, demonstrated that current design provisions, which allow inelastic deformations, but do not require a nonlinear analysis, may not be sufficient to prevent collapse (Villaverde, 1991). The Northridge earthquake of January 17, 1995, has also called into question the fundamental design of moment-resistant connections in steel structures and their ability to reliably achieve the ductility for which they are designed. Both the building codes of the United States (ICBO, 1988) and Mexico (Garcia-Ranz and Gomez, 1988) incorporate the same ductility based design philosophy to resist collapse. This philosophy dictates that critical areas within a structural member, whose yield strength may be exceeded during a severe earthquake, be detailed to achieve a large level of ductility and produce a stable hysteretic behavior, so that the entire structure will also be ductile and display stable hysteretic behavior (Bertero, 1991). The main

[1] EQE International, 44 Montgomery St., San Francisco, CA 94104
[2] Civil Eng. Dept., University of California, Irvine, CA 92717

concern when designing for inelastic behavior is thus the prevention of brittle fracture or severe buckling in areas that undergo inelastic deformations or areas that are adjacent to these.

In view of the possibility of a partial or total collapse in structures subjected to earthquakes, it becomes necessary to know what is their safety margin against collapse and what are their components whose reinforcement may help to increase this safety margin. At present however, there are no tools or analytical methods whereby one can identify the location of failed members or areas of local instability within a member or connection. Furthermore, there are no tools available for the identification of the failure path that leads to a collapse mechanism. The purpose of this paper is thus to propose an algorithm for the detection of a partial or total collapse in steel frame structures and the identification, through the precise delineation of its unstable degrees of freedom, the factors that lead to such a collapse.

Proposed Algorithm

The proposed algorithm, described in detail by Martin (1995), utilizes (a) a finite element formulation; (b) a detailed modeling with shell elements to capture local buckling, spread of plasticity, and the effect of these two factors on global deformation; (c) a step-by-step numerical solution of the nonlinear equations of motion which result from considering both material and geometric nonlinearities, and (d) an examination of the structure's effective stiffness matrix after it has been triangularized to solve through Gaussian elimination these equations of motion at each time step. The technique used for the detection of a partial or total collapse and the delineation of the regions of the structure that are responsible for this collapse is based on the following four premises:

(1) A structure becomes unstable, that is, experiences a partial or total collapse, when one or more of the diagonal elements (pivots) in its factored effective stiffness matrix become zero or negative, and the time at which such instability occurs is when these zero or negative pivots appear for the first time.

(2) If the first zero or negative pivot encountered in the factorization of the effective stiffness matrix of the structure corresponds to its i-th degree of freedom, then an unstable region of the structure is the region that contains degrees of freedom 1 to i.

(3) A structure in a state of instability possesses a rigid body mode whose values are either zero or unity; the zero values correspond to the stable degrees of freedom and the unit values to the unstable ones.

(4) If a zero or negative pivot corresponds to degree of freedom i, then the last unstable degree of freedom in the unstable portion of the structure is degree of freedom i.

Application of Algorithm

To illustrate its basic aspects, the proposed collapse algorithm is applied to the three dimensional steel frame shown in Figure 1. This frame is a simple two-story, two- bay frame with a setback, typical of some laboratory specimens in shaking table experiments. It is designed according to the 1994 seismic provisions from the American Institute for Steel Construction (AISC, 1994). As such, it meets the requirements in regard to: (a) minimum base shear, (b) column-beam moment capacity ratio, (c) interstory drift limit, and (d) minimum shear strength for the panel zones of the beam-to-column connections. Notwithstanding, these requirements are met only marginally with the purpose of facilitating a buckling-initiated collapse.

In the first step of the analysis, the frame is modeled with a collection of shell elements. These shell elements constitute the flanges and webs of its beams, columns, and connections. 5,300 shell elements with 5,700 nodes and 34,000 degrees of freedom are used in the model considered in this study. Then, a time-history analysis of the model is performed when its base is subjected to a portion of the three components of the ground motion recorded at the Newhall - LA County Fire Station during the Northridge earthquake of January 17, 1994 (Darragh et al., 1994). For the purpose of this analysis, the accelerations in the two horizontal components are multiplied by a factor of 1.25, while those in the vertical component are left unchanged. The finite element modeling and the time-history analysis are carried out with NIKE3D (Maker et al., 1991), a nonlinear finite element computer program. The runs are made in a Cray C90 supercomputer at the San Diego Super Computer Center. The CPU time required is approximately 50 seconds per time step. The analysis is continued step by step until a zero or negative pivot is detected in the updated effective stiffness matrix of the frame. This zero and negative pivot is first encountered after 86 steps, or 1.70 seconds after the start of the time marching scheme, indicating thus that at this time the structure is in an imminent state of collapse. Ten nonlinear iterations and eight stiffness reformations are performed during this last time step. The last two effective stiffness arrays are recorded to use them later on in the identification of the buckled regions of the frame responsible for its collapse.

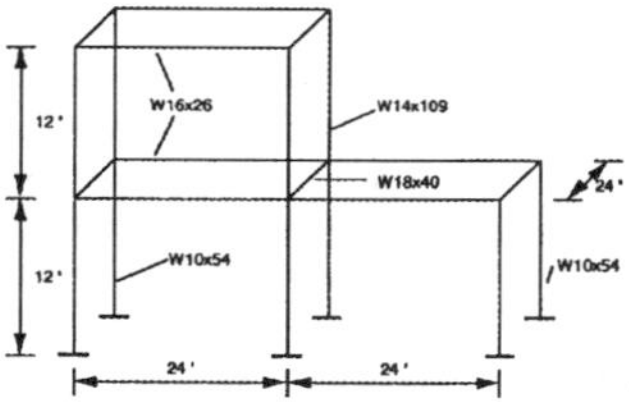

Figure 1. Three-Dimensional Steel Frame (note: 1ft. = 0.3048 m)

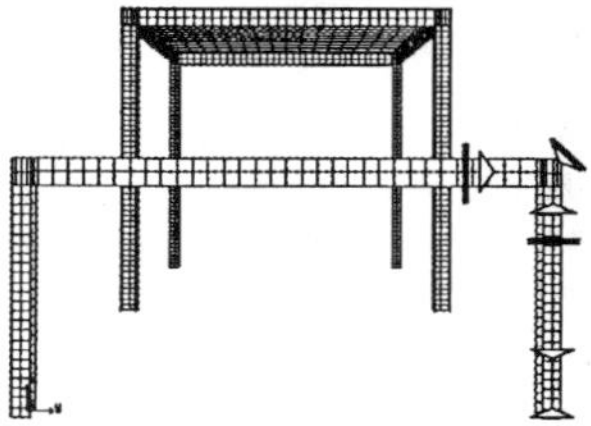

Figure 2. Delimitation of Unstable Degrees of Freedom in 3-D Steel Frame

In the final step of the analysis, the proposed algorithm is used to bracket the regions of the frame that becomes unstable at the time of imminent collapse. Using this algorithm and the saved two stiffness arrays with zero or negative pivots, the regions identified as unstable correspond to those surrounding the connections of the W10x54 column in the first story, rear portion of the frame. These regions are shown in Figure 2, where they are identified by the areas enclosed by the inserted bars and triangles. The first stiffness array with a zero or negative pivot identifies the areas enclosed by the bars. The second identifies those enclosed by the triangles. The buckled regions do not include any of the degrees of freedom within the deck, as all these remain stable at all times. They do include, however, unstable degrees of freedom within the panel zone and continuity plates at the beam-column joint. They also include some stable degrees of freedom within the beams and columns since the bars and triangles in Figure 2 are drawn at the outermost degrees of freedom within each of the unstable regions.

To verify that the regions identified to be the regions where the collapse of the frame originates are indeed so, these regions are reinforced, the frame is analyzed again under the same ground motion, and its response is monitored to see if a collapse occurs again. The reinforcement is limited to the connections of the W10x54 columns and consists of an increase in the thickness of the continuity plates and panel zones to match that of the column flanges. As expected, this time the frame resists the entire ground motion without a partial or total collapse.

Conclusions

The main conclusion that can be drawn from this work is that it is possible to detect with the proposed algorithm the onset of a partial or total collapse in steel frames and delineate the areas of instability where this collapse is initiated. This algorithm may be thus conveniently used to check a structure's safety margin against collapse and to identify critical factors that may affect this safety margin.

Acknowledgments

This research was supported in part by the University of California, Irvine, and the San Diego Supercomputer Center through a generous allocation of computer resources. The authors also wish to express their deep appreciation to B.N. Maker of the Methods Development Group at Lawrence Livermore National Laboratory, who graciously facilitated the use the computer program NIKE3D for this study.

References

AISC (1992). Seismic Provisions for Structural Steel Buildings,. American Institute of Steel Construction, Inc., Chicago, Ill.

Bertero, V. (1991). Design Guidelines for Ductility and Drift Limits: Review of State-of-the-Practice and State-of-the-Art on Ductility and Drift-Based Earthquake-Resistant Design of Buildings," Report No. UCB/EERC-91/15, University of California, Berkeley..

Darragh, R., Cao, T., Cramer, C., Huang, M., and Shakal, A. (1994), "Processed CSMIP Strong-Motion Records from the Northridge, California Earthquake of January 17, 1994," Report No. OSMS 94-06B, California Division of. Mines and Geology.

ICBO (1988), *Uniform Building Code*, International Conference of Building Officials, Whittier, Calif.

Garcia-Ranz, F., and Gomez, R. (1988), "The Mexico earthquake of September 19, 1985 - Seismic design regulations of the 1976 Mexico building code," *Earthquake Spectra*, 4: 427-439.

Maker, B. N., Ferencz, R M ., and Hallquist, J. O. (1991), NIKE3D: A Nonlinear, Implicit, three-dimensional Finite Element Code for Solid and Structural Mechanics: User's Manual, Rept. UCRL-MA-105268, University of California, Lawrence Livermore National Laboratory,.

Martin, S. C. (1995). An Algorithm for the Detection of Collapse in Steel Frame Structures Subjected to Earthquakes and Identification of Causative Factors, Ph.D.Thesis, University of California, Irvine.

Villaverde, R. (1991), "Explanation for the numerous upper floor collapses duriing the 1985 Mexico city earthquake," *Earthq. Eng. Struc. Dyn.*,20, 223-241.

Buckling of Composite Panels with Central Holes

David H. Farnham[1]
Walter J. Horn[2]

Abstract

The buckling of composite plates with circular holes has been conducted An analysis procedure, utilizing the Ritz Method, has been developed to predict the onset of buckling for a wide range of boundary conditions, inplane loads and typical cutout geometries that might be encountered in aircraft applications of composite plates with holes. A computer code was developed to incorporate the method. Results of computational analyses using this code have been compared with those computed with a general purpose finite element code and those obtained by testing of composite plates with circular holes at their center. The results of the comparison indicated good agreement with the finite element code and marginal agreement with the test results.

Introduction

Many aircraft parts such as spars, ribs, frames, bulkheads and floor beams have been designed for fabrication from advanced composite materials because they offer the potential for significant improvements in structural efficiency at reduced cost. Cutouts in the webs of these parts are often necessary to allow for access to internal components and to save weight. These cutouts, however, alter the internal stress field and affect the stability of the web. The objective of this work was to develop an analysis method to predict the onset of buckling for a wide range of boundary conditions, any combination of inplane loadings, and typical cutout geometries that might be encountered in aircraft applications.

Ritchie and Rhodes's paper on the buckling and post-buckling behavior of rectangular isotropic plates with holes (Ritchie and Rhodes, 1975) noted that all previous analysis methods developed to predict buckling loads had followed the same approach of finding the pre-buckling inplane stresses from a plane stress analysis of an infinite plate and then using the Rayleigh-Ritz method to predict the critical buckling load. The deficiency of this method is that the influence of finite plate widths are not accounted for in the inplane pre-buckling stress distributions. They used the finite element method to determine the inplane pre-buckling stress distribution for both the uniform end displacement and the uniform end stress cases. The critical buckling load in each case was determined by minimizing the total strain energy stored in the plate.

Nemeth provided an analytical approach for the buckling of composite plates with cutouts (Nemeth, 1983). He investigated the onset of buckling for a special set of materials, loading, and boundary condition parameters. Composite layups were limited to a class of symmetric layups, known as specially orthotropic layups, for

[1] Senior Engineer, Boeing Commercial Airplane Group, Wichita Kansas, 67277-7730
[2] Associate Professor, Wichita State University, Wichita Kansas, 67260-0044

which $D_{16}=D_{26}=0$. He analyzed uniaxial loading with a constant applied stress along the loaded edges and uniaxial loading with a uniform applied displacement along the loaded side for various boundary conditions on the unloaded edges. Again the buckling problem was divided into two parts; the determination of the prebuckling stresses, and the determination of the buckling load using the prebuckling stresses.

Approach

The approach taken in this study was to expand the analysis method of Ritchie and Rhodes by using the constitutive relations for orthotropic materials. In this two step approach, a finite element analysis was used to predict the stress distributions in the composite plate, and this stress state was then used in a Ritz buckling analysis, to predict the initial buckling load of the composite plate.

Finite Element Analysis

A FORTRAN program was written to generate finite element models of rectangular, composite plates with centrally located elliptical cutouts. This program, created the entire model using the four noded, two dimensional CQUAD4 plate elements. NASTRAN solution 101, a linear static solution, was used to calculate the element normal stresses, σ_x and σ_y, and shear stress, τ_{xy}. Using the appropriate parameters these stresses were output in the material coordinate system instead of an element coordinate system.

Buckling Solution

The strain energy of a plate with negligible transverse normal and shear strains can be wtitten as:

$$U=\frac{1}{2}\int\int\int_V\left(\sigma_x\varepsilon_x+\sigma_y\varepsilon_y+\sigma_{xy}\varepsilon_{xy}\right)dV$$

A laminated composite plate is composed of n layers of elastic orthotropic materials, where each layer obeys a constitutive relationship of the form

$$\begin{bmatrix}\sigma_x\\ \sigma_y\\ \sigma_{xy}\end{bmatrix}_k=\begin{bmatrix}Q_{11} & Q_{12} & Q_{16}\\ Q_{21} & Q_{22} & Q_{26}\\ Q_{61} & Q_{62} & Q_{66}\end{bmatrix}_k\begin{bmatrix}\varepsilon_x\\ \varepsilon_y\\ \varepsilon_{xy}\end{bmatrix}_k.$$

This allows the volume integral to be seperated into piecewise integration through the thickness of the plate, invoking the constitutive relationship for each layer.

The strain energy due to bending can be expressed in terms of the displacements of the plate mid-surface and the integrals can be evaluated through the thickness yielding an expression for the strain energy as an integral over the planar area of the plate.

The theorem of stationary potential energy can be expressed as

$$U+\overline{Q}+Q=\text{stationary value}$$

where U is the strain energy, Q is the potential of external transverse loads and $\overline{Q}$ is the potential energy of the inplane loads. This can be cast in the following form for the case of a symmetrical laminated plate:

$$\frac{1}{2}\iint_A\Big\{D_{11}\left(\frac{\partial^2 w}{\partial x^2}\right)^2+2D_{12}\frac{\partial^2 w}{\partial x^2}\frac{\partial^2 w}{\partial y^2}+D_{22}\left(\frac{\partial^2 w}{\partial y^2}\right)^2+4D_{16}\frac{\partial^2 w}{\partial x^2}\frac{\partial^2 w}{\partial x\partial y}+4D_{26}\frac{\partial^2 w}{\partial y^2}\frac{\partial^2 w}{\partial x\partial y}$$

$$+4D_{66}\left(\frac{\partial^2 w}{\partial x\partial y}\right)^2+N_x\left(\frac{\partial w}{\partial x}\right)^2+N_y\left(\frac{\partial w}{\partial y}\right)^2+2N_{xy}\frac{\partial w}{\partial x}\frac{\partial w}{\partial y}-2qw\Big\}\,dA=\text{stationary value}.$$

where $$D_{ij} = \frac{1}{3}\sum_{k=1}^{n} Q_{ij}^{k}(h_k^3 - h_{k-1}^3),$$

N_x, N_y and N_{xy} are the internal running loads, w is the transverse displacement and q is the applied transverse surface traction.

The criterion for the onset of instability is taken to be the existence of equilibrium solutions other than $w = 0$ for zero applied out-of-plane loading. When such solutions exist, an infinity of non-trivial equilibrium configurations exist..

The Ritz method can be used to reduce the task of finding the extremum of the functional form of the problem to that of finding the minimum of a function. The stability problem is governed by an energy condition of the following form:

$$\Pi\left(u^0, v^0, w\right) = \text{stationary value}$$

where Π is a functional of u^0,v^0 and w. Introducing the following assumed series into this energy expression reduces Π to a function and thus the problem to that of finding the minimum of Π with respect to the generalized coordinates u_i,v_i, and w_i:

$$u^0 = \sum_{i=1}^{n} u_i \Phi_i(x,y), \qquad v^0 = \sum_{i=1}^{m} v_i \Theta_i(x,y), \qquad w = \sum_{i=1}^{p} w_i \Psi_i(x,y)$$

where the functions Φ_i, Θ_i, and Ψ_i are chosen such that they satisfy the kinematic boundary conditions on u^0,v^0 and w respectively.

For the formulation presented, Π becomes quadratic in u_i,v_i, and w_i, and the equations above form a system of linear simultaneous equations in terms of u_i,v_i, and w_i. Thus the problem of determining the functions u^0,v^0 and w which make Π stationary is reduced to the solution of a set of linear algebraic homogeneous equations for the parameters a_{ij}.

The eigenvalue problem described in the preceeding section was implemented in a computer code to generate an eigenvalue problem of the form:

$$[D]\{a_{ij}\} = \lambda[N]\{a_{ij}\}$$

where λ is the eigenvalue, or buckling load and $\{a_{ij}\}$are the eigenvectors, or when used with the displacement equations, w, the buckling mode shapes.

Test Program

Testing was done to verify the analytical development. A test fixture was designed with a Linear Variable Differential Transducer (LVDT) to measure end shortening in the direction of applied load. A shadow Moire system was used to visually determine the onset of buckling and the buckled shape. A system consisting of a mobile LVDT placed perpendicular to the test panel which could be positioned virtually anywhere on the test panel was used to measure the out-of plane displacements of the panels. For each test the mobile LVDT was placed at the point on the test panel calculated to have the greatest transverse displacement at buckling. After buckling the plate, this LVDT was used to map the out-of-plane displacements.

Twenty-four 305mm by 305mm test specimens were fabricated for testing. All panels were made from AS4/PEEK composite lamina of nominal ply thickness of 0.137mm. Fifteen of the panels were sixteen piles thick with a layup of $[45/90/-45/0/(45/-45)_2]_S$. The remaining nine panels were thirty-two piles thick with a layup of $[45/90/-45/0/(45/-45)_2]_S$. The sixteen ply panels were divided into five groups of three with each group having a different size circular hole at its center. These hole sizes were 25.4, 50.8, 101.2 and 127 mm in diameter. One group of panels, used as a control group, had no hole cut in them. The 32 ply panels were also divided into groups of three panels each. The hole sizes for the 32 ply plates were 50.8 and 101.2 mm in diameter. A control group of three panels with no holes were also tested. All panels were fitted with two pairs of back to back strain gauges. The effective test area of the panels, reduced by the presence of the edge supports, was 254mm by 254mm.

The strain reversal method was used to identify the load at the onset of buckling. In this method the strain signals from each strain gage of a back-to-back pair was plotted on the same axes and the load at which one of the strain signals reversed and indicated a decrease in compressive strain for increased load was designated as the buckling load.

Results

Preliminary computer calibration runs provided information to determine that a finite element mesh of 576 elements, zero order Gaussian integration, and eight term displacement series could be used for all analyses. The results from the Ritz buckling solutions using these parameters were compared to the NASTRAN solutions and to the test results in Figure 1.

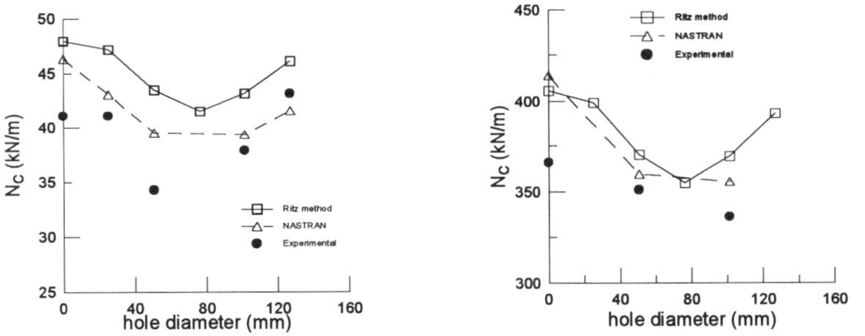

1a. Layup number 1. 1b. Layup number 2

Figure 1. Comparison of the solution scheme with a NASTRAN solution and test data.

Conclusions

1. The Ritz buckling solution for uniaxial loading of composite plates with holes compared well with the NASTRAN buckling solution. The Ritz buckling solution varied from -1.6% less to 10.8% greater than the NASTRAN buckling solutions.
2. The Ritz buckling solution did not compare as well to the uniaxial buckling tests. The Ritz buckling solution varied from 5.4% greater to 26.9% greater than the uniaxial buckling test results.
3. The Ritz buckling solution converged when using zero order Gaussian integration as the number of elements in the finite element mesh was increased. The Ritz buckling solution using first and second order Gaussian integration did not converge when a fine mesh was used.

References

Nemeth M., *Buckling Behavior of Orthotropic Composite Places With Centrally Located Cutouts*, PhD Dissertation, (Virginia Polytechnic Institute and State University, May 1983).

Ritchie D. and Rhodes J., "Buckling and Post-Buckling Behavior of Plates with Holes," *Aeronautical Quarterly* (November 1975): 281-296.

Inelastic Response of Columns After Sudden Loss of Bracing

Rae-Hak Yoo[1] and Raymond H. Plaut[2], Member, ASCE

Abstract

Columns are often braced at one or more locations between their ends. Sometimes this bracing can suddenly be lost, e.g., due to material failure, accident, or explosion. The compressed column then may exhibit large motions and may collapse. The inelastic behavior of columns following a sudden loss of bracing is investigated here. The spread of plasticity during the dynamic response is observed, and the collapse load is computed.

Analysis

A pinned-pinned column of length L with an internal brace at x=a is considered, as shown in Fig. 1. The column is assumed to be elastic, perfectly plastic, and the brace is modeled as an elastic translational spring with stiffness k. The column has an initial deflection $w_0(x)$ (Fig. 1a). Static compressive loads Q_1 and Q_2 are applied at x=a and x=L, respectively, and the resulting equilibrium configuration is denoted $w_s(x)$ (Fig. lb). Suddenly the internal brace and load Q_1 are removed, and the column exhibits a dynamic response with deflection y(x,t) (Fig. 1c). Cases in which the column remains in the elastic range were considered by Plaut and Yoo (1996).

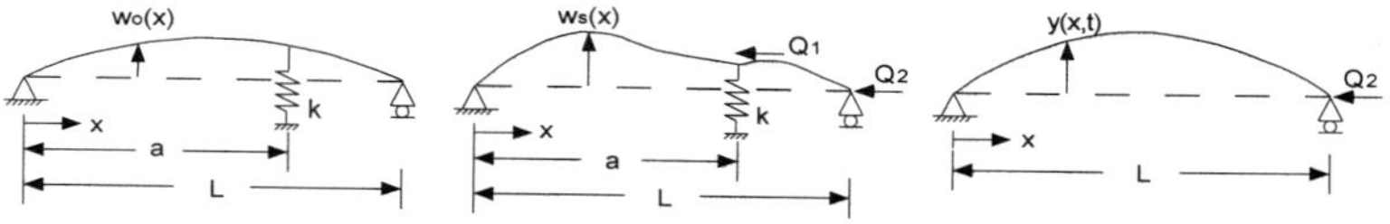

(a) (b) (c)

[1] Grad. Student, [2]Daniel H. Pletta Prof. of Engrg., Charles E. Via, Jr. Dept. of Civ. Engrg., Virginia Polytech. Inst. and State Univ., Blacksburg, VA 24061-0105.

For the numerical results, bending of a wide flange column (W8x40) in the weak direction is considered. The length is L=6.1 m. (20 ft.), the width of the cross section is 20.5 cm. (8.07 in.), the depth is 21.0 cm. (8.25 in.), the web thickness is 0.91 cm. (0.36 in.), and the thickness of the flanges is 1.42 cm. (0.56 in.), giving a moment of inertia about the weak axis of 2,040 cm^4 (49.1 in^4). The mass per unit length is 58.4 kg/m (0.102 slug/in), the yield stress is $2.0x10^{11}$ Pa ($29x10^6$ p.s.i.), and the stiffness of the brace is $k=1.80x10^6$ N/m (10,300 lb/in). This leads to a slenderness parameter of $\lambda_c=1.31$ and an Euler load of $Q_2=1.09x10^6$ N ($2.44x10^5$ lb) for elastic buckling if $w_o(x)=k=Q_1=0$. Residual stresses are included and are given by the piecewise-linear distribution proposed by Galambos and Ketter (1959), with a compressive stress of $7.45x10^7$ Pa (10,800 p.s.i.) at the tips of the flanges and a tensile stress of $4.75x10^7$ Pa (6,890 p.s.i.) in the web.

The initial deflection is assumed to be $w_o(x)=0.001L \sin(\pi x/L)$. The loads are such that the column remains elastic before the bracing is removed, and the linear equilibrium equation is solved analytically for $w_s(x)$, as in Plaut and Yoo (1996). Then the computer program ABAQUS is utilized to determine the dynamic response (Hibbitt et al., 1994). The length of the column is divided into twenty B33 elements of equal length, with eleven integration points used on each flange and three on the web at the end of each element. Newmark's integration method is used with $\gamma=0.5$, $\beta=0.25$, and a time step of 0.002 seconds. If $Q_1=0$, the maximum dynamic response occurs when a=0.5L.

In order to investigate collapse, the loads are increased with a fixed ratio until the response of the column grows indefinitely. The resulting value of Q_2 is plotted as a function of the bracing location in Fig. 2 for three cases: $Q_1=0$, $Q_1=0.25Q_2$, and $Q_1= 0.5Q_2$. When there is no load at the brace, the collapse load is symmetric with respect to the bracing location. For the other two curves, the

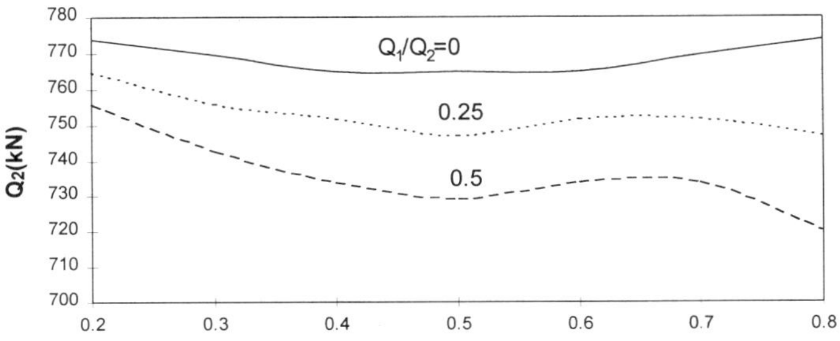

FIG. 2. Effect of Bracing Location on Collapse Value of Q

lowest collapse value of Q_2 occurs when the brace is near the top (i.e., right end) of the column.

The spread of plasticity is studied for two cases with a=0.2, and Q_1=0. The end load is $Q_2=7.56\times10^5$ N (1.7×10^5 lb) in the first case (Fig. 3a) and 8.45×10^5 N (1.9×10^5 lb) in the second (Fig. 3b). At each time step, stresses are checked at the eleven integration points of the flanges at each of the 21 sections along the column, and any point where the stress has reached the yield stress is depicted as a dot in Fig. 3 (which shows a flange at a few selected times).

During the first cycle of motion for the case in Fig. 3a, yielding in compression begins at t=0.038 sec. and is last seen at t=0.104 sec. From the results presented, it is seen that the yielding portion of the column tends to oscillate between the left and right parts of the figure. A small amount of yielding also occurs at t=0.254 sec. and t=0.256 sec. during the second cycle of motion, but otherwise the behavior is elastic. In Fig. 3b, the load is higher than the collapse value. Compressive yielding begins at t=0.026 sec. on the right part of the figure, and yielding in tension begins at t=0.104 sec. The compression and tension zones of plasticity approach each other and the column collapses after t=0.118 sec.

In conclusion, loss of bracing for an imperfect column leads to a dynamic response and possibly to collapse. This study was motivated by the bombing of the World Trade Center complex in New York City in 1993, in which several columns suddenly lost some of their bracing when floors of the parking garage were partially destroyed. Luckily, those columns did not collapse, but a similar situation could lead to a failure. Therefore the type of problem considered here and in Plaut and Yoo (1996) is important.

References

Galambos, T. V., and Ketter, R. L. (1959). "Column under combined bending and thrust." *J. Engrg. Mech. Div., ASCE,* 85, 1-30.

Hibbitt, Karlsson and Sorensen, Inc. (1994). "ABAQUS User's Manual," version 5.4, Pawtucket, Rhode Island.

Plaut, R. H., and Yoo, R.-H. (1996). "Elastic response of columns after sudden loss of bracing." *J. Engrg. Mech.*, ASCE, 122, to appear.

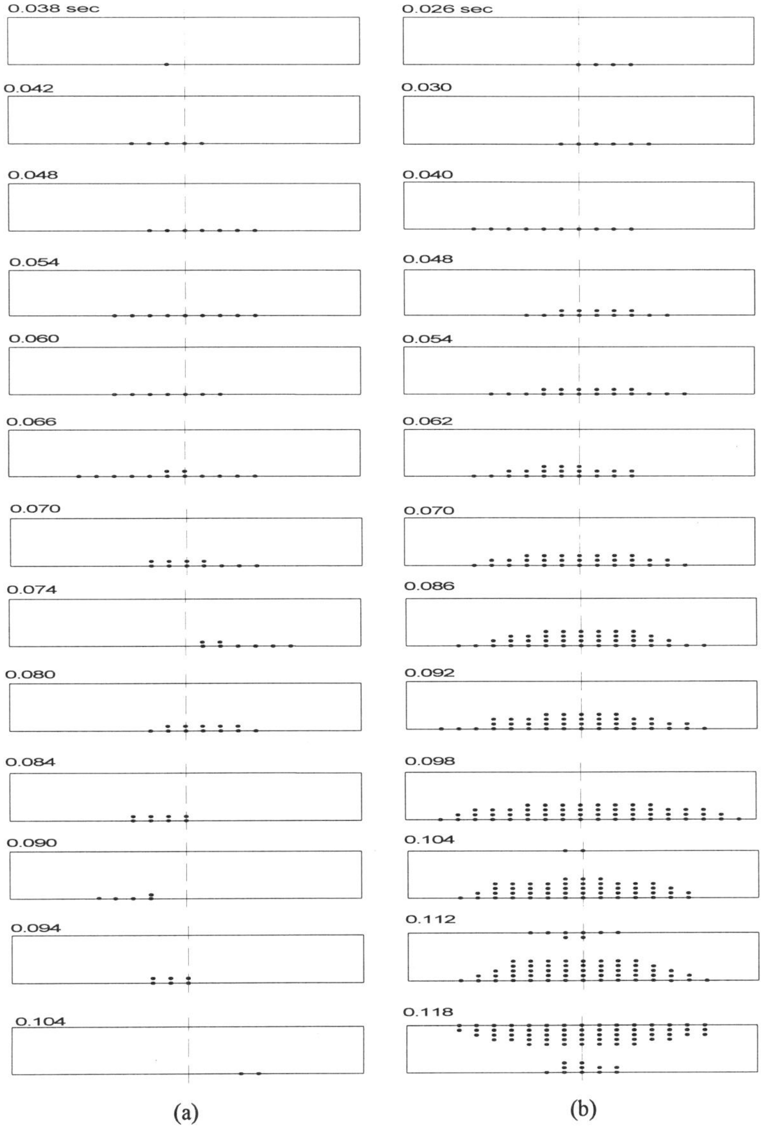

FIG. 3. Spread of Plasticity During Motion for Two Cases

On Buckling Analysis of Beams and Frame Structures by the Differential Quadrature Element Method

Xinwei Wang[1], Huizhi Gu[2] and Bin Liu[3]

Abstract

In this paper, differential quadrature Element Method, developed recently by the present authors, is used to obtain buckling loads of beams and beam structures. The extension of the application range of the differential quadrature method is demonstrated.

Introduction

Recently, the differential quadrature (DQ) method has been successfully used to analyze the deflection, buckling and free vibration problems of structural components. However, there is a certain lack of flexibility when applying this method to real life structural analysis. Additionally, difficulties arise when the method is used to analyze structures or structural elements with discontinuous geometries or discontinuous distributed loads. To overcome the difficulties and to extend the applicability range of the DQ method to these problems, Striz et. al. proposed a method called the quadrature element method (QEM) (Striz, Chen & Bert, 1994). The QEM was used to perform static analyses of truss and beam structures under various loads. However, it is not convenient and accurate enough in the analysis due to introduction of two points representing one end point. The present authors (Wang and Gu, 1995) recently modified the DQ method, called the differential quadrature element method (DQEM), which overcomes the shortcomings of the QEM and successfully performed deflection and free vibrational analyses of frame structures and circular plates with stepped thickness over a concentric region. It is noted that if a uniform beam is

[1] Professor, [2] Associate Professor, [3] Graduate Student, Department of Aircraft Engineering, Nanjing University of Aeronautics & Astronautics, Nanjing 210016, People's Republic of China.

considered and no internal grid points are present in a differential quadrature beam element, the differential quadrature equations are exactly the same as the standard finite element equations for deflection analyses of beams. However, equations are quite different from the conventional finite element method for free vibration and buckling analyses.

In view of the fact that the previous paper (Striz et al. 1994, Wang et al. 1995) have not analyzed the buckling load of frame structures, the title problem is studied. In this investigation, the differential quadrature equation for the beam element with five grid points is derived for illustration. Then the DQEM is used to obtain buckling loads of beams and beam structures. Accurate results are obtained for cases considered as compared with existing analytical solutions.

Differential Quadrature Element Method

Consider an Euler beam subjected to an axial compressive force, the governing differential equation can be expressed as (Timoshenko and Gere, 1961)

$$EI\frac{d^4w}{dx^4} = -P\frac{d^2w}{dx^2} \tag{1}$$

where ***E, I,*** and ***P*** are the Young's modulus, the principal moment of inertia about the y-axis, and the axial compressive force, respectively.

In terms of the DQ element method, Equation (1) can be written as

$$[k]\{\delta\} = \{F\} \tag{2}$$

where

$$\{\delta\} = \left[w_1, w_1', w_2, \ldots, w_{N-1}, w_N, w_N'\right]^T \tag{3}$$

and

$$\{F\} = \left[\,Q_1, M_1, q_2, \ldots, q_{N-1}, Q_N, M_N\,\right]^T \tag{4}$$

are generalized displacement and load vectors, respectively. w_i (i = 1,2,3..., N) are the deflection at the i-th grid point, w'_i. Q_i and M_i (i=1,N) are the rotation, shear force and bending moment at the i-th grid point (end points only), and q_i (i = 2,3,..., N-1) are the transverse load related to axial compressive force p $(-P\,d^2w/dx^2)$ for buckling analysis. [k] is called the weighting coefficient matrix. For the simplest case, i.e., N=2 and EI is a constant, [k] is exactly the same as the stiffness matrix of the finite beam element with four degrees of freedom. In other words, the positive direction of w_i, w_i', Q_i, and M_i are the same as a finite beam element. The essential difference of DQEM from

QEM is that two degrees of freedom is used at each end in DQEM, instead of that two points are used to represent each end point in QEM if a forth-order differential equation is involved. The procedure to determine of the weighting coefficients is similar to the DQ method.

In the case of a five-point DQ beam element, i.e., **N=5**, the coordinates are **[0, L/4, L/2, 3L/4, L]**. Assume EI is a constant, the weighting coefficient matrix is given by Eq.(5) and the part of the load vector is given by Eq. (6) and the other four elements are generalized forces at the end point, depending on either the applied generalized loads or the prescribed generalized displacements at that point. Therefore, applying the boundary conditions is simple.

$$[k] = \frac{EI}{L^4}\begin{bmatrix} \frac{4420L}{3} & 190L^2 & \frac{-8192L}{3} & 2112L & \frac{-4096L}{3} & \frac{1532L}{3} & -50L^2 \\ \frac{998L^2}{9} & \frac{56L^3}{3} & \frac{-512L^2}{3} & 96L^2 & \frac{-512L^2}{9} & \frac{62L^2}{3} & -2L^3 \\ \frac{-7232}{3} & -320L & 4096 & -2304 & \frac{2048}{3} & -64 & 0 \\ \frac{3008}{3} & 80L & \frac{-8192}{3} & 3456 & \frac{-8192}{3} & \frac{-3008}{3} & -80L \\ -64 & 0 & \frac{2048}{3} & -2304 & 4096 & \frac{-7232}{3} & 320L \\ \frac{1532L}{3} & 50L^2 & \frac{-4096L}{3} & 2112L & \frac{-8192L}{3} & \frac{4420L}{3} & -190L^2 \\ \frac{-62L^2}{3} & -2L^3 & \frac{512L^2}{9} & -96L^2 & \frac{512L^2}{3} & \frac{-998L^2}{9} & \frac{56L^3}{3} \end{bmatrix} \tag{5}$$

and

$$\begin{Bmatrix} q_2 \\ q_3 \\ q_4 \end{Bmatrix} = P\begin{bmatrix} \frac{22}{3} & \frac{7L}{16} & -\frac{124}{9} & \frac{15}{2} & -\frac{4}{3} & \frac{5}{18} & -\frac{L}{48} \\ -\frac{10}{9} & \frac{-L}{12} & \frac{64}{9} & -12 & \frac{64}{9} & -\frac{10}{9} & \frac{L}{12} \\ \frac{5}{18} & \frac{L}{48} & -\frac{4}{3} & \frac{15}{2} & -\frac{124}{9} & \frac{22}{3} & \frac{-7L}{16} \end{bmatrix}\{\delta\} \tag{6}$$

To allow for arbitrary orientation of the DQ beam element in a plane, a transformation is performed, which is similar to the QEM (Striz et al. 1994) and the finite element method. However, transformation is not performed at all inner grid points. The procedures for structural analysis are similar to the FEM or QEM and are omitted here.

Examples

Example 1 A column of variable cross section is subjected to the action of a distributed axial load. The column is free at x=0 and fixed at **x=*l***. The flexural rigidity and the intensity of the distributed load can be represented by the following equations (Timoshenko & Gere, 1961),

$$EI = EI_0\left(\frac{x}{l}\right) \qquad q(x) = q_0\left(\frac{x}{l}\right)^{\alpha} \tag{7}$$

The critical buckling load is defined as $p_{cr} = \int_0^l q_0 (x/l)^{\alpha} dx$. For **α=0, 1, 2, 3** and **4**, exactly the same results as reported by Timoshenko and Gere (1961) are obtained with one seven-point DQ element.

Example 2 Consider a portal frame, which has two vertical columns and one horizontal beam. The bottom end of each column is fixed and the top end of each column is subjected to a compressive force. For simplicity, assume that all three beams have the same Yang's modulus, the principal moment of inertia, and the length.

Three seven-point DQ beam elements are used in the analysis. This example is cited from reference 4 (Example 8.6 , Yang,1986). The analytical result is 7.344 EI/L^3. The DQEM result is 7.381EI/L^3. The relative error is only 0.5 per cent.

Conclusion

The new version of DQ method extends the application range of the original DQ method to analyze buckling of frame structures. Further study is needed to extend the DQ method to analyze structures with other structural elements, such as plates and shells.

Acknowledgements--Support for the work was provided by the Jiangsu Natural Science Foundation and by the Science Foundation of Aeronautics of China.

References

1. A. G. Striz, W. Chen and C. W. Bert (1994). " Static analysis of structures by the quadrature element method (QEM)." *International Journal of Solids & Structures*, **31**(30), 2807-2818.
2. S.P. Timoshenko and J.M. Gere (1961). Theory of Elastic Stability, McGraw-Hall Book Company, Inc., New York.
3. X. Wang and H. Gu (1995). " Static analysis of frame structures by the differential quadrature element method." *Unpublished manuscript.*
4. T. Y. Yang (1986). Finite Element Structural Analysis, Prentice-Hall, Inc., Englewood Cliffs, NJ.

Segregation in Hopper Flows

Masami Nakagawa*, Xiaoshan Lin** and G.G.W. Mustoe**

Abstract

We investigate the effects of the particle shapes and sizes on dry granular flows by simulating mass- and funnel-flow hoppers. Our preliminary results based on particular combinations of particles seem to indicate that there is a competing effect between shape and size.

Introduction

Hoppers are commonly used for many particle handling processes such as storing and mixing. Achievement of an optimal design of this process equipment requires fundamental knowledge of the flow behavior of particles. While most existing theories and modeling are applied to flows of idealized particles, often consisting of cohesionless spherical particles, it has been recognized that actual industrial particulate flows differ considerably from these idealized flows due to complex patterns of particle interactions (Behringer and Baxter, 1994). To model more realistic particulate flows, we employ 2D DEM (Discrete Element Method) simulation scheme and introduce particle slenderness. Hoppers with different converging angles of 70 and 20 degrees are employed to simulate mass-flow and funnel- flow in hoppers, respectively. Elongated particles are modeled using ellipses. Flow rate, velocity and rotation of each particle, and interacting forces are observed by discharging the particulate masses of different mixtures from hoppers,. We show the results on segregation obtained from different combinations of particle sizes and shapes.

DEM Simulations

In the DEM simulations, the dynamics of all particles are monitored by applying the Newton's Law directly to each individual particles. Motion of the particles are calculated using an explicit finite difference scheme. Collision detection of the ellipses is

---

*Sandia National Laboratories, P.O. Box 5800, Albuquerque, New Mexico
**Division of Engineering, Colorado School of Mines, Golden, Colorado

performed numerically. Normal interaction forces (particle-to-particle and particle-to-wall) are determined by Herstzian elastic contact theory given by,

$$f_n = \frac{G_s}{3(1-\nu)}\sqrt{8R_e\delta_n^{\ 3}} \qquad (1)$$

where G_s ,ν , and δ_n are the shear modulus, Poisson's ratio of the elastic particles, and the normal overlap of rigid particles, respectively. R_e is defined as R_e=$2R_1R_2/(R_1+R_2)$ and indicates the equivalent radius of curvature at contact point. The tangential interaction between contacting particles is based on a simplified Mindlin's solution for elastic contact, and can be expressed as

$$\Delta f_t = k_t \Delta\delta_t, \quad \text{with } k_t = \frac{4G_s r}{2-\nu} \text{ and } r = \left\{\frac{3(1-\nu)f_n R_e}{8G_s}\right\}^{1/3} \qquad (2)$$

where Δf_t, $\Delta\delta_t$, and k_t are incremental tangential force, relative tangential displacement, and the tangential stiffness, respectively. More detailed discussions on the force model, contact detection scheme and other related issues can be found elsewhere (for example, Lin, 1995; Ting. J.M. et al., 1993).

Hoppers with different converging angles of 70 and 20 degrees are used to simulate mass- and funnel-flows. Elongated particles are modeled using ellipses with an aspect ratio of 2. Combinations of large and small circular particles (mixture:LC-SC), large elliptical and small circular particles (mixture:LE-SC), and large circular and small elliptical particles (mixture:LC-SE) are used to investigate the effects of shapes and sizes on segregation. There are 300 large and 600 small particles in this simulation. Initially, these particles are randomly generated and packed under the gravity in the hopper. Then they are discharged into a partitioned bin to study the spatial distribution of different particles in the flow. The length of the bin is about 60 diameter of (a smaller particle) long and the gap between each partition is 4.5 diameter wide.

Results

Figure 1 shows the effect of particle shapes on the angle of repose. As expected, due to its rolling resistance, the angle of repose of elliptical particles is about 15 degrees higher than that of circular particles. Similar observations were previously made by Walton(1993) and Yamane et al.(1995). Figure 2a, b, and c show snapshots at the same intermediate stage of the discharge for the mixtures (LC-SC), (LE-SC), and (LC-SE), respectively. The hopper contained the (LE-SC) was emptied first. The number of particle contacts at this instant of discharge are 667, 923, and 906, for (LC-SC), (LE-SC), and (LC-SE), respectively. The introduction of slender particles increases the number of contacts by almost 50%. Figure 3 shows a statistics of the spatial distribution of discharged particles in the bin for (LC-SC),(LC-SE), and (LE-SC) mixtures. When larger particles are circular as in the LC-SC and LC-SE mixtures, the stream in the middle tend to contain a larger fraction of small particles. In contrast, this trend is less noticeable when larger particles are elliptical. Elliptical particles seem to flow along the faster stream line in the middle and so do the smaller particles. Although we are not able to make any conclusive remarks based on our small population size of statistics, it seems we are capturing a competing effect on segregation between size and shape.

REFERENCE

•Behringer, R. P. and Baxter, G. W. (1994), "Pattern Formation and Complexity in Granular Flows," *Granular Matter-An Interdisciplinary Approach*, A. Mehta ed., Springer-Verlag New York, Inc., pp.85-119.
•Lin, X. (1995),"Numerical Study of Granular Soil Behavior Using Arrays of Elastic Ellipsoids," Ph.D. Dissertation, University of New Mexico, Albuquerque, New Mexico.
•Ting, J.M. and et al. (1993),"An Ellipse-based Discrete Element Model for Granular Materials," *International Journal for Numerical and Analytical Methods* in Geomechanics.
•Walton, O. R. and Braun, R.L.,"Simulation of Rotary-Drum and Repose Tests for Frictional Spheres and Rigid Sphere Clusters," *Joint DOE/NSF Workshop on Flow of Particulates and Fluids*, Sept. 29- Oct. 1, Ithaca, New York.
•Yamane, K. and et al. (1995),"DEM and MRI Studies of Particulate Flows in a Rotating Cylinder," *Joint ASME/JSME Fluids Engineering Conference*, August 13-18, Hilton Head, South Carolina.

Acknowledgment

This work is, in part, supported by the U.S. Department of Energy, Pittsburgh Energy under contract number DE-AC04-94AL85000. We would also like to acknowledge partial support for this work by Dr. R.S. Miller of section of the mechanics division, office of Naval Research.

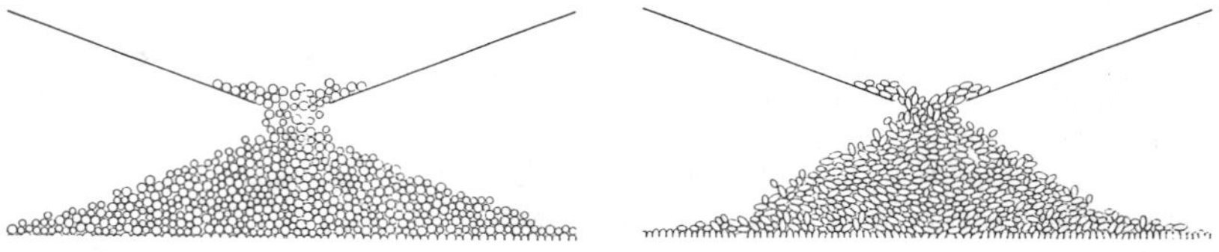

Figure 1. Angle of Repose.

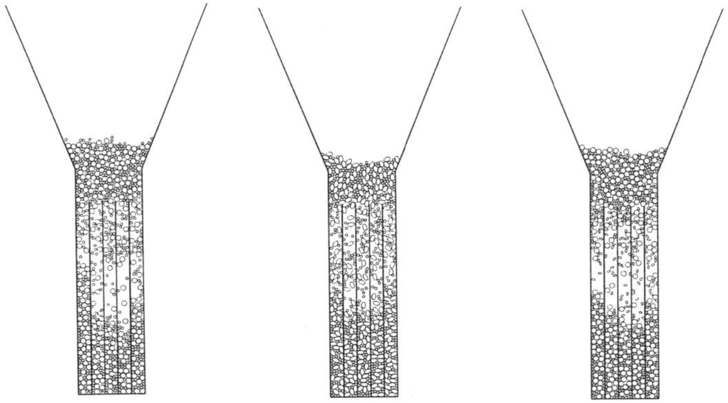

a) LC-SC mixture b) LE-SC mixture c) LC-SE mixture

Figure 2.

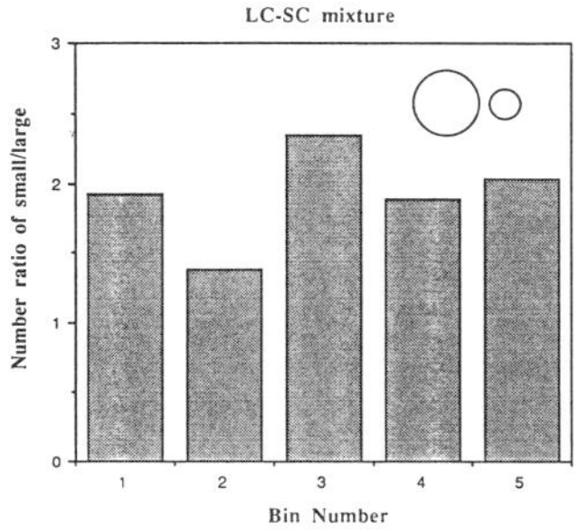

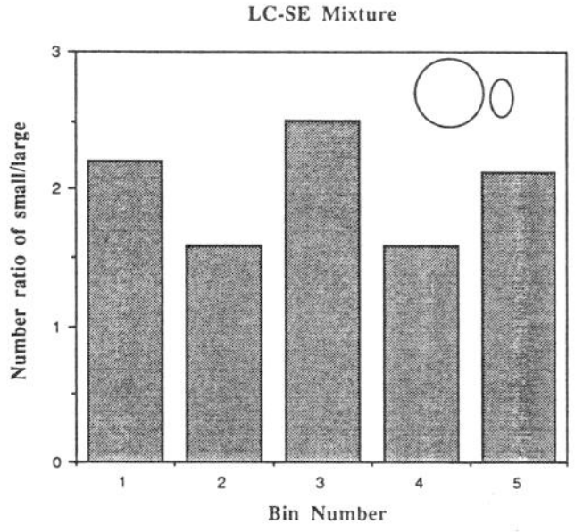

Figure 3. Spatial Distribution of particles in the partitioned bin.

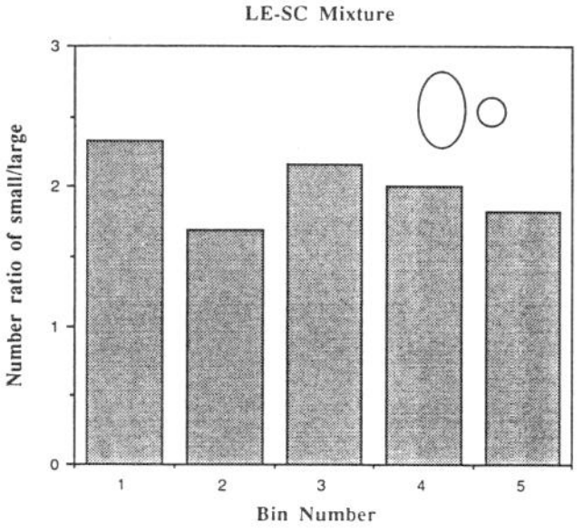

Simulation of Dilute Gas-Solid Flows in Horizontal Channels

Cliff K. K. Lun[1] and Hong S. Liu

Abstract

Reasonable agreement is found between the simulation results and the experimental measurements for mean gas velocity, mean particle velocity and concentration in dilute turbulent gas-solid flows with relatively large particles in a horizontal channel. Interparticle collisions and Magus lift play some important roles in suspending the solid phase. The effects of material collisional properties such as the coefficients of normal restitution and friction are examined. In general, the higher the coefficients of normal restitution and friction, the higher the particle concentration in the upper region of the channel would be resulted.

Introduction

Turbulent two-phase flows are evidently very complex. Experimental and numerical studies of dilute suspensions of relatively large particles had been conducted by researcher such as Lourenco et al. (1983), Tsuji et al. (1982, 1987) and Frank et al. (1993). Measurements of mean fluid velocity in horizontal pipes and channels had shown that the velocity profiles could be significantly modified by the coupling between the fluid phase and the solid phase at dilute systems with bulk solids volume fraction in the order of 10^{-3}. The effect of particle-wall collisions was found to be crucial in wall-bounded suspensions (Lourenco et al. 1983; Tsuji et al. 1987; Frank et al. 1993). On the other hand, the effect of interparticle collisions was often neglected.

A dilute two-phase turbulent model is developed in the present work in which the interparticle collisions and the interfacial coupling effects are incorporated. The effect of material collisional properties on the flow dynamics in the suspension is examined.

----

[1]Department of Engineering, Dalhousie University, Halifax, NS, B3H 3J5 Canada

Two-phase Gas-solid Model

A set of time-averaged equations for continuity, momentum, turbulent kinetic energy (k) and rate of energy dissipation (ε) for the fluid phase of a dilute turbulent fluid-solid system was derived by following the basic conservation principles as those in a single-phase turbulent fluid flow. The anisotropic Reynolds stresses and turbulence intensities model proposed by Nisizima and Yoshizawa (1987) for single-phase fluid flow is adopted in the present study. The sticking-sliding collision model for inelastic, frictional materials presented in Lun and Bent (1994) is employed here for particle-particle collisions and particle-wall collisions. The collision model is simple and can produce result reasonably close to a number of experimental measurements. The collisional process is characterized by three coefficients; namely, the coefficient of normal restitution, e_p, the coefficient of tangential restitution at contact point for sticking contact, β_o, and the friction coefficient μ_p.

The equation of translational particle motion can be written as

$$m_p \frac{d\mathbf{v}}{dt} = m_p \mathbf{g} + \mathbf{F}_{\mathrm{D}} + \mathbf{F}_{\mathrm{LS}} + \mathbf{F}_{\mathrm{LM}}$$

where $\mathbf{F}_{\mathrm{D}}$, $\mathbf{F}_{\mathrm{LS}}$ and $\mathbf{F}_{\mathrm{LM}}$ are the drag force, Saffman lift and Magnus lift respectively. The rate of momentum exchange between the phases can be readily computed by summing the interfacial forces that act on each particle. The rate of angular momentum change of a spherical particle interacting with a viscous fluid may be written as

$$\frac{m_p d^2}{10} \frac{d\omega}{dt} = -\pi\mu\, d^3(\Omega_f - \omega)$$

where $\Omega_f = 0.5\nabla\times U$ is the local mean angular velocity of the fluid.

The set of partial differential equations deduced from the conservation and constitutive equations for the fluid phase are integrated numerically by using a finite-difference method to yield the turbulence flow field in a horizontal channel. The fluid fluctuation velocity and the eddy lifetime at each node of the finite-difference scheme are sampled randomly from Gaussian distributions based on the r.m.s. of the fluid fluctuation velocity and the mean Lagrangian integral time scale respectively. The trajectory and velocity of each particle are determined by integrating the equations of particle motion. The fluid-solid coupling effects due to solids volume fraction and interfacial momentum interactions are incorporated in the simulation.

Result

A computer program is developed to simulate the turbulent air-solid flows in the horizontal channel experiment of Tsuji et al. (1987). The channel height H was 25mm. The solid phase is composed of polystyrene beads with 1mm diameter,

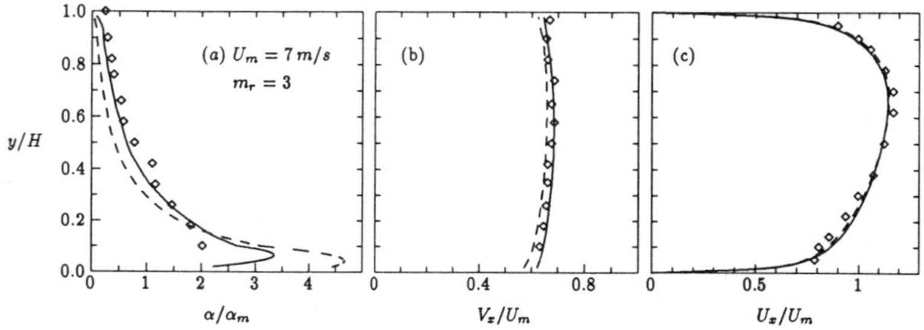

Figure 1. Simulations: – – – – , $e_p = 0.8$; ——— , $e_p = 0.9$.
Data of Tsuji et al. (1987): ◊ , $m_r = 3$.

mass density of 1000 kg/m^3 and particle-wall friction coefficient of 0.4. Normalized (a) volume fraction of the solid phase (α/α_m), (b) mean particle velocity (V_x/U_m), and (c) mean air velocity (U_x/U_m) for bulk fluid velocity $U_m = 7$ m/s and mass loading ratio $m_r = 3$ are plotted in figure 1. The bulk solids volume fraction α_m is 0.0036. Since there is no detailed information about the materials, for simplicity e_p, β_o, and μ_p are assumed to be identical for all collisions. A typical value of $\beta_o = 0.4$ is used in the simulation. Tsuji et al. (1987) suggested an e_p of 0.8 for the polystyrene beads colliding with the solid walls.

Simulation results using e_p of 0.8 and 0.9 are shown in figure 1. In general, the numerical predictions for e_p of 0.9 agree quite well with the experimental measurements. Due to gravity, the solids volume fraction increases from the top to the bottom as shown in figure 1(a). The particle concentration in the upper region increases with increasing e_p while the concentration near the bottom decreases. The mean particle velocity in figure 1(b) is found to increase with increasing e_p. The mean air velocity in figure 1(c) is asymmetric due to the interfacial coupling effects. The relatively higher solids concentration near the bottom causes higher resistance to the air motion than those in the upper region. As a result, air moves relatively slower near the bottom than near the top.

Computations using e_p of 0.9 and μ_p of 0.2 and 0.4 are shown in figure 2. The solids concentration near the bottom is found to increase significantly with decreasing friction from 0.4 to 0.2. The reason is that *on average*, a lower friction coefficient causes lower particle spin in the clockwise direction which in turn generates lower lift forces on the particles, as a result the particles can not rise as high as those with a higher μ_p. In figure 2(b), the mean particle velocity near the top wall is notably reduced due to lowered friction. The mean air velocity profile is slightly modified by the reduction in friction as shown in figure 2(c).

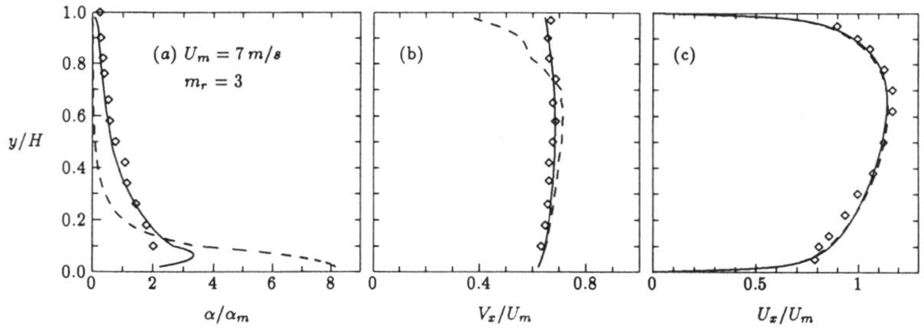

Figure 2. Simulations: – – – – , $\mu_p = 0.2$; ——— , $\mu_p = 0.4$.
Data of Tsuji et al. (1987): ◊ , $m_r = 3$.

Conclusion

The material collisional properties in interparticle collisions and particle-wall collisions play a crucial role in turbulent gas-solid flows. The Magnus lift due to particle spin helps to suspend the particles. The interfacial coupling effects due to solid volume fraction and momentum interactions are responsible for the asymmetry in the mean air velocity distribution. Extension of the present simulation to pipe-flows is now in progress.

Acknowledgment

This work was supported by a research grant from the Natural Science and Engineering Council (**NSERC**) of Canada.

References

Frank, T., Schade, K. P., and Petrak, D. (1993). "Numerical simulation and experimental investigation of gas-solid two-phase flow in a horizontal channel," *Int. J. Multiphase Flow*, 19, 187-198.

Lourenco, L., Reithmuller, M. L., and Essers, J. A. (1983). "The kinetic model for gas particle flow and its numerical implementation," in *Proc. Intl Conf. on the Physical Modeling of Multiphase Flow*, Cranfield, UK, 501-525.

Lun, C. K. K., and Bent, A. A. (1994) "Numerical simulation of inelastic frictional spheres in simple shear flow," *J. Fluid Mech.*, 258, 335-353.

Nisizima, S., and Yoshizawa, A. (1987). "Turbulent channel and Couette flows using an anisotropic k-ε model," *AIAA J.*, 25, 414-420.

Tsuji, Y., and Morikawa, Y. (1982). "LDV measurements of an air-solid two-phase flow in a horizontal pipe," *J. Fluid Mech.*, 120, 385-409.

Tsuji, Y., Morikawa, Y., Tanaka, T., Nakatsukasa, N., and Nakatani, M. (1987). "Numerical simulation of gas-solid two-phase flow in a two-dimensional horizontal channel," *Int. J. Multiphase Flow*, 13, 671-684.

Granular Flow Based on Non-Newtonian Fluid Mechanics

Sergio A. Elaskar[1] , Luis A. Godoy[1] and Alejandro T. Brewer[1]

ABSTRACT

Pressures due to bulk materials on the walls of silos are very important for the silo design. A new model constitutive to simulate discharging processes in mass-flow silos is presented in this paper. The constitutive model is viscoplastic and rigid-viscoplastic. The material is considered incompressible. The walls of the silo are rigid during the simulation of discharge. The problem is governed by continuity, momentum, and constitutive equations; and is dicretized by the finite element method. Velocity and stress fields are calculated as well as wall pressures.

INTRODUCTION

There is a very variety of constitutive models to represent the granular flow. In the context of silo analysis and design, some of the models have been proposed by Jenike and Shield (1959), Hausler and Eibl (1984), Runesson and Nilsson (1986), Schimidt and Wu (1989), and Diez and Godoy (1992). Recent work by the authors (Elaskar, Godoy and Brewer, 1995) shows problems of convergence an satisfaction of constitutive equations in stationary flow of rigid-viscoplastic materials. This work represents an effort to improve the modelling of granular flow based on non-Newtonian fluid mechanics.

CONSTITUTIVE MODEL

The behavior of material is assumed to be isotropic and incompressible, and the material is cohesive and frictional. The constitutive theory presented here considers a viscoplastic and rigid-viscoplastic material (viscoplastic and newtonian-viscoplastic fluid) to represent the behavior of granular solids during steady state discharge.

Rigid-viscoplastic behavior

For a viscoplastic material, the general constitutive equation following Perzyna (Perzyna, 1966) may be written as in which is the plastic discontinuity condition, and is the plastic potential

$$\varepsilon_{ij} = \gamma\left\langle \beta(F)^n \right\rangle \frac{\partial Q}{\partial \sigma_{ij}} = \gamma\langle A\rangle \frac{\partial Q}{\partial \sigma_{ij}} \tag{1}$$

The plastic discontinuity condition of Drucker-Prager is adopted in the form (Zienkiewicz and Taylor, 1991)

---

[1] Departamento de Estructuras. Universidad Nacional de Córdoba. Casilla de Correo 916. (5000) Córdoba. Argentina.

$$F = (3J_2)^{\frac{1}{2}} - k - \alpha' p \tag{2}$$

where J_2 is the second invariant of the deviatoric stress tensor S_{ij}, p is the mean pressure, and

$$k = \frac{6c\cos\phi}{3 - sin\phi} \qquad \alpha' = \frac{6sin\phi}{3 - sin\phi} \tag{3}$$

c is the cohesion and ϕ is the internal angle of friction of the material.
The plastic potential is the von Mises criterion (non-associative plasticity)

$$Q = (3J_2)^{\frac{1}{2}} - k \tag{4}$$

Substitution of these functions in to the constitutive equation (1) results in

$$\varepsilon_{ij} = \gamma \left\langle \beta \left[(3J_2)^{\frac{1}{2}} - (k + \alpha' p) \right]^n \right\rangle \frac{\sqrt{3}S_{ij}}{2\sqrt{J_2}} \tag{5}$$

The general constitutive equation for incompressible viscous fluid is

$$\varepsilon_{ij} = \frac{1}{2\mu_{nv}} S_{ij} \tag{6}$$

where μ_{nv} is the equivalent viscosity of an isotropic fluid.
Comparison of the last two equations leads to the definition of a coefficient of equivalent viscosity

$$\mu_{nv} = \left[\left(\frac{\Sigma}{\beta\gamma} \right)^n + k + \alpha' p \right] \frac{1}{3\Sigma} \tag{7}$$

Is clear that to obtain the equation (7) we lose the rigid condition. We have this condition in consideration when

$$\left\{ \beta \left[\sqrt{3J_2} - (k + \alpha' p) \right]^n \right\} \to 0 \Rightarrow \mu_{nv} \to \infty \tag{8}$$

We then make

$$if \quad \mu_{nv} \geq \mu_{lim} \quad \Rightarrow \quad \mu_{nv} = \mu_{lim} \tag{9}$$

Viscoplastic behavior

The viscoplastic viscosity is

$$\mu_v = \frac{1}{3\gamma} + \frac{6(c_3 + c_1 c)\cos\phi + 6pc_2 sin\phi}{(3 - sin\phi)3\Sigma} \tag{10}$$

where the c_1, c_2, and c_3 are constants.
If

$$h = \frac{6c_3\cos\phi}{3 - sin\phi} \tag{11}$$

and with equations (3), the viscosity results in

$$\mu_v = \frac{1}{3\gamma} + (h + c_1 k + c_2 \alpha' p)\frac{1}{3\Sigma} \tag{12}$$

Constitutive Model: Viscoplastic and rigid-viscoplastic behavior

The rate of deformation tensor is decomposed in one viscoplastic part and another rigid-viscoplastic (newtonian-viscoplastic) part

$$\left(\varepsilon_{ij}\right)_t = \left(\varepsilon_{ij}\right)_v + \left(\varepsilon_{ij}\right)_{nv} \tag{13}$$

and with equations (6), (7), and (12), we get

$$\mu_t = \frac{1}{3\gamma} + \left\{ \left(h + c_1 k + c_2 \alpha' p\right) + \left[\left(\frac{\Sigma}{\beta\gamma} \right)^n + k + \alpha' p \right] \right\} \frac{1}{3\Sigma} \tag{14}$$

where, if

$$\left[\left(\frac{\Sigma}{\beta\gamma} \right)^n + k + \alpha' p \right] \geq \mu_{lim} \quad \Rightarrow \quad \left[\left(\frac{\Sigma}{\beta\gamma} \right)^n + k + \alpha' p \right] = \mu_{lim} \tag{15}$$

and equation (14) becomes

$$\mu_t = \frac{1}{3\gamma} + \left(h + c_1 k + c_2 \alpha' p\right) \frac{1}{3\Sigma} + \mu_{lim} \tag{16}$$

The modelling of a steady-state flow of an incompressible fluid is a standard topic in finite element analysis. In this work a velocity element has been implemented with a 9 node Lagrangean interpolation. A penalty parameter is required to enforce incompressibility, and selective integration of the matrices has been adopted. We calculate one viscosity per element, and the penalty parameter is proportional to the viscosity of each element.
The non-linearity of the systems arises because that equivalent viscosity μ_t is a function of ε_{ij} and p (14). To obtain the solution, an algorithm for direct iteration has been implemented.

NUMERICAL RESULTS

Numerical simulation of steady state flow of granular material due to gravity in a silo structure is present here. The side walls of silo of Fig. 1 (bidimensional silo) were made of steel plates while front walls were made of perspex. The silo has been tested (Kmita, 1991), with pressure measured on side walls at several positions. The material stored inside the silo has $c = 0$, $\phi = 33°$, $\phi_w = 33°$, and $\rho = 16 kg/m^3$. In Fig. 2 we present our numerical results and experimental results (Kmita, 1991).

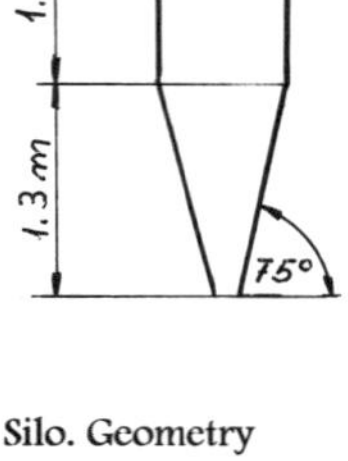

Figure 1: Silo. Geometry

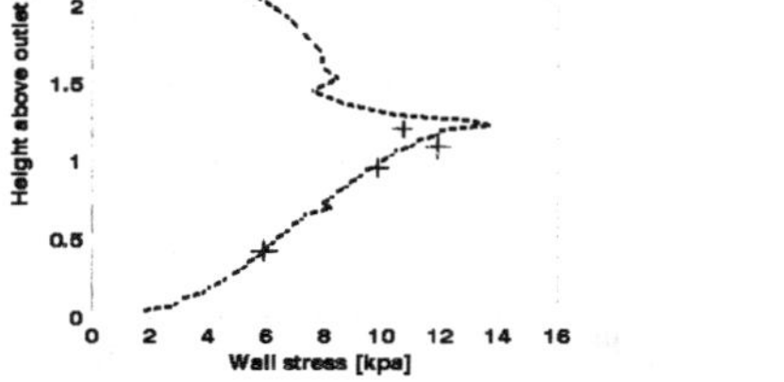

Figure 2: Silo. Pressures on the wall

Conical hopper of Fig. 3 has been tested by Walker and Blanchard, (1967). The material stored inside the hopper has $c = 0$, $\phi = 25°$, $\phi_w = 15°$. In Fig. 4 we present our numerical results, Walker's theory and theoretical analyses of Nguyen, Brennen, and Sabersky, (1979) and experimental results (Walker and Blanchard, 1967). In Fig. 2 and Fig. 4 the constants are

$$c_1 = 0, \quad c_2 = 0.420, \quad c_3 = 0.175 \tag{17}$$

In the silo and the conical hopper, the friction along the walls is simulated by a thin layer of elements with the apropiate parameters. In both, the finite element mesh considered has 130 elements and 583 nodes. In both cases our numerical results show agreement with experimental results.

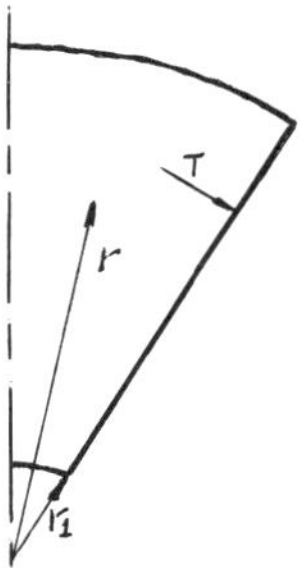

Figure 3: Hopper. Geometry

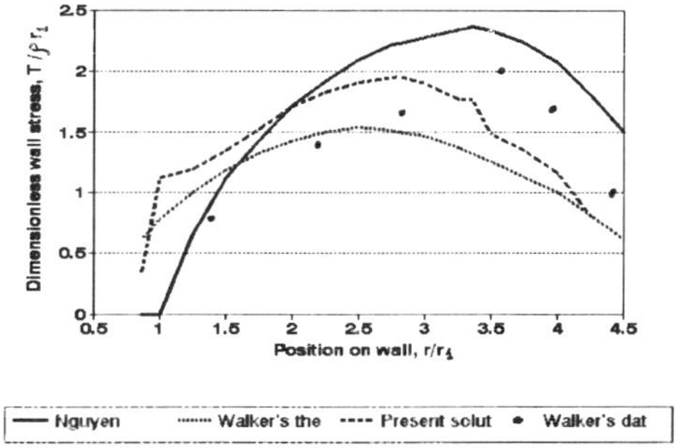

Figure 4: Hopper. Pressures on the wall

REFERENCES

Diez, M. and Godoy, L. 1992, Viscoplastic incompressible flow of frictional cohesive solids, *International Journal of Mechanic Sciences*, 34, 1992, 395-408.

Elaskar, S.; Godoy, L. and Brewer, A., Análisis de la convergencia en fluidos viscoplástico y newtoniano-viscoplástico. In *Mecánica Computacional. AMCA*. Santa Fé. Argentina, 15, 1995, 253-263.

Haussler, U. and Eibl, J., Numerical investigations on discharging silos, *J. of Engineering Mechanics. ASCE*, 110, 1984, 957-971.

Jenike, A. and Shield, R., On the plastic flow of Coulomb solids beyond original failure, *Journal of Applied Mechanics, ASME*, 81(E), 1959, 599-602.

Kmita, J., An experimental analysis of internal silo loads, *Bulk solids handling*, 11(2), 1991, 459-468.

Nguyen, T., Brennen, C., and Sabersky, R., Gravity flow of granular materials in conical hoppers, *Journal of Applied Mechanics, ASME*, 45(3), 1979, 529-535.

Perzyna, P., Fundamental problems in viscoplasticity. In *Recent Advances in Applied Mechanics*, Academics Press. New York, 9, 1966, 243-377.

Runesson, K. and Nilsson, L, Finite element modelling of the gravitational flow of a granular material, *Bulk Solids Handling*, 6(5), 1986, 241-248.

Schmidt, L. and Wu, Y, Prediction of the dynamic wall pressures on silos, *Bulk Solids Handling*, 9(3), 1989, 333-338.

Walker, D. and Blanchard, M., Pressures in experimental coal hoppers, *Chemical Engineering Science*, 22, 1967, 1706-1713.

Zienkiewicz, O. and Taylor R., *The Finite Element Method. Vol 2*. Mc. Graw Hill. London. 1991

FLEXURAL CHARACTERISTICS OF TWO-DIMENSIONAL ADVANCED COMPOSITE GRID REINFORCED CONCRETE

David W. Jensen (Member, ASCE)[1] and Craig W. Smart (Student Member, ASCE)[2]

Abstract

An experimental investigation was conducted to quantify the flexural characteristics of concrete structures reinforced with advanced composite grids. Concrete beams reinforced with 5.1 cm (2 in.) square orthogonal grids composed of non-corrosive composites (carbon/vinyl-ester and fiberglass/vinyl-ester) were compared with standard steel reinforced concrete beams. The flexural strength, stiffness, and failure characteristics were evaluated by monitoring the load, deflection, and strain under four-point bending. The internal strain was measured with both electrical resistance strain gages and fiber-optic strain sensors. The results show that the lower effective modulus of elasticity of composite grids requires more cross- sectional area of axial reinforcement to be employed than standard steel reinforcement to achieve the same deflections under a given load. Higher strength is achieved in the composite grid reinforced concrete, although at the cost of reduced ductility. The embedded fiber-optic strain sensors provided a good indication of damage accumulation.

Introduction

In order to capitalize on aerospace materials technology, researchers have been looking for commercial applications of high strength and stiffness advanced composite materials [Grande]. Meanwhile, civil engineers have been searching for an answer to the corrosion problem in steel reinforced concrete. Recent surveys on the problems of standard steel reinforcement and the potential for composite materials indicate that over 200,000 of the approximately 575,000 bridges in the U.S. are structurally deficient [Dunker and Rabbat]. One of the leading causes of decay in reinforced concrete structures is corrosion of the reinforcing steel [Taggart]. Composites are used extensively in high-corrosive applications such as chemical plants with little to no damage from corrosion, making composites an obvious choice for concrete reinforcement [Liskey].

The orthotropic grid design takes advantage of the unidirectional strength characteristics of composites. The orthogrid is comprised of 0 and 90 degree ribs spaced at 5.1 cm (2.0 in.). This structure functions well as a two-dimensional concrete

[1]Associate Professor, Advanced Composites Manufacturing and Engineering Center, Brigham Young University, 368 Clyde Building, Box 24066, Provo, UT 84602.

[2]Graduate Research Assistant, Department of Civil and Environmental Engineering, Brigham Young University, 368 Clyde Building, Provo, UT 84602.

reinforcement. Furthermore, shipping costs are minimized and high labor costs associated with placement and tying of rebar can be eliminated. In this study, beams reinforced with fiberglass/vinyl-ester and carbon/vinyl-ester grids were compared against beams reinfored with standard steel rebar (see Figures 1 and 2).

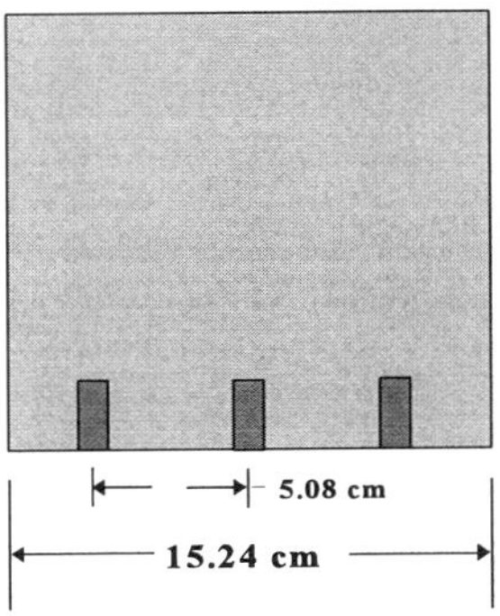

Figure 1. End view of composite grid reinforced beam.

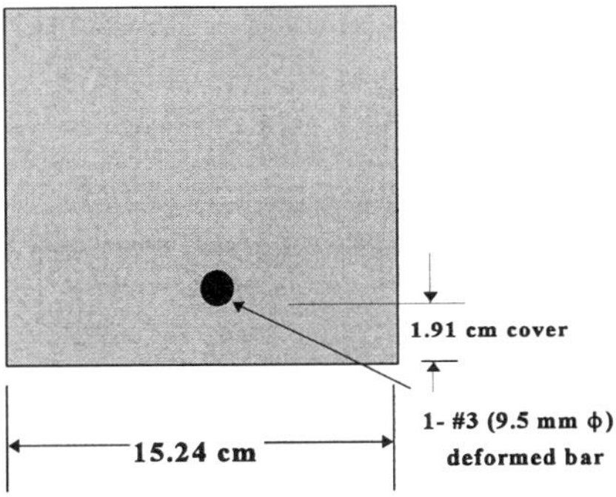

Figure 2. End view of standard steel reinforced beam.

Experiment

Four-point bend tests were performed on 15.24 cm by 15.24 cm by 76.2 cm (6 in. by 6 in. by 30 in.) beams in order to compare the composite grid reinforced concrete with standard steel reinforced concrete. The beams were cast using a 34.5 MPa (5000 psi) concrete mix and cured in a fog room for 14 days. Reinforced beams were cast in four different configurations (see Table 1).

Table 1. Different test configurations.

Type of Reinforcement	Location	# of Specimens
1 #3 steel bar (Fy=414 MPa)*	1.9 cm from bottom surface	3
Fiberglass grid	Flush with bottom surface	3
Carbon grid	Flush with bottom surface	3
2 Carbon grids	Flush w/ bottom and top surfaces	3

*Minimum Allowable Reinforcement [ACI Building Code and Commentary, 318-89, American Concrete Institute, Revised 1992, §10.5.)

The grids were composed of ribs approximately 0.51 cm by 1.5 cm on 5.1 cm centers (0.2 in. by 0.6 in. on 2 in. centers). The volume fraction of the composite grids employed in this study was only 14 percent which is much lower than the 65 percent in typical aerospace structures.

Prior to casting the concrete beams, the internal reinforcement was instrumented with two electrical resistance strain gages and a fiber-optic strain sensor. The strain gages were placed at midspan and under one loading point. The fiber optic strain sensor was mounted on the center longitudinal rib on the top side of the composite grids and on the bottom side of the steel rebar.

The beams were loaded using a 1330 kN (300 kips) Baldwin testing machine. The load, mid-span deflection and strain were monitored and recorded on a Compaq® processor using AutoNet® data aquisition software.

Results

Load versus deflection curves are portrayed in Figures 3 and 4. These graphs show that the composite grid reinforced beams exhibit lower ductility than the steel reinforced beams. Due to the low fiber volume fraction, the carbon grids allowed more deflection at a given load after the first crack. Despite the low modulus of the grids, however, they achieved similar working strengths compared to the steel reinforced beams.

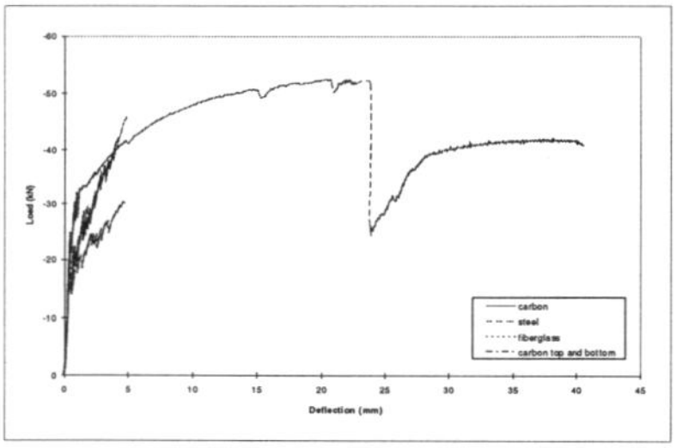

Figure 3. Typical load vs. deflection curves for the four different configurations tested.

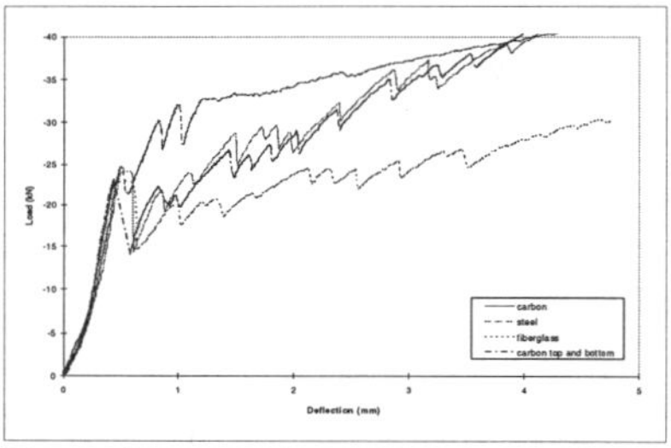

Figure 4. Comparison of first crack load and cracked section deflections.

The first crack loads are compared in Figure 4. It should be noted that the decrease in load upon first crack is significantly greater for the composite reinforced beams than the steel reinforced beams. Also, it is apparent that the deflection of the beam following the first crack is proportional to the modulus of elasticity of the reinforcing material, the steel allowing the least amount of deflection and the fiberglass allowing the most for a given load.

Figure 5 depicts the data gathered from the fiber optic strain sensors and shows the damage accumulation in the beams. The difference in crack patterns between the steel and grid reinforced beams suggests that the load is more evenly distributed in the steel reinforced beam and is highly localized in the grid reinforced beam (see Figure 6).

Conclusions

The results of this preliminary study demonstrate that significant potential exists for two-dimensional advanced composite grid-type reinforcement in concrete beams. By varying the design parameters, the stiffness of the composite grids can be made to match or exceed that of steel. Likewise, the strength of the composite reinforced beams could

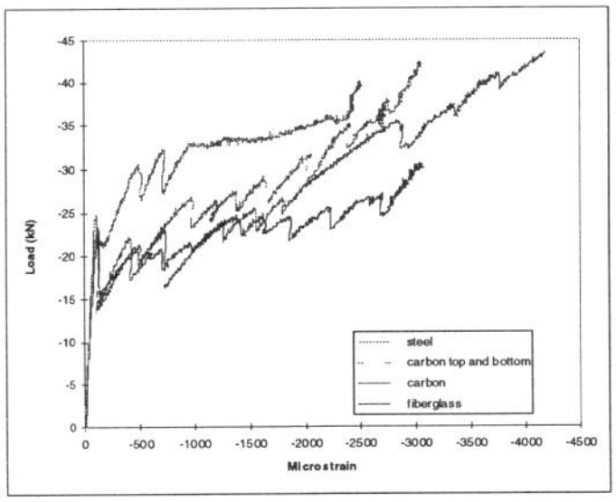

Figure 5. Load vs. microstrain data obtained from fiber-optic strain sensor.

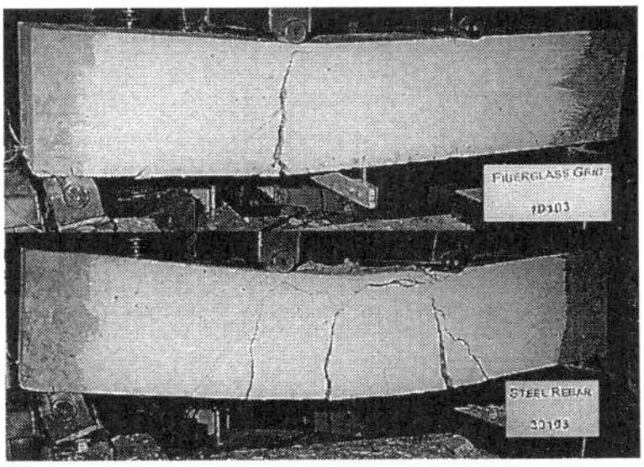

Figure 6. Flexural crack patterns in concrete beams reinforced with fiberglass grid (top) and standard steel bar (bottom).

be increased even further. Finally, the ductility could be improved by adding a secondary reinforcement (e.g., fibrous concrete) [Balaguru and Shah]. The embedded fiber-optic strain sensors provided a good indication of damage accumulation.

Acknowledgements

The authors would like to gratefully acknowledge support from the State of Utah and the Advanced Composites Manufacturing and Engineering Center at the Brigham Young University; the donation of graphite fibers from AKZO Nobel Fortafil Fibers, Inc.; and, the manufacturing of the grids by Fibergrate. Also appreciated is technical support received from Dr. Stephen W. Tsai of the Department of Aeronautics and Astronautics at Stanford University, and Dave Bailey and John Hayes from the U.S. Army Construction Engineering Research Laboratories.

References

Balaguru, P. and Shah, S., *Fiber Reinforced Cement Composites*, McGraw-Hill Inc., 1992.

Dunker, K. F. and B. G. Rabbat, "Assessing Infrastructure Deficiencies: The Case of Highway Bridges," *J. Infrastructure Systems*, June 1995, pp. 100-107.

Grande, J. A., "Advanced Composites Sector Heads for Diversification," *Modern Plastics*, July 1995, pp. 36-43.

Liskey, K., "Structural Applications of Pultruded Composite Products,"*Advanced Composite Materials in Civil Engineering Structures*, ASCE, 1991, pp. 182-193 .

Taggart, H., "Infrastructure Applications Key to Continued Growth," *Composites Design and Application*, Composites Institute of the Society of the Plastics Industry, Inc. (SPI), Winter 1995, pg. 20.

An Experimental Investigation of
Sandwich Flat Panels Under Low Velocity Impact

Anthony N.Palazotto[†], Eric J. Herup*, and Timberlyn Harrington**

Abstract

The results of a study related to the initiation of damage in laminated composite sandwich plates under low-velocity impact is presented. Sandwich structures composed of 8- and 16-ply laminated composite facesheets and 1.27 cm thick honeycomb core have been impacted using drop weight and pendulum impactors and the plate response correlates with the internal damage observed after impact. For impact energies above but near the damage initiation impact energy this system exhibits a linear relationship between incident and absorbed energy. Damage, when present, was in each of the forms of matrix cracking, delamination, and core crushing.

Introduction

This study considers an experimental investigation of low-velocity impact damage to sandwich plates composed of graphite-epoxy facesheets and nomex honeycomb core. Plate specimens of constant core thickness with 8- and 16-ply symmetric crossply facesheets were impacted by both free-falling and pendulum impactors with a hemispherical-nose (the tup). Contact force was monitored throughout the impact event through the use of an instrumented impactor. In addition, plates of each configuration were subjected to static indentation simulating impact loads. After impact/static loading, specimens were nondestructively and destructively analyzed to determine the damage modes present and the extent of damage by each mode. A range of impact energies chosen to bracket the initiation of damage provides a characterization of the damage as a function of impact energy. Comparison of the static and impact damages for these facesheet thicknesses provides insight into the applicability of the quasi-static assumptions commonly made in thin monolithic

---

[†]Professor, Fellow of ASCE, Air Force Institute of Technology, WPAFB, OH 45433
*Ph.D. Candidate
**Former Graduate Student

laminate low-velocity impact analyses to sandwich constructions. This information fills in voids in the published data (Harrington, 1994), particularly at small impact energies.

Specimens and Method

The specimens tested in this study were flat sandwich panels fabricated from AS4/3501-6 graphite/epoxy facesheets secondarily bonded to HRH-10-1/8-9 Nomex honeycomb core using FM 300-2 film adhesive. The core cell size was 0.32 cm providing a core density of 145 kg/m^3. The core thickness was 1.27 cm for all specimens. Facesheets were fabricated from unidirectional tape material with a thickness of 0.0127 cm in a balanced symmetric crossply stacking sequence of either $[0/90]_{2S}$ (sandwich panels using this facesheet are designated 8-ply) or $[0/90]_{4S}$ (sandwich panels using this facesheet are designated 16-ply). These stacking sequences were chosen to minimize the number of independent variables (ply orientation angles) in order to enhance the understanding of damage initiation and progression while exhibiting damage modes typical of sandwich constructions used in current and anticipated aircraft structures. All specimens were 17.78 cm long and 17.78 cm wide and were restrained in a 12.7 cm by 12.7 cm fixture providing near simple support on all edges (Harrington, 1994) of the 12.7 cm by 12.7 cm effective specimen size.

Two different devices were used to impact the specimens, an instrumented vertical drop weight testing machine and an instrumented pendulum testing system. In both cases the same 15 kN load cell and 1.27 cm radius spherical nose were used. Impact energy was regulated by controlling the drop height (or pendulum swing) of the instrumented impactor (hereafter called the tup) and resolved by measuring the mass of the tup and its velocity just prior to impact. In the case of the drop weight tests, force data was recorded at 10 μsec intervals throughout the event duration with a digital data acquisition system. Due to limitations in the drop weight testing system, the pendulum was required in order to provide impact energies low enough to resolve damage initiation in the thinner facesheet specimens. A digital storage oscilloscope was used to record the force data from the pendulum tests at 5 μsec intervals. The tup mass for all drop weight tests was 3.49 kg and the tup mass for the pendulum tests was 3.48 kg. No specimens were impacted more than once. All specimens were visually and ultrasonically (C-scan) inspected after impact to determine the existence of damage in the form of compression failure at the surface under the tup, and interply delamination in the top facesheet. Several specimens were sectioned and photomicrographed after C-scanning to reveal the through-the-thickness distribution of delamination as well as the additional damage in the form of matrix cracking and core crushing.

Results and Discussion

The proportion of the incident energy not returned to the tup at the end of the impact event (assumed to be the energy absorbed by the plate in nonconservative processes such as damage) is presented as a function of the impact energy in figure 1 for 8-ply specimens and in figure 2 for 16-ply specimens. Both sandwich constructions show a jump in absorbed energy at what may be considered a damage initiation energy. The energy at which this jump happened was the same energy level (~0.5 joule in figure 1 and ~1.0 joule in figure 2) that produced the initiation of delaminations and core crushing as found by C-scans and photomicrography. This impact energy level is also the same as that which produced the first significant load drop in the force verses time curves (figures 3 and 4). With high-fidelity time-dependent force data, force measurement and energy calculated from it appear to be reliable indicators of damage in this material system. The energy level that first exhibits a significant proportion of the impact energy absorbed by the specimen is the damage initiation impact energy.

Another more surprising result from these tests is also observed in figures 1 and 2; increasing the impact energy above the damage initiation impact energy increased the absorbed energy, but the proportion of the impact energy which the specimen absorbed was nearly constant (a constant percentage in figures 1 and 2). Clearly, this trend will not continue indefinitely as new damage processes such as fibre tensile failure and matrix-fiber debonding come into play at higher impact energies, but for impact energies above but near the damage initiation impact energy the absorbed energy was proportional to the impact energy. That is to say, in this impact energy region this system exhibits a linear relationship between impact and absorbed energy.

This linear relationship between incident and absorbed energy suggests that for the material, geometric configurations, and energies here examined the damage modes may not be changing as the energy increases. It stands to reason that if there were a switching of damage modes within this impact energy range a linear relationship is highly unlikely. This idea is supported by the post-impact inspections of the specimens which indicated that for these two facesheet thicknesses the same damage modes were present in each of the damaged specimens.

Up to the resolution of this data the impact energy that first causes damage in these specimens causes damage by matrix cracking, delamination, and core crushing. It should be noted that all post impact inspections only show us the result of a complete impact event; a single point in time is represented. The processes that form the damage we see occurred over a period of time and it is considered likely that one damage mode begins first and the load shedding associated with that damage initiates the other damage modes.

The only insight into the time-dependence of the damage process available from these tests comes from the load histories (figures 3 and 4). In figures 3 and 4 the contact force histories are shown with time on the horizontal axis. The individual event histories are shown offset in time in proportion to the impact energy for the particular test. Impact energy increases to the right. The first three histories in figure 3 and the first two histories in figure 4 are for tests in which no damage was found in post-impact inspections. For these events the loading/unloading is a smooth process in which the total energy of the system is conserved. When damage initiates, (as indicated by the onset of high-frequency variation in the contact force history) energy is lost to the damage process and the equilibrium is no longer existent. The first major damage (indicated by the first major load drop in the force history) reduces the stiffness of the plate abruptly, and the localized failure allows the portion of the plate which is in contact with the tup to move momentarily faster that the tup resulting in a brief drop in load. This non-equilibrium motion is of very short duration so that soon the tup catches up and the load may continue to rise as the undamaged portion of the plate resists the motion of the tup. If there remains sufficient energy in the tup, the load may grow again beyond the first failure load.

One interesting feature of these histories is that load at which first major load drop is seen (1200 N in figure 3 and 2700 N in figure 4) seems to be relatively constant for the particular facesheet thickness, and this load is approximately the same as the peak load associated with the threshold impact energy. As the impact energy increases, the first failure load does not also increase. This supports the idea that the damage in these tests is load and not time driven since the first damage occurs at a given load independent of the impact energy. Thus, a static indentation test may produce damage similar to that of an impact test provided both tests reach the same peak load.

Another trend that stands out from these data is that the peak load after the first load drop appears to be linearly related to the impact energy. This supports the idea that the load drop is associated with an energy absorbing process (damage) and that after that process the load will only build up again if there still remains enough energy in the system (i.e. enough remaining tup kinetic energy) to support it. In this way, the load drop is itself an indicator of damage in this (figure 1) as well as previous work (Harrington, 1994). With the more reliable force histories available as a result of reduced tup tip mass, this energy is expected to be a more reliable indicator and quantifier of damage.

All specimens tested were subjected to ultrasonic C-scan inspections before and after impact. The c-scans before the tests revealed no significant flaws, while those after revealed multiple delaminations within the top facesheets. No damage was ever observed in the bottom facesheet. In contrast to this, monolithic (other than sandwich) composite plates generally exhibit damage both at the impact site and at the surface opposite the impact site (Ambur etal., 1995). Even though the time-of-flight scans could indicate the approximate depth of the reflecting surface (a delamination surface) only the first return could be captured, so any delaminations below the uppermost one were effectively hidden by the return above it. Since the specimens had multiple interfaces at which delamination could be present, C-scans could only give a overview of the damage and not detailed information about which interfaces had which damages. In addition, matrix cracking and core crushing could not be picked up by C-scans.

Conclusions

The force histories of damaged specimens generally show a smooth increase in force from the time of impact until a load drop occurs. The load at which this drop takes place is independent of the impact energy and is approximately the same as the peak load seen in a specimen receiving the damage initiation impact energy. This suggests that the process initiating damage may be load and not time driven, so that static indentation tests may provide good simulations of low velocity impact to these structures.

The lowest impact energy at which a significant proportion of the impact energy is absorbed by the specimen is the damage initiation impact energy.

For impact energies above but near the damage initiation impact energy this system exhibits a linear relationship between incident and absorbed energy.

No "first" failure mode was observed in these tests, instead, damage, when present, was in the forms of matrix cracking, delamination, and core crushing.

Acknowledgment

The authors would like to thank William Baron, WL/FIBA for his support under contract F33601-95.

Appendix: References

Ambur, D.R., Starnes, J.H. Jr., and Prasad, C.B., 1995, Low-speed impact damage-initiation characteristics of selected laminated composite plates. The impact resistance of composite materials. *AIAA Journal*, Vol. 33, No. 10.

Harrington, T.M., 1994, An Experimental Investigation of Sandwich Panels Under Low Velocity Impact, AFIT/GAE/ENY/94D-22.

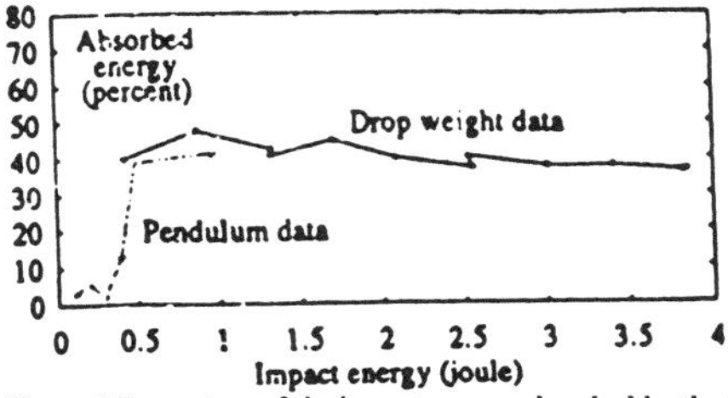

Figure 1 Percentage of the impact energy absorbed by the 8-ply specimens vs. impact energy.

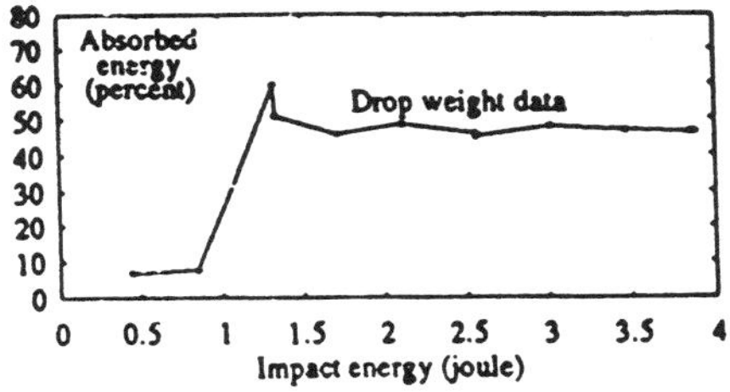

Figure 2 Percentage of the impact energy absorbed by the 16-ply specimens vs. impact energy.

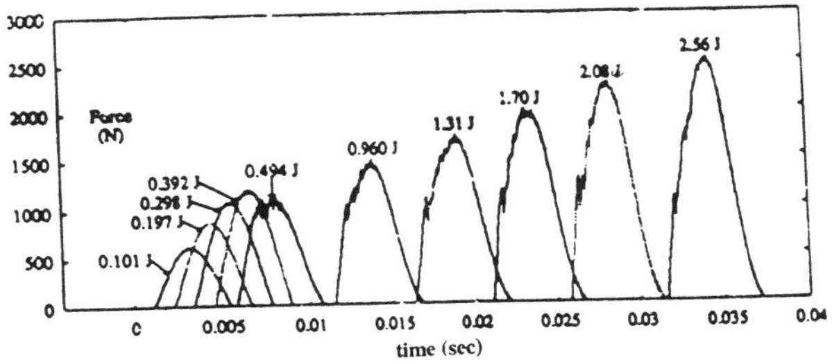

Figure 3 Force histories for 8 ply facesheet specimens (offset by impact energy).

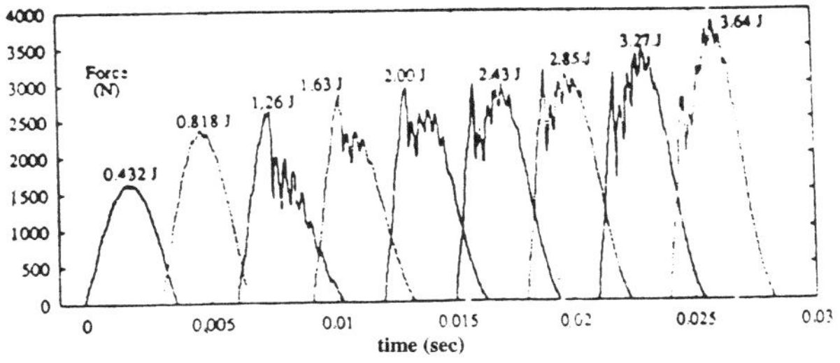

Figure 4 Force histories for 16 ply facesheet specimens (offset by impact energy).

Complex Crack Interaction in Composite Plate

Wieslaw K. Binienda[1]

Abstract

Strain Energy Release Rates (SERR, G_T) are obtained using a rigorous solution for the generalized anisotropic plate with multiple cracks. The solution is formulated using the perturbation mixed boundary-value problem and solved using the Lobatto-Chebyshev collocation technique. Influence of normalized shear (G_{LT}/E_{TT}) and longitudinal Young's (E_{LL}/E_{TT}) moduli on the magnitude of the SERR is examined numerically for two, three and four crack problems.

Introduction

Crack propagation in a composite plate under uniform tensile stress was examined experimentally and documented in the literature. It was shown that a crack path is ragged, branched, and that visible microcracks are formed in front of the major crack (Han and Suresh, 1989 and Han, Warren and Suresh, 1992). A mathematical model for such a class of problems requires consideration of fully interacting multiple cracks in an anisotropic plate, Fig. 1.

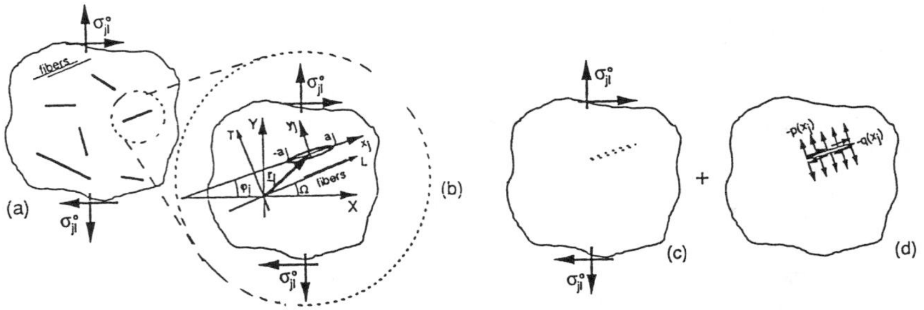

Figure 1. Multicracked plate geometry. (a) Multicracked plate, (b) Fundamental problem for j-th crack, (c) Undamaged plate, (d) Perturbation problem.

[1] Dept. of Civil Engr., The University of Akron, Akron OH 44325-3905, USA

The first step is to derive local stress equations for *j-th* crack in its local coordinate system, x_j-y_j, which is defined with respect to the global coordinate system, *X-Y,* using the position vector, $\underline{r}_j$, and the rotation angle, ϕ_j. The crack is symmetrically situated within its own coordinate system and is $2a_j$ long as shown in Fig. 1(b). The stress equations for the local crack can be obtained by decomposition of the problem into two subproblems: the problem of the undamaged plate under the far field loading, Fig. 1(c) and the perturbation problem, Fig. 1 (d), of a plate with the *j-th* crack subjected to appropriate crack surface tractions $p_j(x_j)$ and $q_j(x_j)$, which are found from the solution of the complementary undamaged problem.

The mixed boundary conditions for the perturbation part, Fig. 1(d), are expressed in terms of normal and shear stress components along the crack surface and in terms of the continuity displacement outside the crack. The analysis of the perturbation problem leads to the identification of singular stresses that govern the local behavior in front of the crack tip.

The second step is to formulate the total stress equations for each crack. The total stress equations include the interaction of all cracks through the summation of the transformed local stresses of all remaining cracks into the local coordinate system for each crack. In order to apply Lobatto-Chebyshev collocation technique and to solve the integral stress equations for the mode-I and mode-II of the Stress Intensity Factors (SIF, k_1 and k_2), the total stress equations are normalized with respect to the local crack length.

The SIF are determined using the discrete auxiliary functions that represent crack opening displacement. Finally, SERR are calculated in terms of the SIFs. The details of the solution for isotropic and anisotropic plates, the accuracy and sensitivity studies were reported by Binienda (Binienda et. al. (1993) and Binienda and Arnold (1995)). It was found that in the case of anisotropic materials the best fracture parameter was the total SERR.

Numerical applications

The focus of the following parametric studies will be limited to investigating the influence of the crack geometry configuration and the strength of anisotropy in an unidirectional fiber reinforced plate subjected to far field unit tensile stress. SERRs for the inner crack tips are reported herein, because they display the largest interaction. Total SERR is calculated using *30* collocation points. The normalized shear modulus (G_{LT}/E_{TT}) is varied from *0.05* to *10* for three cases of the normalized Young's modulus, specifically $(E_{LL}/E_{TT}) = 40, 20$ and 1. For all applications Poisson's ratio is $\nu_{LT}=0.25$, so the isotropic case can be identified for $(E_{LL}/E_{TT}) = 1$ and $(G_{LT}/E_{TT}) = 0.4$.

The case of two collinear cracks of the same length is shown in Fig. 2. The angle of fiber orientation measured from the horizontal axis *X* is $\Omega = 22°$, and the crack tip distance is $0.1a$. The normalized SERRs calculated for this configuration are shown in Fig. 3. The highest values of G_T are obtained for $(E_{LL}/E_{TT}) = 1$ and the smallest for $(E_{LL}/E_{TT}) = 40$. SERR curve for $(E_{LL}/E_{TT}) = 20$ is

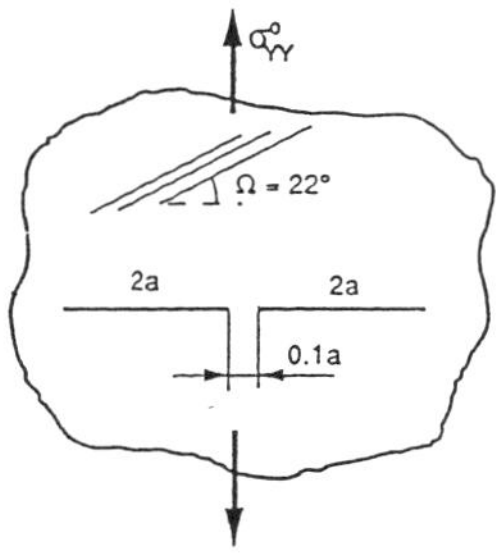

Figure 2. Geometry of two collinear cracks problem.

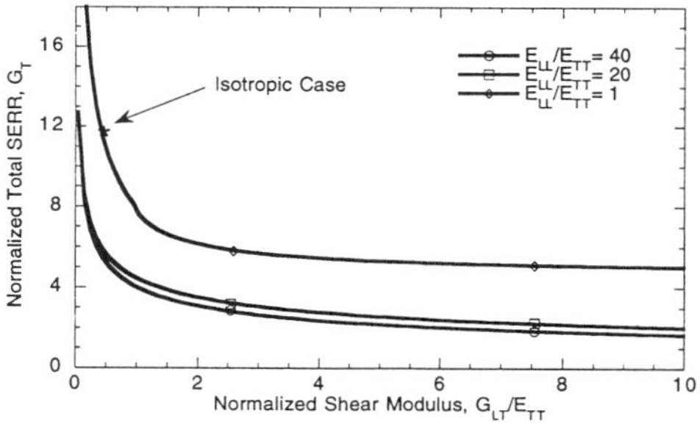

Figure 3.Normalized G_T versuls normalized shear modulus for two collinear cracks.

very close to the one with anisotropy *(E_{LL}/E_{TT}) = 40.* The isotropic case is located in the vicinity of the steepest portion of the SERR curve. So, even very small variation from the isotropic conditions can produce high change in the SERR magnitude. For applications with (E_{LL}/E_{TT}) > *20* and (G_{LT}/E_{TT})> *2* the change of G_T is smaller.

The geometry to the case of two horizontal, equal cracks is shown in Fig. 4. The angle of fiber orientation with respect to *X* axis is now $\Omega = 45°$. Cracks configuration is defined by the angle between the global *X* axis that coincides with the base crack and the line connecting the inner crack tips of the base crack with the second crack that is shifted above the *X* axis. In the numerical application under consideration, this angle is the same as the fiber angle, i.e., $\alpha = 45°$. The distance between the inner crack tips is also kept constant *0.1a*. The behavior of the SERR curves is shown in Fig. 5, and it is similar to the one obtained for the collinear cracks, but the magnitudes of G_T are about 50% higher. Again, any small variation from the isotropic case produces very large changes of the SERR magnitude. Hence, crack propagation prediction for the realistic materials requires an anisotropic problem solution to establish accurately the magnitude of the driving forces.

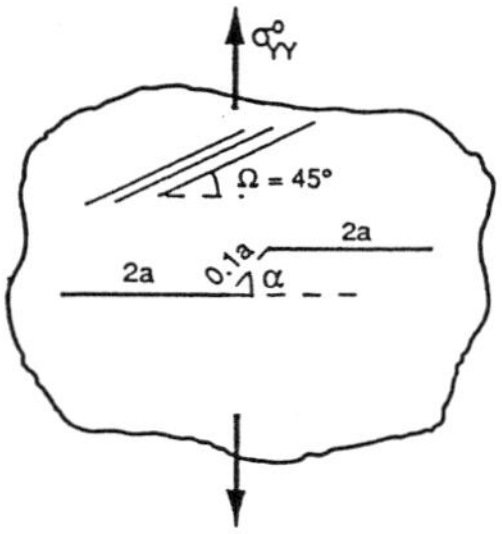

Figure 4. Geometry of two parallel cracks problem.

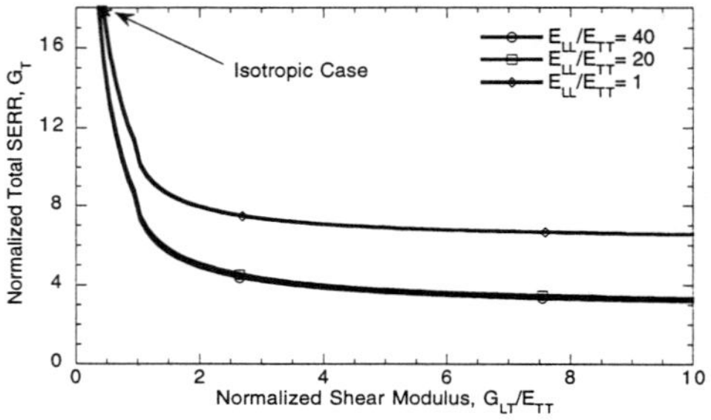

Figure 5.Normalized G_T versuls normalized shear modulus for two parallel cracks.

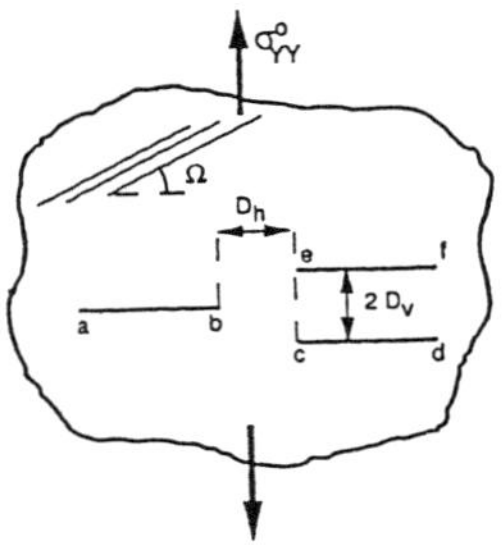

Figure 6. Geometry of three parallel cracks problem.

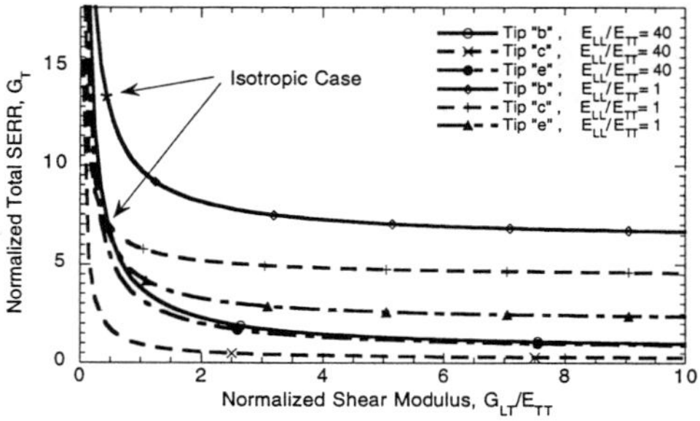

Figure 7. Normalized G_T versuls normalized shear modulus for three parallel cracks.

If another parallel crack is added symmetrically with respect to the X axis to the shifted crack, the three crack configuration with the base crack and two parallel crack is obtained, see Fig. 6. The fiber orientation is $\Omega = 60°$. It can be noted from Fig. 7 that the magnitude of G_T curve for $(E_{LL}/E_{TT}) = 1$ at the base inner crack tip remains almost the same as in the case of two cracks. However, when $(E_{LL}/E_{TT}) = 40$ there is a significant reduction of the SERR magnitude at the same base crack tip "b". In addition one can notice significant reduction of the SERR magnitude at the parallel crack tips because of the shielding effect present between them. The SERR at the parallel crack tip along fiber direction, tip "e", is much higher than it is at the symmetric crack tip below, tip "c" for the anisotropic cases. Obviously this difference is produced because the stresses are channeled along the fiber direction, and they are dissipated in the transverse direction making the interaction between crack tips "b" and "e" much stronger than between the tips "b" and "c".

The final case considered in this paper is produced using four crack geometry, see Fig. 8. The base crack is considered to be a major crack of the length 2a which is surrounded by three microcracks of the length 0.2a that are equally distant from the inner tip "b", 0.1 a. The orientations of the microcracks are 0, 45, and 90 degrees. The fiber orientation in this case is assumed to be very small, $\Omega = 15°$. All the SERR curves continue to display similar behavior as in all previous cases, see Fig. 9. It should be noticed that the magnitudes are much smaller for all three crack tips, and the SERR for the vertical crack tip is close to zero, so it is not shown. All the previously isolated phenomena, i.e., the shielding effect between the microcracks, the stress channeling effect, and the material anisotropy effect play significant role in this case. In addition, one can note that the crack length effect is also present in this solution making the SERR in the base crack the largest for all examined normalized shear and Young's moduli.

Future work will focus on the development of the solution to the multipled kinked and branched crack problems.

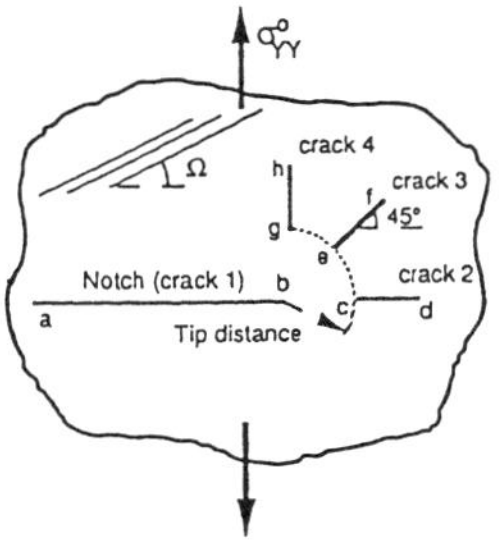

Figure 8. Geometry of a major crack with three microcracks problem.

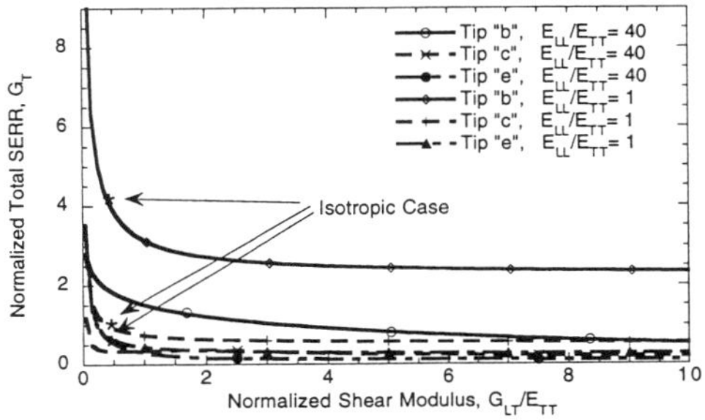

Figure 9.Normalized G_T versuls normalized shear modulus for the major crack and three microcrack.

References

Binienda W.K., Arnold S.M., Tan H.Q. and Xu M.H. (1993) "Stress Intensity Factors in a Fully Interacting, Multicracked, Isotropic Plate" Comp. Mech., v. 12, no. 5, pp. 297-314.

Binienda W.K. and Arnold S.M. (1995) "Driving Force Analysis in an Infinite Anisotropic Plate with Multiple Crack Interaction" Int. J. Fract., v. 71, pp. 213-245.

Han L.X. and Suresh S. (1989) "High Temperature Failure f an Alumina-Silicon Carbide Composite under Cyclic Loads: Mechanisms of Fatigue Crack-Tip Damage" J. Am. Ceram. Soc., v. 72, no. 7, pp. 1233-1238.

Han L. X., Warren R. And Suresh S. (1992) "An Experimental Study of Toughening and Degradation due to Microcracking in a Ceramic Composite" Acta Metall., v. 40, no. 2, pp. 259-274.

Response of MMC Tubes with Internal Fiber Cracks

Sarah C. Baxter[1] and Marek-Jerzy Pindera[2]

Abstract

The effects of partial fiber splits within the silicon carbon fibers of a SiC/Ti composite on the mechanical response of a cylindrical tube are investigated. The fiber cracks are defined as imperfect interfacial subcell bonds within the *repeating unit cell* of the micromechanical model known as the *Generalized Method of Cells.* Results are presented for a single layer, 0 degree fiber orientation, composite tube subjected to an axial torque. These results show significant degradation of the tube's response due to pre-existing fiber splits in both the elastic and inelastic regions.

Introduction

Metal matrix composites (MMCs) are currently being considered for high temperature applications in the aerospace industry. Damage can occur in these composite materials due to either the processing conditions and/or the thermomechanical loading to which the materials are subjected. Specific examples of damage in fibrous MMCs are matrix cracks, transverse fiber fractures, radial fiber cracks or fiber-matrix debonds. In particular, experimental evidence has shown the presence of diametral fiber cracks situated within silicon carbide fibers in SiC/Ti composites.

The laminated tube geometry is of particular interest in modeling a variety of structural elements e.g., pressure vessels or shafts. The possible loads to which the tube may be subjected include axial loads, thermal loading, internal and external pressure and axial torque.

An *exact, implicit, analytic* solution to the inelastic boundary value problem, posed for an arbitrarily laminated composite tube, does exist on the

[1]Postdoctoral Research Associate

[2]Associate Professor, Department of Civil Engr. and Applied Mech., University of Virginia, Charlottesville, VA 22903

global level (Salzar et al., 1996), but no previous attempt has been made to incorporate the combined effects of previously existing damage and inelasticity at the micromechanical level. Due to the lack of a generally accepted inelastic macromechanical constitutive model for heterogeneous materials, the micromechanical model known as the generalized method of cells, GMC (Paley and Aboudi, 1992), is used to predict the inelastic response of the individual layers of a composite tube. The discrete modeling approach of GMC's subcell construction allows specification of the initial position of a pre-existing crack.

Analytical Formulation

The general boundary value problem is that of a metal matrix composite tube, comprised of an arbitrary number of layers, which is subjected to axisymmetric thermo-mechanical loads. Individual layers can be homogeneous or inhomogeneous, where the inhomogeneous layers are metal matrix materials reinforced with continuous fibers. The orientation of the fibers in these layers, with respect to the tube axis, can be varied by layer. The material properties of the constituent materials may be temperature dependent.

A solution to the inelastic boundary value problem described above is obtained by using a displacement formulation under the assumption of generalized plane strain. Within each layer, the displacement field $\mathbf{u} = (u, v, w)$, with respect to a cylindrical coordinate system, x, θ, r, is given by

$$u = \epsilon_0 x, \qquad v = \gamma_0 x r, \qquad w = w(r), \tag{1}$$

where ϵ_0 is the uniform longitudinal strain for all layers, and γ_0 is the angle of twist per unit length of the tube. The x-axis of this coordinate system corresponds to the axis of the tube. The resulting strain field is then

$$\epsilon_{xx} = \frac{du}{dx} = \epsilon_0, \quad \epsilon_{\theta\theta} = \frac{w(r)}{r}, \quad \epsilon_{rr} = \frac{dw(r)}{dr}, \quad \gamma_{x\theta} = \frac{dv}{dx} = \gamma_0 r. \tag{2}$$

The two remaining strains are identically zero. This strain field satisfies the compatibility equations, as well as two of the three equilibrium equations. The surviving equilibrium equation is

$$\frac{d\sigma_{rr}}{dr} + \frac{\sigma_{rr} - \sigma_{\theta\theta}}{r} = 0, \tag{3}$$

where the stresses are at most functions of the radial coordinate r.

If stress-strain equations, valid in the presence of thermal loading and plastic effects, are known then the governing differential equation for the radial displacement of, for example, a transversely isotropic layer is given by

$$\frac{d^2 w}{dr^2} + \frac{1}{r}\frac{dw}{dr} - \frac{w}{r^2} = \frac{1}{r}\frac{(C_{3i} - C_{2i})}{C_{33}}\epsilon_i^p(r) + \frac{C_{3i}}{C_{33}}\frac{d\epsilon_i^p(r)}{dr}. \tag{4}$$

The repeated index i indicates summation over $1, 2, 3, 6$. The quantities ϵ_i^p, where $(\epsilon_1,\ \epsilon_2,\ \epsilon_3,\ \epsilon_6) = (\epsilon_{xx},\ \epsilon_{\theta\theta},\ \epsilon_{rr},\ \epsilon_{x\theta})$ are the plastic strains.

The solution to (4) is then given by

$$w(r) = A_1 r + \frac{A_2}{r} + \frac{1}{2r}\int_{r_{k-1}}^{r} \frac{(C_{3i}+C_{2i})}{C_{33}}\epsilon_i^p(t)dt + \frac{r}{2}\int_{r_{k-1}}^{r} \frac{1}{t}\frac{(C_{3i}-C_{2i})}{C_{33}}\epsilon_i^p(t)dt. \tag{5}$$

A local/global stiffness matrix formulation, GSM (Pindera, 1991), which reformulates the problem in terms of the displacements between the layers, enforces boundary conditions for the problem, considering both the boundaries of the tube structure and the interfaces between the layers. The construction of the GSM requires knowledge of the macroscopic elastic properties of each layer. In this work, the generalized method of cells is used to develop effective properties for a composite material from the properties of its constituents, the layer and fiber-matrix geometry, and the angle of fiber orientation.

Results and Conclusions

The example presented herein is for a single layer tube of 0 degree fiber orientation. The tube is loaded by an axial torque. The composite material is a SiC/Ti system where the titanium matrix is modeled as a nearly elastic-perfectly plastic material. The SCS-6 silicon carbide fiber consists of a carbon core surrounded by a silicon carbide sheath. The silicon carbide fiber occupies 40% of the total composite volume, and the carbide core is 5.4% of the fiber volume. The inner radius of the tube is taken to be 1 m, the outer radius, r_N is 1.2 m. The response of a tube made of the undamaged composite material is compared to that of a tube in which the fibers have radial cracks. These radial cracks extend from the outer edge of the core through half the distance to the outer edge of the fiber. The direction of the cracks corresponds to the radial direction of the tube. Figure (1) plots characteristic shear stress-strain curves for the damaged and undamaged composite tubes. In the presence of damage the elastic modulus is substantially degraded and the apparent initial yield stress is reduced.

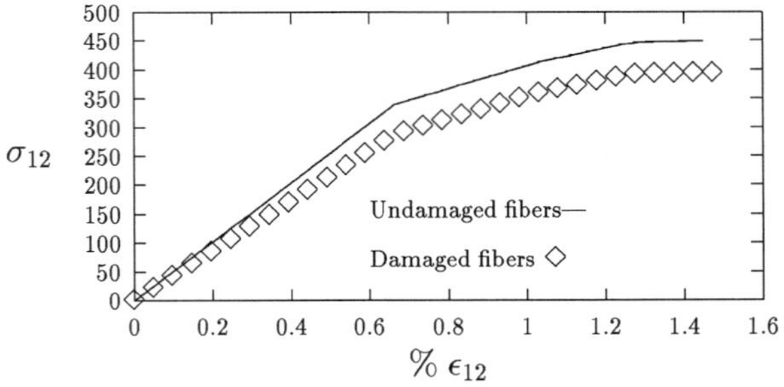

Figure 1: Shear stress versus shear strain

Figure (2) shows the shear stress versus a non-dimensional radial coordinate. The stress levels plotted were chosen from the undamaged tube results. They correspond to a region of elastic deformation and the approximate onset of plastic yielding at the outer surface, the mid-plane and the inner surface, respectively.

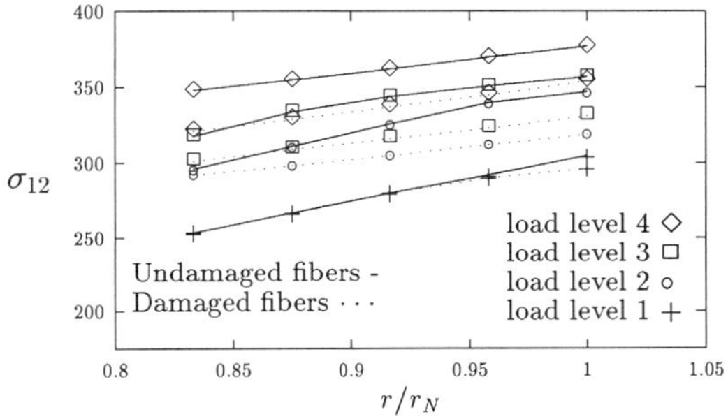

Figure 2: Shear stress distribution versus radial position

The damaged composite tube results were then plotted for these same load values. The stress levels at each load step are smaller in the damaged composite tube. At the first load level, which is in the elastic range for the undamaged tube, yielding has begun at the outer radius of the damaged tube. At the second load level the undamaged composite tube has just begun plastic deformation at its outer radius, while plastic deformation is present throughout the thickness of the damaged composite tube. Although the stress levels are greater in the undamaged composite tube, the stress gradient throughout the thickness of the tube is not substantially different from that of the damaged tube, in either the elastic region or fully yielded plastic region. These results indicate that fiber damage, in the form of pre-existing radial cracks, accelerates yielding and the progression of the yield zone throughout the tube's thickness.

Future work will involve considerations of crack orientations with respect to load and the extent of fiber damage, and parameter studies on crack length.

Reference

Paley, M. and Aboudi, J. (1992) "Micromechanical analysis of composites by the generalized cells model", *Mechanics of Materials*, Vol. 14, 127-139

Pindera, M-J. (1991) "Local/Global Stiffness Matrix Formulation for Composite Materials and Structures", *Composite Engineering*, Vol. 1, 69-83

Salzar, R.S., Pindera, M-J. and Barton, F.W. (1996) "Elastoplastic Analysis of Layered Metal Matrix Composite Cylinders. Part I Theory", *Journal of Pressure Vessel Technology*, February, in press.

PROPERTY DETERIORATION OF COMPOSITES

Charles E. S. Ueng[1], Member ASCE

Abstract

Low cost glass-fiber/polymer-matrix type composites used for civil infrastructures usually exhibit significant amount of strain change when the load is applied for a long period of time, or due to temperature increase during the service time. For the purpose to obtain results on stress analysis of such a structure, it is important to know the variation of Young's modulus as a function of time and temperature. Therefore, a reasonable estimate can be made for the prediction of its long term response and behavior. In this paper, it reports a method on how to conduct a number of tests so a set of curves can be plotted from which the Young's modulus of such a composite can be determined as a function of time and temperature.. These results will be extremely useful in the design process for the prediction of long term deformation behavior of such low cost composites which are widely used in an increasing fashion for civil infrastructures.

Introduction

The versatility of composite materials makes it one of the most widely used construction materials in many engineering applications today. In addition to aerospace and automotive structures, the newest applications are high voltage electric insulation, antenna enclosures, chemical plant structures and cooling systems. Unidirectional composite rods often are used as tension members for earth anchors and prestressing elements in prestressed concrete.

---

[1] Professor of Civil Engineering, and Director, Composites Education and Research Center, School of Civil and Environmental Engineering, Georgia Institute of Technology, Atlanta, GA 30332-0355.

However, such low cost composites which are made of less expensive constituent materials, often exhibit significant amount of strain change when the load is applied for a long period of time, or due to temperature increase during the service time. This type of composites usually contains a significant amount of glass fibers for the purpose to provide necessary strength and resistance for the applied load. Because of this, it is quite sensitive to the temperature change and duration of the loading. A thorough understanding of such behavior is an important factor in the design process so it will make sure that the structure can provide an acceptable serviceability and a useful life span as planned without any unexpected failure due to excessive deformation.

For the purpose to obtain results on stress analysis of such a structure, it is important to know the variation of Young's modulus as a function of time and temperature. Therefore, a reasonable estimate can be made for the prediction of its long term response and behavior. Up to now, very little data or guidance is available in the literature.

In this paper, it reports a method on how to conduct a number of tests so a set of curves can be plotted from which the Young's modulus of such a composite can be determined as a function of time and temperature. These results will be extremely useful in the design process for the prediction of long term deformation behavior of such low cost composites which are widely used in an increasing fashion for civil infrastructures.

Proposed Tests

A workable, systematic approach is proposed here for the experimental determination of the material property parameter including the consideration of creep effect. A number of tests need to be conducted in order to obtain such information.

A simple method of creep testing involves the application of a constant axial load to a bar made of the material of interest. The load is held constant. The creep strain is measured with respect to time at a given temperature, and the rupture time is also recorded if it does occur during the test. Such tests are then repeated at different stress and temperature levels. In the strain-time diagram, it usually exhibits three stages of deformation: (1) Primary, (2) Secondary, and (3) Tertiary. The slope of such a curve gives the strain rate which changes in a decreasing fashion during the primary stage, remains approximately a constant during the secondary stage, and finally increases during the third, or tertiary stage.

In the stress analysis investigation when creep is included in the consideration, a set of stress-strain curves for various constant values of time, called *isochronous* stress-strain curves, are often necessary. Such curves can be constructed from strain versus time data as obtained from different stress levels. First, a set of curves are obtained such as shown in Fig. 1. Then using these curves, one picks up a particular value of t, say $t = t_1$. This vertical line intersects at a number of points with those curves shown. All intersecting points such as $(\varepsilon_1, \sigma_1)$, $(\varepsilon_2, \sigma_2)$, $(\varepsilon_3, \sigma_3)$ and $(\varepsilon_4, \sigma_4)$ can be plotted in the stress-strain plane or σ versus ε axes such as shown in Fig. 2 in which more curves are obtained by letting $t = t_2$, t_3, ---, etc. Such curves are useful for the purpose to provide the needed Young's modulus corresponding to a given strain, say $\varepsilon = \varepsilon'$. For instance, by replotting the three curves from Fig. 2 to Fig. 3, then draw a vertical line for $\varepsilon = \varepsilon'$ which intersects at three points. Connecting those three points with the origin, thus the slope of each line represents E_s , the so called secant modulus at a given value of time. Using such results, one can plot out a curve for E_s versus time t, or log t, in order to describe the material behavior. Picking up another value of strain, say $\varepsilon = \varepsilon''$, ---, etc., one can eventually obtain a set of such curves such as shown in Fig. 4. Note that all these figures are referred to a given constant temperature T. The tests mentioned here can be repeated for a higher temperature level by increasing a given amount each time so another set of curves will become available corresponding to a new temperature. For practical design purposes, it may be necessary to carry out the entire range of temperature for which the structure will experience during the service on the step by step or finite increment basis.

Discussion and Conclusions

From Fig. 3, it indicates that as ε increases, the corresponding value of E_s decreases. Furthermore, the slope of the straight lines is flattened as time t increases from t_1 to t_2, from t_2 to t_3, etc. Physically, this also indicates a deterioration of material stiffness. Similarly, Fig. 4 shows that at a given strain ε, the value of E_s drops slowly as time t increases. As ε increases from ε' to ε'', a lower value of E_s is obtained. In other words, the material will become softer and softer as the strain gets larger.

It is known that the creep related phenomenon of glass fibers will become more noticeable once the service temperature is getting closer and closer to the glass transition temperature. Based on this, it is anticipated that as the temperature level is raised, the curves obtained from these tests will show further reduction of the stiffness as represented by E_s here.

The testing methods proposed here are quite simple and easy to follow through. They can generate the necessary information needed by an engineer for the purpose to carry out his long term prediction of a given structural component which is made from such a composite.

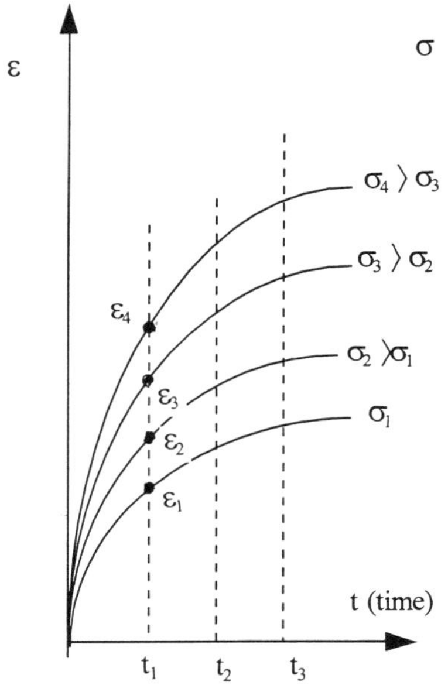

Fig. 1 ε-t curves at different stress levels but a given temperature

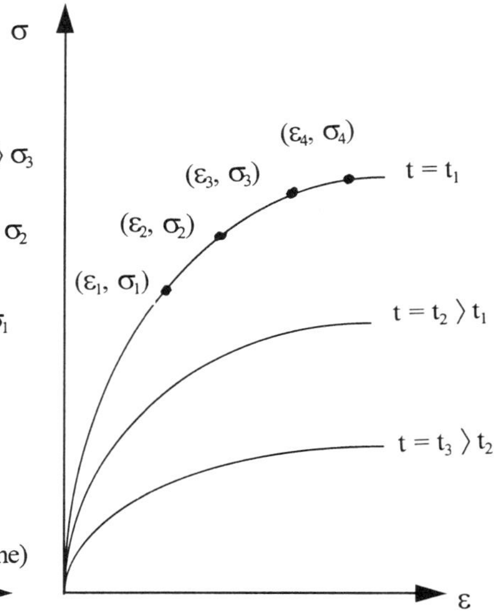

Fig.2 Isochronous stress-strain curves

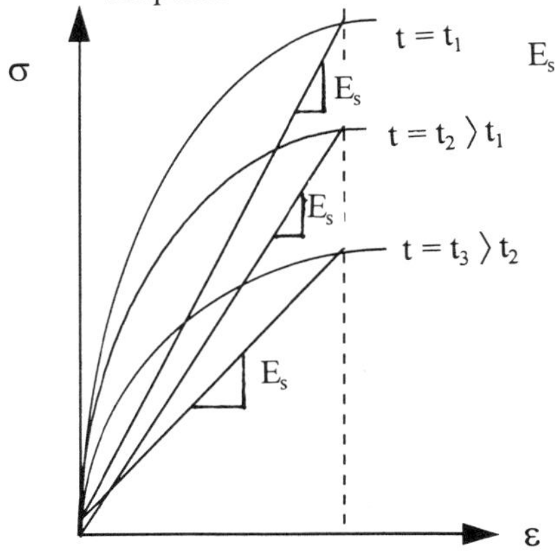

Fig.3 Secant modulus E_s for a given strain $\varepsilon = \varepsilon'$

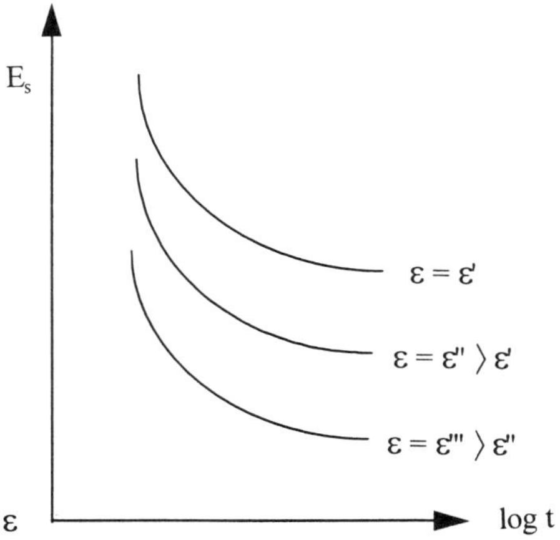

Fig. 4 E_s versus log t for a given strain, $\varepsilon = \varepsilon'$, etc.

MULTIPHASE FLOW IN DEFORMING POROUS MEDIA BY THE FINITE ELEMENT METHOD

Pedro Arduino, ASCE SM [1]
Emir J. Macari, ASCE AM [2]

Abstract

The mechanical behavior of saturated and semisaturated geomaterials is governed largely by the interaction of their solid skeleton with the fluids present in the pore structure. This interaction is particularly strong in dynamic problems and may lead to the catastrophic loss of strength known as liquefaction which frequently occurs under earthquake loading in the case of saturated granular materials. Hence, a multiphase theory becomes necessary to account for a realistic behavior of porous media. In this work the governing equations of a porous media interacting with immiscible porous fluids are derived in the light of the theory of mixtures. A generalized Galerkin procedure is devised to stablish the coupled mixed finite element equation set with u-p-U form. An unconditionally stable implicit solution procedure is used for the time domain numerical solution. Examples of saturated granular deposits under transient loads are presented to illustrate the performance and capabilities of the proposed solution.

1 Introduction

Many engineering problems involve the analysis of the response of porous media. These problems include the settlement of underlaying soil deposits, the dissipation of excess-pore-water pressures resulting from foundation loading, as well as more complex cases such as the design of foundations for vibrating machines, the analysis of offshore structures under wave loading, and the study of propagating earthquake impulses in geologic materials.

Porous media (the pore spaces of which filled with fluids) are characterized by the fact that each one of the materials occupies its own region (i.e one assumes the fluids to be immiscible). The difficulties which arise in describing porous media are evident. Since, in general, the geometrical structure of the interfaces between the different materials is unspecified, one does not know which spatial point is occupied by which particular material. For this reason substitute continuum models (mixture theories) have to be employed, where each material is assumed to occupy the total region.

[1] Ph.D. Research Assistant, Georgia Institute of Technology
[2] Associate Professor, Georgia Institute of Technology

In this context, the theory of porous solids with void pores has been the subject of research since the 1920's. The Terzaghi's effective stress concept and 1-D consolidation theory [1924, 1925] as well as Biot's 3-D consolidation theory and wave propagation analysis of saturated porous media [1941, 1956] may be considered some of the pioneering efforts for the development of an appropriate continuum theory for porous media (i.e. a theory of mixtures). Several other researchers have also focused their attention on the static and dynamic behavior of porous media leading to different theoretical solutions.

In recent years, several numerical approaches have been presented in the literature. Two such works are the finite difference approach given by Garg et al. [1974] and the finite element procedure presented by Gaboussi and Wilson [1972]. Subsequently, other finite element applications have been developed for two and three-field formulations based on a suitable choice of solid displacement, fluid displacement and pore pressure as nodal variables. Zienkiewicz et al. [1984] discussed the limitations and the range of applicability of the various assumptions on which these formulations are based. Simon et al. [1986] presented an accuracy analysis of different formulations by comparing numerical and analytical results of one-dimensional propagation.

Until now, the three-field formulation has not been well studied and only few applications have been presented in the literature. Although, Zienkiewicz and Shiomi [1984] were the first to develop the so called *u-p-U* formulation (i.e. absolute displacement of solid and fluid phases and pore pressures as nodal variables), they did not utilize this method. The main characteristic of the *u-p-U* formulation is the possibility of representing virtually incompressible pore fluid and solid grains. Furthermore, since fluid inertial forces are not neglected, this formulation is applicable to any range of frequency for which the hypotesis of constant permeability is valid, i.e. to any problem in which wavelengths are greater than pore size [Kubik, 1986]. Herein, we present the finite element model and numerical results of the *u-p-U* formulation for one and two dimensional problems.

2 Field Equations

In the light of the theory of mixtures the considered problems have a rather complex character, because the thermodynamical behavior of the different media as well as the transport processes of fluid, and heat have to be analysed in general, leading to geometrically and physically nonlinear relations. In order to obtain a possible simple and practical theory several simplifications need to be introduced.

Assumptions

- A multicomponent model using volume fractions is adopted.
- The different components share the same temperature.
- The compressibility of the grain particles is much lower than that of the skeleton
- The stress deviator in the fluid is assumed negligible.
- Small deformations theory is assumed.

By means of the principles of consevation of mass, linear and angular momentum, the field equations that govern the problem reduce to the following equations:

$$\begin{aligned}(1-n)\,\rho_s\,\ddot{x}_i^{(s)} + (1-n)\,\rho_s\left(\dot{x}_{i,j}^{(s)}\,\dot{x}_j^{(s)}\right) - (1-n)\,\rho_s\,b_i^{(s)} &\;+\\ n^2\,\rho_f (K^{sf})_{ij}^{-1}\left(\dot{x}_j^{(s)} - \dot{x}_j^{(f)}\right) - \sigma_{ij,j}^{'(s)} + (1-n)\,p_{,i} &= 0 \qquad (1)\end{aligned}$$

$$\begin{aligned}n\,\rho_f\,\ddot{x}^{(f)} + n\,\rho_f\left(\dot{x}_{i,j}^{(f)}\,\dot{x}_j^{(f)}\right) - n\,\rho_f\,b_i^{(f)} &\;-\\ n^2\,\rho_f\,(K^{sf})_{ij}^{-1}\left(\dot{x}_i^{(s)} - \dot{x}_i^{(f)}\right) + n\,p_{,i} &= 0 \qquad (2)\end{aligned}$$

$$\frac{n}{K_f}\,\dot{p} + (1-n)\,\dot{x}_{i,i}^{(s)} + n\,\dot{x}_{i,i}^{(f)} = 0 \qquad (3)$$

Equations (1-3) are the equations of motion of the solid and fluid phases, and the continuity equation for the mixture, respectively.

3 Weak Form-Discrete Equations

The weak formulation associated with the set of equations stated in the previous section may be obtained by proceeding along standard lines [Bathe, 1995; Hughes, 1987]. To obtain a solution in terms of solid and fluid displacements and pore pressures, a spatial semi-discretization of the domain Ω into non-overlapping finite element is used followed by an appropriate time stepping process applicable to the final solution.

Spatial Semi-Discretization

If a Galerkin finite element formulation is applied the semi-discretized system takes the form

$$\begin{bmatrix}\mathbf{M}^{ss} & \mathbf{0} & \mathbf{0}\\ \mathbf{0} & \mathbf{0} & \mathbf{0}\\ \mathbf{0} & \mathbf{0} & \mathbf{M}^{ff}\end{bmatrix}\begin{Bmatrix}\ddot{\bar{\mathbf{x}}}^{(s)}\\ \ddot{\bar{\mathbf{p}}}\\ \ddot{\bar{\mathbf{x}}}^{(f)}\end{Bmatrix} + \begin{bmatrix}\mathbf{C}^{ss} & \mathbf{0} & -\mathbf{C}^{sf}\\ \mathbf{0} & \mathbf{0} & \mathbf{0}\\ -\mathbf{C}^{fs} & \mathbf{0} & \mathbf{C}^{ff}\end{bmatrix}\begin{Bmatrix}\dot{\bar{\mathbf{x}}}^{(s)}\\ \dot{\bar{\mathbf{p}}}\\ \dot{\bar{\mathbf{x}}}^{(f)}\end{Bmatrix} + \qquad (4)$$

$$\begin{bmatrix}\mathbf{K}^{ss} & -\mathbf{S}^{sp} & \mathbf{0}\\ -\mathbf{S}^{ps} & -\mathbf{R} & -\mathbf{S}^{pf}\\ \mathbf{0} & -\mathbf{S}^{fp} & \mathbf{0}\end{bmatrix}\begin{Bmatrix}\bar{\mathbf{x}}^{(s)}\\ \bar{\mathbf{p}}\\ \bar{\mathbf{x}}^{(f)}\end{Bmatrix} = \begin{Bmatrix}\mathbf{F}^{s}\\ \mathbf{0}\\ \mathbf{F}^{f}\end{Bmatrix}$$

where each term results from the numerical integration of the above governing equations. Note that the displacement vector term contains displacement and pressure quantites.

Time Integration

The spatially discrete finite element euqations are integrated in time using an implicit predictor corrector method based on the α-method proposed by Hughes et al. [1977]. This work follows the approach suggested by Miranda et al. [1989].

4 Numerical Examples

In order to illustrate the performance of the proposed formulation, results for dilatational wave propagation in saturated porous media are presented. To this end, a simple

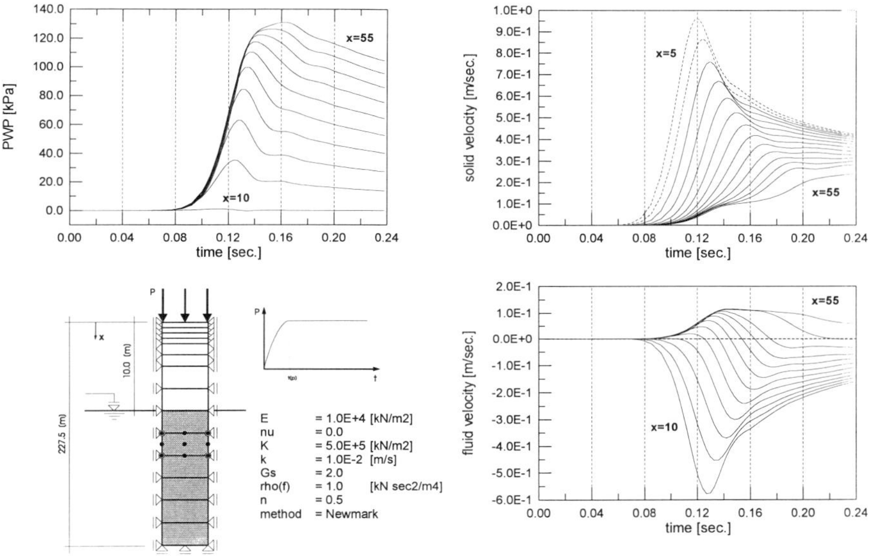

Figure 1: PWP, Solid, and Fluid Velocity Records for 1-D Wave Propagation

one dimensional column model fixed at one end and subjected to an impulsive load at the other end is used. The selected mesh figure 1, consist of 14 2-D elements with a total length of 230 mts. Upon impact at the free end, a compressive wave is sent through the medium, hits the water table, gets partly reflected and partly transmitted through the saturated porous medium. Figure 1 shows the computed solid and fluid velocities at distances ranging from $x = 5.0$ to 55.0 meters away from the free end. Note the distinct two wave fronts in the response evaluated in the near field of the saturated medium. The second wave is more diffusive than the first one and eventually disappears in the far field, where only the wave of the first kind prevails.

5 Summary

In this work the governing equations of porous media interacting with immiscible porous fluids were derived in the light of the theory of mixtures. A generalized Galerkin procedure is devised to establish the coupled mixed finite element equation set with u-p-U form. An unconditionally stable implicit solution procedure was used for the time domain numerical solution. Finally, the essential characteristics of the propagation of dilatational waves were obtained using a one dimensional column model illustrating the performance and capabilities of the proposed solution.

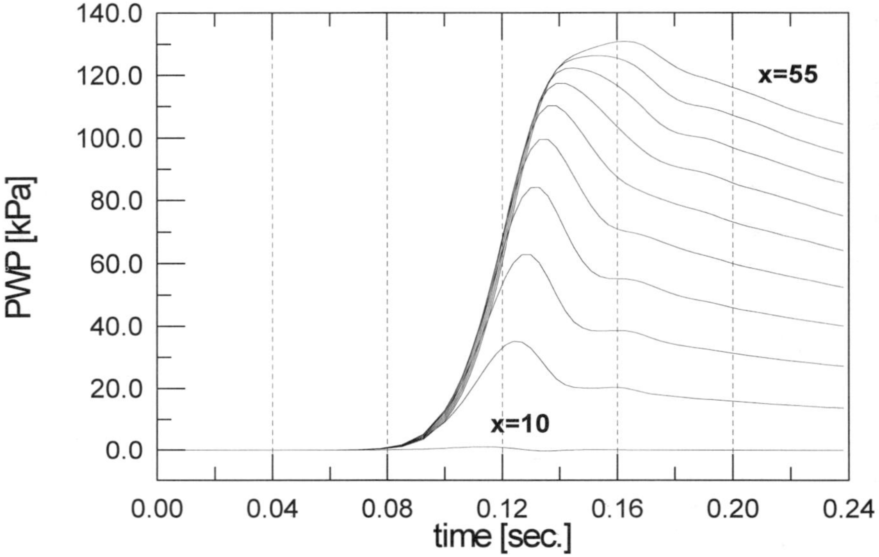

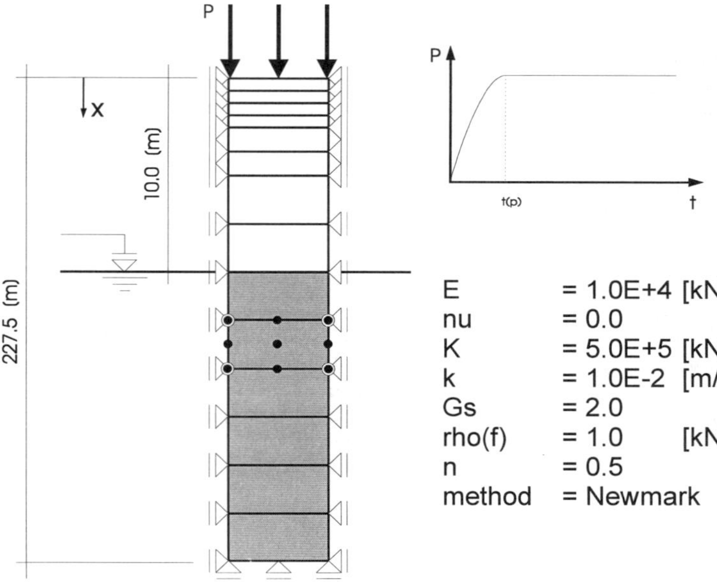

E = 1.0E+4 [kN/m2]
nu = 0.0
K = 5.0E+5 [kN/m2]
k = 1.0E-2 [m/s]
Gs = 2.0
rho(f) = 1.0 [kN sec2/m4]
n = 0.5
method = Newmark

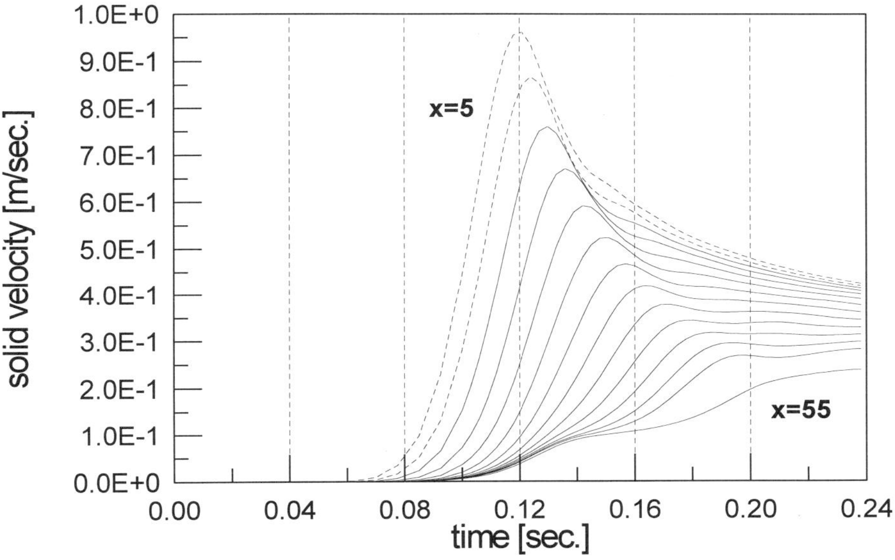
solid velocity [m/sec.]
1.0E+0
9.0E-1
8.0E-1
7.0E-1
6.0E-1
5.0E-1
4.0E-1
3.0E-1
2.0E-1
1.0E-1
0.0E+0
x=5
x=55
0.00
0.04
0.08
0.12
0.16
0.20
0.24
time [sec.]

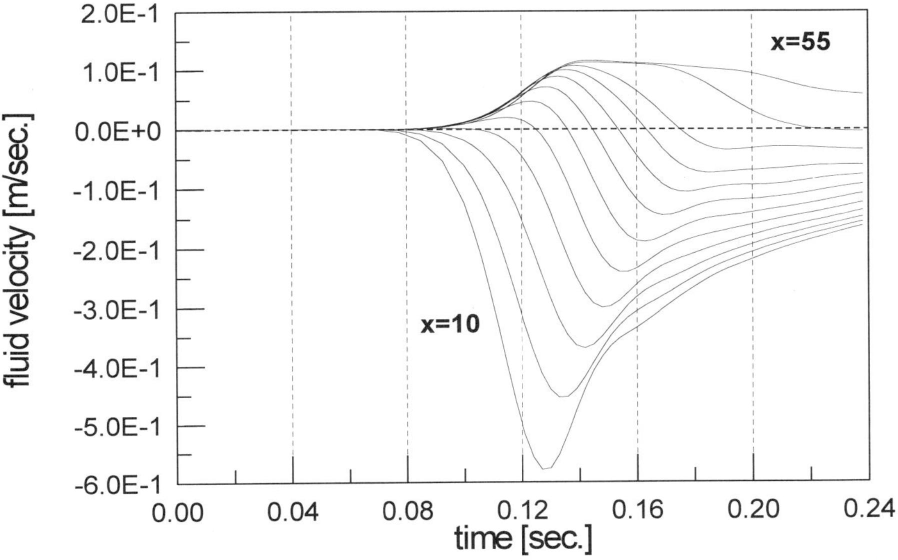
fluid velocity [m/sec.]
2.0E-1
1.0E-1
0.0E+0
-1.0E-1
-2.0E-1
-3.0E-1
-4.0E-1
-5.0E-1
-6.0E-1
x=55
x=10
0.00
0.04
0.08
0.12
0.16
0.20
0.24
time [sec.]

Shock Compression in Granular Media Using DFEM

Abdolreza Joghataie[1] and Jamshid Ghaboussi[1]

Abstract

Effect of discontinuity, porosity and damping due to plasticity of granular materials subjected to shock is investigated. To this end, a new version of Discrete Finite Element codes has been developed to consider the plastic behavior of materials. Shock compression and wave propagation in HMX explosive material has been analyzed numerically and comparison has been made to the available experimental results.

Introduction

A series of experimental studies has been conducted in the Los Alamos and the Sandia National Laboratories to determine the characteristics of shock compression in granular explosives including low density HMX (Sheffield et al.). The problem of shock compression in granular materials has been studied in this paper and numerical simulations of the experiments have been performed. To this end, a Discrete Finite Element code with plasticity of materials has been developed where the Cap plasticity model (Sandler et al. 1976) has been implemented in the computer program. By using Discrete Finite Elements it is possible to include the effects of plasticity, voids, and stress concentration in the granular materials such as HMX. The granular material is simulated as an assemblage of elasto–plastic discrete finite elements. This paper contains some results from the first phase of this on going research.

Equations of Motion of Discrete Finite Elements

Each of the discrete bodies, simulated by a finite element mesh, has the following equations of motion:

$$\boldsymbol{M}\ddot{\boldsymbol{U}} + \boldsymbol{D}\dot{\boldsymbol{U}} = \boldsymbol{P} - \boldsymbol{I} \tag{1}$$

where $\boldsymbol{M}$ and $\boldsymbol{D}$ are the mass and damping matrices, $\boldsymbol{U}$, $\dot{\boldsymbol{U}}$ and $\ddot{\boldsymbol{U}}$ are the displacement, velocity and acceleration, and $\boldsymbol{P}$ and $\boldsymbol{I}$ are the external and the internal force vectors respectively. Contact forces are included in the external force vector $\boldsymbol{P}$. Integration of equations of motion for each block is done by the second central difference method (Barbosa and Ghaboussi 1990).

Cap Plasticity Model

Some of the plastic materials such as soils show strain hardening behavior during their loading. Cap model is one of the proposed models for these kind of materials. In this paper

1. Dept. of Civil Engineering, University of Illinois at Urbana–Champaign, Urbana, Illinois 61801

an algorithm similar to the algorithm proposed by Sandler and Rubin (1979) for the numerical implementation of the cap model has been developed. The cap model is comprised of a fixed failure envelope f_1 and a moving cap f_2 as shown in Figure 1, where their equations can be found too. In these equations I and J are the first and the second stress invariants, A , B , C , and R are material properties. Also k is a parameter that can be a more complicated function of the first invariant of plastic strain ϵ^p , D is a material constant and W is the maximum possible plastic strain of the material.

Simulation Problem

The experimental set up (Sheffield et al. 1993) shown in Figure 2a has been simulated for this study where HMX is confined in a chamber and a hammer with velocity V_0 hits the material and shock wave propagates in the medium and compresses it. HMX has been considered as a discontinuous plastic material as shown in Figure 2b, where a portion of the experimental set up has been studied. First the initial conditions for grains were generated and then they were hit by the hammer.

Generation of initial conditions Grains with elliptic shapes but of slightly different and random sizes and orientations have been generated and a finite element mesh of six elements have been assigned to each of them. The hammer and the bottom of the chamber have been simulated with three elements and one stress gage and one velocity gage have been considered attached to each of them as shown in Figure 2b. During the movement of the hammer, it compresses the material and local stresses are developed both in the hammer and in the material. Since the stress and velocity gauges are in contact with a layer of grains, they provide information about the average values of pressure and velocity of these grains. Hence, to simulate the gauges, it seemed more appropriate to consider for the hammer and the bottom, elements that cover the whole width of the specimen, and the stress at one of the integration points and velocity at one of the corner points have been taken as the gauge measurements as shown in Figure 2b. To shape the initial arrangement of the grains, first they were considered as rigid bodies. They were poured from a height under their own weight and compacted by the hammer and a surcharge, falling under their own weights too. A large contact repulsion force has been assigned to the bodies and the hammer to keep the bodies as separate as possible, until the grains and the hammer have reached an equilibrium and the chamber has been filled.

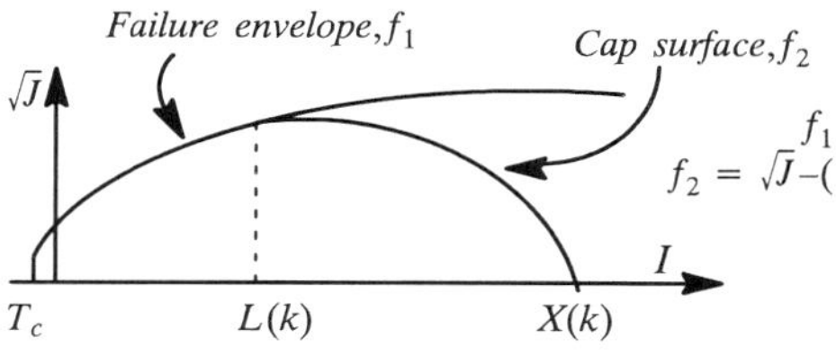

$$f_1 = \sqrt{J} - (A - C \exp(BI)) = 0$$

$$f_2 = \sqrt{J} - (1/R)[(X(k) - L(k))^2 - (\sqrt{J} - L(k))^2]^{1/2} = 0$$

$$k = \epsilon^p$$

$$X(k) = -1/D \, ln(1 - k/W)$$

Figure 1. Cap model

Compression by the hammer In this phase, the grains, the hammer and the bottom of the chamber have been considered plastic and with similar properties as specified in the above mentioned reports. Material properties have been identified through a procedure of trial and error so that material Hugoniot similar to that of HMX, as observed during the experimental investigations, has been obtained. Also similar to the above mentioned experiment, at the Los Alamos National Lab. shot 912, hammer initial velocity of

$V_0 = 288\ m/sec$ and coarse grain HMX with density $\varrho = 1.24\ gr/cm^3$ have been used. As can be seen in Figures 2b to 2e, the material has been compressed and shock wave has been propagated in the material. The shock front, moving faster than the hammer speed, has crushed the material and reached the bottom. Average velocity of shock wave V_s can be calculated from the chamber length L and the time the wave front has reached the bottom T_s as follows:

$$V_s = L\ /\ T_s \approx 0.4\ cm\ /\ 8\mu s = 500\ m/sec \tag{2}$$

which is about twice the hammer velocity.

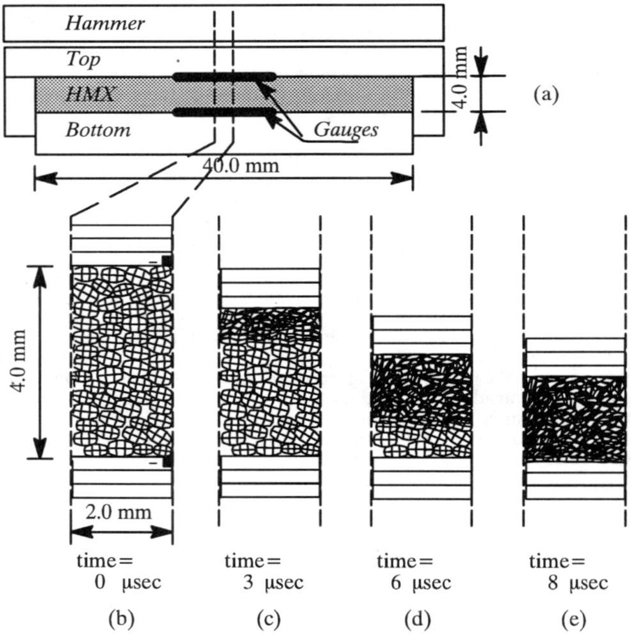

Figure 2 (a) Experimental set up based on (Sheffield, et al. 1993) (b) Initial arrangement of the grains. Velocities have been recorded at the corner points marked by small full squares and stresses have been recorded at the integration points marked by small solid bars. (c), (d) and (e) Three stages of material compression.

Discussion of Results and Comparison to the Experiments

Figures 3a, 3b and 3c represent the history of velocity and stress measurements by the gauges. Solid lines show the numerical results obtained in this study and dashed lines show the approximate experimental results which are extracted from Sheffield et al (1993). While differences can be seen, there are also significant points of similarity. A difference of 2.0 μ seconds is observed between the arrival times of numerical and experimental results. However the shape, the rise times and the final values of velocities and stresses are approximately the same. Void and plasticity have provided damping to the system and slowed down the compression as can be seen in the experiments too. Material has been deformed and filled the voids, and strain hardening, as modeled by the Cap plasticity model, has resulted in the Hugoniot of Figure 3c. As can be seen, material properties and contact forces, which determine the characteristic of shock, have been selected to provide the same bulk modulus as observed in the experiments.

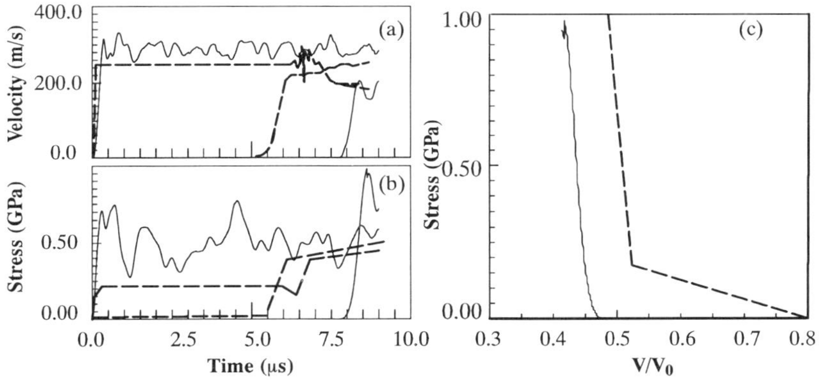

Figure 3 Solid and dashed lines represent results from numerical and experimental studies respectively. (a) Time history of the hammer and the bottom velocities (b) time history of the stress on the hammer and the bottom (c) Hugoniot of the material. $\mathbf{V_0}$ and **V** are the volumes of the uncompressed and the compressed material.

Acknowledgement

This research was partially supported by a research contract between the Los Alamos National Laboratories and University of Illinois at Urbana–Champaign. Support of the Program manager, Dr. P. M. Howe, is greatly acknowledged.

References

1. Barbosa R. and Ghaboussi J., (1990),'Discrete Finite Element Method for Multiple Deformable Bodies', Finite Elements in Analysis and Design, v.7, 145–158.

2. Sandler I.S., DiMaggio F. L. and Baladi G.Y., (1976), 'Generalized Cap Model for Geological Materials', ASCE, J. Geotechnical Engineering, v.102, n.GT7, 683–699.

3. Sandler I. S. and Rubin D., (1979), 'An Algorithm and A Modular Subroutine For the Cap Model', Int. J. for Numerical and Analytical Methods in Geomech., v.3, 173–186.

4. Sheffield S. A., Gustavsen R. L. and Alcon R. R., (1993), 'Shock Initiation Studies of Low Density HMX Using Electromagnetic Particle Velocity and PVDF Stress Gauges', Tenth International Detonation Symposium, Boston, MA, July 12–16, 1993.

Numerical Simulation of Permanent Deformation in Flexible Pavement Systems Subjected to Moving Loads

David J. Kirkner[1], Weixin Shen[2], Michael I. Hammons[3], Donald M. Smith3

Abstract

The structural analysis of a pavement system is an extremely complex problem: the system is layered; the loads are moving; the material behavior may be nonlinear; the domain is unbounded. This paper discusses a finite element modeling methodology which allows all of the key features of the problem to be included. The technique is based on a steady-state assumption and the use of a moving coordinate system. This permits the inclusion of elastic-plastic or elastic-viscoplastic constitutive behavior and easily incorporates the effects of multiple wheel loads. A novel technique for imposing boundary conditions on the face of the domain corresponding to a plane perpendicular to the pavement centerline and located far behind the load eliminates, to a great extent, the end effects which would be present with the use of simple roller boundary conditions. Sample simulations demonstrate the applicability of the technique to model realistic flexible pavement systems. Permanent deformation, including rut profiles and residual stress states are shown.

Introduction

The ability to predict the amount and growth of permanent deformation in pavement systems is an important aspect of pavement design. This paper discusses several issues important for calculating the permanent deformation in three-dimensional, layered, elastic-plastic half spaces. This method is particularly suitable for the analysis of flexible pavement systems, where the permanent deformation may be due in large measure to the failure of the granular base material.

Good reviews of the state-of-the-art of predicting permanent deformations in pavement systems are given in Sousa *et al.* (1991), Zaniewski (1992) and Majidzadeh *et al.* (1992) and the references contained therein.

[1]Associate Professor, Member ASCE and [2]Research Assistant, Department of Civil Engrg. and Geological Sciences, Univ. of Notre Dame, Notre Dame, IN 46556-0767
[3, 4] Research Civil Engineer, Pavement Systems Division, U.S. Waterways Experiment Station, 3909 Halls Ferry Road, Vicksburg, MS 39180-6199

Modeling Strategy

Consider a problem domain consisting of a half-space defined by the region $x_2 < 0$ relative to the fixed x_1, x_2, x_3 coordinate system. This region is subjected to a uniform pressure of magnitude p distributed over a square $a \times a$ which is moving with a constant velocity, c in the x_1 direction. Quasi-static conditions are assumed, *i.e.* inertial effects are neglected. Steady state conditions are assumed, that is, the load began an infinite time in the past. If it is also assumed that the domain is homogeneous in the x_1 direction, a moving coordinate system can be used to advantage. Let $y_1 = x_1 - ct, y_2 = x_2, y_3 = x_3$. Using the moving coordinate system (y_1, y_2, y_3) rather than the fixed reference frame (x_1, x_2, x_3) allows time to be removed as an explicit variable in the problem. Integration in time is thus replaced by integration in y_1. For the work considered herein the constitutive equations will take the form of rate independent plasticity. The Drucker-Prager model is used in the simulations to follow. For details of the finite element scheme used the reader may consult Kirkner (1994).

For problems of infinite extent it is usual to employ roller boundary conditions on the end-faces perpendicular to the x_1 direction. This assumes shearing stress components are small and may be neglected and displacements in the x_1 are also small. However in this traveling load problem residual stresses and displacements exist on the end-face behind the load, *no matter where the domain is truncated.* However if it is assumed that the transition from loading to unloading occurs on the plane defined by $x_1 = 0$ (an approximation), then it can be shown that $u_i(0, x_2, x_3) - u_i^e(0, x_2, x_3) = u_i^R(x_2, x_3), i = 1, 2, 3$. In this equation $u_i(0, x_2, x_3)$ is the displacement in coordinate direction i on the plane defined by $x_1 = 0$. The superscripts e and R mean elastic component and residual displacement respectively. The left hand side of the equation is easily calculated from the current iteration and the value of $u_i^R(x_2, x_3)$ so obtained is used as a specified boundary condition on the left end-face for the next iteration. The result of this approximate boundary condition greatly reduces end effects in the numerical simulation.

Results

A model problem is analyzed to demonstrate the applicability of the methodology to realistic flexible pavement systems. In this sample problem, the pavement consists of three layers, *i.e.*, asphalt concrete (AC), granular subbase (GS) and soil subgrade (SS). Table 1 gives the thickness and material properties of each layer. E is Young's Modulus, μ is Poisson's ratio, c is the cohesion and ϕ is the internal friction angle. The load is applied to a 40 cm by 40 cm square with an uniform pressure of 1 MPa moving left to right.

Table 1: Model Flexible Pavement System

Layer	Thickness	Material Properties
1 (A.C.)	15 cm	Linear Elastic, $E = 2000MPa$, $\mu = 0.45$

Table 1: Model Flexible Pavement System

Layer	Thickness	Material Properties
2 (G.S.)	25 cm	Elastic, $E = 300MPa$, $\mu = 0.30$ Plastic, $c = 0.69MPa$, $\phi = 20°$
3 (S.S.)	100 cm	Elastic, $E = 30MPa$, $\mu = 0.40$ Plastic, $c = 0.28MPa$, $\phi = 10°$

A non-uniform mesh is adopted, so that the mesh is more refined under or near the load. A total of 9,000 elements is used in this simulation. Figure 1 gives the displacement profile along the centerline of the pavement. In the case of a moving load, a displacement profile can also be viewed as a displacement time history that every point on the centerline will follow. Thus, the displacement at the very left (far behind the load) is the rut depth. As indicated by the solid and dotted lines in the figure, the boundary effects which we usually encounter are eliminated by using the iterative technique of imposing displacement boundary conditions.

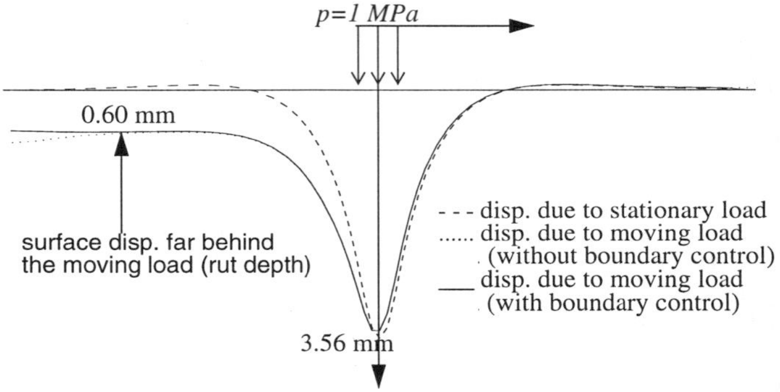

FIGURE 1. Displacement Profile along the Centerline of the Pavement

With the passage of the moving load, the pavement system is left with not only permanent displacements but also residual stresses. Figure 2 gives a stress profile along the centerline of the pavement in the asphalt concrete layer, where the maximum stress usually occurs. Similarly, the stress field far behind the moving load is the residual stress field. Figure 3 shows the rut depth versus pressure. The rut depth grows rapidly with an increase in the surface pressure above 0.32 MPa, which is approximately the load to cause first yield. The difference between rut depths caused by a stationary load and a moving load of the same magnitude is small enough to be neglected for small yielding, but increases with growing load. Within the practical range of tire pressures (0.7-1.2 MPa), the stationary load leads to a rut depth 10~20% greater than that produced by a moving load.

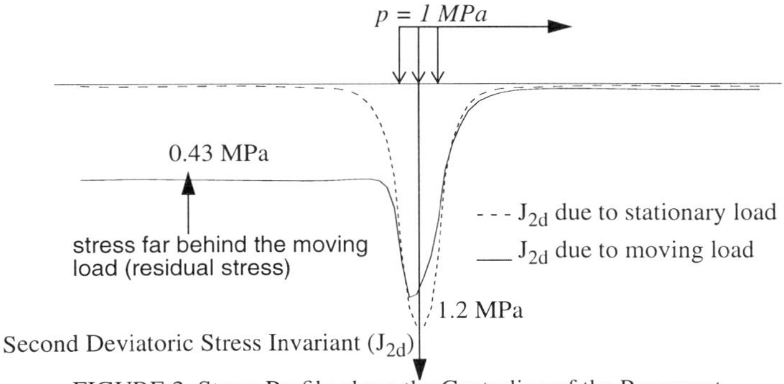

FIGURE 2. Stress Profile along the Centerline of the Pavement

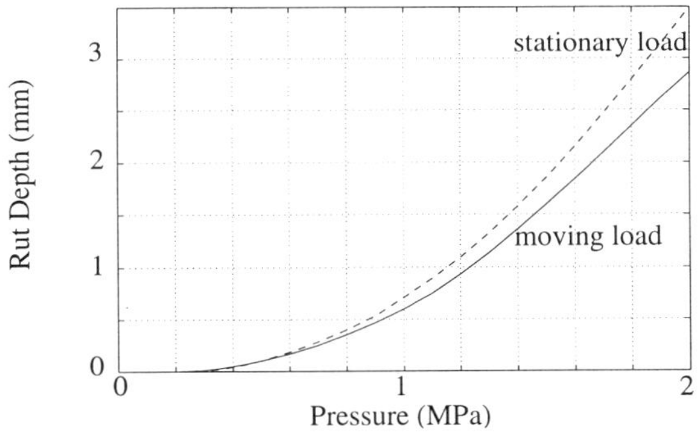

FIGURE 3. Rut Depth vs. Pressure

Appendix: References

Kirkner, D. J., C3ent Deformations in Flexible Pavement Systems," *Transportation Research Record,* TRB, National Research Council, 1448, 34-39.

Majidzadeh, K., Saraf, C. L., Mesarovic, S., and Ives, G. J. (1992) "State of the Art Review of Rutting and Cracking in Pavements." *Proceedings, Unified Airport Pavement Design and Analysis Concepts Workshop*, DOT/FAA/RD-92/17.

Sousa, J. B., Craus, J., Monismith, C. L. (1991) *Summary of Report on Permanent Deformation in Asphalt Concrete*, Strategic Highway Research Program, National Research Council, SHRP-A/IR-91-104.

Zaniewski, J. (1992) "State of the Art of Airport Pavement Analysis and Design." *Proceedings, Unified Airport Pavement Design and Analysis Concepts Workshop*, DOT/FAA/RD-92/17.

Analysis Tools of Transitioning and Turbulent Flows

M. R. Hajj[1]

Introduction

Most flows encountered in nature and in technological applications are turbulent. These flows are unsteady and exhibit high levels of fluctuating velocity. These fluctuations are random in space and in time. In flows that are originally laminar, turbulence takes place via a transition process that is a characterized by a sequence of instabilities. Both turbulence and the closely related field of transition have been the subject of scientific research for over a century. This fact is a proof of the difficulties involved in studying transitioning and turbulent flows.

Over the past six years I have been involved in an investigation of the nonlinear dynamics of transitioning wakes and mixing layers. The major emphasis has been to gain new insight into the transition process by quantifying the role of nonlinear wave-wave interaction mechanisms in the different stages of the transition leading to the breakdown to turbulence. Quantification of these mechanisms is an important step in developing an understanding of the transition process and in attempting to control it. More recently, I have been involved in an investigation aiming at improving the simulation and analysis of extreme wind loads on low-rise structures. Important components of this work have been to characterize the turbulent atmospheric wind and examine the relation between velocity fluctuations and low-pressure peaks on full-scale and model low-rise structures.

The objective of this paper is to present and discuss some of the analysis tools that are used to examine transitioning and turbulent flows. More specifically, the discussion is concerned with the use of polyspectral analysis to analyze the transition of free shear flows such as plane wakes and mixing layers and the use of wavelet analysis to characterize turbulence structures in atmospheric wind.

---

[1] Assistant Professor, Department of Engineering Science and Mechanics, Virginia Tech, Blacksburg, VA 24061-0219

Analysis of Transitioning Flows Using Polyspectral Moments

A common element of the transition of shear flows is the evolution of the velocity fluctuations field from a spectrum of weak random background noise in the initial laminar state to an intense spectrum in the turbulent state. Such a transformation takes place via vorticity restructuring which can be described by a sequence of instabilities. Each of these instabilities is characterized by growth of certain unstable modes and their interactions. In the primary instability stage, the unstable modes grow exponentially and nonlinear interactions are very weak. Further downstream, in the secondary and tertiary instability stages, the transition is dominated by energy transfer among spectral components via nonlinear wave-wave interactions. Estimates of polyspectral moments of velocity fluctuations have been shown to provide new physical insight into these interactions.

One way to introduce polyspectral moments is by forming the analogy with the well-known second-order correlation and power spectral density functions. For a real-valued zero-mean stationary fluctuations, $x(t)$, that represents a random velocity or pressure waveform, one can express the power spectral density function as:

$$P(f) = \lim_{T\to\infty}\frac{1}{T} < X(f)X^*(f) > \qquad (1-a)$$

where $< ... >$ denotes an expected value and $X(f)$ is the Fourier transform of $x(t)$. The next higher-order spectrum, namely, the auto-bispectrum is expressed as:

$$B(f_1, f_2) = \lim_{T\to\infty}\frac{1}{T} < X(f_1 + f_2)X^*(f_1)X^*(f_2) > \qquad (1-b)$$

The phase of auto-bispectrum is given by $\theta(f_1 + f_2) - \theta(f_1) - \theta(f_2)$. Clearly, the auto-bispectrum measures the degree of coherence among the modes with frequencies f_1 and f_2 and $f_1 + f_2$. If these modes are independently excited, their phases are statistically independent and, subsequently, the phase of the auto-bispectrum is randomly distributed. The averaging, indicated in equation (1-b), will then yield a zero value for the auto-bispectrum. On the other hand, if the three modes are coupled, a phase coherence exists among them and the averaging will not give a zero value for the auto-bispectrum. A normalized value of the auto-bispectrum, namely, the auto-bicoherence, can be defined as

$$b^2(f_1, f_2) = \frac{|B(f_1, f_2)|^2}{< |X(f_1)X(f_2)|^2 >< |X(f_1 + f_2)|^2 >} \qquad (2)$$

It can be shown by Schwartz inequality that the auto-bicoherence, $b^2(f_1, f_2)$, is bounded by zero which indicates no coupling, and by one, which indicates high quadratic coupling. In Hajj et al. (1992, 1993) it is shown how the auto-bispectrum was used to (1) detect the highly coupled modes in the different stages of the transition of mixing layers, (2) establish the important role of the low-frequency modes in the transition process, and (3) quantify the role of the phase relation between the fundamental and its subharmonic in the transition process of mixing layers. Recently, estimates of nonlinear energy transfer rates among spectral components in the transition of plane wakes were presented by Hajj and Janajreh (1995).

Analysis of Atmospheric Wind Using Wavelets

Characterizing atmospheric wind is an essential step towards accurate simulation of wind loads on low-rise structures in wind tunnels. Wind velocity components are unsteady and exhibit high levels of fluctuations, which implies the presence of all dynamic scales. All of these scales could have an effect on the separating flow and/or on the vortex formation. Subsequently, the energy distribution and time-varying characteristics of these scales will affect the suction pressure on the roofs of low-rise structures. To date, frequency decomposition has been the main tool used to represent the energy content of turbulence scales. However, one may be interested in the time-varying parameters of these scales. Under these conditions, time-frequency representation should be used. In the case of atmospheric wind turbulence, time-frequency analysis can be used to describe localized behavior in time, such as intermittent events.

Several time-frequency analysis techniques are available. The short-time Fourier transform and the Gabor transform have been used to examine time-varying signals by examining signals through windows over which signals are approximately stationary. Similarly, the wavelets representation allows one to capture time variations of the scales contained in time-varying signals. Its basic advantage over other time-frequency domain analysis is in the resulting resolution. In the short-time Fourier transform or Gabor transform, a single analysis window is used. In contrast, the wavelet transform uses short windows at high frequencies and long windows at low frequencies in a mathematically elegant fashion. The application of wavelet analysis to turbulent flows has been presented by Farge (1992). While frequency analysis is performed by projecting a signal onto a number of sinusoids that are infinite in extent, the wavelet analysis is performed by projecting the signal onto a set of highly localized basis functions. The basis functions are called wavelets and are obtained from a single "mother" wavelet by dilations and translations. Thus, in wavelets, the notion of a scale replaces the notion of frequency which leads to the so-called time-scale representation.

The wavelet transform can either be continuous or discrete. The discrete wavelet transform (DWT) decomposes the signal onto a series of orthogonal basis functions. The resulting decomposition can then be used to reconstruct the time signal without any redundancy. For a signal, $x(t)$, the DWT is defined such as

$$x(t) = a_o + \sum_{j=1}^{N} \sum_{k=0}^{2^j-1} a_{2^j+k} \psi(2^j t - kT), \; 0 \leq x \leq T \qquad (3-a)$$

where

$$a_o = \int_0^T x(t)dt \qquad (3-b)$$

and

$$a_{2^j+k} = 2^j \int_o^T x(t)\psi(2^j t - kT)dt \qquad (3-c)$$

where $\psi(2^j t - kT)$ represents the wavelet functions. For a time series with n samples, the upper limit of j, N, is given such that $2^N = n$.

The continuous wavelet transform (CWT) is defined by the function

$$W(a,\tau) = \int_{-\infty}^{+\infty} x(t)a^{-\frac{1}{2}}\psi((t-\tau)/a)dt \qquad (4)$$

where ψ is the wavelet function, τ and a are respectively, the translation and dilation parameters. The CWT is a non-orthogonal transform algorithm. There is redundant information in the CWT. In our work, we have applied both types to atmospheric wind. The results show that wavelets are better suited than Fourier analysis for the representation of the turbulence scales contained in the atmospheric wind. In particular, one can qualify the intermittency character of all turbulence scales, Hajj et al. (1996).

Conclusions

The analysis tools, polyspectral moments and wavelets, provide useful insight into the characteristics of transitioning and turbulent flows. Polyspectral analysis is a nice method for capturing nonlinear couplings among various modes of a system as it departs a linear stage and moves into nonlinear stages. One can, by polyspectral analysis, identify the way in which identifiable modes interact and establish a measure of the strength of these interactions. The usefulness of such analysis in studying transition to turbulence has been established. Wavelets are nicely adapted to describe localized behavior in time such as intermittent events, local structures, or transients, such as the atmospheric wind.

References

Farge, M. (1992) Annu. Rev. Fluid Mech., Vol. 24, pp. 395-457.

Hajj, M. R. & Janajreh, I. M., (1995), Proc. 10th ASCE/EMD Conference, Vol. 2, pp. 851-854, Boulder, CO.

Hajj, M. R., Miksad, R. W., & Powers, E. J., (1992), J. Fluid Mech. Vol. 236, pp. 385-413.

Hajj, M. R., Miksad, R. W., & Powers, E. J., (1993), J. Fluid Mech., Vol. 403, pp. 403-426.

Hajj, M. R., Tieleman, H. W., & Bikdash, M. (1996), Proc. 11th ASCE/EMD Conference, Ft. Lauderdale, FL.

DOMINANT EDDY SIMULATION IN TURBULENT FLOW

J.-B. Zhang[1], and V. H. Chu[2], Member, ASCE

Abstract

The normal-mode procedure in the classical hydrodynamic stability theory is used to simulation turbulent flow. In the simulation, the turbulent flow is represented by a WKB series. The dominant eddy of the turbulent flow is determined as the lowest order of approximation in the series. It is argued that the mean and fluctuation motions of the dominant eddy are parts of the same entity, and therfore must have the same length scale and the same growth rate. This similarity assumption allows the calculation of the turbulent flow to the lowest order of approximation as an eigenvalue problem. The calculations conducted for the turbulent mixing layer have produced realistic results that are consistent with the experimental observation.

Introduction

The classical hydrodynamic stability theory was developed with the primarily goal of studying the transition from laminar to turbulent flow. The analysis for the stability of parallel shear flow begins with a basic mean flow, $U(y)$. A perturbation proportional to $a\exp[i\alpha(x-ct)]$ is then added to the basic flow. The solution for the linearized equation for the perturbation is an eigenvalue problem. The eigenvalue, c, has a real part and an imaginery part. The real part, c_r, is the phase speed of the wavelike perturbation,

$$a\exp[i\alpha(x-ct)] = A\exp[i\alpha(x-c_r t)]. \tag{1}$$

The amplitude of this purtubation, $A = a\exp(-\alpha c_i t)$, grows exponential in time, and the growth rate is

$$\frac{1}{A}\frac{dA}{dt} = -\alpha c_i. \tag{2}$$

The linear theory is for the onset of the instability and is not suppose to be valid in the later stage of transition when the basic mean flow is significantly modified by the exponential growth of the disturbance. However, many fully developed

[1]Grad. Student, Dept. Civil Engrg., McGill Univ., Montreal H3A 2K6, Canada
[2]Prof., Dept. Civil Engrg., McGill Univ., Montreal H3A 2K6, Canada.

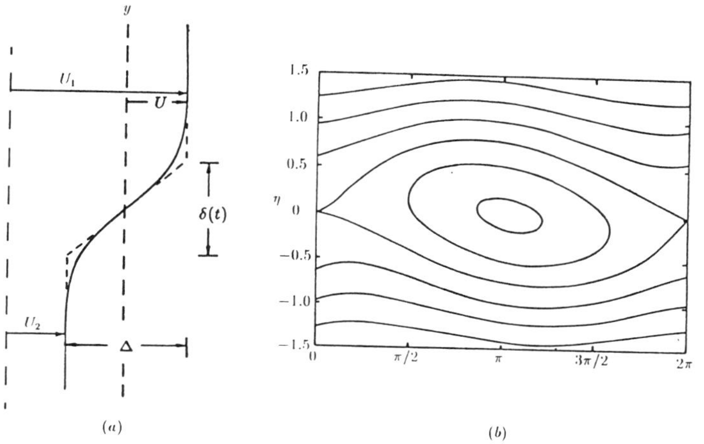

Figure 1: The unsteady mixing layer; (a) mean velocity profile, and (b) the streamline pattern produced by the composite of the mean and the fluctuation parts.

turbulent shear flows are observed to exhibit instability behaviour predicted by the linear stability analysis. For example, the calculation of Michalke (1964) for the most unstable wave in a mixing layer has produced a maximum growth rate

$$\frac{1}{A}\frac{dA}{dt} = -\alpha c_i = 0.19\,\frac{\Delta}{\delta} \tag{3}$$

where Δ is the velocity difference across the mixing layer and δ the thickness of the layer (see Figure 1). If the mean flow and the disturbance are assumed to be parts of the same turbulent structure, the thickness of the mixing layer would increase at a rate proportional to the amplitude of the wave. Assuming that $\delta \sim A$, the relation given by Equation 3 would give a spreading rate

$$\frac{d\delta}{dt} = U_c\frac{d\delta}{dx} = 0.19\,\Delta, \tag{4}$$

in agreement with the rate observed in a forced turbulent mixing layer (Gaster, Kit and Wygnanski, 1985). The success of this re-interpretation of results obtained from a linear theory is surprising, but similarily correct results have been obtained for stratified shear flows (Chu and Baddour, 1980, 1984).

The theoretical basis for a stability theory of the free turbulent shear flow is introduced in this paper. Calculations have been conducted for the development of the turbulent mixing layer.

Dominant Eddy Hypothesis

Following the classical procedure of Reynolds (1894) the flow is separated into mean and fluctuating parts:

$$\begin{aligned} u &= U(y,t) + u'(x,y,z,t) \\ v &= v'(x,y,z,t) \\ w &= w'(x,y,z,t) \\ p &= P(y,t) + p'(x,y,z,t) \end{aligned} \tag{5}$$

However, this separation of the flow into parts is viewed as an *artificial* procedure. As the mixing layer rolls up to form eddies and grows in thickness, the mean and the fluctuation parts must increase with the same rate since the mean and the fluctuation are integral parts of the same structure. The re-unification of the parts would be possible if both parts are defined by the same similarity variable, $\eta = y/\delta(t)$. In terms of the similarity variable, the mean velocity is given by $\Delta U(\eta)$ while the fluctuation part by the WKB expansion:

$$\begin{aligned} u' = \; & \Delta \;\; [\hat{u}_1(\eta)e^{i\theta} + \hat{u}_1^*(\eta)e^{-i\theta} + \hat{u}_2(\eta)e^{2i\theta} + \hat{u}_2^*(\eta)e^{-2i\theta} \\ & + \;\; \hat{u}_3(\eta)e^{3i\theta} + \hat{u}_3^*(\eta)e^{-3i\theta} + \;\ldots\ldots.] \end{aligned} \tag{6}$$

$$\begin{aligned} v' = \; & \Delta \;\; [\hat{v}_1(\eta)e^{i\theta} + \hat{v}_1^*(\eta)e^{-i\theta} + \hat{v}_2(\eta)e^{2i\theta} + \hat{v}_2^*(\eta)e^{-2i\theta} \\ & + \;\; \hat{v}_3(\eta)e^{3i\theta} + \hat{v}_3^*(\eta)e^{-3i\theta} + \;\ldots\ldots.] \end{aligned} \tag{7}$$

$$\begin{aligned} w' = \; & \Delta \;\; [\hat{w}_1(\eta)e^{i\theta} + \hat{w}_1^*(\eta)e^{-i\theta} + \hat{w}_2(\eta)e^{2i\theta} + \hat{w}_2^*(\eta)e^{-2i\theta} \\ & + \;\; \hat{w}_3(\eta)e^{3i\theta} + \hat{w}_3^*(\eta)e^{-3i\theta} + \;\ldots\ldots.] \end{aligned} \tag{8}$$

$$\begin{aligned} p' = \; & \Delta^2 \;\; [\hat{p}_1(\eta)e^{i\theta} + \hat{p}_1^*(\eta)e^{-i\theta} + \hat{p}_2(\eta)e^{2i\theta} + \hat{p}_2^*(\eta)e^{-2i\theta} \\ & + \;\; \hat{p}_3(\eta)e^{3i\theta} + \hat{p}_3^*(\eta)e^{-3i\theta} + \;\ldots\ldots.] \end{aligned} \tag{9}$$

where $*$ denotes the complex conjugates, θ the phase function, and $\hat{u}_i, \hat{v}_i, \hat{w}_i, \hat{p}_i$ are dimensionless amplitudes of the fluctuations. The wave number, (α, β), in the x- and z-direction and wave frequence, ω, are defined by the phase as follows:

$$\theta_x = \frac{\alpha}{\delta}, \quad \theta_z = \frac{\beta}{\delta}, \quad \theta_t = -\frac{\omega\Delta}{\delta} = -\frac{\sqrt{\alpha^2+\beta^2}\, c\Delta}{\delta}. \tag{10}$$

The solution for the various harmonics in the WKB expansion may be obtained using the aysmptotic method of Stuart (1960) and Watson (1960).

The flow corresponding to the maximum growth is two-dimensional. The mixing layer with the three-dimensional turbulent motion is observed in the experiments to have a smaller growth rate, $\overline{\delta}_t \simeq 0.09$. With the growth rate $\overline{\delta}_t = 0.09$, motions with a range of wave number varying from $\overline{\alpha} = 0.2$ to 1.6 are possible. The superposition of the harmonics can lead to a distribution of Reynolds stresses that are consistent with the mean flow. Figure 2 shows

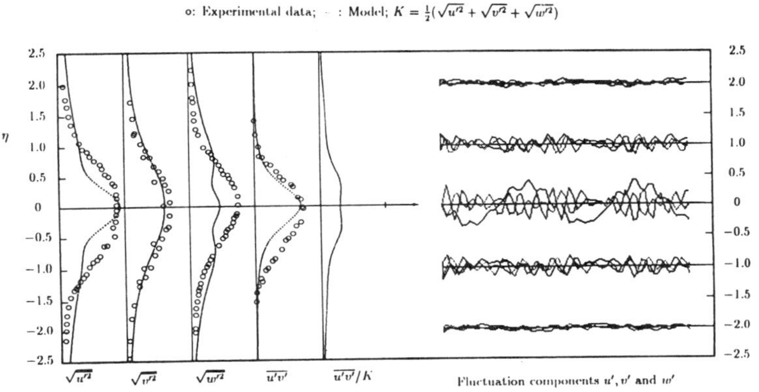

Figure 2: The simulation of the turbulent mixing layer by superposition of the harmonics.

the results of the simulation obtained from superposition of the harmonics at wave number $\alpha = 0.2, 0.4, 0.8$, and 1.6. The results for $\sqrt{\overline{u'^2}}$, $\sqrt{\overline{v'^2}}$, $\sqrt{\overline{v'^2}}$, $\overline{u'v'}$ and $\overline{u'v'}/K$ are in close agreement with the experimental data (shown as circles in the figure). Perhaps the most remarkable result is the correlation coefficient $\overline{u'v'}/K$, which is found to have a maximum value of 0.35 in very close agreement with experimental observation.

References

Chu, V.H. and Baddour, R.H. (1980) "Stabilization of turbulence in plane shear layers," *Proceedings of 2nd International Symposium on Stratified Flows,* Trondheim, Norway, Vol. 1, pp. 367-377.

Chu, V. H. and Baddour, R. E. (1984) "Turbulent stratified shear flows," *J. Fluid Mechanics,* Vol. 132, pp. 353-378.

Gaster, M., Kit, E. and Wygnanski, I. (1985) "Large-scale structure in a forced turbulent mixing layer," *J. Fluid Mech.,* Vol. 150, pp. 23-39.

Reynolds, O. (1894) "On the dynamic theory of incompressible viscous fluids and the determination of the criterion," *Phil. Trans. Roy. Soc. London,* Vol. 186, pp. 123-161.

Stuart, J. T. (1960) "On the non-linear mechanics of wave disturbances in stable and unstable parallel flow, Part I. The basic behaviour in plane Poiseuille flow," *J. Fluid Mech.,* Vol. 9, pp. 353-370.

DEPOSITION OF PARTICLES FROM A VERTICAL JET

M.J. Neves[1], H.J.S. Fernando[2], A.A. Neves[3]

Abstract

An experimental study was carried out to investigate the deposition of heavy particles from a particle-laden momentum jet discharging in the vertical direction. Some practical applications related to this problem can be referred, for example, geophysical phenomena such as eruption of volcanoes and sedimentation in lakes, pollution created by industrial chimneys, sedimentation in the near field of sea outfalls, etc. Solutions are presented to evaluate the maximum height of rise of the particles and the rate of deposition on the ground.

Analysis

The situation is represented in Figure 1. After a theoretical analysis inspired by Tchen (1947) equation, Neves and Fernando (1994) suggested the use of two length scales: (i) *the momentum length scale*

$$l_m = \frac{M_0^{1/2}}{w_s} \qquad (1)$$

where M_0= kinematic momentum flux = $\pi R^2 w_0^2$; R = port radius; w_0 = initial jet velocity; w_s= settling velocity = $-g' d^2 / 18\nu$; ν = viscosity; g' = reduced gravity = $g(\rho_p - \rho_f)/\rho_f$; ρ_p= density of the particle; ρ_f = density of the fluid. This parameter is related to the height at which the single particle settling velocity balances the local velocity of a single-phase jet; (ii) *the buoyancy length scale,*

----

[1] University of Porto, Rua dos Bragas, 4099 Porto Codex, Portugal. [2] Arizona State University, Tempe, USA. [3] Instituto Superior de Engenharia do Porto, Portugal

$$l_b = \frac{w_s d^2}{\nu} = 18 \frac{w_s^2}{g'} \tag{2}$$

This parameter is related to the length over which a particle, starting from the rest, would accelerate before achieving the terminal settling velocity.

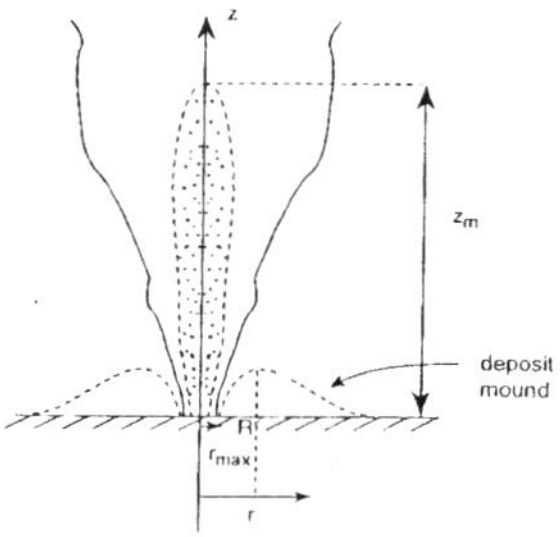

Figure 1 - Sketch of a vertical particle-laden jet

The experimental procedure is described in detail by Neves and Fernando (1994). Three different sizes of polystyrene particles were used, with diameters of 868, 799 and 530 microns and densities of 1.0445, 1.0445 and 1.0251 g/cm3. The volume fraction of the particles varied in the range $4.5 \times 10^{-5} < c_0 < 2.3 \times 10^{-3}$.

Figure 2 shows that the *maximum height of rise of the particles* depends slightly on the concentration, but at low values,

$$z_m \cong 6.9\ l_m \tag{3}$$

For practical applications one of the most important points is to evaluate the *rates of deposition on the ground.* A special catchment device was used, made out of

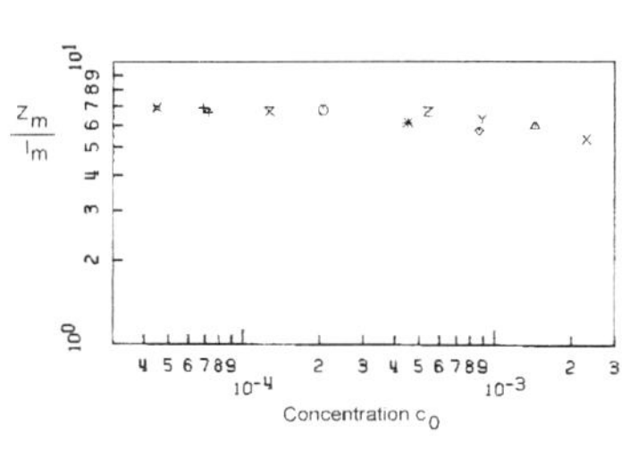

Figure 2 - Maximum height of rise of the particles

a flat sheet of metallic mesh, onto which thin round cylindrical elements were attached, originating several annular regions at different distances from the centre of the discharging orifice. The particles were allowed to accumulate in this catchment device for a period of 3-10 minutes, and then the number of particles in each annular region was counted manually, which was thought to be the most accurate technique.

Figure 3 shows that the deposition rate $\dot{N}$ (number of particles per unit time and unit area) can be expressed by

$$\frac{\dot{N}}{\dot{N}_{\max}} = 65\frac{r'}{r'_{\max}}\exp\left[-0.054\left(\frac{r'}{r'_{\max}}+7.8\right)^2\right] \tag{4}$$

where $r' = r - R$ is the distance from the edge of a jet with radius R. $\dot{N}_{\max}$ represents the maximum value of $\dot{N}$, and $r'_{\max}$ is the distance where it occurs. For using this equation, means of evaluating $\dot{N}_{\max}$ and r'_{max} have to be given. With respect to the former, its behaviour is shown in Figure 4. When l_b / l_m is small, say less than 0.15,

$$\dot{N}_{\max} = 1.4\frac{n_0 Q_0}{l_m^2} \tag{5}$$

which means that, in this case, the deposition rates only depend on the momentum length scale. On the contrary, at larger values of l_b / l_m it can be shown after simple transformations that

$$\dot{N}_{\max} = 0.03\frac{n_0 Q_0}{l_b^2} \tag{6}$$

which means that, in these circumstances, the governing parameter is the buoyancy length scale. n_0 represents the number of particles discharged per unit volume, and Q_0 is the volume flux.

A similar trend can be observed with respect to $r'_{\max}$, the distance where the rate of deposition reaches its maximum. Figure 5 shows that at low values of l_b / l_m,

$$r'_{\max} \cong 0.15\, l_m \tag{7}$$

whereas at large values, it can be shown that

$$r'_{\max} \cong l_b \tag{8}$$

confirming that in this regime the phenomenon is dominated by l_b, which takes into account the relaxation of particles under gravitational effects.

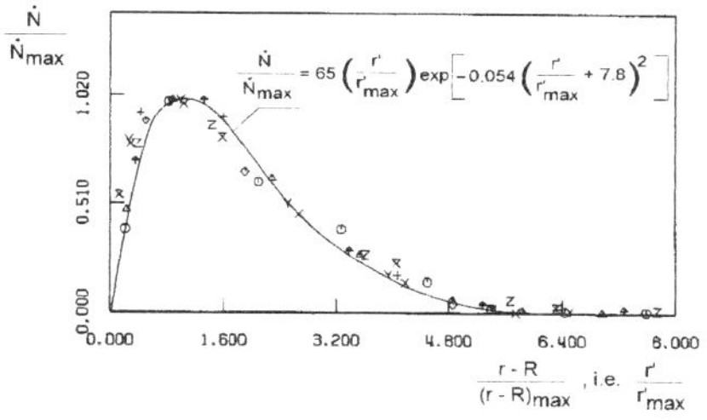

Figure 3 - Rate of deposition on the ground

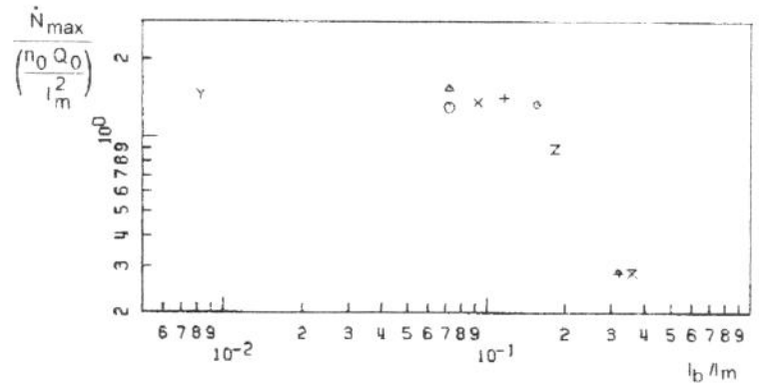

Figure 4 - Maximum rate of deposition

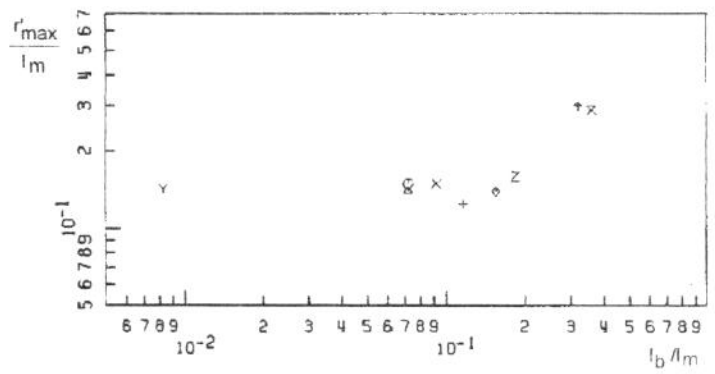

Figure 5 - Distance where the rate of deposition reaches the maximum

References

Neves, M. and Fernando, H., 1994, Particle Dispersion in Turbulent Flows, *Report for NATO Scientific Affairs Division*, Arizona State University, USA.

Neves, M. and Fernando, H., 1995, Sedimentation of Particles from Jets Discharged by Ocean Outfalls: A Theoretical and Laboratory Study, *Water Science and Technology, Vol. 32, No 2, pp 133-139.*

Tchen, C.M., 1947, Mean Value and Correlation Functions Connected With the Motion of Small Particles Suspended in a Turbulent Fluid, *PhD Thesis, Delft.*

BIFURCATION OF LINE THERMALS

M. Dehghani[1], and V. H. Chu[2], Member, ASCE

Abstract

Line thermals experiments were performed in the laboratory by releasing saline solution from a line source in a large tank of water. The salinity concentration distribution in the thermals was determined by video imaging technique. The concentration field provided a precise description of the process that was not possible in the previous studies of the line thermals. Due to unstable convolution of the density interface at the front of the thermal, and the formation of a stable interface at the back, the entrainment of the surrounding fluid is concentrated along an intense shearing passage between the vortex cores. The asymmetry of the mixing process led to a lopsided density distribution and occasional overturning of the thermal.

Introduction

The turbulent motion of a line thermal is produced by instantaneous release of buoyant fluid along a line source. Through a process known as entrainment, the surrounding irrotational fluid is incorporated into the turbulent region of the thermal. The result is increase in the cross-sectional area of the thermal with distance from the source. Figure 1 shows the development of the thermal with distance from a source located at the water surface of a large tank of water. Dimensional analysis shows that the radius of the thermal cross section increases with distance from the source following a linear spreading relation

$$R = \beta z, \tag{1}$$

and the distance from the leading edge of the advancing front increases with time following a two-third power law relation

$$\zeta_f = K_f \left(\frac{B_o}{\rho}\right)^{\frac{1}{3}} t^{\frac{2}{3}} \tag{2}$$

where B_o is the buoyancy force per unit length of the line thermal.

[1]Grad. Student, Dept. Civil Engrg., McGill Univ., Montreal H3A 2K6, Canada
[2]Prof., Dept. Civil Engrg., McGill Univ., Montreal H3A 2K6, Canada.

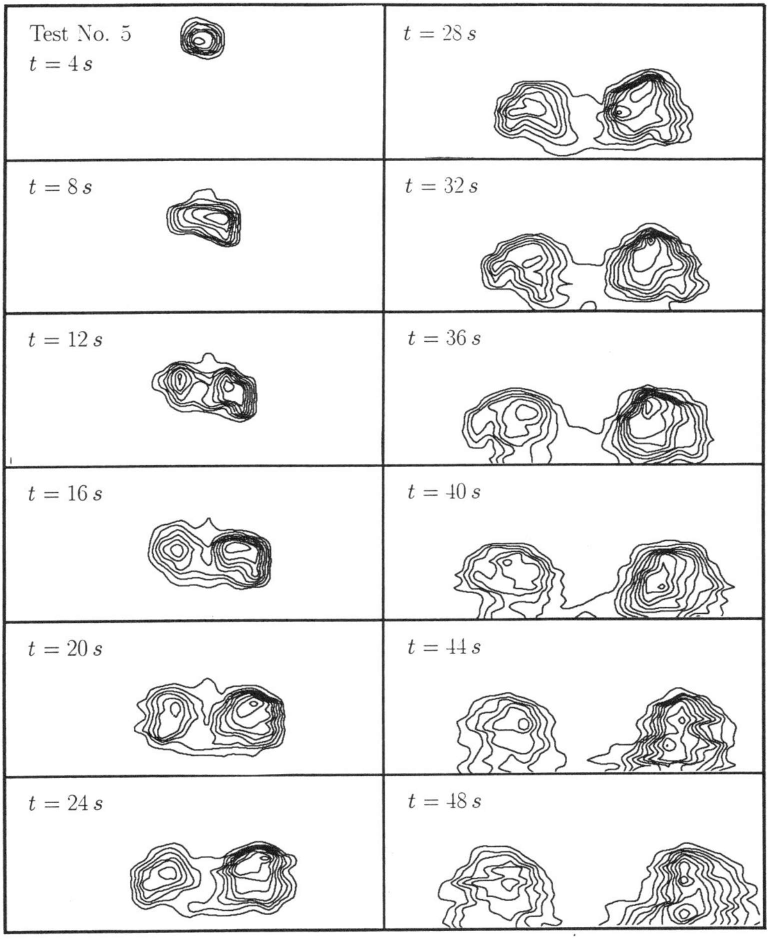

Figure 1: Concentration contours of the thermal cross sections obtained from a sequence of video images at 4 second intervals; test no.5.

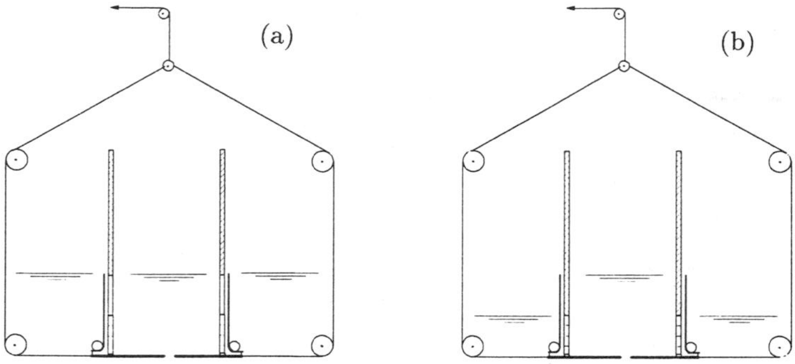

Figure 2: Two arrangements of release of the source fluid from a 5.2 cm x 6.2 cm enclosure; (a) source fluid at the same level as the water surface in the tank, (b) source fluid at a level 2.6 cm above the water surface.

Much of the preocupation in the previous experimental studies was to evaluate the spreading coeffcient, β, and the path constant, K_f. Through video imaging, the saline concentration distribution in the thermal are determined for the first time from the present series of experiments. The video images have also provided a detailed description of the entrainment process. Phenomena such as overturning and bifurcation are observed to play a significant role in the entrainment of irrotational fluid into the thermals.

Experiments

The experiments were conducted in a 2.4 m wide, 1.2 m high and 0.6 m thick tank. The thermal was initiated by sudden opening of a set of gates releasing the saline solution from a 5.2 cm x 6.4 cm rectangular box as shown in Figure 2. The entire gate assembly was attached to a weight through a system of strings and pullys. The slide opening of the gates on the side, and on the bottom, is symmetrical and were activated by droping of the weight. Opening of the gates exposed an unstable gravity-stratified interface at the bottom of the box. The convolution of this unstable interface from rest is not a deterministic process. To avoid uncertain action of the release, two of the experiments (test no.3 and no.4) were performed with the solution in the box initially at a level 2.6 cm higher than the water surface in the tank as shown in Figure 2b. The drop of the block over a distance of 2.6 cm produced a non-zero initial velocity at the source. This small initial impulse led to a more symmetrical action and hence more reproducible results. The initial velocity due to the impulse was estimated to be 35.7 cm/s, giving an initial Reynolds number of 22,000.

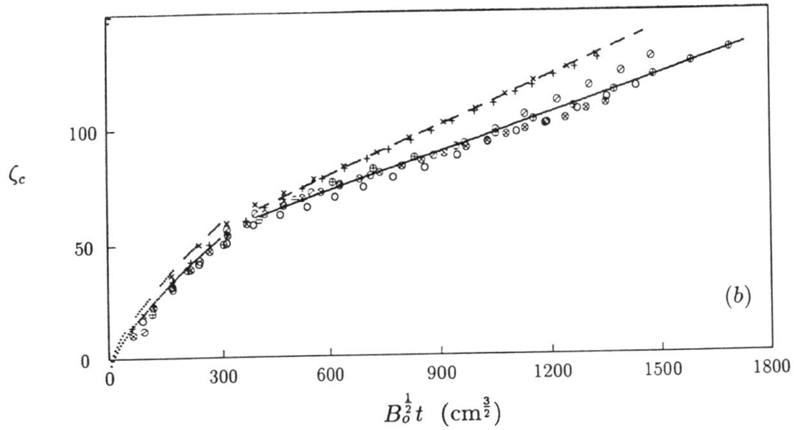

Figure 3: Two stage development of the line thermal; the data of the tests with an initial impulse (no.3 and no.4) are denotes by the + and × symbols.

Results

The development of the thermals were recorded by a video camera. Figure 1 shows the concentration contours obtained from one sequence of the video images. Figure 3 shows the trajectory of the mass center; the distance of the mass center from the source, ζ_c, is plotted with time, t, in this figure. The initial development is relatively unaffected by the bottom of the tank. Hence, the data is fitted with the two-third power law. The thermal becomes bifurcated in the later stage of the development. The movement of the bifurcated element along the floor is characterized by a constant velocity. The impulse (in tests no.3 and no.4) has almost no effect on the initial development of the thermal, but is found to have a huge effect (30% increase) in the velocity along the floor. The impulse delayed the overturning and bifurcation of the thermal. The effect was the reduction in mixing and consequently the increase in the floor velocity.

References

Richards, J. M. (1963) "Experiments on the motions of isolated cylindrical thermals through unstratified surroundings," *Int. J. Wat. Poll.*, Vol. 7, pp. 17-34.

Tsang, G. (1971) "Laboratory study of line thermals," *Atmos. Environment*, Vol. 5, pp. 445-471.

CHAOTIC ADVECTION IN A BIOENGINEERING SYSTEM

Ahmet C. Omurtag[†], Victor Stickel[†], and Rene Chevray[†]

Abstract

Chaotic mixing is applied to the problem of the recovery of fetal cells from the maternal circulation. This is intended as part of a novel method in the prenatal detection of genetic and physical abnormalities in infants. The extremely low concentration of infant cells in the maternal circulation, the fragility of the cells, and the fact that the cells adhere to collecting surfaces only at very low shear rates contribute to the difficulty of this problem. In the fetal cell collector investigated here, a laminar chaotic flow is maintained to increase collection efficiency by continually randomizing the positions of the cells. Numerical simulations show that the infant cells in the chamber come into contact with the collectors (coated with an adhesive antibody in the actual system) and are collected at reasonably high rates.

Introduction

In recent years there have been improvements in the basic understanding of mixing and stirring as a result of the introduction of concepts from dynamical systems theory. After Aref's initial insight[1] into the interrelation between incompressible two dimensional flows and Hamiltonian dynamical systems, many numerical and experimental studies have been performed on chaotic mixing. These investigations have established that the mixing properties of laminar chaotic flows, referred to as chaotic advection, parallel some of those attributed so far only to turbulence[2]. Some of the basic developments are ready for practical implementation, especially as mixing devices to be used in the polymer industry or in biomedical applications. The requirements of rapid homogenization and low hydromechanical stress represent two opposing demands on a biomedical process of mixing. The former necessitates some form of mechanical agitation. In the absence of agitation, mixing would be confined to diffusion and consequently be very slow. High levels of agitation, however, can be disruptive for the organic ingredients or for the intended result. Shear strain rates as low as 10 s^{-1} are known to severely retard adhesion to collecting surfaces[3]. Chaotic advection provides a suitable tool to handle these conflicting desiderata. It involves laminar flow therefore the shear strain levels can be kept extremely low. In addition, the exponentially rapid separation of nearby elements and the continual stretching and folding deformations throughout the flow, which are characteristic of chaotic behavior, provide a high mixing capability, despite the laminar flow regime.

[†] Department of Mechanical Engineering, Columbia University New York, New York 10027

The Cell collection process

The model used in the present study consists of a small cylindrical chamber (approximately 10 ml) eccentrically protruded by a smaller inner cylinder. This geometry is chosen due to its facility in generating laminar chaotic flows and the large number of previous studies performed on it (for references, Dutta and Chevray[2]). The liquid sample containing infant cells fills the annular region. A number of vertical fibrous collectors are fixed in this region and the infant cells are initially randomly distributed in the fluid as in Fig.1. The fluid in the eccentric annular region is stirred by performing a large number of *cycles.* Each cycle contains a rotation, with a prescribed slow speed, of the inner cylinder for time Δt_i and one rotation of the outer cylinder for time Δt_o. The ratio of the cylinder radii, the eccentricity, and the amounts of rotation are adjusted to create the largest extent of chaotic region in the chamber. The rotation speed is low enough to maintain Stokes regime where an analytic solution for the velocity field is available[4]. The extent of chaos is monitored by means of a Poincare section generated by recording the position of several particles at the end of each cycle for a large number of cycles. The Poincare section corresponding to the stirring procedure used in this study is shown in Fig.2. Chaos is almost global for these parameters as seen by the points scattered nearly over the entire region. Chaotic advection is employed in the infant cell recovery process in order to maximize the concentration of cells in the flow presented to each collector. This aim is best achieved if the positions of the cells are continually rerandomized in the chamber by an efficient mixing process. In order to establish an upper bound for the collection rate we assume that the concentration in the flow near every collector equals the global concentration in the medium. Then the number of infant cells in the chamber, *n(t)*, is governed by the relation $dn/dt = -[E\,Q(t)/V]n$, where *Q(t)* is the flow rate of fluid swept by the projected or frontal area of the entire set of collectors, *V* is the chamber volume, and *E* is the fiber efficiency to be explained below. Integrating this relation and rearranging, one obtains the *fraction of cells collected:*

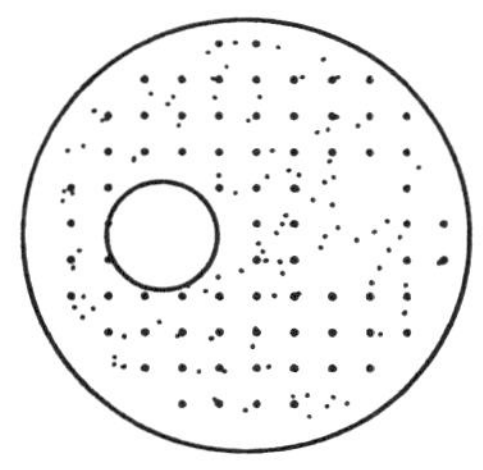

Fig.1. Approximately 80 infant cells (small dots) are randomly initialized among fixed fibrous collectors

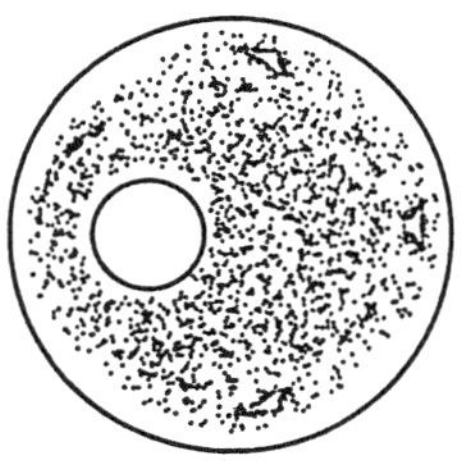

Fig.2. The Poincare section generated by the stirring procedure used in this study

$$f_N = 1 - \exp\{-E\, r_f\, \alpha\, N\} \qquad (1)$$

where $\alpha = \frac{2}{A}\sum_k (u_i \Delta t_i + u_o \Delta t_o)_k$, the index *k* representing individual collectors. *A* is the area of the eccentric annular region. The undisturbed flow velocity at the fiber center is represented by u_i and u_o for the inner and outer cylinders, respectively. *N* is the number of cycles and r_f is the fiber radius. α is essentially a parameter indicating the efficiency of the mixing process per unit fiber length per cycle. It depends on the position and number of collectors as well as on the stirring procedure.

The Fluid mechanical efficiency of fibers

Due to the deflection of streamlines by the fibers, only a fraction of the cells in the undisturbed flow (the flow that would exist if no fibers were there) swept by the collector area are actually collected. This fraction is expressed by the fluid mechanical efficiency, E. In order to determine E we construct a model of the flow around a cylindrical collector based on Happel's solution of Stokes flow around a cylinder[5]. The solution assumes that the radial velocity and the shear strain rate vanish at the boundaries of a larger cylindrical region of radius R surrounding the fibrous cylinder. It was found that $R \approx 10\, r_f$ yields realistic results in the calculation described below. There are three basic mechanisms of collection of particles suspended in a moving fluid: (i) inertial, (ii) interceptive, and (iii) diffusive. The Stokes number of an infant cell is low enough to make (i) negligible in the present system. The remaining mechanisms are studied as follows. Fig.3 schematically illustrates the interceptive collection process. Interceptive efficiency is defined as $E_i = (r_f + r_c)/r_o$ where r_o is simply found by integrating the trajectory of a cell backward from its position of grazing contact with the cylinder until it reaches the boundary of the region of radius R. The overall efficiency $E = E_i + E_d$, where E_d represents the diffusive efficiency, is obtained by determining the fraction of diffusing-advecting cells originally within the projected fiber area that eventually come into contact with the fiber. The motion of the cells is governed by the Langevin equation,

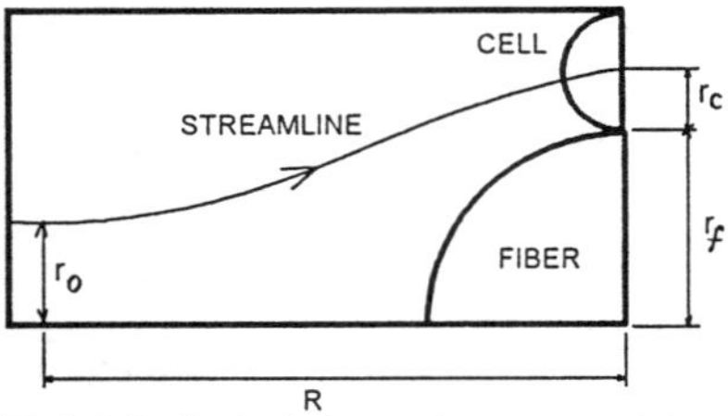

Fig.3. Cell collection by interception

$$\dot{\mathbf{x}} = \mathbf{u}(\mathbf{x}, t) + \mathbf{G}(t). \qquad (2)$$

Here, $\mathbf{u}(\mathbf{x},t)$ is the fluid velocity at the particle position $\mathbf{x}$ and $G_i(t)$ is a Brownian velocity component satisfying the conditions $\langle G_i(t)\rangle = 0$ and $\langle G_i(t)G_j(t')\rangle = 2D\delta_{ij}\delta(t-t')$. Ensemble average has been shown as $\langle \bullet \rangle$ and D is the diffusivity of an infant cell given by the Stokes-Einstein relation. The results are shown in Fig.4. For fiber radii smaller than the critical value $r_f \approx 100\mu m$, interceptive collection is dominant. Diffusive collection dominates for larger fibers. These results are consistent with those found from the approximate analytical solution of the steady diffusion equation in van de Ven[6] with negligible interception. The results for a spherical collector reported by Yang et al.[7] are plotted as dashed lines in Fig.4 to allow comparison with the cylindrical case.

Numerical results

In the interception dominant region, E varies approximately as the inverse of the fiber radius. Hence for small collectors, Eq.1 clearly indicates that the collection rate depends only weakly, if at all, on the fiber radius. In order to minimize the disruption of the flow field by collectors, however, fairly small collectors are needed. We choose $r_f = 50\ \mu m$. The fibers are fixed and the infant cells ($r_c = 10\ \mu m$) are initialized randomly in the chamber as in Fig.1. The cells are then advected in accordance with the Langevin equation with the local fluid velocity

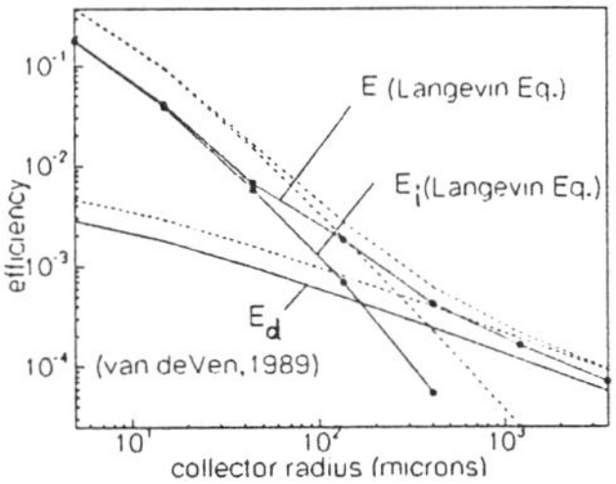

Fig.4. The fluid mechanical efficiency as a function of fiber radius. The dashed lines represent the corresponding efficiencies of a spherical collector.

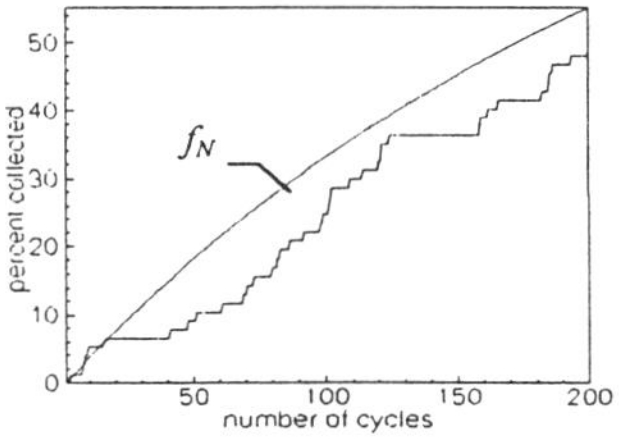

Fig.5. The fraction of cells collected as a function of the number of cycles. The smooth curve is the upper bound from Eq.1.

associated with the stirring procedure in Fig.2. Whenever a cell approaches a fiber to within a collision radius, r_f+r_c , it is collected with a probability equal to the efficiency *E*. The resulting value of the fraction of cells collected is plotted in Fig.5 along with the theoretical upper bound from Eq.1. The closeness of the simulated result to the upper bound indicates the effectiveness of chaotic motion in randomizing the locations of the infant cells.

Experimentation

As a means of validating our theoretical models we designed and fabricated a fetal-cell separation experiment. The experimental apparatus consists of two eccentric cylinders (with eccentricity $e = 2.88mm$) of radii, $r_i = 2.3mm$, and, $r_o = 9.2mm$, respectively, with a holding volume of approximately 10 mL between the cylinders. The cylinders are rotated independently by two high resolution stepper motor/driver units at speeds from 0.06-6.0 rpm. System control is achieved with an IBM 386 compatible PC. The collecting fibers are supported from above the cylinders. The outer cylinder is transparent and flow visualization data are recorded with a high resolution video capture system. This data is compared to graphical theoretical predictions using image processing algorithms. After the stirring cycles have been completed, the collecting fibers are removed and the number of captured cells is quantified via fluorescence microscopy.

Acknowledgement. The support of the NSF/Whittaker Foundation under Grant No. CUO1627201 is gratefully acknowledged.

[1] Aref, H. 1984, 'Stirring by chaotic advection', *J. Fluid Mech. 143:* 1.

[2] Dutta, P. & Chevray, R. 1995, 'Inertial Effects in Chaotic Mixing with Diffusion', *J. Fluid Mech.285:* 1

[3] Mandrusov, E. , Houng, A., Klein, E., & Leonard, E.F. (in press) 'Membrane-based cell affinity chromatography to retrieve viable cells', *Biotech Prog*

[4] Ballal, B.Y. & Rivlin, R.S. 1977 Flow of a Newtonian fluid between eccentric rotating cylinders: inertial effects. *Arch. Rat. Mech. Anal., 62:* 237.

[5] Happel, J. 1959 *AIChE J 5*:174

[6] Van de Ven, T. 1989 *Colloidal Hydrodynamics* (Academic Press:New York)

[7] Yang S.M., Han S.P., & Hong J.Jet al. 1995 'Capture of Small particles on a bubble collector by Brownian Diffusion and Interception' *Jl Colloid and Interface Sc 169*:125

Thermohaline Buoyancy Effects on Turbulent Flows

R. E. Baddour, M. ASCE[1]

Abstract

Nonlinear effects of thermohaline buoyancy on turbulent flows are described. The buoyancy fields of turbulent flows are examined on the basis of a thermohaline buoyancy equation which was derived from an equation of state which is nonlinear in temperature and linear in salinity. Numerical calculations and experimental results have indicated that linear thermohaline buoyancy could alter significantly the development, mixing and stability of turbulent flows.

Introduction

Buoyancy has important effects on the development, mixing and stability of turbulent flows. Large scale circulations in lakes and oceans are also influenced by spatial and temporal variations of the buoyancy field. The buoyancy fields in natural bodies of water are primarily caused by variations in temperature and salinity. This study examines the effects of temperature and salinity gradients on turbulent flows.

Thermohaline buoyancy equation

The buoyancy acting on a fluid is proportional to the density difference between the fluid and the surrounding environment. Such density anomalies are produced in water by both temperature and salinity gradients. *Equations of state* which define the density of water as a function of temperature and salinity can be traced back to the work by *Knudsen, Forch and Sörensen* in 1902. A more recent equation for seawater by *Millero and Poisson (1981)* has the form:

$$\rho(T,S) = f_1(T) + f_2(T)S + f_3(T)S^{\frac{3}{2}} + CS^2 \qquad (1)$$

[1] Professor, Dept. of Civil Engineering, The University of Western Ontario, London, Ontario, Canada N6A 5B9

where ρ is the density in kg/m^3, $-2<T<40$ the temperature in °C, $0<S<40$ the salinity in ppt, C is a constant and f_1, f_2, and f_3 are nonlinear functions of temperature. The function f_1 is the density of pure water as determined by *Bigg (1967)*. The two terms in Equ 1 which are proportional to $S^{3/2}$ and S^2 are, however, less than 2% of the linear salinity term. Hence without loss of accuracy the equation of state for seawater can be assumed to have the simpler form

$$\rho(T,S) = f_1(T) + f'_2(T)\,S \tag{2}$$

The function f'_2 was adjusted to minimize the error of omitting the nonlinear salinity terms. With f_1 and f'_2 expressed as polynomials we have

$$\rho = \sum_{i=0}^{n} a_i T^i + S \sum_{i=0}^{m} b_i T^i \tag{3}$$

Accurate estimates of density throughout the range of temperature and salinity of interest can be obtained with n = 5 and m = 2. The polynomial constants are a_0= 999.842594, a_1= 6.793952 x 10^{-2}, a_2= - 9.095290 x 10^{-3} a_3= 1.001685 x 10^{-4}, a_4= - 1.120083 x 10^{-6}, a_5= 6.536332 x 10^{-9}, b_0= 0.806924, b_1= - 3.042878 x 10^{-3} and b_2= 3.280458 x 10^{-5}.

It is desirable to relate the buoyancy, which is proportional to density difference, $\Delta\rho$, to temperature and salinity differences, ΔT and ΔS. This is achieved by transforming the coordinates of the equation of state from (T,S,ρ) to $(\Delta T, \Delta S, \Delta\rho)$. This transformation requires moving the origin of Equ 3 from $(0,0,0)$ to (T_a, S_a, ρ_a), where subscript a refers to a reference state for the water environment. This transformation gives, *Baddour, 1994*

$$B = \frac{-g}{\rho_a}\left[\sum_{i=1}^{n} \alpha_i \Delta T^i + S_a\left(\sum_{i=1}^{m} \beta_i \Delta T^i\right) + \Delta S\left(\beta_0 + \sum_{i=1}^{m} \beta_i \Delta T^i\right)\right] \tag{4}$$

$$\alpha_i = \sum_{j=i}^{n} \frac{j(j-1)(j-2).....(i-j+1)}{i!} a^j T_a^{j-i} \quad i=1,n \tag{5}$$

$$\beta_0 = \sum_{j=0}^{m} b_j T_a^j \tag{6}$$

$$\beta_i = \sum_{j=i}^{m} \frac{j(j-1)(j-2)...(i-j+1)}{i!} b^j T_a^{j-i} \quad i=1,m \tag{7}$$

where B = -g $\Delta\rho/\rho_a$ is the buoyancy per unit mass of fluid and g = gravity constant.

Surface jets and internal hydraulic jumps

Plane surface thermal jets were investigated by *Farghaly and Baddour (1991)* and *Zaghloul et al(1995)*. In the latter study the jet was discharged in a weak co-current. Temperature and buoyancy profiles obtained when T_a was 4.6°C are shown in Fig. 1. Note that the buoyancy profile coincides on the temperature profile when Equ 4 is linearized. This is clearly not the case in this experiment. Nonlinear mixing increased the buoyancy gradient near the surface and decreased the buoyancy gradient near the interface. The latter effect destabilized the interfacial region of the jet and enhanced mixing. Similar observations were made by *Baddour (1991)* who compared the behaviour of linear and nonlinear thermal hydraulic jumps. Fig 2 illustrates effects of nonlinear buoyancy on mixing in thermal hydraulic jumps.

Vertical jets

The buoyancy fields of thermohaline jets were also studied using Equ 4. Typical buoyancy profiles associated with Gaussian temperature and salinity distributions in vertical jets can be seen in Fig 3. The buoyancy B is normalized by the absolute value of the centerline buoyancy $|B_m|$ to identify the positive and negative buoyancy regimes of jets. Gaussian buoyancy profiles occurs only when Equ 4 is linearized. Deviations from linear mixing conditions are quite significant when salinity differences are small. Of particular importance in modelling turbulent jets is the integral buoyancy force. The buoyancy force is obtained by integrating Equ 4 over the cross section area of the jet. As an example, for a two-dimensional jet, the buoyancy force is

$$G = -\frac{\sqrt{\pi}\, bg}{\rho_w}\left[\sum_{i=1}^{n} \frac{\alpha_i \Delta T_m^i}{\sqrt{i/\lambda_T^2}} + S_w \sum_{i=1}^{m} \frac{\beta_i \Delta T_m^i}{\sqrt{i/\lambda_T^2}} + \Delta S_m \left(\frac{\beta_0}{\sqrt{1/\lambda_S^2}} + \sum_{i=1}^{m} \frac{\beta_i \Delta T_m^i}{\sqrt{1/\lambda_S^2 + i/\lambda_T^2}} \right)\right] \tag{8}$$

where b, λ_Tb, and λ_Sb are the characteristic widths of the velocity, temperature and salinity profiles of the jet. *Jones and Baddour (1991)* applied this method to study positively-buoyant thermal jets, and more recently *Zhang and Baddour (1995)* investigated thermohaline dense jets. Experimental observations of maximum jet penetrations in nonlinear mixing regimes were found in good agreement with numerical predictions. Fig. 4 shows typical calculations of momentum and volume fluxes in plane thermohaline dense jets. These results show that nonlinear buoyancy modelled using Equ 8 alter the mixing characteristic, and drastically reduce the penetration of vertical dense jets.

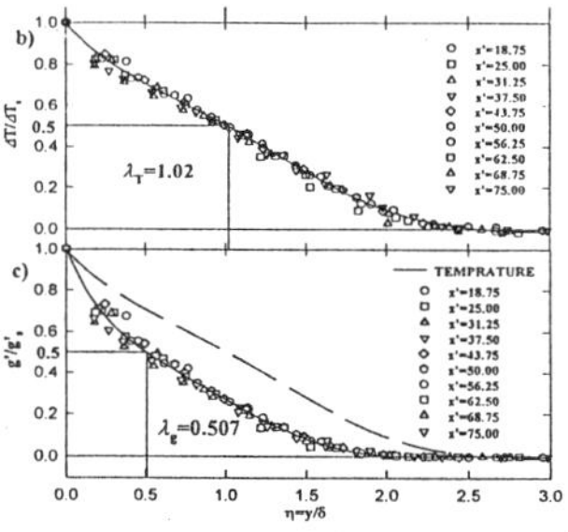

Fig. 1 Temperature and Buoyancy Profiles in a surface thermal jet, *Zaghloul et al, 1996*
$Fr = 9.35$, $U_a' = 0.16$, $Re = 6350$, $T_a = 4.6$

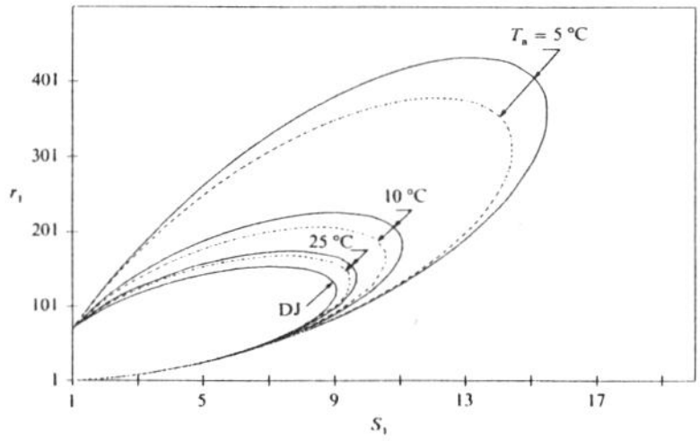

Fig 2. Nonlinear mixing properties of thermal hydraulic jumps, *Baddour, 1991*
$Fr = 50$, $\Delta T_0 = 10$ & 15°C, DJ = density jump

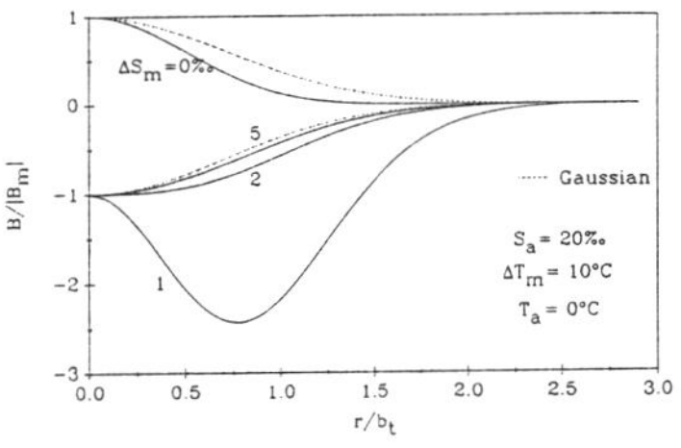

Fig 3. Buoyancy profiles in thermohaline jets

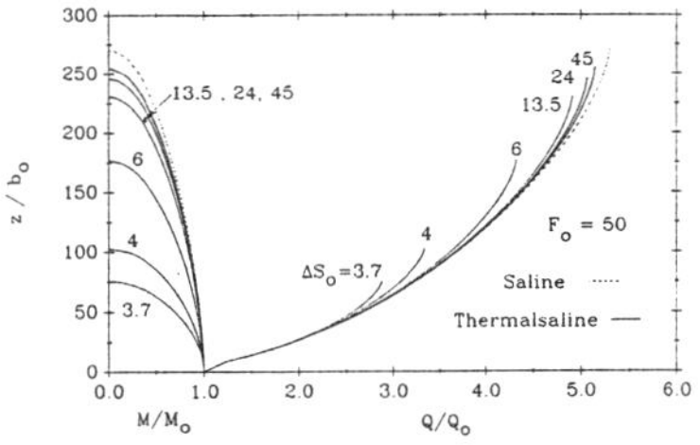

Fig 4. Predictions of momentum and volume fluxes in plane thermohaline dense jets
$Fr=50$, $\Delta T_0 = 18.5$°C, $S_a=0$ ppt, $T_a = 5$°C

ACKNOWLEDGEMENT

The research was supported by grants from the Natural Sciences and Engineering Research Council of Canada.

REFERENCES

Baddour, R.E. 1991 *J. Fluid. Mech.* 226, 243-256
Baddour, R.E. 1994 *NATO ASI Series*, 255, 117-130
Bigg, P.H. 1967 *Brit. J. Appl. Phys.* 18, 521-525
Farghaly, H.A. & Baddour R.E. 1991 *Proc. CSCE Conf.* Hamilton, 164-177
Jones, D.G. & Baddour, R.E. 1991 *J. Hydr. Res.* 29, 601-613
Millero, F.J. & Poisson, A 1981 *Deep Sea Res.* 28A, 625-629
Zaghloul, A.M, Martinuzzi, R. & Baddour, R.E. 1996 *Proc. ASCE EM Conf.*
Zhang, H. & Baddour, R.E. 1995 *Proc. ASCE EM Conf.* Boulder, 1010-1013.

Magnetic Fluid Dynamics of Blood Flow

Yousef Haik§ , Ching Jen Chen+ and Vinay Pai§

Abstract

This paper presents the biomagnetic fluid dynamics of human blood. The human blood was found to posses the property of paramagnetic material when deoxygenated (vein) and diamagnetic when oxygenated (artery). This unique property of the blood can be utilized to develop medical applications that can be utilized for enhancing the human health. In this paper the experimental results on the effect of the magnetic field on a diamagnetic human blood flow is presented. The mathematical model to describe the behavior of the blood flow under the influence of magnetic field is also presented.

Introduction

The recent increase in the frequency of the human exposure to magnetic fields such as the magnetic resonance image diagnostic technique (MRI) and passenger transport systems based on the magnetic levitation lead to many studies initiated to investigate the effect of the magnetic fields on the human body. In 1992, The Committee on Interagency Radiation Research and Policy Coordination (CIRRPC) [1], published a report to summarize the studies on the health effect from the exposure to electric and magnetic field. The conclusion indicates that there is no convincing evidence to support the contention that exposure to low frequency magnetic fields will lead to health hazards.

Studies of magnetic effect on human blood was conducted by many researchers. Pauling and Coryell (1936) [2] reported that the oxygenated hemoglobin (artery blood) is diamagnetic while the deoxygenated hemoglobin (vine) is paramagnetic. Okazaki et al (1986) [3] experimentally demonstrated that a drift of the human red blood cells flow line due to inhomogenous magnetic field in a laminar flow. Shiga et al (1993) [4] reported that the attraction of a narrow stream of red blood cells suspension towards the stronger magnetic field in a wide laminar flow was found to be proportional to the magnetic field susceptibility of the red blood cells, the product of the field strength and its spatial gradient, the reciprocal of the flow velocity and the hematocrit. Their experiment was carried on a venous blood which has a paramagnetic characteristics.

---

§ Graduate Students

+ Dean of FAMU-FSU College of Engineering, 2525 Pottsdamer Rd, Tallahassee, Fl 32310

In this study a preliminary experiment was conducted on human blood flow against the magnetic field direction in the National High Magnetic Field Laboratory. The mathematical model is also presented.

Experimental Setup and Procedure

The experimental study was carried out at the National High Magnetic Field Laboratory. The blood for the experiment was drawn from the vein of a thirty year old male volunteer. The hematocrit value for the blood was 0.51. The experimental setup is shown in figure 1. The blood which was collected in a blood bag was allowed to flow in a plastic tubing of internal diameter 3 mm and total length 800 mm, while the tube was placed inside the 50 mm core of a 20 tesla resistive magnet. The blood from the tubing was allowed to flow and collected in another blood bag. The temperature in the setup was monitored by means of thermocouples to ensure that the temperatures remained constant throughout the experiment. Care was taken to center the apparatus in the magnet in order to cancel out the magnetic gradient effects. The 20 tesla hybrid magnet at the National High Magnetic Field Laboratory has a positive magnetic gradient going from the center upward and negative magnetic field gradient going from the center downward.

The blood flow in the tubing was carried out first in the absence of magnetic fields. The Reynolds number in this case was approximately 390. The flow was then subjected to various magnetic field strengths up to 10 tesla. At different field strengths, the time for the flow of blood from the blood bag above the magnet to the one below the magnet was carefully measured. The experimental results are shown in table 1. A reduction of 30% in the flow rate for the blood was detected for 10 tesla field.

Mathematical Formulation

The mathematical model for the biomagnetic fluid dynamics is based on the modified Stokes principles stated below:

1. The stress in the biomagnetic fluid is a continuous, linear function of the velocity gradient, the three thermodynamics properties (p, ρ, T), space (x_i), time (t), and the magnetization property (M)
2. The biofluids are homogeneous.
3. The stress in the biomagnetic fluid is an addition of fluid stress and the magnetic stress created by the magnetization process.
4. In the presence of the magnetic field, the stress of the fluid at rest is equal to the hydrostatic and magnetic pressure.
5. If the motion is pure dilatation, the stress due to the rate of rotation vanishes and the normal stress is equal to the hydrostatic and magnetic pressure.

From the modified principles of constitutive relations, the equations of motion are derived for incompressible flow [5] and they are

Continuity Equation

$$\nabla \bullet \overline{V} = 0 \tag{1}$$

Linear Momentum

$$\rho \frac{D\overline{V}}{Dt} = -\nabla p + \rho \overline{F} + \eta \nabla^2 \overline{V} + \xi\left(\nabla^2 \overline{V} + 2\nabla \times \overline{\omega}\right) + \mu_o\left(\overline{M} \bullet \nabla\right)\overline{H} \tag{2}$$

Equation (2) states that the time rate change of momentum is equal to the pressure force, the body force, the viscous symmetric and antisymmetric shear force and the magnetic force (the last two terms). The magnetic force term consists of two parts: the magnetic force due to polarization and the Lorentz force due to the induced current. Equation (2) is similar to the Navier-Stokes equations with three additional terms due to the antisymmetric part of the stress tensor and the magnetic stress. The term

$$\xi\left(\nabla^2 \overline{V} + 2\nabla \times \overline{\omega}\right)$$

models the antisymmetric part of the stress tensor. This is because when the biomagnetic fluid is subjected to a magnetic field the action of magnetization will introduce a rotational motion to orient the magnetic fluid particles with the magnetic field. This action of orientation is directional dependent and will introduce the antisymmetric stress tensor.

The magnetic stress is modeled based on the Maxwell stress tensor and is generated from the polarization force and the Lorentz force effects. The term

$$\mu_o\left(\overline{M} \bullet \nabla\right)\overline{H}$$

represents the magnetic force due to polarization. This stress depends on the existence of the magnetic gradient. When the magnetic gradient is absent, this force vanishes.

Angular Momentum

$$\rho I \frac{D\overline{\omega}}{Dt} = \mu_o \overline{M} \times \overline{H} + + \eta' \nabla^2 \overline{\omega} + 2\xi\left(\nabla \times \overline{V} - 2\overline{\omega}\right) . \tag{3}$$

The time rate change of the rotational motion of the magnetic particles is due to the magnetic torque, the exchange between the internal and external angular momentum and the surface couple. The Angular momentum equation is no longer a simple algebraic equation of stress symmetry as in the Navier-Stokes equations but rather a non linear differential equation that specifies the interaction of the internal angular momentum due to the particle spin and rotationality of the fluid element. The term

$$\mu_o \overline{M} \mathrm{x} \overline{H}$$

represents the magnetic torque exerted on the magnetic particle. The term

$$2\xi\left(\nabla \mathrm{x} \overline{V} - 2\overline{\omega}\right)$$

represents the interaction between the internal and external angular momentum in the fluid element, and

$$\eta' \nabla^2 \overline{\omega}$$

represents the diffusion of fluid rotation. The diffusion term was modeled on the assumption that the surface couple is similar to the surface stress due to the rotational motion of the fluid element.

Conclusions

The study of the effect of static magnetic field on the blood flow is an important but unexplored area for research. This study presents the mathematical model of biomagentic fluid dynamics and two experimental results of the influence of magnetic field on diamagnetic blood. It was found that a 30% reduction in the blood flow rate when subjected to 10 tesla field.

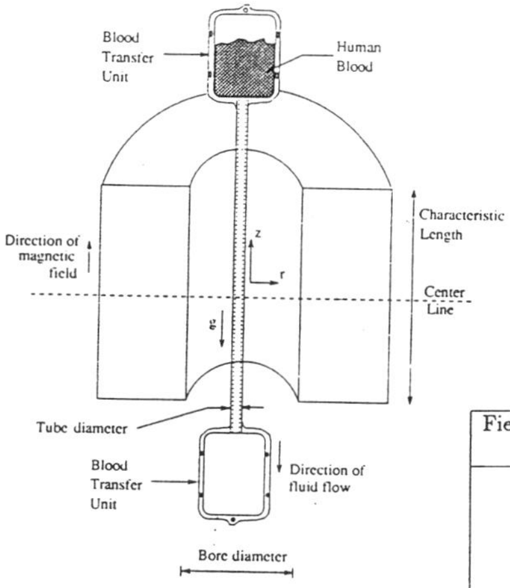

Figure 1: Experimental Setup

Field Strength (tesla)	Time (s)	
	Volume: 300 cc	Volume: 100 cc
0.0	115	--
1.0	--	912
3.0	125	1113
5.0	137	1185
10.0	147	--

References

1. Oak Ridge Associated Universities, "Health Effects of Low-Frequency Electric and Magnetic Fields," ORAU 92/F8, June 1992.

2. Pauling, L. and Coryell, C. D., "The Magnetic Properties and Structure of Hemoglobin, Oxyhemoglobin and Carbonmonoxy Hemoglobin," Proceedings of the National Academy of Science, USA, Vol. 22, pp. 210 - 216, 1936.

3. Okazaki, M., Maeda, N., and Shiga, T., "Drift of an Erythrocyte Flow Line Due to the Magnetic Field," Experientia, Vol. 42, pp. 842-843, 1986.

4. Shiga, T., Okazaki, M., Seiyama, A., and Maeda, N., "Paramagnetic Attraction of Erythrocyte Flow Due to an Inhomogenous Magnetic Field," Bioelectrochemistry and Bioenergetics, Vol. 30, pp. 181-188, 1993.

5. Haik, Y., Chen, C. J. and Pai, V. "Development of Biomagnetic Fluid Dynamics," FAMU-FSU College of Engineering Technical Report, Tallahassee, Florida, 1995.

Simulation of Suspended Particles Transport in the Entrance Region of Tube Flow

Shi-kang Wang* N.H.C. Hwang**

Abstract

The behavior of suspended particles in the entrance region of tube flow can provide a useful hydrodynamic model in understanding cell hemodynamics in the intensely branching vascular trees. A quasi-multi-particle simulation based on the "inertia-vorticity" model without particles interaction is proposed to simulate the transport of suspended particles in the entrance region of tube flow. The results show that the peak location of particle concentration can be adjusted by changing either the initial concentration distribution at the inlet or the entry velocity of the particles.

Introduction

Segre and Silberberg (1962a, 1962b) investigated the motions of an initially uniform, dilute suspension of neutrally buoyant rigid spheres transported in a Poiseuille flow through a circular tube having an internal diameter of 11.0mm. At low Reynolds numbers, the spheres were carried to an equilibrium annular region at about 0.6 tube radius from the axis, irrespective of the position at which the spheres enter the tube.

Blood cells flowing in small arteries and arterioles have been observed to exhibit much larger

* Visiting professor, on leave from Tianjin University, Tianjin, China
** Knight Professor, Dept. of Biomedical Engineering, University of Miami, Coral Gables Florida 33124

radial dispersive motions than predicted by the Stokes-Einstein formula characterizing Brownian diffusion. Wang and Hwang (1992) have shown that the often cited convective-diffusion theory may not be an adequate model for describing the transverse migration of suspended cells in blood flow. However, the behavior of suspended particles in the entrance region of tube flow can provide an important model in understanding cardiovascular hemodynamics in the intensely branching vascular trees.

Based on our preliminary experimental observations (Wang et al, 1991; Wang and Hwang, 1994),we suggesed an "inertia-vorticity" (IV) model to describe the radial migration of suspended particles in the entrance region of a tube.

This paper presents a quasi-multi-particle simulation based on the "inertia-vorticity" model without particle interaction. The results show that the peak locaton of particle concentration depends on the initial concentration distribution at the inlet, the size and the entry velocity of neutrally buoyant particles.

Single-Particle Simulation---IV Model

In the entrance zone fluid flow may be categorized into two regions: in the core region, fluid flows with zero radial velocity gradient, while in the boundary layer, the fluid experiences shear stress due to the existing velocity gradient in the radial direction. In the region of $\eta<0.55$ ($\eta=r/R$ is the dimensionless radial position), the fluid accelerates for a certain distance then decelerates due to the wall influence. Around $\eta=0.55$, the flow accelerates to reach a constant value downstream (Wang and Hwang, 1994). In the core region neutrally buoyant particles in suspension move in parallel with the tube wall. Suspended particles may not have the same velocity as the surrounding fluid, depending largely on the size and density of the particle, the particle entry velocity and its entry position at the inlet. The fluid/particle velocity diffrence can be obtained numerically using the unsteady motion eqautions for a small rigid particle. When the particle enters into the wall shear region with a velocity different from that of the surrounding fluid medium, it immediately experiences a lift force due to the existing vorticity in the shear region. The forces that dominate the particle motion in the shear region are the Stokes drag force and the lift force.

Quasi-Multi-Particle Simulation

For low-concentration particle/fluid flow, we can assume the particle motion to be unaffected by the presence of neighboring particles. Based upon the above computational results of the IV model, a quasi-multi-particle simulation was carried out under the following two simplified conditions:
1. The initial distribution of particle at the inlet is uniform
2. Neglecting the interaction between particles, even if the particles travel to the same location.

Three dimensional arrays of the concentration distribution for two different particle sizes (a=1.0 and 0.85mm) and the entry velocity $\omega_p = 1$, are shown in Figure(1a,b). The decreasing of the particle number in the tube cross-section with increasing x* illustrated that a portion of entry particles had already deposited on the tube wall. In Figure (1a) the dimensionless number of moving particles decreases to 0.7 at cross section of x*=0.025, and continuously decreases to 0.64. The presence of the peak concentration means that a portion of particles are driven into a narrow annular region due to the effect of the inertia lift force. At the end of the entrance (i.e. x*=0.15), the concentration peak for a=1mm is situated in the range of η between 0 and 0.1. For the particles with radius a=0.85mm, the concentration peak is shifted towards the tube wall centered at η=0.2. Although the quasi-multi-particle simulation neglects the interaction between particles, the obtained results still are an inspiration to us. It is possible that the peak location of particle concentration can be adjusted by changing either the initial concentration distribution or the entry velocity of the particles.

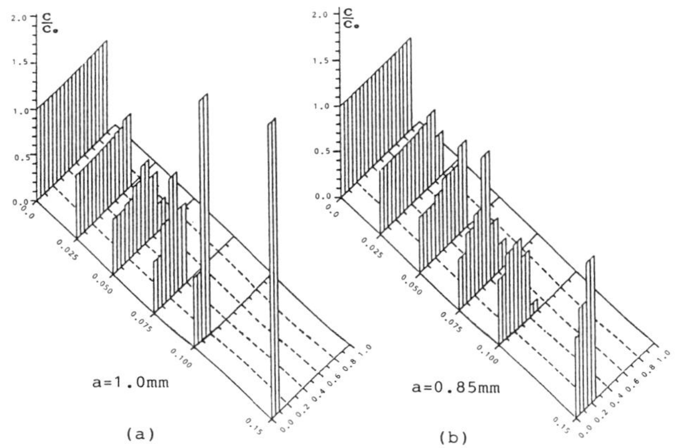

Figure 1 Quasi-Multi-Particle Simulation for $\omega_p = 1$

References

SEGRE, G. and SILBERBERG, A. (1962a) Behavior of macroscopic rigid spheres in Poiseuille flow. Part 1. Determination of local concentration by statistical analysis of particle passages through crossed light beams. J. Fluid Mech., 14, part 1, 115-135.

SEGRE, G. and SILBERBERG, A. (1962b) Behavior of macroscopic rigid spheres in Poiseuille flow. Part 2. Experimental results and interpretation. J. Fluid Mech., 14, part 1, 136-157.

WANG, S.K., LI, W., and HWANG, N.H.C. (1991) Behavior of single rigid particle in the entrance region of tube flow. ASME Advances in Bioengineering BED20, 387.

WANG, S.K. and HWANG, N.H.C. (1992) On transport of suspended particulates in tube flow. Biorheology 29, 353-377.

WANG, S.K. and HWANG, N.H.C. (1994) On the transport of a suspended particle in flow through the entrance region of a tube. Biorhology 31, 549-563

HANDRAIL GRASPABILITY

Donald O. Dusenberry,[1] Member, ASCE and
Howard Simpson,[2] Fellow, ASCE

ABSTRACT: As part of an experimental program to develop design criteria for cross-sectional profiles of stairway handrails, measurements were made of the maximum grasp force (the maximum restraining force exerted by the subject as the rail is moved away from the subject) on two wood handrail designs -- a 6010 decorative ("milled") handrail (the most common residential handrail style in use today) and a 51-mm (2-in.) round handrail. A total of 292 tests were performed with 73 subjects, ranging in age from 10 years to 83 years. It is concluded that the milled shape performs at least as well as the round shape, and that the maximum restraining force is developed with a hook grip (rather than, as is commonly assumed, a "power" grip).

INTRODUCTION

Falls on stairs constitute a leading cause of accidental injury in the United States. While there is some controversy as to the role of handrails in the prevention of accidents, the available data relating the severity of injuries from stairway falls to the presence of handrails indicate that, in general, the availability of a handrail reduces the severity of accidents that do occur (Carson 1978; Templer 1992). It is reasonable to conclude, therefore, that an important function of the handrail is to serve as a grab bar for arresting or mitigating a fall.

Unfortunately, there is very little published information on the relationship between handrail cross-sectional profile and the magnitude of grab force ("graspability") that it is capable of developing. The only significant data that has been available are the results of a limited study sponsored by the Canadian National Research Council (Maki 1985). This study, on ten male and ten female subjects, involved only one decorative ("milled") profile -- a shape not in use in the United States. Of the various shapes tested, the author found a 38-mm (1.5-in.) circular shape to be optimum. Spurred largely by the results of this study, there has been a trend among the model building codes toward mandating all handrails to be approximately circular in cross-section, with a maximum dimension of 51 mm (2 in.). (A 38-mm rail is unacceptably flexible in many applications).

---

[1] Principal, Simpson Gumpertz & Heger Inc., 297 Broadway, Arlington, MA 02174
[2] Senior Principal, Simpson Gumpertz & Heger Inc., 297 Broadway, Arlington, MA 02174

EXPERIMENTAL PROGRAM

We have initiated a comprehensive investigation into the influence of handrail profile on the prevention and mitigation of stair injuries. This new research should help resolve the important issues of handrail shape and safety and result in improvements in code provisions governing handrail design. A major goal of this work is to develop an understanding of the nature of the stairway fall and handrail grab response phenomena. We have commissioned a test program at the University of Toronto to record and analyze the motions and grab responses of subjects during actual stair falls, and the forces applied to handrails in the attempt to arrest a fall.

A major part of our test program is the investigation of the influence of handrail cross-sectional profile on graspability, for the purpose of developing criteria that lead to shapes that are not only safe and functional, but also permit choice, variety, and aesthetic expression. This work is being conducted in our own laboratories at Simpson Gumpertz & Heger Inc.

Our initial test series obtained comparative graspability data for two wood handrail designs - a 6010 milled handrail (the handrail style in most common use for residential construction in the United States) and a 51-mm (two-inch) round handrail (Figure 1). We measured the maximum pull applied by each subject to each handrail as the rail was slowly moved away from the subject. Our test apparatus (see Photo) includes an adjustable, padded table against which the subjects brace themselves, a sliding platform on which the handrail specimen is mounted, an electric motor with speed control for moving the handrail away from the test subject, and a load cell and associated instrumentation for measuring the force exerted by the subject on the handrail.

A total of 73 subjects, ranging in age from 10 years to 83 years, was tested. The distribution of hand sizes for the test subjects closely represents the hand size distribution for the U.S. general population. Our subject group included hands smaller than the 5th percentile woman's hand and larger than the 95th percentile man's hand. (See Figure 2 for age, hand size, and gender distributions.) The maximum grasp force for each subject's left and right hand was recorded, for a total of 292 tests.

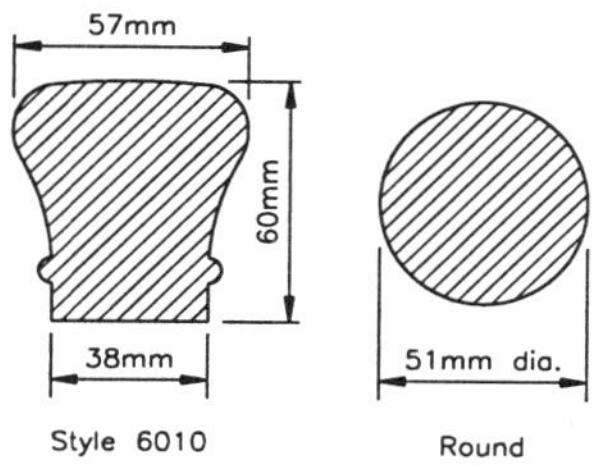

Figure 1. Cross Sections of Handrail Shapes

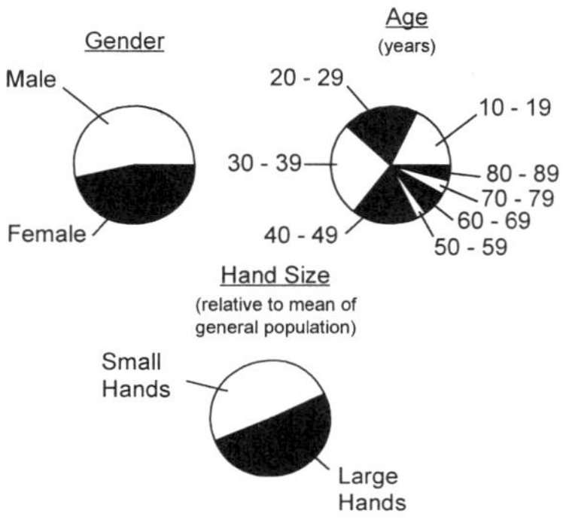

Figure 2. Summary of Tested Individuals

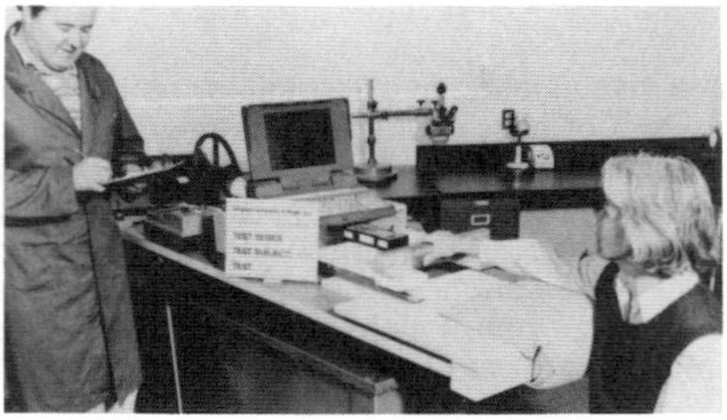

Test Apparatus

TEST RESULTS AND COMMENTS

Test results are plotted in Figure 3, which shows, for each hand size, the mean ratio of milled rail grasp strength to round rail grasp strength. The scatter of test results is such that ratios between 0.9 and 1.1, shown by the shaded zone in the figure, can be considered to approximate equivalence. On this basis, the tests show that the most commonly installed milled handrail is essentially as graspable as the 51-mm-diameter round handrail permitted by all U.S. building codes. In fact, for the two shapes tested, women as a group and men with small hands tend to be able to develop larger forces when grasping the milled handrail.

Our tests revealed the sequence of grasp configuration changes that occurs as subjects attempt to maintain their hold on handrails at high load. At the beginning of the tests, all subjects almost invariably grab the handrail in a "power grip," which allows a strong prehensile clasp on the handrail. However, as the rail is moved away from the subject and maximum grasp force is approached, the thumb is pulled away from the rail. The resulting grip becomes a "hook grip," with the subject's fingers wrapped around the handrail without contact between the thumb and the handrail (Figure 4). Just prior to release, it is the strength of the fingers hooked around the shape that determines the maximum force that each subject can exert. The relatively small radius at the top edges of the tested milled

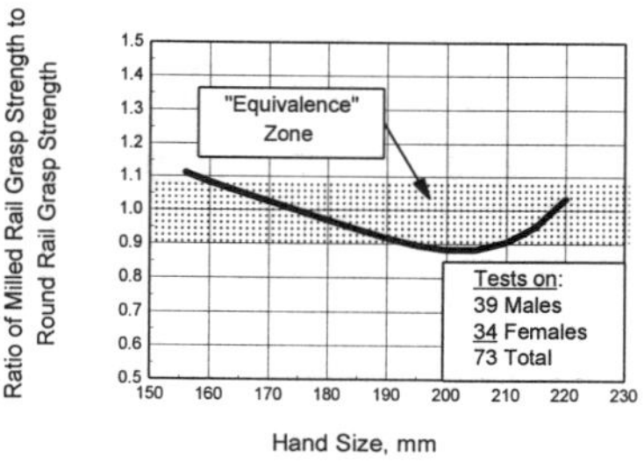

Figure 3. Relationship Between Relative Grasp Strength and Hand Size

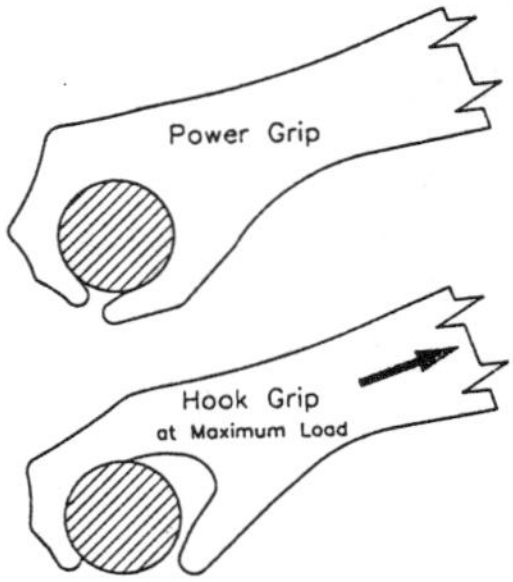

Figure 4. Power and Hook Grips

shape allowed all subjects, and especially subjects with small hands, to maintain firm grip, even after the thumb was pulled away from the handrail.

THEORETICAL STUDIES

It is impractical to test all combinations of hand size and handrail shape to record direct measurements of performance. However, we expect to establish the basic principles of hand biomechanics that come into play as a human hand uses and grasps a handrail, and then to apply the results of our tests to the entire spectrum of user hand sizes and to untested handrail profiles.

Theoretical approaches may best benefit the evaluation of handrail graspability by the aged and the infirmed. These groups have relatively high stairway accident rates but, because of their physical condition, they are the most difficult to use as subjects in comprehensive tests.

FUTURE RESEARCH

Our next series of tests will compare the performance of several new shapes that have been milled for our studies. The purpose of these tests will be to independently evaluate the influence of certain handrail profile dimensions on graspability. The data obtained from our research on stairway fall scenarios and the role of handrails in arresting falls, together with the information we develop on the relationship between the handrail profile and graspability, will be of great assistance in the formulation of rational design rules for handrail shape.

CONCLUSIONS

We have initiated a research program to investigate the issues that affect the functional adequacy and safety of the handrail profile. Upon completion of these studies, we expect to be able to formulate rational code rules for handrail shapes. In the meantime, the tests that we have performed to date and summarized in this paper demonstrate that, for the pull direction tested, the most commonly used milled shape performs at least as well as a 51-mm-diameter round shape.

ACKNOWLEDGMENTS

The research reported herein was sponsored by the Stairway Research Committee of the Stairway Manufacturers Association. The experimental program on stairway falls is being conducted by University of Toronto Associate Professor Brian Maki at the Centre for Studies in Aging, Sunnybrook Health Science Center, Toronto, Canada. The authors gratefully acknowledge the biomechanics expertise provided by Professor Emeritus Robert Mann of the Massachusetts Institute of Technology and the collaboration of orthopedic surgeon Alan Ertel, M.D.

APPENDIX I. REFERENCES

Carson, D. H., Archea, J. C., Margulis, S. T., and Carson, F. E. (1978). "Safety on stairs." *NBS Building Science Series 108*, National Bureau of Standards, Washington, D.C.

Maki, B. E. (1985), "Influence of handrail shape, size and surface texture on the ability of young and elderly users to generate stabilizing forces and moments." Report; Contract No. OSR84-00197 for the National Research Council of Canada, Toronto, Ontario.

Templer, J. (1992). *The Staircase.* The M.I.T. Press, Cambridge, Mass.

Mechanical Stress in Pediatric Heart Disease: Computational Modeling of Associated Defects in Subaortic Stenosis

Michael D. VanAuker[1], Pedro del Nido, Theresa A. Tacy, Gunnlaugur Sigfusson, and Edward G. Cape

Subaortic stenosis (SAS) constitutes up to 20% of all forms of left ventricular outflow in children. It is an acquired and progressive lesion with unknown etiology. An understanding of the etiology of SAS is the first step in improving surgical timing, techniques, and developing alternative therapy. Previous studies have identified an association between (1) SAS and (2) associated cardiac defects and morphologic abnormalities. The most common associated defect is a perimembranous ventricular septal defect (VSD), which is a hole in the interventricular septum which permits shunting of blood from the left to the right ventricle. More subtle morphologic abnormalities have also been observed in association with SAS in the absence of congenital defects, such as a steeper aorto-septal angle. This angle is defined as that formed by the long axis of the ascending aorta and the interventricular septum. These defects and abnormalities may appear disparate when examined superficially, but the principles of fluid dynamics suggest that they may cause a consistent change in shear stress along the interventricular septum in regions in which SAS occurs. Mechanical stress effects on endothelial cells have been widely investigated in vascular disease; however, shear stress effects have not been investigated within the cardiac chambers. This work, therefore, is part of an ongoing study which addresses the hypothesis that congenital heart defects and morphologic abnormalities affect septal shear stress, which acts as an environmental trigger for cellular growth factors. To address this hypothesis, a study of patients with SAS at the Children's Hospital of Pittsburgh was undertaken to identify and confirm the associations between certain defects and abnormalities, and to design a computational model to explore the effects of these abnormal morphologies on the flow fields in the left ventricular outflow tract. Using this model, we have demonstrated several important results: (1) a steepened aorto-septal angle causes important changes in wall shear stress in the regions in which SAS develops; (2) for a given outflow velocity, a larger outflow tract size is associated with relatively greater septal shear stress and greater sensitivity to aorto-septal angle effects; (3) the presence of a VSD can further augment peak shear stress levels. These findings may help identify patients at risk for either development or recurrence of SAS, and may illustrate why SAS is rare in infancy and more commonly found later in childhood.

[1]Cardiac Dynamics Laboratory, Children's Hospital of Pittsburgh, University of Pittsburgh, 3705 Fifth Avenue, Pittsburgh, PA 15213

Anisotropy Effect on One-Dimensional Consolidation

L. Cui[1] Y. Abousleiman[1] A. H-D. Cheng[2] J.-C. Roegiers[1]

Introduction

Starting from Terzaghi (1923), one-dimensional consolidation model had played an important role in soil and geomechanics applications (Schiffman *et al.*, 1969). To take into consideration of coupling and compressibility effects of solid and fluid constituents, the original Terzaghi theory had been modified to the theory of poroelasticity by Biot (1941). However, the typical application is still limited to the assumption of material isotropy. In reality, soil and rock are more likely to be transversely isotropic due to their modes of deposition and formation.

The theory of anisotropic poroelasticity was established by Biot (1955). However, not until lately, detail studies of constitutive equations (Thompson and Willis, 1991; Cui, 1995; Cheng, 1995) and laboratory measurements of anisotropic poroelastic constants (Aoki *et al.*, 1993) were conducted. Recently, the effect of anisotropic material in some poroelastic problems have been investigated (Abousleiman *et al.*, 1995; Cui, 1995).

This paper applies anisotropic poroelasticity to the 1-D consolidation and presents an analytical solution. For a 1-D consolidation in a transversely isotropic material with the isotropic plane perpendicular to the direction of loading, the anisotropy effect is demonstrated.

One-Dimensional Consolidation

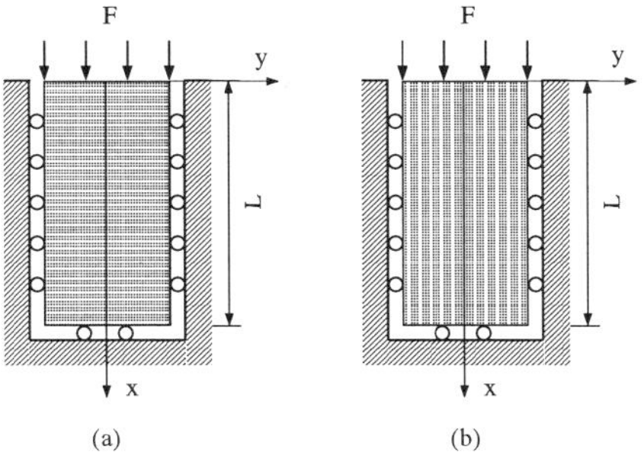

Figure 1: 1-D consolidation in a transversely isotropic material.

The focus of this paper is on the Terzaghi's consolidation problem and its extension to account for material anisotropy. Figure 1 is a schematic of the 1-D consolidation showing typical loading, geometry and boundary conditions. It is assumed that the material is orthotropic, and its axes of symmetry are coincident with the axes of Cartesian coordinates. Utilizing the geometric and

---

[1]Rock Mechanics Institute, University of Oklahoma, Norman, Oklahoma 73019-1014.
[2]Department of Civil and Environmental Engineering, University of Delaware, Newark, Delaware 19716.

physical symmetries, based on anisotropic poroelasticity (Cui, 1995), it can be shown that the constitutive relations and Darcy's law may be written as:

$$\sigma_{xx} = D_{11}\varepsilon_{xx} - \alpha_x p \;, \quad \sigma_{yy} = D_{12}\varepsilon_{xx} - \alpha_y p \;, \quad \sigma_{zz} = D_{13}\varepsilon_{xx} - \alpha_z p \tag{1a}$$

$$p = M(\zeta - \alpha_x\varepsilon_{xx}) \;, \quad q_x = -\kappa_x \partial p/\partial x \tag{1b}$$

where D_{ij} are elements of the drained elastic matrix, M is the Biot's modulus, α_x, α_y, and α_z are anisotropic Biot's effective stress coefficients, and κ_x is the permeability in the x-direction.

Governing Equations and Solutions

Ignoring body forces, the governing equations of 1-D consolidation in an orthotropic material can be derived as:

$$\frac{\partial \sigma_{xx}}{\partial x} = 0 \tag{2a}$$

$$\frac{\partial p}{\partial t} - c_x\frac{\partial^2 p}{\partial x^2} = -\frac{\alpha_x}{D_{11}S_x}\frac{\partial \sigma_{xx}}{\partial t} \tag{2b}$$

where, $c_x = \kappa_x/S_x$ is the consolidation coefficient in the x-direction, and S_x is the storage coefficient in the x-direction, defined as $S_x = 1/M + \alpha_x^2/D_{11}$. The boundary conditions should be given at $x = 0$ and $x = L$, where L is the length of the column (which can be infinity.) They are expressed by:

$$\sigma_{xx}|_{x=0} = -F(t) \;;\; p|_{x=0} = \beta F(t) \;;\; \partial p/\partial x|_{x=L} = 0 \;;\; \text{and,}\; u_x|_{x=L} = 0 \;; \tag{3}$$

where $F(t)$ is the loading applied on the top surface, $\beta = 0$ if the top is drained, or $\beta = 1$ if the loading is applied through fluid pressure.

Herein, the loading is assumed to remain constant (for time-dependent loading, the convolution integral should be applied (Cui, 1995).) Hence, $F(t) = P_0\mathrm{H}(t)$, in which $\mathrm{H}(t)$ is the Heaviside step function and P_0 is the magnitude of the load. In this case, initial conditions are $p|_{t=0} = p^u = \alpha_x P_0/D_{11}S_x$, and $u_x|_{t=0} = u_x^u = P_0(L-x)/D_{11}MS_x$, where p^u and u_x^u are the pore pressure and the axial displacement at the undrained state, respectively. Hence the pore pressure and the displacement can be integrated as:

$$p = p^u(1 - f_1(x,t)) + \beta P_0 f_1(x,t) \tag{4a}$$

$$u_x = u_x^u + \frac{\alpha_x}{D_{11}}(p^u - \beta P_0)f_2(x,t) \tag{4b}$$

in which f_1 and f_2 are two influence functions, and they are given by:

$$f_1 = 1 - \sum_{m=1,3,\cdots}^{\infty} \frac{4}{m\pi}\sin\left(\frac{m\pi}{2L}x\right)\exp\left(-\frac{m^2\pi^2}{4L^2}c_x t\right) \tag{5a}$$

$$f_2 = \sum_{m=1,3,\cdots}^{\infty} \frac{8L}{m^2\pi^2}\cos\left(\frac{m\pi}{2L}x\right)\left[1 - \exp\left(-\frac{m^2\pi^2}{4L^2}c_x t\right)\right] \tag{5b}$$

It is noticed that the anisotropy of the material is brought into the solution, because D_{11} depends on Young's moduli and Poisson's ratios related to all three directions of the material symmetry.

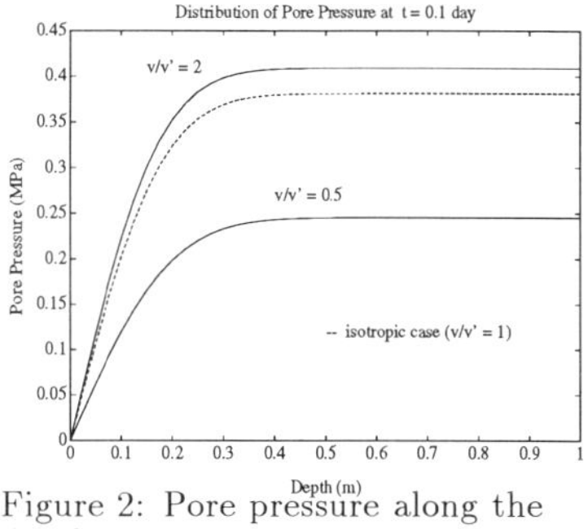

Figure 2: Pore pressure along the depth at $t = 0.1$ day.

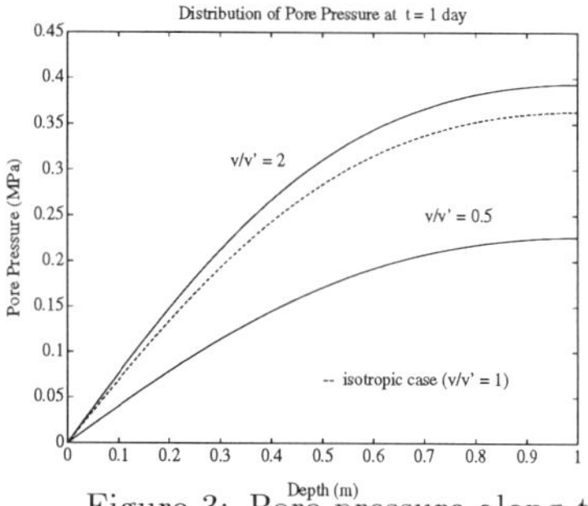

Figure 3: Pore pressure along the depth at $t = 1$ day.

Example

To illustrate the effect of anisotropy, the 1-D consolidation problem in a transversely isotropic material is analyzed. Considering the following data set (Abousleiman *et al.*, 1995): $E = 20.6$ GPa, $\nu = 0.189$, $K_s = 48.2$ GPa, and $M = 15.8$ GPa. For the sake of convenience, only the anisotropy effect caused by anisotropic Poisson's ratios will be investigated in this paper. Therefore, it is assumed that $E = E'$, and $\nu' = \nu/n_\nu$, where n_ν is the ratio of ν/ν'. κ_x is given as 8.64×10^{-6} m^2/MPa $\cdot$ day. For a given n_ν, the transversely isotropic material is fully defined and the other properties of the material can be calculated by the formulae proposed by Cui (1995). Different values of n_ν give different transversely isotropic poroelastic materials. When $n_\nu = 1$, the material is isotropic. By solving the problems with n_ν equal to 1, 2, and 0.5, the effect of anisotropy on the 1-D consolidation problem is demonstrated.

The loading and the geometry of the problem are described as: depth of column is 1 m; normal pressure on top is $P_0 = 1$ MPa; and, the top surface is drained, hence $\beta = 0$. It is supposed that the isotropic plane coincides with the y-z plane (see Figure 1 (a)). The distributions of pore pressure along the depth at two different times are shown in Figures 2 and 3. For the materials defined above, the results show that the distribution of pore pressure varies with the value of n_ν. For $n_\nu = 2$, the pore pressure distribution is not far from the isotropic case. On the other hand, for $n_\nu = 0.5$, the distribution of pore pressure is much different from that for the isotropic case. The material anisotropy also affects the subsidence at the top surface. This effect can be seen in Figure 4. The greater the value of n_ν, the greater the subsidence. The difference is quite significant, especially for the case $n_\nu = 0.5$.

Conclusion

The material anisotropy effect on the 1-D consolidation in a transversely isotropic medium is observed. For the material with $n_\nu = 0.5$, the anisotropy effect is more significant. The pore pressure, displacement, and the rate of subsidence are affected not only by material properties in the isotropic plane, but also the ones related to the direction normal to the plane. Particularly, it should

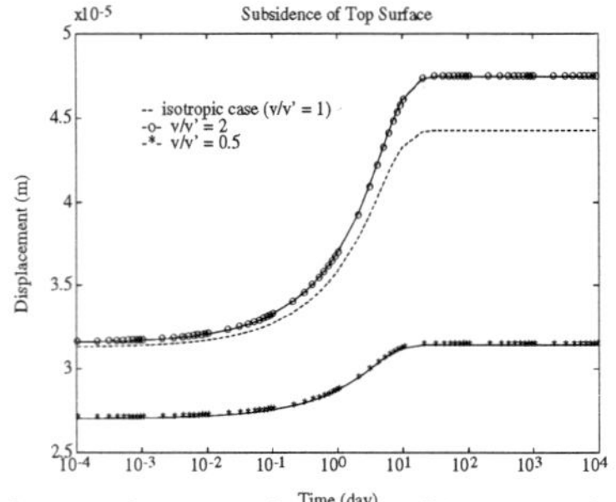

Figure 4: Subsidence of top surface as a function of time.

be pointed out that these anisotropic effects do not exist in Terzaghi's uncoupled consolidation theory. The out-of-plane anisotropic material properties are brought into the analysis only through the coupled nature of poroelasticity theory.

Acknowledgment

The work reported herein was supported by the Rock Mechanics Research Center at The University of Oklahoma, and the Oklahoma Center for Advancement of Science and Technology.

References

Abousleiman, Y., Cheng, A.H-D., Cui, L., Detournay, E., and Roegiers, J.-C., 1995. Mandel problem revisited: Consolidation of a porous transversely isotropic rock. *Géotechnique.* in press.

Aoki, T., Tan, C.P., and Bamford, W.E., 1993. Effects of elastic and strength anisotropy on borehole failures in saturated rocks. *Int. J. Rock Mech. Min. Sci. & Geomech. Abstr.* **30**. 1031-1034.

Biot, M.A., 1941. General theory of three dimensional consolidation. *J. Appl. Phys.* **12**. 155-164.

Biot, M.A., 1955. Theory of elasticity and consolidation for a porous anisotropic solid. *J. Appl. Phys.* **26**. 182-185.

Cheng, A.H-D., 1995. Material coefficients of anisotropic poroelasticity. Submitted to *Int. J. Rock Mech. Min. Sci. & Geomech. Abstr.*

Cui, L., 1995. *Poroelasticity with Application to Rock Mechanics.* Ph.D. Thesis. The University of Delaware.

Schiffman, R.L., Chen, A.T-F., and Jordan, J.C., 1969. An analysis of consolidation theory. *J. Soil Mech. Foundations Div., ASCE.* **95**, SM1. 285-312.

Terzaghi, K., 1923. Die berechnung der durchlassigkeitsziffer des tones aus dem verlauf der hydrodynamischen spannun-gerscheinungen. *Sitzungsber. Akad. Wissen Math. Naturwiss. Kl. Abt. IIa.* **132**. 105-124.

Thompson, M., and Willis, J.R., 1991. A reformulation of the equations of anisotropic poroelasticity. *J. Appl. Mech., ASME.* **58**. 612-616.

Normal Modes of a Poroelastic Soil Layer on a Rigid Bed Rock

Terumi Touhei [1] and Nimal Rajapakse [2] , Member, ASCE

Abstract

Normal modes of a three dimensional poroelastic layer of finite thickness resting on a rigid base are examined. The period equation for normal modes is established by using the transfer matrix method incorporating the Stokes-Helmholtz decomposition and Fourier-Bessel expansion. The roots of the period equation are examined to investigate the influence of poroelastic material properties on the dynamic response.

Introduction

It is important to study the normal modes of a poroelastic layer and examine the influence of porosity, internal friction and interaction between solid matrix and pore fluid on the wave propagation. In this paper, the period equation for normal modes of a three dimensional poroelastic layer is solved. The formulation is based on the transfer matrix method incorporating the Stokes-Helmholtz decomposition and Fourier-Bessel transform. The Newton-Raphson scheme is intoroduced to directly obtain the complex roots of the period equation.

Formulation

The governing equation for wave propagation in a poroelastic medium can be expressed as (Biot, 1962),

$$(\lambda+\mu+\alpha^2 M)\nabla\nabla\cdot\boldsymbol{u}(\boldsymbol{x},t)+\mu\nabla^2\boldsymbol{u}(\boldsymbol{x},t)+\alpha M\nabla\nabla\cdot\boldsymbol{v}(\boldsymbol{x},t)=\partial_t^2[\rho\boldsymbol{u}(\boldsymbol{x},t)+\rho_f\boldsymbol{v}(\boldsymbol{x},t)] \tag{1}$$

$$\alpha M\nabla\nabla\cdot\boldsymbol{u}(\boldsymbol{x},t)+M\nabla\nabla\cdot\boldsymbol{v}(\boldsymbol{x},t)=\partial_t^2[\rho_f\boldsymbol{u}(\boldsymbol{x},t)+m\boldsymbol{v}(\boldsymbol{x},t)]+b\partial_t\boldsymbol{v}(\boldsymbol{x},t) \tag{2}$$

[1] Assistant Professor, Science Univ. of Tokyo, 2641 Yamazaki Noda City 278, Japan
[2] Professor, Univ. of Manitoba, Winnipeg, Canada R3T 5V6

where $\boldsymbol{u}$ is the displacement of the solid matrix, $\boldsymbol{v}$ the average fluid displacement relative to the solid matrix, $\boldsymbol{x}$ the position in the space, t the time, and ∂ the partial differential operator whose subscript refers to the parameter for the differentiation. In addition, the material parameters λ, μ, ρ, ρ_f, m, α and M are defined by Biot (1962).

Based on the Stokes-Helmholtz decomposition and Fourier-Bessel transform, the displacement of the solid matrix and the fluid displacement are expressed as,

$$\boldsymbol{u}(\boldsymbol{x},t) = \sum_{m=-\infty}^{\infty}\sum_{n=0}^{\infty}[U_{zk_n}^m(z,t)\boldsymbol{R}_{k_n}^m(r,\phi)+U_{rk_n}^m(z,t)\boldsymbol{S}_{k_n}^m(r,\phi)+U_{rk_n}^m(z,t)\boldsymbol{T}_{k_n}^m(r,\phi)] \tag{3}$$

$$\boldsymbol{v}(\boldsymbol{x},t) = \sum_{m=-\infty}^{\infty}\sum_{n=0}^{\infty}[V_{zk_n}^m(z,t)\boldsymbol{R}_{k_n}^m(r,\phi)+V_{rk_n}^m(z,t)\boldsymbol{S}_{k_n}^m(r,\phi)+V_{\phi k_n}^m(z,t)\boldsymbol{T}_{k_n}^m(r,\phi)] \tag{4}$$

where (r,ϕ) denote the horizontal cylindrical coordinates, z the vertical coordinate, k_n the discrete wavenumber, m the circumferential order number and $\boldsymbol{R}_{k_n}^m$, $\boldsymbol{S}_{k_n}^m$ and $\boldsymbol{T}_{k_n}^m$ are the surface vector harmonics (Olson et al, 1984). The representations of displacement field shown in Eqs. (3) and (4) are based on the discrete wavenumber method introduced by Olson et al.

The components of a traction vector on the z plane in the wavenumber domain can be expressed as,

$$\begin{aligned} P_{zk_n}^m &= (\lambda+2\mu+\alpha^2 M)\partial_z U_{zk_n}^m - k_n(\lambda+\alpha^2 M)U_{rk_n}^m \\ &\quad +\alpha M\partial_z V_{zk_n}^m - \alpha M k_n V_{rk_n}^m \\ P_{rk_n}^m &= \mu\partial_z U_{rk_n}^m + \mu k_n U_{zk_n}^m \\ P_{\phi k_n}^m &= \mu\partial_z U_{\phi k_n}^m \end{aligned} \tag{5}$$

Likewise, the component of pore fluid pressure can be expressed as,

$$P_{k_n}^m = -\alpha M(\partial_z U_{zk_n}^m - k_n U_{rk_n}^m) - M(\partial_z V_{zk_n}^m - k_n V_{rk_n}^m) \tag{6}$$

A transfer matrix method can be developed for a layered poroelastic medium using the quantities $U_{zk_n}^m$, $U_{rk_n}^m$, $U_{\phi k_n}^m$, $V_{zk_n}^m$, $P_{zk_n}^m$, $P_{rk_n}^m$, $P_{\phi k_n}^m$ and $P_{k_n}^m$ for each layer. Note that the components of relative displacement of fluid, $V_{rk_n}^m$ and $V_{\phi k_n}$ are not required in the transfer matrix method, since these quantities depend on $U_{zk_n}^m$,$U_{rk_n}^m$ and $U_{\phi k_n}^m$ in the frequency domain.

A period equation of the following form can be obtained from the transfer matrix method for a layered poroelastic medium,

$$F(k,\omega) = F_L(k,\omega)F_R(k,\omega) = 0 \tag{7}$$

where k is a wavenumber, ω is a circular frequency, F_L is a characteristic function for the Love waves and F_R is that for the Rayleigh waves. As can be seen in Eq. (7), the period equation can be factored into two equations related to the Love waves and Rayleigh waves.

The solutions of the period equation for a poroelastic medium are complex quantities due to the presence of internal friction which is denoted by coefficient b in Eq. (2). To obtain the complex roots of the period equation, the following Newton-Raphson scheme is introduced in the present study,

$$\omega_{j+1} = \omega_j - \frac{F(k,\omega_j)}{F'(k,\omega_j)} \qquad (j = 1, 2, ...) \tag{8}$$

where ω_j is an approximate complex eigenfrequency for given wavenumber k (real). The derivative of $F(k,\omega_j)$ denoted by $F'(k,\omega_j)$ can be obtained from the following,

$$F'(k,\omega_j) = \frac{1}{2\pi i}\oint \frac{F(k,\omega)}{(\omega-\omega_j)^2}d\omega \tag{9}$$

Note that the integration in Eq. (9) is carried out in the complex ω plane.

Numerical Examples

Solutions of the period equation are examined for the model shown in Figure 1, in which a homogeneous poroelastic layer is overlying a rigid bed rock.

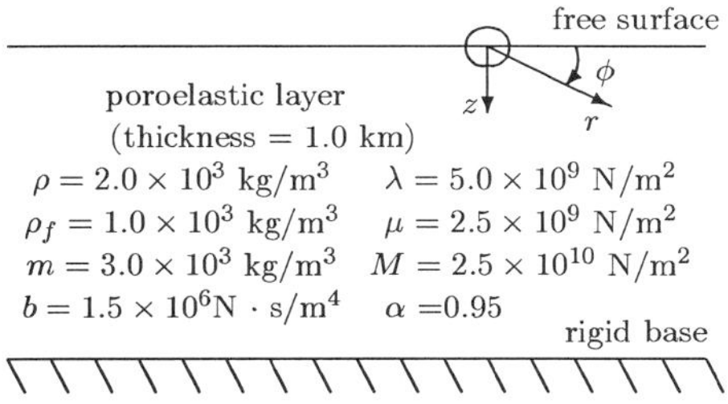

Figure 1. Poroelastic layer resting on rigid bed rock

The dispersion property of the normal modes are initially investigated. Dispersion curves are shown in Figure 2, in which the relationship between phase velocity and frequency of normal modes is described. The phase velocity of the normal modes is equal to $\mathrm{Re}(\omega)/k$ and is normalized so that the normalized unit phase velocity corresponds to the shear wave velocity in an ideal elastic medium, (i.e., $\sqrt{\mu/\rho}$).

As can be seen in Figure 2, the phase velocity of one of the normal modes becomes asymptotic to the normalized phase velocity of unity. This mode corresponds to the fundamental Rayleigh wave, because the phase velocity is lowest when compared to the other modes. Note that the phase velocity of this fundamental Rayleigh wave mode is higher than that of the Rayleigh wave in an ideal elastic medium. It is also found from Figure 2 that the phase velocity of some normal modes including Love waves converges at a normalized velocity of about 1.1 in a certain high frequency range. This indicates that the body waves can be synthesized from the superposition of normal modes. The velocity at which these modes converge is the shear wave velocity of a poroelastic medium, which is larger than the shear wave velocity of an ideal elastic medium. The dispersion curves also show that wave velocity in a poroelastic medium is higher than that in an ideal elastic medium.

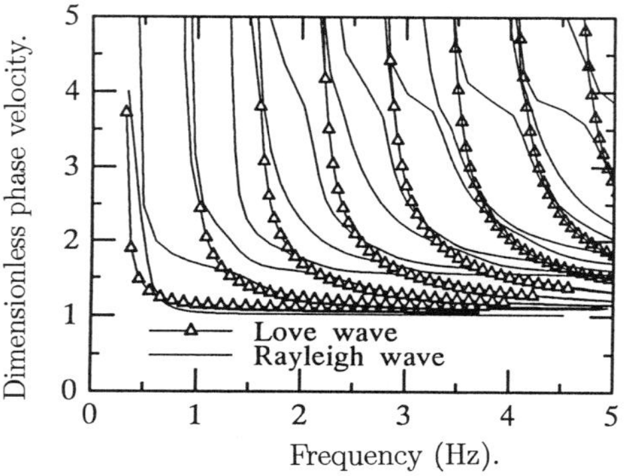

Figure 2. Dispersion property of the normal modes.

Conclusion

In this paper, the properties of normal modes of a poroelastic layer on a rigid base are investigated. The period equation for a poroelastic layered medium can be solved by using the Newton-Raphson scheme. Dispersion curves indicate that the wave velocity of normal modes in a poroelastic medium is higher than that in an ideal elastic medium.

References

Biot, M. A. (1962). "Mechanics of deformation and acoustic propagation in porous media." *J. Appl. Phys.*, **33**, 1482-1498.

Olson, A. H., J. A. Orcutt and G. A. Frazier (1984). "The discrete wavenumber/finite element method for synthetic seismograms." *Geophys. J. R. astr. Soc.*, **77**, 421-460.

FURTHER APPLICATION OF DYNAMIC POROELASTICITY TO GEOTECHNICAL ENGINEERING VIA BEM

Jianming Chen *
Department of Civil Engineering, University of Delaware, Newark, Delaware 19716-3210, U.S.A.

ABSTRACT

Based on the boundary element formulation for Poro-Thermo-Elasticity (Chen, 1992), a purely poroelastodynamic BEM formulation is presented in the Laplace domain. Several geotechnical problems of considerable importance are solved, which includes the transient response of a long cylindrical cavity in a poroelastic medium under different kinds of boundary conditions and vibration analysis of a transient strip load on the surface of a fluid-saturated porous half-space. Time history and spacial variations of displacement, stress, pore pressure and fluid flux in poroelastic medium are shown graphically.

INTRODUCTION

Many practical engineering problems involve dynamic analysis of poroelasticity , which has applications in geotechnical engineering, earthquake engineering, geophysics, acoustics and biomechanics. The first model of dynamic poroelasticity was proposed by Biot (1956a,b). Its essential correctness has been confirmed both theoretically and experimentally. Recently a frequency domain Boundary Element Method for dynamic poroelasticity was attemped by Cheng and Badmus (1991), which is based on fundamental solutions for thermoelasticity rather than poroelasticity. The purpose of this paper is to present a Laplace Transform domain BEM for dynamic poroelasticity and its implementation for the calculation in Geotechnical Engineering.

GOVERNING EQUATIONS

Following Zienkiewicz *et al.* (1984) and Cheng *et al.* (1991), Biot's equations governing the behavior of a poroelastic medium can be written in the Laplace transform domain in the form of the equations of motion and an energy equation, which involve four independent variables u_i and p, as

$$(\lambda + \mu)\tilde{u}_{j,ij} + \mu\tilde{u}_{i,jj} - \alpha_1\tilde{p}_{,i} - \rho_1 s^2\tilde{u}_i + \tilde{f}_i = 0 \qquad (1.a)$$

$$\zeta\tilde{p}_{,ii} - \frac{s}{Q}\tilde{p} - \alpha_2 s\tilde{u}_{i,i} + \tilde{\gamma} = 0 \qquad (1.b)$$

where $i, j = 1, 2, 3$, the tilde denotes the Laplace transformation, $\alpha_1 = \alpha_2 = \alpha - \rho_f s\zeta$, $\rho_1 = \rho - \rho_f^2 s\zeta$, $\zeta = (\frac{1}{\kappa} + ms)^{-1}$ and s is the Laplace transform parameter. Additionally, p denotes the excessive fluid pore pressure, u_i is the displacement of the solid skeleton, the fluid mass flux vector is defined as q_i. The elastic constants λ and μ are termed drained Lamé constants, κ is the permeability coefficient. The quantity ρ_f represents the fluid density, where n is the porosity. ρ is the density of the solid-fluid mixture. Meanwhile, f_i and γ are the body force and fluid source, respectively. In addition α and Q are Biot's parameters accounting for compressibility in the two-phase material.

TRANSFORM DOMAIN INTEGRAL EQUATION AND NUMERICAL PROCEDURE

* **Structural Engineer/ STV Group/ Seelye Stevenson Value Knecht, Inc., 225 Park Ave. Sourth, New Yord, NY 10003-1604, U.S.A.**

The transform domain coupled boundary integral equation by Chen (1992) through the Green's function method are presented as follows:

$$c_{ij}\tilde{u}_i(\xi,s) = \int_S [\tilde{G}^*_{ij}(x,\xi;s)\tilde{t}_i(x,s) - \tilde{F}^*_{ij}(x,\xi;s)\tilde{u}_i(x,s)]dS(x) \tag{2}$$

where c_{ij} denotes the shape coefficient for the boundary at ξ and reduces to $\delta_{ij}/2$ for the smooth surface.

The generalized displacement and traction vectors

$$\tilde{u}_i = (\tilde{u}_1,\ \tilde{u}_2,\ \tilde{u}_3,\ \tilde{p})^\top \quad \tilde{t}_i = (\tilde{t}_1,\ \tilde{t}_2,\ \tilde{t}_3,\ \tilde{q})^\top \tag{3}$$

Displacement, traction and flux kernels are defined by, respectively

$$\begin{gathered} \tilde{G}^*_{\alpha j}(x,\xi;s) = G^*_{\alpha j}(x,\xi;s) \quad \tilde{G}^*_{4j}(x,\xi;s) = -G^*_{4j}(x,\xi;s) \\ \tilde{F}^*_{\alpha j} = [(\lambda G^*_{\gamma j,\gamma} + \alpha s G^*_{4j})\delta_{\alpha\beta} + \mu(G^*_{\alpha j,\beta} + G^*_{\beta j,\alpha})]n_\beta \quad \tilde{F}^*_{4j} = \zeta G^*_{4j,n} - \rho_f s\zeta G^*_{\alpha j} n_\alpha \end{gathered} \tag{4}$$

while surface traction and flux

$$\tilde{t}_\alpha = [(\lambda\tilde{u}_{\gamma,\gamma} - \alpha\tilde{p})\delta_{\alpha\beta} + \mu(\tilde{u}_{\alpha,\beta} + \tilde{u}_{\beta,\alpha})]n_\beta \quad \tilde{q} = -\zeta\tilde{p}_{,n} - \rho_f s^2\zeta\tilde{u}_\alpha n_\alpha \tag{5}$$

In which i,j =1,2,3,4; α,β,γ=1,2,3 for three-dimensional case. $G^*_{\alpha j}$, G^*_{4j} are the fundamental solutions of the adjoint differential equation for eqns.(1a,b). One finds that $G^*_{ij} = s\tilde{G}_{ij}$, $G^*_{4i} = -s\tilde{G}_{i4}$, $G^*_{j4} = -s\tilde{G}_{4j}$, $G^*_{44} = s\tilde{G}_{44}$ with $i,j = 1,2,3$, where $\tilde{G}_{ij}$, $\tilde{G}_{i4}$, $\tilde{G}_{4j}$, and $\tilde{G}_{44}$ are fundamental solotions for eqns.(1a,b), which is given in Chen (1992). For the two dimensional problems the limits on the subscript indices i,j and α,β,γ shrinking by one and with boundary surface S reducing to the boundary curve.

In order to exploit the above integral equations for the solution of boundary value problems, various computational aspects such as spatial and time discretizations, kernel singularity, numerical integration, numerical inversion of Laplace transform, assembly of the system equations and solution algorithms must be introduced. This BEM procedure is currently well known in the literature, therefor we skip the details in current paper.

GEOTECHNICAL APPLICATIONS

Explosions Inside a Cylindrical Cavity : Consider a circular cylindrical cavity of radius $R = 1.0$ embedded in a infinite space. Two kinds of hydraulic boundary conditions at the surface of the cavity are considered. *i.e.* a permeable surface and an impermeable surface. The cavity is subjected to (1) a suddenly applied and maintained internal radial pressure of intensity p_0, (2) triangle pulse with a linear rise and fall for the time range of two units. The maximum strength attained by the explosion is p_0. For this particular problem, the dimensionless variables are $\xi = r/R$, $\tau = \frac{t}{R}\sqrt{\frac{\mu}{\rho}}$ and the following nondimensional material parameter are introduced *i.e.* $\lambda^* = \frac{\lambda}{\mu}$, $\mu^* = 1.0$, $Q^* = \frac{Q}{\mu}$, $\rho^*_f = \frac{\rho_f}{\rho}$, $\rho^* = 1.0$, $m^* = \frac{\rho^*_f}{n}$, $b^* = \frac{\eta R}{k\sqrt{\rho\mu}}$ and used as: $\lambda^* = 2$, $Q^* = 23$, $b^* = 16$, $\alpha = 0.984$, $\rho^*_f = 0.5$, $m^* = 1.25$. It is seen, from Figures (1a) and (1b), that radial displacement and hoops tress show a higher value for the impermeable surface than for the permeable surface. Moreover they reach their maximum value closely at time instants which are nearly independent of the hydraulic boundary condition. It is noted that the radial displacement for loading case (2) decreases rapidly after it reached it's maximum value, and experiences a minor inward displacement before reaching steady state. While the hoop stress is compressive at $\tau = 0$, for loading case (1) it increases with τ. The hoop stress for loading case (2) is zero at $\tau = 0$ and reaches its compressive maximum near $\tau = 1$. Thereafter both of them increase with time and reach their peak tensile values near $\tau = 3, \tau = 2$ respectively then decrease rapidly. Static state is reached near $\tau > 10$ for all types of loading. The pore pressure histories at $r = R$ are shown in Figure (1c). For loading case (1) pore pressure has a finite value at $\tau = 0.$, then decreases rapidly and suction is noted, afterwards a slight change in sign occurs and finally pore pressure is nearly zero. For loading case (2), the initial pore pressure is zero, it increases rapidly from $\tau > 0$ and after reaching the positive maximum value it's behavior is much similar to that for the loading case (1). In Figure (1d), the dynamic fluid flux value at permeable boundary are seen to undergo a pronounced oscillation. For loading case (1), initially the fluid flux is infinite and outward (into cavity). It decreases rapidly with time and changes to inward flux (into media). For loading case (2), initially the flux is outward with finite value and as time goes on, it increases quickly to it's maximum then decreases quickly to it's minimum value (into media flux), then relatively rapidly reaches the quiescent state after one or two oscillations. The Figure (1d) illustrates that the transient effects are more prominent for flux due to loading case (2) than that for the loading case (1).

Transient Strip Load on the Surface of a Half-Space : Consider the plain strain condition, with non-dimensional material constants: $\kappa^* = 1.0$, $\lambda^* = 0.1715$, $\mu^* = 0.3007$, $Q^* = 0.37422$, $\rho^* = 1.0$, $\rho_f^* = 0.4325$, $m^* = 2.3006$, $\alpha = 0.779$. Therefore the non-dimensional propagation speeds of the first compressional wave, second compressional (diffusive) wave and shear wave are 1.0, 0.3682 and 0.5716 respectively. The normal stress of magnitude $\sigma_0 = 1$ uniformly distributed over a length $2b = 2$ over a porous half plane is suddenly applied and maintained. ξ_1 is vertical axis from center of strip into half space while ξ_2 is horizontal axis. Figure (2a) illustrate the vertical displacement at the centerline points ($\xi_1 = 0.0$, $\xi_2 = 0.0$) and one unit below the centerline ($\xi_1 = 1.0$, $\xi_2 = 0.0$) respectively. Next, Figure (2b) contains the pore pressure time history for points on the centerline at points ($\xi_1 = 1.0$, $\xi_2 = 0.0$) and ($\xi_1 = 2.0$, $\xi_2 = 0.0$) below the surface. Surface fluid flux time history under the middle of the strip loading are demonstrated in Figure (2c). While Figure (2d) depicts nondimensional pore pressure versus non-dimensional spatial variation along the center line at various times $\tau = 2.0, 2.4, 3.6$. As mentioned earlier, the evaluation of the accuracy of the method is more difficult due to lack of appropriate analytical solution, but the results seem reasonable.

CONCLUSION

In this paper a transform domain BEM for dynamic poroelasticity is derived in a way that permits an easy discretization and implementation in a numerical code. Next step of current research is to develop an efficient time domain BEM for dynamic poroelasticity.

REFERENCES

Biot. M.A. (1956b,c), 'Theory of Propagation of Elastic Waves in a Fluid-Saturated Porous Solid, I. Low Frequency Range, II. Higher Frequency range', *J. Acoust. Soc. Am.*, V28, pp. 168-178, 179-191

Cheng, A. H-D. and Badmus, T. (1991), "Integral Equation for Dynamic Poroelasticity in Frequency Domain with BEM Solution", *J. Engrg. Mech.*, ASCE, Vol. 117, No.5, pp. 1136-1158

Zienkiewicz, O.C. and Shiomi, T. (1984), 'Dynamic Behavior of Saturated Porous Media; the Generalized Biot Formulation and its Numerical Solution', *Int. J. Num. Meth. Engrg.*, V8, pp. 71-96.

Chen, J., (1992), 'Fundamental Solutions and Boundary Element Formulations for Dynamic Problems of Poroelasticity and Thermoelasticity', Ph.D. Dissertation, State University of New York at Buffalo.

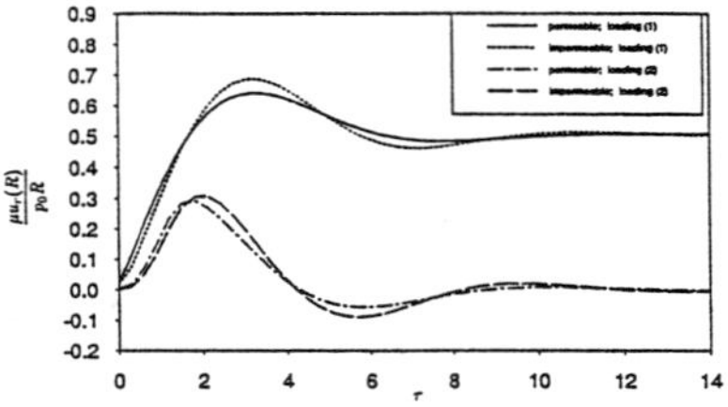

Figure 1a Radial Displacement Time History at Cavity Surface ($R = 1.0$) for $0 \le \tau \le 14$

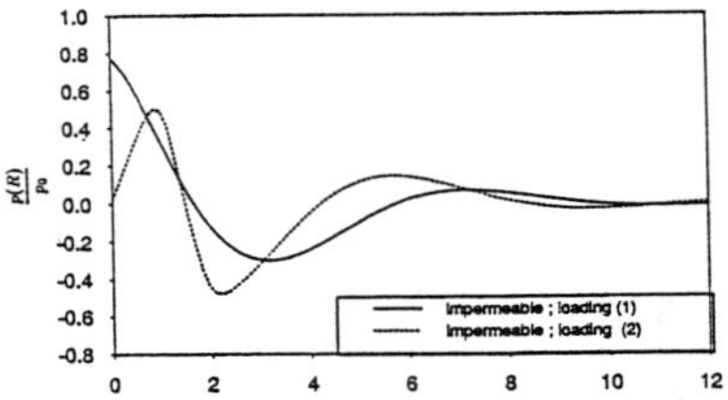

Figure 1c Pore Pressure Time History at Cavity Surface ($R = 1.0$) for $0 \le \tau \le 12$

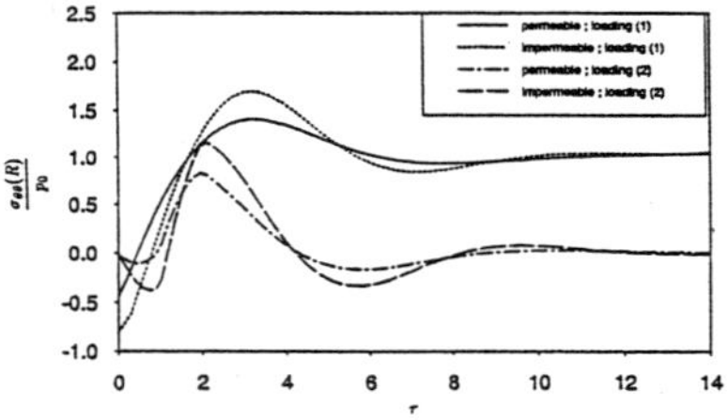

Figure 1b Hoop Stress Time History at Cavity Surface ($R = 1.0$) for $0 \le \tau \le 14$

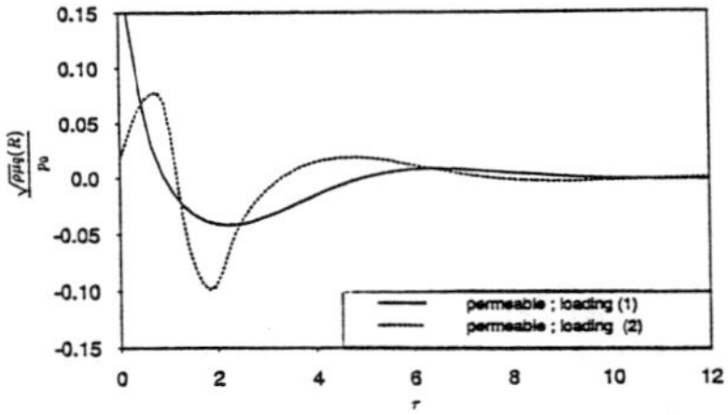

Figure 1d Fluid Flux Time History at Cavity Surface ($R = 1.0$) for $0 \le \tau \le 12$

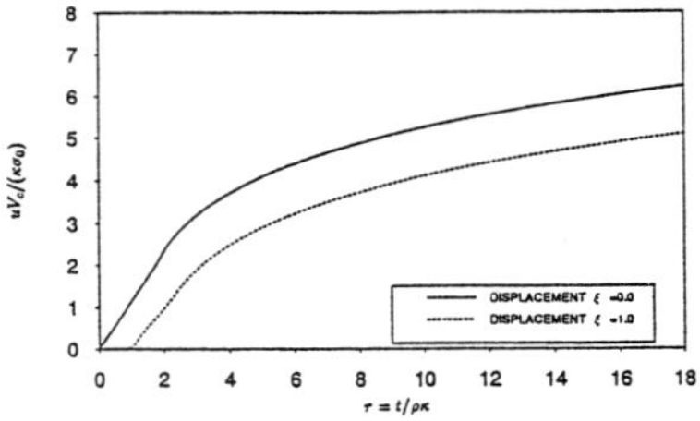

Figure 2a Displacement Time History at $\xi = 0.0, 1.0$ for $0 \le \tau \le 18$

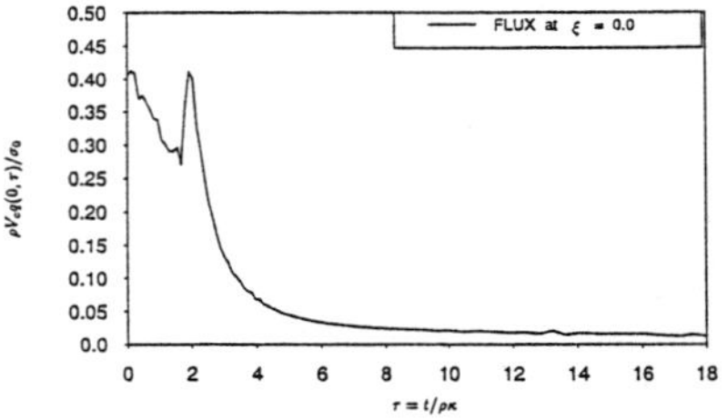

Figure 2c Surface Fluid Flux Time History Under the Middle of the Strip Loading $0 \le \tau \le 18$

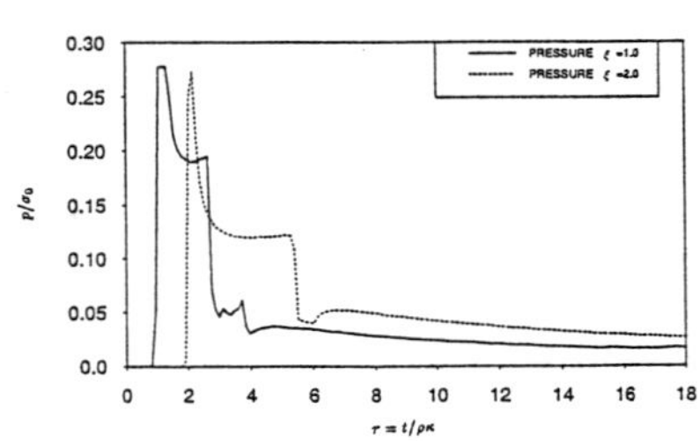

Figure 2b Pore Fluid Pressure Time History at $\xi = 1.0, 2.0$ $0 \le \tau \le 18$

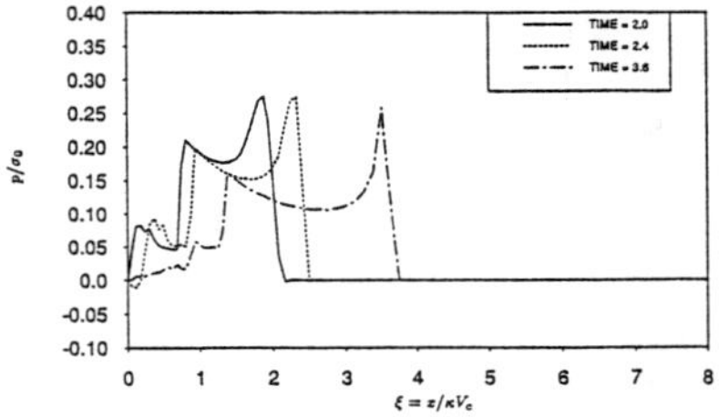

Figure 2d Spatial Variation along Centerline in Pore Fluid Pressure p/σ_0 at τ=2.0,2.4,3.6

Coupled and Uncoupled Poroelastic Solutions to Land Subsidence due to Groundwater Withdrawal

Giuseppe Gambolati[1], Mario Putti[2] and Pietro Teatini[2]

Abstract

Coupled and uncoupled finite element solutions to predict land subsidence due to groundwater withdrawal are presented and discussed for a realistic pumped aquifer system with hydrogeological properties typical of the unconsolidated Quaternary sediments of the Po river plain in Italy. The results show that both pressure drawdown and land settlement are almost insensitive to coupling between the three-dimensional flow and stress fields. By distinction, the horizontal displacements at the ground surface exhibit a limited difference, with the uncoupled solution approaching faster the ultimate steady state value at some distance from the pumping well.

Introduction

The basic linear poroelastic theory controlling pressure drawdown and soil displacement in aquifer systems during unsteady groundwater pumping couples the three-dimensional flow and stress fields through a set of interrelated equations [Biot, 1941; Verruijt, 1969; Detournay and Cheng, 1993]. However, uncoupled equations are generally used to predict the actual land subsidence due to fluid (water, gas, oil) withdrawal [Gambolati and Freeze, 1973; Gambolati *et al.*, 1991; Helm, 1975; Rivera, 1990; van Hasselt, 1992]. Theoretical evidence that coupling between flow and stress is weak in anthropically subsiding basins has also been provided [Gambolati, 1977, 1992; Debbarh, 1988]. In this note we present and discuss the numerical results obtained from both coupled and uncoupled three-dimensional models of land subsidence in aquifer systems subject to a unit pressure decline in the pumping well. It is shown that both pressure drawdown and land settlement are practically the same irrespective of the approach used. A more appreciable difference affects the rate by which the horizontal displacement at the ground surface achieves the steady state value. Due to linearity of the theory, the extension of the results from the single well analysis to any arbitrary distribution of abstraction wells is straightforward.

Coupled and uncoupled poroelastic model of anthropic land subsidence

Following the development of Gambolati *et al.* [1996] the coupled equations of soil

[1]Professor; [2]Researcher. Department of Mathematical Models for Applied Sciences, University of Padova, via Belzoni 7, 35131 Padova, Italy

consolidation in an isotropic medium may be written as:

$$(\lambda + G)\frac{\partial \epsilon}{\partial x_i} + G\nabla^2 u_i = \left(1 - \frac{E}{E_r}\right)\frac{\partial p}{\partial x_i} \qquad i = 1,2,3 \tag{1}$$

$$\nabla\left(\frac{K_{i,j}}{\gamma}\nabla p\right) = \left[n\beta + c_{br}(1-n)\right]\frac{\partial p}{\partial t} + \frac{\partial \epsilon}{\partial t} \tag{2}$$

where x_i, $i = 1,2,3$ are the Cartesian coordinate directions, u_i, $i = 1,2,3$ are the components of the position vector, p is the incremental pore pressure, ϵ is the volume strain or dilatation, λ, G, E are the Lamé constant, and the shear and Young moduli of the porous skeleton, E_r and c_{br} are the Young modulus and the volumetric compressibility of the individual grains , β and γ are the compressibility and the specific weight of water, n is the medium porosity, and $K_{i,j}$ is the medium hydraulic conductivity tensor.

Setting $\epsilon = \alpha(1-\beta')p$, where $\alpha = 1/(\lambda + 2G) = [(1-2\nu)(1+\nu)]/[E(1-\nu)]$ is the vertical medium compressibility, and $\beta' = E/E_r$, leads to a new uncoupled expression of the flow eq. (2):

$$\nabla\left(\frac{K_{i,j}}{\gamma}\nabla p\right) = \left[n\beta + c_{br}(1-n) + \alpha(1-\beta')\right]\frac{\partial p}{\partial t} \tag{3}$$

Eqs. (1) and (3) represent the uncoupled three-dimensional form of the consolidation theory. Both coupled and uncoupled models must be supplemented by appropriate initial and boundary conditions [Gambolati *et al.*, 1996].

Comparison of coupled and uncoupled solutions

The coupled (eqs. (1)-(2)) and uncoupled (eqs. (1)-(3)) models for land subsidence prediction due to groundwater withdrawal are solved in a non-homogeneous aquifer system with soil compressibility α which varies according to sediment depth and medium lithology, as is shown in Gambolati *et al.* [1991]. The constitutive relationship used for α is representative of the Quaternary and Pliocene sediments underlying the Po river plain and the Upper Adriatic Sea. Poisson's ratio is taken equal to 0.25. The basement of the system is set at a depth $H = 20000$ m while the outer boundary is located 30000 m from a fully penetrating well, where an instantaneous pressure drawdown $p = -1$ kg/m^2 is prescribed at time $t = 0$. The pumped aquifer is 20 m thick and is overlain/underlain by an alternating sequence of sandy and clayey formations. We assume $K_{i,j} = K = 10^{-4}$ m/s in the aquifer and the other sandy units, $K = 10^{-7}$ m/s in the clayey units. Two problems are addressed by assuming an average aquifer depth h equal to 100 m and 500 m. Both the coupled and the uncoupled models are solved by finite elements taking advantage of the axi-symmetric nature of the problems and the results are shown and compared in the sequel.

Figures 1 and 2 provide the behavior of p and of land subsidence (u_3 at $x_3 = 0$) for the two h values at different times as obtained from the numerical models. Figure 3 gives the horizontal ground surface displacement consistent with the outcome of Figures 1 and 2. The solid and dashed profiles represent the coupled and uncoupled solutions, respectively. Note that the aquifer pressure drawdown and the land subsidence are practically the same for the two depths of stress source irrespective of the model (coupled or uncoupled) used. By distinction, note in Figure 3 that the uncoupled $u_r = u_1 = u_2$ exhibits some difference from the coupled u_r during an intermediate time at a distance r from the well where u_r achieves the largest values. The difference in the rate by which u_r approaches the steady state is much less pronounced far away from the strength source.

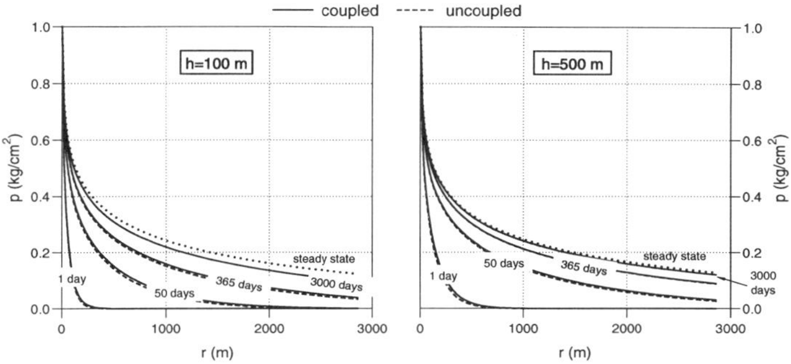

Figure 1: Comparison of coupled and uncoupled aquifer pressure decline p vs distance r from the pumping well.

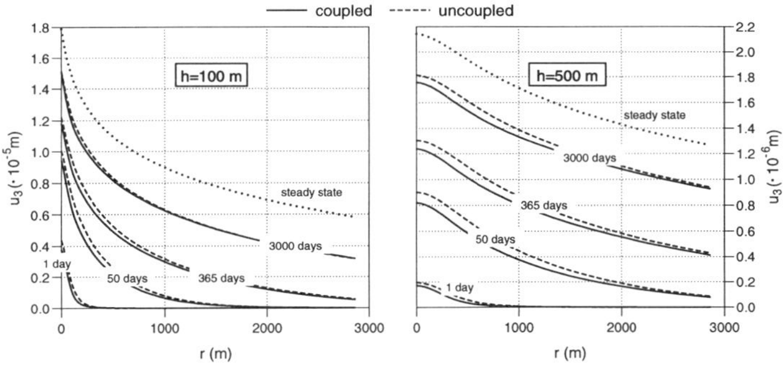

Figure 2: Comparison of coupled and uncoupled land subsidence u_3 (x_3=0) vs distance r from the pumping well.

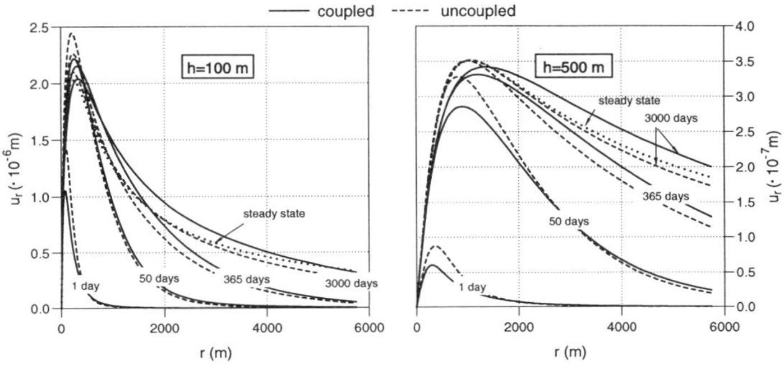

Figure 3: Comparison of coupled and uncoupled horizontal ground surface displacement u_r (x_3=0) vs distance r from the pumping well.

Conclusion

The previous analysis shows that the uncoupled poroelastic model of flow and stress predicts quite satisfactorily the transient aquifer pressure response and the resulting land subsidence over a pumped aquifer basin of realistic size. However, the results also indicate that the use of the coupled approach may be required for the accurate prediction of the time rate by which the surface horizontal displacement achieves the steady state solution.

Acknowledgment This research was supported by the EC Environment Research Programme (contract: EV5V-CT94-0498, Climatology and Natural Hazards).

References

Biot, M. A. (1941). General theory of three-dimensional consolidation. *J. Appl. Phys.*, **12**(2), 155–164.

Debbarh, A. (1988). *Consolidation of Elastic Saturated Soil due to Water Withdrawal by Numerical Laplace-Fourier Inversion Methods.* Ph.D. thesis, Dept. of Civil and Min. Eng., Univ. of Minn., Minneapolis.

Detournay, E. and Cheng, A. H. D. (1993). Fundamentals of poroelasticity. In C. Fairhurst, editor, *Comprehensive Rock Engineering: Principles, Practice & Projects, Vol. II, Analysis and Design Method*, chapter 5, pages 113–170. Pergamon Press.

Gambolati, G. (1977). Deviations from Theis solution in aquifers undergoing three-dimensional consolidation. *Water Resour. Res.*, **13**(1), 62–68.

Gambolati, G. (1992). Comment on "Coupling versus uncoupling in soil consolidation". *Int. J. Numer. Anal. Methods Geom.*, **16**(11), 833–837.

Gambolati, G. and Freeze, R. A. (1973). Mathematical simulation of the subsidence of Venice. 1. Theory. *Water Resour. Res.*, **9**(3), 721–733.

Gambolati, G., Ricceri, G., Bertoni, W., Brighenti, G., and Vuillermin, E. (1991). Mathematical simulation of the subsidence of Ravenna. *Water Resour. Res.*, **27**(11), 2899–2918.

Gambolati, G., Putti, M., and Teatini, P. (1996). Land subsidence. In V. Singh, editor, *Hydrology of Disasters*, chapter 10. Kluwer Academic P.C., Dordrecht. To appear.

Helm, D. C. (1975). One-dimensional simulation of aquifer system compaction near Pixley, California. 1. Constant parameters. *Water Resour. Res.*, **11**(3), 465–478.

Rivera, A. (1990). *Modèle hydrogéologique quasi-tridimensionnel non-linéaire pour simuler la subsidence dans les systèmes aquifères multicouches. Cas de Mexico.* Ph.D. thesis, Ecole des Mines de Paris-CIG.

van Hasselt, J. P. (1992). Reservoir compaction and surface subsidence resulting from oil and gas production. *Geologie en Mijnbouw*, **71**, 107–118.

Verruijt, A. (1969). Elastic storage of aquifers. In R. De Wiest, editor, *Flow Through Porous Media*, pages 331–376, New York. Academic Press.

Fluid Pressure Polarization and Effective Response of Fluid-Saturated Materials with Cavities of Various Shapes

Mark Kachanov[1] and Boris Shafiro[2]

Abstract

Fluid saturated material with cavities of various shapes filled with compressible fluid is considered. Fluid pressures in different cavities induced by externally applied loads, will be different, depending on cavity shapes and on cavity orientations with respect to the applied loads. This phenomenon ("pressure polarization") is coupled with stress interactions between cavities and with the effective elastic properties. Results are obtained in the undrained approximation.

Introduction

We consider a linear elastic solid containing pores of various shapes filled with non-viscous compressible fluid. The externally applied stress $\boldsymbol{\sigma}$ induces fluid pressures q in cavities that depend on cavity orientations with respect to $\boldsymbol{\sigma}$ and cavity shapes. This *pressure polarization* (differences between q in different cavities) is coupled with stress interactions between cavities and with the overall elastic response.

The effective elastic properties of fluid-saturated materials were first examined by (Budiansky and O'Connell 1976). Their analysis was limited to narrow, crack-like cavities and to the case of the isotropic overall properties (random crack orientation); the problem of fluid pressure polarization was not addressed. (Kachanov 1993) analyzed this problem for an arbitrary orientational distribution of crack-like cavities (anisotropic overall properties) and examined the coupling between stress intensity factors at the tips of fluid-filled cracks and the fluid pressure polarization. The present work gives a general 3-D analysis of materials that covers fluid-filled pores of arbitrary ellipsoidal shapes. In particular, the essential for real microstructures case of *mixtures* of cavities of diverse shapes (pores + cracks, etc.) is considered. For a presentation in detail, see (Kachanov *et al.* 1995; Shafiro and Kachanov 1996).

The undrained approximation is assumed in the present work: the fluid mass in each cavity is constant. If the matrix allows fluid diffusion, this approximation corresponds to the "short time" response. (In the opposite, "fully drained", limit, the effective response coincides with that of the dry solid (Kachanov *et al.* 1994).

[1] Professor, Dept. of Mech. Engineering, Tufts University, Medford MA 02155
[2] Gr. student, Dept. of Mech. Engineering, Tufts University, Medford MA 02155

The problem of the overall strain response of a material with fluid-filled cavities to the applied stress $\boldsymbol{\sigma}$ can be represented as a superposition of the subproblems of Fig. 1: (a) matrix under the uniform stress $\boldsymbol{\sigma}$, containing a cavity loaded at the cavity boundary by the traction $\boldsymbol{\sigma}\cdot\boldsymbol{N}$, where $\boldsymbol{N}$ is the outward to the material (inward the cavity) unit normal to the boundary; (b) matrix with a cavity loaded at the cavity boundary by the traction $-\boldsymbol{\sigma}\cdot\boldsymbol{N}$, stresses vanish at infinity; (c) same as (b) but with tractions $-q\boldsymbol{N}$ at the cavity boundary, where q is the (yet unknown) fluid pressure in the cavity (assumed positive when compressive).

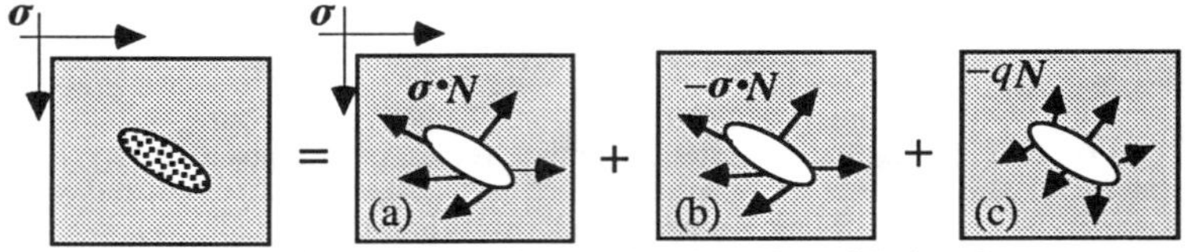

Figure 1. Superposition for a solid with a fluid-filled cavity

The pressure q is to be understood as the response to $\boldsymbol{\sigma}$. If a certain "background" pressure at $\boldsymbol{\sigma}=0$ is present, then q represents its change due to $\boldsymbol{\sigma}$.

We represent the total strain in a dry solid containing a cavity as a sum $\boldsymbol{\varepsilon}=\boldsymbol{s}_0:\boldsymbol{\sigma}+\Delta\boldsymbol{\varepsilon}$ where $\boldsymbol{s}_0$ is the compliance tensor of the matrix; a colon denotes contraction over two indices. Since $\Delta\boldsymbol{\varepsilon}$ is a linear function of $\boldsymbol{\sigma}$, it can be written as $\Delta\boldsymbol{\varepsilon}=\boldsymbol{H}:\boldsymbol{\sigma}$ where the fourth rank tensor $\boldsymbol{H}$ is the *cavity compliance tensor*. The $\boldsymbol{H}$-tensors were found for dry 3-D ellipsoids by (Kachanov *et al.* 1994).

For a *fluid-filled* cavity, the superposition of Fig. 1 reduces the analysis to the case of a *dry* cavity by adding the hydrostatic stress $q\boldsymbol{I}$ to $\boldsymbol{\sigma}$ ($\boldsymbol{I}$ is a unit second rank tensor). The problem is thus reduced to finding the fluid pressure q induced by $\boldsymbol{\sigma}$. In order to find q, we first relate it to the change in fluid density η (with respect to the reference density η_0 at $\boldsymbol{\sigma}=0$) through a constitutive equation for the fluid. We assume that the fluid is linearly compressible (κ is the fluid compressibility):

$$\Delta\eta/\eta_0 \ \left(= -\Delta V/V_{cav}\right) = \kappa q \tag{1}$$

Then, we express the change of cavity volume $\Delta V/V_{cav}$ produced by the total traction $-\boldsymbol{\sigma}\cdot\boldsymbol{N}-q\boldsymbol{N}$ on the boundary of the (dry) cavity in terms of compliance $\boldsymbol{H}$ of the dry cavity. This will yield q as a linear function of $\boldsymbol{\sigma}$.

Fluid pressures induced in 3-D ellipsoidal cavities. Pressure polarization

Being a linear function of $\boldsymbol{\sigma}$, fluid pressure q in a single cavity can be characterized by the second rank *pressure polarization tensor* $\boldsymbol{Q}$: $q=\boldsymbol{Q}:\boldsymbol{\sigma}=Q_{ij}\sigma_{ij}$, where $\boldsymbol{Q}$ depends on the geometry of the cavity and on its orientation with respect to $\boldsymbol{\sigma}$. For a *general ellipsoid* (with the axes $2a_1, 2a_2, 2a_3$ aligned with unit vectors $\boldsymbol{l},\boldsymbol{m},\boldsymbol{n}$), the results of (Kachanov *et al.* 1994) for the $\boldsymbol{H}$-tensor of a dry cavity, combined with the constitutive equation (1) for the fluid, yield

$$\boldsymbol{Q}=-\frac{1}{1+\delta}\frac{1}{R}\left[R_1\,\boldsymbol{ll}+R_2\,\boldsymbol{mm}+R_3\,\boldsymbol{nn}+(1-2\nu_0)\boldsymbol{I}\right] \tag{2}$$

where R_1, R_2, R_3 characterize ellipsoid's geometry; they are combinations of H_{ijkl}, found in terms of Eshelby's tensor by (Kachanov *et al.* 1994): $R_1 = H_{11jj}$, $R_2 = H_{22jj}$, $R_3 = H_{33jj}$ and $R = R_1 + R_2 + R_3$ characterizes the *compressibility* of a dry cavity: its relative volume change under the hydrostatic loading P is PR/E_0 (E_0 and ν_0 are Young's modulus and Poisson's ratio of the matrix).

The dimensionless parameter $\delta = \kappa E_0/R$ plays a key role: it characterizes the matrix/fluid properties and the cavity geometry. It determines the sensitivity q to $\boldsymbol{\sigma}$. In the limit of a highly compressible fluid ("air") or a very stiff matrix, it is large and q induced by $\boldsymbol{\sigma}$ is small in the case of low κ, or a very soft matrix, δ is small and q is maximal. This parameter generalizes the one introduced by (Budiansky and O'Connell 1976) for crack-like slits to the general ellipsoids.

For a *spheroid* $(a_1 = a_2 \equiv a)$, with unit vector $\boldsymbol{n}$ along the axis of symmetry

$$\boldsymbol{Q} = -\frac{1}{1+\delta}\frac{1}{R}\left[\left(\frac{3}{2}A_3 + A_4 + A_5\right)\boldsymbol{nn} + \left(3A_1 + A_2 + \frac{1}{2}A_3 + 1 - 2\nu_0\right)\boldsymbol{I}\right] \tag{3}$$

where A_i, R, are combinations of Eshelby's tensor components.

In the case of a *sphere*,

$$\boldsymbol{Q} = -\frac{1}{1+\delta}\frac{5-7\nu_0}{9(1-\nu_0)}\boldsymbol{I}, \quad R = 9(1-\nu_0)/2, \quad \delta = 2\kappa E_0/(9-9\nu_0) \tag{4}$$

In the limit of a *crack-like* spheroidal cavity (aspect ratio $\zeta = a_3/a << 1$),

$$\boldsymbol{Q} = -\frac{1}{1+\delta}\left[\boldsymbol{nn} + \zeta\beta(1-2\nu_0)\boldsymbol{I}\right], \quad R = 1/(\beta\zeta), \quad \delta = \beta\zeta\kappa E_0 \tag{5}$$

where $\beta = 3\pi/\left[16\left(1-\nu_0^2\right)\right]$ and $\boldsymbol{n}$ is a unit normal to the crack.

In the limit of a *needle-shaped* spheroidal cavity (aspect ratio $\zeta = a_3/a >> 1$):

$$\boldsymbol{Q} = -\frac{1}{1+\delta}\frac{1}{5-4\nu_0}\left[-(1+\nu_0)\boldsymbol{nn} + 3(1-\nu_0)\boldsymbol{I}\right], \; R = 5-4\nu_0, \; \delta = \frac{\kappa E_0}{5-4\nu_0} \tag{6}$$

In a solid with *many* cavities, q induced by $\boldsymbol{\sigma}$ will be different for different cavities. This *pressure polarization* is coupled with the effective elastic response.

Effective elastic properties. Proper parameters of defect density

The compliance tensor $\hat{\boldsymbol{H}}$ of a single fluid-filled cavity is a sum of $\boldsymbol{H}$ for a dry cavity plus a correctional term $\Delta\boldsymbol{H}$ accounting for the fluid presence: $\hat{\boldsymbol{H}} = \boldsymbol{H} + \Delta\boldsymbol{H}$ where $\Delta\boldsymbol{H}$ is found in terms of $\boldsymbol{H}$ and $\boldsymbol{Q}$:

$$\Delta\boldsymbol{H} = (\boldsymbol{QH}:\boldsymbol{I} + \boldsymbol{H}:\boldsymbol{IQ})/2 \quad \text{or} \quad \Delta H_{ijkl} = \left(Q_{ij}H_{klmn}\delta_{mn} + H_{ijmn}\delta_{mn}Q_{kl}\right)/2 \tag{7}$$

We restrict the analysis to the approximation of non-interacting cavities (each cavity is placed in the externally applied stress $\boldsymbol{\sigma}$ and experiences no influence of neighbours). Besides being rigorous for small cavity densities, it constitutes a basic building block for various approximate schemes, that place non-interacting defects into some sort of "effective environment" - effective matrix (self-consistent, differential schemes) or effective stress (Mori-Tanaka's scheme).

The effective properties are best analyzed in terms of elastic potentials. Formulation in potentials yields results for all orientational and aspect ratio distributions of defects in a unified way. Besides, the structure of potentials identifies the *proper parameters of cavity density*, in terms of which the effective properties are expressed in a unified way. Such parameters are non-trivial: for example, porosity (relative volume of cavities) is not a proper density parameter (for the cavity shapes other than spherical). For fluid-filled cavities, the density parameters, in addition to being shape dependent, also depend on the physical parameter δ.

The elastic potential in stresses (complementary energy density) $f(\boldsymbol{\sigma})=(1/2)\,\boldsymbol{\sigma}:\boldsymbol{\varepsilon}$ can be represented as a sum $f(\boldsymbol{\sigma})=f_0+\Delta f$ where $f_0=(1/2E_0)\left[(1+\nu_0)tr(\boldsymbol{\sigma}\cdot\boldsymbol{\sigma})-\nu_0(tr\boldsymbol{\sigma})^2\right]$ is the potential in the absence of cavities and Δf is due to cavities. The problem is reduced to finding Δf as a function of $\boldsymbol{\sigma}$. The effective compliances s_{ijkl} are then found from $\varepsilon_{ij}=\partial f/\partial\sigma_{ij}=s_{ijkl}\sigma_{kl}$.

The potential Δf has the form:

$$\Delta f=\frac{1}{2}\boldsymbol{\sigma}:\sum\Delta\boldsymbol{\varepsilon}^{(k)}=\frac{1}{2}\boldsymbol{\sigma}:\sum\hat{\boldsymbol{H}}^{(k)}:\boldsymbol{\sigma}=\underbrace{\frac{1}{2}\boldsymbol{\sigma}:\sum\boldsymbol{H}^{(k)}:\boldsymbol{\sigma}}_{\Delta f_{dry}}+\frac{1}{2}\boldsymbol{\sigma}:\sum\Delta\boldsymbol{H}^{(k)}:\boldsymbol{\sigma} \qquad (8)$$

It can be specialized for various cavity shapes: narrow crack-like cavities, spheres, needles, etc.(Kachanov *et al.* 1995; Shafiro and Kachanov 1996). In each case, the structure of Δf dictates the proper parameters of cavity density. *Mixtures of pores of diverse shapes*, typical for real microstructures are covered in a unified way.

Acknowledgment: This work was supported by the Army Research Office.

References

Budiansky, B. and O'Connell, R.J. (1976). Elastic moduli of a cracked solid. *Int. J. Solids & Struct.*, **12**, 81-97.

Kachanov, M. (1980). Continuum model of medium with cracks. *J.Eng.Mech.Div.*, **106**(EM5), 1039-1051.

Kachanov, M. (1993). Elastic solids with many cracks and related problems. In *Advances in Applied Mechanics*, **30** (Hutchinson, J.W. and Wu, T.Y., eds), Academic Press, 259-445.

Kachanov, M., Tsukrov, I. and Shafiro, B. (1994). Effective moduli of solids with cavities of various shapes. *Appl. Mech. Reviews*, **47**(1), S151-S174.

Kachanov, M., Tsukrov, I. and Shafiro, B. (1995). Materials with fluid-saturated cracks and cavities: fluid pressure polarization and effective elastic response. *Int. J. Fract. (Reports of Current Research)*, in press.

Shafiro, B. and Kachanov, M. (1996). Materials with fluid-filled pores of various shapes. *Int. J. Solids & Struct.*, submitted.

REDUCTION OF STRUCTURAL RESPONSE BY ENERGY DISSIPATION DEVICES MODELLED ON SHAPE MEMORY MATERIALS

C.W.S. To* and **J.M. Kelly†**

Abstract

A mathematical model based on the dynamic behavior of shape memory materials (SMM) is proposed for the study of a class of energy dissipation restraints (EDR's). A formulation and response analysis for a multi-storey building treated as a shear beam structure with EDR modelled on the SMM in any or every storey is presented. Computed results were obtained for various multi-degree-of-freedom (MDOF) systems excited by deterministic disturbances. Chaotic motions were observed. Of the cases considered in the investigation, results obtained indicated that the EDR's reduced significantly the structural responses at all storey levels.

Introduction

An aseismic design of low to medium-rise buildings primarily involves with the reduction of vibration of their fundamental natural frequency. The latter is within the frequency range where the earthquake energy is highest. The high energy and coincidental frequency of the earthquake set the building in resonance. Thus, the building acts as an amplifier of the ground vibration. The amplified structural vibration can cause severe damage to the contents, occupants and the structure itself.

One class of techniques for treating such amplified vibration is the passive structural vibration controls. Among these techniques base isolation, in which the building is decoupled from the horizontal components of earthquake ground motion such that the lateral structural vibration is greatly attenuated, has received much attention within the earthquake engineering community. Several mechanisms for base isolation have been developed over the last two decades. However, recent developments in rubber technology have made base isolation employing laminated rubber bearing a practical reality (Kelly, 1990).

Application of energy absorbing devices (EAD) in the building provides another technique of passive vibration control. Experimental and model studies involving EAD have recently been carried out by Kelly and associates (Nims and Kelly, 1990). The studies (Lin, et al., 1988) showed that overall structural vibration of the laboratory model under typical

* Department of Mechanical Engineering, The University of Western Ontario, London, Ontario, Canada N6A 5B9.

† Department of Civil Engineering, University of California, Berkeley, California 94720.

horizontal earthquake excitations was reduced very significantly. This finding is consistent with that reported by Nims and Kelly (1990). It is also evident from Lin, et al. (1988) that the efficiency of the viscoelastic dampers is reduced as the temperature is increased.

In the investigation briefly presented here the SMM model of Graesser and Cozzarelli (1989) is examined so as to better understand and identify various model parameters for use in a mathematical model for a class of EDR exemplified by the Fluor Daniel EDR. The EDR model is then incorporated into the derivation of equaitons of motion and response analysis of a n-storey building subjected to a horizontal excitation.

Shape Memory Material (SMM) Model

While various hysteretic models available in the literature have several desirable features, such as description of degrading and pinching (Noori, et al., 1988), no tendency to exhibit displacement drift and does not violate Illyushin's and Drucker's postulates (Thyagarajan and Iwan, 1990), and rate independent hysteretic behavior (Ozdemir, 1976), they are incapable of describing the dynamic behavior of superelastic materials that have zero residual strain. This important feature has been commonly observed in shape memory alloys. A model for describing the SMM was proposed by Graesser and Cozzarelli (1989). Essentially, this SMM model is based on Ozdemir's model. The two basic equations for the SMM model are

$$\dot{\sigma} = E\left[\dot{\varepsilon} - |\dot{\varepsilon}|\left|\frac{\sigma-\beta}{\sigma_y}\right|^{N-1}\left(\frac{\sigma-\beta}{\sigma_y}\right)\right], \tag{1}$$

$$\beta = E\,\alpha\left[\varepsilon^{nl} + \gamma_T |\varepsilon|^c \, erf(\rho\,\varepsilon)\, U(-\varepsilon\,\dot{\varepsilon})\right], \tag{2}$$

where ρ , c , and γ_T are material constants. The material constant γ_T is the one that accounts for the hysteretic behavior affected by varying temperatures. U(.) is the unit step function of the enclosing quantity. This unit step function is included to activate the second term inside the square brackets of equation (2) during the unloading process. Thus, during the loading process the second term inside the square brackets of equation (2) is set to zero. The symbol σ is the one dimensional stress, ε is the corresponding strain, β is the one dimensional backstress, E is the elastic modulus, σ_y is the yield strength, α is the post-yield to pre-yield stiffness ratio, N determines the shape of the hysteresis loop, the over-dot denotes the first derivative with respect to time t, and the superscript "nl" designates the nonlinear component.

Response Analysis of A Multi-Storey Building and Numerical Results

The important self-centering feature of the EDR resembles the superelastic material behavior that is common in SMM's and therefore a mathematical model of EDR is proposed by applying equations (1) and (2). Consider a n-storey building as a shear beam structure whose mass is approximated as n lumped masses, m_1 , m_2 , m_3, ..., and m_n , being located at the floors. The walls of the building and n energy dissipation restraints braced diagonally provide the hysteretic restoring forces. Without loss of generality only motion in the plane of the building is assumed. Thus, the motion can be described in terms of the horizontal displacements of the floors with respect to the stationary ground, x_1 , x_2 , x_3 , ... , and x_n. Let the relative displacement or inter-storey drift of the i'th floor be y_i. That is,

$$y_i = x_i - x_{i-1} \quad i = 1, 2, \ldots, n. \tag{3}$$

Writing the equations of motion in terms of y_i and going through some algebraic manipulations, the matrix equation of motion becomes

$$M\ddot{Y} + C\dot{Y} + K^{(r)}Y + K_h^{(r)}U^{(r)} = F \tag{4}$$

where the symbols are defined by To and Kelly (1996).

Let $Z_1 = Y$, $Z_2 = U^{(1)}$, $Z_3 = U^{(2)}$, $Z_4 = Y^{nl(1)}$, $Z_5 = Y^{nl(2)}$, $Z_6 = \dot{Y}$, and $Z = (Z_1, Z_2, Z_3, Z_4, Z_5, Z_6)^T$, one can show that

$$\frac{dZ}{dt} = AZ + P. \tag{5}$$

Equation (5) is now ready to be numerically integrated by using the fourth order Runge-Kutta algorithm or other similar numerical integration schemes. In the present investigation, equation (5) has been applied to a number of systems under deterministic excitations. For brevity, representative results are included in Figure 1. Of the cases considered in the investigation, results indicated that the addition of EDR's can significantly reduce structural responses. Chaotic motions were also observed.

Concluding Remarks

A mathematical model for a class of EDR exemplified by the Fluor Daniel EDR is presented. The proposed model is based on the dynamic behavior of SMM. One important feature of the SMM model is its ability to describe the dynamic behavior of superelastic materials that have zero residual strain. This, in turn, enables one to model the self-centering characteristic of the EDR. A formulation and response analysis for a multi-storey building treated as a shear beam structure with EDR installed in any or every storey have been briefly outlined. The computed numerical results indicated that the addition of EDR's can significantly reduce structural responses. The latter may exhibit chaotic behavior.

Acknowledgements

The investigation reported above was performed between July 1991 and June 1992 while the first author (CWST) was visiting the senior author (JMK) and Professor J.L. Sackman. It was funded by Fluor Daniel, Inc. and the National Science Foundation. The first author wishes to acknowledge the initial assistance of Dr. D.K. Nims.

Appendix-References

1. Graesser, E.J. and Cozzarelli, F.A., "Multidimensional Models of Hysteretic Material Behavior for Vibration Analysis of Shape Memory Energy Absorbing Devices," Technical Report NCEER-89-0018, *National Center for Earthq. Eng. Res.*, State University of New York at Buffalo, 1989.

2. Kelly, J.M., "Base Isolation: Linear Theory and Design," *Earthquake Spectra*, Vol. 6, No. 2, 1990, pp. 223-244.
3. Lin, R.C., Liang, Z., Soong, T.T. and Zhang, R.H., "An Experimental Study of Seismic Structural Response with Added Viscoelastic Dampers," Technical Report NCEER-88-0018, *National Center for Earthq. Eng. Res.*, State University of New York at Buffalo, 1988.
4. Nims, D. and Kelly, J.M., "Introduction to the Small Scale Seismic Testing of A Self-centering Friction Energy Dissipator for Structures," Draft Report to Sponsor, Dept. of Civil Engineering, University of California, Berkeley, 1990.
5. Noori, M., Davoodi, H., and Saffer, A., "An Ito-Based General Approximation Technique for Random Vibration of Hysteretic Systems, Part I: Gaussian Analysis," *Journal of Sound and Vibration*, Vol. 127, No. 2, 1988, pp. 331-342.
6. Ozdemir, H., "Nonlinear Transient Dynamic Analysis of Yielding Structures," Ph.D. Dissertation, University of California, Berkeley, 1976.
7. Thyagarajan, R.S. and Iwan, W.D., "Performance Characteristics of A Widely Used Hysteretic Model in Structural Dynamics," *Proceedings of Fourth U.S. National Conference on Earthquake Engineering*, Palm Springs, California, May 20-24, 1990, pp. 177-186.
8. To, C.W.S., and Kelly, J.M., "A Mathematical Model for An Energy Dissipating Restraint and Its Application to Mitigate Earthquake-Induced Structural Response," NDRL TR9603, Department of Mechanical Engineering, University of Western Ontario, Canada, April 1996.

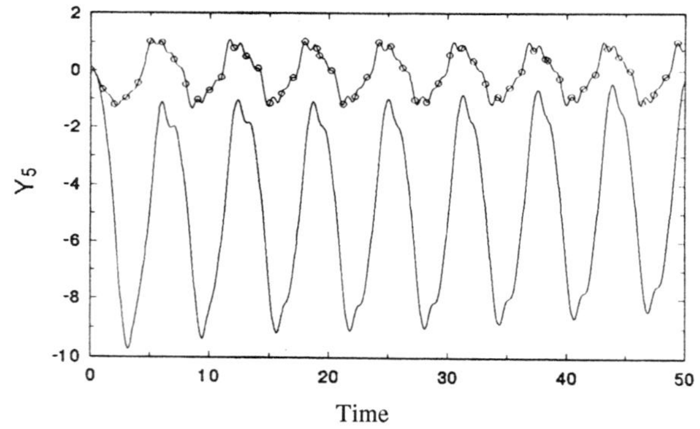

Figure 1. Inter-storey drift of a 5DOF system: ⊖ with EDR, — without EDR.

Vibration Control of Structures Utilizing Architectural Walls

H. Allison Smith[1], Luciana R. Barroso[2] and Vicki L. Vance[3]

Abstract

A study of architectural walls for energy dissipation and vibration control of buildings in seismic zones is presented. Typical architectural walls exhibit inelastic behavior; however, both the stiffness and energy dissipation capacity degrade after the walls are damaged. Integrating passive damping devices within these walls augments their energy dissipative abilities by widening the hysteresis loops and increasing the number of cycles to failure. The paper proposes a method of vibration control that would utilize architectural walls as hysteretic energy dissipators.

Introduction

In the majority of earthquake engineering analyses, architectural walls, which are considered non-structural components, are not modeled explicitly. These walls, however, may contribute significantly to the overall structural stiffness and can be a major source of energy dissipation during seismic events. A prior study by the authors (Smith *et al.*, 1996) involved quantification of the effects of architectural walls on structural behavior and resulted in the development of an analytical model which represents the stiffness and energy dissipative characteristics of typical steel stud architectural walls (Vance *et al.*, 1996).

The objective of this research project is to investigate the potential of utilizing architectural walls as vibration control devices by integrating them with passive control mechanisms, such as viscoelastic and friction-slip dampers. Conventional building design in earthquake engineering relies on the inelastic behavior of structural members to dissipate the input energy from lateral loads. While large amounts of energy may be dissipated through inelastic behavior, damage to structural members is not desirable. Thus, by incorporating another means

[1]Assistant Professor; Dept. of Civil Engineering, Stanford Univ., Stanford, CA 94305.
[2]Graduate Student; Dept. of Civil Engineering, Stanford Univ., Stanford, CA 94305.
[3]Graduate Student; Dept. of Civil Engineering, Stanford Univ., Stanford, CA 94305.

of dissipating energy, additional seismic energy can be dissipated without compromising the load-resisting system. Typical architectural walls exhibit inelastic behavior which corresponds to dissipated seismic energy, but their behavior degrades as the walls are damaged after repeated cycles. Integrating passive damping devices within these walls could augment their energy dissipative abilities by widening the hysteresis loops.

Previous Work

Seismic control utilizing walls has focused on those walls that are already considered part of the lateral load resisting system. A Viscous-Damped Wall (VDW) system has been implemented in a 16-story steel-frame building in Shizuoka, Japan (Miyazaki, 1992). Each wall is essentially a viscous damper, where a plate connected to the top of the story is placed within a viscous fluid contained by two outer plates connected to the floor of the story. As interstory drift occurs, the inner plate moves within the viscous fluid, providing the extra damping to the system. The final structure demonstrated 20% or greater damping.

A. Filiatrault (1990) has developed a method for analytical predictions of seismic response of friction damped timber shear walls. Timber shear walls progressively lose lateral stiffness under repeated loading cycles. Triangular friction devices are placed at each corner of the wall to reduce the pinching behavior of the shear wall's hysteresis curve. The triangular shape of the device maintains the integrity of the corners, while the energy is dissipated at all four corners of the wall structure.

Seismic Energy Dissipation due to Architectural Walls

Interior architectural walls are the focus of this research. Though several different types of steel stud assemblages are used, a typical design is composed of steel studs attached to runners by either friction or welds. A gypsum wallboard facing panel is attached to the studs by drywall screws. Occasionally, a thin diagonal steel plate is attached to gusset plates at opposite corners of the wall. Though each wall configuration exhibits a different response characteristic, they all display the same general characteristics. Typically, the wall response stabilizes in the second cycle of loading at the same level of displacement. These stable cycles display significantly smaller values of stiffness and energy dissipation than the initial cycles. The pinching of the hysteresis loops is the major problem to overcome in the utilization of the walls as vibration control devices. Of the different types of wall assemblages, architectural walls with the diagonal strap appear the most promising since the diagonal strap increases both the strength and stiffness of the wall configuration; however, the strap can cause additional damage if it buckles during a seismic event.

Integration of Passive Control Devices within Architectural Walls

Ease of implementation is one of the main issues in deciding whether a damper integration scheme will be viable. Any major change in the wall design will decrease the chance that the proposed configuration will be accepted and utilized. In studying how the energy dissipative capacities of architectural walls can be enhanced via passive control devices, various configurations of both friction and viscoelastic dampers are being investigated. Viscoelastic dampers reduce the dynamic amplification factor of the system in the frequency range of the system's natural frequency while having little impact outside that range. While the effectiveness of friction-slip dampers does not depend on frequency, energy dissipation only occurs if the forces developed exceed a predetermined *slip load.*

Two different locations for the passive devices were originally considered in this study. The first utilizes the connection between the diagonal strap and gusset plate. That location was discarded for two reasons. The steel strap is extremely thin and buckles easily in compression. In addition, the introduction of a device at that location would be a major design change in the architectural wall as the strap would undoubtedly have to be thicker in order for the mechanism to be effective. The location currently being investigated introduces passive devices at the corners of the walls, similar to those developed by Filiatrault (1990) for timber shear walls. The corner location takes advantage of the large rotations incurred at those points under lateral loads.

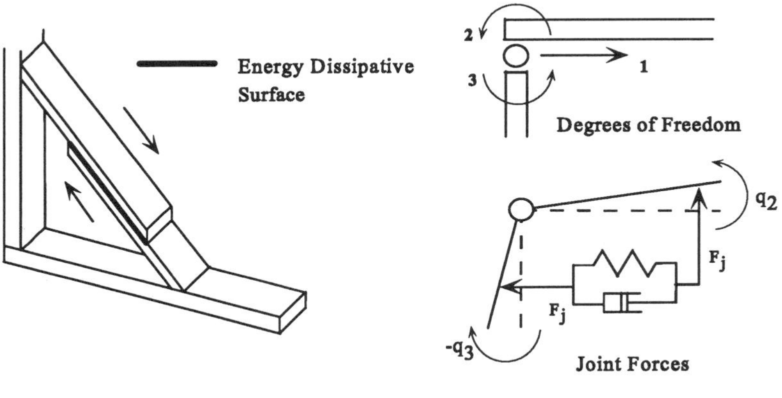

Figure 1: Modeling of Damper at Wall Corner

The passive control devices will be added to the analytical wall model, or 'super-element', previously developed by the authors (Vance *et al.*, 1996). Because the 'super-element' is composed of individual finite elements to represent the different wall components, the dampers are being modeled individually. Each device is modeled as a flexible connection between the exterior studs and runners using rotational stiffness (see Fig. 1). The addition of the connection introduces two additional degree of freedom at each corner of the wall, where these connections were modeled as fixed in the original model. Since the joint now has finite stiffness, the rotation of the elements on either side of the connection is no longer the same. The force in the joint, F_j, can be determined from the rotational degrees of freedom and the appropriate element stiffness and damping. When modeling the friction-slip damper, only the stiffness term is needed. The spring stiffness is bilinear, with the break occurring at a specified slip-load. In the case of the viscoelastic damper, the stiffness term is linear and a linear damping term is also included. The response of the resulting assemblage is then analyzed to determine the effectiveness of the system. Once a specific configuration has been chosen, a single finite element model will be developed to represent the behavior of the entire assemblage and to reduce the total number of degrees of freedom. This model will then be utilized to analyze the impact on overall building response.

Conclusions

The proposed research suggests a method of vibration control that would utilize architectural walls as hysteretic energy dissipators. By integrating passive control devices within the walls' design, the energy dissipative capacity will increase and stabilize. Thus, placement of these walls throughout the structure, would result in a significant impact on dynamic performance. Through careful selection of design parameters for the passive control device, the engineer can tailor the dynamic response of the building. Under strong ground motions, energy dissipation can be distributed throughout the structure without compromising the structural system.

References

Filiatrault, A., (1990),"Analytical Predictions of the Seismic Response of Friction Damped Timber Shear Walls," *Earthquake Engineering and Structural Dynamics* Vol. 19, pp. 259-273.

Miyazaki, M. and Y. Mitsusaka, (1992), "Design of a Building with 20% or Greater Damping," *Proceedings of the Tenth World Conference on Earthquake Engineering*, Madrid, Spain, pp. 4143-4148.

Smith, H.A. and Vance, V.L., (1996), "Model to Incorporate Architectural Walls in Structural Analysis," to appear in the *ASCE Journal of Structural Engineering*, April.

Vance, V.L., Smith, H.A., and Barroso, L.R., (1996), "'Super-Element' to Represent the Behavior of Architectural Stud Partition Walls," *Proceedings of the 11th ASCE Conference on Engineering Mechanics*, Fort Lauderdale, Florida, May.

Hysteretic systems: chaotic region and control

M. Battaini, F. Casciati and L. Faravelli [1]

Abstract

In the study of the possible chaotic motions of a nonlinear dynamic system the problem is to find out, in the space of suitable parameters, the chaotic regions. The continuous curves ("manifolds"), on which the Poincaré maps lie, could be built. Alternatively, the possible chaotic regions can be identified by calculation of the system Lyapunov exponents. The study, already developed in the literature for the Duffing oscillator, was recently extended to hysteretic oscillators. In this paper, it is tried to design an hysteretic oscillator which is operative in its chaotic region: its motion is then controlled by a sequence of small perturbations.

Introduction

Chaotic motion was initially regarded as neither predictable nor controllable. It was just something to avoid in technical practice [1],[5],[10]. Further studies emphasized how a suitable selection of driving terms can lead a chaotic system to behave in a desired manner. In a recent paper [11] the presence of chaos is regarded as a desired feature for an actively controlled system; tiny perturbations on a chaotic state can be used to stabilize a regular dynamic behaviour and to rapidly direct chaotic trajectories toward a desired state. The aim of the present paper is to study the feasibility of a technique to control the response of an hysteretic oscillator that shows a chaotic behaviour. " One can convert the motion of a chaotic dynamical system to periodic motion by controlling the system about one of the many unstable periodic orbits embedded in the chaotic attractor, through only small time dependent perturbations in an accessible system parameters. " [8]

Governing Equations

The system to be investigated is a hysteretic single-degree-of-freedom oscillator whose equation of motion can be expressed as:

$$m\ddot{x} + c\dot{x} + \Phi(x,t) = f(t) \tag{1}$$

where m and c represent the mass and the viscous damping coefficient, respectively; $f(t)$ is the external action, assumed to be periodic of amplitude α and frequency ω, respectively, and Φ is the hysteretic restoring force. In general, the term Φ can be regarded as the sum of a linear part rx and a hysteretic contribution z:

$$\Phi = rx + z \tag{2}$$

[1] Department of Structural Mechanics, University of Pavia, Via Abbiategrasso 211, 27100 Pavia, Italy

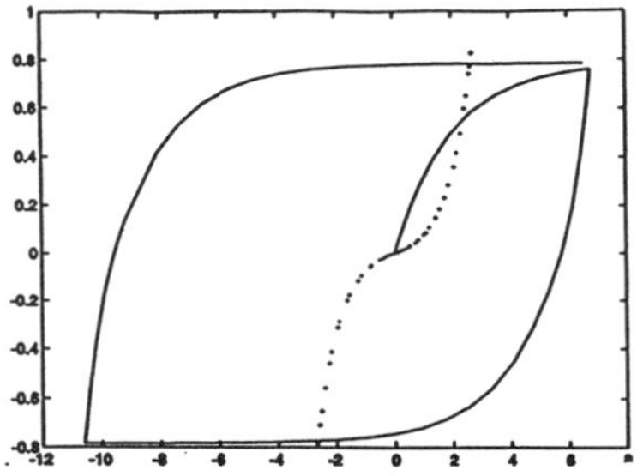

Figure 1: Hysteretic model of Eq. (3), $A = 0.4$, $\gamma = -0.5$ and $\beta = -0.3$ (solid line), $\gamma = -0.01$ and $\beta = -1.2$ (dashed line)

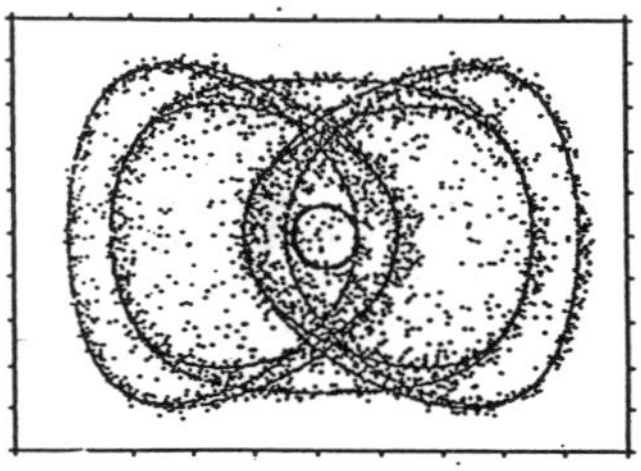

Figure 2: Poincarè map in the displacement - velocity plane. The solid line represents the target trajectory.

In Eq. (2) r represents the linear stiffness term and z is an auxiliary variable. Its first derivative with respect to time t is [4], [6]:

$$\dot{z} = \gamma \mid \dot{x} \mid z + \beta \dot{x} \mid z \mid + A\dot{x} \tag{3}$$

where A is regarded as a fixed parameter and γ and β vary to determine the dissipation regime. For an accurate selection of the parameters γ and β one can obtain a constitutive law with either low or large dissipation (see Figure 1). On the basis of the experience of a previous study [2], the possibility of a chaotic behaviour is investigated using two groups of parameters and performing two parametric analyses. The first group collects the parameters of the external force (represented as a periodic function): $\boldsymbol{p}_1 = \{\alpha, \nu\}^T$; the second set groups the parameters of the hysteretic oscillator defined by Eq. (3): $\boldsymbol{p}_2 = \{\beta, \gamma\}^T$. The chaotic behaviour is investigated by the calculation of the Lyapunov exponents [12]. The presence of chaos requires that the dissipation be very low; if the dissipation term is increased,the system, after an initial transient phase, becomes stable.

For any point for which a chaotic response is detected three sections of the Poincaré map are computed over a time range of 20000 force period and neglecting the initial transient part of 500 period. The three sections in the phase space are: displacement - velocity; displacement - z; velocity - z. An example is shown in Figure 2.

Controlling a chaotic system

Chaotic systems are characterized by extreme sensitivity to tiny perturbations. This characteristic can be used both to stabilize the dynamic behaviour and to direct chaotic trajectories toward a desired state. Thus small, carefully chosen perturbations are able to effect a large beneficial change in the long term system behaviour.

Control algorithm. A possible technique to control a chaotic system stabilizes of the unstable orbits by small parameter adjustments as discussed in [11]. Put Eqs. (1, (2) and (3) in the general form of a system of autonomous ordinary differential equations $d\psi/dt = G(\psi)$, where $d\psi(t)$ is a vector of length $M = 4, (x, \dot{x}, r, t)$ and t is the continuos time variable. By an integration of $d\psi/dt = G(\psi)$ over (t_n, t_{n+1}), the response ψ_{n+1} at time $n+1$ is calculated from the one at time t_n and the map $\psi_{n+1} = F_\psi(\psi_n)$ is built. Let p be the system parameter vector that could be modified to control the response. One writes the map in the form:

$$\psi_{n+1} = F_\psi(\psi_n, p) \tag{4}$$

It is possible to reduce the size M of the map ψ_n by introducing a $M-1$ Poincaré map. Let $Z_n(x, \dot{x}$ and $r)$ be the reduced associated vector. As a consequence map (4) becomes:

$$Z_{n+1} = F(Z_n, p) \tag{5}$$

and the resulting dimension N account for $M-1$ and the dimension of p; in the following p is regarded as a scalar. Let p_0 be any nominal value of p and $Z^*_{n+1} = F(Z^*_n, p_0)$ be the associated trajectory. For values of p close to p_0 the following linear map can be introduced:

$$(Z_{n+1} - Z^*_{n+1}) = \Theta(Z_n - Z^*_n) + \Gamma(p - p_0) \tag{6}$$

where Θ is the $N \times N$ Jacobian matrix, $\Theta = \partial F/\partial Z$, and Γ is the column vector of size N, $\Gamma = \partial F/\partial p$. These partial derivatives are calculated at $Z = Z^*_n$ and $p = p_0$. If one adopts the policy of changing p at each iteration, the last value of the sequence Z_n is found on the basis of small changes of p from p_0 to p_n.

Introduce now the linear control law:

$$(p_n - p_0) = -K^T(Z_n - Z^*_n) \tag{7}$$

which, for given K (a constant column vector of size N), specifies p_n for each iteration. Introducing Eq. (7) into Eq. (6), one obtains:

$$\Delta Z_{n+1} = (\Theta - \Gamma K^T)\Delta Z_n \tag{8}$$

where $\Delta Z_n = Z_n - Z^*_n$.

The trajectory Z_n will be stable if one can take K so that the matrix $(\Theta - \Gamma K^T)$ has all the eigenvalues with modulus lower than one. In this case $\Delta Z_n \to 0$ as $n \to \infty$. Since one knows Z_n and Z^*_n, at time t_n, one can select K^T in such a way that the eigenvalues of the matrix $(\Theta - \Gamma K^T)$ are lower than 1. Then Eq. (7) provides the value of p_n. This strategy has the advantage that the explicit form of F is not required. On the other hand such a procedure is only effective in stabilizing a trajectory near existing periodic orbits.

Technological constrain. Due to practical constrains, one cannot make the difference between p_n and p_0 too large, i.e. $\mid p_n - p_0 \mid$ is bounded in Eq. (7) by an allowed value, say Δp_{max}. If the system would require $\mid p_n - p_0 \mid > \Delta p_{max}$, no perturbation is applied and the control policy is to wait until the state falls near the target trajectory. When this occurs, the system is driven into the desired periodic state by applying the appropriate small perturbation computed by Eq. (7).

Operative aspects. The matrix Θ and the vector Γ can be extracted from observations over a long time period. One has to collect a long sequence of vectors $Z_1, .., Z_n$. If two successive Z_i are close to each other then there will be a stable orbit Z^* nearby. One, now, identifies within the succeeding data other pairs Z_n, Z_{n+1} restricted to belong to a small region. Orbits on a strange attractor are ergodic, so many others pairs will appear if the data string is long enough. Assuming that the restricted region is small, the following linear relation can be used to model the close returning pairs:

$$Z_{n+1} = \hat{\Theta} Z_h + \hat{C} \tag{9}$$

Generally one wants to use as many pairs as possible; this requires to fit the matrix $\hat{\Theta}$ and the vector $\hat{C}$ using a least square fit to the data. The matrix $\hat{\Theta}$ is an approximation of the matrix Θ of Eq. (7).
Analogously one estimates Γ and there compute K. Eq. (7) eventually provides the control strategy.

Conclusions

This paper moves within the framework of a new approach to control the structural response by introducing hybrid control devices with chaotic response. It suggests to take advantage of the extreme sensitivity of chaotic systems to tiny perturbations to stabilize the dynamic behaviour.

Acknowledgement

This paper was supported a grant from the Italian Ministry of University and Scientific and Technological Research (MURST).

References

[1] G. L. Baker and J. P. Gollub, 1990, *Chaotic Dynamics*, Cambridge University Press, Cambridge
[2] M. Battaini and F. Casciati, "Controlling a chaotic response", Euromech 341, 25-27 September 1995, Giens, France, paper submitted for pubbication in the Journal of Structural Control
[3] F. Bontempi and F. Casciati, 1994, Chaotic Motion and Stochastic Excitation, *Nonlinear Dynamics* 6: 179-191
[4] F. Casciati, 1989, Stochastic Dynamics of Hysteretic Media, *Structural Safety*, 6 259-269
[5] F. Casciati et al., 1993, *Dynamic motion: chaotic and stochastic behaviour*, CISM Courses and Lectures No. 340, International Centre for Mechanical Sciences, Springer-Verlag Wien New York
[6] F. Casciati and L. Faravelli, 1991a, *Fragility Analysis of Complex Structural Systems*, Research Studies Press, Taunton
[7] F.C. Moon, M.A. Johnson and W.T. Holmes, Controlling Chaos in a Two-Well Oscillator, To appear in *Journal of Bifurcation and Chaos*, 1996
[8] E. Ott, C. Grebogy and J.A. Yorke, 1990, Phys. Rev. Lett. 64, 1196; *Chaos*, edited by D. K. Campbell (American Institute of Physics, New York, 1990), pp. 153-172
[9] G. Rega, F. Benedettini and A. Salvatori, 1991, Periodic and Chotic Motions of an unsymmetrical oscillator in nonlinear structural dynamics, *Chaos, Solitons and Fractals* 1(1), 39-54
[10] H. Troger and A. Steindl, 1994, *Nonlinear stability and Bifurcation Theory, An Introduction for Engineers and Applied Scientists*, Springer- Verlag Wien New York
[11] T. Shinbrot, C. Grebogy, E. Ott and J.A. Yorke, 1993, Using small perturbations to control chaos, *Nature*, Vol 363, 411-417
[12] A. Wolf and J.B. Swift, H.L. Swinney and J.A. Vastano, 1985, Determining Lyapunov Exponents from a Time Series, *Physica 16 D, 285-317*, North-Holland, Amsterdam

A STRATEGY FOR ACTIVE CONTROL OF TALL CIVIL STRUCTURES USING REGENERATIVE ELECTRIC ACTUATORS

A.C. Nerves* and R. Krishnan*

Abstract: The active control of tall civil structures in the face of environmental forces, such as wind and earthquake, is an emerging area of research. A number of techniques, such as passive and liquid dampers and semi-active systems such as gravity actuation and base isolation techniques, have been studied. This paper considers a novel technique for utilizing the displacement caused by environmental forces on a tall building to generate electrical energy when the building displacement is within the safety limits. This recovered energy is used to reduce the peak oscillations of the building by applying forces counter to the environmental forces. To achieve this, a tuned mass damper is used as the intermediary for the energy transfer between the building and the actuator. The actuator proposed for this study is the permanent- magnet brushless DC (referred hereafter as PMBDA) type. The rationale for the choice of PMBDA for this application is provided. A strategy for the active control of tall buildings with PMBDA is formulated and an appropriate controller is designed to mitigate the peak oscillations of the tall building. Such a control strategy has to take into account the nature of the load it is controlling, i.e. tall buildings, and hence proper design entails a systematic survey of various controllers. These include classical linear, nonlinear, and intelligent controllers. The tall building, with this regenerative four-quadrant electric actuator, is dynamically simulated and analyzed for both wind and earthquake forces.

1. INTRODUCTION

The damping of wind- and earthquake-induced oscillations in tall civil structures is of paramount importance to the national economy as well as for the safety and well-being of the people in those structures. It is assuming national importance in many countries such as USA, Japan, China, Mexico, and the Philippines. Many techniques have been proposed for the damping and control of these oscillations caused by environmental forces of random nature. Many of these techniques in the literature are broadly classifiable into passive, semi-active , and active forms of control (Soong,1990). This paper is concerned with a new form of semi-active control whereby the energy of the environmental forces themselves is harnessed and used for the mitigation and control of the peak magnitude of oscillations in tall buildings. It is named as Regenerative Electric Actuation (REA) for tall civil structures due to the fact that this actuation method both generates electrical energy and uses it to produce a counter-force to the environmental force.

2. PRINCIPLE OF REGENERATIVE ELECTRIC ACTUATION FOR THE MITIGATION OF WIND- AND EARTHQUAKE-INDUCED OSCILLATIONS IN TALL BUILDINGS

Mitigation of wind- and earthquake-induced oscillations in tall buildings is of crucial importance to the safety of the people in the building and to the protection of the structural integrity of the building itself. Passive damping in the form of tuned mass

* The Bradley Department of Electrical Engineering, Virginia Tech, Blacksburg, VA 24061

dampers (TMD) is a widely prevalent practice for tall buildings. Basically, a TMD consists of a mass attached to a structure such that it oscillates at the same frequency as the structure but with a phase shift. It is found that the TMD's have better performance with wind forces than with earthquake excitations. To counter this high frequency excitation of the earthquake, various semi-active control methodologies have come to fore recently. All of these rely on energy for actuation from an external supply which most likely is the utility power supply. As the utility power supply itself may be under threat and therefore unavailable during the wind and earthquake excitations, *it is important that the energy source to counter the oscillations of the building must itself be derived from these oscillations*. Such a fundamental idea was proposed in (Krishnan, 1991). The energy from the environmental forces could be converted into electrical energy and stored for reuse by an actuator which produces a counter-force to the environmental forces. Therefore, the actuator has to be an electrical one as it can handle energy flow in both directions, i.e., it can generate electrical energy when given a mechanical input (hereafter referred to as generation mode) as well as produce mechanical energy when given an electrical input (hereafter referred to as motoring mode). The interconnection between the building and the electric actuator is through the medium of the tuned mass damper. It must be noted that during generating and motoring modes, the energy taken out of the TMD could be precisely controlled. That indirectly amounts to a controlled damping of the building. This is a significant factor due to the fact that controlled damping is able to accommodate most safety criteria for building displacement, velocity, acceleration, inter-story shift and its velocity, etc.

The electric actuation being proposed has the unique advantage of being able to operate even in the absence of a power source such as the utility power supply. This is unlike other or most of the semi-active and active control systems proposed for civil structures. Therefore, the reliability of the control system is increased manifold, hitherto considered impossible both in concept and in practice. Furthermore, the electric actuation has the advantage of controlling the damping during both the energy-acquisition mode and in the force-application mode. Therefore, control is guaranteed at all times, unlike other systems. The space and volume requirements for the electric actuator seem to be feasible and within the constraints of the housing required for most of the other control systems. Hence, it should not be of any concern from the point of view of implementation.

3. SELECTION OF ACTUATOR FOR REA

The use of an AC machine as the actuator is not pursued because it requires external excitation for four-quadrant operation. In the other hand, the use of a DC machine with commutators is limited by saturation and commutation problems due to the high currents. Therefore, a permanent-magnet brushless DC machine is chosen as the regenerative electric actuator because this machine type has a higher peak-torque-to-rated-torque ratio. Furthermore, this type of machine has a more compact frame size compared to other types of electric machines.

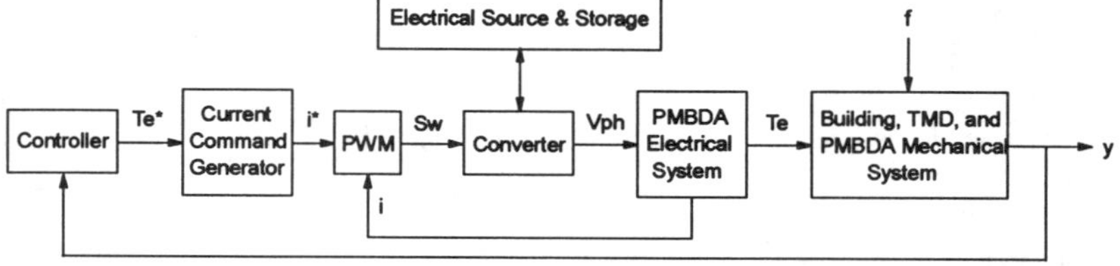

Fig. 1. Basic Control Scheme

4. BASIC CONTROL SCHEME

Fig. 1 shows the basic scheme of the proposed active/semi-active structural control system. The building-TMD system is being subjected to an external excitation ***f***. Sensors measure the response ***y*** of the building-TMD and sends the measurements to the controller. The controller then forms a command input T_e* to the actuator subsystem. The actuator either applies the required control force to the building-TMD (motoring mode) or converts the building-TMD motion to electrical energy for storage (generating mode). A ball-screw mechanism converts the actuator torque into a linear control force (and vice versa).

5. SYSTEM MODELING

The building-TMD model used in (Chang, 1980) is modified by using the Bouc-Wen model for building stiffness (Wen, 1989). After incorporating the PMBDA mechanical equation with the building-TMD equations, the system equations are obtained as

$$m_s \ddot{y}_s + c_s \dot{y}_s + F_{ks} - n^2 J \ddot{z} - (c_d + n^2 B)\dot{z} + k_d z = f - nT_e \quad (1)$$

$$m_d \ddot{y}_s + (m_d + n^2 J)\ddot{z} + (c_d + n^2 B)\dot{z} + k_d z = nT_e \quad ; \quad n = 2\pi e / p \quad (2)$$

where m, c, and k are mass, damping, and stiffness constants with subscripts s and d referring to the corresponding building and TMD parameters, respectively; y_s is the building displacement; z is the relative displacement between the building and the TMD; F_{ks} is the nonlinear stiffness force; n is the linear-to-rotational transformation ratio; J and B are the PMBDA moment of inertia and damping coefficient, respectively; f is the external excitation force; T_e is the PMBDA electromagnetic torque; e is the ball-screw efficiency; and p is ball-screw pitch (m/rev). The PMBDA electrical system is modeled by (Krishnan, 1996)

$$v_j = Ri_j + (L - M)\frac{di_j}{dt} + f_j(\theta_r)K_b\omega_r \qquad j = a,b,c \quad (3)$$

$$T_e = (P/2)K_b[f_a(\theta_r)i_a + f_b(\theta_r)i_b + f_c(\theta_r)i_c] \quad (4)$$

where v is the applied phase voltage; i is the phase current; R is the phase resistance, $(L-M)$ is the phase inductance; f_j is an induced-EMF function which depends on the rotor position θ_r ; K_b is the EMF constant; ω_r is the rotor speed; and P is the number of PMBDA poles.

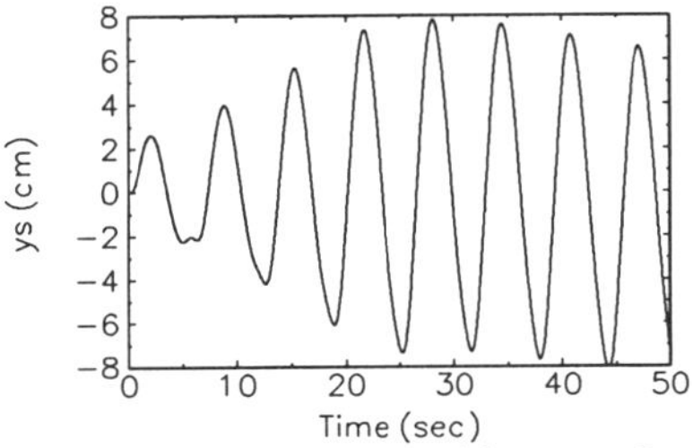

Fig. 2. Displacement without control

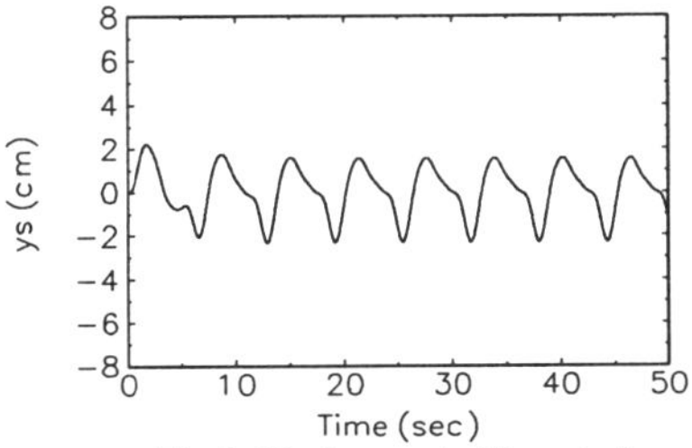

Fig. 3. Displacement with control

6. SIMULATION RESULTS

Numerical simulations are conducted for the whole system to determine its behavior when subjected to wind excitation. The building parameters, TMD parameters, and wind excitation used in [4] are also used in this study. The PMBDA is assumed to have J =0.1 kg-m^2 and B =5.34×10^{-4} Nm/rad/sec. A ball-screw with a pitch of 3.3 mm/rev is also assumed. For Case 1, Fig. 2 shows the building displacement without control, for a wind excitation with parameters p =9.75 and ω=1.0 (Chang, 1980). With the PMBDA activated, the resulting displacement is shown in Fig. 3. For Case 2, Fig. 4 shows the building displacement without control for a wind excitation with parameters p =5 and ω=1.1. With the PMBDA in regeneration mode (|T_e*|=5.65 kN-m), the resulting displacement is shown in Fig. 5. The instantaneous power from the source for both cases are given in Fig. 6 and Fig. 7.

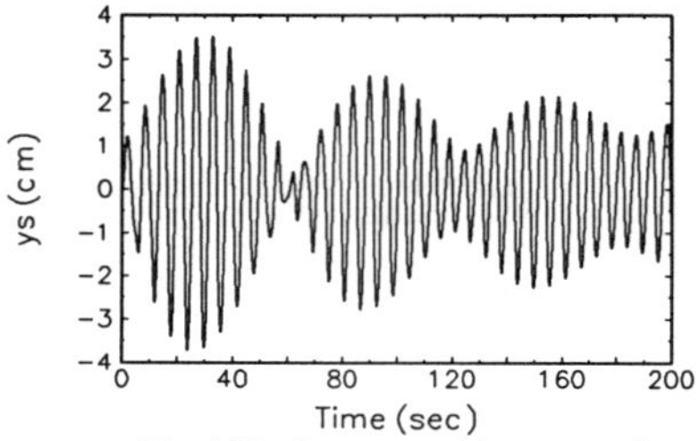

Fig. 4 Displacement without control

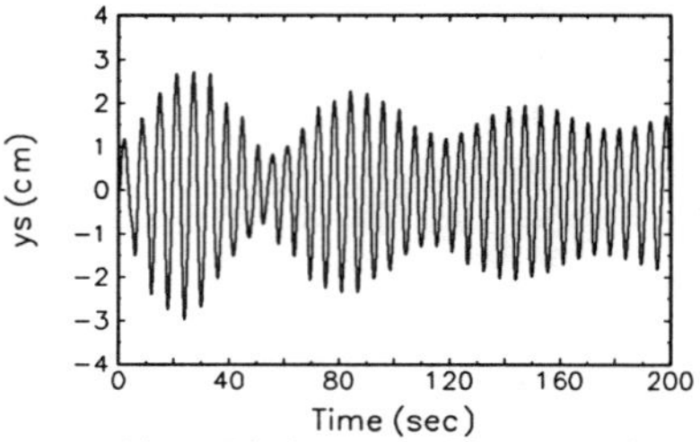

Fig. 5. Displacement with regeneration

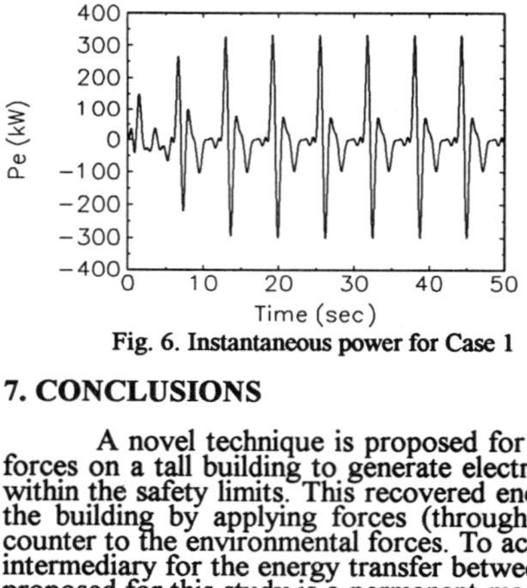

Fig. 6. Instantaneous power for Case 1

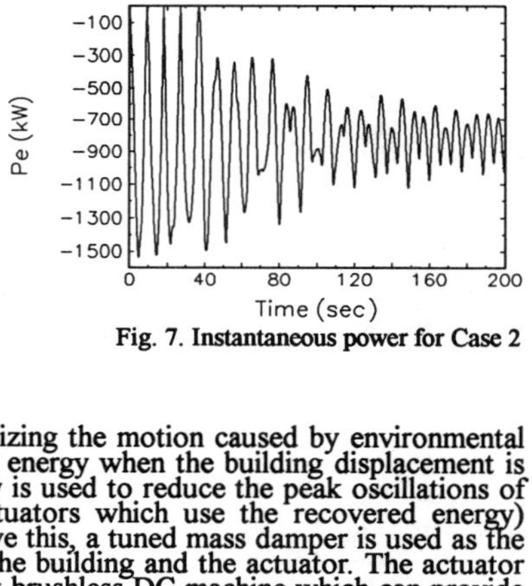

Fig. 7. Instantaneous power for Case 2

7. CONCLUSIONS

A novel technique is proposed for utilizing the motion caused by environmental forces on a tall building to generate electrical energy when the building displacement is within the safety limits. This recovered energy is used to reduce the peak oscillations of the building by applying forces (through actuators which use the recovered energy) counter to the environmental forces. To achieve this, a tuned mass damper is used as the intermediary for the energy transfer between the building and the actuator. The actuator proposed for this study is a permanent-magnet brushless DC machine which can provide a higher instantaneous torque in a compact frame size. An appropriate controller is designed for the proposed control strategy. This study is the first of its kind, to propose and investigate active control of structures using regenerative electric actuators.

Based on the results obtained, it is shown that the proposed strategy, using regenerative electric actuators, is both a viable and a reliable alternative for active control of tall civil structures.

REFERENCES

[1] Chang, J.C.H. and Soong, T.T., "Structural Control Using Active Tuned Mass Damper," ASCE Journal of the Engineering Mechanics Division, Vol. 106, No. EM6, December 1980, pp. 1091-1098.

[2] Krishnan, R. *et al.*, "Active Control of Structures using Sliding Mode Control Techniques with Regenerative Electric Actuator," NSF Grant No. BDS-9301574, October 1991.

[3] Krishnan, R., Electronic Control of Machines, Book to be published by Prentice-Hall.

[4] Soong, T.T., Active Structural Control: Theory and Practice, Longman Scientific and Technical, New York, 1990.

[5] Wen, Y.K., "Methods of Random Vibration for Inelastic Systems," Journal of Applied Mechanics Review, ASME, Vol. .42, No. 2, 1989, pp. 39-52.

OBLIQUE WAVE INTERACTION WITH VERTICAL WALL STRUCTURES

Xugui Ren[1] and Keh-Han Wang[2]

Abstract

The temporal and spatial evolution of an obliquely incident cnoidal wave in a wide channel was modelled using the generalized Boussinesq equations. Waves interaction with vertical walls was examined. An Euler's predictor-corrector finite-difference algorithm was used for numerical computation. A set of two-dimensional open boundary conditions was also incorporated for time-domain simulation. The nonlinear wave interactions and the formation of Mach stem were successfully modelled. The predicted wave profiles are in good agreement with the experimental data.

Introduction

The study of propagation and interaction of nonlinear waves in shallow water has many important engineering applications in harbors and shallow estuaries. Cnoidal waves are typical periodic nonlinear long waves. In an effort to model the three-dimensional nonlinear long waves in a more general way, Wu (1981) presented a set of generalized Boussinesq equations (gB). This set of equations has been successfully applied in modelling wave propagation and wave-structure interaction.

For oblique wave interaction with a vertical wall, Melville (1980) has conducted extensive laboratory experiments to investigate Mach reflection of a solitary wave. The experimental study of oblique interaction of a cnoidal wave was presented by Hammack et al. (1989). In this study, the propagation of cnoidal waves and their interactions with the channel wall were modeled based on the gB model using a predictor-corrector finite-difference scheme (Wang et al. 1992). A solution describing the periodic nonlinear shallow waves for the gB model is derived as the incident wave condition. A set of two-dimensional radiation conditions is also developed to effectively transmit the cnoidal waves out of the computational domain. From numerical results, the nonlinear wave interactions and the formation of Mach stem along the reflecting wall were analyzed. The simulated free-surface profile indicates that the large wave crests are connected by saddle-like regions of small amplitudes to form an array of regular surface pattern. The predicted wave profiles were also compared with laboratory measured data.

---

[1]Former Graduate Student, [2] Assistant Professor, Department of Civil and Environmental Engineering, University of Houston, Houston, TX 77204-4791.

Theoretical Model

The schematic diagram of obliquely incident cnoidal waves propagating in a wave channel is shown in Figure. 1. The theoretical model adopted for this study is the generalized Boussinesq (gB) two-equation model (Wu 1981). Under the conditions of constant water depth, the dimensionless gB two equations are given as

$$\zeta_t + \nabla \cdot [(1+\zeta)\nabla\Phi] = 0 , \tag{1}$$

$$\Phi_t + \frac{1}{2}(\nabla \Phi)^2 + \zeta - \frac{1}{3}\nabla^2\Phi_t = 0 , \tag{2}$$

where subscript t represents time derivative, $\nabla = (\partial/\partial x, \partial/\partial y)$ denotes the horizontal gradient in Cartesian coordinates, Φ represents the depth-averaged velocity potential and ζ is the free-surface elevation. Here, x, y, and ζ are referenced to d, t to d/c_o, and Φ to dc_o, $c_o = \sqrt{gd}$ and d = a representative constant water depth.

Incident Cnoidal Waves and Boundary Conditions

For modelling wave propagation and interaction with vertical walls, a second-order cnoidal wave solution for the gB equations was used as an incident wave condition along the left boundary. From Jiang (1994), the depth-averaged velocity potential for an incident plane cnoidal wave is given in a series form as

$$\Phi = \frac{\varepsilon}{kd}\left\{ \sum_{n=1}^{\infty}\left[\frac{A_n + \varepsilon B_n}{in}\right] e^{inkd(x-ct)} \right\} + \varepsilon^2 \frac{\gamma(-3\gamma-2m+4)-m_1}{6m^2}(x-ct) \tag{3}$$

The coefficient A_n and B_n can be found in Jiang (1994). Here, γ equals to E(m)/K(m). E(m) and K(m) are respectively the complete elliptic integral of the first kind and the second kind with modulus m and $m_1 = 1 - m$. k is the wave number and c (referenced to c_o) is the dimensionless phase velocity. The parameter ε is defined as $\varepsilon = H/d$, where H is a typical wave height. Using equation (3), the incident cnoidal wave profile is determined from equation (2).

The boundary conditions on rigid walls require that the normal fluid velocity vanishes there. The radiation conditions along the downstream open boundary are

$$\frac{\partial G}{\partial t} + c\left[\left(\frac{\partial G}{\partial x}\right)^2 + \left(\frac{\partial G}{\partial y}\right)^2\right]^{1/2} = 0, \quad \text{when } \frac{\partial G}{\partial x} > 0 \tag{4}$$

$$\frac{\partial G}{\partial t} - c\left[\left(\frac{\partial G}{\partial x}\right)^2 + \left(\frac{\partial G}{\partial y}\right)^2\right]^{1/2} = 0, \quad \text{when } \frac{\partial G}{\partial x} < 0 \tag{5}$$

where $G = \{\Phi, \zeta\}$. Here, the second-order cnoidal wave celerity c is used in equations (4) and (5). Similarly, the upstream boundary conditions are formulated as

$$\frac{\partial G}{\partial t} - c_s\left[\left(\frac{\partial G}{\partial x}\right)^2 + \left(\frac{\partial G}{\partial y}\right)^2\right]^{1/2} = \frac{\partial G^I}{\partial t} - c_s\left[\left(\frac{\partial G^I}{\partial x}\right)^2 + \left(\frac{\partial G^I}{\partial y}\right)^2\right]^{1/2}, \text{ when } \frac{\partial G}{\partial x} > 0 \tag{6}$$

$$\frac{\partial G}{\partial t} + c_s\left[\left(\frac{\partial G}{\partial x}\right)^2 + \left(\frac{\partial G}{\partial y}\right)^2\right]^{1/2} = \frac{\partial G^I}{\partial t} + c_s\left[\left(\frac{\partial G^I}{\partial x}\right)^2 + \left(\frac{\partial G^I}{\partial y}\right)^2\right]^{1/2}, \text{ when } \frac{\partial G}{\partial x} < 0 \tag{7}$$

in which $G^I = \{\Phi^I, \zeta^I\}$ represents the layer-mean velocity potential or surface elevation of the incident waves. The dimensionless scattered wave celerity c_s is set to be 1.0.

Results and Discussion

The numerical investigation of the Mach reflection of cnoidal waves was carried out in a computational domain of 64 x 32 (0<x<64, 0<y<32). The incident wave height

$\varepsilon = 0.3$ and the wave length $\lambda = 8.0$. The oblique wave profile is input along the left boundary for time-domain simulation. The angle of incidence is 45°. The free-surface profile at t = 90 is presented in Fig. 2 to show the periodic interactions between the incident cnoidal waves and the reflected ones. The apex of the incident and the reflected cnoidal waves moves away from the wall at a constant angle and is jointed to the wall by a Mach stem of cnoidal waves normal to the x-axis. Since the cnoidal waves are directed into the domain obliquely from the left open boundary, the diffraction effect becomes apparent near the wall opposite to the reflecting wall.

The Mach reflection of the cnoidal waves is more clearly depicted in the gray level contour plot as shown in Fig. 3 (t = 60 and t = 100). The light region represents high elevation whereas the dark region shows the low elevation. Again, the large crests are connected by saddle-like regions of small amplitudes to form an array of regular surface pattern. The formation of Mach stem along the reflecting boundary can be clearly observed.

The model predictions were compared with the experimental data obtained by Hammack et al. and reported by Kirby (1991) in a summary of his discrete angular spectrum model study. The experimental facility, test procedure and data acquisition are described in Hammack et al. (1989). The water depth used for the test is 20 cm. The computational domain is $0< x < 62.75$ and $0< y < 66.3$. Two arrays of wave gauges (designated 9-8-7-6-4-3-2-1 across the basin and 13-14-15 along the reflecting wall) were installed to measure wave elevations. The incident wave height $\varepsilon = 0.25$ and wavelength $\lambda= 9.89$.

The comparison between computed water surface elevations and measured ones shows good agreement for an incident wave angle of 10.13° (Figure 4). Typically, wave runup on the reflecting wall (gauge 13) was predicted well by the model simulations. Comparison was also made for an incident wave angle of 44.32°. The simulated wave elevations are again in good agreement with measured data.

Conclusions

The oblique wave interaction with vertical walls was studied numerically using Boussinesq equations. The nonlinear wave interaction and the formation of Mach stem were successfully modeled. It is found that the large wave crests are connected by saddle-like regions of small amplitudes to form an array of regular surface pattern. The predicted wave profiles are in good agreement with the experimental data.

References

Hammack, J., Scheffner, N., and Segur, H. 1989. Two-dimensional periodic waves in shallow water. J. Fluid Mech., **209**, 567-589.

Kirby, J. T. 1991. A discrete angular spectrum model for nonlinear shallow water waves". Personal communication.

Jiang, L. 1994. Hydrodynamic interactions between nonlinear long waves and cylinder arrays. M.S. thesis, Department of Civil and Environmental Engineering, University of Houston.

Melville, W.K.1980. On the Mach reflection of a solitary wave. J. Fluid Mech., **98**, 285-297.

Wang, K. H. , Wu, T. Y. & Yates, G. T. 1992. Three-dimensional scattering of solitary wave by vertical cylinder. *J. of Waterway, Port, Coastal and Ocean Eng.* , ASCE. **118**, 551-566.

Wu, T. Y. 1981. Long waves on ocean and coastal waters. *J. Eng. Mech. Div.* (ASCE) **107**, 501-522.

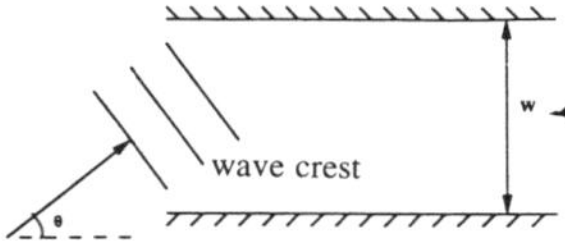

Fig. 1. Schematic diagram of obliquely incident cnoidal waves.

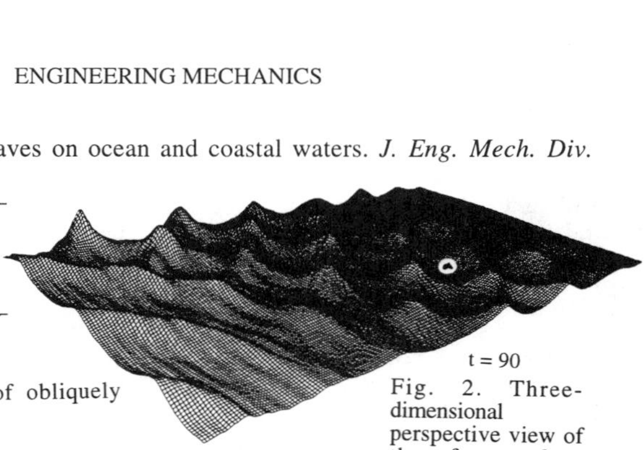

Fig. 2. Three-dimensional perspective view of the free-surface elevation. (Angle of incidence = 45°, ε = 0.3 and λ = 8.0.)

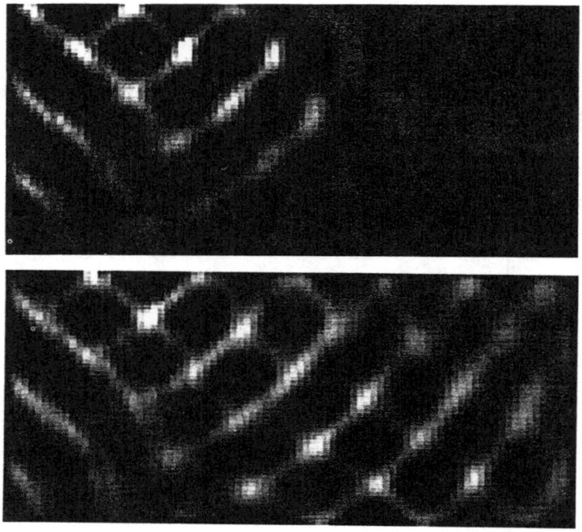

t = 60

t = 100

Fig. 3. Gray level contour plot of oblique wave interaction. (Angle of incidence = 45°, ε = 0.3 and λ = 8.0.)

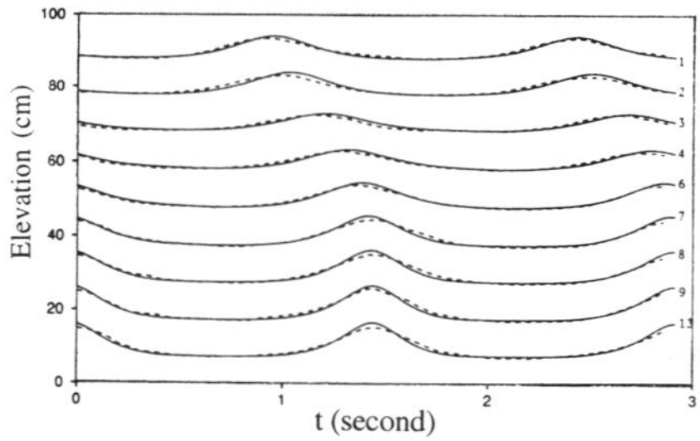

Fig. 4. Comparison of simulated and measured wave profiles. (incident wave angle =10.13°, ——— simulated, -------- measured)

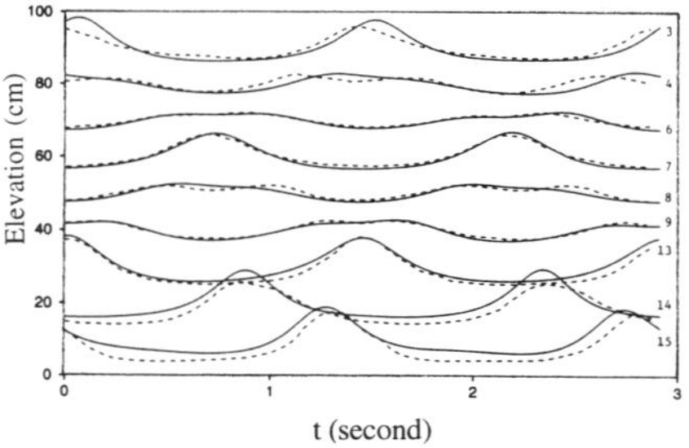

Fig. 5. Comparison of simulated and measured wave profiles. (incident wave angle =45°, ——— simulated, -------- measured)

Dynamic Effects of Sediment and Foundation on Dam Hydrodynamic Pressure Under Vertical Ground Accelerations

Kuo-Chyang Chang[1], Tin-Kan Hung[2], David A. Margo[3] and Jeen-Shang Lin[4]

Reservoir sediment is an important issue in the planning, design, evaluation and management of water resources. The concern of sediment dynamics on dam safety during earthquakes has resulted in several analytical and computational analyses. The concept of damping effect was implemented in the reflection coefficient at the water-sediment interface by Fenves and Chopra (1984), estimating the reduction in hydrodynamic pressure by reservoir bottom soft materials. The dynamic pore pressure in the sediment was not considered in the analysis.

More rigorous analyses were followed by several investigators, using Biot's linearized theory of poroelastic media. Cheng (1986) and Bougaha and Tassoulas (1991) reported that for saturated pore water the reduction of dynamic pressure by sediment was rather negligibly small. The damping was mainly contributed by the elastic foundation. However, a significant increase in hydrodynamic pressure by the sediment was found possible by Chen and Hung (1990 & 1993) even though the pore water was fully saturated. The increase in hydrodynamic force on dams was attributed to the inertial force of the sediment. Further justification of the augmentation of hydrodynamic pressure by the sediment dynamics appears necessary. Numerical solutions of the velocity and pore pressure of the sediment during ground excitations require that both the mass and momentum conservations of the sediment (pore water and solids) be satisfied in the pressure wave equation (Hung and Chang 1995). Similar to the numerical procedures of solving the Navier-Stokes equations, the pressure wave equation replaces the equation of mass conservation of pore water and sediment solids. The pressure field thus obtained is used to calculate velocity from the momentum equations which are parabolic in nature. In our previous study (Chen and Hung 1993), the dynamic expressions of the velocity of pore water and sediment solids are coupled with the storage equation. This formulation appears to underestimate the dynamic pressure in pore water because the specific gravity of sediment solids is not properly implemented in the pressure wave equation (Hung and

----

[1] Postdoctoral Fellow; presently senior engineer, Department of Water Resources, R.O.C.
[2] Professor, Department of Civil and Environmental Engineering, University of Pittsburgh, Pittsburgh, PA 15261; [3] Research Assistant; [4] Associate Professor.

Chang 1995). Another formulation of the pressure wave equation could result in a much lower and incorrect dynamic pressure when the equation is derived from the coupling of the storage equation with the momentum equations of pore water. It was caused by the absence of the inertia of sediment solids in the pressure wave equation.This wave equation is applicable, however, for cases where the solid skeleton of the poroelastic medium is motionless. Conversely, an anomalous high pore pressure could be obtained numerically when the pressure is calculated by the coupling of the momentum of sediment solids with the storage equation, demonstrating also a misleading solution from the incorrect momentum in the pressure wave equation (Hung and Chang 1995).

In the present study, the analytical and computational procedures of Hung and Chang (1995) is applied for studying the dan-water-sediment interaction and the dam-water-foundation interaction. The results in Fig. 1a show the transient dynamic pressure distributions on a rigid dam face for vertical harmonic ground acceleration: $a_y = \alpha g \cos \omega t$ (α= 0.2 and ω = 62.83 rad/sec or 10 Hz). The horizontal dash line marks the water-sediment interface. The solid curve represents the peak pressure profile when the solution becomes periodic. The original reservoir depth (h) is 61 m (200 ft) which is reduced to 45.75 m (i.e., d_o = 150 ft) by a layer of sandy sediment with the specific weight of 2.5 and the porosity of 0.3. The temporal variation of the dimensionless hydrodynamic force **F** (= $F/0.5\rho g h^2$) shown in Fig. 2 corresponds to a ground excitation at the resonant frequency of the reservoir (ω_o = 45.53 rad/sec) with two intensities (α = 0.2 and 0.3). The dynamic force oscillations reflect that the increase in the inertia effect is balanced by the damping force. Curve A in Fig. 3 shows the amplitude of the periodic force $\mathbf{F}^*$ as a function of the dimensionless frequency $\omega h/C$ (where the celerity of saturated water C = 4,720 ft/sec). Curve B represents the contribution of dynamic force by the impounded water alone. The corresponding analytical solution of Cheng (1986) is plotted here as Curve C. The gap between Curves A and B is contributed by pore pressure. The elastic force is insignificant in this case. The disagreement between Curves B and C is due to the application of different Neumann boundary conditions. In our analysis, the inertia of pore water and sediment solids at the sediment-foundation interface is implemented (Chen and Hung 1993) while in the analytical solution of Cheng (1986) only the inertia of pore water is considered. A correction factor for Curve C is proposed for the rigid foundation; it is simply the ratio of the combined inertia of pore water and sediment solids to the inertia of pore water. With this factor, Curve C can be modified and replaced by the hollow points in the figure. They are in good agreement with Curve B.

Further compliments to Cheng's solution are reflected in Fig. 4 for the water-foundation interaction without the sediment. The finite element solutions of the dimensionless periodic force, **F***, are in good agreement with Cheng's for a stiff foundation (E_f =7x10^{10} N/m^2). A smaller time step and refined elements are necessary when the frequency of the ground motion is near the natural frequency of the reservoir. Fig. 5 shows a good agreement between the finite element solutions with those calculated by using the radiation boundary condition at the water-foundation interface (Hall and Chopra 1982) for a soft foundation with E_f = 7x10^9 N/m^2. They are also in good agreement with the analytical solution of Cheng (1986).

The study was support by NSF grant CMS-9319341 from the Earthquake Hazards Mitigation Program.

REFERENCES

Bougaha, S., and Tassoulas, J.L. (1991). J of Engrg Mech., Vol. 117, 1839-1850.
Chen, B.F., and Hung, T.K. (1990). Hydraulic Engineering, Proc. Of 1990 National Conference, ASCE, Vol. 1, 109-114.
Chen, B.F., and Hung, T.K. (1993). J. of Engineering Mechanics, Vol.119, 1411-1433.
Cheng, A.H.D. (1986), J of Engineering Mechanics, Vol.111, 654-663.
Fenves, G., and Chopra, A.K. (1984). Earthquake Engrg. Struct. Dyn., Vol.11, 663-680.
Hall, J.F., and Chopra, A.K. (1982). Earthquake Engrg. Struct. Dyn., Vol.10, 305-332.
Hung, T.K., and Chang, K.C. (1995). Proc. of Interna. Conf. On Math. Modelling, Universiti Brunei Darussalam, 86-100.

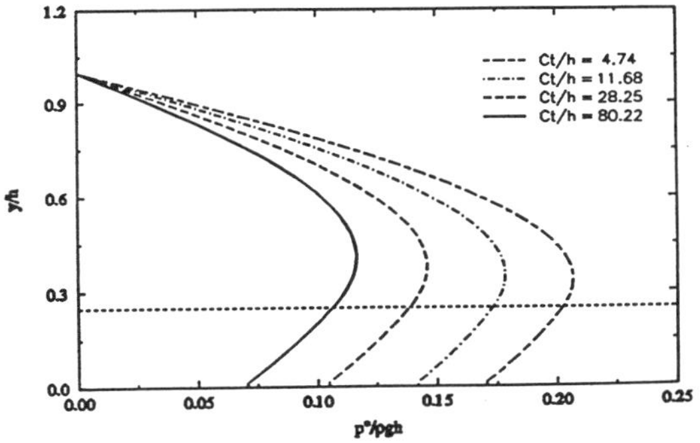

Fig. 1. Dynamic pressure distributions on rigid dam for ω = 62.83 rad/sec

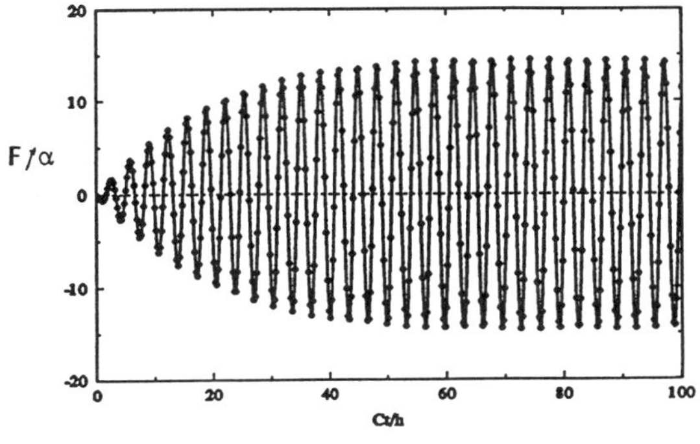

Fig. 2. Dimensionless hydrodynamic force **F** at natural frequency ω_o.

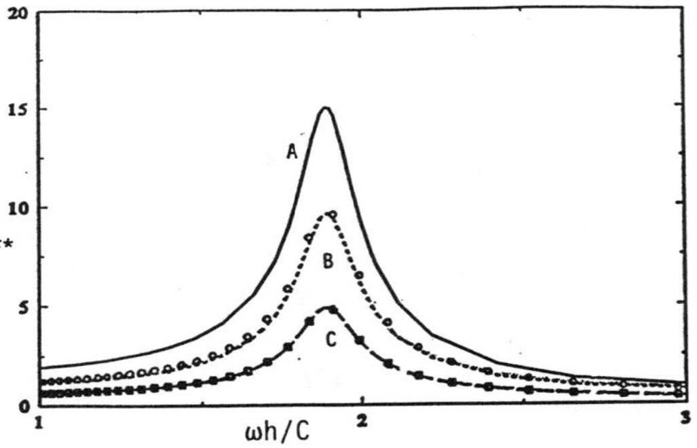

Fig. 3. Dimensionless periodic force **F*** as function of frequency parameter ωh/C.

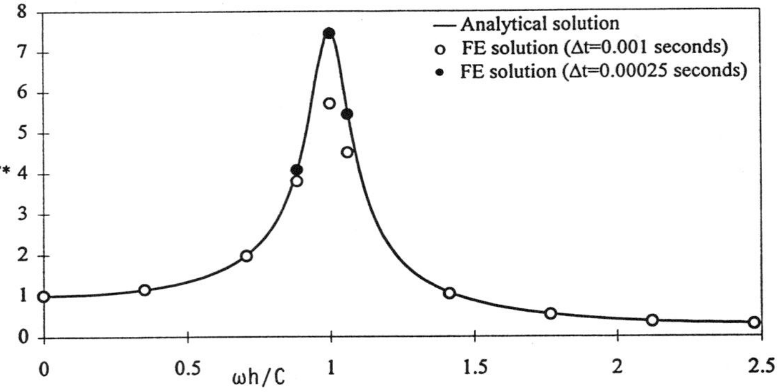

Fig. 4. Dimensionless periodic force **F*** for stiff foundation without sediment.

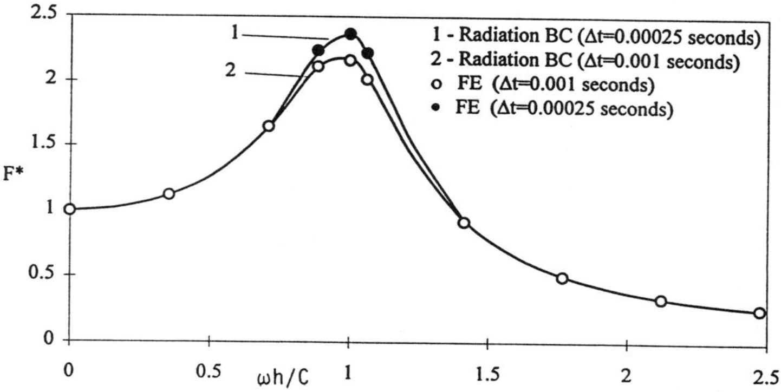

Fig. 5. Dimensionless periodic force **F*** for soft foundation without sediment.

Viscous Ship Waves on Water of Finite Depth

Andy T. Chan and Allen T. Chwang[1]

Abstract

The wave generation by a submerged point force in a viscous fluid of finite depth is analyzed. In the first part of this work, the image system of an Oseenlet near a no-slip wall is derived. The second part deals with the generation of waves by these solutions. The wave generated is shown to exhibit the Kelvin ship wave pattern.

Introduction

When a body cruises with constant velocity in a fluid, it induces on the free-surface the Kelvin ship wave pattern. In this paper, we investigate the waves generated by a moving body simulated by an Oseenlet with a uniform velocity in a viscous fluid of finite depth. The solutions for the case of infinite depth have been obtained by Chan & Chwang (1996). The solutions, which satisfy the linearized free-surface conditions, possess many characteristics of Kelvin's ship wave pattern except that they also include a viscous decay term. They will be used as the building blocks for the present study.

In the first part of this paper, we shall derive the image system of an Oseenlet near a no-slip boundary. The second part will extend the first part of the work to derive the solutions for the wave generated by an Oseenlet.

The image system of an Oseenlet near a no-slip boundary

We first consider the problem of a steady three-dimensional Oseen flow past a horizontal no-slip boundary in the presence of a singular force acting at the opposite direction to the flow. We want to remove the physical boundary by finding the complementary terms to the Oseenlet situated at a distance h from a stationary plane boundary, such that the no-slip boundary condition can be satisfied on the boundary.

The viscous fluid occupies the upper half-space $z > 0$ with an infinite plate, $-\infty < (x,y) < \infty$, located at $z = 0$, and a singular force moving towards the negative x-direction with constant velocity U. The non-dimensional Oseen equations are

[1] Department of Mechanical Engineering, The University of Hong Kong, Hong Kong.

$$\nabla \cdot \mathbf{u} = 0 \text{ and } \frac{\partial \mathbf{u}}{\partial x} = -\nabla p + \varepsilon \nabla^2 \mathbf{u}, \tag{1}$$

where $\mathbf{u}$ = (u,v,w) is the velocity vector and p is the pressure. The dimensionless parameter ε is the reciprocal of the Reynolds number and will be used as the perturbation parameter in the asymptotic expansion of wave amplitude.

For a singular force located at (0,0,h) with magnitude $4\pi F$ along the negative x-direction, the velocity solution to (1) for an unbounded fluid is given by

$$\mathbf{u}_u[x,y,z-h] = -\frac{F}{\varepsilon}\frac{e^{\lambda(x-r)}}{r}\mathbf{e}_x + F\nabla[\frac{e^{\lambda(x-r)}-1}{r}], \tag{2}$$

where $\lambda = 1/(2\varepsilon)$ and $r = \sqrt{x^2+y^2+(z-h)^2}$. (3)

We now impose the no-slip boundary condition on the horizontal plane boundary. Similar to Blake (1971), we locate an Oseenlet of equal magnitude but opposite in sign at the mirror image point (0,0,-h) with respect to the plane wall. Therefore the resultant velocity field can be written as

$$\mathbf{u} = \mathbf{u}_u[x,y,z-h] - \mathbf{u}_u[x,y,z+h] + \mathbf{u}_b, \tag{4}$$

where $\mathbf{u}_b$ are unknown function introduced to satisfy the no-slip boundary conditions.

Applying the no-slip boundary conditions at the wall (z = 0), it can be shown that

$$\mathbf{u}_b = \frac{F}{2\pi}\int_{-\infty}^{\infty}\int_{-\infty}^{\infty} e^{i(k_1x+k_2y)}\left[\frac{ik_1B\Omega}{A}(e^{-Bz}-e^{-Az})\mathbf{e}_x + \frac{ik_2B\Omega}{A}(e^{-Bz}-e^{-Az})\mathbf{e}_y \right.$$
$$\left. + AB\Omega(\frac{e^{-Bz}}{B}-\frac{e^{-Az}}{A})\mathbf{e}_z\right]dk_1dk_2, \tag{5}$$

where the functions A, B and Ω are defined as

$$A = \sqrt{k_1^2+k_2^2},\ B = \sqrt{k_1^2+k_2^2+\frac{ik_1}{\varepsilon}} \text{ and } \Omega = -\frac{2(e^{-Bh}-e^{-Ah})}{B-A}. \tag{6}$$

It then follows from Olmstead (1975) that a solution of the Oseen equations (1) can be given by a distribution N[x,y] of vertical Oseen doublets in the form of

$$\mathbf{u}_b = -\int_{-\infty}^{\infty}\int_{-\infty}^{\infty}\frac{\partial \mathbf{u}_u}{\partial z}[x-x', y-y', z]N[x',y']dx'dy', \tag{7}$$

where N[x,y] has the expression

$$N[x,y] = \frac{32\varepsilon}{\pi}\int_0^{\frac{\pi}{2}}\int_0^{\frac{\pi}{2}} e^{-xC}\cos[yC\cos\lambda\tan\vartheta]\frac{\cos[hC]-\cos[hC\frac{\cos\lambda}{\cos\vartheta}]}{\cos\vartheta-\cos\lambda}\cot\lambda\, d\lambda\, d\vartheta, \tag{8}$$

where $C = \cot\lambda/(\varepsilon\sin\lambda\cos\vartheta)$. (9)

<u>Surface wave generation by an Oseenlet with a horizontal no-slip boundary</u>

We now consider the general problem of finding the wave amplitude generated by an Oseenlet submerged at a depth h_2 beneath the free surface (z = 0) in a viscous fluid of depth h_1. As mentioned earlier, the no-slip boundary can be removed by locating an Oseenlet of opposite sign at $(0,0,-2h_1 + h_2)$ and a planar distribution N[x,y] of vertical Oseen doublets at $z = -h_1$. The linearized free surface conditions represent the vanishing of stresses at $z = 0$,

$$\frac{\partial u}{\partial z}+\frac{\partial w}{\partial x}=0,\ \frac{\partial v}{\partial z}+\frac{\partial w}{\partial y}=0 \text{ and } \frac{\partial p}{\partial x}-2\varepsilon\frac{\partial^2 w}{\partial x\partial z}-w=0. \tag{9}$$

Applying conditions (9), we have the final asymptotic solutions for the wave amplitude generated by an Oseenlet with magnitude $4\pi F$ at depth z_0 from the free surface as:

$$\eta \approx \eta_o+\eta_v = \eta_o + O[\sqrt{\varepsilon}] \tag{10}$$

where

$$\eta_o[z_0]=4F\left(\frac{2\pi}{R\cos\theta}\right)^{\frac{1}{2}}(1-8\tan^2\theta)^{-\frac{1}{4}}\left\{\left(\frac{m^+ +1}{2}\right)^{\frac{3}{4}} e^{-\frac{z_0}{2}(1+m^+)-\frac{4\varepsilon R}{m^+}\left(\frac{m^+ +1}{2}\right)^3\cos\theta}\right.$$
$$\times\cos\left[R\left(\frac{m^+ +1}{2}\right)^{\frac{1}{2}}\left(\cos\theta-\left(\frac{m^+ -1}{2}\right)^{\frac{1}{2}}\sin\theta\right)-\frac{\pi}{4}\right]+\left(\frac{m^- +1}{2}\right)^{\frac{3}{4}}\times$$
$$\left. e^{-\frac{z_0}{2}(1+m^-)-\frac{4\varepsilon R}{m^+}\left(\frac{m^- +1}{2}\right)^3\cos\theta}\cos\left[R\left(\frac{m^+ +1}{2}\right)^{\frac{1}{2}}\left(\cos\theta-\left(\frac{m^+ -1}{2}\right)^{\frac{1}{2}}\sin\theta\right)+\frac{\pi}{4}\right]\right\}, \tag{11}$$

and
$$m^{\pm}=\frac{\cot^2\theta}{4}\left(1\pm\sqrt{1-8\tan^2\theta}\right). \tag{12}$$

Thus the resultant wave amplitude η_B for the finite depth system can be expressed as

$$\eta_B \approx \eta_o[h_2]-\eta_o[2h_1-h_2]+\int_{-\infty}^{\infty}\int_{-\infty}^{\infty}\eta_o[x-x',y-y',h_1]N[x',y']dx'dy', \tag{13}$$

Since N[x,y] is of the order of $\varepsilon\exp[-x/\varepsilon]$ and that of the viscous part of the wave amplitude is $\sqrt{\varepsilon}$, the dominant terms in the wave amplitude for small ε are thus the inviscid part of the Oseenlet solution and its mirror image. The characteristics of the viscous ship wave pattern described by Chan & Chwang (1996) are thus retained, except that the wave amplitude is reduced for small depth h_2. Figure 1 shows the free surface profile for an Oseenlet of magnitude $F = 1/4$ at $\varepsilon = 0.001$, $h_1 = 10$ and $h_2 = 1$.

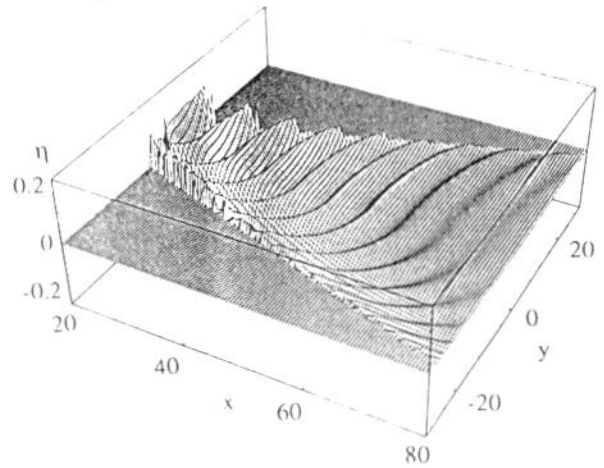

Fig. 1. Free surface profile generated by an Oseenlet of strength $F = 1/4$ at $\varepsilon = 0.001$, $h_1 = 10$ and $h_2 = 1$.

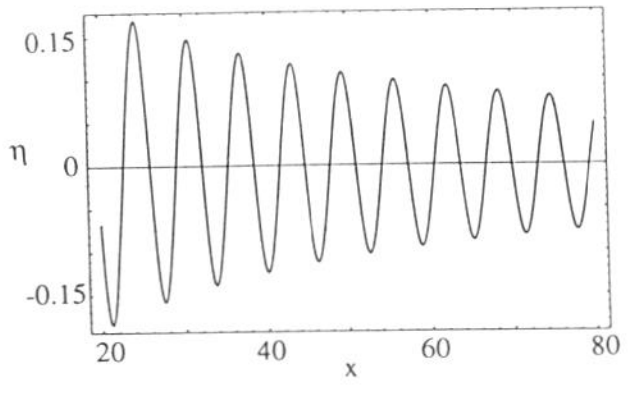

Fig. 2. Wave amplitude along the line of motion of an Oseenlet of strength $F = 1/4$ at $\varepsilon = 0.001$, $h_2 = 1$ and $h_1 = 10$.

The wave amplitude along the line of motion of an Oseenlet is plotted in Fig. 2 for $h_1 = 10$, $\varepsilon = 0.001$ and $h_2 = 1$. Because the no-slip boundary reduces the velocity at its near field, wave amplitude increases with h_1 but remains small for small h_1. The maximum wave amplitude along the line of motion is plotted in Fig. 3 against the normalized depth h_1 at $\varepsilon = 0.001$ and $h_2 = 1$.

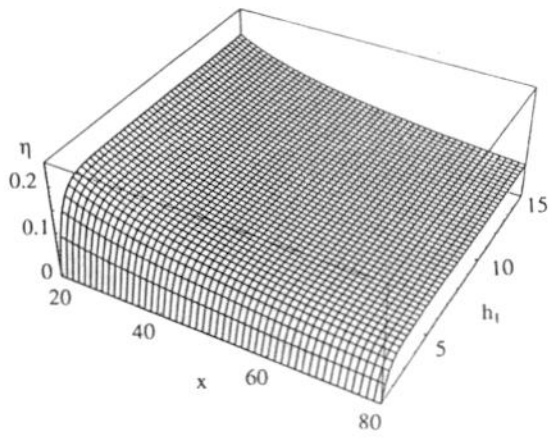

Fig. 3. Maximum wave amplitude along the line of motion of an Oseenlet versus water depth h_1 at $\varepsilon = 0.001$ and $h_2 = 1$.

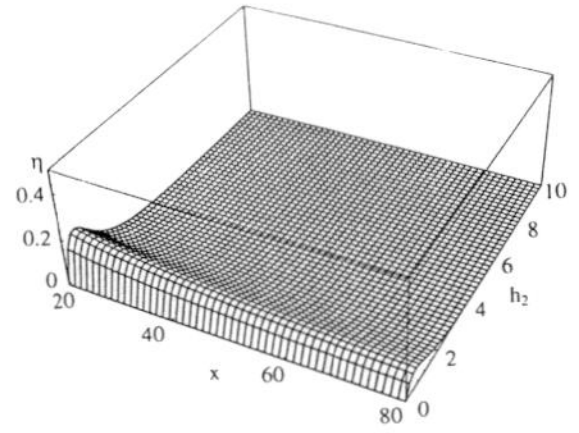

Fig. 4. Maximum wave amplitude along the line of motion of an Oseenlet versus the submergence depth h_2 at $\varepsilon = 0.001$ and $h_2/h_1 = 0.5$.

Fig. 4 shows the variation of maximum wave amplitude against the submergence depth h_2 with the relative submergence depth $h_2/h_1 = 0.5$ at $\varepsilon = 0.001$. The wave amplitude first rises with the water depth to a maximum value and then drops to zero. It has been shown that the wave amplitude increases with water depth. However, as h_2/h_1 is kept constant, the submergence of the Oseenlet also increases, and it reduces the wave amplitude.

Conclusions

The wave amplitude due to an Oseenlet moving in water of finite depth is derived. The wave amplitude can be expressed as the wave generated by the Oseenlet minus that of its mirror image plus the contribution due to a planar distribution of vertical Oseen doublets. The dominant terms of the wave amplitude are those of the Oseenlet and its mirror image only. The presence of the no-slip boundary is found to reduce the wave amplitude for small depth because it reduces the velocity of the system.

Acknowledgements

This work was sponsored by the Committee on Research and Conference Grants, The University of Hong Kong, under Grant No. 335/064/0071.

References

Chan A.T. & Chwang A.T. (1996), "Interaction of laminar far-wake with a free surface," *Phys. Fluids* (accepted for publication).

Blake J.R. (1971), "A note on the image system for a stokeslet in a no-slip boundary," *Proc. Camb. Philo. Soc.*, Vol. 70, pp. 303-310.

Olmstead W.E. (1975), "A homogeneous solution for viscous flow around a half plane," *Quart. Appl. Math.*, Vol. 33, pp. 165-169.

A Complete Three Dimensional Analysis of Pressures on a Vertical Cylinder by Earthquakes Including Fluid-Structure Interaction

Bang-Fuh Chen[1], member

Abstract

A complete three dimensional analysis was made, that is, three dimensional equations of motion and the simultaneous action of three components of ground acceleration were included in the analysis. The flexibility of the offshore cylinder could significantly increase the hydrodynamic pressures acting on cylinder faces and the fluid-structure interaction is necessary in offshore cylinder analysis. For a solid and stubbier circular cylinder, the vertical component of ground acceleration may be negligible.

Keywords: nonlinear hydrodynamic pressure, flexible offshore cylinder, earthquake, three dimensional analysis.

Introduction

The analysis of surface flow around an offshore cylinder has been studied in the past decade. The generation of the corresponding surface flow could be resulted from the cylinder motion, such as Wang and Chwang (1989), and Chen (1995a), or from the assulating waves pass the cylinder, such as Wang et. al. (1992). Except Chen 1995a, all the analyses adopted three-dimensional wave equations but used only one fixed-directional ground motion or one-way assulating wave as the examples. They claimed the results of multiple exciting directions can easily be obtained by linear superposition. However the principle of the superposition is invalid when the flow field presents high nonlinearity. In order to fully understand the detail insight of the nonlinear effect on offshore cylinder hydrodynamics, both three-dimensional momentum equations and simultaneous action of three-dimensional ground acceleration should be included in the analysis. Besides, we could assume the fluid-structure interaction effect on cylinder hydrodynamics be significiant. In the present analysis, a developed finite-difference approximation of three-dimensional Euler's equations coupled with the nonlinear free-surface boundary condition is associated with the finite element approximation of cylinder structure to evaluate the nonlinear hydrodynamic pressures, the free surface profile and dynamic response of a deformable vertical cylinder. The cylinder is exicted simultaneously by a three-dimensional ground motion.

[1] Associate Professor, Department of Marine Environment, National Sun Yat-Sen University, Kaohsiung,Taiwan, R. O. C.

Basic Equations

The 3-D Euler's equations in cylindrical coordinates can be written as

$$u_t + uu_r + vu_\theta/r + wu_z - v^2/r = p_r/\rho \tag{1}$$

$$v_t + uv_r + vv_\theta/r + wv_z + uv/r = -p_\theta/\rho \tag{2}$$

$$w_t + uw_r + vw_\theta/r + ww_z = -p_z/\rho - g \tag{3}$$

and the corresponding continuity equation of incompressible fluid is

$$u_r + v_\theta/r + w_z + u/r = 0 \tag{4}$$

in which, u, v and w are, respectively, the velocity components of fluid in r, θ and z directions, p the total pressure, ρ the fluid density and g the gravitational acceleration. The terms with subscripts indicate the partial derivatives with respect to the subscripts.

The sea water depth, $h(r, \theta, t)$, are determined from the kinematic free surface boundary condition

$$h_t + uh_r + vh_\theta/r = w \tag{5}$$

The definition sketch of the problem is shown in Figure 1 where b_2 is the distance between origin and upstream end, d_0 the undisturbed sea depth, a the radius of the cylinder and $b_1(\theta, z, t)$ the profile of cylinder face. The following functions are used to remove the dependence of b_1 and h to time.

$$r^* = (r - b_1(\theta, z, t))/(b_2 - b_1(\theta, z, t)) \quad , \quad z^* = 1 - (z + d_0)/h(r, \theta, t) \tag{6}$$

The coordinates (r^*, z^*) can be further transformed so that the layer near the free surface, the sea bottom, the cylinder face and the upstream end will be stretched (Chen, 1995a).

The isoparametric eight-node prismatic finite element (linear finite element) is used to model the cylinder. The matrix equation of a linear structural dynamic system can be written as the form:

$$M\ddot{X} + C\dot{X} + KX = Q \tag{7}$$

where M is the lump mass matrix, C the viscous damping matrix, K the stiffness matrix, Q the force matrix, X, $\dot{X}$, and $\ddot{X}$ are, respectively, the displacement, velocity and acceleration vectors. The mode-superposition technique is used to solve for the dynamic response of the cylinder. The detail algorithm and iteration procedures were the same as reported earlier in (Chen 1995a,b).

Results and Discussion

The mesh sizes ΔR, $\Delta\Theta$ and ΔZ are set to be equal to 0.1, 0.05 and 0.1 respectively. The dimensionless hydrodynamic coefficients is defined as $C_f = \mathbf{P}/\rho \mathbf{g} \mathbf{d}_0^2$, where $\mathbf{P}$ is the resultant hydrodynamic force acting on the cylinder.

The offshore cylinder was assumed to be made of concrete. The radius (r_0) and height (l_0) of the cylinder are 10 feet and 100 feet respectively. The undisturbed sea depth is 80 feet. The Young's modulus, Poisson ratio and the unit weight were, respectively, 3.44 x 10^7 kPa, 0.2 and 24330 N/m^3. The first 15 fundamental modes and corresponding frequencies were calculated by the subspace iteration method and the leading five are 20.61, 20.61, 115.90, 115.90 and 116.77 (rad/sec). By the action of the two components of horizontal ground acceleration of El-Centro 1979 earthquake, Fig. 2 shows the comparison of the resultant hydrodynamic force coefficients acting on cylinder due to rigid and deformable cylinder motions. The hydrodynamic forces acting on a deformable cylinder are much larger than those on a rigid cylinder and the fluid-structure interaction could significantly augment the hydrodynamic pressure acting on the cylinder. Due to the simultaneous action of three components of ground acceleration, Fig. 3 shows the time history of the hydrodynamic pressures at $\theta = 0^0$ and $z = 0.5d_0$. The results of two horizontal ground acceleration were also plotted in the figure. The corresponding hydrodynamic force coefficient of both cases are presented in Fig. 4. Since the dominant vertical component of the ground acceleration, the addition of the vertical exciting significantly increases the hydrodynamic pressures acting on cylinder face. However, the exciting of the vertical ground acceleration to the fluid is basically the change of the acceleration of gravity and the hydrodynamic pressure around the cylinder is pseudo-static and isotropic and the corresponding contribution to hydrodynamic force coefficient is actually zero (all be balanced in radial direction). For this stubbier cylinder ($r_0 = 10$ feet and $l_0 = 100$ feet), the calculated hydrodynamic force coefficeints due to only two horizontal components exciting are much larger than those due to three components exciting simutaneously.

Conclusions

The flexibility of the offshore cylinder could significantly increase the hydrodynamic pressures acting on cylinder faces, that is, the fluid-structure interaction is necessary in offshore cylinder analysis. For a solid circular cylinder, the vertical component of ground acceleration may be negligible unless the exciting frequency is close to the natural frequence of the cylinder structure. But the other structure types, such as slender cylinder or hollow circular cylinder with thin wall thickness, the significant structural dynamic response and the compressive hydrodynamic pressure (or force) may be essential and further studies should be made.

Acknowledgement

This study is partially supported by the National Science Council of R. O. C. under the grant NSC 83-0209-E-110-006.

Reference

1. Chen, B. F., (1995a), "A complete three dimensional nonlinear hydrodynamic analysis of a vertical cylinder during earthquakes, Part I: a rigid cylinder motion", Submitted to J. Ergrg. Mech.
2. Chen, B. F., (1995b), " A comprehensive study of nonlinear hydrodynamic pressure on concrete gravity dam", to be issued in Engineering Structures.
3. Wang, K. H. and Chwang, A. T., (1989), "Nonlinear free-surface flow around an impulsively moving cylinder", J. Ship Research, 33(3), 194-202.
4. Wang, K. H., Wu, T. Y., and Yates, G. T., (1992), "Three-dimensional scattering of solitary waves by vertical cylinder", J. Wtrways Port. Coast. and Oc. engrg. , Vol. 118, (5), pp. 551-566.

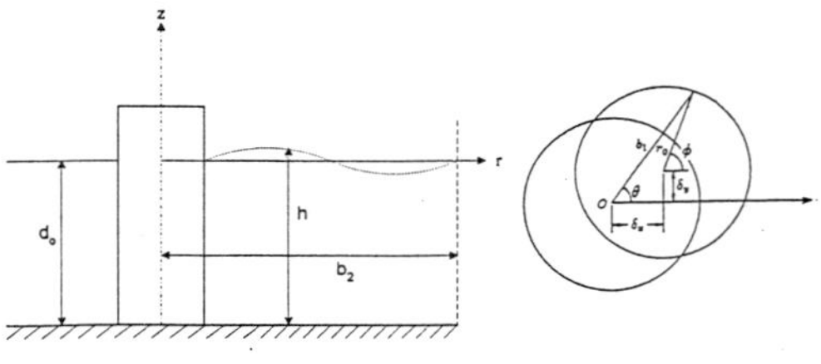

Figure 1: The definition sketch

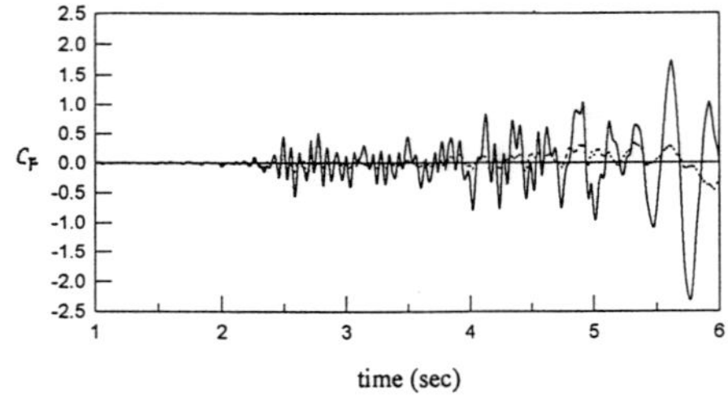

Figure 2: Due to two horizontal ground exciting, the comparison of hydrodynamic force coefficients due to rigid (dot line) and deformable (solid line) cylinder motions.

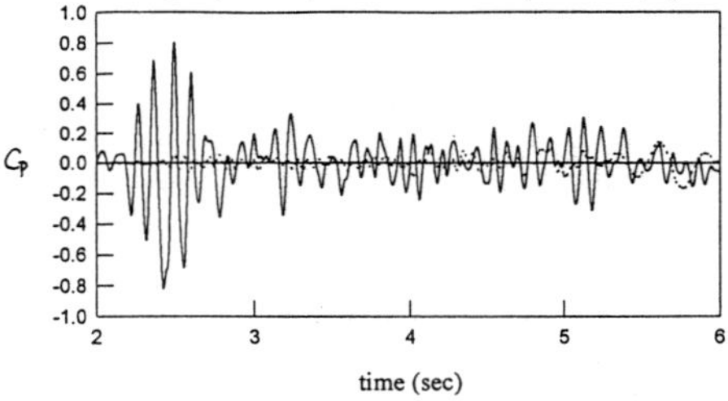

Figure 3: For a deformable cylinder, the comparison of the time history of the hydrodynamic pressures at $\theta = 0^0$ and $z = 0.5d_0$, due to simultaneous action of three (solid line) components and two horizontal (dot line) components of ground acceleration.

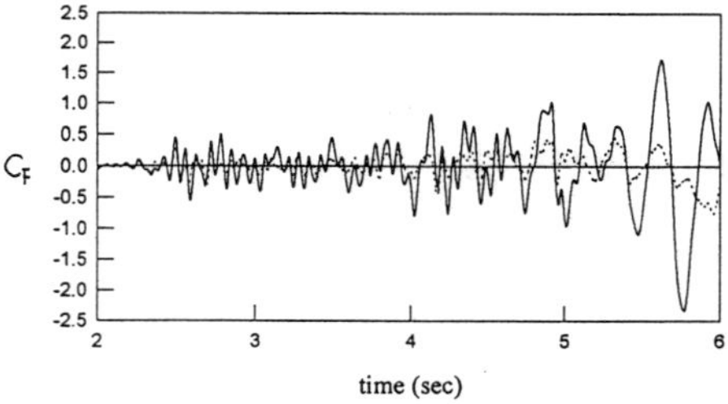

Figure 4: For a deformable cylinder, the comparison of the time history of the hydrodynamic force coefficients, due to simultaneous action of three (dot line) components and two horizontal (solid line) components of ground acceleration.

First Exit Times in Non-Linear Dynamical Systems by Advanced Monte Carlo Simulation

H.J. Pradlwarter[1] and W. Kliemann[2]

Abstract

Dealing with nonlinear dynamical systems, the system dynamics, the excitation dynamics and the underlying noise process, respectively, are all nonlinear, and the noise statistics is non-Gaussian. Hence one can not expect to obtain explicit expressions for failure probabilities, failure distributions, etc.. The reliability of non-linear dynamical systems can for such general cases only be assessed quantitatively by numerical Monte Carlo simulation. Monte Carlo simulation, however, is not well suited to assess exit probabilities less than 10^{-4} or to estimate the low probability tail of the exit time. For such cases, Advanced Monte Carlo simulation is suggested. The proposed procedure, is utilized to estimate exit times in the low probability range.

Introduction

The dynamics of nonlinear systems has attracted considerable interest in the last two decades in wide field such as mathematics, natural science, engineering and economy. For more realistic modeling of observed natural phenomena nonlinear stochastic systems might be required, where random disturbances are included in the mathematical representation. These disturbances can be external (environmental) or internal (parameter excitation) leading to multiplicative terms affected by their own dynamics. Such systems are generally modeled as random differential equation of the form

$$\dot{x} = X_0(x) + \sum_{i=1}^{m} \xi_i(t)X_i(t) \qquad (1)$$

where x is a state vector of the system, $\dot{x} = X_0(t)$ is the undisturbed deterministic system, $\xi_i(t)$,i=1...m are the random perturbations and $X_i(x)$ define the perturbation

---

1) Institute of Engineering Mechanics, University of Innsbruck, Innsbruck, AUSTRIA

2) Department of Mathematics, Iowa State University, Ames, IA, USA

dynamics. These functions simplify to $X_i(x) \equiv 1$ for the case of external noise. The characteristics of the random process $\xi_i(t)$ are assumed to be known from observations.

The deterministic system $\dot{x} = X_0(t)$ itself might have in general a rather complex behavior. Besides equilibrium points, periodic orbits and homoclinic orbits also chaotic behavior might occur in higher dimensions. The stochastic processes $\xi_i(t)$ are in general non-Gaussian and the perturbation dynamics nonlinear. Some analytical techniques are available to study the long term behavior of systems of low dimensions. However, it is not possible to obtain analytical solution of the statistics of the nonlinear system in finite time as one is concerned in real life and hence in reliability studies. Instead of analytical methods one use efficient numerical techniques as Monte Carlo simulation to compute the random orbits following the stochastic differential equation (1).

"Double & Clump II"

Monte Carlo simulation (MCS) is not well suited to assess exit probabilities less than 10^{-4} or to estimate the low probability tail of the exit time. Only a very small fraction of all realization will exit by chance to the absorbing boundary while most of the realizations do not contribute to the exit probability.

For such cases, straight, i.e. direct MCS is quite inefficient and to increase the efficiency Advanced Monte Carlo simulation procedures have been suggested [1,2,3,4]. The references [1,2] discuss the "Double & Clump" (D&C) procedure as proposed originally, and [3] deals with a similar approach denoted by "Russian Roulette & Splitting" and in [4] Genetic Algorithms are utilized. These procedures are independent from the integration scheme and operate at subsequent (frozen) time steps. All these approaches redirect the realizations on the basis of a heuristic in particular an energy criterion. Unfortunately, the selection criterion depends on the specific problem and no clear guidance how to establish such a criterion could be given. For this reason, a new approach has been developed, for such selection criteria are no longer needed. The approach has some similar features as the original D&C and will be called "Double & Clump II" (D&C II).

The key to this newly developed approach is to distribute dynamically the sample density (realizations with different probabilities) approximately uniform within the state space. The tendency to a uniform density of realizations in the state space is achieved by clumping subsequently the closest state pair to a single realization until the number of realization is reduced to half of n, where n is the sample size. The closest pair (k1,k2) is determined dynamically by evaluating the smallest distance among all remaining samples (including previously clumped realizations)

$$d_{k1,k2} = \|y_{k1} - y_{k2}\| = \sqrt{\sum_{i=1}^{m} (y_{k1,i} - y_{k2,i})^2} \qquad (2)$$

in a normalized state space, given by

$$y_{k,i} = \frac{x_{k,i} - \hat{x}_i}{\sigma_i} \tag{3}$$

where m is the dimension of the state space, $\hat{x}_i$ denotes the sample mean, σ_i the sample standard deviation of the i-th component, respectively. Clumping means that two realizations are merged into a single realization. The single state vector x_{k0} resulting from clumping the pair of state vectors x_{k1} and x_{k2} is selected as the mean, i.e.

$$x_{k0} = \frac{1}{w_{k0}}(x_{k1}w_{k1} + x_{k2}w_{k2}) \; ; \; w_{k0} = w_{k1} + w_{k2} \tag{4}$$

where w_k represents the weight of the k-th realization or the associated probability which is set to 1/n before D&C is utilized the first time. The main effort is spent for determining subsequently the closest pairs (k1,k2), where special techniques and data structures are utilized for efficiency resulting in an computational effort proportional to n·ln(n). Clumping introduces a bias in the distribution, which is balanced by local adjustments of clumped realizations such that the marginal distributions are retained.

After clumping, all remaining n/2 realizations (denoted as x_{j0}) are doubled ($x_{j0} \rightarrow x_{j1}, x_{j2}$), keeping the sample size constant, i.e.

$$x_{j1} = x_{j2} = x_{j0} \; ; \; w_{j1} = w_{j2} = \frac{w_{j0}}{2} \tag{5}$$

Note that doubling does not introduce any bias to the distribution of the samples.

Numerical example

In order to study the efficiency and applicability of the proposed D&C II procedure, the Takens-Bogdanov oscillator as utilized in [5] is examined. This oscillator is characterized by the two dimensional differential equation

$$\begin{aligned} \dot{x} &= y \\ \dot{y} &= \lambda_1 + \lambda_2 x + x^2 + xy + \xi(t) \end{aligned} \tag{6}$$

is examined with bifurcation parameter $(\lambda_1, \lambda_2) = (-0.3, -1.0)$ and external random excitation $\xi(t) = \rho \cdot \sin(\eta(t))$ where $\eta(t)$ is the one-dimensional Wiener process. This system has been investigated in [5] for an excitation range $\rho \in [0, 0.3]$ with respect to the first passage time of the threshold set

$$L = [-1.5, 1.5] \text{ x } [-100, 100] \tag{7}$$

Starting at time t=0 at the stable operating point (focus) (x,y) = (-0.2416,0.0) it has been shown that the system has a finite exit time for an excitation range $\rho > \rho_c = 0.924$. However, very few samples passing the threshold could be generated by straight MCS within the time range [0,1000] for $\rho_c < \rho < 0.20$. Thus, the proposed Advanced MCS D&C II is utilized to investigate this excitation range. The results of this numerical investigation are shown in Fig.1. It is to be noted how fast the exit times increase with decreasing excitation parameter ρ.

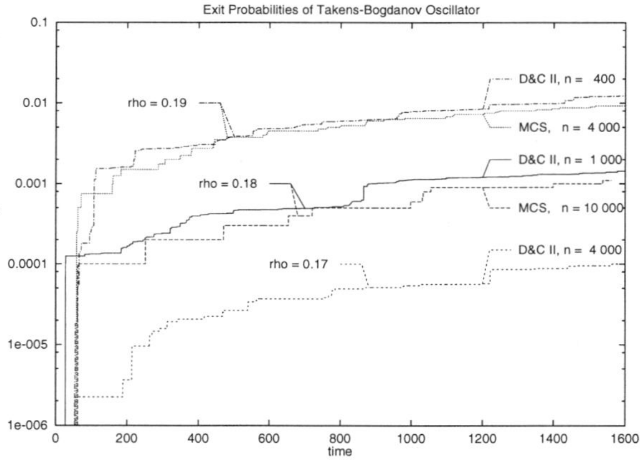

Fig.1: Low exit probabilities of the Takens-Bogdanov oscillator

Concluding Remarks

An extended version of Advanced Monte Carlo simulation procedure, i.e. D&C II has been applied to access the low probability range of exit times of the nonlinear Takens-Bogdanov oscillator. This version does not need a specific selection criterion as required in all other procedures.

Acknowledgment

This work was supported by the Austrian Industrial Research Promotion Fund (FFF) under project N0. 6/795, which is gratefully acknowledged by the authors.

References

[1] Pradlwarter, H.J.: "A Selective MC Simulation Technique for Nonlinear Structural Reliability Assessment", Proc. ASCE Specialty Conference, Denver, Colorado, 1992, ASCE, N.Y., 1992, pp.451-454.

[2] Pradlwarter, H.J., Schuëller, G.I., Melnikov, P.M.: "Reliability of MDOF-Systems", J. *Probabilistic Engineering Mechanics*, 9 (1994), pp.235-243.

[3] Pradlwarter, H.J., Schuëller, G.I.: "On Advanced MCS Procedures in Stochastic Structural Dynamics", *to appear* Aug.96, *Int. J. of Non-Linear Mechanics*.

[4] Johnson, E.A., Bergmann, L.A.: "Monte Carlo Simulation of Dynamical Systems of Engineering Interest in a Massively Parallel Computing Environment", Proc. IUTAM-Symposium on "Advances in Nonlinear Stochastic Mechanics", Trondheim, Norway, July 3-7, 1995, Eds: A.Naess and S.Krenk, Kluwer Academic Publisher, Dordrecht, The Netherlands, to appear, 1996.

[5] Colonius, F., Häckl, G, Kliemann, W.: "Dynamic Reliability of Nonlinear Systems under Random Excitation", DE-Vol.84-1, 1995 Design Engineering Techn. Conf., Vol.3-Part A, ASME 1995, pp.1007-1024.

Statistical Seismic Responses of Structures using Response Spectrum Matching Technique

Ruichong Zhang[1] and Masanobu Shinozuka[2]

Abstract

The response spectrum matching technique is used in this study to investigate statistical seismic peak response characteristics of high-rise structures. As an example, the seismic responses of 310 m Nanjing TV tower at Nanjing, China is carried out for illustration. Numerical results show that the variation of the peak responses are much small compared with their corresponding mean values, especially for the displacement and velocity responses, implying the method used in this study can predict the peak responses quite well only using a few response spectrum-compatible earthquake ground motions.

Introduction

The acceleration input data in the seismic structural analysis can be taken from either a real earthquake record or a simulated motion using spectral representation technique (Shinozuka and Deodatis, 1991) with given power spectrum of the acceleration. However, such ground motion may not have the consistent response spectrum as the target one. Consequently, the structural responses so obtained may deviate significantly from the seismic design requirement, leading to the uncertainty of the earthquake-resistant capability of the structures under consideration. Instead, the responses spectrum matching technique (Lilhanand and Tseng, 1988 and Abrahamson, 1993) is used in this study. Specifically, an

[1]Research Assistant Professor, Department of Civil Engineering, University of Southern California, Los Angeles, CA 90089-2531.

[2]Fred Champion Professor in Civil Engineering, Department of Civil Engineering, University of Southern California, Los Angeles, CA 90089-2531.

earthquake record or a simulated non-stationary ground motion is first selected as an input seed. The selected ground motion is then scaled with respect to the peak value and adjusted so that the modified motion will have the response spectrum consistent with the target one within acceptable error. The statistics of peak structural responses to such modified earthquake motion, which is often crucial to the earthquake-resistant structural design, can thus be investigated.

Response Spectrum Matching Technique

Following Abrahamson (1993), the earthquake input seed is first selected from either real records or simulated motions. The selected ground motion is then scaled and used as the input data to compute the response spectrum. The response spectrum difference (ΔR) between target or design and computed using the earthquake input is matched at natural frequency ω_i and damping ratio ξ_i by a small adjustment time history ($\delta a(t)$), which is assumed to be composed of a series of fundamental functions ($f_j(t)$). Mathematically, it can be represented by

$$\Delta R_i = \int_0^\infty \delta a(\tau) h_i(t_i - \tau) d\tau \tag{1}$$

where h_i is the acceleration impulse response of the oscillator at frequency ω_i and damping ratio ξ_i, and

$$\delta a(\tau) = \sum_{j=1}^{N} b_j f_j(t) \tag{2}$$

in which b_j is the set of coefficients to be determined and N denotes the number of spectral points to match. By solving equation (1), one may obtain a set of coefficients b_j and consequently the earthquake motion is modified by adding the adjustment $\delta a(t)$. Such a procedure is repeated until the response spectrum difference ΔR_i is within the prescribed level.

Numerical Example

The response spectrum matching technique is first demonstrated using El Centro earthquake record. Fig. 1 presents the scaled acceleration record (one fifth of original peak value), modified response spectrum-compatible motion as well as their difference. As seen from Fig. 1, the modified earthquake motion does not significantly change the original record, indicating the modified earthquake motion retain the nature of real earthquake, such as non-stationarity, dominant frequency, *etc.* Fig. 2 depicts the comparisons among target response spectrum, response spectrum due to the scaled acceleration and response spectrum due to modified acceleration time history. It is quite apparent from Fig. 2 that the response spectrum matching technique works out efficiently.

As an example, the response of 310 m high Nanjing TV tower at Nanjing, China with the given response spectrum on the basis of Chinese Seismic Design Code is carried out for illustration. Based on previous investigations and measured data (Cheng, et al. 1993, and Ding and Ren, 1994), the Nanjing TV tower is modeled as 16 degrees of freedom system. The stiffness matrix is obtained by modeling the TV tower as a set of segments with approximately uniform properties (Reinhorn, et al, 1995). 50 acceleration time histories of Loma Prieta earthquake records, collected from the sites located between 10 and 100 miles around the epicenter, are used as input seeds and scaled on the basis of design requirement. They are then modified using the response spectrum matching technique and used as the input to calculate the corresponding responses of the TV tower at 240 m. The statistics of peak responses are finally found in table 1.

Table 1: Mean and standard deviation of structural peak responses

	Displ. (cm)	Velocity (cm/sec)	Accel. (cm/sec^2)
Mean	7.44	13.5	0.93
Standard Deviation	0.001	0.013	0.219

The numerical results in table 1 show that the deviation of peak responses is much small compared with their corresponding mean values, especially for the displacement and velocity peak responses, implying the method used in this study can predict the peak responses quite well only using few earthquake input data.

References

Shinozuka, M. and Deodatis, G. (1991). "Simulation of stochastic processes by spectral representation," *Applied Mechanics Review*, Vol. 44, pp. 191-203.

Lilhanand, K. and Tseng, W. (1988). "Development and Application of realistic earthquake time histories compatible with multiple damping design spectra", *Proceedings of World Conference on Earthquake Engineering*, Tokyo and Kyoto, Vol. II, pp. 819-824.

Abrahamson, N.A. (1993) "Nonstationary spectral matching program RSP-MATCH".

Cheng, W.R. et al. (1992) "The analysis of dynamic characteristics of Nanjing TV tower," *Southeast University*

Ding, Dajun and Ren, Zhenhua (1994) "Design testing and analyses of the Nanjing TV Tower," *Concrete International*, pp. 42-44.

Reinhorn, A.M., Soong, T.T. and Cao, H. (1995) "Preliminary evaluation of wind induced response of Nanjing TV tower", Report, US/PRC Cooperation Program.

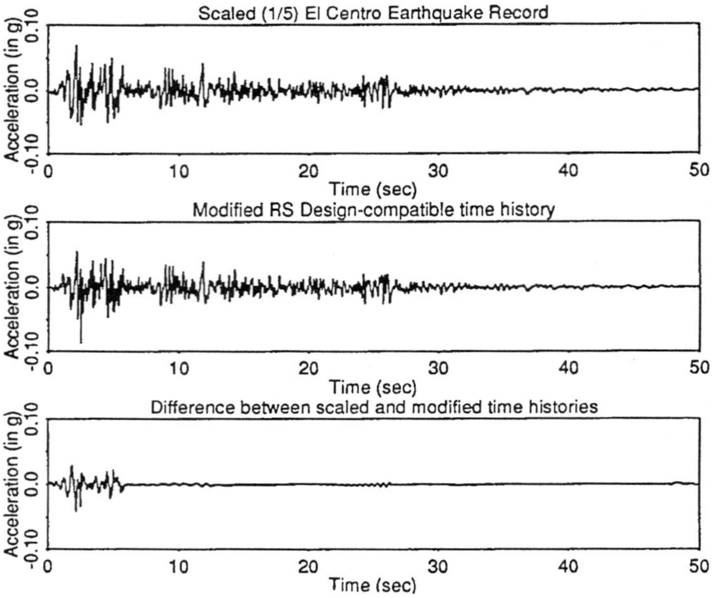

Fig. 1 Scaled and modified El Centro earthquake acceleration time histories as well as their difference.

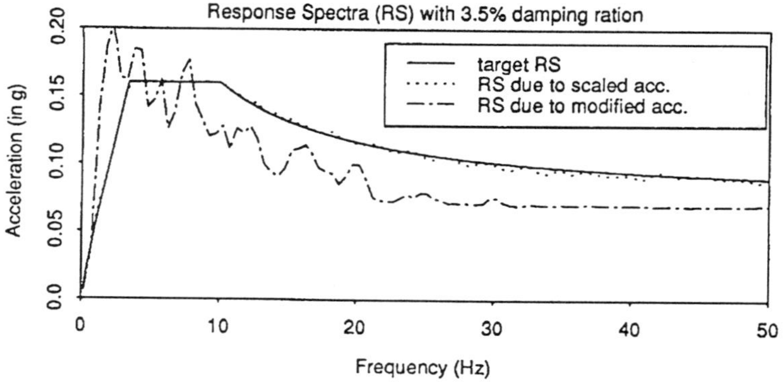

Fig. 2 Comparison between target acceleration response spectrum and those due to scaled and modified acceleration time histories.

Reliability - Based Structural Optimization - Software Development

M. Gasser * and G.I. Schuëller**

Abstract:

A software environment specifically designed for *Computational Stochastic Structural Analysis* (COSSAN) and reliability-based optimization (RBO) of realistic structures is presented. This software concentrates on the treatment of the randomness of the loads and structural parameters as extension of the usual design variables in structural design decision making. A general optimization procedure and the command structure used for RBO are presented. The example of an offshore platform is utilized for showing the software environment and engineering application of RBO.

1. Introduction

Already in 1956 *A.M. Freudenthal* discussed the problem of specification of an acceptable risk *"...on the basis of economic balance between the cost of increasing the safety and the cost of failure"* [1]. There it is claimed that the optimal *economic* probability of failure should make of the sum of all anticipated costs a minimum. Increasing attention has been focused on reliability-based structural optimization during the last two decades (see e.g. [2-4]). Although considerable progress has been made in the fields of structural optimization and reliability theory, *practical applications* of reliability-based optimization procedures to realistic structures are rarely presented in the literature. This may on one hand be due to the complexity of the problem, but on the other hand also due to the lack of availability of a easy-to-use software. Therefore, it is necessary to develop a software platform which enables the design engineer to access most efficiently all the modern computational procedures of structural and advanced reliability analysis in combination with optimization algorithms.

2. Methods of Analysis

Several reliability-based optimization formulations have been suggested in recent studies. The most commonly applied criteria in RBO utilize approaches in terms of constrained single objective functions. This may be expressed in terms of minimization of weight or lifetime costs subject to reliability constraints. However, significant structural optimization can only be performed within a framework of potential consequences of structural failure. Therefore it is necessary to introduce reliability measures in the objective function. In order to evaluate an objective in terms of costs, the objective function has to include three general types of costs, the initial construction costs and consequence costs related to partial and total failure

* Engineer, Kennelbach, Austria; formerly Res. Assoc. Inst. Engr. Mech., Univ. Innsbruck, Austria
** Chair, Institute of Engineering Mechanics, University of Innsbruck, Austria

respectively. Thus, a general reliability-based optimization problem can be defined as follows:

$$\mathrm{E}[C(\mathbf{x},\mathbf{Y})] = c_I(\mathbf{x}) + \sum_j \mathrm{E}\left[C_{f,j}(\mathbf{x},\mathbf{Y})\right] \rightarrow \min \tag{1}$$

where $c_I(\boldsymbol{x})$ denotes the initial costs, and $C_{f,j}(\mathbf{x},\mathbf{Y})$ are the expected failure costs related to partial and total system failure. The vector **Y** contains the random variables which are used to describe the statistical uncertainties of the structural system. The estimation of expected failure consequence costs involve usually the determination of corresponding failure probabilities which further requires the calculation of multi-dimensional probability integrals. Considering one single partial failure criterion related to a specific structural damage state the objective function i.e. equ. (1) rewrites

$$\mathrm{E}[C] = c_I + c_s\left(1 - p_{fc}\right)p_{fs} + c_c p_{fc} \rightarrow \min \tag{2}$$

where p_{fs} and p_{fc} are the occurrence probabilities with respect to partial and total failure respectively. The factors c_s and c_c are the respective consequence cost factors. At this point it should be noted that the failure probabilities depend on the respective design state and therefore, they have to be evaluated repeatedly during the iterative design optimization process. For the highly nonlinear optimization problem as defined by eqs. (1), (2), the well known NLPQL-algorithm [5] is used to control the optimization procedure. Furthermore, the necessary tools, such as structural analysis, reliability analysis and optimization are implemented in the interactive controllable multi-purpose software environment COSSAN [6].

3. COSSAN - Software Environment

It is well known that the practical use of information processing requires not only efficient computational procedures, but also the respective software. The most general and practical approach to reliability-bases structural optimization is the direct connection of structural analysis, reliability analysis, and optimization algorithms (see e.g. [7]). In the first step the decision model has to be developed and the problem functions have to be defined, e.g. the objective function determines the total expected costs including failure consequence costs subject to reliability constraints. In addition to that the reliability model has to cover all uncertainties of the decision system so that the reliability measures can be estimated utilizing an efficient reliability analysis method such as the RSM. After all definitions and having the calculation tools available, the problem functions can be determined. This is the point now where the optimization algorithm can be executed. During the iterative procedure the problem functions and their gradients are evaluated according to the directives of the mathematical programming algorithm. External or third-party software can be utilized to calculate the problem functional (e.g. FE-software). The mechanical system and/or the reliability model are modified by iterative change of the design. Subsequently, the structural response quantities are calculated (use of FE) to determine the limit state points respectively the response surface and other functional required. With the calculation of reliability measures, e.g. failure probabilities (use of RSM and advanced MCS), the problem function values are determined. Iteration and/or other parameters may be controlled interactively which is required when the optimization procedure does not show convergence.

4. Numerical Example

A template, steel made platform (Fig. 1) has been selected. It has a quadratic cross-section and consists of elements with tubular cross-sections.

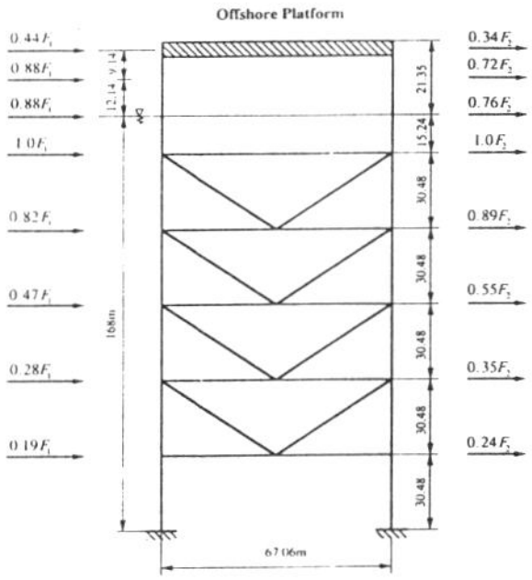

Fig. 1: Model of fixed offshore structure

Uncertainties in the structural resistance are taken into account by a fully correlated random yield strength of the structural elements. A coefficient of variation of 0.05 is assumed for the used steel type (St52-3). First yielding (partial failure) in any structural component and complete plastification (collapse) are the considered two failure criteria.It is assumed that the main part of the loading to the offshore structure results from extreme waves due to severe storms. For details on the load analysis it is referred to [7].

Carrying out the RBO, first, an objective formulation may be considered where the production costs are formulated as a function directly dependent on the probability of failure (mostly due to collapse). Now structural costs and expected failure consequence costs have to be defined as functions of the structural design variables which is achieved by introducing failure probabilities and cost occurrence probabilities in the problem formulation. Optimal structural design variables can be calculated and optimal failure probabilities may be deduced subsequently.

Equ. (2) is subjected to side constraints: $0.05 \leq x_1 \leq 0.1$[m], $0.20 \leq x_2 \leq 0.5$[m] where production and expected consequence costs due to partial and collapse failure are to be minimized. The design parameter x_1 denotes the wall thickness of the piles and x_2 is the mean radius of the tubular stiffeners. The wall thickness of the bottom horizontal elements are considered as linearly dependent on x_1 respectively. An operation period of T=10 years and a mean rate of occurrence of storms of 0.2 per year are assumed. In this context it should be emphasized that these failure probabilities are dependent on the structural design variables. The structural costs are guided by the mass m dependent on the design vector $\mathbf{x}$ and the cost factor c_0 (100 cost units per unit mass). In this example the optimization procedure is carried out nine times for combinations of several values of the cost consequence factors c_s and c_c. For solution of the optimization problem formulated in equ. (3) the NLPQL-algorithm, as implemented in the interactive software environment COSSAN, is used. It should be noted that in each iteration step the response surfaces (polynomials of second order) with respect to first yielding and collapse have to be determined.

The required limit state points are calculated by nonlinear FE-analyses. It is found that for all combinations of failure consequence costs c_s and c_c the optimal wall thickness of the piles is equal to its lower bound. The reason for this circumstance is that the initial structural costs $c_0 m(\mathbf{x})$ are guided almost completely by the dimension of the piles (x_1). But the probabilities of failure p_{fs} and p_{fc} are primarily determined by the cross-sectional dimension of the stiffening elements (see Fig. 3).

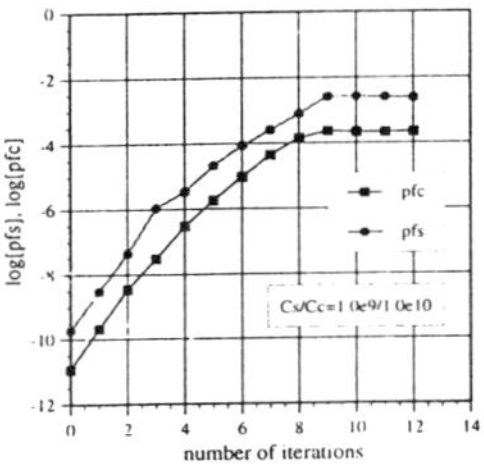

Fig.2: Representative iteration history of partial and total failure probabilities

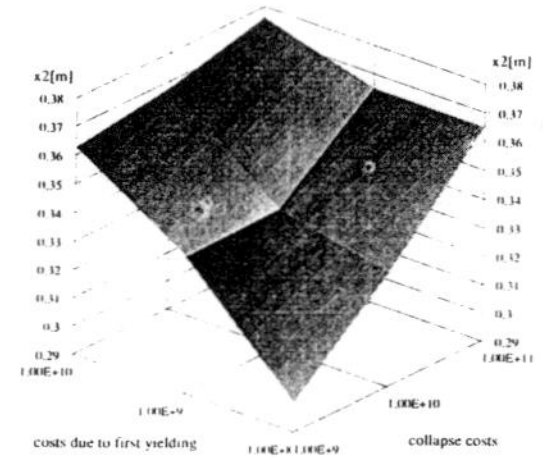

Fig.3: Optimal radius of horizontal and diagonal stiffeners dependent on failure consequence costs

The strong influence of the individual cost factors respectively the failure criteria on the structural design becomes apparent considering Fig. 3. It should be noticed that partial as well as total failure are of almost equal importance in reliability-based decision making. Both, increasing collapse costs and partial failure costs lead to larger cross-sectional dimensions. Therefore, and with respect to the design problem formulation given in equ. (2), neglecting the partial failure criterion would lead to a weaker structural design.

5. Conclusion

The realization of COSSAN [6] as a software package based on an object oriented programming language provides an interactively tool for stochastic structural analysis and optimization. State-of-the-art FE-codes, data analysis procedures, advanced simulation and reliability analysis methods, structural identification tools, random vibration methods, and optimization methods are easily accessible to the practitioner. However, if desired, the possibility for full utilization of an user-preferred third-party analysis software (e.g. FE-systems) is provided too. This allows reliability-based structural analysis and optimization in a most flexible form on the basis of highly accurate and numerically efficient procedures. There are generally no restrictions w.r.t. of problem formulations and design criteria. Additional optimization algorithms and other analysis tools can easily be implemented in the software environment so that COSSAN can kept up-to-date at any time.

References

1. Freudenthal, A.M.: "Safety and the Probability of Structural Failure", *Transactions* ASCE, Vol. 121, 1956, pp. 1337-1397.
2. Thoft-Christensen, P.: "On Reliability-Based Structural Optimization", in Der Kiureghian, A., Thoft-Christensen, P. (eds), Reliability and Optimization of Structural Systems, *Proceedings* of the 3rd IFIP WG 7.5 Conference, Berkeley, March 26 - 28, 1990, pp. 387-402.
3. Frangopol, D.M.: "Reliability-Based Structural Optimization Research at the University of Colorado: A Brief Retrospective 1983 - 1991", in Grierson, D.E. et. al. (eds), Progress in Structural Engineering, Kluwer Academic Publishers, Netherlands, 1991, pp. 301-322.
4. Cohn, M.Z., Dinovitzer, A.S.: "Application of Structural Optimization", *J. Struct. Engineering,* Vol. 120, No. 2, Feb. 1994, pp. 617 - 650.
5. Schittkowsky, K.: "Theory, Implementation, and Test of a Nonlinear Programming Algorithm", *Proceedings* of Euromech-Colloquium 164 on "Optimization Methods in Structural Design", Universität Siegen, 1982, Bibliographisches Institut, Mannheim, 1983, pp. 122 - 132.
6. COSSAN "Computational Stochastic Structural Analysis", User Manual, draft, Institute of Engineering Mechanics, University of Innsbruck, 1995
7. Gasser M.: "Structural Optimization Based on Reliability Criteria Utilizing Approximate Methods", submitted as Dissertation, Inst. Engr. Mech. University of Innsbruck, 1995.

A Comparative Analysis of FORM/SORM and Polynomial Chaos Expansions for Highly Nonlinear Systems

R. Ghanem [1], Member, ASCE

D. Ghiocel [2], Member ASCE.

Abstract

The concept of most probable design point (MPP) in structural reliability has dominated engineering practice for over 3 decades. This concept is based on linearization or quadratization of the response surface at some appropriate point from which approximations to the probability of failure are obtained.

In view of its pointwise nature, the MPP approach cannot be expected to handle response surfaces that feature singularities or that are highly nonlinear. Such response surfaces are common in applications involving bifurcation-type and nonlinear behavior. Typical applications are buckling under random properties, and softening Duffing-type oscillators featuring more than one attractor.

In this paper, the deficiencies of the MPP method are highlighted by comparing their predictions to those obtained by Monte Carlo simulations. It is also demonstrated that the spectral stochastic, whereby the response surface is expanded globally, as opposed to locally around the MPP, can substantially improve the accuracy of the reliability results.

Snap Through Bifurcation

The normalized load-deflection curve for a an arch subjected to a normalized transverse load H/EA, where E denotes the modulus of elasticity and A denotes the cross-sectional area, is given by the following equation

$$P = 2\left[-1 + \frac{1}{\sqrt{1-2D\sin\alpha + D^2}}\right](\sin\alpha - D) \ . \tag{1}$$

This plot of this relationship is shown in Figure (2). The value of the deflection at which bifurcation occurs is denoted by, D_{cr}, and is obtained as the smallest real solution to the equation,

$$\frac{\partial P}{\partial D} = 0 \ . \tag{2}$$

[1]The Johns Hopkins University, Baltimore, MD 21218.

[2]Stevenson & Associates, Cleveland, OH.

This equation can be solved analytically, resulting in the following expression for D_{cr},

$$D_{cr} = \sin\alpha - \sqrt{\theta^{2/3} - \theta}, \tag{3}$$

where

$$\theta = (1 - \alpha^2) \, . \tag{4}$$

The critical value of the load associated with this critical value of deflection is denoted by P_{cr} and is given by the following equation,

$$P_{cr} = \left(\theta^{2/3} - \theta\right)^2 \left[\frac{1}{\theta^{1/3}} - 1\right] \, . \tag{5}$$

Let the load P be normally distributed according to $\mathcal{N}(\frac{4}{7}10^{-2}; 0.3)$. Given that the initial load is random, a jump will occur whenever the applied load exceeds P_{cr}, and the portion of the curve corresponding to a decrease in the applied load will not be followed. Thus, in the course of the numerical solution of Equation (2) for the value of D associated with a specified value of P, the search for a solution is restricted to values of D that are less than D_{cr} whenever the simulated load is smaller than P_{cr}. This observation resolves the nonuniqueness issue associated with the solution to this problem. For values of P that are greater than P_{cr}, no constraints are applied, since in this range, a unique solution does exist.

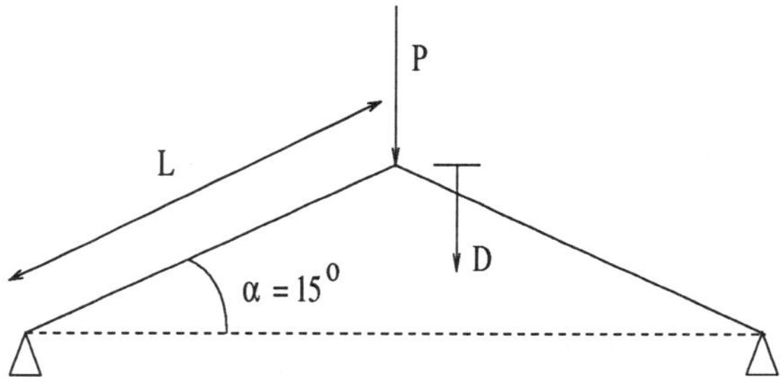

Figure 1: Physical Setup.

MPP Approach

The FORM/SORM methods have long been accepted as efficient probabilistic algorithms for reliability analysis. They have also been known to be limited in their applicability to cases involving response surfaces that are lightly nonlinear and convex (Ebber et.al, 1994). Generally, SORM results

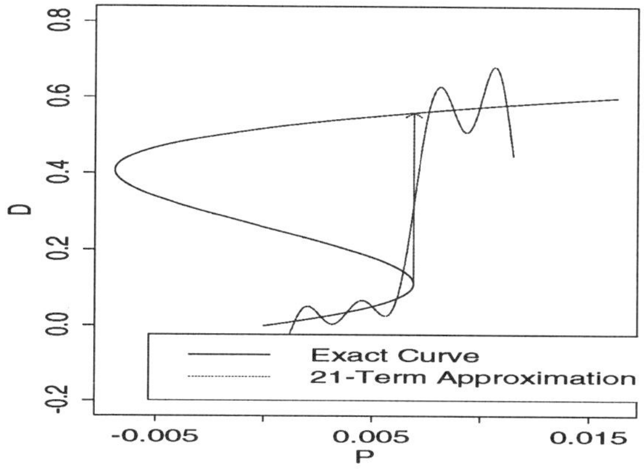

Figure 2: Exact Load Deflection Curve.

are more sensitive to the nonconvexity issue than the FORM. The problem considered in this paper features a nonconvex response surface that includes a jump. As expected, the SORM procedure failed to converge in all cases considered, while the FORM procedure converged in few of the cases.

Polynomial Chaos Expansion

Let $\{\Psi_i\}$ denote the set of polynomial chaoses (Ghanem and Spanos, 1991) that are orthogonal with respect to the Gaussian measure. Moreover, let the random load P be defined through its mean value $< P >$ and its variance v. Then, P can be represented in terms of the standard Gaussian variable, X,as

$$P = X < P > v + < P > \; . \tag{6}$$

The deflection of the arch from its initial position, d, is a function of X, through its dependence on P, and can thus be represented by its expansion along the Polynomial Chaos set,

$$d(P(X)) \; = \; \sum_{i=0}^{N} d_i \Psi_i(X) \; . \tag{7}$$

Explicitly, the first few terms in the above expansion have the form,

$$d(X) = = < d > + d_1 X + d_2(X^2 - 1) + d_3(X^3 - 3X) + \dots \; . \tag{8}$$

In the view of the orthogonality of the Polynomial Chaos terms, the coordinates in the above expansion are given by the following simple formula,

$$d_i \; = \; \frac{< d(X)\Psi_i(X) >}{< \Psi_i^2 >} \; . \tag{9}$$

The denominator in this equation can be readily evaluated and tabulated, while the numerator is easily calculated using the following integral,

$$< d(X)\Psi_i(X) > = \frac{1}{\sqrt{2\pi}} \int_{-\infty}^{\infty} d(x)\Psi_i(x)e^{-\frac{x^2}{2}} dx \ . \qquad (10)$$

Figure (2) compares the approximation obtained from a 21-term expansion with the exact expression. It is noted that the jump is well captured. Once this approximation has been obtained, simulation can be conducted in a very expeditious fashion. Of course, in the event that such a closed form expression is not avaibale apriori, other procedures must be followed (Ghanem and Spanos 1991) in order to obtained the coefficients in the Polynomial Chaos expansion.

Comparative results of FORM/SORM and Polynomial Chaos expansion versus Monte Carlo simulation are given in Table 1. Observe that for most cases considered, corresponding to points near the bifurcation point, neither the FORM nor the SORM approach yields meaningful results. On the other hand, the Polynomial Chaos expansion results in values of similar order of magnitude as the Monte Carlo simulation. The discrepancy is attributed to the overshoot in the approximation in the neighborhood of the bifurcation.

Applied Load Mean; Covariance	Normalized Critical Displacement	FORM		SORM		MCS	Series Expansion, P_f
		β	P_f	β	P_f	P_f	P_f
5.7×10^{-3}; 0.3	0.57	-	-	-	-	2.4×10^{-2}	9.96×10^{-2}
5.7×10^{-3}; 0.3	0.56	-	-	-	-	2.4×10^{-1}	1.0×10^{-1}
6.6×10^{-3}; 0.4	0.57	0.88	0.19	0.88	-	0.18	0.21
6.6×10^{-3}; 0.4	0.56	-	-	-	-	0.44	0.24

Table 1: Comparison of numerical results associated with the model problem.

References

Ebbeler, D.H., et.al, "Computational methods for probabilistic failure assessment." 12th *U.S. National Congress of Applied Mechanics,* Seattle, WA, June, 1994.

Ghanem, R., and Spanos, P., *Stochastic Fnite Elements: A Spectral Approach,* Springer-Verlag, 1991.

MULTI-COUPLED DISORDERED PERIODIC SYSTEMS

Cindy X. Qiu[1]

Abstract

An analytical procedure is described to obtain the localization factor, namely, the average exponential decay rate per cell of wave propagation in a disordered periodic chain. The procedure is based on the multiplicative ergodic theorem of Oseledec, which justifies the replacement of a spatial average by an ensemble average.

Introduction

Many structures are designed to be spatially periodic, but disorder from ideal periodicity always occurs, because of material, geometric, and manufacturing variabilities. Most early studies (e.g. Kissel, 1988; Cai and Lin, 1991, 1992; Ariaratnam and Xie, 1991, 1994) were restricted to single-channel systems, in which only one type of motion can be transmitted from a cell unit to the next in a system. The objective of these studies was to determine the localization factor, namely, the rate of exponential decay of wave propagation due to disorder. In this paper, localization factors are obtained analytically for multi-channel systems by ensemble-averaging.

Analysis

Shown in Figure 1 is a structure of infinite total length, which is composed of identical cell-units connected end-to-end to form a spatially periodic pattern. The state vectors at the two sides of cell n are related as follows

---

[1]Research Assistant, Center for Applied Stochastics Research, Florida Atlantic University, Boca Raton, FL 33431-0991.

$$\begin{Bmatrix} w(n_r) \\ f(n_r) \end{Bmatrix} = [T(n)] \begin{Bmatrix} w(n-1_r) \\ f(n-1_r) \end{Bmatrix} \tag{1}$$

where each $w(\,)$ is a d-dimensional vector of generalized displacements, each $f(\,)$ is a d-dimensional vector of generalized forces, and $[T(n)]$ is a $2d \times 2d$ transfer matrix. Denoting the eigenvalues of $[T(n)]$ by $\upsilon_j = \exp(\eta_j)$ and making use of the fact that eigenvalues of a transfer matrix are reciprocal pairs, the characteristic equation for the eigenvalues may be cast in the form of

$$\cosh(d\eta) + a_1 \cosh[(d-1)\eta] + \ldots + a_{d-1} \cosh\eta + a_d/2 = 0 \tag{2}$$

Note that $\cosh\eta = (e^{\eta} + e^{-\eta})/2 = (\upsilon + \upsilon^{-1})/2$. Eq.(2) is an equation for $\cosh\eta$, which is of a lower order than that of an equation for υ. Once it is solved, we can obtain

$$\upsilon = \cosh\eta \pm\sqrt{\cosh^2\eta - 1} = \exp(\pm\eta) \tag{3}$$

The localization factor is equal to the smallest positive Lyapunov exponent. According to the multiplicative ergodic theorem of Oseledec(1968), this can be obtained from an ensemble average, namely,

$$\gamma = \ln\left|\int\int\left[\cosh\eta + \sqrt{\cosh^2\eta - 1}\right] p(\alpha)\, p(w_d \mid \alpha)\, dw_d\, d\alpha\right| \tag{4}$$

where $p(w_d|\alpha)$ is the induced stationary probability density of a unit vector w_d in the canonical sub-space associated with the dth Lyapunov exponent. Unfortunately, the determination of $p(w_d|\alpha)$ remains an unsolved problem. Instead, we propose an approximation

$$\gamma \approx \left\{\ln\left|\int\left[\cosh\eta + \sqrt{\cosh^2\eta - 1}\right] p(\alpha)\, d\alpha\right|\right\}_{\min} \tag{5}$$

Eq.(5) is adopted in the numerical example to be described below.

Numerical Example

Consider a multi-span beam. We regard the ideal periodic structure as being composed of periodic units, each of which consists of a beam segment between two neighboring elastic supports, plus the elastic support at the right end of the beam segment. Only the variations in the spring constants of elastic support are taken into account. Specifically, the spring constant of the nth cell-unit is denoted by $k[1 + X(n)]$, where k is a constant, and $X(n)$ for different n are assumed to be independent and uniformly distributed in $\left[-\sqrt{3}\,\sigma_k, \sqrt{3}\,\sigma_k\right]$.

The computed localization factors are plotted in Figure 2, along with the simulation results, represented by diamonds. The coupling parameter $\delta = EI/(\ell^3 k) = 0.01$ is selected in the computation. The first four frequency bands in the unit of rad/s, $0 < \omega < 141.0$, $141.0 < \omega < 151.0$, $151.0 < \omega < 261.0$, and $261.0 < \omega < 605.0$, referring to the corresponding ideally periodic structure (e.g. Signoreli and von Flotow, 1988), were found to be a complex band, a wave-passage band, a wave-stoppage band, and an other wave-passage band, respectively. The computed results compare reasonably well with the simulation ones, except near the boundaries of each wave-passage band. The simulation results were obtained using the procedure proposed independently by Ariaratnam and Xie(1994) and by Castanier and Pierre(1995) based on an orthogonalization scheme due to Wolf, *et al*(1985).

Concluding Remarks

An approximate analytical procedure is given to compute the localization factor of a multi-coupled disordered chain. Numerical example is provided of a dual-coupled chain, permitting the propagation of two types of waves. These numerical results are compared with those obtained by Monte Carlo simulations. The comparison shows that the approximate procedure is quite adequate except at the close proximity of a boundary of a wave-passage frequency band.

Acknowledgements

The work reported in this paper is supported by the Office of Naval Research under Grant N00014-93-0879. Guidance provided by Professor Y.K.Lin is gratefully acknowledged.

References

Ariaratnam, S. T. and Xie, W. C.(1991). "On the Localization Phenomenon in Randomly Disordered Engineering Structures." Nonlinear Stochastic Mechanics, eds. N. Bellomo and f. Casciati. Springer-Verlag, Berlin, 13-24.

Ariaratnam, S. T. and Xie, W. C.(1994). "Localization of Stress Wave Propagation in Disordered Multi-Wave Structures." *Structural Safety and Reliability*, 1, eds. G. I. Schuëller, M. Shinozuka, and J. T. P. Yao. Balkema, Rotterdam, 77-83.

Ariaratnam, S. T. and Xie, W. C.(1995). "Wave Localization in Randomly Disordered Nearly Periodic Long Continuous Beams." *J. of Sound and Vibration*, 181(1), 7-22.

Cai, G. Q. and Lin, Y. K.(1991). "Localization of Wave Propagation in Disordered Periodic Structures." *AIAA Journal*, 29(3), 450-456.

Cai, G. Q. and Y. K. Lin (1992). "Statistical Distribution of Frequency Response in Disordered Periodic Structures." *AIAA Journal*, 30(5), 1400-1407.

Castanier, M. P. and Pierre, C. (1995). "Lyapunov Exponents and Localization

Phenomena in Muti-Coupled Nearly Periodic Systems." *J. of Sound and Vibration*, 181(3), 493-515.

Signoreli, J. and von Flotow, A. H.(1988). "Wave Propagation, Power Flow, and Resonance in a Truss Beam." *J. of Sound and Vibration*, 126(1), 127-144.

Kissel, G. J.(1988). "Localization in Disordered Periodic Structures." *Ph.D Dissertation, Massachusetts Inst. of Technology, Cambridge, MA.*

Oseledec, V. I.(1968). "A Multiplicative Ergodic Theorem Lyapunov Characteristic Numbers for Dynamical Systems." *Trans. of Moscow Mathematical Society*, 19, 197-231.

Wolf, A., Swift, Jack B., Swinney, Harry L.(1985). "Determining Lyapunov Exponents from a Time Series." *Physical* 16D, 285-317.

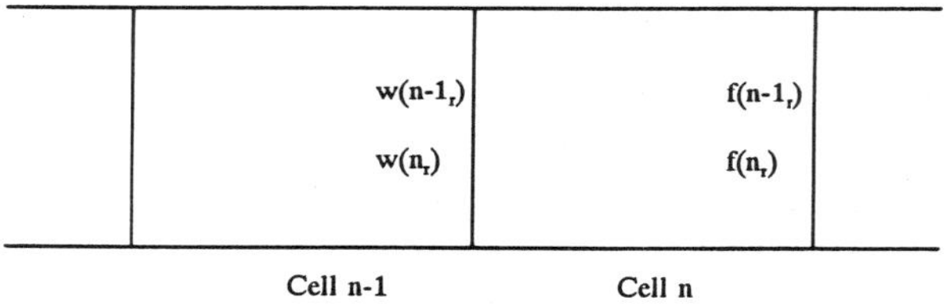

Figure 1. Typical Cell-Units of Periodic Structures

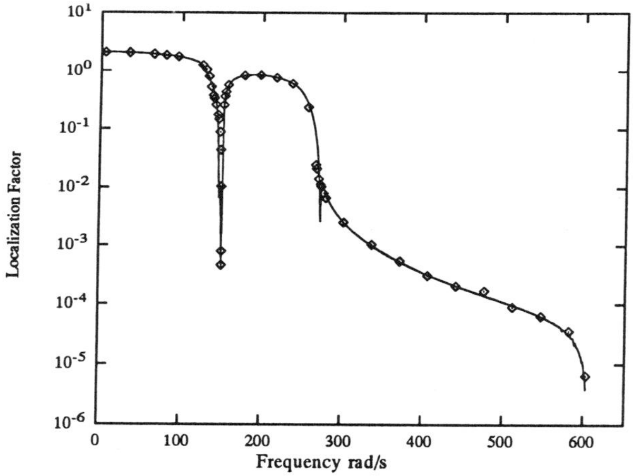

Figure 2. Localization Factor of A Dual-Coupled Disordered Periodic Beam: $\delta=EI/(\ell^3 k)=0.01$, $\sigma=\ell(\omega^2 m/EI)^{1/4}=0.1$

Analytical Study and Verification of a Coupled Theory of Mixtures for Application in Cone Penetration and Tunnel Boring in Soils

George Z. Voyiadjis[1], F.ASCE, Murad Y. Abu-Farsakh[1], and Mehmet T. Tumay[1], F.ASCE

Abstract

Elasto-plastic coupled equations are formulated in order to describe the time-dependent deformation for saturated soils. Formulation of these equations is based on the theory of mixtures for inelastic porous media proposed by Prevost (1980) and Kiousis and Voyiadjis (1985). The saturated soil is considered to be a mixture of two deformable media, the solid grains and the water, each of which is regarded as a continuum. The coupled equations as well as the numerical simulation of both the continuous penetration of the piezocone and the tunnel boring in soil are implemented in a finite element program. The results of the numerical simulation are compared with experimental laboratory measurements.

Introduction

The soil skeleton consists of solid grain particles and voids that are filled with water and/or air. Soil can be considered as a mixture of multiphase deformable medium of solid grains, water, and air, each of which is regarded as a continuum and follow its own motion. The theory of mixtures for a small strain linear elastic porous media was first developed by Biot (1955). An extension of Biot's theory for large strain nonlinear inelastic porous media was presented by Prevost (1980). In this work, an analytical model is formulated for a cohesive soil skeleton based on the principle of virtual work and the theory of mixtures for inelastic porous media as proposed by Prevost (1980) and Kiousis and Voyiadjis (1985). The model is used for the analysis of the piezocone penetration problem in cohesive soils. The results of the numerical simulation are compared with experimental laboratory measurements.

[1]Dept. of Civil & Environmental Engineering, Louisiana State University, Baton Rouge, LA 70803

Piezocone Penetrometer Analysis

In this work, the penetration of a piezocone penetrometer in cohesive soils is numerically simulated assuming that the penetrometer is infinitely stiff and the soil-penetrometer interface friction is negligible. The continuous penetration of the piezocone is simulated by applying an incremental vertical movement of the cone tip boundary as shown in Figure 1. During the incremental penetration, the nodes are checked for possible changes in boundary conditions. When tensile stresses occur in the node immediately below the lower end of the cone at location (1), the node is released as indicated by location (2) and allowed to move freely. Once the free node at location (2) reaches the cone tip boundary, its movement will be restricted along the skew surface of the cone tip as indicated by (3). When the horizontal coordinate of the nodes at location (3) exceed the penetrometer radius, these nodes are released at location (4) until they return to the boundary of the shaft. They are then restricted from further movement in the horizontal direction as shown in location (5).

The numerical simulation is compared with the piezocone penetrometer tests (PCPT) conducted by Kurup, et al. (1994) on cohesive soil specimen prepared in a calibration chamber at LSU. The projected cone area is 100 mm^2 with a cone apex angle of 60°. The soil specimen used contains 33% kaolinite and 67% fine sand. The specimen is consolidated to an isotropic stress of 207 kPa. The filter used to measure the porewater pressure is located either in the lowest 1/4 of the cone tip (U1 configuration) or starting 0.5 mm above the base of the cone with 2 mm vertical height (U2 configuration).

The modified Cam clay model is adapted in this work to describe the plastic behavior of soil. The soil parameters used are $\kappa = 0.024$, $\lambda = 0.11$, $\Gamma = 2.162$, and $M = 1.2$. The coefficient of permeability used is 0.5×10^{-5} m/s.

The results of the numerical simulation are compared with the experimental results obtained by Kurup, et al. (1994). Figure 2 shows the cone resistance profile with the penetration depth, while Figure 3 shows the excess pore pressure profile with the penetration depth. It can be seen from the numerical simulation that the steady state conditions are approached at about 30 mm which is faster than those obtained by the experiments. Similar results are obtained by Kiousis, et al. (1988) in the simulation of cone penetrometer (10 cm^2) where the pore pressure distribution reaches steady state condition at about 25 mm.

Conclusion

Although the development of the cone resistance and excess pore pressure obtained by finite element is faster than the experimental results, it is closer to the U2 configuration. The final values obtained as the steady state condition is

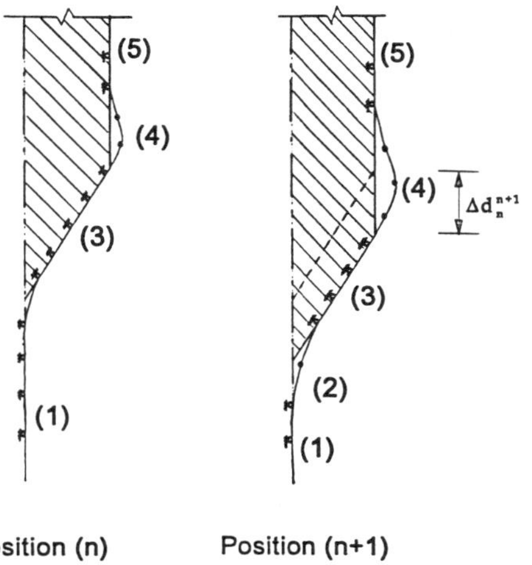

Figure 1. Incremental Penetration of Piezocone Penetrometer

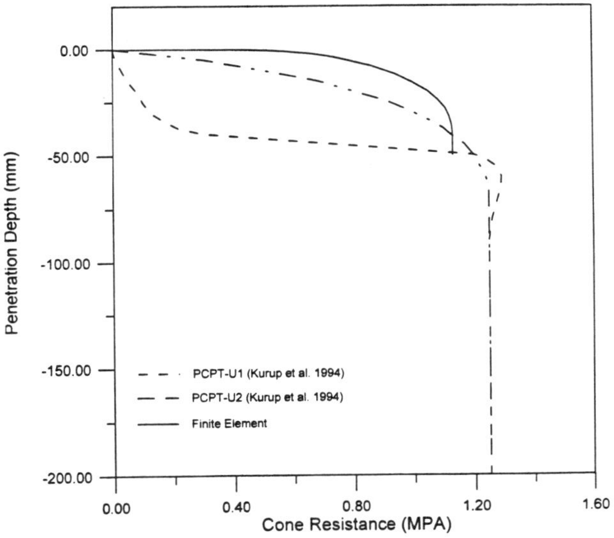

Figure 2. Cone Resistance Profiles

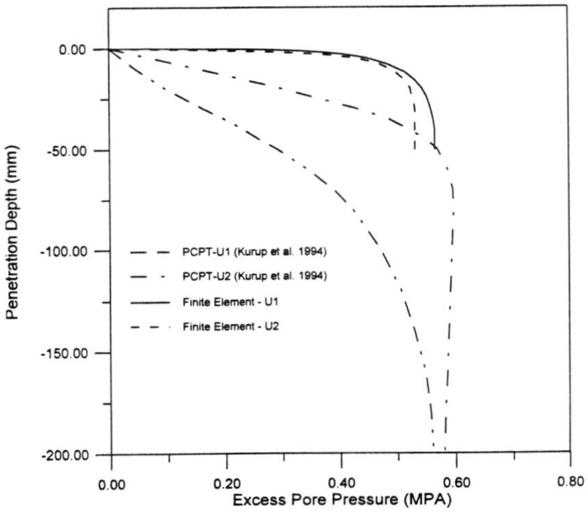

Figure 3. Excess Pore Pressure Profiles

approached are quite close to the experiments. The current difference at the start of penetration is due to some errors usually encountered at the initial stage of the experimental testing that need to be corrected. This can be seen from the difference between the U1 and U2 configuration test results at the initial stages while the only difference is the filter location between the two configurations.

Appendix -- References

Biot, M. A., 1955, "Theory of Elasticity and Consolidation for a Porous Anisotropic Solid," Journal of Applied Physics, Vol. 26, pp. 182-185.

Kiousis, P. D. and Voyiadjis, G. Z., 1985, "Lagrangean Continuum Theory for Saturated Porous Media," Journal of Engineering Mechanics, Vol. 111, No. 10, pp. 1277-1288.

Kiousis, P. D., Voyiadjis, G. Z. and Tumay, M. T., 1988, "A large Strain Theory and its Application in the Analysis of the Cone Penetration Mechanism," International Journal for Numerical and Analytical Methods in Geomechanics, Vol. 12, pp. 45-60.

Kurup, P. U., Voyiadjis, G. Z. and Tumay, M. T., 1994, "Calibration Chamber Studies of Piezocone Test in Cohesive Soils," Journal of Geotechnical Engineering, Vol. 120, No. 1, pp. 81-107.

Prevost, J. H., 1980, "Mechanics of Continuous Porous Media," International Journal of Engineering Science, Vol. 18, pp. 787-800.

Elasto-Plasticity of Sand Deformation

Eqramul Hoque[1] *and Fumio Tatsuoka*[2]

Abstract

Elastic and plastic components of sand deformation were precisely separated and evaluated. The importance of the above was demonstrated by comparing the deformation of a cyclically prestrained triaxial specimen with that of a virgin one. As a result of cycic prestraining, the deformation was dominated largely by elastic strain components and dilatancy rate decreased substantially.

Introduction

Precise separation of elastic and plastic components of deformation is essential for proper elasto-plastic modeling of sand having a wide variety of strain history. In this study, the importance of the above was confirmed by comparing the elasto-plastic deformation characteristics of a prestrained specimen with those of a virgin one. Elastic components were measured frequently during shearing by applying very small unload/reload cycles. Effects of cyclic prestraining (CP) on the deformation of soil is very important also for the design of structures subjected to heavy traffic, seismic activity, machine vibration, and other dynamic loading such as off-shore structures. A large number of loading cycles of the order of single amplitude axial strain $(\varepsilon_a)_{SA}$=0.02~0.1% on dry sand has been observed to increase the elastic limit strain and tangent stiffness both in cyclic and monotonic loadings, and to decrease significantly damping at relatively large strain levels (say, $(\varepsilon_a)_{SA}\geq$0.01%), with relatively smaller effects on the elastic Young's modulus (E^e) (Teachavorasinskun, 1992). Therefore, the deformation characteristics including dilatancy of a prestrained specimen should differ considerably from that of a virgin one.

Testing method

Triaxial tests were performed in which axial load was measured by using a very sensitive and rigid load cell located inside the triaxial cell. Axial deformations were obtained by bedding-error-free 'on-sample' local measurements using a pair of Local Deformation Transducers (LDT, Goto et al., 1991). Local radial deformations were measured by using proximeters at three heights (2.5, 7.5 and 12.5 cm from the bottom). A dry specimen (15 cm high and 7.5 cm in diameter) of Hostun sand (French sand;

[1] Graduate student, IIS, Tokyo University
[2] Professor, ditto

U_c=1.94, G_s=2.65, D_{50}=0.31 mm, e_{max}=0.95, and e_{min}=0.55) was reconstituted by pluviating through air. Porous stones were used at the ends of the specimen. After consolidation isotropically at a partial vacuum of σ_c=78.4 kPa, vertical sinusoidal cyclic deviator load (CL) at constant confining stress (σ_c) equal to 78.4 kPa was applied. Responses of a virgin specimen to cyclic loading at a loading frequency (f) of 0.1 Hz along the predetermined CP stress path with an axial stress range of nominally σ_a=78.4~196 kPa (CP_{str}) was obtained first. Cyclic prestraining of 40,000 cycles was then applied in the stress amplitude CP_{str} at f=0.25 Hz. Figs. 1a and b show the relationships between axial stress (σ_a) and axial strain (ε_a) and radial strain (ε_r), respectively, for the virgin and prestrained specimens. The ε_a and ε_r axes for the prestrained specimen were shifted arbitrarily. The actual values at the bottom ends of the curves for the prestrained specimen (denoted as S_p) are ε_a=0.012% and ε_r=-0.008%. Cyclic prestraining of 40,000 cycles was applied in the stress amplitude CP_{str} at f=0.25 Hz. Before and after CP, the elastic Young's modulus (E^e) defined for $(\varepsilon_a)_{SA}$≤0.001% was evaluated from ten very small unload/reload cycles with f=0.1 Hz applied at different stress states along the CP stress path.

Test Results

The relationship between the elastic Young's modulus E^e and σ_a obtained during loading ($\Delta\sigma_a$>0) and unloading ($\Delta\sigma_a$<0) before and after CP are shown in Fig. 2. The E^e values have been divided by a void ratio function f(e) to account for the effects of a slight variation in e during the test. It can be seen that E^e can be modeled as $E^e=E_1.f(e).\sigma_a^m$ (m<1), when $E_1=E^e$ at σ_a=98 kPa and f(e)=1. It is seen that the E^e~σ_a relations of the virgin specimen along the CP stress path is very similar to that obtained at isotropic stress states (i.e., $\sigma_a=\sigma_r$). This result indicates that the E^e value of a granular material depends essentially on the principal stress in the direction of the major principal strain increment for which the E^e value is defined. By this property, the elasticity of granular material becomes anisotropic under anisotropic stress conditions (i.e., stress state-induced anisotropy; Hoque et al., 1995). Differences in the values of E_1 and m (Fig. 2)

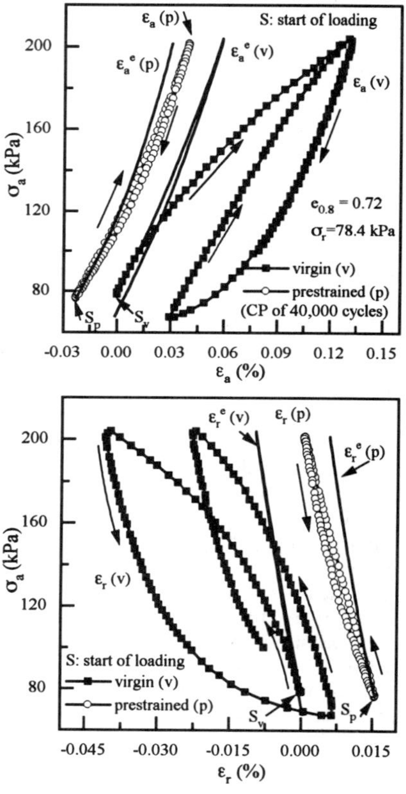

Fig. 1: (a) σ_a~ε_a and (b) σ_a~ε_r relationships of virgin and prestrained states of specimen in CP stress path.

between loading and unloading paths before CP is due to effects of preloading to σ_a=156.8

kPa. Due to CP, the exponent m increased noticeably with an overall increase in E^e at higher stress levels and the corresponding decrease at lower stress levels. That is, the specimen became more anisotropic in the small strain stiffness due to CP, which may be called the 'cyclic prestrain-induced anisotropy.'

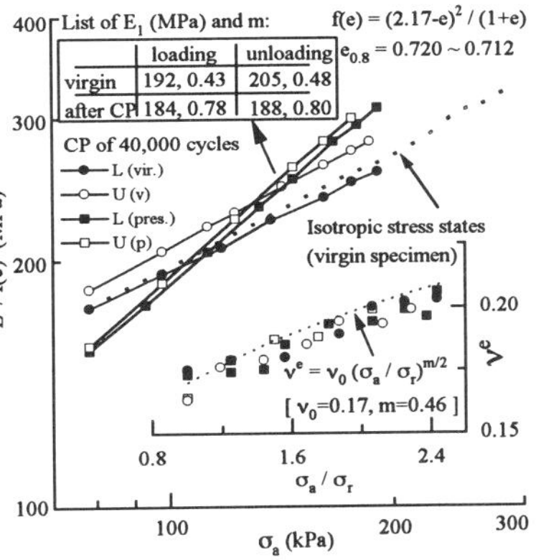

Fig. 2: Relationships of elastic parameters (E^e, v^e) with stress states.

In Fig. 1, the relationship between the elastic Poisson's ratio v^e obtained from the small unload/reload cycles and σ_a/σ_r is also shown. The differences in v^e at different states (mentioned earlier) were negligible. The relation can be fitted reasonably by, $v^e=v_0.(\sigma_a/\sigma_r)^{m/2}$, where $v_0=v^e$ at $\sigma_a/\sigma_r=1$ and m=0.46 (Hoque et al., 1995).

Elastic strains for the whole stress range CP_{str} were obtained by integrating $d\varepsilon_a^e=d\sigma_a/E^e$ and $d\varepsilon_r^e=-v^e.d\varepsilon_a^e=-v^e.d\sigma_a/E^e$ by using the relations for E^e and v^e (mentioned earlier) together with the respective elastic parameters (E_1, m) listed in Fig. 2. The relationships between σ_a and the elastic axial and radial strains (ε_a^e, ε_r^e) were arbitrarily plotted with the origins denoted as S_v and S_p at σ_a=0.8 kgf/cm² of the respective σ_a~ε_a and ~ε_r relation (Fig. 1). A drastic change in the σ_a~ε_a relation was caused by a substantial reduction in the plastic strain components (ε_a^p, ε_r^p) due

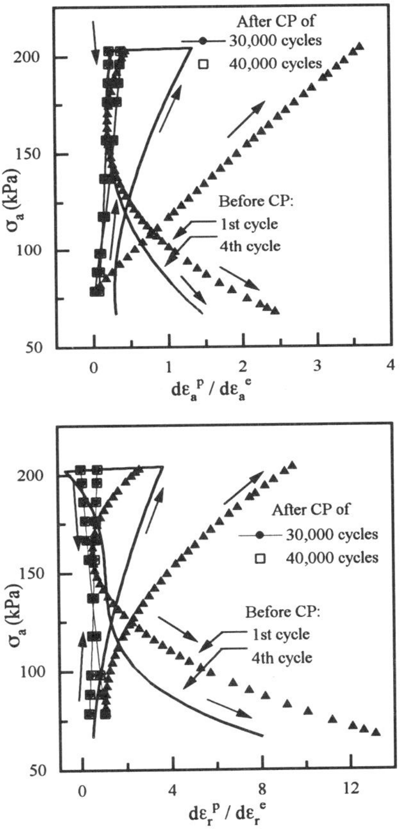

Fig. 3: Variations of (a) axial plastic-to-elastic, (b) radial plastic-to-elastic strains increments with σ_a along CP stress path.

to CP, while the σ_a~ε_a^e relation changed very slightly. Now we can understand why the stress-strain relation after CP is concave upward with a small hysteresis area; this is due to the intrinsic elastic property of sand. A considerable decrease in the plastic-to-elastic strain increment ratio due to CP can also be seen from Fig. 3. Before CP, the ratio varied largely and nonlinearly with σ_a, whereas after CP, it varied slightly and linearly with σ_a during loading and unloading. It is also important to note that the deformation for the range of σ_a which was applied as CP has not become totally elastic even after this large amount of CP. Cyclic prestraining by 30,000 and 40,000 cycles did not show any difference.

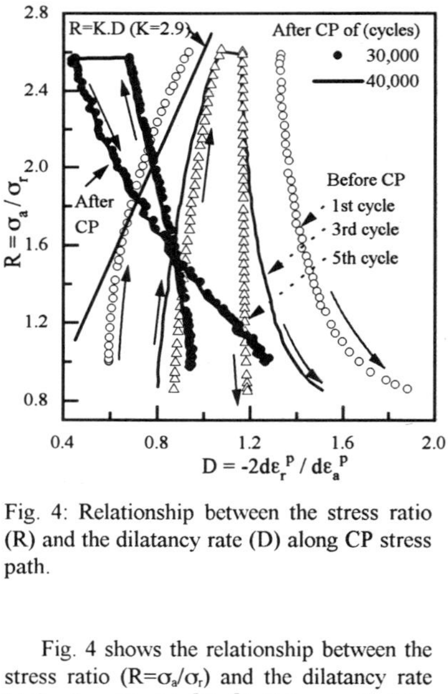

Fig. 4: Relationship between the stress ratio (R) and the dilatancy rate (D) along CP stress path.

Fig. 4 shows the relationship between the stress ratio ($R=\sigma_a/\sigma_r$) and the dilatancy rate (defined as $D=-2d\varepsilon_r^P/d\varepsilon_a^P$). It is seen that the relation during loading at the virgin states follow closely the Rowe's stress-dilatancy relation R=KD with K=2.9, but the other relations do not. It is seen that before CP, the relations approaches to the non-volumetric strain increment condition (i.e., D=1.0) for both loading ($d\sigma_a>0$) and unloading ($d\sigma_a<0$). The relations after CP show a different trend; during loading more contractive behavior (D>1.0) with the increase in $R=\sigma_a/\sigma_r$, whereas during unloading more dilative behavior (D<1.0). This behavior is much more complicated than that predicted by Pradhan and Tatsuoka (1989).

Conclusions

The elastic Young's modulus for the axial direction increases with the increase in the axial stress. The plastic strain increment decreases substantially with cyclic loading along a fixed stress path. The deformation does not become totally elastic even after applying a large amount of CP with a relatively large amplitude of axial stress cycle. The elastic deformation characteristics are very stable, but they changes slightly showing cyclic strain-induced anisotropy. These results suggest that the deformation of sand cannot be modeled precisely by a simple elasto-plastic model having a fixed range of elastic zone. The stress-dilatancy relation is changed by cyclic loading in a complicated manner.

Appendix I

(1) Goto, S., Tatsuoka, F., Shibuya, S., Kim, Y.-S. and Sato, T. (1991), "A Simple Gauge for Local Small Strain Measurements in the Laboratory," Soils and Foundations, Vol. 31, No. 1, pp. 169-180.
(2) Hoque, E., Tatsuoka, F., Sato, T., and Kohata, Y. (1995), "Inherent and Stress-Induced Anisotropy in Small Strain Stiffness of Granular Materials," Proc. of 1st Int. Conf. on Earthquake Geotechnical Engg., IS-TOKYO, Vol. 1, pp. 277-282.
(3) Pradhan, T.B.S. and Tatsuoka, F. (1989), "On Stress-Dilatancy Equations on Sand Subjected to Cyclic Loading," Soils and Foundations, Vol. 29, No. 1, pp. 65-81.
(4) Teachavorasinskun, S. (1992), "Deformation Characteristics of Sands at Small Strains,' Doctoral thesis, University of Tokyo.

Micromechanical Modelling for Granular Materials

Ching. S. Chang[1]

Abstract

Granular material is analyzed as an assembly of particles. Under applied loads, micro-slip occur at inter-particle contacts due to sliding and separation of particles. The micro-slip at inter-particle contacts contribute to the deformation of the packing. A number of approaches have recently been used for modelling the stress-strain behavior considering the micro-slip at inter-particle contacts. This paper gives a brief overview of these approaches.

Introduction

In a micromechanical approach for constitutive modelling of granular material, the kinematic hypothesis of uniform strain has often been made to describe the displacement field of discrete particles (e.g., Ref. 1, 2). The kinematic hypothesis furnishes a simple and useful relationship for estimating the movement of particles. However, the approach leads to an upper bound solution due to the kinematic constraint of the system. Studies have shown that the assumption is reasonably good for sands under certain loading conditions (Ref. 3), in which inter-particle contacts experience mostly partial slip with very few sliding. Uniform strain is not a good assumption for loading conditions with high deviatoric stress in which large shear deformation occurs with a large number of particle sliding. Thus the strain exhibits non-uniformity within the material sample and the overall behavior is no longer a straight volume average of the local behavior (Ref. 3). To account for the strain heterogeneity, three models have been considered, namely, (1) cell model, (2) high-gradient model, and (3) statically-based model. An overview of our work on these three models is given below.

[1] Professor, University of Massachusetts, Amherst, MA 01003

Cell Model

To model the heterogeneity, material is divided into cells (i.e. micro-elements). Each cell represents a particle group thus has a different configuration. A statistical set of cells is used to characterize the packing structure of the granular material. Under a deformation of the material, each cell is allowed to have a different value of strain depending on the stiffness of each cell. Thus the uniform strain assumption is removed.

The key issue of this model is a reliable method to correctly determine the strain for each cell under the application of a given macro-scale strain. After conceptually viewing the granular material as a cell structure, the material in certain respects is analogous to the polycrystalline material. The self-consistent method used in polycrystalline was applied in the granular material to relate the micro-element strain and the macro-scale strain. Stress-strain relationship was derived based on the self-consistent method, for a two dimensional granular packing considering plastic sliding (Ref. 4). The predicted stress-strain curves and the predicted volume change behavior show great improvement over the results from uniform strain approach. The self-consistent method can not be readily extended to a three-dimensional problem because it requires a Green's function for general anisotropic media which is not available for three-dimensional condition in the current literature. For this purpose, the explicit series expression of a Green's function for a general three-dimensional anisotropic media may be considered. (Ref. 5).

Other than the self-consistent method, a canonical ensemble average technique used in statistical mechanics was adopt for developing the stress-strain relationship. A partition function was used to relate the micro-element strain and the macro-scale strain. A constitutive theory was derived considering the strain fluctuation and particle sliding (Ref. 3, 6). Application of this method to an idealized simple packing show that the yielding function and the non-associated flow behavior for this packing resemble those observed from experiments on sand.

The cell model provides a more flexible deformation mode allowing strain fluctuation to take place within a representative volume of granular media under an applied macro-scale strain. The model greatly improves its performance compared to the model with uniform strain assumption (Ref. 4, 6). It is noted that, although the strains are different for all cells, the strain of each cell is still assumed to be uniform. Further investigation is needed to release the constraint within a cell.

Higher-order Model

The higher-order model utilizes higher order derivatives of displacement and rotation to represent the deformation of the granular media. Therefore the deformation field is no longer modeled as uniform strain within the representative volume. From previous study, it has shown that the deformation of a packing can be in general represented by continuum variables of higher order derivatives of displacement and rotation (Ref. 7, 8). These continuum variables can be used to retrieve and keep track the movement of each discrete particles, at least in an approximate manner. By selecting suitable continuum variables, the equivalent continuum model for the discrete system can be represented by different types of continua, for example, the micro-polar continuum, the high gradient continuum, the non-local continuum, etc. All types of these equivalent continua are suitable for representing granular media to some degree. More complex continua capture more features of the deformation behavior for granular material. So far, we have only studied the elastic behavior of high-gradient model. Plastic behavior for such continua has not yet been studied. However, it is expected that the continua are useful in modelling localized deformation in plastic region due to microstructure effect.

Statically-based Model

The assumption opposite from the kinematic hypothesis of uniform strain is the static hypothesis of uniform stress. The statically-based model utilizes the uniform stress assumption, consequently leads to a non-uniform strain field. The kinematic hypothesis provides a method of determining particle displacements from the macro-scale strain. On the other hand, the static hypothesis provides a method of determining contact forces from the macro-scale stress. The stress-strain relationships of granular material can be derived from both approaches.

Very little attention has been paid in the literature to the approach of employing a static hypothesis for granular systems. To this end, we propose a static hypothesis that delineates the distribution of forces at inter-particle contacts in a system of discrete particles (Ref. 7, 9). Following the approach of a static hypothesis, elastic moduli are derived in closed-form for assemblies of spheres with isotropic and anisotropic packing structures (Ref. 10). The stress-strain behavior derived from static hypothesis is quite different from that of the kinematic hypothesis especially for particles with small tangential contact stiffness. In elastic conditions, for particles with very small tangential contact stiffness, the shear modulus is approach to zero and the Poisson's ratio is approach to 0.5. On the other hand, the kinematic hypothesis predicts a Poisson's ratio of 0.25 and a considerably high shear modulus.

In the formulation of static hypothesis, the derived stress-strain relationship is corresponding to a lower bound solution, in contrast to the upper bound solution from the kinematic hypothesis. Therefore, although the assumption of uniform stress is as inaccurate as that of uniform strain, the development of a static hypothesis is useful in providing a set of bounds for estimating the exact solution. The bounds are particularly useful when the microstructure information is limited so that an exact solution cannot be determined.

Acknowledgement

This work was sponsored by the Air Force Office of Scientific research, USAF, under grant/contract number F49620-95I-0117.

References

1. Walton, K. (1987), "The Effective Elastic Moduli of a Random Packing of Spheres", *Journal of Mechanics and Physics of Solids*, Vol. 35, No. 3, 213-226.
2. Jenkins, J.T. (1988), "Volume Change in Small Strain Axi-symmetric Deformations of a Granular Material", *Micromechanics of Granular Materials*, Eds. M. Satake, and J.T. Jenkins, Elsevier, 143-152.
3. Chang, C. S., Kabir, M., and Chang, Y. (1992), "Micromechanics Modelling for the Stress-Strain Behavior of Granular Soil - II: Evaluation," *Journal of Geotechnical Engineering*, ASCE, Vol. 118, No. 12, 1975-1994.
4. Chang, C. S., Misra, A., and Acheampong, K. (1992), "Elastoplastic Deformation of Granulates with Frictional Contacts," *Journal of Engineering Mechanics*, ASCE, Vol. 118, No. 8, 1692-1708.
5. Chang, C. S., and Chang, Y. (1995), "Green's Function for Elastic Medium with General Anisotropy," *Journal of Applied Mechanics*, ASME, Vol. 62, No. 3, 573-579.
6. Chang, C. S. (1993), "Micromechanics Modelling for Deformation and Failure of Granulates with Frictional Contacts," *Mechanics of Material*, Elsevier Science Publishers, Amsterdam, Vol. 16, No. 1-2, 13-24.
7. Chang, C. S., and Liao, C. L. (1994), "Estimates of elastic modulus for media of randomly packed granules," *Applied Mechanics Reviews*, ASME, Vol. 47, No. 1, Part 2, 197-206.
8. Chang, C. S., and Gao, J. (1995), "Second-Gradient Constitutive Theory for Granular Material with Random Packing Structure," *International Journal of Solids and Structures*, Vol. 32, No. 16, 2279-2293.
9. Chang, C. S., and Gao, J. (1996), "Kinematic and Static Hypotheses for Constitutive Modelling of Granulates Considering Particle Rotation," *Acta Mechanica*, Springer-Verlag (in print).
10. Chang, C.S., Chao, S. C., and Chang, Y. (1995), "Estimates of Mechanical Properties of Granulates with Anisotropic Random Packing Structure," *International Journal of Solids and Structures*, Pergamon Press, Vol. 32, No. 14, 1989-2008.

Refined Finite Element Analysis of Geomaterials

Boris Jeremić,[1] **Stein Sture,**[2]

Abstract

Derivation of a *Hyper-Spherical Arc-Length* type solution procedure for the finite element analysis of nonlinear material (hardening or softening)is presented. All relevant parameters are variable, resulting in a rather general path–following solution procedure with possible specializations to load (λ) control or displacement control procedures. Various solution techniques, used in non–linear finite element analysis of highly nonlinear constitutive behavior observed in the analysis of cohesionless granular materials subjected to low effective stresses, are discussed.

Residual Force Equations

Incremental equilibrium of the stationary body can be expressed through the *principle of virtual displacements* as:

$$\int_V \tau_{ij}\ \delta e_{ij}\ dV = \int_V f_i\ \delta u_i\ dV \quad \text{where} \quad e_{ij} = \frac{1}{2}\left(u_{i,j} + u_{j,i}\right) \tag{1}$$

The internal work given on the left side is equal to the actual stresses τ_{ij} going through the virtual strains δe_{ij} that corresponds to the imposed virtual displacements. The external work on the right side is equal to the actual forces f_i going through the virtual displacements δu_i. With the classical finite element approximation of displacement field $u \approx \hat{u}_a = H_I \bar{u}_{Ia}$ and after some algebra the equilibrium state can be expressed as:

$$\bigcup_m \int_{V^m} H_{I,b}\ {}^{tan}E_{abcd}\ H_{J,d}\ dV^m\ \bar{u}_{Jc} = \bigcup_m \int_{V^m} f_a\ H_I\ dV^m \quad \text{or} \quad \left(f_{Ia}\left(\bar{u}_{Jc}\right)\right)_{int} = \lambda\left(f_{Ia}\right)_{ext} \tag{2}$$

where ${}^{tan}E_{abcd}$ is the *consistent tangent stiffness tensor* defined in Jeremić and Sture [4]. It should be noted that the previous formulae have been derived for one increment, thus rendering explicit notation of incremental values unnecessary. By defining a tensor of residuals $r_{Ia}(\bar{u}_{Jc}, \lambda)$, or $\mathbf{r}(\mathbf{u}, \lambda)$ in vector form, equilibrium is obtained when:

$$r_{Ia}(\bar{u}_{Jc}, \lambda) = \left(f_{Ia}\left(\bar{u}_{Jc}\right)\right)_{int} - \lambda\left(f_{Ia}\right)_{ext} = 0 \quad \text{or} \quad \mathbf{r}(\mathbf{u}, \lambda) = \mathbf{f}_{int}(\mathbf{u}) - \lambda \mathbf{f}_{ext} = 0 \tag{3}$$

where $\mathbf{f}_{int}(\mathbf{u})$ are the internal forces which are functions of the displacements, $\mathbf{u}$, the vector $\mathbf{f}_{ext}$ is a *fixed external loading vector* and the scalar λ is a *load–level parameter* that multiplies

[1]Graduate Research Assistant,

[2]Professor, Department of Civil, Environmental, and Architectural Engineering, University of Colorado, Boulder, Colorado 80309-0428.

$\mathbf{f}_{ext}$. Switch from the tensorial to matrix notation is useful when it comes to solving a system of equations. Equation (3) describes the case of *proportional loading* in which *the loading pattern is kept fixed.*

Constraining the Residual Force Equations

Various forms of *path following methods* have stemmed from the original concept of the arc–length method[3]. The aim is to find the intersection of the residual equation (3) with $s = constant$, where s is the arc-length, approximated using the incremental form:

$$a = (\Delta s)^2 - (\Delta l)^2 = \left(\frac{\psi_u^2}{u_{ref}^2} \Delta\mathbf{u}^T \mathbf{S} \Delta\mathbf{u} + \Delta\lambda^2 \psi_f^2 \mathbf{f}_{ext}^T \mathbf{f}_{ext} \right) - (\Delta l)^2 = 0 \qquad (4)$$

where Δl is the *radius* of the desired intersection and represents an approximation to the incremental arc length. The essence of the arc-length methods is that the load parameter λ

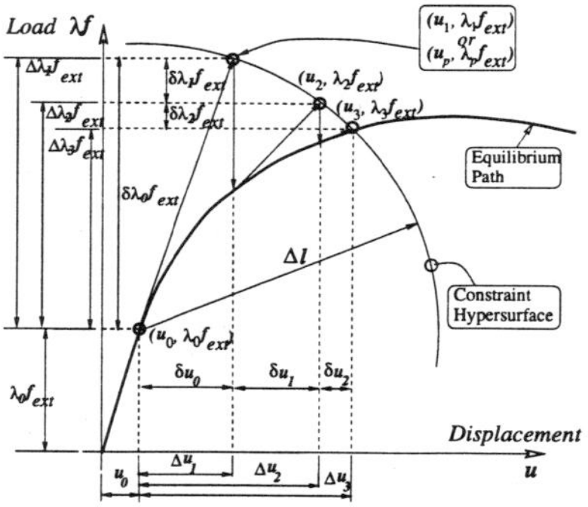

Figure 1: Spherical arc-length method and notation for one DOF system.

becomes a variable, in which case we are dealing with $n+1$ unknowns[4]. In order to solve this problem we have n equilibrium equations (3) and one constraint equation (4). We can solve the augmented system of $n+1$ equations iteratively by applying the *Newton – Raphson* scheme to equation (3):

$$0 = \mathbf{r}^{new}(\mathbf{u}, \lambda) = \mathbf{r}^{old}(\mathbf{u}, \lambda) + \frac{\partial \mathbf{r}(\mathbf{u}, \lambda)}{\partial \mathbf{u}} \delta\mathbf{u} + \frac{\partial \mathbf{r}(\mathbf{u}, \lambda)}{\partial \lambda} \delta\lambda = \mathbf{r}^{old}(\mathbf{u}, \lambda) + \mathbf{K}_t \, \delta\mathbf{u} - \mathbf{f}_{ext} \, \delta\lambda \qquad (5)$$

where $\mathbf{K}_t = \frac{\partial \mathbf{r}(\mathbf{u},\lambda)}{\partial \mathbf{u}}$ is the *tangent stiffness matrix*. It is convenient to solve equation (5) for $\delta\mathbf{u}$. The iterative displacement $\delta\mathbf{u}$ is split into two parts:

$$\Delta\mathbf{u}^{new} = \Delta\mathbf{u}^{old} + \delta\mathbf{u} = \Delta\mathbf{u}^{old} - \mathbf{K}_t^{-1}\mathbf{r}^{old} + \delta\lambda \mathbf{K}_t^{-1}\mathbf{f}_{ext} = \Delta\mathbf{u}^{old} + \delta\bar{\mathbf{u}} + \delta\lambda\delta\mathbf{u}_t \qquad (6)$$

where $\delta\lambda$ is the only unknown. The constraint from equation (4) can be used here, then by substituting $\Delta\mathbf{u}^{new}$ from equation (6) and by recalling that $\Delta\lambda^{new} = \Delta\lambda^{old} + \delta\lambda$ one ends up with the following quadratic scalar equation from which $\delta\lambda$ is solved:

$$\frac{\psi_u^2}{u_{ref}^2} \left(\Delta\mathbf{u}^{old} + \delta\bar{\mathbf{u}} + \delta\lambda\delta\mathbf{u}_t\right)^T \mathbf{S} \left(\Delta\mathbf{u}^{old} + \delta\bar{\mathbf{u}} + \delta\lambda\delta\mathbf{u}_t\right) + \left(\Delta\lambda^{old} + \delta\lambda\right)^2 \psi_f^2 \mathbf{f}_{ext}^T \mathbf{f}_{ext} = (\Delta l)^2$$

[3] See Felippa [2] and Crisfield [1].

[4] n unknown displacement variables and one extra unknown in the form of load parameter λ.

The constraint equation is given in a rather general form. Some further comments and observations are in order. By assigning various numbers to parameters ψ_u, ψ_f, $\mathbf{S}$ and u_{ref} one can obtain different constraint schemes from (4).

Coefficients ψ_u and ψ_f may not be zero simultaneously. Useful choices for the scaling matrix $\mathbf{S}$ are $\mathbf{I}$, $\mathbf{K}_t$ and $diag\,(\mathbf{K}_t)$. If $\mathbf{S} = \mathbf{I}$ and $u_{ref} = 1$ the method is the classical *arclength method.* If we choose $\mathbf{S} = diag\,(\mathbf{K}_t)$ nice scaling is achieved, but otherwise no physical meaning can be attributed to this category. With $\mathbf{S} = \mathbf{K}_t$ and $\psi_f \equiv 0$ one ends up with something very similar to the *external work constraint.* A rather general equation (4) can be further specialized to load (λ) control with ($\psi_u \equiv 0; \psi_f \equiv 1; \mathbf{S} \equiv \mathbf{I}$) or state control[5] with ($\psi_u \equiv 1; \psi_f \equiv 0; \mathbf{S} \equiv \mathbf{I}$). In the finite element literature, the term *displacement control* has been traditionally associated with the case in which only one of the components of the displacement vector $\mathbf{u}$ is specified. This may be regarded either as a variant of state control, in which a norm that singles out the ith component is used, or as a variant of the λ control, if the control parameter is taken as $\lambda \mathbf{u}_i$. One important aspect of scaling constraint equations by using $\mathbf{S} = diag\,(\mathbf{K}_t)$ or $\mathbf{S} = \mathbf{K}_t$ is the possibility of encountering a non–positive definite stiffness matrix $\mathbf{K}_t$. It often happens that after the limit point is passed, at least one of the eigenvalues of $\mathbf{K}_t$ is non–positive, thus rendering the constraint hypersurface non–convex.

Application to Geomaterials

The generalized constraint was tested using FEI[6]. A simple triaxial drained test on a loose sand was modeled, using MRS–Lade non–linear material model (Sture et al. [5]). Vertical displacements and vertical forces were followed. The numerical experiment was conducted in two stages, the first being isotropic compression $0.0 \rightarrow 10.0 kN/m^2$ and the second a standard triaxial compression test $q/p = 3/1$ under displacement control. Figure 2 presents typical load–displacement response for both loading stages of such an experiment. Specimen was $0.15m$ long with a $0.075m$ diameter, so the end displacement of $0.025m$ equals 16.7% of axial strain. In Figure (3) we follow one increment through three different types of control: mixed

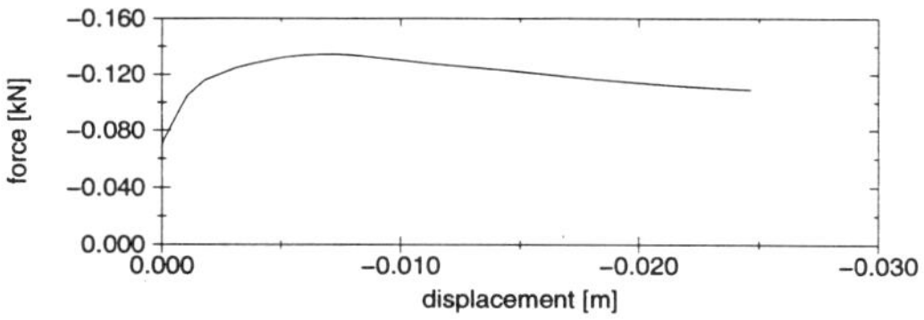

Figure 2: Non–linear response for the dry triaxial sand specimen.

(FU) control, λ (F) control and displacement (U) control. Since this was a displacement driven numerical experiment, λ control actually mimics displacement control. Figure (3) also shows that the use of the *consistent tangent stiffness tensor* in building the system stiffness matrix together with a full Newton – Raphson iterative solution scheme at the global level results in a quadratic convergence rate for the residual norm $\|\mathbf{r}\|$. At the end, converged points are close[7] to the actual equilibrium path. As for all path dependent problems, actual difference in the step length introduces additional error.

Conclusions

A derivation of a *Hyper-Spherical Arc-Length* type solution procedure for the non–linear

[5]The cylindrical constraint, or general displacement control.

[6]Finite Element Interpreter, our testbed program, under development using **nDarray** and FEMtools class libraries written in C++ programming language.

[7]Close here means within convergence tolerance.

$\downarrow$iter.,$\|\mathbf{r}\|$ $\rightarrow$	FU	F	U
prev.(1)	$0.00e+0$	$0.00e+0$	$0.00e+0$
pred.(2)	$8.05e-4$	$2.11e-3$	$1.03e-3$
(3)	$1.91e-4$	$7.70e-4$	$2.29e-4$
(4)	$3.83e-5$	$2.79e-4$	$3.93e-5$
(5)	$6.76e-6$	$9.95e-5$	$5.55e-6$
(6)	–	$3.39e-5$	–
(7)	–	$9.90e-6$	–

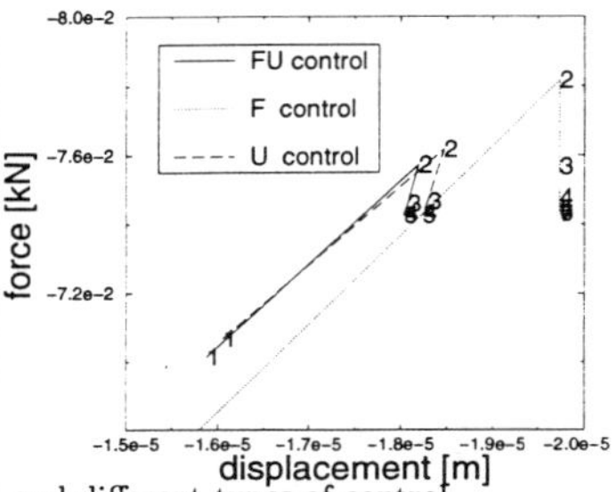

Figure 3: Iterative paths for one increment and different types of control.

finite element analysis was presented. Relevant parameters were kept variable, thus rendering a rather general path–following solution procedure. It seems that a full Newton–Raphson iterative scheme together with generalized constraint equations provides necessary flexibility and robustness in dealing with non–linear problems.

This endeavor represent a further step in the direction of creating robust finite element analysis applications for granular materials at low effective stresses, with high dilatancy rates, strong pressure sensitivity and highly nonlinear constitutive behavior seen at low effective stress levels.

Acknowledgment

The authors gratefully acknowledge support through NASA grant NAS8-38779 from Marshall Space Flight Center.

References

[1] Crisfield, M. A. *Non - Linear Finite Element Analysis of Solids and Structures Volume 1 : Essentials.* John Wiley and Sons, Inc. New York, 1991.

[2] Felippa, C. A. Dynamic relaxation under general incremental control. In *Innovative Methods for Nonlinear Problems*, W. K. Liu, T. Belytschko, and K. C. Park, Eds. Pineridge Press, Swansea U.K., 1984, pp. 103–133.

[3] Jeremić, B., and Sture, S. Finite element implementation of a elasto plastic material model. Report to: NASA, Marshall Space Flight Center, Contract: NAS8-38779, University of Colorado at Boulder, 1995.

[4] Jeremić, B., and Sture, S. Implicit integrations in geoplasticity. In *Proceedings of 10th Conference* (May 1995), S. Sture, Ed., Engineering Mechanics Division of the American Society of Civil Engineers, pp. 1099–1102.

[5] Sture, S., Runesson, K., and Macari-Pasqualino, E. J. Analysis and calibration of a three invariant plasticity model for granular materials. *Ingenieur Archiv 59* (1989), 253–266.

Non-Linear Models for Resilient Modulus Characterization of Granular Soils

Anand J. Puppala, A.M.[1], Louay N. Mohammad, M.[2], Aaron Allen[3]

Abstract: Non-linear resilient modulus models are generally used for accurate characterization of the granular subgrade soils. This paper evaluates three characterization models in their predictability to the resilient modulus of sandy soil. This analysis shows that the model which uses triaxial test stresses provides reasonable predictions of resilient properties. This model also adequately accounts for the moisture content and dry density effects.

Introduction

The 1986 AASHTO Guide for Design of Pavement Structures recommended using resilient modulus, M_r as a definitive property to characterize flexible pavement materials (Barksdale et al. 1990). The resilient modulus is a dynamic response defined as the ratio of the repeated axial deviator stress to the recoverable axial strain. AASHTO has developed separate laboratory test methods for granular and cohesive soils. The bulk stress model, recommended in AASHTO test methods, is generally used to characterize granular soils. This characterization is essential in flexible pavement design to estimate the structural capacity of flexible pavements in the mechanistic based design. One major limitation of the bulk stress model is that it provides identical resilient properties for identical bulk stresses obtained either from high confining and low deviatoric stresses or from low confining and high deviatoric stresses. This indicates that this model does not distinguish the individual/isolated effects of confining and deviatoric stresses. As a consequence, several non-linear regression models have been developed for better characterization. These models account for both confining and deviatoric stress effects. This paper presents a comparison of resilient modulus predictions by three non-linear regression models, including the bulk stress model. Repeated load triaxial resilient modulus test results on a granular sand are used in the model analysis.

[1]Research Associate, [2]Assistant Professor, [3]Research Assistant, Louisiana Transportation Research Center, Louisiana State University, Baton Rouge, LA70808

Regression Models

Three regression models are used in present material characterization methods. These are the bulk stress model, the octahedral stress model and the triaxial stress model. The bulk stress model uses bulk stress as the sole attribute. The bulk stress, θ, is defined as the summation of three principal stresses ($\sigma_1+\sigma_2+\sigma_3$). The equation of the bulk stress model is:

$$M_r = a\ \theta^b \quad (1)$$

where a and b are model constants and M_r is the resilient modulus.

The octahedral stress model can be expressed as:

$$\frac{M_r}{\sigma_{atm}} = k_1 \left(\frac{\sigma_{oct}}{\sigma_{atm}}\right)^{k_2} \left(\frac{\tau_{oct}}{\sigma_{atm}}\right)^{k_3} \quad (2)$$

where M_r is resilient modulus, σ_{oct} is octahedral normal stress ($\sigma_1 + \sigma_2 + \sigma_3 / 3$), τ_{oct} is octahedral shear stress (⅓ $((\sigma_1 - \sigma_2)^2 + (\sigma_2 - \sigma_3)^2 + (\sigma_1 - \sigma_3)^2)^{1/2}$), σ_1, σ_2, σ_3, and σ_{atm} are major, intermediate, minor principal stresses, and atmospheric pressure, respectively.

The third model, referred to as the triaxial stress model, uses confining and deviatoric stresses of a triaxial test as attributes. This model is described as:

$$\frac{M_r}{\sigma_{atm}} = k_4 \left(\frac{\sigma_3}{\sigma_{atm}}\right)^{k_5} \left(\frac{\sigma_d}{\sigma_{atm}}\right)^{k_6} \quad (3)$$

It should be noted that this model's attributes are not influenced by the multicollinearity effects since the attributes are direct triaxial test related stresses. The constants of each model can be determined by performing multiple regression analysis on resilient modulus triaxial test results. It should be noted that the last two models are modifications of models developed by Uzan (1985).

Analysis

Resilient modulus tests conducted at the Asphalt Institute in Lexington, Kentucky have provided the experimental data required for the characterization

(Hadley, 1983). Sand specimens were compacted at different moisture contents and dry densities and were tested in a repeated load triaxial test setup. Five different moisture content and dry density levels were used in testing. These moisture content and dry density levels represent dry of optimum, optimum and wet of optimum ranges of a standard Proctor test. Tests were conducted at three sets of confining pressures, 14, 28 and 42 kPa (2, 4 and 6 psi). At each confining pressure, four sets of deviatoric stresses of magnitudes varying from 14 to 112 kPa (2 to 16 psi) were applied. The stresses and the sequence order in which they were applied are different from the current AASHTO testing procedures.

Table 1: Model Correlations

Bulk Stress Model	**Octahedral Stress Model**	**Triaxial Stress Model**
log a = -0.85 + 0.06* γ_d - 0.27 * w (0.98)	log k_1 = 2.56 + 0.013 * γ_d - 0.08 * w (0.96)	log k_4 = -9.61 + 0.12 * γ_d - 0.08 * w (0.69)
b = - 1.23 + 0.002 * γ_d + 0.11 * w (0.96)	k_2 = - 34.9 + 0.31 * γ_d - 0.003 *w (0.72)	k_5 = -19.6 + 0.17 * γ_d + 0.05 *w (0.69)
	k_3 = 28.1 - 0.25 * γ_d + 0.07 * w (0.68)	k_6 = 15.2 - 0.14 * γ_d + 0.06 * w (0.68)

Note: γ_d is dry density in pcf; w is moisture content; R^2 value is given in parenthesis

The model analysis is conducted by first analyzing the regression model and then correlating the model constants with moisture content and dry density levels. The model correlations are presented in Table 1. Evaluation of regression models is attempted by performing a comparison analysis with the experimental results. The developed correlations of each model are used to determine the model constants at the optimum moisture content and dry density level. These constants are then used with its respective model to predict the resilient moduli at different confining and deviatoric stress levels. The predictions are compared with the corresponding experimental test results in Figure 1.

Comparisons indicate that the triaxial stress model provides predictions closer to measured data. The octahedral and bulk stress model predictions are different from the measured data, indicating that these models may not provide reasonable predictions. Better predictions by the triaxial model can be attributed to the absence of multicollinearity problems in the model.

Summary and Conclusion

Three regression models are evaluated with respect to their ability to reliably predict resilient moduli. The triaxial model which uses test stresses as attributes provided the best predictions. The bulk stress model predictions are different from the measured data, which indicate that this model may not be reliable in providing reasonable predictions. Based on the comparisons, the triaxial model is recommended for future use.

References

1. Barksdale, R.D., Rix, G.J., Itani, S., Khosla, P.N., Kim, R., Lambe, P.C., and Rahman, M.S. (1990) "Laboratory Determination of Resilient Modulus For Flexible Pavement Design," *NCHRP Report 1-28*, Georgia Institute of Technology, Atlanta, Georgia.
2. Hadley, W. O. (1983) "Experimental Base Project Reports 78 - 1 to 78 - 6," Louisiana Tech University, Ruston, Louisiana.
3. Uzan, J. (1985) "Characterization of Granular Materials," *Transportation Research Record 1022, TRB,* National Research Council, Washington, D.C., pp.52-59.

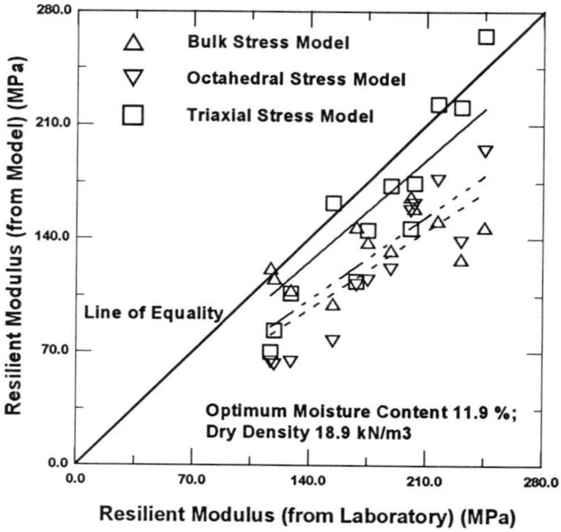

Figure 1: Comparison of Predicted and Measured Resilient Properties

Fatigue Model of Asphalt Concrete

Jian Zhou[1] and Robert Y. Liang[2]

Abstract

The decrease of fracture energy per cycle under a displacement-controlled fatigue testing is calculated based upon the two-stage line zone model. A parameter α is introduced in the fracture energy to take into account the amount of the damage incurred for each cycle. The parameter α can be obtained from a regression analysis of the calculated fracture energy and the test results from overlay tester. The shift factors based on the developed two-stage fatigue model for different rest times are calculated.

Introduction

In general, the fatigue data obtained from laboratory tests, which are determined under the conditions of constant load or constant displacement, underpredict the field fatigue life. Finn et al (1983) demonstrated that the laboratory-derived phenomenological fatigue relationships for the asphalt concrete used at the AASHTO Road Test required a shift factor of 13 in fatigue life to match actual fatigue cracking data derived from AASHTO field sections. This difference between laboratory and field fatigue curves may be attributed to the difference of loading conditions between the laboratory and the field.

In the field, there are intervals between successive applications of load; the length of such intervals depends upon axle spacing, vehicle speed and the distance between vehicles. It is conceivable that during these "no loads" intervals or rest periods as they are commonly called, incurred damage in the pavement is recovered. The issue is complicated by the fact that the healing mechanism during the rest periods are not fully understood and the degree of damage incurred during a load pulse is not well defined.

[1] Graduate Assistant, Dept. of Civil Eng., The Univ. of Akron, Akron, OH44325
[2] Professor and Chair, Dept. of Civil Eng., The Univ. of Akron, Akron, OH 44325

According to the research work provided by Kim and Little(1989), there are two different mechanisms occurring in a partially cracked asphalt concrete during the rest period. One is the relaxation of stresses in the material due to the viscoelastic nature of asphalt concrete after removing the applied load, and the other one is the chemical healing across microcrack and macrocrack face. Both of these mechanisms enhance the fatigue life of AC pavement and occur simultaneously during rest periods. These two mechanisms have been evaluated separately(Kim and Little, 1989) by applying the correspondence principle of the theory of nonlinear viscoelasticity developed by Schapery (1975).

Tseng and Lytton (1990) presented an analytical technique that may be used to estimate a shift factor between laboratory and field fatigue life. The shift factor consists of two components: shift factor caused by the relaxation of residual stress SF_r and shift factor caused by the healing process SF_h. Thus, the shift factor SF is

$$SF = (SF_r)(SF_h) \tag{1}$$

Tseng and Lytton (1990) reported an increase in work done to open the crack after rest periods in controlled-displacement crack propagation testing in asphalt concrete mixes modified with various additives. Based on this fracture energy concept, the shift factor caused by healing was obtained.

In this paper, the decrease of fracture energy per cycle under the displacement-controlled fatigue testing is calculated based upon the two-stage line zone model. A parameter α is introduced in the fracture energy to take into account the amount of the damage incurred for each cycle. The amount of fracture energy decrease after rest period can be obtained in terms of the varying parameter α for different rest times. The parameter α, as a function of rest time, can be obtained from a regression analysis of the calculated fracture energy and the test results from overlay tester. Finally, the fatigue lives based on the developed two-stage fatigue model for different rest times are calculated and compared to the results provided by Tseng and Lytton (1990).

Analysis of Fracture Energy

In displacement-controlled fatigue testing, the strain energy decreases with the increase of crack length (Aglan and Figueroa, 1993), therefore the energy release rate always increases with increasing crack length. The strain energy change ΔU per cycle of load is calculated by

$$\Delta U = -\frac{dU}{dN} = -\frac{dU}{da}\frac{da}{dN} = -G\frac{da}{dN} \tag{2}$$

where a is the crack length; ΔU is the strain energy change per cycle; G is energy release rate; da/dN is the rate of crack growth.

As it was well known, the Dugdale's plastic line zone model assumed that the fracture stress was constant in the fracture zone. This model was widely used to characterize the fracture zone size and fracture behavior of quasi-brittle materials. According to the two-stage plastic line zone model proposed by Williams (1977) for fatigue crack growth, the fracture zone can be divided into two zones: the damage zone incurred by loading and unloading at the crack tip, and the plastic zone which was the same as that in the Dugdale model. The failure stress in the damage zone was characterized by a parameter α which can account for the amount of damage incurred by a cycle of loading-unloading. The interpretation of parameter α as an amount of damage per cycle was critical in our analysis of the rest period effects on the fatigue life.

The strain energy change per cycle ΔU can be obtained using the two-stage model as follows:

$$\Delta U = -\sigma_c u_c [\alpha \frac{da}{dN} + \frac{1-\alpha}{\delta_1}\left(\frac{da}{dN}\right)^2] \tag{3}$$

where σ_c is the fracture stress in the fracture zone; u_c is the crack tip opening displacement; δ_1 is fracture zone length which can be derived by the analysis of two-stage plastic zone model

The two-stage fatigue crack growth equation derived by Williams (1977) provides the crack growth rate da/dN.

$$\frac{da}{dN} = \frac{\delta_c}{(1-\alpha)^2}\left[\left(\frac{K}{K_c}\right)^2 - \alpha\right] \tag{4}$$

where K is the stress intensity factor; K_c is the critical value of K.

The Rest Period Effects on the Strain Energy Change

As discussed in the previous sections, the healing of microcracks in the damaged material after the rest periods tends to enhance the fatigue resistance of the material. Thus, the rest periods result in an increment of fatigue life, compared to that without the rest periods. Several researchers (Kim and Little, 1989; Tseng and Lytton, 1990) have noticed this type of mechanism and studied the beneficial change of microstructure in the damaged material. Therefore, the ratio of energy release rate after rest period is

$$\frac{\Delta U_r}{\Delta U} = \frac{\left\{\alpha^*\left(\frac{da}{dN}\right)^* + \frac{(1-\alpha^*)}{\delta_1^*}\left(\frac{da}{dN}\right)^{*2}\right\}}{\left\{\alpha\left(\frac{da}{dN}\right) + \frac{(1-\alpha)}{\delta_1}\left(\frac{da}{dN}\right)^2\right\}} - 1 \tag{5}$$

where $\Delta U_r = \Delta U^* - \Delta U$; ΔU^* is the fracture energy after rest period. Tseng and Lytton (1990) provided some data about the relationship between ratio $\Delta U_r / \Delta U$ and rest time for overlay test.

Let us assume the initial value of parameter α as 0.05 which can be used to determine the fatigue life without rest periods N_0. The relationship between α^* and rest time t_r / t_0 is derived by a regression procedure and is shown in Fig. 1:

$$\alpha^* = 0.05 + 8.12 \times 10^{-3} (t_r / t_0)^{0.2945} \qquad (6)$$

Once we have established the regression equations for α^*, it is possible to calculate the fatigue life for different values of rest periods. Here, the overlay tester is used to study the effects of rest on the fatigue life. An initial notch is introduced in the specimen. A shift factor SF is defined as $SF=N_f/N_0$, where N_f is the number of cycles to failure with rest periods; N_0 is the number of cycles to failure without rest periods. The fatigue life is predicted based on two-stage crack growth law as follows

$$N_f = \frac{(1-\alpha)^2 (a_f - a_0)}{2\delta_c} \int_{-1}^{1} \frac{d\xi}{\left[\frac{\sigma \sqrt{H}}{K_c} \sqrt{\frac{H}{\pi a}} F\left(\frac{\xi (a_f - a_0) + (a_f + a_0)}{2H} \right) \right]^2 - \alpha} \qquad (7)$$

where F is the geometric function related to crack length.

Tseng and Lytton (1990) proposed a shift factor which is a function of rest time.

$$SF = 1 + 0.3432 (t_r / t_0)^{0.427} \qquad (8)$$

The relationship between the predicted shift factor($SF=N_f/N_0$) based on the two-stage model and the rest time is shown in Fig. 2 The calculated shift factors based on Eq. (8) are also shown in Fig. 2. It is shown that the shift factors based on the two-stage model accurately reflect the observed trend.

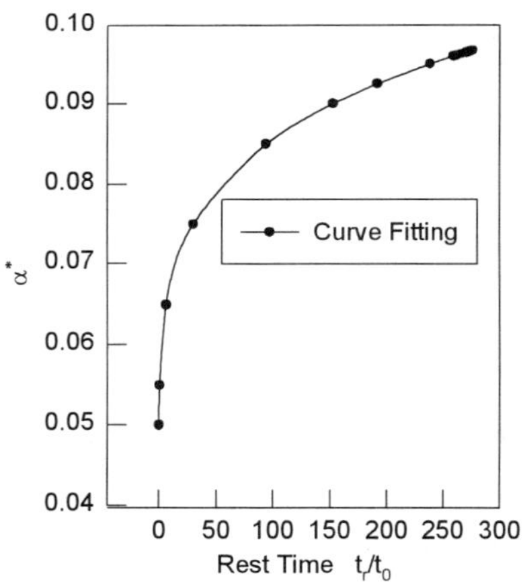

Fig.1 Parameter α^* as a function of rest time

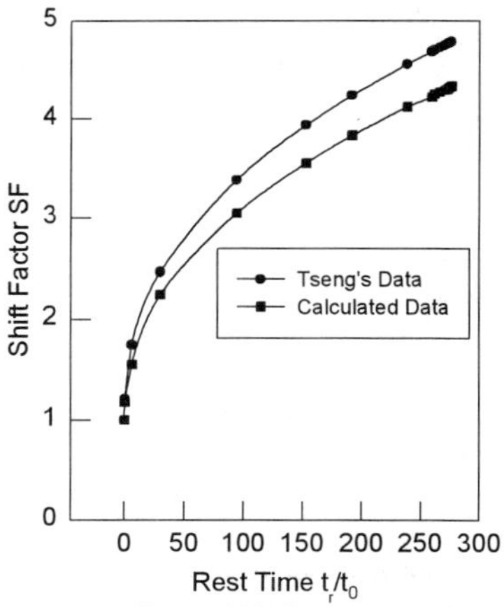

Fig. 2 Shift Factor as a Function of Rest Time

Summary

A simple method to study the effects of the rest period on the fatigue life of asphalt concrete is introduced based on the considerations of the energy release rate during cyclic loading. The basic idea is to consider the healing of microcracks in front of a main crack which results in more energy needed to propagate the crack after the rest periods. The two-stage plastic zone model is used to study the fatigue crack growth of asphalt concrete. The parameter α introduced in the model can be identified as the damage incurred per cycle. It is reasonable to assume that the damage per cycle decreases after the rest period due to the healing process of microcracks.

The healing is a very complicated process during the rest periods. There are many factors which affect the healing such as the rest time, temperature, test frequency and asphalt component. We did not get into the healing mechanism in detail. The study is based on the continuum fracture mechanics concept.

References

1. Aglan, H. A. and Figueroa, J. L., "Damage-Evolution Approach to Fracture Cracking in Pavements", Journal of Engineering Mechanics, Vol. 119, No. 6, 1993, pp. 1243-1259.
2. Finn, F. N., Monosmith, C. L., Markvich, N. J., "Pavement Performance and Asphalt Concrete Mix Design", Proc., Association of Asphalt Paving Technologists, Vol. 52, 1983, pp. 121-144.
3. Kim, Y. R. and Little, D. N., "Evaluation of Healing in Asphalt Concrete by Means of the Theory of Nonlinear Viscoelasticity", Transportation Research Record, 1228, 1989, pp. 198-210.
4. Schapery, R. A., "A Theory of Crack in Initiation and Growth in Viscoelastic Media, I. Theoretical Development", Int. J of Fracture, Vol. 11, No. 1, 1975, pp. 141-159.
5. Tseng, K-H, and Lytton. R. L., "Fatigue Damage Properties of Asphaltic Concrete Pavements", Transportation Research Record, 1286, 1990, pp. 150-163.
6. Williams, J. G., "A Model of Fatigue Crack Growth in Polymers", Journal of Materials Science, Vol. 12, 1977, pp. 2525-2533.

DEFORMATION PATTERNS IN BIAXIAL SHEAR OF PARTICULATES

Anil Misra[1] and Hongjun Jiang[2]

ABSTRACT

Biaxial experiments are performed with rod assemblies to study the micromechanical deformation behavior of granular materials. The focus in this paper is on the micromechanical behavior under mixed boundary conditions, with stress-controlled lateral boundaries and displacement controlled vertical boundaries. Particle motions, i.e. displacements and rotations, are measured during the test. The particle motions are analyzed to study deformation patterns, including shear localization. Deformation patterns in a number of rod assemblies with regular as well as random packing structures are investigated.

INTRODUCTION

The mechanical properties of dense particulate materials are significantly influenced by the micromechanical phenomena that occur at particle level. In recent years, efforts have been made to study these particle-level phenomena from theoretical as well as experimental view points. Along the theoretical methods, two approaches have been followed, namely: (1) the micro-mechanical modelling method (see Misra and Chang 1993), and (2) the computer simulation method, such as the discrete element method (see Cundall and Strack 1979).

Experimentally, the behavior of particulate materials from the micro-mechanical viewpoint has been mainly studied using simulated materials, such as rod assemblies used by Schneebli (see Drescher and DeJosselin DeJong 1972). Several researchers have since performed experiments on rod assemblies to study the mechanical behavior of granular materials. These experiments have provided a slew of fundamental results useful for understanding mechanisms of deformation and force transference in granular media. However, the particle displacement and rotation fields, especially under mixed boundary conditions, have not been investigated. In the present paper, we perform mixed boundary biaxial tests on rod assemblies with stress-controlled lateral boundaries and displacement controlled vertical boundaries. Particle motions, i.e. displacements and rotations, are measured during the test. The particle motions are analyzed to study micromechanical

[1] Assistant Professor and [2] Research Assistant, Department of Civil Engineering, University of Missouri, 5100 Rockhill Road, Kansas City, MO 64110.

phenomena in rod assemblies with regular as well as random structures.

DESCRIPTION OF THE BIAXIAL APPARATUS

A general view along with a schematic plan of the biaxial apparatus used in this study is shown in Figure 1. The axial load is applied using a pneumatic cylinder with a 4" diameter piston bolted to the top bar of the frame assembly, such that a uniform displacement is applied across the top of the rod assembly. The lateral load is applied from both sides using bicycle tubing enclosed in special cloth sleeves designed to prevent tube bulging. The tube-sleeve bladder assemblies provides a uniform pressure distribution to lateral side of the rod assembly. To avoid friction between the top bar slot and bladder assembly, a special trap is used which makes the shape of bladder assembly fit the top bar slot. Two separate air supplies, with regulators and air pressure gages, are used for independent loading in the lateral and the axial directions. Both bladder assemblies are connected to the same regulator so that the air pressure in each tube-sleeve bladder remains equal. The undeformed dimensions of a typical rod assembly are 16"x10".

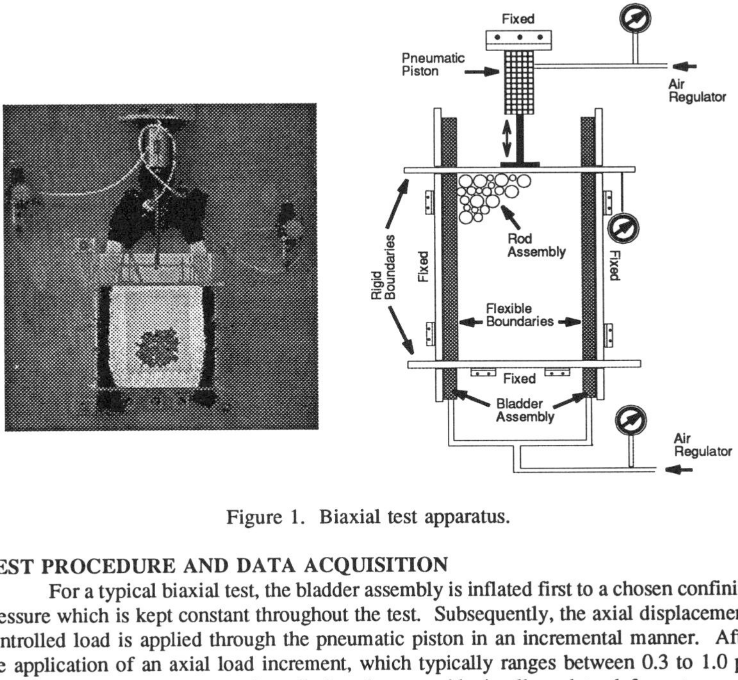

Figure 1. Biaxial test apparatus.

TEST PROCEDURE AND DATA ACQUISITION

For a typical biaxial test, the bladder assembly is inflated first to a chosen confining pressure which is kept constant throughout the test. Subsequently, the axial displacement-controlled load is applied through the pneumatic piston in an incremental manner. After the application of an axial load increment, which typically ranges between 0.3 to 1.0 psi of air pressure in the pneumatic cylinder, the assembly is allowed to deform to a new equilibrium configuration. To ensure that the loading process is quasi-static, a time lapse of up to 15 minutes is allowed between two load increments. The dimensions of the deformed rod assembly is measured before the application of a new load increment. The axial stress on the assembly is calculated from the air pressure in the pneumatic cylinder and the average lateral dimension of the rod assembly. The lateral strain is also based upon the average lateral dimension of the rod assembly. The lateral stress is, of course, the air pressure in the bladder assembly, while the axial strain is calculated from the change in the axial dimension. The deformed rod assembly is also photographed before the application of a new load increment. The digitized images of these photographs are analyzed to obtain the particle locations and orientations at the beginning of each load increment. To facilitate the image analysis, each rod has been marked with two dots. Also the rod sizes are distinguished by colors painted on rod ends. A typical digitized photograph is shown in Figure 2. The particle locations and orientations obtained from this digitized image are used to plot the rod assembly in Figure 2.

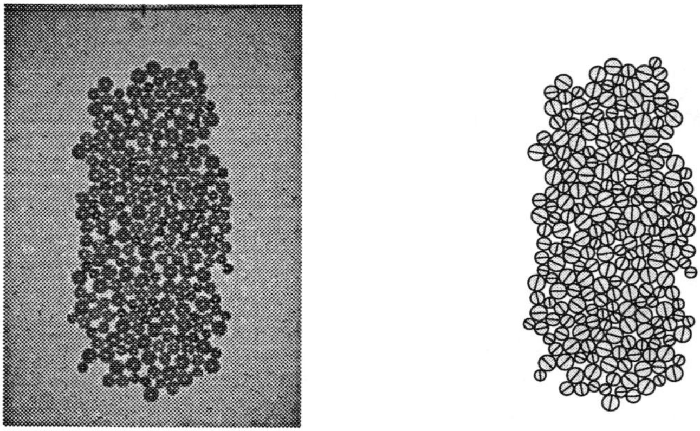

Figure 2. Particle level data acquisition.

RESULTS OF BIAXIAL TEST

A series of experiments were performed using the above biaxial apparatus. The confining stress was varied from 2 psi to 7 psi for these tests. Figure 3 shows the typical stress-strain curves obtained for plastic rod assemblies for a confining stress of 4 psi. As evident from Figure 3, the tests are fairly repeatable and the stress-strain results almost coincide. During the test, the rod assemblies first contracted up to an axial strain of 0.05. The rod assemblies then dilated, displaying the classical sand-type behavior.

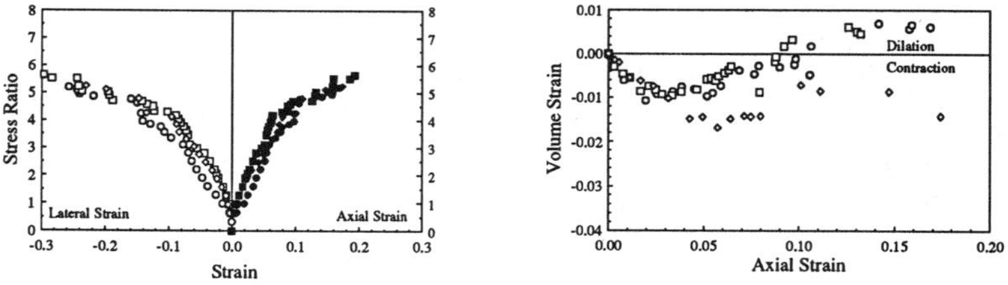

Figure 3. Stress-strain behavior of rod assemblies.

Examples of measured displacement and rotation fields are given in Figure 4. As illustrated in Figure 5, the displacement field is not linear and interesting deviation patterns are observed.

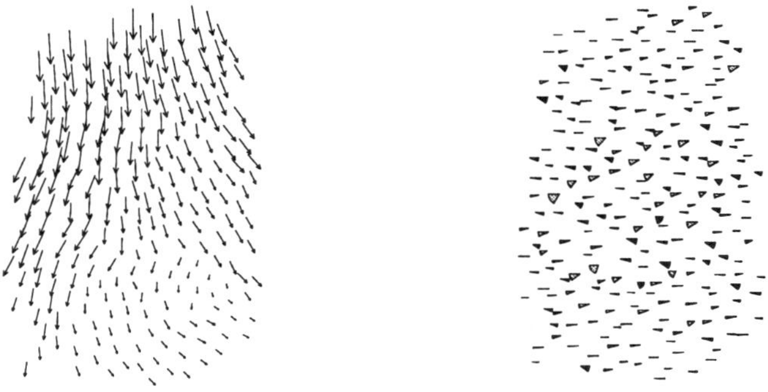

Figure 4. An example of measured displacement and rotation fields.

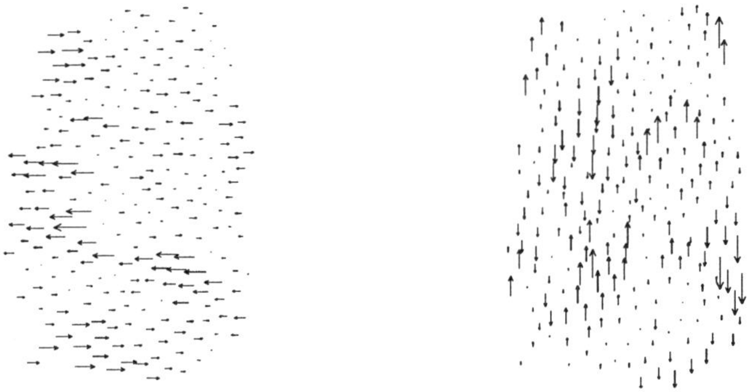

Figure 5. Deviation of particle displacements from a linear displacement field.

References

Cundall, P.A. and Strack, O.D.L., "A discrete numerical model for granular assemblies," *Geotechnique*, Vol. 29, No. 1, 47-65, 1979.

Drescher, A. and DeJosselin DeJong, G., "Photoelastic verification of a mechanical model for the flow of a granular material," *Journal of Mechanics and Physics of Solids*, Vol. 20, 337-351, 1972.

Misra, A. and Chang, C.S., "Effective elastic moduli of heterogeneous granular solids," *International Journal of Solids and Structures*, Vol. 30, No. 18, 2547-2566, 1993.

MRI Studies of Direct Shear Tests on Round Particles

Tang-Tat Ng[1], Marlene Kelley[2] and James Sampson[3]

Abstract

In order to study the microscopic behavior of granular soils, magnetic resonance imaging (MRI) technology is used to detect round particle locations and movements during loading. The results will be used to verify computer simulation results based upon Discrete Element Method (DEM). The use of MRI technology will allow the researchers to non-invasively visualize the interior and behavior of a suitable granular mass. Pharmaceutical pills act as idealized granular soil particles because of their strong MRI signal. The focus of these experiments will be on the stress-strain behavior, shear strength, volumetric change, initial packing of the particles, and the fabric evolution during shear.

Introduction

In natural granular soils, both stiffness and strength are usually anisotropic. The main reason for this structural anisotropy is that soils contain a wide range of elongated shapes which tend to lie horizontally when deposited in a gravitational field. As the particles line up, their contact normals group around the vertical direction. The three main elements which affect the geometric orientation of the soil are the shape and size of the particles, the distribution of contact normals, and the volume of the voids. The number of contact points and the distribution of contact normals are important factors in determining the anisotropy and macromechanical behavior of the granular material.

By using the MRI to image the evolution of fabric development under loading, we will be able to visually observe the interaction between individual particles. From these images, the changes in geometric orientation can be determined.

1. Assistant Professor, Department of Civil Engineering, University of New Mexico.
2. Student, Department of Civil Engineering, University of New Mexico.
3. Student, Department of Mechanical Engineering, University of New Mexico.

Two computer programs, ELLIPSE2 and ELLIPSE3, were developed by the first author (Ng, 1992; Lin and Ng, 1994). The programs simulate the stress-strain response of ellipsoidal particles. Three-dimensional images of particles and micromechanical statistics can also be generated by the programs. The numerical simulations are an attempt to establish a connection between the parameters affecting the micromechanical response and the macromechanical behavior of soil particles. From this correlation, a constitutive relation between micro and macromechanical behavior of soil can be determined. Due to the page limitation, this paper will only present the MRI and the nonmetal direct shear device.

Magnetic Resonance Imaging (MRI)

The energy level of some atoms is split in the presence of an external magnetic field due to the interaction of the atomic magnetic moments and the external magnetic field. When a sample is irradiated with photons of energy, some of the nuclei translate from a lower energy level to a higher energy level by absorption. These nuclei then decay down to a lower energy level, emitting photons of energy. The rate at which these photons are absorbed and emitted is called Radio Frequency radiation. The measurement of this resonant frequency can be used to determine the magnetic moment of the nuclei, and can give information about the magnetic field seen by the molecule. This process is called Nuclear Magnetic Resonance (NMR) or Magnetic Resonance Imaging (MRI).

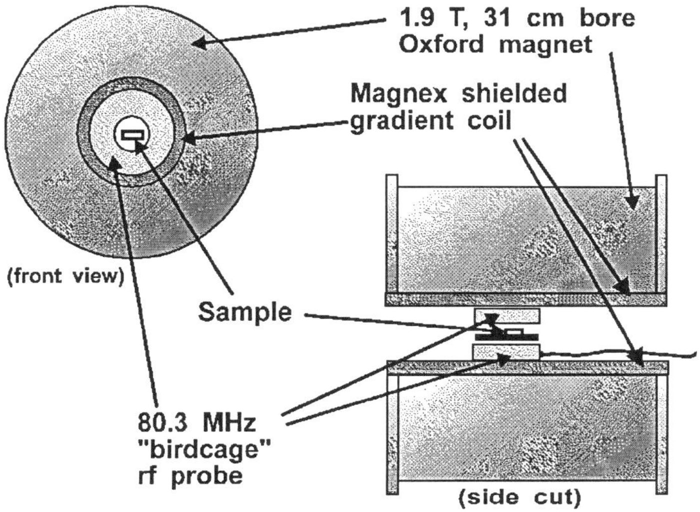

Figure 1. Magnetic Resonance Imaging Setup

Typical magnets used in MRI are superconducting magnets with field strengths up to 12 Tesla for small bore (~0.1m), 2-4.7 T for medium bore (~0.3m), and 1-1.5 T for large bore (~1m). The magnet used at the Lovelace Research Institute is of medium bore (1.9T), and has a precession frequency of approximately 80 MHz. A schematic diagram of the MRI is given in Figure 1.

Design of Shear Test Box

The direct shear test is an inexpensive, fast, and a simple strength test for soils. This device is so common that description of the test can be obtained in any soil mechanics text books.

In order to image the vitamin pills (particles with strong signals) undergoing compression and shear loads within a 14.5 cm diameter "birdcage", a shear box must be fabricated. Design limitations for the shear box included space considerations, nonmagnetic materials, and acceptable loading techniques. The use of nonmagnetic metals was discouraged due to possible interference with the uniform magnetic field within the bore of the MRI.

A lot of tentative designs have been tested and the final design chosen incorporates a fitted direct shear apparatus with the normal load being applied pneumatically. The device shown in Figure 2. consists of four major components: the fitted base frame, top shear box, pneumatic plunger assembly, and upper frame. The materials of choice for the shear box were Plexiglas, Polycarbon and Teflon.

The fitted base frame slides into and is supported by the "birdcage" within the MRI and remains fixed throughout the testing process. Through a regulated air system and internal air bladder, the pneumatic plunger assembly applies a constant pressure to the vitamin pill sample while allowing for volumetric changes throughout the shearing process. The upper frame is fixed to the fitted base frame and insures the pneumatic plunger assembly freedom of movement. Teflon strips between the upper frame and the pneumatic plunger assembly provide for a nearly frictionless surface during the horizontal displacement of the top shear box. The sample has a cross sectional area of 25 cm^2.

The only nonmagnetic metallic portion of the designed shear box was the valve stem of the air bladder used to provide the vertical load on the specimen. It is generally felt that the presence of a small amount of metal will not significantly disturb the imaging ability of the magnetic flux.

Figure 2. Nonmagnetic Direct Shear Device

Acknowledgments

The authors would like to thank Dr. Masami Nakagawa (Sandia National Laboratories), Dr. Eiichi Fukushima (Lovelace Research Institute), Dr. Steve Altobelli (Lovelace Research Institute), and Mr. Clifford Bussell (UNM) for their support, and encouragement. This project was supported by the Minority Engineering Mathematics and Science Program (University of New Mexico) and the National Science Foundation under Grant CMS-9505338 and the cognizant program director was Dr. Priscilla P. Nelson. This support is gratefully acknowledged.

References

Lin, X. and Ng, T.-T. (1994) "Numerical Modeling of Granular Soil Using Random Arrays of Three-Dimensional Elastic Ellipsoids," *Computer Methods and Advances in Geomechanics*, Vol. 1, 605-610.

Ng, T.-T. (1992) "Numerical Simulations of Granular Soils Using Elliptical Particles," *Microstructural Characterization in Constitutive Modeling of Metals and Granular Media*, MD-Vol. 32, 95-117.

EFFECT OF MAXWELL BINDER ON TWO-PHASE MATERIALS

Han Zhu[1], Jeff W. Rish, III[2] and William C. Dass[1]

Abstract

This study presents some results from an investigation of the constitutive behavior of a two-phase (the binder and particles) particulate material. In this simplified analysis, the particles behave elastically and the binder is treated as a Maxwell visco-elastic solid. The particles are hexagonally arrayed in a binder matrix, and the basic-cell technique is employed to determine the constitutive relationships for the cell. The constitutive equations are derived in both integral and differential representations, and they show the over-all visco-elastic response of this two-phase material can be represented by a model of generalized multi-axial Maxwell-elements.

Introduction

In this study, we aim to investigate asphaltic concrete (AC), a major building material used in construction of roads and airfields. Asphalt concrete consists of mineral aggregates, air voids and asphalt binder. Aggregates are very rigid relative to the binder, and are usually assumed to be elastic. On the other hand, the physical properties of the asphalt binder are much more complicated. A few recent studies (Little, et al., 1993) show that, as far as asphalt binder modeling is concerned, the visco-elastic nature of asphalt binder can be fairly represented by using a system of generalized Maxwell elements.

Granting that aggregates are elastic and asphalt binder is visco-elastic with Maxwell behavior, the question becomes one of how to describe the constitutive behavior of the mixture, e.g., asphaltic concrete? More importantly, can we develop analytic models that will reasonably predict the behavior of asphaltic concrete quantitatively under the practical road or airfield service environment based solely on the physical properties of aggregates and asphalt binder and the load-time history for the material? This question is being addressed in this study.

---

[1] Applied Research Associates, Inc., Tyndall AFB, Florida 32403
[2] Wright Laboratory, WL/FIVCO, Tyndall AFB, Florida 32403

Aggregates constitute about 95 percent of AC by weight, and the asphalt binder makes up the other 5 percent. Air voids occupy 5-8 percent of the total volume of asphaltic concrete. Since the presence of aggregates in AC is very high both in terms of weight and volume, it can be easily seen that when traffic loads are imposed on AC, load transfer mainly takes place through the aggregate skeleton. Analytical description of this process is complicated by the particle-binder-particle interactions that occur when the load is applied. This indicates that, to find an answer to the question proposed in this study, particle-binder-particle contact laws must be developed.

Contact laws are used to characterize the force-displacement relationships between two particles. Obviously, these local relationships must be extended to the level where the packing of aggregates can be accounted for, i.e., an extension from micro to macro with respect to constitutive behavior of the material. Therefore, we pursue the analysis of AC by formulating the approximated contact laws and using micromechanics to establish the stress-strain relationships.

Development of binder-contact laws

The particle geometry used to develop the contact laws is shown in Fig. 1. It is a system of two-particles with a binder in between. The particles are elastic and the binder can be either elastic or visco-elastic. The governing equation for the normal contact interfacial pressure distribution p(r), total force applied to the system P_z and the relative approach δ_z are given by (elastic binder case):

$$\delta_z = h(r)\frac{p(r)}{E_b} + \frac{(1 - \nu_1^2)}{\pi E_p}\int_0^a \int_0^{2\pi} \frac{p(\rho)\ \rho\ d\rho\, d\theta}{\sqrt{r^2 + \rho^2 - 2r\rho\cos\theta}} \tag{1}$$

$$P_z = 2\pi \int_0^a r p(r)\, dr \tag{2}$$

The exact analytical solution of p(r) to Eqs. (1) and (2) is very difficult to obtain. More importantly, considering the large numbers of particles/inclusions that are present in any engineering structures made of real two-phase materials, contact-law compliances should be sought in terms of explicit representations in order to facilitate the general applicability of those compliances. This means that a simple and closed-form relationship between P_z and δ_z is needed. Two upper and lower bounds and two physically based approximations for P_z-δ_z relationships are successfully derived by utilizing the monotonic behavior of the kernel and *Chebyshev's inequality for integrals*. The results are given by (Zhu et al., 1995):

$$(b_2C_1+C_2)P_z < \delta_z < (C_1+b_1C_2)P_z, \ (upper,\ lower\ bounds) \tag{3}$$

$$\delta_z = (C_1+C_2)P_z, \ (physical\ approximations) \tag{4}$$

$$b_2 < 1 < b_1, \quad C_1 = \frac{h_0}{\pi a^2 E_b X}; \quad C_2 = \frac{1-\nu_p^2}{2aE_p}, \ X = \frac{\ln(1+d)}{d} \tag{5}$$

where Eq. (4) is a "best estimate" of the compliance. The visco-elastic normal compliance for the case of a Maxwell binder is obtained in a similar fashion. The derived relationship is rate-dependent, and is presented in Eq. (6).

$$\dot{\delta}_z(t) = (C_1+C_2)\dot{P}_z(t) + C_1E_bP_z(t)\,/\,\eta \tag{6}$$

The quantity, η, in Eq. (6) is the binder viscosity constant. Tangential and rolling compliances for elastic and Maxwell binders have also been derived.

Micromechanical Analysis

The following analysis presents results obtained as a consequence of studying the stiffness of the previously described two-phase particulate materials from a micromechanics viewpoint. A simple geometry is used in which the particles are hexagonally arrayed. The basic cell technique is employed for both on-axis and off-axis cell arrangements, and a detailed static and kinematic analysis is carried for the cell under compressive and shear forces. The derived integral and differential relationships relating the stress and strain components for a Maxwell binder at the cell level, (σ_x, σ_y, σ_{xy}) and (ϵ_x, ϵ_y, ϵ_{xy}), are obtained in a matrix form. For a special case of only non-vanishing σ_x and ϵ_x, the relationship between σ_x and ϵ_x is given by:

$$\begin{aligned}
\overline{\sigma_x} = {} & \frac{\sqrt{3}5K_n}{8}[\overline{\epsilon_x}(t) - g_n\int_0^t e^{-g_n(t-t_1)}\overline{\epsilon_x}(t_1)dt_1] \\
& + \frac{\sqrt{3}K_s}{5}[\overline{\epsilon_x}(t) - g_s\int_0^t e^{-g_s(t-t_1)}\overline{\epsilon_x}(t_1)dt_1] \\
& + K^*[\overline{\epsilon_x}(t) - g^*\int_0^t e^{-g^*(t-t_1)}\overline{\epsilon_x}(t_1)dt_1] \\
& + 0.5Q'_{66}[\overline{\epsilon_x}(t) - g'\int_0^t e^{-g'(t-t_1)}\overline{\epsilon_x}(t_1)dt_1]
\end{aligned} \tag{7}$$

It can also be shown that Eq. (7) represents a visco-elastic model which can be graphically described in Fig. (2).

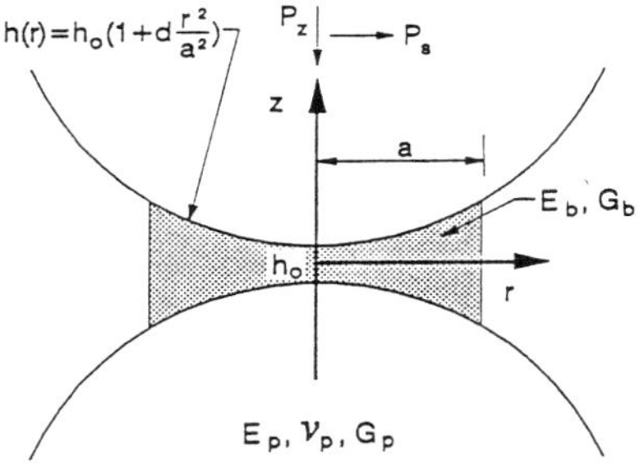

Fig. 1

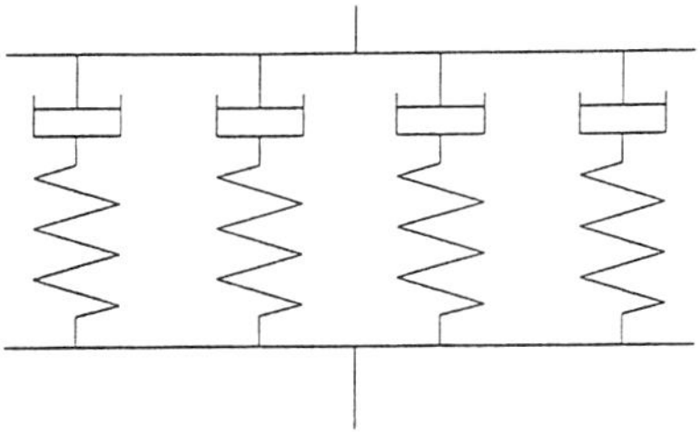

Fig. 2

Summaries

The idea of employing binder-contact laws and micromechanics to develop the constitutive models of two-phase materials with elastic or Maxwell type binders is presented in this study. The results presented here are limited in scope; however, more in-depth studies are currently in progress. It is interesting to know that many phenomenological visco-elastic models proposed for asphalt concrete (Bouldin et al., 1993; SHRP-A-415) are equivalent to the models derived here, in which all parameters are determined by the asphalt binder and aggregates' properties.

Acknowledgement

This work was sponsored by the Air Force Office of Scientific Research, and performed by Wright Laboratory, Tyndall AFB FL, under project No. 2302CW. Contact laws were developed through collaboration with Professor C.S. Chang of the University of Massachusetts at Amherst.

References

1. Bouldin M.G., Rowe G.M., Sousa J.B., and Sharrock M.J., "Mix rheology - a tool for predicting the high temperature performance of hot mix asphalt". *A.A.T.P.* 1993.
2. Little D.N., Prapnnachari S., Letton A., and Kim Y.R., "Investigation of the microstructural mechanisms of relaxation and fracture healing in asphalt". AFOSR-89-0520, 1993.
3. Zhu H., Chang C.S., and Rish III J.W., "Normal and Tangential Compliance for Conforming Binder Contact I: Elastic Binder; II: Visco-elastic Binder", to appear in *International Journal of Solids and Structures.*

Dynamic Brittle Material Response Based on a Nonlocal Damage Model

E. P. Chen[1]

Abstract

A nonlocal analysis of the dynamic damage accumulation processes in brittle solids is presented in this investigation. The analysis is based on a microcrack based continuum damage model which has been implemented into a transient dynamic finite element computer code. Numerical study of the damage accumulation process in a concrete plate with a central hole and subjected to the action of a step tensile pulse applied at opposite edges of the plate was carried out. Comparisons between calculated results based on local and nonlocal formulations are made and nonlocal effects are discussed.

Introduction

Continuous demands on safety and efficient design have placed increasing emphases on fracture and failure analyses of engineering materials and structures. Continuum damage mechanics have been applied to study the phenomenon of brittle fracture in solids under dynamic loads with varying degrees of success. Because of the complex geometries and loading conditions involved, numerical simulation techniques have become the tools of choice. Two common problems associated with fracture analyses by classical continuum damage mechanics are softening and localization. The consequences of failing to take account of these two problems properly usually manifested in solutions which are dependent on the discretization size. A potential remedy is to cast the field equations in a nonlocal setting such that the high gradients associated with the field parameters in these problems can be captured. The purpose of this research is to investigate the appropriateness to apply nonlocal damage models to analyze dynamic brittle fracture. The study is based on the numerical results obtained from a nonlocal damage model which is a modified version of a damage model devel-

----

[1]Material and Structural Mechanics Department, Sandia National Laboratories, Albuquerque, New Mexico 87185-0437

oped by the author (Chen 1986, 1995; Taylor et al. 1986) to simulate rock fragmentation. The nonlocal model (Chen 1995) utilizes a simple nonlocal formulation (Bazant and Lin 1988) by applying nonlocality only to those internal state variable involved with material damage while strains remain local. The model has been implemented into the transient finite element code PRONTO 2D (Taylor and Flanagan 1987) for numerical analysis. An example problem has been selected and results have been obtained from both local and nonlocal calculations. Comparisons between these results have been made. Based on these findings, conclusions on the appropriateness of nonlocal damage models have been made.

A numerical example

General nonlocal effects are examined based on the comparisons of results between local and nonlocal calculations. The example problem selected is the same one used in (Chen 1995) which involves the sudden stretching of a plate with a centrally located hole. Consider the 0.2 m by 0.4m rectangular plate with a 0.1 m diameter hole in the center in Fig. 1. A step tensile pulse is applied symmetrically to the left and right edge of the plate. The plate consists of an Oil Shale material whose properties are given in Taylor et al. (1986). For numerical calculations, the pulse is given a strength of 10.0 MPa. Both the nonlocal and local damage models as implemented in the finite element code PRONTO 2D are used to obtain the dynamic response of the plate. The plane strain condition is assumed to prevail. Because of symmetry, only one quarter of the geometry in Fig. 1 needs to be included in the finite element mesh. Four meshes with varying degree of fineness are chosen to investigate mesh size effects. The most coarse mesh has 75 elements and 96 nodes which divides the boundary of the hole into 10 elements. The second mesh is constructed by simply doubling the numbers of divisions in both the x and y directions from the first one. Thus, the mesh consists of 300 elements and 341 nodes. In a similar expansion, the third mesh is constructed with 1,200 elements and 1,281 nodes. To avoid an excessive number of elements, the fourth mesh only doubles the number of divisions in the third mesh in the lower right part of the plate and consists of 2,800 elements and 2,921 nodes. This is justified because with the applied tensile pulse, the high stress gradients are anticipated to concentrate around the lower right part of the plate and it is not necessary to place fine mesh in other parts of the plate.

Calculations were carried out on the Cray YMP computer at Sandia National Laboratories. The results of the local continuum model are presented first. The evolution of damage focused on the lower 0.09-by-0.09 meter area of the plate is depicted at four time intervals at 0.175, 0.2, 0.215 and 0.225 miliseconds after the tensile pulse has been applied. The results are given in Figs. 2 for the finest mesh. It is clear that damage localizes near the lower right edge of the hole. The fact that damage occurs at a direction deviating from the bottom edge of the hole is due to the effect of wave propagation and the strain rate dependence of the damage model. A more detailed explanation is given in (Chen 1995) and will not be repeated here.

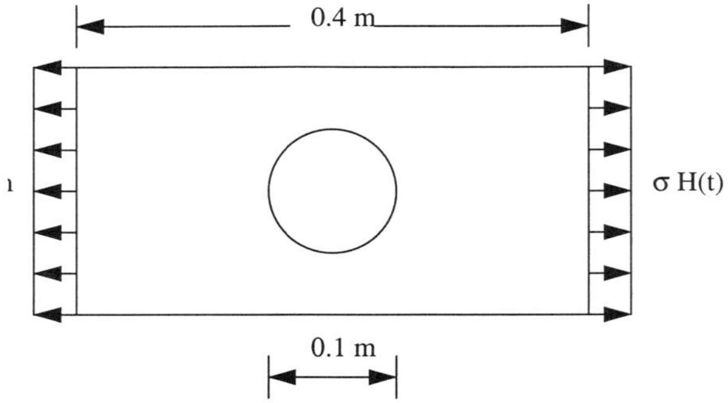

Figure 1. Schematic drawing of the example problem.

The corresponding nonlocal results to those in Fig. 2 are shown in Fig. 3. The smearing effect of the nonlocal results becomes more clear. It is observed that the narrow band of localized damage in Fig. 2 is replaced by lower damage values spread over a wider area near the same location, Fig. 3.

Acknowlegement

This work was funded by the Laboratory Directed Research and Development Program, Sandia National Laboratories, under the auspices of the U. S. Department of Energy under Contract Number DE-AC04-94AL85000.

References

Bazant, Z. P. and Lin, F.B., "Non-Local Yield Limit Degradation," International Journal for Numerical Methods in Engineering, Vol. 35, pp. 1805-1823, 1988.

Chen, E. P. and Taylor, L. M., "Fracture of Brittle Rock Under Dynamic Loading Conditions," in *Fracture Mechanics of Ceramics*, Volume 7, R. C. Bradt, A. G. Evans, D. P. H. Hasselman, and F. F. Lange, Editors, Plenum press, New York, pp.175-186, 1986.

Chen, E. P., "Dynamic Brittle Material Response Based on a Continuum Damage Model," in, R. C. Batra, A. K. Mal, And G. P. MacSithigh, Editors, AMD-Vol. 205, American Society of Mechanical Engineers, New York, pp. 21-34, 1995.

Taylor, L. M., Chen, E. P., and Kuszmaul, J. S., "Microcrack-Induced Damage Accumulation in Brittle Rock under Dynamic Loading," Journal of Computer Methods in Applied Mechanics and Engineering, Vol. 55, pp. 301-320, 1986.

Taylor, L. M., and D. P. Flanagan, "PRONTO 2D - A Two-Dimensional Transient Solid Dynamics Program," Sandia National Laboratories Report SAND86-0594, Albuquerque, New Mexico, 1987.

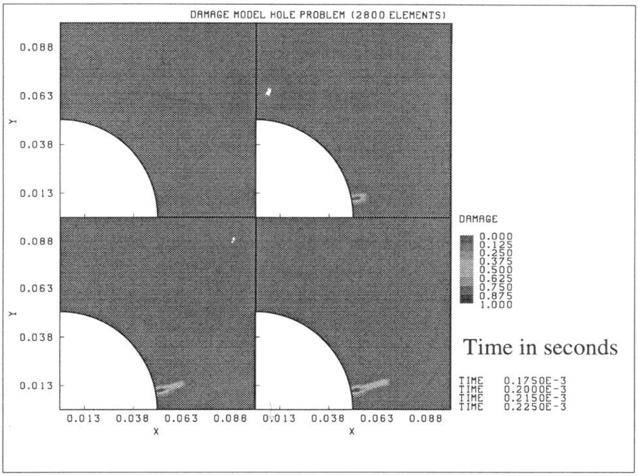

Figure 2. Damage evolution for the finest mesh.

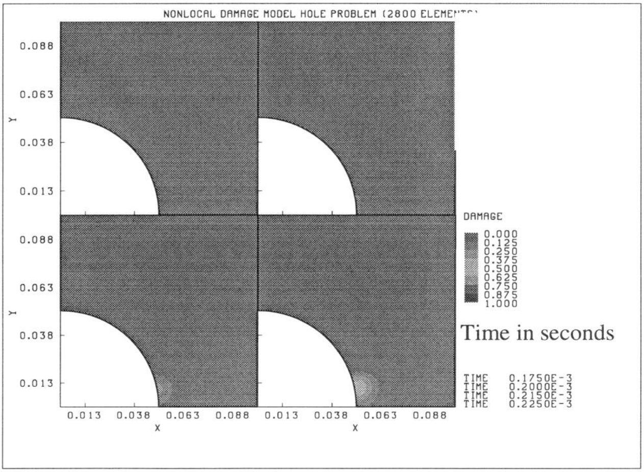

Figure 3. Nonlocal damage evolution for the finest mesh.

Peak-Load Method for Fracture Parameters of Two-Parameter Fracture Model

Tianxi Tang[1], Chengsheng Ouyang[2], and Surendra P. Shah[3]

Abstract

Peak-load method is a test method to determine material fracture parameters K^S_{Ic} and $CTOD_c$ of the two-parameter fracture model for quasi brittle materials. It calculates K^S_{Ic} and $CTOD_c$ from the measured peak loads of three or more distinct specimens. The method has been verified by a series of tests including beam bending, eccentric compression and split tension.

Introduction

The two-parameter fracture model (TPFM) (Jenq and Shah, 1985) introduces two material fracture parameters K^S_{Ic} and $CTOD_c$ to characterize the fracture properties of quasi brittle materials (such as concrete and ceramics). A three-point bend beam has been suggested for measuring values of K^S_{Ic} and $CTOD_c$ by RILEM (1990). In the test the initial compliance and the compliance at the peak load must be measured. Therefore a closed-loop testing machine is required to record load versus crack mouth opening displacement relation in loading and unloading. Tang, Ouyang and Shah (1994) proposed a simple procedure to determine K^S_{Ic} and $CTOD_c$ from peak loads of three or more distinct specimens. This method is named peak-load

---

[1]Engineering Research Associate, Texas Transportation Institute, College Station, TX 77843-3135.

[2]Engineer, Iowa Department of Transportation, 800 Lincoln Way, Ames, IA 50010.

[3]Professor and Director of NSF Center for Advanced Cement-Based Materials, Northwestern University, Evanston, IL 60208-4400.

method to distinguish it from the test method recommended by RILEM, which is called compliance method. With the peak-load method, the fracture parameters can be determined in laboratories and work sites where a closed-loop testing machine is not available. Distinction among specimens can be estimated by a quantity named distinction number (Tang, Yang, Ouyang, Libardi and Shah, 1995). When the distinction numbers of the specimens are farther apart, the more confident solution for K^S_{Ic} and $CTOD_c$ are obtained.

Peak-load method

TPFM provides the failure criterion for a structure or component of a quasi brittle material in terms of simultaneous satisfaction of the following two equations:

$$K_I(\sigma_{Nc}, a_c) = K_{Ic}^S , \tag{1}$$

$$CTOD(\sigma_{Nc}, a_c) = CTOD_c \tag{2}$$

Solution of K^S_{Ic} and $CTOD_c$ from measured σ_{Nc} values through Eqs. (1) and (2) is the basis of the peak-load method. If one applies Eqs. (1) and (2) to two different specimens, four equations are obtained. If critical nominal stresses σ^1_{Nc} and σ^2_{Nc} for two different specimens are measured, these equations involve four unknowns: a^1_c, a^2_c, K^S_{Ic}, and $CTOD_c$. (The superscript denotes the specimen.) With σ^1_{Nc} known, Eqs. (1) and (2) can be written as

$$K_{Ic}^S = K_I^1(a_c^1) , \tag{3}$$

$$CTOD = CTOD(a_c^1) \cdot \tag{4}$$

Combination of Eqs. (3) and (4) can be considered the parametric form of a function for $CTOD_c$ in terms of K^S_{Ic} with a^1_c as the parameter for specimen 1. A similar relation of $CTOD_c$ versus K^S_{Ic} for specimen 2 can also be established. Intersection of the curves of $CTOD_c$ versus K^S_{Ic} for the specimens gives the solution of K^S_{Ic} and $CTOD_c$.

Since test data for quasi brittle materials are usually scattering, at least three specimens are suggested. Based on the $CTOD_c$ - K^S_{Ic} curve of each specimen, an average of $CTOD_c$ values can be calculated. Then solution can be found by applying the least-squares error criterion, that is, by minimizing the following function:

$$f(K_{Ic}^S) = \sum_{i=1}^{n} (CTOD_c^{avg} - CTOD_c^i)^2 \tag{5}$$

where n is the number of specimens, $CTOD^{avg}_c$ is the average value of $CTOD_c$ of all n specimens, and $CTOD^i_c$ is the $CTOD_c$ value for the ith specimen. In the function f

in terms of K^S_{Ic}, the value of K^S_{Ic} at which the minimum f occurs is the solution of material parameter K^S_{Ic}. By substituting the obtained value for K^S_{Ic} into the average $CTOD_c$ - K^S_{Ic} curve, the solution of $CTOD_c$ is obtained.

The test specimens should be so different that these $CTOD_c$ - K^S_{Ic} curves are significantly different in the slope at the solution. Thus a high Δf is achieved for a given small ΔK^S_{Ic} from the solution of K^S_{Ic}. Accordingly the determination of the two parameters would be of high confidence. The following quantity β can be used to measure the confidence for the determined K^S_{Ic} and $CTOD_c$:

$$\beta = \sum_{i=1}^{n} (k^{avg} - k^{i})^2 \tag{11}$$

$$k = \left[\frac{dCTOD_c}{dK^S_{Ic}} \right]_{K^S_{Ic}=K^S_{Ic}*} \tag{12}$$

because $\Delta f = \beta(\Delta K^S_{Ic})^2$. The quantity β, equal to half the curvature of the f versus K^S_{Ic} curve at the minimum f, is named confidence level for K^S_{Ic} and $CTOD_c$ determined by the peak-load method. Thus specimens can be approximately distinguished by the value of k, in the sense of the peak-load method. The quantity k is named specimen distinction number. By comparing LEFM formulas for K_I and $CTOD$, one can find how k depends on specimen geometry, specimen size and notch length. With these two proposed quantities, specimens can be well designed so as to enhance the confidence for K^S_{Ic} and $CTOD_c$ determined by the peak-load method.

Experimental verification of peak-load method

To verify the proposed peak-load method, a series of tests have been performed including beam bending, eccentric compression, and split tension. Among these tests, eight three-point bend beams and six split tension cylinders from the same batch of mortar mix were prepared. Proportion of water, cement and aggregate in the mix was 1 : 2.5 : 5. The maximum size of the aggregate was 3 mm. The notch in the beam specimen was formed by sawing, whereas it was precast for cylinder specimens. All the beams were 50.8 mm deep and 25.4 mm wide with the ratio of the support span to the specimen depth of 4:1. All the cylinder specimens were 102 mm in diameter and 31.8 mm in thickness. All the beams were tested according to the compliance method to obtain K^S_{Ic} and $CTOD_c$. Table 1 contains the results obtained with the compliance method and the peak load method, which match each other very well. When the peak-load method is applied to both the beams and split tension cylinders, confidence of the determination, β, is high, because different types of specimens provide largely different specimen distinction numbers.

Table 3. Values of K^S_{Ic} and $CTOD_c$ obtained with different methods

test method	specimens	K^S_{Ic} (MPa·m$^{1/2}$)	$CTOD_c$ (mm)	β (10^{-4} GPa^{-2}·m)
compliance	beams	0.883	0.00754	N/A
peak load	beams	0.929	0.00820	0.135
peak load	cylinders	0.921	0.00709	0.177
peak load	beams and cylinders	0.881	0.00739	1.40

Tests have shown that a hole drilled at the axis of the split tension cylinder changes the specimen distinction number largely. As for a rectangular prism specimen, different loading conditions, beam bending and eccentric compression, can also provide specimens largely different distinction numbers.

Conclusions

The peak-load method of determining K^S_{Ic} and $CTOD_c$ from the measured peak load of distinct specimens is a simple and reliable method. The values of K^S_{Ic} and $CTOD_c$ obtained by the peak-load method and obtained by the method recommended by RILEM match each other very well. By using the specimen distinction number and the confidence level as the criterion, specimens can be well designed to achieve confident determination..

References

Jenq, Y. S. and Shah, S. P. (1985). "A two parameter fracture model for concrete." *J. of Engrg. Mech.*, ASCE, 111(4), 1227-41.

RILEM Committee on Fracture Mechanics of Concrete - Test Methods (1990). "Determination of the fracture parameters (K^S_{Ic} and $CTOD_c$) of plain concrete using three-point bend tests." *Materials and Structures*, RILEM, 23(138), 457-460.

Tang, T., Ouyang, C., and Shah, S. P. (1994). "A simple method for determining material fracture parameters from peak loads." *ACI Materials J.*, accepted for publication.

Tang, T., Yang, S., Ouyang, C., Libardi, W., and Shah, S. P. (1995). "Determination of K^S_{Ic} and $CTOD_c$ from peak loads and relationship between two-parameter fracture model and size effect model." *Fracture Mechanics of Concrete Structures*, ed., F. H. Wittmann, AEDIFICATIO Publishers, Freiburg, Germany, Vol. 1, 135-144.

A Damage Mechanics-based Approach to Structural Deterioration

Baidurya Bhattacharya[1] and Bruce Ellingwood[2]

Abstract

Structural deterioration in the form of inelastic deformation, high- temperature creep or fatigue, depends on the operating environment and loading conditions, and often occurs without visible manifestation. Continuum damage mechanics (CDM) makes it feasible to predict effects of such processes and to estimate residual strength or service life of existing structures. From first principles of thermodynamics, closed-form solutions are obtained for isotropic damage as a function of uniaxial plastic strain, which, in turn, can be related to time. The model is validated with laboratory tests by other authors.

Introduction

CDM deals with the characterization and analysis of growth of strength-reducing microstructural defects with the help of macroscopic state variables. Damage is quantified by the surface density of cracks and cavities on an elemental cross-sectional area weighted by their average stress-raising effects (Lemaitre, 1985). If damage is independent of the orientation of the cross-section, it is said to be *isotropic* and is expressed by a scalar variable, D, between zero and one. If σ is the nominal stress, the *effective stress* in a damaged body is

$$\tilde{\sigma} = \frac{\sigma}{1-D} \tag{1}$$

Damage growth is a dissipative process governed by the laws of thermodynamics. The existing methods (eg, Hansen and Schreyer, 1994) express the rate of intrinsic dissipation as $\Phi = -\rho(\partial\psi/\partial\mathbf{H}) \cdot \dot{\mathbf{H}} \geq 0$, where $\mathbf{H}$ denotes the "dissipation variables", ψ is the free energy per unit volume and ρ is the density. For isotropic damage, the "damage energy release rate", Y, is defined as $Y = -\rho(\partial\psi/\partial D)$. As a duality, the damage growth rate is $\dot{D} = \partial\Phi/\partial Y$. It has been customary to propose expressions for Φ in terms of Y, and then to differentiate them in order to obtain equations for $\dot{D}$ (eg Lemaitre 1985; Chaboche,

[1]Graduate Student, Department of Civil Engineering, The Johns Hopkins University, Baltimore, MD 21218.

[2]Professor and Chairman, Department of Civil Engineering, The Johns Hopkins University, Baltimore, MD 21218.

1988b; Hansen and Schreyer, 1994). The necessity to conceive dissipation functions that do not descend from first principles for linking damage growth laws to thermodynamics is a handicap of the existing methods. Moreover, the proliferation of undetermined "material parameters" spawned by such models has been criticized (Krajcinovic and Mastilovic, 1995).

Thermodynamic Foundations of Damage

Let us consider a system $\mathcal{R}$ defined by the closed boundary $\partial\mathcal{R}$. Throughout the following development, we shall assume that the system is in diathermal contact with a heat reservoir at the (constant) absolute temperature θ. The rate of heat flow into the system from the reservoir is $\dot{Q}$, causing an increase in the entropy of $\mathcal{R}$ at the rate $\dot{S}$. The rate of increase of entropy of the universe (the system and the reservoir) is positive by the second law of thermodynamics:

$$\dot{S} - \frac{\dot{Q}}{\theta} \geq 0 \tag{2}$$

Let the Helmholtz Free Energy, defined as $\Psi = U - \theta S$, be a function,

$$\Psi = \Psi(\theta, \mathbf{x}, D) \tag{3}$$

where U is the internal energy and $\mathbf{x}$ is the set of internal variables that contribute to work. The second law inequality (2) can be modified with the help of the first law (which states $\dot{K} + \dot{U} = \dot{Q} + \dot{W}$), by differentiating eq (3) and noting that $\dot{\theta} = 0$:

$$\Gamma \equiv -\dot{K} + \dot{W} - \frac{\partial\Psi}{\partial\mathbf{x}} \cdot \dot{\mathbf{x}} - \frac{\partial\Psi}{\partial D} \cdot \dot{D} \geq 0 \tag{4}$$

where K is the kinetic energy and W is the net work done on the system. We denote the above expression as the dissipation rate, Γ.

The work done on the system (adjusted for the increase in kinetic energy), in a dissipative process at constant temperature is greater than the increase in Ψ. Therefore the free energy can be expressed as the difference:

$$\Psi = \int_{t_1}^{t_2} (\dot{W} - \dot{K})\, dt - \int_{t_1}^{t_2} \Gamma\, dt \tag{5}$$

where t_1 is the initial time and $t_2 > t_1$ is an arbitrary instant of time.

A system in diathermal contact with a heat reservoir is in equilibrium (not necessarily stable) if the first variation of its Helmholtz free energy is zero (McLellan, 1980). For the purpose of modeling accumulation of damage in deformable continua, we assume that (i) the initial state of the system at time t_1 is one of thermodynamic equilibrium, and (ii) the damage-causing processes are slow processes and, though irreversible, proceed sufficiently close to equilibrium at every t_2 under consideration. Under these assumptions:

$$\delta\Psi = \delta\left[\int_{t_1}^{t_2}\left(\dot{W} - \dot{K} + \frac{\partial\Psi}{\partial D}\dot{D}\right) dt\right] - \delta\left[\int_{t_1}^{t_2}\left(\dot{W} - \dot{K} - \frac{\partial\Psi}{\partial\mathbf{x}}\dot{\mathbf{x}}\right) dt\right] = 0 \tag{6}$$

For a deformable body occupying $\mathcal{R}$ bounded by the surface $\partial\mathcal{R}=\partial\mathcal{R}_1\cup\partial\mathcal{R}_2$, $\mathbf{x}$ (eq 3) is equivalent to the symmetric strain tensor $\epsilon_{ij}=\frac{1}{2}(u_{i,j}+u_{j,i})$, where $u_{i,j}$ is the partial derivative of the deformation u_i w.r.t. the displacement j. We define the stress tensor $\sigma_{ij}=\partial\psi/\partial\epsilon_{ij}$. Applying a variation δu_i on $\partial\mathcal{R}_1$ consistent with the boundary conditions $\delta u_i=0$ on $\partial\mathcal{R}_2$ $\forall t\in[t_1,t_2]$, and using the symmetry of the stress tensor, the second term in eq (6) can be shown to vanish, provided that $\mathcal{R}$ is in equilibrium. Hence the variation of the first integral in eq (6) is zero:

$$\delta\left[\int_{t_1}^{t_2}\left(\dot{W}-\dot{K}+\frac{\partial\Psi}{\partial D}\dot{D}\right)dt\right]=0 \tag{7}$$

Damage due to Ductile Deformation

Let us assume that damage can be expressed as a function of the strain tensor. Then, on applying a variation consistent with the boundary conditions, and proceeding as above, we arrive at the partial differential equations

$$F_i-\rho a_i-(\psi_D D'_{ij})_{,j} = 0 \quad \text{on } \mathcal{R} \tag{8}$$

$$T_i+\psi_D D'_{ij}n_j = 0 \quad \text{on } \partial\mathcal{R}_1 \tag{9}$$

where F_i are body forces, T_i is surface traction, a_i is acceleration, ψ_D is the partial derivative of ψ wrt D and D'_{ij} denotes the derivative of D with respect to the strain ϵ_{ij}. For uniaxial straining under a far field stress, σ_∞, acting normal to the surface, eq (9) can be simplified to

$$\sigma_\infty+\psi_D D'=0 \tag{10}$$

where D' is the derivative of damage with respect to the axial strain.

The general form of ψ for ductile deformation, creep and fatigue is

$$\psi=\int\sigma_{ij}\,d\epsilon_{ij}-\gamma \tag{11}$$

where γ is the surface energy of voids that arise due to damage growth.

Applying the principle of strain equivalence (Chaboche, 1988a) to the elastic and plastic strains separately (assuming that the strain-hardening exponent M remains unaffected by damage) and using eq (1), we obtain the damaged moduli of elasticity and hardening, respectively: $\tilde{E}=E(1-D)$ and $\tilde{K}=K(1-D)$, where E and K are the undamaged moduli. Moreover, $D=0$ as long as the plastic strain $\epsilon_p<\epsilon_0$. The first term in eq (11) therefore equals:

$$\int_0^{\epsilon}\sigma d\epsilon'=\int_0^{\epsilon_{e0}}E\epsilon'_e\,d\epsilon'_e+\int_{\epsilon_{e0}}^{\epsilon_e}E(1-D)\epsilon'_e\,d\epsilon'_e+\int_0^{\epsilon_0}K\epsilon_p'^{\frac{1}{M}}\,d\epsilon'_p+\int_{\epsilon_0}^{\epsilon_p}K(1-D)\epsilon_p'^{\frac{1}{M}}\,d\epsilon'_p \tag{12}$$

Further to our assumption of isotropic damage, let us assume that the only defects within the material are spherical voids which are distributed uniformly inside the material. Within a representative volume element, V_0, in the shape of a sphere of radius u, there are n voids of radius a_i. Ignoring stress-concentration effects, it can be shown that $D = \frac{1}{u^3}\sum_{i=1}^{n} a_i^3$. The energy required for unit growth of void surface area, assuming a linear force-displacement relation at the microscale, is $\sigma_c \pi a_i^3$ (σ_c is the local failure stress). Summing over i and normalizing by V_0, we find that $\gamma = \frac{3}{4}\sigma_c D$. Using this and eq (12), we can obtain ψ (eq 11).

For ductile materials, the plastic strain dominates so that $D' = dD/d\epsilon_p = -\sigma_\infty/\psi_D$ (eq 10), and usually $K/(2E) \sim 0$. This leads to the closed-form solution

$$D = 1 - \frac{C_2}{\epsilon_p^{1+\frac{1}{M}} + C_1} \tag{13}$$

where C_1 and C_2 are the constants: $C_1 = \frac{3}{4}(1+\frac{1}{M})\frac{\sigma_c}{K} - \epsilon_0^{1+\frac{1}{M}}$ and $C_2 = C_1 + \epsilon_0^{1+\frac{1}{M}}$. This approximates to a linear relation between D and ϵ_p in the strain range $\epsilon_p < C_1$, for alloys with large M.

A comparison of the solution of eq (10) with zero initial condition to the results of Lemaitre (1985) is presented in the figure. The plot of eq (13) is virtually identical to that of eq (10). No new or undetermined material parameters need be introduced in the current approach. The material is equivalent to SAE 1035 steel (E=180 Gpa, K=871 Mpa, M=3.4, σ_c=1200 Mpa and $\epsilon_0 = 0$).

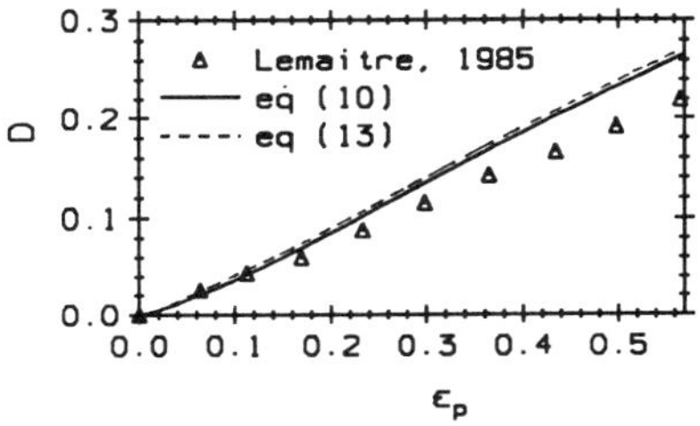

Conclusion

The present approach starting from the first principles of thermodynamics, under the assumption of damage processes taking place close to equilibrium, isotropic damage and uniformly distributed spherical voids, is able to derive the phenomenological model of Kachanov (1986) (for creep), the arbitrary dissipation potential model of Lemaitre (1985) (for ductile flow) and a general damage model for fatigue. Predictions show excellent agreement with experimental results. Research to extend the above approach to a stochastic formulation of damage growth is under way.

References

1. Chaboche, J. L.(1988a,b).*J. of Appl. Mech.*,vol 55, pp 59-72.
2. Chow, C.L. and Wei, Y. (1991). *Int'l. J. of Fract.*, vol 50, no 4, pp 301-316.
3. Hansen, N.R. and Schreyer, H.L. (1994). *Int'l J. of Solids and Struct.*, vol 31, no 3, pp 359-389.
4. Kachanov, L.M. (1986). *Introduction to Continuum Damage Mechanics*, Martinus Nijhoff.
5. Krajcinovic, D. and Mastilovic, S. (1995). *Mechanics of Materials*, vol 21, no 3, pp 217-230.
6. Lemaitre, J. (1985). *J. of Engrg Mat'ls and Tech.*,vol 107,no 1, pp 83-89.
7. McLellan, A.G. (1980). *The Classical Thermodynamics of Deformable Materials*, Cambridge Univ. Press.

Interaction between the Crack Tips of a Circular Arc Crack

Y. C. Shiah[1] and Y. M. Tsai[2]

Abstract

A new conformal mapping is devised to transform the contour of a circular arc crack into a unit circle. By this mapping, the interaction between both tips of the crack is studied, which exhibits different characteristics from those of a pair of collinear cracks. The solution has the particular feature that without any limiting process, it can be continuously reduced to the particular case of the straight crack problem simply by setting a mapping variable equal to zero.

Introduction

Most of the work done so far for the two dimensional case was dealing with the crack ideally modelled as a perfectly straight slit for simplicity. Following his own work (Muskhelishvili in 1942 [1]), where the problem of straight cut was solved by treating the crack as the discontinuity of linear relationship (the Hilbert problem), Muskhelishvili in 1975 [2] derived the complex stress functions for an elastic isotropic plate weakened by a circular arc crack under biaxial tension at infinity. Further, Sih [3] in 1962 derived the stress-intensity factor for such a problem based upon Muskhelishvili's solution.

As the subtending angle of the circular arc crack gets larger, the interaction between the crack tips gets more intense. For the case of a pair of collinear cracks, as

[1]Graduate Student, 2019 Blacking Engineering Building, Aerospace Engineering and Engineering Mechanics Department, Iowa State University, Ames, Iowa 50011

[2]Professor, 2019 Black Engineering, Aerospace Engineering and Engineering Mechanics Department, Iowa State University, Ames, Iowa 50011

the distance between the crack tips approaches zero, the stress intensity factor of the inner tips would go to infinity (Sih in 1964 [4]). In this paper, the interaction between the crack tips of a circular arc crack is studied as a function of the distance between the crack tips (see figure 1).

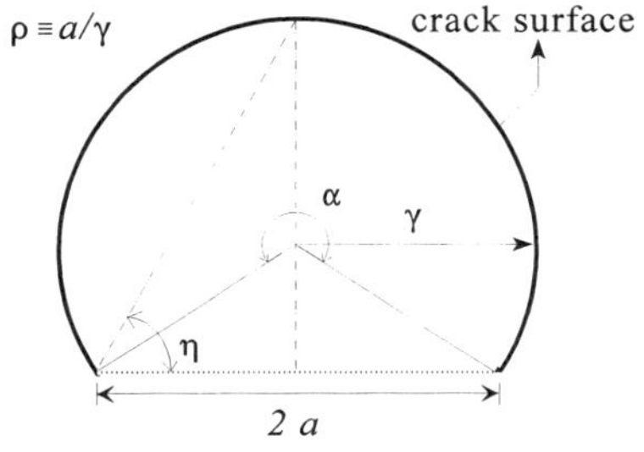

Figure 1. Variable Definitions

Conformal Mapping Scheme

In calculation of stress functions, it is expedient for the purpose of evaluating Cauchy integral to transform the boundary contour into a unit circle. For this purpose, a mapping function is devised as follows:

$$\omega(\zeta) = R\left(\frac{\zeta^2+1}{\zeta+ic}\right) \quad \textbf{(1)}$$

where $R = a\,(1\text{-}c^2)^{-1/2} / 2$. By suitable choice of c and a, one may obtain circular cracks of any dimension. Particularly, if the constant c is chosen 0, the above mapping equation is identical with the equation for the straight crack problem. This transformation, an extension of the particular form for straight crack problems, actually transforms the unbounded plane outside a circular arc crack with radius $\gamma = R/c$ and chord length $2a$ onto the unbounded plane outside the unit circle. It is essential to note that, through this transformation equation (1), the tips of the circular arc crack are mapped to

$$\sigma_{(L\backslash R)} = (-\backslash+)\sqrt{1-c^2} - ic \quad \textbf{(2)}$$

where the subscripts L and R represent the left crack tip and the right crack tip respectively. By setting ζ equal to $\sigma_{(L\backslash R)}$ in equation (1), it can be easily shown that

$$c = \sin\eta \quad \textbf{(3)}$$

where η is within the region $[0,\pi/2]$.

Derivation for the Complex Stress Function

A complex variable technique was developed by Muskhelishvili (see Muskhelishvili in 1975 [2] for details) to solve the boundary value problems governed by the biharmonic equation. The boundary equation of the first fundamental problem is

$$\phi(\sigma)+\frac{\omega(\sigma)}{\omega'(\sigma)}\overline{\phi'(\sigma)}+\overline{\psi(\sigma)} = f \tag{4}$$

where primes denote the differentiation with respect to ζ in the mapped plane and σ is the boundary value on the unit circle in the ζ-plane. In equation (4), the quantity f is defined on the boundary by

$$f = i\int_0^s (X_n+iY_n)ds \tag{5}$$

where X_n and Y_n denote outward traction components applied to the boundary along arc s. Let the uniform traction F be composed of the outward normal stress $-P$ (compression) and the tangential stress Q (clockwise direction) written as $-P + i\,Q$ act upon the whole edge of the circular arc crack. Then the quantity f, according to its definition, can be written as $f = F\ \omega(\sigma)$. Consequently, after Cauchy integrals are evaluated for the boundary equation (4), the general expression for the complex stress function ϕ can be obtained as following.

$$\phi(\zeta) = -\mathscr{F}(\zeta)-\frac{1-c^2}{1+c^2}\cdot\frac{c^2\mathscr{F}'(-ic^{-1})-\overline{\mathscr{F}'}(ic^{-1})}{\zeta+ic} \tag{6}$$

where $\mathscr{F}$ is the Cauchy integration of the function f. For the special case when uniform stresses are applied upon the whole crack edge, the following nice expression for the function $\phi(\zeta)$ can be obtained.

$$\phi(\zeta) = \frac{R(F-c^2\overline{F})}{(1+c^2)(\zeta+ic)} \tag{7}$$

It is noteworthy that if the mapping variable c is set equal to zero, the above stress function is actually identical with the one for the straight crack problem. By use of the approach established by Sih in 1962 [3], the stress intensity factors of the deeply curved crack ($\pi/4\leq\eta<\pi/2$) are written as

$$K_{I_{(L\backslash R)}} = \sqrt{\pi a}\left(\frac{\sqrt{2}\cdot\sqrt{1-\sqrt{1-\rho^2}}}{3+\sqrt{1-\rho^2}}\cdot P(-\backslash+)\frac{\sqrt{2}\cdot\sqrt{1+\sqrt{1-\rho^2}}}{1-\sqrt{1-\rho^2}}\cdot Q\right)$$

$$K_{II_{(L\backslash R)}} = \sqrt{\pi a}\left(\frac{(+\backslash-)\sqrt{2}\cdot\sqrt{1+\sqrt{1-\rho^2}}}{3+\sqrt{1-\rho^2}}\cdot P+\frac{\sqrt{2}}{\sqrt{1-\sqrt{1-\rho^2}}}\cdot Q\right) \quad \textbf{(8)}$$

where ρ is the non-dimensional curvature defined by a/γ.

Conclusion and Discussion

The interaction of the crack tips of a circular arc crack exhibits different characteristics from those of a pair of collinear cracks. For the collinear cracks (Sih in 1964 [4]), as the distance between the inner tips approaches zero, divergence with an order of $a^{-1/2}$ will happen for the stress intensity factors of both fracture modes. In contrast, for the circular arc crack on which only uniform normal stress is applied, its stress intensity factors of both fracture modes will converge to zero as the distance a goes to zero. Instead, if only shear stress is applied on the crack surface, the similar divergence with an order of $a^{-1/2}$ will happen for K_{II}. From the analysis in this paper, the conclusion can be drawn that as the distance between the crack tips approaches zero, the stress intensity factors will be dominated by the applied shear stress only.

References

1. Muskhelishvili, N. I. "The Fundamental Boundary Problems of the Theory of Elasticity for the Plane with Straight Cuts" in Russian, *Soobshcheniya A. N. Gruz. S.S.R. Сообщения, А. Н., Груз,* 1942.

2. Muskhelishvili, N. I. "Some Basic Problems of the Mathematical Theory of Elasticity" 4th Ed. English translation, Noordhoff International Publishing, Leyden, p.542-545, 1975.

3. Sih, G. C. "Crack-Tip, Stress-Intensity Factors for Plane Extension and Plate Bending Problems" Transactions, Am. Soc. Mechanics, Vol. 29, p. 306-310, 1962.

4. Sih, G. C. "Boundary Problems for Longitudinal Shear Cracks" Proceedings, Second Conference on Theoretical and Applied Mechanics, Pergamon Press, p. 117, New York, N. Y. 1964.

Modeling Rotation of Principal Load Axes in Brittle Solids with Damage

S. Karnawat[1] and S. Yazdani[2]
Members, ASCE

Abstract

A phenomenological constitutive model, motivated by the directionality of the progressive failure and its consequences on tensile and compressive strength of brittle solids which were preloaded in orthogonal directions, is developed within the framework of continuum damage mechanics theories. A separate elastic damage surface is postulated for the tensile regime and the compressive regime. The use of effective damage parameters instead of cumulative ones is proposed to model the preloading effects. The concept of current state of damage planes is used to capture the oriented feature of damage for certain loading paths.

Introduction

The overall response of brittle solids to external loads is generally governed by the kinetics of microdefects in its microstructure. For example, it is well recognized that initiation, growth, and coalescence of microcracks and microvoids control the response of most cement-based solids. The continuum damage mechanics theories are proven to be useful to model this feature of brittle material behavior (Chaboche, 1993; Yazdani, 1993; Ortiz, 1985). A phenomenological theory based on micromechanical aspects is commonly used in reproducing the observed stress-strain response of brittle solids under proportional and non-proportional loading paths. In general, a set of damage variables that are related to internal changes are utilized in thermodynamics framework of internal variables to construct a constitutive model (Lubliner, 1972; Coleman and Gurtin, 1967).

Two distinct microcracking patterns are observed in the mechanical behavior of brittle solids (Ortiz, 1985). Under the application of increasing tensile stress, mode-I type cracks are formed which are perpendicular to the direction of applied stress. The failure of solid takes place when these cracks merge into a single large crack that

[1] Engineering Instructor, Turtle Mountain Community College, Belcourt, ND 58316.
[2] Associate Prof., Dept. of Civ. Engrg., North Dakota State University, Fargo, ND 58105.

split the specimen. On the other hand, under compressive stress, failure of the solid occurs due to mode-II type cracks which are parallel to the loading axis (Wastiels 1979). To address these two internal mechanism, two separate elastic damage surfaces are proposed. The damage surfaces are allowed to evolve isotropically with corresponding internal damage variables.

Stankowski (1990) has simulated the effect of preloading on the stiffness and strength of square composite specimens in which the direction of loading is rotated by 90° at different state of damage. The failure simulation indicated the degradation in strength for both tensile and compressive loading histories. Similar findings are reported by van Mier (1984) which are based on the triaxial compression experiments on a cubical concrete specimens. van Mier has pointed out that the carrying capacity of the solid is affected by cracking due to the first loading. This issue is investigated and modeled in this study. Due to space restriction, only elastic damage is considered in this paper. In a future publication, inelastic damage, mixed-mode preloading effects, and crack closure issues will be presented as an extension to the proposed theory and validated through available experimental data.

Continuum Damage Model

The proposed phenomenological model can be cast by employing the principles of irreversible thermodynamics. To conserve space, this exercise is omitted in this paper, but one can easily build a two-surface model following the steps presented in Yazdani and Schreyer (1990). The stress-space formulation is chosen to describe dissipative surfaces and damage rules. As suggested by Ortiz (1985), the fourth-order elastic compliance tensor is no longer constant but evolves with damage. Thus, the load-induced anisotropy is easily captured in the material stiffness. It is assumed that the elastic compliance takes on the following form of the additive decomposition:

$$\mathbf{C}(\mathrm{r}) = \mathbf{C}^0 + \mathbf{C}^+(\mathrm{r}) + \mathbf{C}^-(\mathrm{r}) \tag{1}$$

in which $\mathbf{C}^0$ is the original compliance tensor, $\mathbf{C}^+$ and $\mathbf{C}^-$ are the added flexibility tensors due to mode-I and mode-II type cracks, respectively. The internal variable r represents a set of damage variables that describe internal dissipative processes due to microcracking.

Two scalar parameters, $\tilde{k}^+$ and $\tilde{k}^-$, are chosen to describe the damage in mode-I and mode-II, respectively (similar arguments are made by Stevens and Liu, 1992, and Berthaud et al., 1990). Both $\tilde{k}^+$ and $\tilde{k}^-$ are assumed to be independent from the outset, so Equation (1) takes on the following form:

$$\mathbf{C}(\tilde{k}^+, \tilde{k}^-) = \mathbf{C}^0 + \mathbf{C}^+(\tilde{k}^+) + \mathbf{C}^-(\tilde{k}^-) \tag{2}$$

Due to the nonlinear structure of Equations, the time derivative of the total flexibility tensor is considered as:

$$\dot{\mathbf{C}}(\widetilde{k}^+,\widetilde{k}^-)=\dot{\mathbf{C}}^+(\widetilde{k}^+)+\dot{\mathbf{C}}^-(\widetilde{k}^-). \tag{3}$$

The rate of the added flexibility in mode-I and mode-II is obtained by postulating the associated damage rules in which the damage response tensors are obtained from the damage functions as follows:

$$\dot{\mathbf{C}}^+(\widetilde{k}^+)=\dot{\widetilde{k}}^+\frac{\partial\Psi^+}{\partial\sigma^+\otimes\sigma^+},\qquad \dot{\mathbf{C}}^-(\widetilde{k}^-)=\dot{\widetilde{k}}^-\frac{\partial\Psi^-}{\partial\sigma^-\otimes\sigma^-} \tag{4}$$

where

$$\Psi^+(\sigma^+,\widetilde{k}^+)=\frac{1}{2}\sigma^+:\mathbf{R}^+:\sigma^+-\frac{1}{2}\left[t^+(\widetilde{k}^+)\right]^2=0, \tag{5}$$

and

$$\Psi^-(\sigma^-,\widetilde{k}^-)=\frac{1}{2}\sigma^-:\mathbf{R}^-:\sigma^--\frac{1}{2}\left[t^-(\widetilde{k}^-)\right]^2=0. \tag{6}$$

In the above formulations, Ψ^+ and Ψ^- are the elastic isotropic damage surfaces with its associated damage hardening-softening functions, t^+ and t^-, designated for mode-I and mode-II, respectively. The positive and negative parts of the stress tensor are denoted by σ^+ and σ^- (Ortiz, 1985). The symbol "$\otimes$" stands for the tensor product and the symbol ":" denotes inner product. The fourth order response tensors $\mathbf{R}^+$ and $\mathbf{R}^-$ are given as follows:

$$\mathbf{R}^+=\mathbf{R}^{+a}+\mathbf{R}^{+h}=\mathbf{S}^+\otimes\mathbf{S}^++\mathbf{R}^{+h}\text{ where }\mathbf{S}^+=\frac{\dot{\sigma}^+}{(\dot{\sigma}^+:\dot{\sigma}^+)^{1/2}}, \tag{7}$$

and

$$\mathbf{R}^-=\mathbf{R}^{-a}+\mathbf{R}^{-h}=\mathbf{S}^-\otimes\mathbf{S}^-+\mathbf{R}^{-h}\text{ where }\mathbf{S}^-=\frac{\dot{\tilde{\sigma}}}{(\dot{\tilde{\sigma}}:\dot{\tilde{\sigma}})^{1/2}}. \tag{8}$$

The lateral response of solid is modeled by $\mathbf{R}^{+h}$ and $\mathbf{R}^{-h}$, which are kept the same as proposed by Yazdani (1993), and $\dot{\tilde{\sigma}}=\dot{\sigma}^- - \lambda \mathrm{i}$ with λ being the maximum eigenvalue of σ^- and $\mathbf{i}$ representing the second order identity tensor.

The substitution of Equations (5) and (6) into Equation (4) yields:

$$\dot{\mathbf{C}}^+(\widetilde{k}^+)=\dot{\widetilde{k}}^+\mathbf{R}^+=\dot{\mathbf{C}}^{+a}+\dot{\mathbf{C}}^{+h}\quad\text{and}\quad\dot{\mathbf{C}}^-(\widetilde{k}^-)=\dot{\widetilde{k}}^-\mathbf{R}^-=\dot{\mathbf{C}}^{-a}+\dot{\mathbf{C}}^{-h}. \tag{9}$$

To capture damage in various directions, the effective damage parameters, $\widetilde{k}^+$ and $\widetilde{k}^-$, are related to current state of damage planes as follows:

$$\tilde{k}^{+} = \mathbf{S}^{+} : \mathbf{C}^{+a} : \mathbf{S}^{+}, \qquad \tilde{k}^{-} = \mathbf{S}^{-} : \mathbf{C}^{-a} : \mathbf{S}^{-} \tag{10}$$

where $\mathbf{C}^{+a} = \tilde{k}^{+} \mathbf{R}^{+a}$ and $\mathbf{C}^{-a} = \tilde{k}^{-} \mathbf{R}^{-a}$.

Finally, to accommodate the preloading effects, the damage functions for both modes need to be identified. The cumulative damage parameters, $k^{+} = \sum \tilde{k}^{+}$ and $k^{-} = \sum \tilde{k}^{-}$, as a measure of total dissipation need to be introduced in damage functions to complete the formulation. Here, the summation of the effective damage parameters is taken over all the loading paths. Interested readers are referred to a paper by Yazdani and Karnawat (1996) in which the orthogonal preloading issues related to mode-I are modeled.

References

Berthaud, Y., Laborderie, C., and Ramtani, S. (1990). "Damage modeling and crack closure closure effect." *Damage mechanics in engineering materials*, 109, J. W. Ju, D. Krajcinovic, H. L. Schreyer, eds., ASME, New York, N.Y. 263-276.

Chaboche, J. -L. (1993). "Development of continuum damage mechanics for elastic solids sustaining anisotropic and unilateral damage." *Int. J. Damage Mech.*, 2, 311-329.

Coleman, B. D. and Gurtin, M. E. (1967). "Thermodynamics with internal state variables." *J. Chem. Phys.*, 47(2), 597-613.

Ortiz, M. (1985). "A constitutive theory for the inelastic behavior of concrete." *Mech. of Mat.*, 4, 67-93.

Lubliner, J. (1972). "On the thermodynamic foundations of non-linear solid mechanics." *Int. J. Non-Linear Mech.*, 7(3), 237-254.

Stankowski, T. (1990). "Numerical simulation of progressive failure in particle composites." *Ph. D. Dissertation*, University of Colorado, Boulder.

Stevens, D. J. and Liu, D. (1992). "Strain-based constitutive model with mixed evolution rules for concrete." *J. Engrg. Mech.*, ASCE, 118(6), 1184-1200.

van Mier, J. G. M. (1984). "Strain-softening of concrete under multiaxial loading conditions." *Ph. D. Dissertation*, De Technische Hogeschool Eindehoven, The Netherlands.

Wastiels, J. (1979). "Behavior of concrete under multiaxial stresses - a review." *Cement and Concrete Research*, 9, 35-44.

Yazdani, S. and Karnawat, S. (1996). "Mode I damage modeling in brittle solids with rotation of principal load axis." *Int. J. Damage Mech.*, to be published.

Yazdani, S. (1993). "On a class of continuum damage mechanics theories." *Int. J. Damage Mech.*, 2, 162-176.

Yazdani, S. and Schreyer, H. L. (1990). "Combined plasticity and damage mechanics model for plain concrete." *J. Engrg. Mech.*, 116(7), 1435-1450.

MODAL IDENTIFICATION OF A CABLE-STAYED BRIDGE

W-H. P. Yen[1], M. ASCE, T. T. Baber[2], M .ASCE, F. W. Barton[2] , F. ASCE

Abstract

Transient vibration tests were conducted to obtain the bridge response under the effects of two measured trucks traversing the bridge with near steady speed. With the input and output of the bridge known, a system identification method for experimental determination of the modal parameters was developed and applied to investigate the bridge's dynamic properties. This method was applied with symmetric characteristics to reduce the number of parameters in identification procedure, and to eliminate the error effects from modal interference between symmetric and antisymmetric modes.

Introduction

A long-span cable-stayed bridge structure possesses a complex dynamic response. The bridge deck dynamic behavior often consists of vertical bending, torsion, horizontal bending and transverse bending modes. These modes usually interact with each other complicating the analytical or computational prediction of the dynamic response. Therefore, a field test of the bridge is desirable to calibrate an analytical model and improve the prediction accuracy. Although ambient vibration tests have addressed finding the natural frequency, damping and mode shapes, some limitations, such as modal interference or response amplitude at specific modes, still exist because the inputs are unknown.

In this study, forced vibration tests were performed on the I-295 James River Bridge. The excitation used a transient forcing by two trucks of known weight traversing the bridge side by side with a steady speed. With the forces and output response known, modal parameters were identified by a system identification method. This method was applied with the symmetry property of the bridge to separate closely spaced modes and also reduce the number of parameters to be estimated. A computational analysis was conducted using a finite element model in order to obtain the a priori parameter estimates and the mode shapes.

---

[1] Structures Division, FHWA R&D 6300 Georgetown Pike, McLean, VA 22101

[2] Civil Engineering Department, University of Virginia, Thornton Hall, Charlottesville, VA 22901

Descriptions of the Bridge

The I-295 James River Bridge is a modern precast, segmentally erected, post-tensioned, cable stayed box girder bridge located at east of Richmond, Virginia. An elevation drawing of the seven continuous central spans of the bridge is shown in Figure 1. The bridge deck consists of twin box girders.

Identification Method with an Incomplete Modal Model

Often, the common practical problems of system identification using frequency domain methods in a large structure such as a bridge or a high rising building are 1) incomplete field data, 2) closely spaced modes and 3) too many parameters to be estimated. This study develops a system identification method modified from a generalized multiple inputs and multiple outputs (MIMO) modal parameter estimation algorithm written by Craig and Blair (1985) for solving the problems mentioned above. The detail of the method can be found in Yen's (1992) study.

Experimental Procedure and Input Response Modeling

The experiments were conducted at the same time as the ambient vibration test. The accelerometers were located as shown in Figure 2. The data acquisition system began to record the data when two trucks, weighing 50 kips and 51 kips respectively, ran across the expansion joints (the ends of the bridge). The two trucks traversed the bridge at the nearly steady speed of 35 mile/hour. The vertical acceleration response was measured at 9 points for each loading sequence, including three symmetric pairs with respect to the center line of the deck. Horizontal accelerations were recorded at the same time but the parameter estimation utilized only the vertical responses. Station 5 was taken as the reference station.

The input response was calculated by simulating each truck as two concentrated forces of known magnitude a constant distance apart and moving along a straight line across the bridge with a constant speed (See Figure 3). The route of the truck moving was idealized to travel over the nodes of the computational model to facilitate computation of the response using the modal model.

Results Comparison

The modal stiffness and damping obtained from the forced vibration test are compared to the results which were obtained from the ambient vibration test in Table 1. C_i is the ith modal damping parameter estimated, while K_j is the jth modal stiffness parameter. The results from the system identification method indicate that in the higher modes the damping values are larger than the values which were estimated from ambient vibration tests. Damping parameters showed a large variation in the estimated values. This is consistent with the practical problem of measuring damping.

Conclusion

This paper describes a forced vibration test of a cable-stayed bridge, and application of the symmetrical characteristics to an identification method to solve an incomplete model with closely spaced modes. Identified parameters show consistency with ambient vibration data in natural frequencies. The input data was obtained by making some assumptions to idealize the actual conditions, and was less accurate in the magnitudes and phase data than those data measured from input transducers. Obviously the controllability of this form of input is rather limited. The accuracy could be improved by changing the excitation type to multiple shakers if physically possible.

Acknowledgment

Partial support for the work was provided by Grant Number CES-8715463 from the National Science Foundation. Additional support was provided through the Virginia Transportation Research Council. Assistance in the field tests by W. T. McKeel, Jr., James W. French, Jean-Marie Roque and Paul Duemmel is greatfully acknowledged.

Appendix I- References

Craig, Roy R. Jr., Blair Mark A. 1985 " A Generalized Multiple-Input, Multiple-Output Modal Parameter Estimation Algorithm" AIAA Journal Vol. 23, No. 6, pp 931-937. June 1985.

Ewins, D. J. 1984. " Modal testing: Theory and Practice " John Wiley & Son INC. N.Y.

Fritzen, C-P,1985 "Identification of Mass, Damping, and Stiffness Matrices of Mechanical Systems.",Transaction of ASME June 1985 Paper No. 85-DET-91.

Yen, W-H P. (1992), "Modal Identification of Bridge Structures," Ph. D. Dissertation, University of Virginia, Charlottesville, VA, January.

Table 1. Modal Parameters Comparison

Modal Stiffness	Ambient	Sys. Id.	Modal Damping	Ambient	Sys. Id.	Modes
K1	0.566	0.572	C1	0.71	0.865	1st V.B.
K2	1.085	1.065	C2	0.51	0.822	1st T.
K3	1.175	1.212	C3	0.42	0.722	2nd V.B.
K4	2.112	2.212	C4	0.52	1.023	3rd T.
K5	2.210	2.228	C5	0.52	0.467	2nd T.
K6	3.275	3.195	C6	0.64	1.976	4th V. B.
K7	3.425	3.303	C7	0.52	2.817	3rd T.

Note: V. B. = Vertical Bending, T. = Torsion

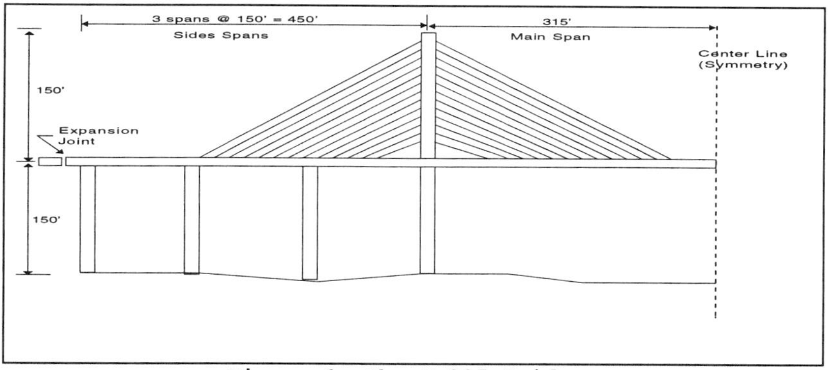

Figure 1. The I-295 Bridge

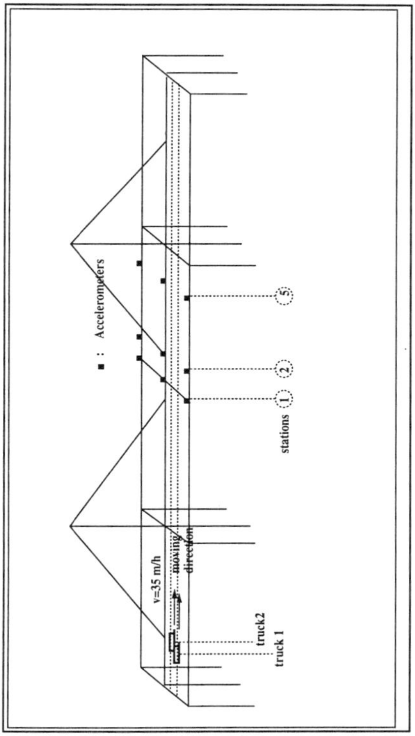

Figure 2. Experiment Setup

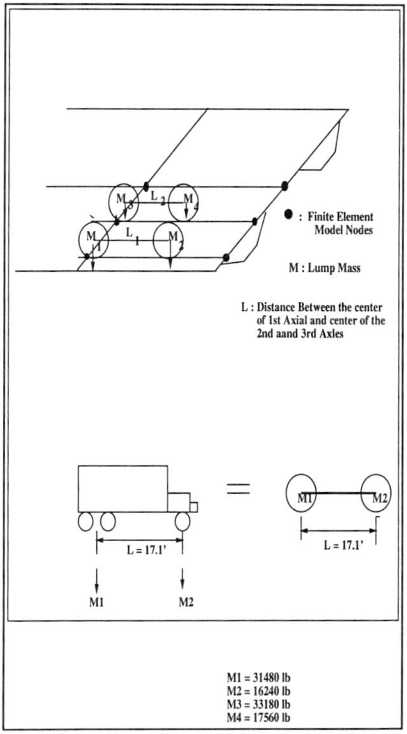

Figure 3. Idealized Input

NDE of Distributed Cracking in Concrete

Scott F. Selleck[1], Eric N. Landis[2], Michael L. Peterson[3], and Surendra P. Shah[4]

Abstract

An experimental program was conducted to evaluate ultrasonic techniques for evaluating distributed cracking in concrete structures. Distributed cracking here refers primarily to microcracking and other high porosity zones which typically precede large scale cracking. Several mechanisms were identified as primary causes for distributed cracking in concrete. These mechanisms include freeze-thaw, salt scaling, and mechanical stresses. A series of specimens were cast and subjected to these effects. Immersion scanning techniques were used to measure ultrasonic properties (pulse velocity and peak-to-peak amplitude). The changes in ultrasonic properties with increasing degrees of damage were compared to corresponding changes in material properties. Amplitude attenuation was found to be a much more sensitive indicator of damage than pulse velocity. The results of the experiments show that NDE techniques can be developed which use relative ultrasonic attenuation measurements as an indicator of damage and structural integrity in field structures.

Introduction

A serious maintenance problem with bridges is concrete deck deterioration due primarily to mechanical stresses, thermal loading, chemical attack, and reinforcing steel corrosion. This deterioration may take the form of distributed microscopic damage which can evolve into large defects such as cracks, delaminations, spalling, and scaling. Robust nondestructive testing techniques such as impact-echo effectively locate large cracks and delaminations. However mature techniques do not exist to evaluate distributed microscopic damage. The focus of this work was to evaluate whether ultrasonic waves can detect early, microscopic damage in concrete.

[1] Bessir, Gibble, Norden Consulting Engineers Inc., Old Saybrook, CT.
[2] Deptartment of Civil & Env. Engineering, University of Maine, Orono, ME.
[3] Deptartment of Mechanical Engineering, Colorado State University, Fort Collins, CO.
[4] Center for Advanced Cement-Based Materials, Northwestern University, Evanston, IL.

Experimental Program

Concrete specimens were subjected to damage that simulates environmental deterioration of bridge decks in northern climates. Two types of damage were induced: freeze-thaw and salt scaling. Freeze-thaw cycles were assumed to produce a relatively uniform distribution of microscopic damage through the specimen, while salt scaling tests were assumed to produce a gradient of damage through the specimen thickness. The damaged concrete specimens were insonified by ultrasonic pulses in the form of P-waves. The time record of the specimen response, known as a time domain signal, was analyzed. Characteristics of the time domain signal vary with changes in the microstructure of the material through which they propagate. The characteristics of the ultrasonic signals investigated were the amplitude and pulse velocity in the time domain, and peak frequency in the frequency domain. The most sensitive time domain signal may be the best for field testing of concrete. The trends observed in data obtained from ultrasonic testing were compared to the trends observed in data obtained from microscopic tests. Microscopic investigation was used to quantify the type, and amount, of microscopic damage that occurred.

The concrete used for this work had a mix proportion of 1:2.25:2.25:0.5, by weight respectively, Type I portland cement: 3/8 in. pea gravel: sand: water. The mix had a 28 day strength of 6500 psi.

Eleven specimens were prepared: six 16 x 4 x 3" specimens for freeze-thaw tests, and five 16 x 4 x 7 3/4" specimens for salt scaling tests. Freeze-thaw tests were performed in accordance with ASTM C666, and salt-scaling tests were performed in accordance with ASTM C672.

The ultrasonic testing method used was immersion scanning which is illustrated in Figure 1. The specimen is immersed in a water bath. The ultrasonic signal passes from the source transducer through water, through the specimen, and through water again before it reaches the receiving transducer. The water acts as a uniform couplant between the specimens and the transducers. Through a waveform analysis of the received signal, the signal's time-of-flight and peak-to-peak amplitude may be recorded. The time-of-flight is of course dependent on the pulse velocity of an ultrasonic signal in the material, and the peak-to-peak amplitude is dependent on the attenuation of the signal as it passes through the material.

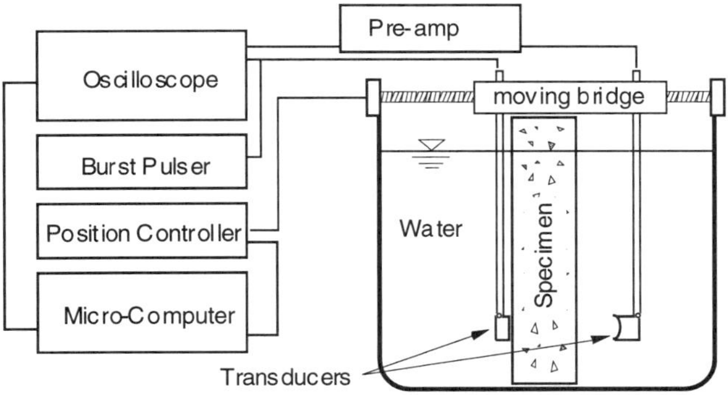

Figure 1. Ultrasonic Test Configuration

Ultrasonic scans were conducted as follows. The position controller moved the transducer pair to a location of interest. An ultrasonic signal was sent through the material, and the measured waveforms were recorded. Two types of tests were conducted: Line-scans and Area-scans. The Line-scan is a 1-D map composed of a series of time domain signals along a line. Each time domain signal was analyzed for a specific characteristic (either time-of-flight or amplitude). This characteristic was plotted against the position where the time domain signal was acquired along a specified height. The Area-scan is a 2-D map of a scanned area where a characteristic of each time domain signal was plotted against the coordinates where the time domain signal was acquired. At each point of the 2-D map, the transducer pair stopped moving and data was acquired. The two methods produce data plots for which variations in ultrasonic properties may be easily distinguished.

The freeze-thaw specimens were subjected to ultrasonic tests at 14 days of age. One specimen was kept in a water bath and included in all ultrasonic tests to monitor curing effects. The remaining specimens were subjected to increasing numbers of rapid freeze-thaw cycles as described in ASTM C666. At a certain number of freeze-thaw cycles each specimen was subjected to ultrasonic scanning.

The salt scaling specimens were subjected to ultrasonic measurements at approximately 28 days of age. One specimen was placed in a water bath to monitor curing effects. The remaining four specimens were subjected to consecutive cycles of salt scaling as specified by ASTM C672. Two specimens were removed from salt scaling cycling after 25 cycles, and two more were removed after 40 cycles. At the end of salt scaling cycling, each specimen was subjected to ultrasonic scanning.

Results and Implications for NDE of Concrete

The ultrasonic test results for a specimen subjected to freeze-thaw cycles are illustrated in Figures 2 and 3. Line scans are shown for undamaged and damaged specimens. The specimen was subjected to 33 freeze-thaw cycles which resulted in a 25% reduction in its dynamic elastic modulus. In each figure, the heavy line is an average of three line scans made at three different locations on the specimen. These figures show that pulse velocity is a fairly consistent measurement whereas peak to peak amplitude is highly erratic. However, it is also shown that the pulse velocity is relatively insensitive to damage as compared to peak to peak amplitude. That is, there is only a marginal change in pulse velocity in the damaged versus undamaged specimen, whereas there is a significant change in average signal amplitude. These differences are even more pronounced in the plots of Figures 4 and 5. This specimen had a 50% reduction in dynamic elastic modulus. The reduction of pulse velocity due to damage was approximately 10%, whereas the reduction of amplitude was approximately 80%. These figure show clearly that ultrasonic attenuation (as manifested in the peak-to-peak amplitude measurements) is much more sensitive to changes in material microstructure than pulse velocity.

The implications for nondestructive evaluation of concrete structures are as follows. Pulse velocity is a poor indicator of damage because small variations in pulse velocity can represent large variations in material properties. This is significant since much effort has gone in to developing pulse velocity-based NDE methods for concrete (Popovics and Rose, 1994). Ultrasonic attenuation is much more sensitive to damage, but at a cost of much more noisy data. The conclusions of this work are that ultrasonic attenuation shows promise as a damage indicator in concrete structures. However,

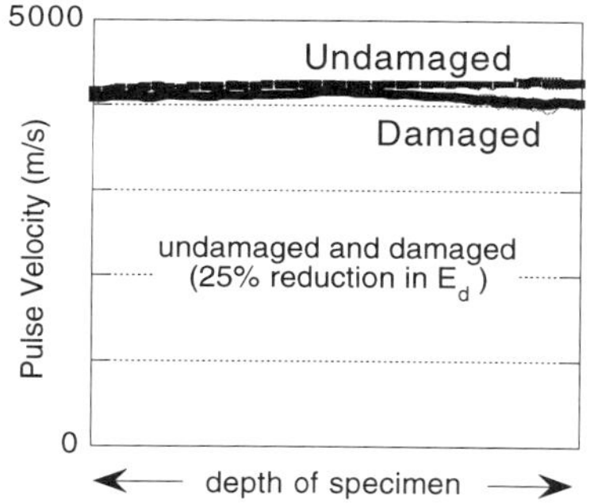

Figure 2. Variation in Ultrasonic Velocity vs. Position

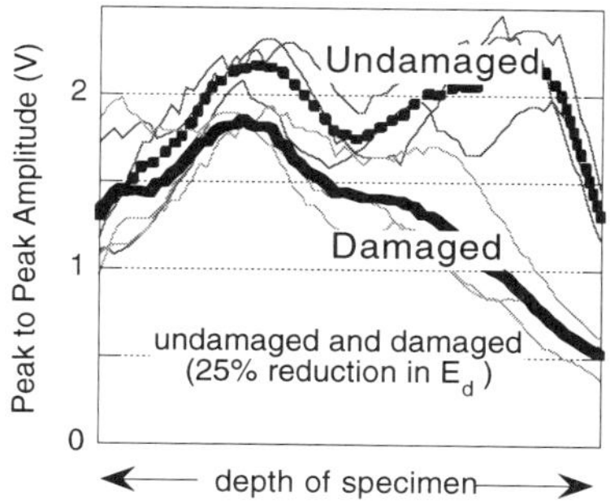

Figure 3. Variation in Ultrasonic Amplitude vs. Position

substantial development time would be required for the signal processing and data interpretation routines which would be necessary for a robust field implementation.

Acknowledgments

This work is supported in part by Infrastructure Technology Institute of Northwestern University. Additional support was provided by the NSF Center for Advanced Cement-Based Materials, and the Center for Quality Engineering and Failure Prevention, both at Northwestern University.

Reference

Popovics, J. S., and Rose, J. L. (1994). "A survey of developments in ultrasonic NDE of concrete." *IEEE Transactions on Ultrasonics, Ferroelectrics, and Frequency Control*, 41(1), 140-143.

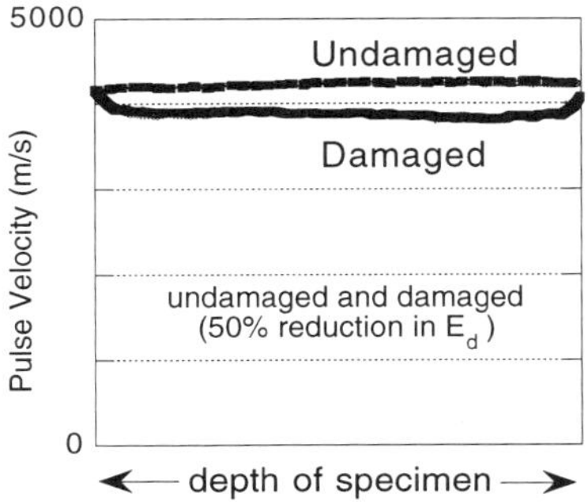

Figure 4. Variation of Ultrasonic Pulse Velocity vs. Position

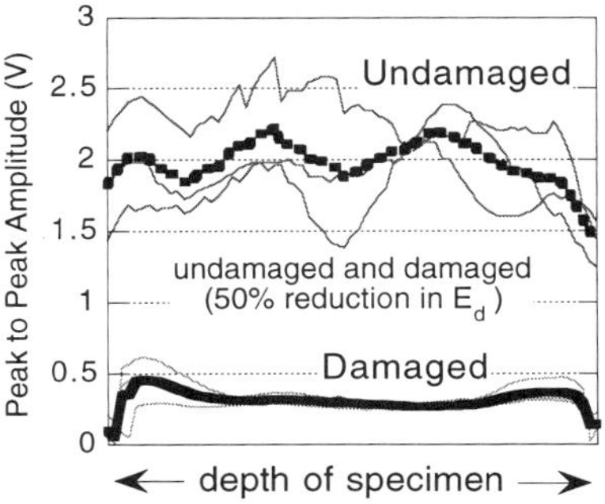

Figure 5. Variation in Ultrasonic Amplitude vs. Position

Intelligent Bridge Monitoring System

Pei-Ling Liu[1], Member, ASCE,
Yun-Fu Luo[2], and Shyh-Jang Sun[2]

Abstract

The objective of this study is to develop an artificial neural network system to monitor the safety of a bridge structure. A truck is driven at constant speed through the target bridge. The response of the bridge is then processed by a backpropagation neural network to evaluate the condition of the bridge. Finite element analysis is carried out to compute the response of the bridge under various damage conditions. The neural network is trained by the simulation data. The monitoring system is evaluated in terms of its reliability in damage detection.

Introduction

Bridges play an indispensable role in land transportation. The collapse of a bridge often results in tremendous losses. Therefore, the safety assessment of bridges is an important and challenging task for civil engineers. The conventional way of evaluating the integrity of a bridge is by visual inspection. However, bridge structures are usually very large. Therefore, such inspection is costly and time-consuming. Furthermore, the inspection may not be complete because the structure is not totally accessible.

This paper proposes an intelligent inspection system for bridges. A truck of constant mass is driven at constant speed through the target bridge. The response of the bridge is then processed by an artificial neural network (ANN) to evaluate the condition of the bridge automatically.

Artificial Neural Network

The ANN is a powerful tool in the areas of pattern recognition, signal processing, control, and so on. Recently, several researchers have applied ANN in the damage assessment of structures, for example, Wu et al. (1990), Elkordy and Chang (1992), and Tsou and Shen (1994).

---

[1] Associate Professor, Institute of Applied Mechanics, National Taiwan University, Taipei, Taiwan.
[2] Graduate student, Institute of Applied Mechanics, National Taiwan University, Taipei, Taiwan.

The ANN adopted in this study is a multi-layered backpropagation network (Zurada, 1992). The backpropagation network constitutes of an input layer, an output layer, and several hidden layers. Each layer is composed of several neurons, and each neuron is linked to every neuron of the adjacent layers.

The input layer reads the input data and passes it to the first hidden layer. The neurons in the first hidden layer take weighted sums of the input data and evaluate the activation function f (a sigmoid function in this study):

$$y_k = \mathrm{f}\left(\sum_{j=1}^{p} w_{kj} x_j\right) \tag{1}$$

The output of the activation function, y_k, is then passed forward to the next layer. The neurons in the next layer again take weighted sums of the incoming data, evaluate the activation function, and pass the function values forward. The same process is repeated until the output layer sends out the results.

The weights of the links, which characterize the network, are determined by training. To train a backpropagation network, a number of samples must be provided. Suppose the target output of a sample is $\mathbf{d} = \begin{bmatrix} d_1 & d_2 & \cdots & d_r \end{bmatrix}^T$ and the network output is $\mathbf{o} = \begin{bmatrix} o_1 & o_2 & \cdots & o_r \end{bmatrix}^T$. A sum-squared error can be defined as follows:

$$E = \frac{1}{2}\sum_{k=1}^{r} (d_k - o_k)^2 \tag{2}$$

In each training cycle, the weights are modified along the direction of negative gradient of the error function. The weight corrections are performed layer by layer, starting from the output layer backward to the input layer.

The training data are best obtained from full-scale tests. However, this is not possible because the bridge must be in various damaged conditions. Therefore, the training data are generated by numerical simulations in this study.

Dynamic Analysis of a Bridge Subjected to a Moving Truck

A truck moving at constant speed across a bridge can be modeled by a moving mass on a continuous beam. The dynamic response of a beam subjected to a moving mass is as follows:

$$\frac{\partial^2}{\partial x^2}\left(EI\frac{\partial^2 w(x,t)}{\partial x^2}\right) + \rho A\frac{\partial^2 w(x,t)}{\partial t^2} + M\delta(x-\eta)\frac{\partial^2 w(\eta,t)}{\partial t^2} = Mg\delta(x-\eta) \tag{3}$$

where w is the transverse displacement of the beam, EI is the flexural rigidity, ρA is the mass density per unit length, M is the mass of the truck, g is the gravitational acceleration, δ is the Dirac delta function, and η is the position of the truck.

Finite element formulation of the above equation leads to

$$\mathbf{KU}+\mathbf{M}\ddot{\mathbf{U}}=\mathbf{F} \tag{4}$$

where **K** , **M**, **U**, and **F** are the stiffness matrix, mass matrix, nodal displacement vector, and external load vector, respectively. Notice that **M** and **F** are dependent on the truck position. The equation can be integrated by the Newmark scheme to obtain the beam response.

Numerical Example

Consider a three-spanned bridge, each span being 60m long. The original properties of the beam are: $EI = 7.34\times10^{10}$ N·m^2 and ρA=1570 kg/m. The mass of the test truck is 20000 kg, and the speed is 60 km/hr. The bridge is discretized into 30 uniform beam elements. Accelerometers are mounted on nodes 2, 6, 10, 12, 16,20, 22, 26, and 30 to measure the vertical accelerations at the left, middle, and right of each span.

In this study, damage of the bridge is modeled by stiffness reduction. Nine damaged locations are considered: 1L, 1M, 1R, 2L, 2M, 2R, 3L, 3M, and 3R, where 1, 2, and 3 denote the span number, and L, M, and R denote the left, middle, and right of the span. Each damaged location constitutes of two elements. For example, 1L means that elements 1 and 2 are damaged, 2M means that elements 15 and 16 are damaged, and 3R means that elements 29 and 30 are damaged. At each location, 18 damage levels are simulated, the ratio of the damaged and original stiffness being 0.1, 0.15, 0.2, ..., and 0.95. Finite element analysis was carried out for each of these cases, and the results were used as the database for ANN training and testing. The training database has 82 samples, including the undamaged case and the cases with stiffness ratios 0.1, 0.2, ..., and 0.9. The test database has 81 samples, including the cases with stiffness ratios 0.15, 0.25, ..., and 0.95.

While designing an ANN, the most crucial step is the selection of input parameters. The input parameters must be chosen such that they are representative of the output. Hence, the input vector must be able to reflect the current condition of the bridge in this application.

Examining the acceleration curves, one can see that the wave forms indeed vary with the damage condition. However, it is not appropriate to use these curves directly as the input because they contain too many data points. Therefore, a lot of efforts have been made in this study to extract useful features from these curves. Finally, 81 input parameters are defined as follows:

$$x_{ij}=\frac{\sum_{k=(j-1)m+1}^{jm}\tilde{a}_i^2(k\Delta t)-a_i^2(k\Delta t)}{\sum_{k=(j-1)m+1}^{jm}a_i^2(k\Delta t)} \qquad i,j=1,\ \dots,\ 9 \tag{5}$$

in which $\tilde{a}_i$ and a_i denote the accelerations of the current and original bridges, respectively, measured by the ith accelerometer, Δt is the sampling period, and m= T/9Δt, where T is the passing time of the truck. Notice that each input parameter can reflect the change of kinetic energy of a part of the bridge in a certain time interval.

The output layer of the ANN has 9 neurons, representing the ratios between the current and original stiffness of the 9 locations. The output value is between 0 and 1. The number of hidden layers and neurons are determined by trial runs. The 81-60-50-9 ANN turned out to perform best, where 60 and 50 respectively denote the numbers of neurons in the first and second hidden layers.

Using the ANN to analyze the test data, the damaged locations are found accurately. However, the maximum error of the output ratios is 30.7%. In order to improve the system, an 81-50-30-1 ANN is introduced as a postprocessor. This ANN determines the damage level when the damaged location is known. With the help of this postprocessor, the maximum error of the stiffness ratios is reduced to 5.6%.

Since noise always presents in practice, random noise was superposed to the test data to see if the system is vulnerable to noise. As the noise level increases from 5% to 30% of the standard deviation of the acceleration curve, the maximum error increases from 7.3% to 23.6%. Apparently, this system is insensitive to noise.

Conclusions

An intelligent monitoring system for a bridge is developed in this paper. The back-propagation neural network is adopted in the system for data processing. Numerical tests show that the system can detect the location and level of damage accurately. Furthermore, the system is insensitive to noise. Therefore, the monitoring system has a high potential in real applications.

Acknowledgment

This work partly supported by the National Science Council of the Republic of China under Grant NSC85-2211-E-002-017.

References

1. Elkordy, M. F. and Chang, K. C. (1993) "Neural networks trained by analytical simulated damage state", *Journal of Computing in Civil Engineering*, **7**(2), 130-145.
2. Tsou, P. and Shen M.-H. H. (1994) "Structure damage detection and identification using neural networks," *AIAA Journal*, **32**(1), 176-183.
3. Wu, X., Ghaboussi, J., and Garrett, J. H. (1992) "Use of neural networks in detection of structure damage," *Computer and Structures*, **42**(4), 649-659.
4. Yao, G. C., Chang, K. C., and Lee G. C. (1992) "Damage Diagnosis of Steel Frames Using Vibrational Signature Analysis," *Journal of Engineering*, ASCE, **118**(9), 1949-1961.
5. Zurada, M. J. (1992) *Introduction to Artificial Neural Systems*, Info Access & Distribution Pte Ltd., Singapore.

Evaluation of Microcrack Damage Growth and Healing of Asphalt Concrete Pavements Using Stress Wave Method

Yongon Kim[1] and Y. Richard Kim[1], Member, ASCE

Abstract

This paper presents a nondestructive technique based on the stress wave propagation method as a means of evaluating the change in the "effective" modulus of asphalt concrete layer due to the fatigue crack growth and healing. This method has been applied to pavements in service that have been loaded by Accelerated Loading Facility (ALF) in FHWA Highway Research Center. The results indicate that the wave speed, and therefore, the "effective" modulus decreases as the number of load cycles increases and increases after a rest period is introduced.

Test Procedure

A series of stress wave tests have been completed for two of the FHWA ALF sections with 102 mm (4 in) thick AC5 and AC20 layers. The data acquisition and the signal analysis were performed using a computer program, developed in North Carolina State University. The program is capable of performing Fast Fourier Transform, Fourier Phase Method, and Short Kernel Method (Holt, et al 1994).

The ALF applies cyclic loading with frequency of 10 seconds per cycle. The wheel used in this study was 318 mm (12.5 in) wide with a load of 53.4 kN (12,000 lb) and tire pressure of 689.5 kPa (100 psi). The position of wheel load was controlled to produce normally distributed wandering transverse to the traffic direction. The resulting width of wheel path was 1.07 m (42 in).

The pavements were subjected to repetitions of a group of loading cycles followed by a 24-hour rest period while the pavement temperature was maintained at 20 C (68 F). Number of loading cycles in each group of loading changed from 8000 cycles in early days of testing to 25,000 cycles in later part of testing.

One section was tested for one week, and the ALF was moved to the other section at the end of the week. This alternating test program was repeated until the

[1]Graduate Research Assistant and Associate Professor, Department of Civil Engineering, North Carolina State University, Raleigh, NC 27695-7908

two sections failed. Surface wave data was acquired at the following times; before any loads were applied, immediately after each group of loading ends, and at the end of 24-hour rest periods. Wave speeds were measured in 6 different positions in the middle area of each section. Positions 1 and 5 were on the center line of the wheel path. Positions 2, 3, and 4 were located at 152 mm (6 in), 305 mm (12 in), and 457 mm (18 in) apart from the center line. One more position, named control, was selected at the outside of the wheel path and tested to monitor the effects of environmental variables, if there is any.

Detailed visual surface condition surveys were made periodically so that cracks on the pavement surface were mapped and crack lengths were measured after each survey.

Test Results

In general, the wave speeds in the AC20 section were higher (about 1250 m/sec) than those of the AC5 section (about 1100 m/sec), which was expected because of higher stiffness of AC20 pavement. Figure 1 shows the change of wave speed at Positions 1, 3, 4, and control in AC20 section. In Positions 1 and 5 with greatest number of load applications, the major portion of modulus reduction occurred within 25,000 passes of loading and then the curve became relatively flat afterwards. This decrease in wave speed appears at a later time with a more gradual solpe as the offset from the center increases. In most cases except for Position 4 and Control, that are on the edge of the wheel path and outside of the wheel path respectively, the wave speed decreased after loading periods and increased after the 24-hour rest period. The beneficial effects could be attributed to microcrack healing or steric hardening or both. In this paper, these beneficial effects are categorized under the term "healing". The healing rate is defined:

$$\text{Healing Rate} = (V_b - V_a) / V_a \times 100\ (\%)$$

where V_a=wave speed after completion of each group of loading
V_b=wave speed after a 24-hour rest period

Figure 2 shows the rate of healing in each test position for each section. It is noticeable that Position 2, under the edge of the tire, showed the greatest healing rate, instead of Position 1 or 5 at the center of the wheel path. The type of damage in Position 2 is thought to be mostly due to shear stresses. This finding suggests that the microcrack damage is more pronounced at the edge of a tire (due to shear stresses) than at the center of the wheel path (due to tensile stresses at the bottom of the AC layer). According to this observation, the traditional approach of using the tensile stress and strain as a design criterion for pavement design needs to be carefully reconsidered. AC20 showed about twice higher healing rate than AC5. It may take longer for microcracks in the AC20 layer to heal completely because of its higher viscosity, but the higher healing rate of the AC20 pavement suggests greater strength regain in the AC20 pavement after a 24-hour rest period.

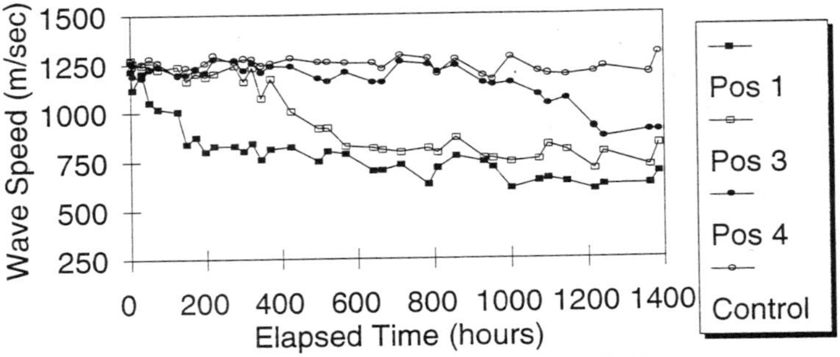

Fig. 1. Wave Speed Change of AC 20

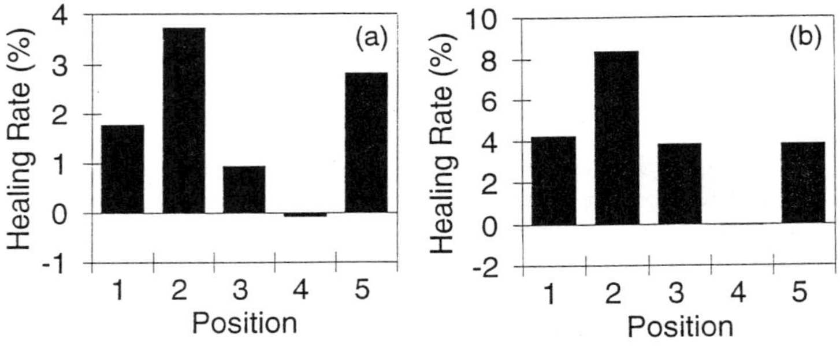

Fig. 2. Average Healing Rate of (a) AC5 and (b) AC20

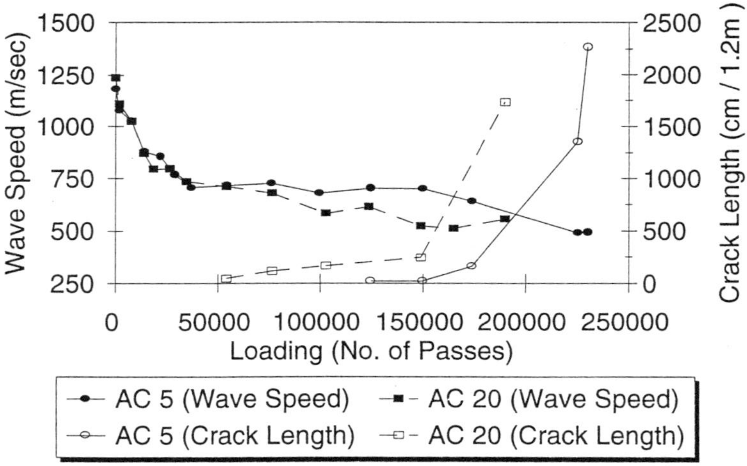

Fig. 3. Wave Speed & Crack Length

The AC5 pavement failed after 250,000 cycles, whereas the AC20 section failed after 200,000 cycles. Although the AC20 showed a higher wave speed at the beginning because of the higher stiffness, it showed visual cracks earlier than the AC5 and its wave speed decreased more than the AC5 at the later period of loading. Both the wave speeds calculated from the stress wave analysis in Position 1 and the measured crack lengths are plotted against the number of loading cycles in Fig. 3. The wave speed decrease in the first stage of loading is considered due to microcrack initiation and propagation. Therefore, even long time before visual cracks occur on the pavement surface, microcracks seem to occur and they have a great role in the reduction of structural integrity of pavements. Another observation to be made from this figure is that the dramatic increase in crack length, which indicates the failure of pavements, coincides with additional reduction of wave speeds.

Conclusions

1. Stress wave test and analysis methods used in this study successfully detected microcrack growth and healing of AC5 and AC20 pavements.

2. The stress wave analysis of the data obtained from the ALF pavements demonstrates the importance of microcrack growth and healing on the performance of asphalt concrete pavements.

3. Increase in surface crack length coincides well with the reduction of wave speeds.

4. The AC5 pavement outlasted the AC20 pavement, i.e., the considerable decrease of wave speed and development of visible cracks in the AC5 pavement were observed later than AC20.

Acknowledgements

The experimental data used in this paper were generated from a research project sponsored by Texas A&M Research Foundation, Western Research Institute, and the Federal Highway Administration. The authors are grateful for these organizations' support. The technical support and advices provided by Mr. Ramon Bonaquist is greatly appreciated.

Appendix : References

Holt, J.D., S. Chen, and R. A. Douglas. *Determining Lengths of Installed Timber Piles by Dispersive Wave Propagation*, Transportation Research Record 1447, TRB, National Research Council, Washington, D.C., 1994, pp. 110-115

Surface Response of a Cracked Layered Half-space Subjected to an Antiplane Impact

S. W. Liu[1], J. C. Sung[2], and M. S. Lin[2]

Abstract

The surface response of a cracked layered half-space subjected to an antiplane impact is investigated by using a hybrid method which combines the finite element method with the boundary integral representation. Both surface-breaking and sub-surface cracks are considered in this study. Numerical results show that not only the propagation of reflected, diffracted, and direct impact waves can be clearly observed, but also the crack can be easily detected by the surface response.

Introduction

The antiplane shear waves are created while the free surface of the layered half-space is subjected to a time dependent antiplane impact loading. It is of importance to be able to detect the cracks in the layered solid, and one of the most practical and efficient methods for accomplishing this is ultrasonic nondestructive technique which utilizes the scattering of elastic waves and the subsequent detection of these scattered waves by appropriate transducers. In recent years the scattering of elastic waves by cracks in multilayered solid materials have received considerable attention in the literature devoted to nondestructive testing and fracture mechanics in view of the rapid increase of coated and multilayered plates in engineering applications. The scattering of Loves waves by an interface crack and a normal surface-breaking crack have been investigated by Neerhoff (1979) and Angel (1986,1994), respectively. The method adopted in this study is similar to that of Liu and Datta (1993).

Formulation

Consider a layer which is perfectly bonded to a half-space of a different material as shown in Fig. 1. The surface of the layered half-space is subjected to an antiplane shear impact at $x = l, z = 0$. The out-of-plane displacement is denoted by u_j (j=1 for layer and j=2 for half-space). The two-dimensional antiplane wave motion is governed by

- -

[1]Liu, Department of Physics and Mathematics, Chinese Air Force Academy, P.O. Box 90277-4, Kangsan, Kaohsiung, Taiwan 82000, R.O.C.

[2]Sung and Lin, Department of Civil Engineering, National Cheng Kung University, Tainan, Taiwan 70101, R.O.C.

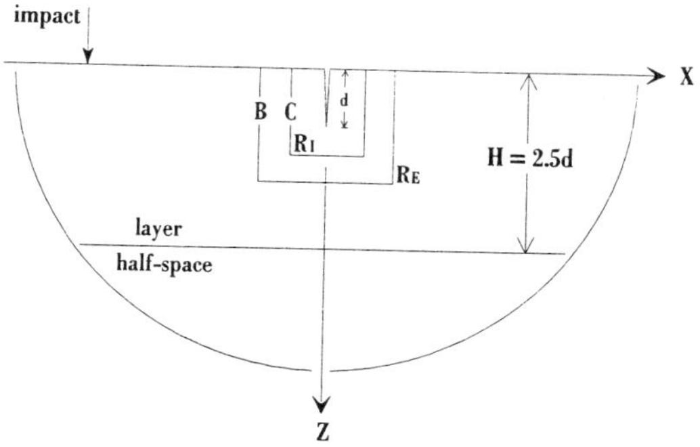

Figure 1. Geometry of the problem.

$$\beta_j^2\left(\frac{\partial^2 u_j}{\partial x^2} + \frac{\partial^2 u_j}{\partial z^2}\right) = \frac{\partial^2 u_j}{\partial t^2} - \frac{p(t)}{\rho_j}\delta(x-l)\delta(z), \tag{1}$$

where δ and $p(t)$ are Dirac delta function and impact loading function, respectively. The Fourier transform of Eq.(1) is given by

$$\beta_j^2\left(\frac{\partial^2 U_j}{\partial x^2} + \frac{\partial^2 U_j}{\partial z^2}\right) = -\omega^2 U_j - \frac{P(\omega)}{\rho_j}\delta(x-l)\delta(z), \tag{2}$$

where ω is the circular frequency. The problem is solved by a hybrid method which combines the finite element discretization of the near field with the boundary integral representation of the far field. We define the near field R_I to be bounded outside by an imaginary boundary B. The far field R_E is bounded inside by the contour C. The near field is discretized into a set of finite elements. The equation of motion can be written in the following form:

$$\begin{bmatrix} \mathbf{S}_{BB} & \mathbf{S}_{BI} \\ \mathbf{S}_{IB} & \mathbf{S}_{II} \end{bmatrix} \begin{Bmatrix} \mathbf{d}_B \\ \mathbf{d}_I \end{Bmatrix} = \begin{Bmatrix} \mathbf{f}_B \\ \mathbf{0} \end{Bmatrix}, \tag{3}$$

where $\mathbf{S}$ is the impedance matrix, $\mathbf{f}_B$ denotes the nodal force at B, and $\mathbf{d}_B$ and $\mathbf{d}_I$ are the nodal displacements at B and interior nodes, respectively. In the far field, based on the Betti's reciprocal theorem, the total displacements at the nodal points on boundary B can be expressed as

$$U(x_s, z_s) = \int_C (T^G U - T U^G)dC + U^F, \tag{4}$$

where T is the surface traction, the supercript G and F denote the Green's function and free field, respectively.

The integral in Eq.(4) can be discretized by using the finite element formulation and combined with the second equation of Eq.(3),

$$\begin{bmatrix} \mathbf{I} - \mathbf{A}_{BB} & -\mathbf{A}_{BI} \\ \mathbf{S}_{IB} & \mathbf{S}_{II} \end{bmatrix} \begin{Bmatrix} \mathbf{d}_B \\ \mathbf{d}_I \end{Bmatrix} = \begin{Bmatrix} \mathbf{d}_B^F \\ \mathbf{0} \end{Bmatrix}. \tag{5}$$

Once the nodal displacements are solved from Eq.(5), the displacements in the near field and far field can be easily computed by using finite element method and boundary integral representation, respectively. The transient response is then obtained from the spectrum by using fast Fourier transform.

Numerical Results

The thickness H of the layer is equal to $2.5d$. The material properties are normalized with respect to those of the layer. Thus the normalized shear wave velocity in the layer is equal to 1. The nondimensional time is defined as $\bar{t} = \frac{t\beta}{d}$, where β is the shear wave velocity in the layer. The impact loading is exerted on the free surface at $x = -0.95d$.

The surface responses from $x/d = -2.0$ to 2.0 on the layered half-space without any crack are calculated analytically and shown in Fig.2. After the direct impact wave, two reflected waves are observed in the time window. Figure 3 shows the surface displacements on a layered half-space containing a surface-breaking crack. The direct impact wave first arrives at the crack mouth at $\bar{t} = 0.95$, then a reflected wave is generated by the left crack face and reaches x/d=-2 at $\bar{t} = 2.95$. The direct impact wave is then diffracted by the crack-tip at $\bar{t} \approx 1.4$ and an anti-symmetrical cylindrical scattered wave is created.

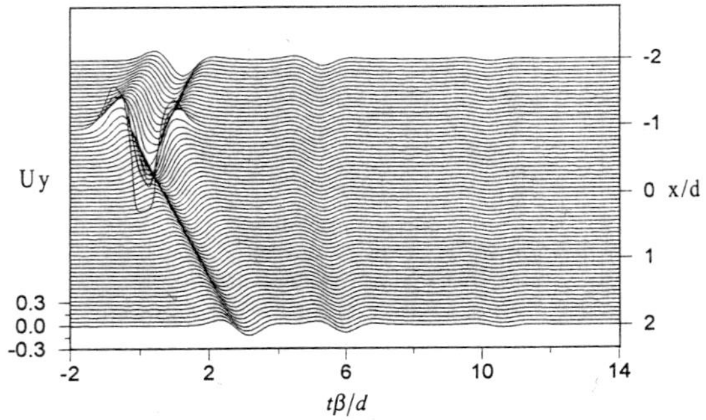

Figure 2. Surface displacements (without any crack).

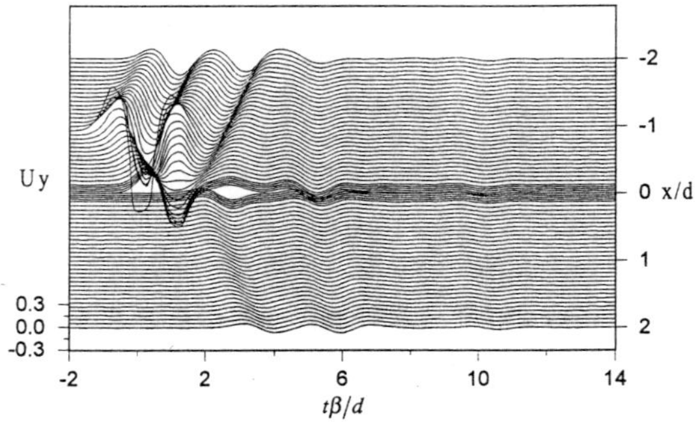

Figure 3. Surface displacements (with a surface-breaking crack).

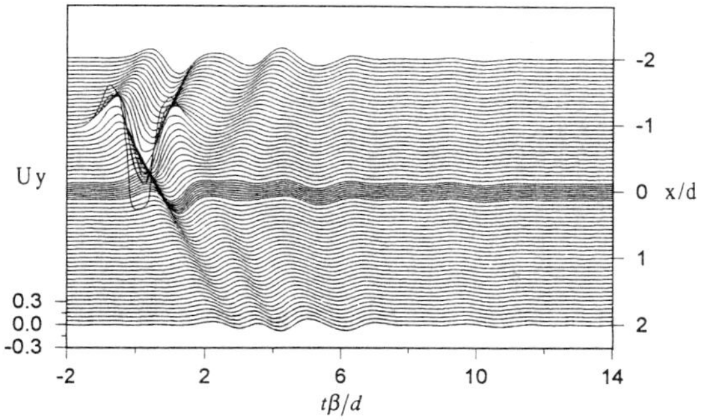

Figure 4. Surface displacements (with a sub-surface crack).

Finally, the direct impact wave travels to the interface and a reflected wave is created by the interface. Except these major diffracted and reflected waves mentioned above, a series of minor scattering waves continue to be created but their amplitudes decay gradually. Figure 4 shows the surface displacements of a layered half-space containing a sub-surface crack. It is observed that two diffracted waves appear on the free surface. The first one is caused by the diffraction of the direct impact wave by the upper tip at $\bar{t} \approx 1.0$. The second one originates from the diffraction occured at $\bar{t} \approx 1.6$ when the impact wave interacts with the lower tip. These scattered waves can be used to find the location and length of the crack. The multi-reflections in the layer are also observed on the free surface.

Conclusion

The surface responses can be used successfully to detect the thickness of layer and the location, length, and depth of the crack. Without mode conversion in the interaction of shear waves with cracks and other boundaries, the propagation of reflected, diffracted, and direct impact wave can be clearly observed and the mechanism of the impact wave scattering by cracks can be understood. This study is an important step for the investigation on more complicated scattering problems as well as inverse problems.

Acknowledgements

The present work is supported by the National Science Council, Republic of China under Grants NSC 84-0410-E-006-014.

References

Neerhoff, F.L., Diffraction of Love waves by a stress-free crack of finite width in the plane interface of a layered composite, *Appl. Sci. Res.* **35** (1979) 265-315.
Angel, Y.C., Scattering of Love waves by a surface-breaking crack, *ASME J. Appl. Mech.* **53** (1986) 587-592.
Angel, Y.C., Reflection and transmission of antiplane surface waves by a surface-breaking crack in a layered elastic solid, *Wave Motion* **20** (1994) 371-383.
Liu, S.W. and Datta, S.K., Scattering of ultrasonic wave by cracks in a plate, *ASME J. Appl. Mech.* **60** (1993) 352-357.

DETECTION OF CRACKS IN CONCRETE USING THE IMPACT RESPONSES

H. L. (Roger) Chen[1] and Lianfeng Pei[2]

Abstract

An impact response study was conducted for the identification of cracks in model concrete bridge decks and pavements. Concrete specimens were cast to simulate bridge deck and pavement slabs. These specimens contain horizontal cracks of different sizes and locations. The output responses contain frequencies of free vibration and are picked up by an accelerometer. The input and output signals are recorded and the transfer functions and their Power Spectral Density (PSD) are analyzed. The difference of the frequency response functions between specimens with or without cracks is evaluated to determine the presence of cracks or reinforcements and the position of cracks. The response frequencies are verified with finite element calculation. The results show that the impact technique is an effective nondestructive testing method for concrete structures.

Introduction

Impact response technique has been applied extensively for nondestructive testing of concrete structures (Carino and Sansalone, 1988). The dynamic response of a concrete structure under mechanical impact can be analyzed to reflect the damage condition in the structure. Due to the high attenuation in concrete, a steel impactor which can easily generate an impact frequency response up to 5 kHz was selected for this study. The measurement of the frequency response function can be used in the model analysis. The magnitude of change in natural frequencies is a function of the severity and of the location of deterioration in the structure. Ratios of changes in natural frequencies normalized with respect to the largest frequency change can serve to indicate the location of deterioration directly (Hearn

---

1. Associate Professor, 2. Graduate Research Assistant, Department of Civil and Environmental Engineering, West Virginia University, Morgantown, WV 26506.

and Testa 1991). This helps to explain the usefulness of impact response in the localization of damage in practical situations (Morassi 1993).

Instrumentation and Test Setup

The impact response test system used is composed of an instrumented impact-hammer, an accelerometer, and a data acquisition system including a digital oscilloscope and a personal computer. The instrumented PCB impact-hammer has a steel tip with an impact duration of about 0.4 to 0.5 ms. The accelerometer (PCB Model 2215) has a frequency range up to 6 kHz. The data acquisition system has a sampling frequency of 500 kHz, and a two-channel digital oscilloscope was used to record and to store the input and output responses. Beewax was used to adhere the accelerometer on the concrete surface. By dividing the Fourier transforms of the output response by the impact force function, the transfer function is calculated in the frequency domain. The transfer function was expressed in terms of PSD.

Test Specimens

Three 61x61x15cm concrete specimens with different internal conditions were cast to simulate bridge deck, and three 60x60x25cm concrete specimens were cast to simulate highway pavement. Specimen B11, B12, and B13 are 15cm thick specimens while B21, B22, and B23 are 25cm thick specimens. B11 and B21 are plain concrete. B12 and B22 have a 18x18x0.5cm plain-water crack located at 6.5cm and 12.5cm below surface, respectively. B13 and B23 have a 18x18x0.5cm saline-water crack located at 7.5cm and 12.5cm below surface, respectively. The embedded cracks were placed horizontally in the middle of the concrete slabs.

Experimental Results

Three test cases corresponding to three different testing locations (points) on the specimen's surface are presented here. Point A is at the center; point B is 10cm away from A; point C is 25cm from A. (1) Case 1: The accelerometer is placed at A, and the impact-hammer is 1.25cm away from point A. Figure 1 shows the test results of PSD of the specimens B21 and B22. Specimen B21 has distinct frequency components at 2075, 3925, and 5500 Hz. In B22 the frequency of 4800 Hz has a large amplitude appeared which clearly indicates the embedded crack. (2) Case 2: Also for specimens B21 and B22, the accelerometer is placed at point A, and the impact is at point B. Test result is shown in Figure 2. The impact point was just above the edge of the plain-water crack, and except the reduction of the amplitude, no major changes had occurred in the transfer function. (3) Case 3: The accelerometer is at point A, and the impact-hammer is at point C. Result is plotted in Figure 3. It shows that the amplitudes of all frequency components continuously decreased while the impact location was moving away. However, the crack is still detectable (at 4800 Hz), even when the impactor was already 25cm from the accelerometer and 15cm away from the edge of the embedded crack.

Analysis and Discussion

Dynamic finite element analysis using three-dimensional brick elements was performed to analyze the free vibration modes of specimens B11 and B21. Ultrasonic pulse-velocity tests were conducted to measure the longitudinal velocity and the Young's modulus of concrete. The FEM results show that the frequencies of the first 3 symmetric bending modes are 1516, 3120 and 5079 Hz for specimen B11, when comparing with the experimental results (Table1), the differences are 4.5%, 1.0% and 1.6%, respectively. For specimen B21, the FEM results are 2083, 3634, and 5499 Hz, and the difference in comparison with the experimental results are 0.2%, 7.0%, and 0.02%. The test results are summarized in Table 1 for all the specimens. In Table 1, F1, F2, and F3 represent the first three symmetrical bending modes of the specimens, and F4 represents the frequency caused by the embedded cracks. Results indicate that the crack-induced changes in frequency components are very distinct, and can be used to quantify the cracks.

Summary

Impact response tests were performed on six model concrete specimens. The test results indicate that the impact response is capable of detecting subsurface water-filled concrete cracks. Similar results were obtained by exchanging the positions of the accelerometer and impact-hammer. The location and size of the crack can be evaluated by analyzing the frequency response function at several different measuring positions.

Acknowledgements

The authors acknowledge the support from the West Virginia Department of Highway, especially Mr. Gary Robson and Mr. Aaron Gilisby of the Material and Soil Testing Division. The assistance from the Constructed Facilities Center at WVU is also acknowledged.

References

Carino, N.J., and Sansalone, M. (1988). "Impact-Echo Method." **Concrete International**, April, pp. 38-46.

Hearn, G., and Testa R.B. (1991). "Modal Analysis for Damage Detection in Structures." **J. of Structure Engineering**, Vol. 117, No. 10, October, pp. 3042-3063.

Morassi, A. (1993). "Crack-Induced Changes in Eigenparameters of Beam Structures." **J. of Engineering Mechanics**, Vol. 119, No. 9, September, pp. 1798-1803.

Specimen	F1 (Hz)	F2 (Hz)	F3 (Hz)	F4 (Hz)
B11	1450	3150	5000	
B12	1450	3125	4775	3725
B13	1500	3200	5000	4100
B21	2075	3925	5500	
B22	1975	3775	5275	4800
B23	1975	3775	5300	4925

Table 1: Test Results in Case 1

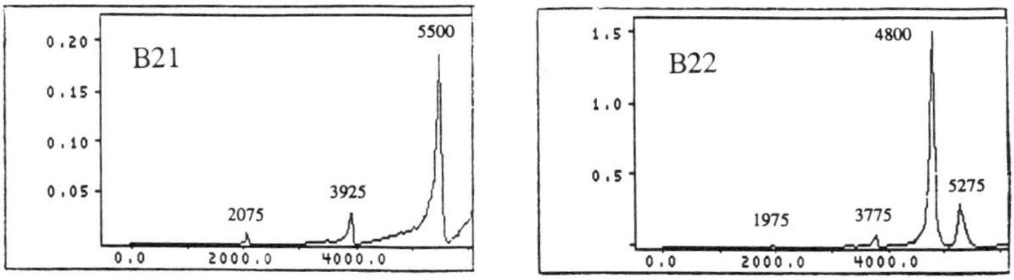

Figure 1: Test Results in Case 1

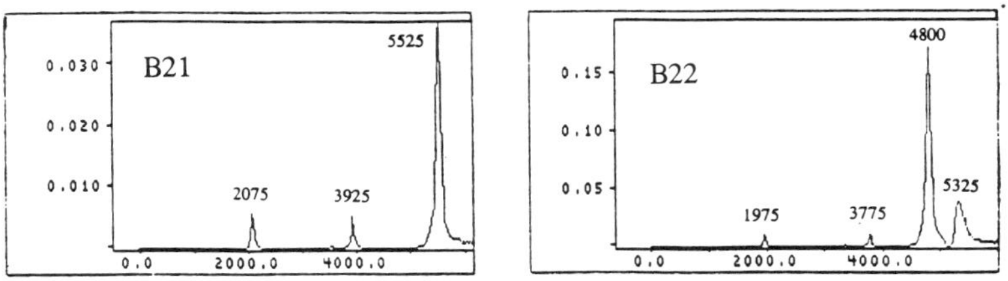

Figure 2: Test Results in Case 2

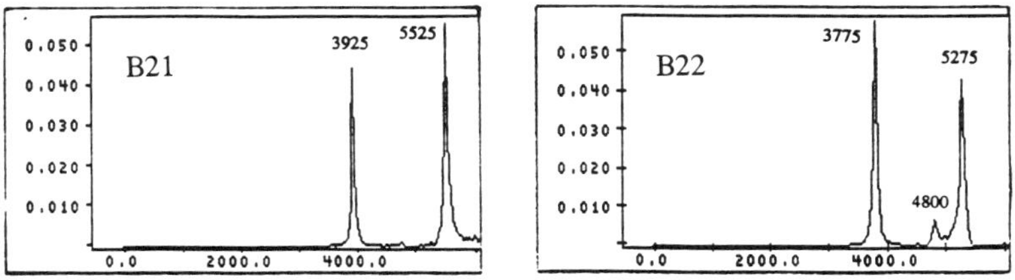

Figure 3: Test Results in Case 3

Crack Growth in Uniaxially Aligned Fiber Reinforced Mortar

Mohsen A. Issa[1], Member, ASCE, and A. B. Shafiq[2]

Abstract

The influence of spacing between the reinforcing fibers on the fracture energy, crack tip stresses and fatigue crack growth characteristics is presented in this paper. Experimental results indicate an increase in the fracture energy as the size of the reinforcing fibers and the corresponding spacing between the fibers is reduced. A large scatter in the fatigue data was observed. A significant increase in the lifetime was observed as the stress level was reduced from 80% to 65%. Fatigue life was found to be unlimited at 50% stress level. The present study indicates that the lifetime and fatigue crack growth characteristics are independent of the spacing between the reinforcing fibers.

Introduction

The effect of fiber size and spacing between the reinforcing fibers in concrete on the fracture behavior under quasi-static loading conditions has been rigorously studied (Issa et al., 1996; Romualdi and Batson, 1963; Shah and Rangan, 1971). Results of static analysis generally indicate an increase in the fracture properties of the material when the size or the corresponding spacing between the fibers is reduced, however, a wide scatter in the data has led to conflicting conclusions. The effect of reinforcement size and spacing in concrete under fatigue loading conditions has attracted very little attention to date. As most structural components experience cyclic loading during their life-time, it is imperative to conduct fatigue testing to understand the material behavior under service conditions.

Experimental Program

Specimens were prepared with spacings of 37.5, 15 and 5.8 mm (1.5 , 0.6

----

1 Associate Professor, CME Dept., UIC, Chicago, Illinois
2 Graduate Student, CME Dept., UIC, Chicago, Illinois

and 0.23 in.) between the fibers. The average compressive strength of the matrix was 51.7 MPa (7.5 ksi) and the yield strength of the steel reinforcing fibers was 380 MPa (55 ksi). The volume percentage of the reinforcement was kept constant. Testing was performed in three-point bending. Displacement control mode at a rate of 0.075 mm/min (0.003 in./min) was used for the quasi-static cyclic testing. Closed loop load control mode was used for fatigue testing at stress levels of 80%, 65% and 50% at a constant frequency of 2 Hz.

Fracture Energy (G_F)

Fracture energy is a critical material property as it is the materials' resistance to crack growth and it depends on the interlocking of the fracture surfaces, crack bridging and strength of the concrete and reinforcement. Fracture energy was determined from the loading-unloading technique used by Issa et al. (1996) and was calculated as:

$$G_F = \frac{W_o}{A_{lig}} \tag{1}$$

where "W_o" is the strain energy release rate obtained from the area under the load-CMOD diagram, and "A_{lig}" is the area of the cracked ligament. Average values along with the standard deviation of fracture energies of different specimens are plotted against the crack advance (Δa) as shown in Fig. 1. An increase in fracture energy was observed as the number of fibers in the wake of the crack tip increased. This increase in fracture energy can be attributed to the increased surface area, better bond and increased stress transfer capability. As the crack grew, G_F for plain mortar decreased whereas, G_F in the fiber reinforced mortar exhibited a parabolically increasing trend until the load carrying capacity was transferred from the matrix to the fibers. Fracture energy had a decreasing trend once the stresses were transferred to the fibers.

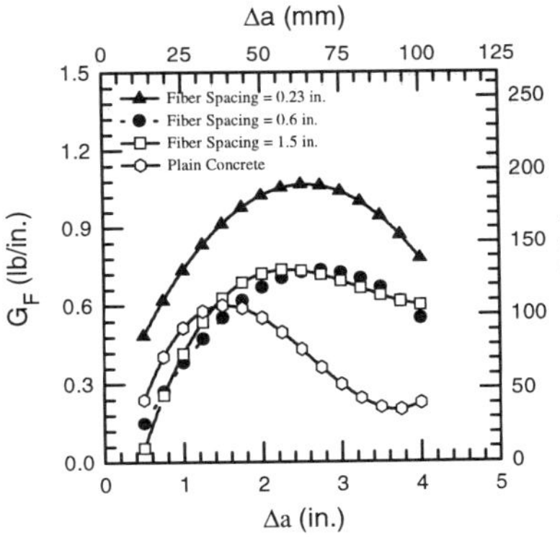

Fig. 1 Fracture energy.

In general, fracture energy increased as the number of fibers increased, however, there seems to be a threshold range of the fiber spacing above which the influence is not well pronounced. Similar results have been reported by Romualdi and Batson (1963) who concluded that the effect of reinforcement is not significant until the fiber spacing is reduced to approximately 7.5 mm (0.3 in.). On the contrary, Shah and Rangan (1971) concluded that the effect of spacing is insignificant below a fiber spacing of 25 mm (1 in.).

Fatigue Crack Propagation: Spacing Effect

Under fatigue loading the spacing between the reinforcing fibers played an insignificant role in the crack growth characteristics and lifetime distribution of the composite. The strain rate in fatigue testing was approximately 15 to 20 times higher than that of static testing. Although classical fracture mechanics is a rate-independent theory, the fracture properties of all materials depend upon the loading rate. An increase in strain rate generally shifts the failure from ductile to brittle mode as it affects the constituents that comprise the composite differently. The micro-yielding, micro-cracking and almost the entire process zone ahead of the crack tip undergoes mechanisms that ignore the spacing between the reinforcing fibers. In addition, an increased strain rate yields in greater dissipation of energy due to the loss of creep effects, larger debonding due to losses in the two phase action and inertial effects, etc. Finally, the effect of large scatter in the data cannot be ignored as the influence of fiber spacing may be of the same order as the scatter itself.

Fatigue Life Distribution

A large scatter in the fatigue data is typically observed in concrete specimens. A two parameter Weibull distribution was utilized to assess the fatigue life data. The survival probability (or the reliability function) for the two parameter Weibull probability distribution function can be written as

$$SF\ (n) = \exp\left[-\left(\frac{n}{\nu_s}\right)^{\alpha_s}\right] \tag{2}$$

where SF(n) is the probability that life will be longer than n, α_s is the shape parameter, ν_s is the scale parameter. Therefore, the parameters α_s and ν_s can be directly obtained from Eq. 2. Non-linear regression analysis was performed to obtain the coefficients α_s and ν_s which are used to create a plot of "ℓn(N)" versus the survival probability "ℓn(ℓn(1/SF))" (Fig. 2). If all the data falls along a straight line, it indicates that a good agreement with the chosen two parameter Weibull probability density function has been obtained. Fig. 2 shows the plots at various stress levels and fiber spacings of fatigue data versus the Weibull distribution. To check the validity, Kolmogorov-Smirnov goodness-of-fit tests were performed and the two parameter Weibull distribution was found to be acceptable, i.e., within 5% accuracy.

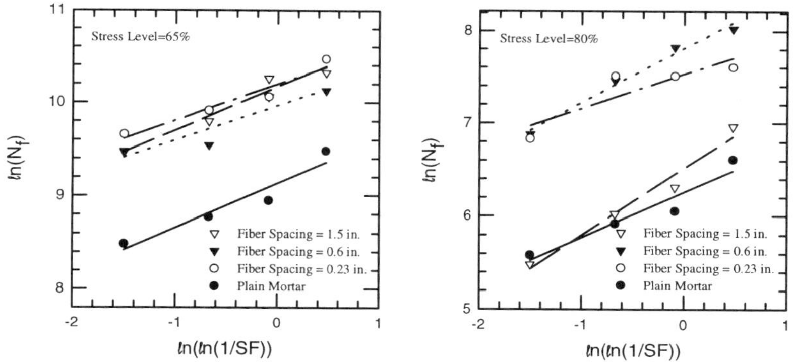

Fig. 2 Survival probabilities vs the number of cycles to failure.

Conclusions

1. Reducing the spacing between the fibers increased the fracture energy of the mortar reinforced with long aligned fibers, however, the effect was not significant until the spacing was reduced to 6.35 mm (0.25 in.).
2. The fatigue crack growth characteristics of mortar reinforced with continuous fibers were independent of the size and spacing between the fibers.
3. The crack growth characteristics of fiber reinforced mortar appeared to be dependent upon the strain rate and the stress level.
4. A two parameter Weibull distribution function compares well with the lifetime fatigue data.
5. Increasing the stress level from 65% to 80% reduced the lifetime of the composite by one order of magnitude.

References

Issa, M., Shafiq, A. B. and Hammad, A.: " Crack arrest in mortar reinforced with continuous aligned fibers," Cem. and Conc. Res., submitted for publication.

Romualdi, J. and Batson, G.: "Mechanics of crack arrest in concrete", J. Eng. Mech., Proc. Am. Soc. Civil Engineers, 1963, pp. 147-169.

Shah, S. and Rangan, V.:"Fiber reinforced concrete properties," ACI J., 1971, pp. 126-135.

Mechanical Response of Woven Graphite/Copper Composites

Brett A. Bednarcyk[1], Christopher C. Pauly[2], and Marek-Jerzy Pindera[3]

Abstract

Tension, compression, combined tension-compression, and Iosipescu shear tests were performed on 8-H woven Gr/Cu-alloy composites. This woven metal matrix composite (MMC) is a candidate material for high heat flux aerospace applications. The material's response exhibited significant plasticity, characterized by low yield stress, kinematic hardening, and significant stiffening in tension. Typical test results are presented with particular emphasis on the ultimate strength and failure characteristics.

Introduction

Woven composites are rapidly gaining popularity. These materials consist of a reinforcement phase made of a weave of fiber bundles, called yarns, in a matrix phase. By incorporating a woven reinforcement phase into a composite, rather than utilizing unidirectional fibers only, several benefits are realized. A single ply of a woven composite can have balanced longitudinal and transverse thermo-mechanical properties, unlike a unidirectional ply which often has poor thermo-mechanical properties transverse to the fiber direction. Woven composites have superior out-of-plane properties with regard to impact and crack resistance. Perhaps most importantly, a woven reinforcement phase offers superior stability during composite manufacturing compared to unidirectional fibers.

The advantages of a woven reinforcement phase have been realized for polymer matrix composites (PMCs) for quite some time, but only recently has the concept been extended to MMCs. One such MMC is graphite/copper (Gr/Cu). Gr/Cu is a candidate material for high heat flux aerospace applications (such as space craft radiator panels) because of the high thermal conductivity offered by the copper matrix. A good deal of work has been done on unidirectional Gr/Cu composites (Ellis, 1992; Bednarcyk and Pindera, 1994; DeVincent, 1995) though production of the material continues to be problematic. In order to facilitate the manufacturing of Gr/Cu, an 8-harness satin woven

[1] PhD. Candidate, [2]Research Assistant, [3]Associate Professor
Dept. of Civil Engr. & Applied Mech., University of Virginia, Charlottesville, VA 22903, USA

graphite reinforcement phase has recently been proposed. While the literature is replete with experimental and modeling investigations for woven PMCs and woven ceramic matrix composites (CMCs), there has been relatively little work done on woven MMCs. The present investigation seeks to characterize the room-temperature mechanical response of 8-harness woven Gr/Cu-alloy composites using tension, compression, combined tension-compression, and Iosipescu shear tests.

Experimental Investigation

The Gr/Cu-alloy composites used in this investigation were produced by Metal Matrix Castings, Inc. under contract with NASA Lewis Research Center. The reinforcement phase, in each case, consisted of three plies of 8-harness satin woven VCX-11 graphite cloth purchased from Amoco. Three different matrix alloys were used to produce eighteen composite plates: pure Cu, Cu-0.7 weight percent Cr, and Cu-0.5 weight percent Ti. The small amounts of the alloying metals are meant to improve fiber-matrix bonding rather than to alter the bulk properties of the mostly copper matrix. Tension and Iosipescu shear specimens were prepared and tested in the fill direction (0) and warp direction (90) of the woven reinforcement. Only fill direction specimens were prepared for the compression and combined tension-compression tests.

All tests were run using an MTS 810 servo-hydraulic testing apparatus. In addition, special specimen fixtures were used for each type of test. For tension testing, low-shear bell-housing grips were used that allow the test specimen to rotate in one plane (Pindera and Herakovich, 1986). For compression and combined tension-compression testing, a buckling-preventive fixture was used (Lin and Pindera, 1988). For shear testing, an Iosipescu fixture was used. Strain measurements were taken using an extensometer and unidirectional and rosette strain gauges mounted on either side of each specimen.

Results and Discussion

Figure 1a shows typical results of the cyclic tension and cyclic compression tests. In tension, the woven Gr/Cu exhibits early yielding with permanent strain appearing almost immediately. This seems to be attributable to local stresses, possibly at yarn cross-over areas, which are high enough to yield the copper-alloy matrix. At higher tensile strains the composites exhibit stiffening. It is believed that this effect is due to movement or rotation of the woven fabric when the yarns straighten out. As this straightening occurs, the yarns bear more of the load, and the composite response is stiffer.

In compression, the initial modulus of the composite is similar to that in tension, but the onset of permanent strain is more gradual. Stiffening does not occur since the yarns do not straighten, and the compressive strength is much lower (see Figure 4).

The hysteresis loops evident in both tension and compression are believed to be due to matrix plasticity rather than damage. The changes in the unloading modulus are most likely due to kinematic hardening effects. This hypothesis is supported by the cyclic evolution of the initial tensile modulus of the material, Figure 1b. The initial modulus upon loading for each cycle is nearly constant, degrading only slightly. The initial modulus upon unloading for each cycle increases slightly as a result of the stiffening. If damage were responsible for the hysteresis, degradation of the composite modulus would be expected.

Failure of the specimens in tension was typified by fracture of the copper-alloy matrix, with a large number of fibers remaining intact across the failure surface. In compression, the composite micro-buckled, usually near one grip, also leaving a great deal of fibers intact in the failure region. In both cases an audible cracking or tearing sound indicated failure.

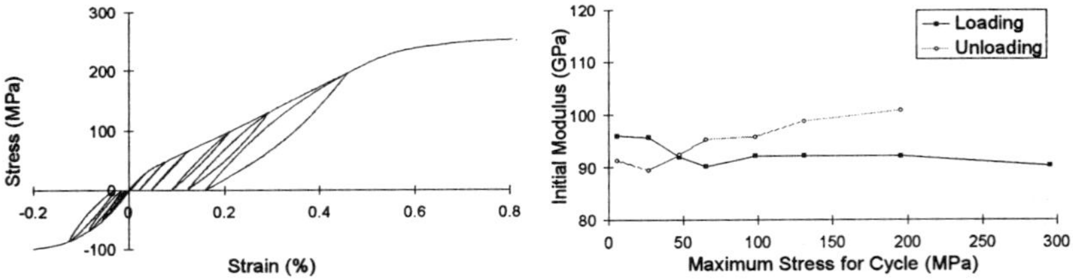

Figures 1a and 1b. Typical cyclic tension and compression behavior of 8-H satin woven Gr/Cu, and corresponding initial loading and unloading moduli for tensile cycles.

Figure 2 shows the behavior of Gr/Cu in combined tension-compression. Once again, composite yielding occurred at low stresses, but the composite typically failed in compression before reaching tensile strains sufficient to induce stiffening. Each tensile loading cycle passed through the point of maximum stress reached by the previous cycle. This was not the case for the compressive loading cycles. This again points to kinematic hardening in the copper-alloy matrix. Compressive failures were similar to monotonic and cyclic compression failures. Ultimate strengths for the combined tension-compression tests were included in the compression failure data (see Figure 4).

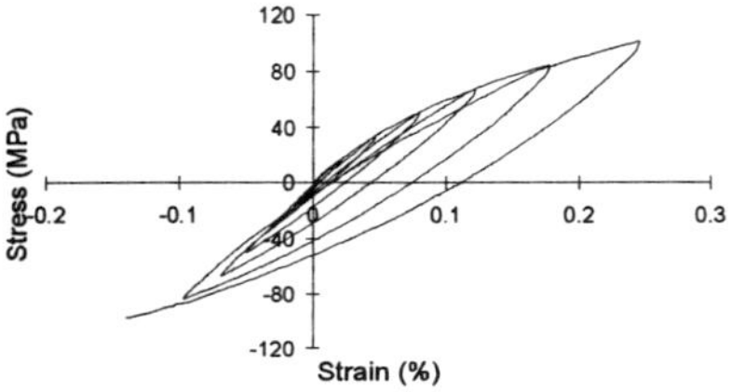

Figure 2. Typical tension-compression behavior for 8-H satin woven Gr/Cu.

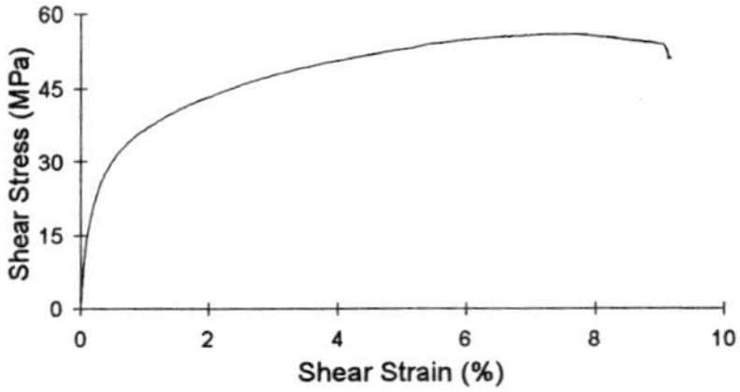

Figure 3. Typical shear behavior for 8-H satin woven Gr/Cu.

Figure 3 shows the shear stress-strain response for woven Gr/Cu. The composite exhibits substantial shear strain before failure, though failure occurred at relative low shear stresses (see Figure 4). The shear specimen failures were unique from tension and compression failures in that there were no distinct events to indicate failure. There was no audible indication or noticeable instability, rather the load began to decrease upon further application of deformation. Post-test inspection of the shear specimens revealed the presence of micro-buckling and yarn rotation in the test section.

Figure 4 shows average ultimate strength results for the Gr/Cu-alloy composites in tension, compression, and shear. The error bars indicate the high and low strength values. Compressive results are limited to fill-direction specimens.

The specimen orientation (fill or warp) seems to have had little effect on the ultimate strengths. Tensile and shear strengths were comparable for both testing directions for each matrix alloy, and no trend based on orientation is apparent. The ultimate tensile strength of the pure copper matrix composites was greatest, followed by the Cr-alloy matrix composites and the Ti-alloy matrix composites. Thus the composite with the poorest fiber-matrix bond exhibited the highest tensile strength. A possible explanation is that poor bonding in the pure copper matrix composites allowed the fiber yarns to rotate most freely thus delaying fracture of the matrix that typified composite failure. The compressive and shear strengths of the different matrix alloy composites were similar, and matrix alloying seems to have had little effect on these strengths.

Future work will focus on analytical modeling of 8-H woven Gr/Cu. The experimental results generated in this investigation will be used for model comparison and verification.

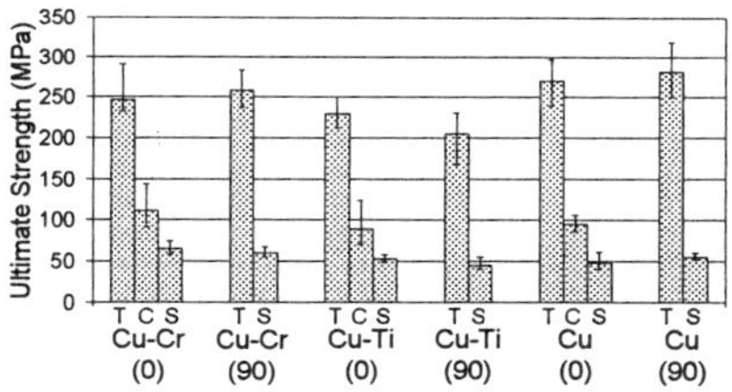

Figure 4. Ultimate strengths of 8-H satin woven Gr/Cu-alloy specimens in fill (0) and warp (90) directions in tension (T), compression (C), and shear (S).

References

Bednarcyk, B.A. and Pindera, M.-J. (1994) "Micromechanical Modeling of the Thermal Expansion of Graphite/Copper Composites With Nonuniform Microstructure" *NASA-CR-195368*.

DeVincent, S.M. (1995) "Interfacial Effects on the Thermal and Mechanical Properties of Graphite/Copper Composites" *NASA-CR-198370*.

Ellis, D.L. (1992) " Properties of Graphite Fiber Reinforced Copper Matrix Composites for Space Power Applications" *NASA-CR-191026*.

Lin, M.W.Y. and Pindera, M.-J. (1988) "Elastoplastic Response of Unidirectional Graphite/Aluminum under Combined Tension-Compression Cyclic Loading" *CCMS Report 88-01*, Virginia Polytechnic Institute and State University.

Pindera, M.-J. and Herakovich, C.T. (1986) "Shear Characterization of Unidirectional Composites with the Off-Axis Tension Test" *Experimental Mechanics*. Vol. 26, No. 1.

Constitutive Modeling
of Composites in Opto-Mechatronics

Tau C. Fan[1], M. ASCE

Abstract

Opto-mechatronics is a specialized treatment of mechanical systems in the electro-magnetic field in which the incident waves are derived from the quantized electro-magnetic waves. This happens in thermal disturbance of lattice when photons in electro-magnetic waves are converted into phonons of light waves. Wave guides in telecommunication in polymer fabrics are one of the applications.

Attentions are called in this study for the design of a wave guide with suitable constitutive modeling in appropriate strain matrices and the consideration of the impact of electro-magnetic and light waves in the generalized Hooke's law. Also cited in the failure criteria are the noneigen value branch of instability in a nonconservative case and the potential presence of dynamic stress concentration in an inhomogeneous solid. In addition, the energy integrals in fracture criteria consist of J-, G-, and C-integrals (Fan 1987) to include the strain, thermal, dynamic and quantum effects as a result of the new strain matrix.

Introduction

Well known is the advantage of telecommunication today, yet the efficiency of transmission is still to be addressed because of inconsistency and poor quality beyond certain threshold of distance. Primarily it is due to the loss of power, geometrical imperfection and/or materials deficiency. It is postulated that with adequate power, the design criteria of configuration and materials selection are met. Then the failure criteria may be in question, notably the back scattering, $\bar{E} \times (v \times \bar{E})$, electrostatic parametric instability, $\bar{E}.v\bar{E}$, and dynamic instability in a

[1]Professor, Materials Science and Engineering, World Open University, P.O. Box 955, Palos Verdes Estates, CA 90274

noneigen value problem in a nonconservative system. (Fan 1988) provides the list of failure criteria.

The potential loss of signal power may be caused by (1) conversion of light into heat, (2) absorption of light due to electronic transition between the valence and conduction bands, (3) mechanical vibrations. Unfortunately, none of the cases states the dynamic instability such as a noneigen value problem in a nonconservative system. For comparison, a panel flutter is an eigen value problem but in a nonconservative system, while a static buckling is an eigen value problem in a conservative system usually. More discussion will be in the following section.

Analytically, the governing compatibility equations to determine the required displacements with associated boundary and initial conditions are derivable from the appropriate strain matrices suitable for the prescribed environment with mechanical, electro-magnetic and optical influences which are to be introduced into the subject constitutive modeling.

The study of phonons in light waves leads to durality of wave-particle theory which deals with corpuscular theory in mass and wave diffractions in wave theory. The light wave is transversely applied to components, thus the shear wave will be considered normally. The phonon energy can be traced to the energy of a lattice vibration in molecular mechanics. When the lattice vibration is quantized, the quantum of energy is called phonon, similar to photon in an electro-magnetic wave. The elastic wave of phonons is closely associated with photons in the electro-magnetic wave during the thermal vibration of crystals on lattice. Consequently, the phonon energy will be considered in the optical event.

Strain Matrices

The key component is the wave guide between the transmitter and a receiver in telecommunication. Its configurations vary from a box beam to a cylinder with cavity. A circular cylinder with cavity is considered in this note. It is treated as a plane strain problem, because of the large ratio of its length to the diameter. It is further regarded as transversely isotropic, (Fan 1993) and axisymmetrical with respect to geometry, loading and materials system. Excluding the hygrothermal influence, the compliance index, $a_{i,j}$ in this case, (i=1,2,3; j=1,2,...9) are used in the generalized Hooke's law.

$$\varepsilon_{rr} = a_{11}\,\sigma_{rr} + a_{12}\,\sigma_{\theta\theta} + a_{13}\,\sigma_{zz} + a_{14}\,E_{rr} + a_{17}E_{rr}$$

$$a_{11} = 1/E_{rr},\ a_{12} = -\,\nu_{r\theta}/E_{rr},\ a_{13} = -\,\nu_{rz}/E_{rr} = a_{31} = a_{23} = a_{32}$$

$$a_{14} = A/[1/2\,(L^{1}\,I^{2}) + Id \times \chi_{M} H],\ a_{33} = 1/E_{zz},\ a_{14} = a_{25} = a_{36}$$

$$a_{17} = \hat{h}\,\omega\, d/W^{2} = a_{28} = a_{39}$$

Similarly,

$$\varepsilon_{\theta\theta} = a_{21}\sigma_{rr} + a_{22}\sigma_{\theta\theta} + a_{23}\sigma_{zz} + a_{25}E_{\theta\theta} + a_{28}E_{\theta\theta}$$

$$\varepsilon_{zz} = a_{31}\sigma_{rr} + a_{32}\sigma_{\theta\theta} + a_{33}\sigma_{zz} + a_{36}E_{zz} + a_{39}E_{zz}$$

Prescribed conditions lead to

$$a_{11} = a_{22} = 1/E,\ a_{12} = a_{21} = -\nu_{1/E},\quad \nu_{rz} = \nu_{\theta z} = \nu_2$$

$$\nu_{r\theta} = \nu_{\theta r} = \nu_1$$

It follows that the new strain matrix explicitly considers the influence of electro-magnetic and optical effects. The governing equations can be constructed per (Fan 1993) with infinitesimal strain for strain displacement relations, and stress functions, , compatible with the strains in a plane strain such as

$$r\ \partial\varepsilon_{\theta\theta}/\partial r = \varepsilon_{rr} - \varepsilon_{\theta\theta}\quad ,\quad \varepsilon_{zz} = 0$$

The resulting stresses are for the responses in an equilibrium case. The critical frequency in particular for the cylinder in dynamic instability will be examined in the next section.

Dynamic Stability

In failure criteria on instability, there are eigen value problems in a conservative system such as buckling in a static case, eigen value problems in a nonconservative system such as flutter in a dynamic case. In addition there is dynamic instability, a noneigen value problem in a nonconservative system. In molecular mechanics, the vibration of lattice due to thermal disturbance is postulated to be the case of dynamic instability.

In the conservation of crystal momentum, hK, the wavevector of scattering photon is equal to the sum of the wavevector of incident photon and a vector of reciprocal lattice (Lin 1989). If phonons interact with neutrons in scattering, there is either emission or absorption of phonon in this inelastic scattering. It follows that the above equilibrium will be changed when the phonon energy term is added or subtracted from the incident photons. This is a nonconservative case as it differs from that in a conservative case where the energy is derivable from its potential, and the independent variable t (time) is not explicitly expressed in a Lagrangian. In an eigen value problem such as flutter, its solution is a product of $F(r)e^{\omega t}$. F(r) is to be determined and the critical flutter is then found by considering the small oscillations of elastic body from its position of equilibrium. In dynamic stability, the solution is a product of F(r)g(t) where r is a spatial coordinate, and F(t) is to be determined.

The treatment of a cylinder subject to radial loads has been studied in (Fan 1966). The governing dynamic equations are transformed to be a Mathieu type which defines the regions of instability. Their presence definitely impacts the efficiency of signal transmission as expected. This deficiency related to light energy can be alleviated by geometrical optimization and suitable materials selection in the electro-magnetic properties.

Other Dynamic Considerations

Studies in (Pao & Mow 1971, Fan & Pongracz 1986) show the influence of wave propagations in solids. The local stresses and strains can be greatly increased with material inhomogeneity, surface imperfection, etc. The related stress concentration factor can be greater than 4.0 which is beyond the limit in a static case. As expected, the local displacement will be unfavorably modified. It translates into the deficiency in the function of a wave guide. Recent study (Chen and He 1996) shows the influence of a composite system on waveguide.

In fracture mechanics, the stress intensity factors and the energy integrals are derivable from the suitable strain matrices for the pertinent generalized Hooke's law. It follows that the proposed strain matrix is admissible in the study when there are electro-magnetic and optical fields as associated environment besides the mechanical effects.

Conclusion

In the design of wave guides, it is essential to examine the strain matrices of composites compatible with environment in application. Besides the geometrical variation, the selection of materials system jointly in the transfer function to provide the potential optimum responses is necessary. This process can be illustrated in the additional analytical procedure in the noneigen value branch and local stress concentration besides the newly proposed energy integrals in fracture mechanics. A specific design is planned in the near future to demonstrate the expected influence.

Appendix 1. References

Chen, Hung-Liang & He, Yidong (1996). "Analysis of Liquid-Core Cylindrical Acoustic Waveguides Embedded in Solid Media", J. Engineering Mechanics, ASCE 122 (1), 1

Fan, T.C. (1966). Dynamic Stability of Thin Cylinder Under Radial Pressure. RM-4583-PR, The RAND Corporation, Washington, D.C.

Fan, T.C. & Pongracz, H. (1986). "Responses of Particulate Media Under Intermittent Impulses", Proc. of 10th U.S. National Congress of Applied Mechanics, Univ. of Texas

Fan, T.C. (1987). "Rupture of Composites in Quantum Mechanics", Proc. of 20th Mid-Western Mechanics Conf., Purdue Univ., p. 957

Fan, Tau-Chao (1988). "Strain Matrices of Composites", SAMPE Quarterly, 20 (4), 38

Fan, T.C. (1993). "Strain Matrices of Composites in Magneto-Elasticity", Proc. of 38th SAMPE International Symposium, Anaheim, California, p. 868

Lin, Chinlon, Editor (1989). Optoelectronic Technology and Lightwave Communications Systems. Van Nostrand Reinhold, New York

Pao, Yihhsing & Mow, Chaochow (1971). Diffraction of Elastic Waves and Dynamic Stress Concentrations. Crane, Russak, New York

Appendix 2. Notations

d = Diameter of cylinder
E = Electric field intensity; Modulus of elasticity
H = Magnetic field intensity
$\hat{h}$ = Planck's constant
I = Current
K = Wavevector , L == Inductance
r,θ, z = Polar coordinates
w = weight
ε = Strain
σ = Stress
ν = Poison's ratio
Φ = Stress function
ω = Circular frequency
μ == Magnetic Permeability

Observations of Internal Fracture in Mortar using X-Ray Microtomography

Eric N. Landis[1], Edwin N. Nagy[2], Denis T. Keane[3], and Nhan Huynh[4]

Abstract

The internal cracking patterns of a series of small mortar specimens were observed using a technique called x-ray microtomography. The experimental arrangement was such that three dimensional images of internal cracks could be made at several different points during the load history of the sample. This arrangement allowed progressive crack growth to be observed at micron-scale resolution in a single specimen. The specimens were loaded in compression. The data obtained will be used in true three dimensional measurements of work of fracture. The effects of particle packing and to a limited extent, pore structure may be observed.

Introduction

The fracture and failure mechanisms of concrete are dictated by its heterogeneous microstructure. Inhomogenieties at a wide range of length scales contribute to the observed roughness in fracture surfaces, and to the increased toughness as demonstrated by its post-peak load-displacement response. Although it is widely recognized that concrete fracture has a strong three dimensional dependence, experimental observation of fracture in three dimensions is a difficult problem. Most observations of fracture and damage progression have only been made in two dimensions. High resolution observations have been limited to either the specimen surface, or to thin cross sections examined after a test, the latter being a "destructive" test such that damage progression cannot be observed on a single specimen. Nondestructive tests which can be applied in three dimensions, such as acoustic emission and acoustic tomography lack spatial resolution necessary to resolve microstructural features. X-ray microtomography (XMT) is a technique by which three dimensional measurements of internal cracking may be made in situ, at micron

[1] Assistant Professor, Dept. of Civil & Env. Engineering, University of Maine, Orono, ME 04469.
[2] Graduate Research Assistant, Dept. of Civil & Env. Engineering, University of Maine.
[3] Senior Scientist, DND-CAT Synchrotron Research Center, Argonne National Laboratroy, IL 60439.
[4] Undergraduate Student, Dept. of Civil Engineering, Temple University, Philadelphia, PA 19122.

resolution. Previous applications of more traditional CAT-scans to concrete have been primarily exploratory (Morgan *et al*, 1980, Martz *et al*, 1993). These applications involved locating defects, voids, and reinforcing steel. Spatial resolutions of about 1 mm were typical. For the XMT experiments described here, the resolution was about 10 microns.

X-Ray Microtomography

X-ray microtomography is a technique which measures the 3-D structure of a material's x-ray absorption. Radiographs of a sample are taken at hundreds of different viewing angles. A computer algorithm inverts this data to produce the 3-D absorption map. Microtomography can resolve volume elements within a sample down to approximately two cubic microns.

Conventional x-ray computed tomography, e.g. medical applications, achieves a spatial resolution on the order of hundreds of microns. Microtomography achieves a resolution of 2 to 15 microns by combining synchrotron radiation with a high resolution x-ray detector. Synchrotron radiation provides x-rays of much greater brightness than conventional x-ray sources. The higher flux and better collimation of this source provide more x-rays per unit area on the sample, allowing for better spatial resolution. The higher flux and continuous spectrum allow researchers to choose a monochromatic beam with a small bandpass, which greatly improves the sensitivity of microtomography to smaller x-ray absorption variations within the sample. The use of a high resolution CCD based detector allows the greater spatial resolution of microtomography to be realized.

Experimental Program

The goal of the experiments was to examine internal cracking at high resolution while the specimen was under load. A series of small (4 mm diameter) mortar cylinders were cast for the experiments. The mix proportion was 1:2:0.5 cement to sand to water, respectively. The intent was to mimic the "mortar phase" of a typical concrete mix.

The XMT experiments were conducted at the National Synchrotron Light Source (NSLS), beamline X2B at Brookhaven National Lab. The operational configuration was as follows. The x-ray beam from the synchrotron was monochromated by a Si crystal to select an energy of 32 keV. The detector is a phosphor screen with high spatial resolution which converts x-rays to visible light. A microscope objective lens focuses the light onto a CCD chip to produce 13.5 micron pixels. The 512 X 512 format of the CCD gave a field of view of approximately 6.5 X 6.5 mm. The specimens were exposed to the beam at 900 different angles, where absorption radiographs were taken. The data set was then sent through the Exxon reconstruction algorithm, which is based on Direct Fourier Reconstruction methods.

A special loading stage was constructed so that specimens could be examined while under load. The requirements of the stage were that it needed to be small enough to fit into the limited space at the beamline, and it had to load the specimen in such a way that the frame did not interfere with any of the scans. Specifically, the entire specimen and surrounding loading pins had to be in the field of view at all times during the x-ray scans. The loading stage is illustrated in Figure 1.

In the experiments described here, two specimens were tested. Four tomographic scans were made of each specimen. The first scan was of the sample with little or no damage, and the remaining three at progressively higher levels of damage. Typical reconstruction images are shown in Figure 3 of scans made at points labeled (a) through (c) on the load-displacement curve of Figure 2. The three images of Figure 3 represent roughly the same plane through the sample at three different levels of damage. It should be noted that for each XMT scan, 320 such "slices" are produced. These slices may be stacked to produce true three dimensional renderings of the specimen's internal structure.

The present analysis of the XMT data can be summarized as follows. Each reconstruction image is processed to evaluate the total length of cracks in the cross section. This is done through a combination of thresholding and feature analysis. The totals for each slice are then summed to produce the total crack area in the specimen. Ultimately this crack area may be compared to the work of the external load to produce a three dimensionally-based work of fracture estimate.

Implications for Fracture Research

The significance of these experiments is that quantitative measurements of internal cracking have not been possible at the kind of spatial resolution which is achieved here. In addition, the nature of the digital reconstructions of the internal structure are such that analysis is not limited to planar cracks or any other particular geometry. The tortuous, three dimensional nature of internal cracking in cement-based materials may be examined with relatively high resolution. Effects of grain-size distribution and particle packing may be observed and should be important in development of high performance concrete.

Acknowledgments

This work is supported in part by the National Science Foundation (Grant # CMS-9522984) under a program directed by Dr. Ken P. Chong. Part of this research was conducted at the National Synchrotron Light Source, Brookhaven National Laboratory, which is supported by the U.S. Department of Energy, Division of Materials Sciences and Division of Chemical Sciences (DOE contract number DE-AC02-76CH00016). Nhan Huynh was supported by the McNair Scholars summer program at the University of Maine.

References

Morgan, I. L., Ellinger, H., Klinksiek, R., and Thompson, j. N. (1980). "Examination of Concrete by Computerized Tomography." *ACI Journal*, 77(January/February), 23-27.

Martz, H. E., Schneberk, D. J., Roberson, G. P., and Monterio, P. J. M. (1993). "Computerized Tomography Analysis of Reinforced Concrete." *ACI Materials Journal*, 90(3), 259-264.

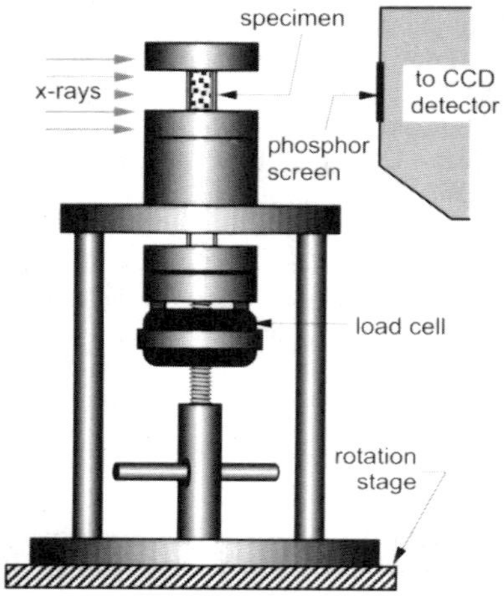

Figure 1. Loading stage and experimental setup.

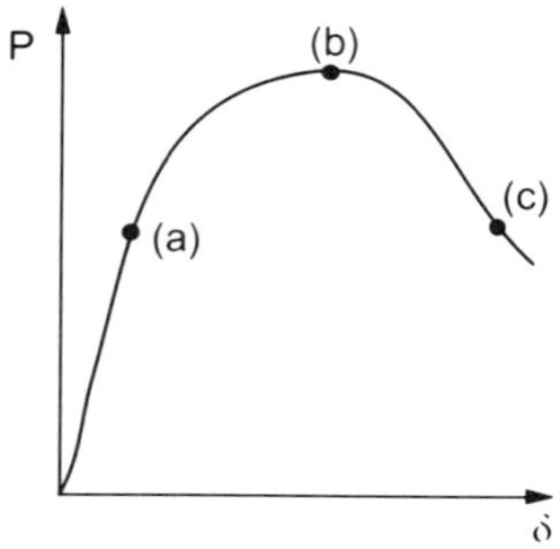

Figure 2. Points on load vs. displacement diagram where XMT scans were made.

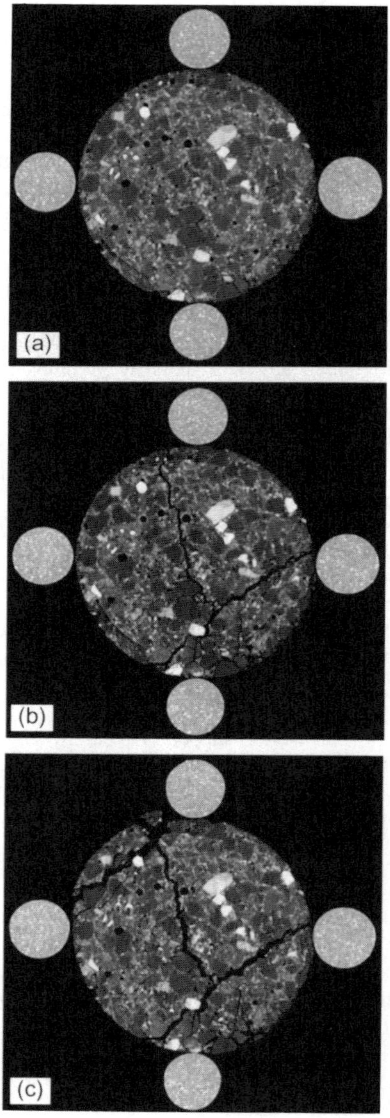

Figure 3. Reconstruction images of the same plane taken at three different loads.

Structural Redundancy and Fracture in Composite Lattice/Skin Structures

A. K. Maji[1] member ASCE, E. Fosness[2] and D. Satpathi[3]

Abstract

Graphite/Epoxy composites are being investigated as an alternative to aluminum (or aluminum honeycomb material) in a number of space structures applications ranging from solar array panels to payload shrouds and motor cases. High modulus graphite fiber and toughened epoxy matrices have extended the scope of composite structures. Lattice/skin structures called 'isogrids' were investigated (NASA, 1973). These structures consist of a system of stiffening ribs in three directions, connected to a pseudo-isotropic skin consisting of a $[0/+60/-60]_S$ lay-up sequence. Experiments have demonstrated that failure initiates at the bond between the rib system and the skin both due to stress concentration and material weakness introduced during the manufacturing process.

Fracture mechanics based analyses using a commercial finite element code (with material anisotropy) were used to study the initiation and propagation of cracks along the rib/skin interface. The energy release rate criterion was used along with aspects of nonlinear fracture. The consequence of disbond was investigated with conventional mechanics of anisotropic materials.

Background

Rhodes and Mikulas (1975) explored the use of composite materials for manufacturing the next generation of isogrid structures. Reddy et al. (1985) compared the isogrid with other lattice reinforcement schemes (orthogrid and generalized orthogrid), under various loading combinations, and concluded that the isogrid was the most attractive due to its damage tolerant characteristics. The redundancy of the rib stiffening system was experimentally proven by cutting individual ribs and exploring buckling loads, the redistribution of stresses and load paths, etc. The design was governed primarily by the local buckling of the ribs and skin pockets, and global buckling of the structural system. A detailed description of the research reported here will be available in Maji et al., (1996).

1. Senior Research Scientist, Air Force Phillips Laboratory, PL/VTSC, Albuquerque, NM 87117-5776. ph : (505)-846-9334.
2. Research Engineer, Air Force Phillips Lab., PL/VTSS, NM 87117-5776
3. Graduate Research Assistant, U. of New Mexico, NM 87131

Research Objectives

Damage tolerance of isogrids was discussed by Ambur and Rehfield (1991) based upon tests on isogrid panels with the ribs cut at various locations. However, most investigators have reported that the separation of the rib/skin connection, followed by the local buckling of individual ribs is the critical failure mechanism in isogrids. The origin of this problem is easily understood in isogrids where the skin is separately attached (epoxied) to the lattice rib structure subsequent to curing. The more common manufacturing method involves co-curing the skin (prepreg lay-ups) and the ribs (prepreg tows hand-laid or filament-wound). Even then, the rib/skin bond is a matrix dominated interface which consequently forms a weak link in the isogrid structure. Disbonds can develop during manufacture, or due to in-service loads. Our visual observations showed that the strength of the rib/skin interface is reduced by the thinning of the ribs at that critical location, due to overcompaction of the ribs during cure. In addition, resin-rich bands devoid of fibers (observed under a scanning electron microscope) further compromise the strength.

The issues addressed in this paper are : i) when is an interface crack likely to propagate ? and ii) what is the effect of complete disbond on local buckling ?

Fracture Mechanics

The Energy Release Rate G was used as the fracture parameter. According to linear elastic fracture mechanics, a crack propagates when $G = G_C$, which is an experimentally determined material property. The finite element method (FEM) using an energy derivative approach was used to determine G :

$$G(a) = \frac{\partial \Pi}{\partial a} = -\frac{1}{2}\frac{\partial\{[u][K][u]\}}{\partial a} \tag{1}$$

where a is the crack length (times specimen thickness), Π is the total strain energy of the specimen obtained from the displacement [u] and stiffness [K] matrices. Therefore the change in the total strain energy of the specimen is to be determined at two slightly different crack lengths. The total strain energy Π can be determined directly from the software output, or evaluated indirectly from the external work :

$$\Pi = \frac{P\delta}{2} \tag{2}$$

Finite Element Analyses

A commercially available finite element software, 'IDEAS' (SDRC, 1993) was used because of its ability to incorporate an orthotropic material model and its postprocessing capabilities. Since this software did not have any specific fracture mechanics capability (such as singular elements), G was evaluated from the energy derivative approach discussed earlier .

The T-beams tested in our laboratory (discussed later) were modeled next using a 2-dimensional mesh with a different thickness for the rib and the skin sections. Orthotropic material properties were used, different for the rib (web) and the skin (flange), as discussed in the next section. A convergence study was performed by progressively refining the mesh from 48 to 2048 elements. Four noded rectangular 'thin shell' elements were used to generate the initial mesh. The T-beam geometry with cracks of 2.5cm (1") and 1.27cm (1/2") length (Figure 1) at the rib-skin interface were analyzed to determine the load-point displacements (Figure 2). From this data, G was evaluated using equations 1 and 2. The orthotropic material properties used in these analyses are discussed next.

Experimental Program

The specimens were cut out of a 46cm. square isogrid panel made of IM7 graphite fiber and 977-2 epoxy prepregs and tows (ICI Fiberite, Tempe, AZ). Three point bending tests were conducted. Artificial cracks, 2.54 and 1.27cm (1" and 1/2") long were introduced at the rib-skin connection by a rotary cutting blade with a diamond abrasive wheel. Test results of the tests are shown in Figure 3.

G values were determined for a load of 4.448N (1lb) for the two fractured specimens. These were 0.0347 and 0.1138 N/cm. respectively (note : thickness of material at the interface was 0.0762cm). Prepreg manufacturer provided a G_C value of 367.7 Nm/m^2 for the resin. Based on the LEFM energy release rate criteria (crack propagates when $G = G_C$), the failure loads for the 1.27cm and 2.54cm specimens were calculated to be 472N and 143N respectively. While this agrees reasonable well with the test data (Figure 3), the smaller slope of the experimental data (load vs. crack length) demonstrates 'nonlinear fracture' behavior.

Consequence of Rib/Skin Disbond :

Under axial compression disbond is of no consequence, except to reduce the local buckling load of both the rib and the skin. If the rib is considered simply supported at the three edges where it is connected, its buckling load is given by :

$$N_x = D_{11}\frac{\pi^2}{a^2} + 12\frac{D_{33}}{d^2} \text{ where } D_{11} = \frac{E_x b^3}{12(1-\nu_{xy}\nu_{yx})}, D_{33} = \frac{G_{xy}b^3}{12} \qquad (3)$$

d is the rib height (=1.27cm), b is rib width (0.127cm) and a is the unsupported length of the rib between nodes. When the rib separates from the skin, the buckling load reduces to only the first term of equation 3. For the isogrid studied here (a = 9.14cm), the second term is twice as large as the first. Therefore rib/skin debonding will reduce the rib's buckling load to 1/3 of its original value.

Skin buckling varies inversely as the square of the largest circle that can be inscribed within the simply supported boundaries. The largest circle inside the triangular ribs has a diameter of $a\tan 30^0$. When a rib is separated, the largest circle that can be inscribed inside two adjacent triangles has a diameter of $a\cos 30^0$. The buckling load varies inversely with the square of the circle diameter. Therefore rib/skin disbond will decrease the skin buckling load to 44% of the original value.

Acknowledgement

D. Satpathi was supported by NSF grant # MSS 9212733.

References

1. Ambur D.R. and Rehfield L.W., Effect of Stiffness Characteristics on the Response of Composite Grid-Stiffened Structures", AIAA # 91-1087-CP, 1991.
2. Maji et al., "Evaluation of Rib/Skin Fracture In Composite Isogrid", to appear, ASCE J. of Engineering Mechanics, 1996.
3. NASA report # CR124075, by McDonnell Douglas Astronautics Co., 1973.
4. Reddy A.D., Rao V.R. and Rehfield L.W., "Continuous Filament Wound Composite Concepts for Aircraft Fuselage Structures", J. Aircraft, V22, No. 3, March 1985, pp. 249-255.
5. Rhodes M. and Mikulas M.M., "Composite Lattice Structures", NASA Technical Memorandum, TM X-72771, 1975.

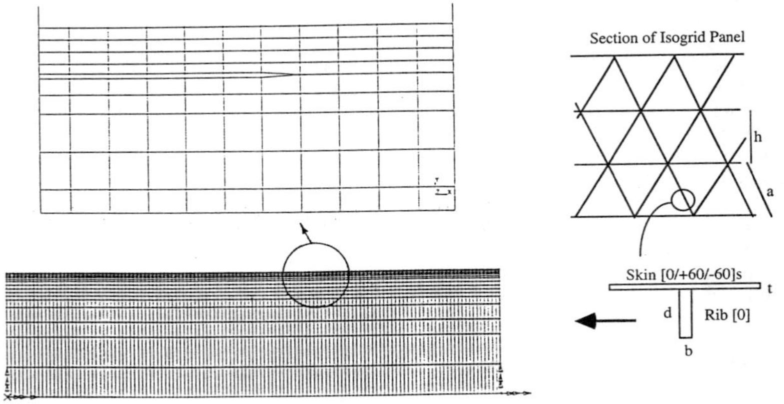

Figure 1. Isogrid Panel and FEM Mesh of the Interface

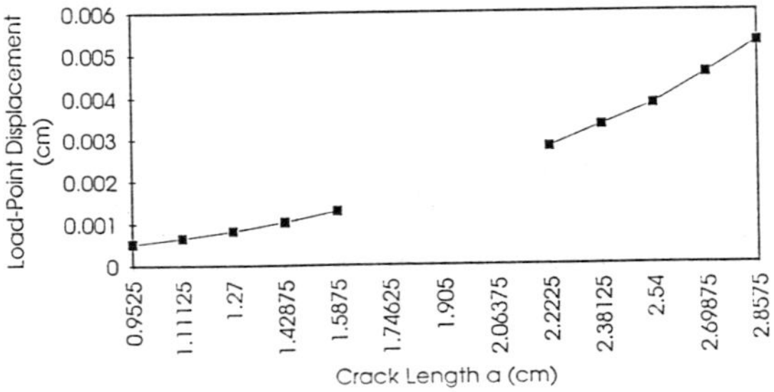

Figure 2. Load-Point Displacement vs. Crack Extension

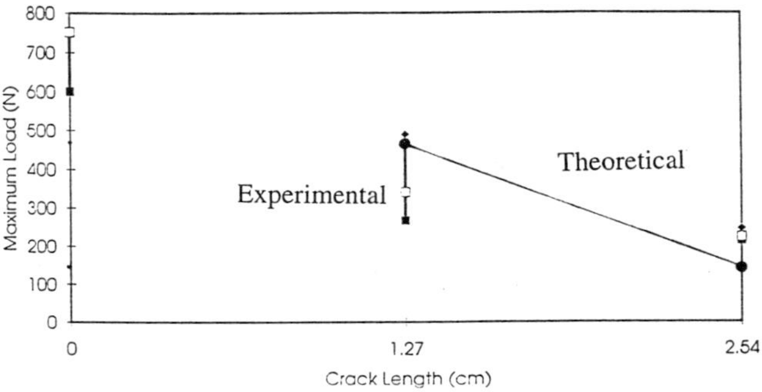

Figure 3. Failure Load for Various Specimens (Experimental vs. Theoretical)

SUBJECT INDEX

Page number refers to first page of paper

AUTHOR INDEX

Page number refers to first page of paper